An Introduction to General Relativity
and Cosmology

General relativity is a cornerstone of modern physics, and is of major importance in its applications to cosmology. Experts in the field Plebański and Krasiński provide a thorough introduction to general relativity to guide the reader through complete derivations of the most important results.

An Introduction to General Relativity and Cosmology is a unique text that presents a detailed coverage of cosmology as described by exact methods of relativity and inhomogeneous cosmological models. Geometric, physical and astrophysical properties of inhomogeneous cosmological models and advanced aspects of the Kerr metric are all systematically derived and clearly presented so that the reader can follow and verify all details. The book contains a detailed presentation of many topics that are not found in other textbooks.

This textbook for advanced undergraduates and graduates of physics and astronomy will enable students to develop expertise in the mathematical techniques necessary to study general relativity.

An Introduction to General Relativity and Cosmology

Jerzy Plebański

Centro de Investigación y de Estudios Avanzados
Instituto Politécnico Nacional
Apartado Postal 14-740, 07000 México D.F., Mexico

Andrzej Krasiński

Centrum Astronomiczne im. M. Kopernika,
Polska Akademia Nauk, Bartycka 18, 00 716 Warszawa,
Poland

CAMBRIDGE
UNIVERSITY PRESS

CAMBRIDGE UNIVERSITY PRESS
Cambridge, New York, Melbourne, Madrid, Cape Town,
Singapore, São Paulo, Delhi, Mexico City

Cambridge University Press
The Edinburgh Building, Cambridge CB2 8RU, UK

Published in the United States of America by
Cambridge University Press, New York

www.cambridge.org
Information on this title: www.cambridge.org/9781107407367

First published 2006
First paperback edition 2012

A catalogue record for this publication is available from the British Library

ISBN 978-1-107-40736-7 Hardback
ISBN 978-0-521-85623-2 Paperback

Cambridge University Press has no responsibility for the persistence or
accuracy of URLs for external or third-party internet websites referred to in
this publication, and does not guarantee that any content on such websites is,
or will remain, accurate or appropriate. Information regarding prices, travel
timetables, and other factual information given in this work is correct at
the time of first printing but Cambridge University Press does not guarantee
the accuracy of such information thereafter.

An errata list for this publication is available at
www.cambridge.org/9781107407367

Contents

Figures

The scope of this text

General relativity is the currently accepted theory of gravitation. Under this heading one could include a huge amount of material. For the needs of this theory an elaborate mathematical apparatus was created. It has partly become a self-standing sub-discipline of mathematics and physics, and it keeps developing, providing input or inspiration to physical theories that are being newly created (such as gauge field theories, super-gravitation, and, more recently, the brane-world theories). From the gravitation theory, descriptions of astronomical phenomena taking place in strong gravitational fields and in large-scale sub-volumes of the Universe are derived. This part of gravitation theory develops in connection with results of astronomical observations. For the needs of this area, another sophisticated formalism was created (the Parametrised Post-Newtonian formalism). Finally, some tests of the gravitational theory can be carried out in laboratories, either terrestrial or orbital. These tests, their improvements and projects of further tests have led to developments in mathematical methods and in technology that are by now an almost separate branch of science – as an example, one can mention here the (monumentally expensive) search for gravitational waves and the calculations of properties of the wave signals to be expected.

In this situation, no single textbook can attempt to present the whole of gravitation theory, and the present text is no exception. We made the working assumption that relativity is part of physics (this view is not universally accepted!). The purpose of this course is to present those results that are most interesting from the point of view of a physicist, and were historically the most important. We are going to lead the reader through the mathematical part of the theory by a rather short route, but in such a way that the reader does not have to take anything on our word, is able to verify every detail, and, after reading the whole text, will be prepared to solve several problems by him/herself. Further help in this should be provided by the exercises in the text and the literature recommended for further reading.

The introductory part (Chapters 1–7), although assembled by J. Plebański long ago, has never been published in book form.[1] It differs from other courses on relativity in that it introduces differential geometry by a top-down method. We begin with general manifolds,

---

[1] A part of that material had been semi-published as copies of typewritten notes (Plebański, 1964).

on which no structures except tensors are defined, and discuss their basic properties. Then we add the notion of the covariant derivative and affine connection, without introducing the metric yet, and again proceed as far as possible. At that level we define geodesics via parallel displacement and we present the properties of curvature. Only at this point do we introduce the metric tensor and the (pseudo-)Riemannian geometry and specialise the results derived earlier to this case. Then we proceed to the presentation of more detailed topics, such as symmetries, the Bianchi classification and the Petrov classification.

Some of the chapters on classical relativistic topics contain material that, to the best of our knowledge, has never been published in any textbook. In particular, this applies to Chapter 8 (on symmetries) and to Chapter 16 (on cosmology with general geometry). Chapters 18 and 19 (on inhomogeneous cosmologies) are entirely based on original papers. Parts of Chapters 18 and 19 cover the material introduced in A. K.'s monograph on inhomogeneous cosmological models (Krasiński, 1997). However, the presentation here was thoroughly rearranged, extended, and brought up to date. We no longer briefly mention all contributions to the subject; rather, we have placed the emphasis on complete and clear derivations of the most important results. That material has so far existed only in scattered journal papers and has been assembled into a textbook for the first time (A. K.'s monograph (Krasiński, 1997) was only a concise review). Taken together, this collection of knowledge constitutes an important and interesting part of relativistic cosmology whose meaning has, unfortunately, not yet been appreciated properly by the astronomical community.

Most figures for this text, even when they look the same as the corresponding figures in the papers cited, were newly generated by A. K. using the program Gnuplot, sometimes on the basis of numerical calculations programmed in Fortran 90. The only figures taken verbatim from other sources are those that illustrated the joint papers by C. Hellaby and A. K.

J. Plebański kindly agreed to be included as a co-author of this text – having done his part of the job more than 30 years ago. Unfortunately, he was not able to participate actively in the writing up and proofreading. He died while the book was being edited. Therefore, the second author (A. K.) is exclusively responsible for any errors that may be found in this book.

Note for the reader. Some parts of this book may be skipped on first reading, since they are not necessary for understanding the material that follows. They are marked by asterisks. Chapters 18 and 19 are expected to be the highlights of this book. However, they go far beyond standard courses of relativity and may be skipped by those readers who wish to remain on the well-beaten track. Hesitating readers may read on, but can skip the sections marked by asterisks.

Andrzej Krasiński
Warsaw, September 2005

Acknowledgements

We thank Charles Hellaby for comments on the various properties of the Lemaître–Tolman models and for providing copies of his unpublished works on this subject. Some of the figures used in this text were copied from C. Hellaby's files, with his permission. We are grateful to Pankaj S. Joshi for helpful comments on cosmic censorship and singularities, and to Amos Ori for clarifying the matter of shell crossings in charged dust. The correspondence with Amos significantly contributed to clarifying several points in Section 19.3. We are also grateful to George Ellis for his very useful comments on the first draft of this book. We thank Bogdan Mielnik and Maciej Przanowski, who were of great help in the difficult communication between one of the authors residing in Poland and the other in Mexico. M. Przanowski has carefully proofread a large part of this text and caught several errors. So did Krzysztof Bolejko, who was the first reader of this text, even before it was typed into a computer file. J. P. acknowledges the support from the Consejo Nacional de Ciencia y Tecnología projects 32427E and 41993F.

1

How the theory of relativity came into being
(a brief historical sketch)

1.1 Special versus general relativity

The name 'relativity' covers two physical theories. The older one, called special relativity, published in 1905, is a theory of electromagnetic and mechanical phenomena taking place in reference systems that move with large velocities relative to an observer, but are not influenced by gravitation. It is considered to be a closed theory. Its parts had entered the basic courses of classical mechanics, quantum mechanics and electrodynamics. Students of physics study these subjects before they begin to learn general relativity. Therefore, we shall not deal with special relativity here. Familiarity with it is, however, necessary for understanding the general theory. The latter was published in 1915. It describes the properties of time and space, and mechanical and electromagnetic phenomena in the presence of a gravitational field.

1.2 Space and inertia in Newtonian physics

In the Newtonian mechanics and gravitation theory the space was just a background – a room to be filled with matter. It was considered obvious that the space is Euclidean. The masses of matter particles were considered their internal properties independent of any interactions with the remaining matter. However, from time to time it was suggested that not all of the phenomena in the Universe can be explained using such an approach. The best known among those concepts was the so-called Mach's principle. This approach was made known by Ernst Mach in the second half of the nineteenth century, but had been originated by the English philosopher Bishop George Berkeley, in 1710, while Newton was still alive. Mach started with the following observation: in the Newtonian mechanics a seemingly obvious assumption is tacitly made, namely that all the space points can be labelled, for example by assigning Cartesian coordinates to them. One can then observe the motion of matter by finding in which point of space a given particle is located at a given instant. However, this is not actually possible. If we accept another basic assumption of Newton, namely that the space is Euclidean, then its points do not differ from one another in any way. They can be labelled only by matter being present in the space. In truth, we thus can observe only the motion of one portion of matter relative to another portion of matter. Hence, a correctly formulated theory should speak only about relative motion (of matter relative to matter), not about absolute motion (of matter relative to space). If this is so, then the motion of a single particle in a totally empty Universe would not be detectable. Without any other matter we could not establish whether the lone particle is at rest, or is moving or experiencing acceleration. But the reaction of matter to acceleration is the only way to measure its inertia. Hence, that lone particle would have zero inertia. It follows then that inertia is, likewise, not an absolute property of matter, but is relative, and is induced by the remaining matter in the Universe, supposedly via the gravitational interaction.

One can question this principle in several ways. No-one will ever be able to find him/herself in an empty Universe, so any theorems on such an example cannot be verified. It is possible that the inertia of matter is a 'stronger' property than the homogeneity of space, and would still exist in an empty Universe, thus making it possible to measure absolute acceleration. Criticism of Mach's principle is made easier by the fact that it has never been formulated as a precise physical theory. It is just a collection of critical remarks and suggestions, partly based on calculations. It happens sometimes, though, that a new way of looking at an old theory, even if not sufficiently well justified, becomes a starting point for meaningful discoveries. This was the case with Mach's principle that inspired Einstein at the starting point of his work.

1.3 Newton's theory and the orbits of planets

In addition to the above-mentioned theoretical problem, Newton's theory had a serious empirical problem. It was known already in the first half of the nineteenth century that the planets revolve around the Sun in orbits that are not exactly elliptic. The real orbits are rosettes – curves that can be imagined as follows: let a point go around an ellipse, but at the same time let the ellipse rotate slowly around its focus in the same direction (see Fig. 1.1). Newton's theory explained this as follows: an orbit of a planet is an exact ellipse only if we assume that the Sun has just one planet.[1] Since the Sun has several planets, they interact gravitationally and mutually perturb their orbits. When these perturbations are taken into account, the effect is *qualitatively* the same as observed.

However, in 1859, Urbain J. LeVerrier (the same person who, a few years earlier, had predicted the existence of Neptune on the basis of similar calculations) verified whether the calculated and observed motions of Mercury's perihelion agree. It turned out that they do not – and that the discrepancy is much larger than the observational error. The calculated velocity of rotation of the perihelion was smaller than the one observed by $43''$ (arc seconds) per century (the modern result is $43.11 \pm 0.45''$ per century (Will, 1981)). Astronomers and physicists tried to explain this effect in various simple ways, e.g. by assuming that yet another planet, called Vulcan, revolves around the Sun

---

[1] More assumptions were actually made, but the other ones seemed so obvious at that time that they were not even mentioned: that the Sun is exactly spherical, and that the space around the Sun is exactly empty. None of these is strictly correct, but the departures of observations from theory caused by the non-sphericity of the Sun and by the interplanetary matter are insignificant.

Fig. 1.1.   Real planetary orbits, in consequence of various perturbations, are not ellipses, but non-closed curves. The angle of revolution of the perihelion shown in this figure is greatly exaggerated. In reality, the greatest angle of perihelion motion observed in the Solar System, for Mercury, equals approximately 1.5° per 100 years.

inside Mercury's orbit and perturbs it; by allowing for gravitational interaction of Mercury with the interplanetary dust; or by assuming that the Sun is flattened in consequence of its rotation. In the last case, the gravitational field of the Sun would not be spherically symmetric, and a sufficiently large flattening would explain the additional rotation of Mercury's perihelion. All these hypotheses did not pass the observational tests. The hypothetical planet Vulcan would have to be so massive that it would be visible in telescopes, but wasn't. There was not enough interplanetary dust to cause the observed effect. The Sun, if it were sufficiently flattened to explain Mercury's motion, would cause yet another effect: the planes of the planetary orbits would swing periodically around their mean positions with an amplitude of about 43″ per century, and that motion would have been observed, but wasn't (Dicke, 1964).

In spite of these difficulties, nobody doubted the correctness of Newton's theory. The general opinion was that Mach's critique would be answered by formal corrections in the theory, and the anomalous perihelion motion of Mercury would be explained by new observational discoveries. Nobody expected that any other gravitation theory could replace Newton's that had been going from one success to another for over 200 years. General relativity was not created in response to experimental or observational needs. It resulted from speculation, it preceded all but one of the experiments and observations that confirmed it, and it became broadly testable only about 50 years after it had been created, in the 1960s. So much time did technology need to catch up and go beyond the opportunities provided by astronomical phenomena.

1.4 The basic assumptions of general relativity

It is interesting to follow the development of relativistic ideas in the same order as that in which they actually appeared in the literature. However, this was not a straight and smooth road. Einstein made a few mistakes and put forward a few hypotheses that he had to revoke later. He had been constructing the theory gradually, while at the same time learning the Riemannian geometry – the mathematical basis of relativity. If we followed that gradual progress, we would have to take into account not only some blind paths, but also competitors of Einstein, some of whom questioned the need for the (then) new theory, while some others tried to get ahead of Einstein, but without success (Mehra, 1974). Learning relativity in this way would not be efficient, so we will take a shortcut. We shall begin by justifying the need for relativity theory, then we shall present the basic elements of Riemann's geometry, and then we will present Einstein's theory in its final shape. The history of relativity's taking shape is presented in Mehra's book (Mehra, 1974), and its original presentation is to be found in the collection of classic papers (Einstein *et al.*, 1923).

Einstein's starting point was a critique of Newton's theory based on Mach's ideas. Newtonian physics said that, in a space free of any interactions, material bodies would either remain at rest or would move by uniform rectilinear motion. Since, however, the real Universe is permeated by gravitational fields that cannot be shielded, all bodies in the Universe move on curved trajectories in consequence of gravitational interactions.

There is a problem here. When we say that a trajectory is curved, we assume that we can define a straight line. But how can we do this when no actual body follows a straight line? The terrestrial standards of straight lines are useful only because no distances on the Earth are truly great, and at short distances the deformation of 'rigid' bodies due to gravitation is unmeasurably small. Maybe then the trajectory of a light ray would be a good model of a straight line?

To see whether this could be the case, consider two Cartesian reference systems K and K', whose axes (x, y, z) and (x', y', z') are, respectively, parallel. Let K be inertial, and let K' move with respect to K along the z-axis with acceleration $g(t, x, y, z)$. Let the origins of both systems coincide at $t = 0$. Then

$$x' = x, \qquad y' = y, \qquad z' = z - \int_0^t d\tau \int_0^\tau ds\, g(s, x, y, z).$$

Hence, the equations of motion of a free particle, that in K are

$$\frac{d^2x}{dt^2} = \frac{d^2y}{dt^2} = \frac{d^2z}{dt^2} = 0,$$

in K′ assume the form

$$\frac{d^2x'}{dt^2} = \frac{d^2y'}{dt^2} = 0, \qquad \frac{d^2z'}{dt^2} = -g(t, x, y, z).$$

The quantity that we interpreted in K as acceleration would be interpreted in K′ as the intensity of a gravitational field (with opposite sign). The gravitational field can thus be simulated by accelerated motion, or, more exactly, the gravitational force is simulated by the force of inertia. If so, then light in a gravitational field should behave similarly to when it is observed from an accelerated reference system.

How would we see a light ray in such a system? Imagine a space vehicle that flies across a light ray. Let the light ray enter through the window W and fall on a screen on the other side of the vehicle (see Fig. 1.2). If the vehicle were at rest, the light ray entering at W would hit the screen at the point A. Since the vehicle keeps flying, it will move a bit before the ray hits the screen, and the bright spot will appear at the point B. Now assume that the light ray indeed moves in a straight line when observed by an observer who is at rest. Then it is easy to see that the path WB will be straight when the vehicle moves with a constant velocity, whereas it will be curved when the vehicle moves with acceleration. Hence, if the gravitational field behaves analogously to the field of inertial forces, then the light ray should be deflected also by gravitation. Consequently, it cannot be the standard of a straight line.

If we are unable to provide a physical model of a fundamental notion of Newtonian physics, let us try to do without it. Let us assume that no such thing exists as 'gravitational forces' that curve the trajectories of celestial bodies, but that the geometry of space is

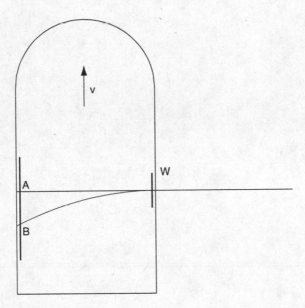

Fig. 1.2. A space vehicle flying across a light ray. See the explanation in the text.

modified by gravitation in such a way that the observed trajectories are paths of free motion. Such a theory might be more complicated than the Newtonian one in practical instances, but it will use only such notions as are related to actual observations, without an unobservable background of the Euclidean space.

A modified geometry means non-Euclidean geometry. A theory created in order to deal with broad classes of non-Euclidean geometries is differential geometry. It is the mathematical basis of general relativity, and we will begin by studying it.

Part I

Elements of differential geometry

2

A short sketch of 2-dimensional differential geometry

2.1 Constructing parallel straight lines in a flat space

The classical Greek geometric constructions, with the help of rulers and compasses, fail over large distances. For example, if we wish to construct a straight line parallel to the momentary velocity of the Earth that passes through a given point on the Moon, compasses and rulers do not help. What method might work in such a situation? For the beginning, let us assume that great distance is our only problem – that we live in a space without gravitation, so we can use a light ray or the trajectory of a stone shot from a sling as a model of a straight line.

Assume that an observer is at the point A (see Fig. 2.1) on the straight line p, and wants to construct a straight line through the point B that would be parallel to p. The

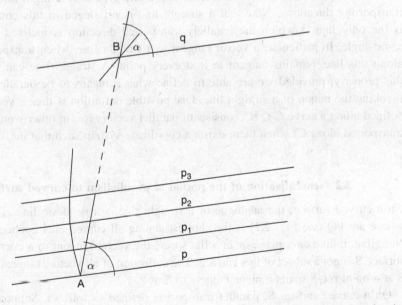

Fig. 2.1. Constructing parallel straight lines at a distance in a flat space. See the explanation in the text.

following programme is 'technically realistic': we first determine the straight line passing through both A and B (for example, by directing a telescope towards B), then we measure the angle α between the lines p and AB, then, from B, we construct a straight line q that is inclined to AB at the same angle α *and lies in the same plane as* p *and* AB. The second condition requires that we can control points of q other than B, and it can pose some problems. However, if our observer is able to construct parallel straight lines that are not too distant from the given one, he/she can carry out the following operation: the observer moves from A to A_1, constructs a straight line $p_1 \parallel p$, then moves on to A_2, constructs a straight line $p_2 \parallel p_1$, etc., until, in the nth step, he/she reaches B and constructs $q = p_n \parallel p_{n-1}$ there.

This construction can be generalised. The observer does not have to move from A to B on a straight line. He/she can start from A in an arbitrary direction and, at a point A_1, construct a straight line parallel to p; it has to lie in the plane pAA_1 and be inclined to AA_1 at the same angle as p. Then, from A_1 the observer can continue in still another arbitrary direction and at a point A_2 repeat the construction: a straight line p_2 has to lie in the plane $p_1A_1A_2$ and be inclined to A_1A_2 at the same angle as p_1 was. When the broken line he/she is following reaches B, the last straight line will be the one we wanted to construct.

We can imagine broken lines whose straight segments are becoming still shorter. In the limit, we conclude that we would be able to carry out this construction along an arbitrary differentiable curve. The plane needed in the construction will be in each step determined by the tangent vector of the curve and the last straight line we had constructed.

In this way, we arrived at the idea of constructing parallel straight lines by parallely transporting directions. Note that a straight line is privileged in this construction: this is the only line to which the parallely transported direction is inclined always at the same angle. In particular, a vector tangent to a straight line, when transported parallely along this line, remains tangent to it at every point. A straight line can be defined by this property, provided we are able to define what it means to be parallel without first invoking the notion of a straight line. One possible definition is this: a vector field $\mathbf{v}(x)$ defined along a curve $C \subset \mathbb{R}^n$ consists of parallel vectors (or, in other words, is parallely transported along C) when there exists a coordinate system such that $\partial v^i / \partial x^j \equiv 0$.

2.2 Generalisation of the notion of parallelism to curved surfaces

On a curved surface, the analogue of a straight line is a geodesic line. This is a curve whose arc PQ (see Fig. 2.2) is the shortest among all curved arcs connecting P and Q. Note that, unlike on a plane or in a flat space, the vector tangent to a curve on a curved surface S is not a subset of this surface. The collection of all vectors tangent to the surface S at a point $p \in S$ spans a plane tangent to S at P.

On a curved surface S, parallel transport is defined as follows. Suppose that we are given the pair of points P and Q, an arc of a curve C connecting P and Q and a vector tangent to S at P that we plan to parallely transport to Q. If C is a geodesic, then we

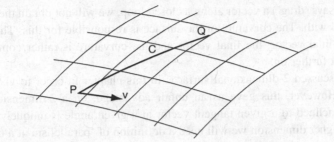

Fig. 2.2.   Parallel transport of vectors on a curved surface. See the explanation in the text.

transport the vector **v** along it in such a way that it is everywhere inclined to the tangent vector of C at the same angle. If C is not a geodesic, then we proceed as follows:

1. We divide the arc PQ into n segments.
2. We connect the ends of each arc by a geodesic.
3. We transport **v** parallely along each geodesic arc.
4. We calculate the result of this operation as $n \to \infty$.

It is easy to note that the parallel transport thus defined depends on the curve along which the transport was carried out. For example, consider a sphere, its pole C and two points A and B lying on the equator, 90° away from each other (Fig. 2.3). Let **v** be the vector tangent to the equator at A. Transport **v** parallely to C along the arc AC, and then again along the arcs AB and BC. All three arcs are parts of great circles, which are geodesics, so **v** makes always the same angle with the tangent vectors of the arcs. The first transport will yield a vector at C that is tangent to BC, while the second one will yield a vector at C perpendicular to BC. In consequence, if we transport (in differential

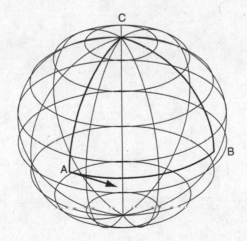

Fig. 2.3.   Parallel transport of vectors on a sphere. See the explanation in the text.

geometry one says 'drag') a vector along a closed loop, we will not obtain the same vector that we started with. The curvature of the surface is responsible for this. The connection between the initial vector, the final vector and the curvature is rather complicated; we will come to it further on.

We have discussed 2-dimensional surfaces in this chapter in order to visualise things more easily. However, this gave us an unfair advantage: on a 2-dimensional surface, the direction inclined to a given tangent vector at a given angle is uniquely determined. In spaces of higher dimension we will need a definition of 'parallelism at a distance' that will be analogous to $\partial v^i / \partial x^j = 0$ that we used in a flat space.

3

Tensors, tensor densities

3.1 What are tensors good for?

In Newtonian physics, a preferred class of reference systems is used. They are the inertial systems – those in which the three Newtonian principles of dynamics hold true. However, it may be difficult in practice to identify the inertial systems. As we have seen in Chapter 1, the inertial force imitates the gravitational force, so it may not be easy to make sure whether a given object moves with acceleration or remains at rest in a gravitational field. Hence, the laws of physics should be formulated in such a way that no reference system is privileged. The choice of a reference system, even when it is evidently convenient (e.g. the centre of mass system), is an act of human will, while the laws of physics should not depend on our decisions.

Tensors are objects defined so that no reference system is privileged. For the beginning, we will settle for a vague definition that we will make precise later. Suppose we change the coordinate system in an n-dimensional space from $\{x^{\alpha}\}$, $\alpha = 1, 2, \ldots, n$ to $\{x^{\alpha'}\}$, $\alpha' = 1, 2, \ldots, n$. A tensor is a collection of functions on that space that changes in a specific way under such a coordinate transformation. The appropriate class of spaces and the 'specific way' in which the functions change will be defined in subsequent sections.

3.2 Differentiable manifolds

As already stated, in relativity we will be using non-Euclidean spaces. The most general class of spaces that we will consider are **differentiable manifolds**. This is a generalisation of the notion of a curved surface for which a tangent plane exists at every point of it. An n-dimensional differentiable manifold of class p is a space M_n in which every point x has a neighbourhood $\mathcal{O}_x$ such that the following conditions hold:

1. There exists a one-to-one mapping κ_x of the neighbourhood $\mathcal{O}_x$ onto a subset of $\mathbb{R}^n$, called a **map** of $\mathcal{O}_x$. The coordinates of the image $\kappa_x(x)$ are called the coordinates of $x \subset M_n$.
2. If the neighbourhoods $\mathcal{O}_x$ and $\mathcal{O}_y$ of $x, y \in M_n$ have a non-empty intersection ($\mathcal{O}_x \cap \mathcal{O}_y \neq \emptyset$), κ_x is a map of $\mathcal{O}_x$ and κ_y is a map of $\mathcal{O}_y$, then the mappings $(\kappa_y \circ \kappa_x^{-1})$ and $(\kappa_y \circ \kappa_x^{-1})^{-1}$ are mappings of class p of $\mathbb{R}^n$ into itself.

13

A **tangent space** to the manifold M_n at the point x is a vector space spanned by vectors tangent at x to curves in M_n that pass through x.

If $\kappa_x(x) = \{x^1, \ldots, x^n\}$ are the coordinates of the point x, then the equation $x^i =$ constant, where $i \in \{1, \ldots, n\}$ is a fixed index, defines a hypersurface in $\mathbb{R}^n$ and thus also a hypersurface H in M_n. A coordinate system is thus a set of n one-parameter families of hypersurfaces, x^i being the parameter in the ith family.

Now let the manifold be $\mathbb{R}^n$. Each hypersurface H defines then a family of vectors: if $\Phi(x) = C$ (where C is an arbitrary constant) is the equation of H, then $\partial\Phi/\partial x^\alpha$, where $\{x^\alpha\}$ are the coordinates of the point x, is an equation of a family of vectors attached to H and orthogonal to it. Taking all the hypersurfaces of the $\Phi = C$ family, we obtain a family of curves tangent to the vectors $\partial\Phi/\partial x^\alpha$. Thus, each coordinate system in $\mathbb{R}^n$ defines a family of curves.

The converse is not true: an n-parameter family of curves C_x in $\mathbb{R}^n$ defines a family of hypersurfaces orthogonal to C_x only when the vectors tangent to C_x have zero rotation (to be defined later).

The reason why, for this example, we had to take the special case of $M_n = \mathbb{R}^n$ is that, as we shall see later, in a general vector space vectors like the gradient of a function (called *covariant vectors*) and vectors like a tangent vector to a curve (called *contravariant vectors*) are unrelated objects of different kinds. A relation between them exists only in spaces tangent to such manifolds in which a *metric* is defined, see Chapter 7. $\mathbb{R}^n$ is one of them. Without a metric, a covariant vector cannot be converted into a contravariant one, and a curve tangent to a field of covariant vectors cannot be constructed.

Let $U \subset M_n$ be an open subset. Suppose we are given a collection of n families of curves, each of $(n-1)$ parameters such that n curves pass through each point $x \in U$. Suppose that the tangent vectors to these curves are linearly independent at every $x \in U$. Then the tangent vectors to these curves at the point x, $e_a(x)|_{a=1,\ldots,n}$ are a basis of the space tangent to M_n at x. Let $v(x)$ be an arbitrary vector tangent to M_n at x. Then

$$v = \sum_{a=1}^n v^a(x) e_a(x).$$

The coefficients $\{v^a\}$ are called the components of the vector v in the basis $e_a(x)$. The mapping $x \to v(x)$ assigns to each point $x \in U \subset M_n$ a vector tangent to M_n at x and is called a **vector field** on U.

Note that the vectors of a vector field are defined on tangent spaces to the manifold, while the components of vector fields are functions on the manifold. In particular, the vectors $e_a(x)$ can be identified with directional derivatives, and can be defined by a coordinate system as $e_a(x)\Phi(x) = (\partial/\partial x^a)\Phi(x)$ (that is, the ath vector in the basis is the directional derivative down the ath family of hypersurfaces in the coordinate system.) Then the v^a are components of the vector **v** in the coordinate system $\{x\}$. Again, they are functions on the manifold.

We adopt three conventions.

1. If any index in a formula appears twice, a sum over all of its values is implied. Hence, if α changes from 1 to n, while $\{V_\alpha\}$ and $\{U^\alpha\}$, $\alpha = 1,\ldots, n$ are collections of functions labelled by α, then

$$U^\alpha V_\alpha \equiv \sum_{\alpha=1}^n U^\alpha V_\alpha.$$

2. The collection of all coordinates $\{x^\alpha\}$, $\alpha = 1,\ldots, n$ will be denoted $\{x\}$, and a function $f(\{x\})$ will be denoted $f(x)$.
3. The derivative with respect to x^α will be denoted by a comma followed by a subscript: thus

$$f,_\alpha \equiv \frac{\partial f}{\partial x^\alpha}.$$

3.3 Scalars

The simplest tensor is a scalar. It is a function on a manifold whose value, when the coordinate system is transformed, changes simply by substituting the transformation in its argument:

$$\varphi'(x'(x)) = \varphi(x). \tag{3.1}$$

Examples of scalars are physical constants, rest masses of elementary particles, their electric charges, a density distribution in a continuous medium. The coordinates of a point in a manifold are scalars, too, since their values transform by the simple law (3.1) when coordinates are changed from $\{x'\}$ to $\{x''\}$: $x^\alpha(x''(x')) = x^\alpha(x')$.

3.4 Contravariant vectors

The functions $v^\alpha(x)$, $\alpha = 1,\ldots, n$ are said to be coordinates of a contravariant vector field when, by a change of coordinates on M_n, they transform by the law

$$v^{\alpha'}(x'(x)) = x^{\alpha'},_\alpha v^\alpha(x). \tag{3.2}$$

Examples are vectors tangent to curves. Suppose that a curve C is given by the parametric equations $t \to x^\alpha(t)$, $\alpha = 1,\ldots, n$, $t \in [a, b] \subset \mathbb{R}^1$, where x^α is the value of a coordinate of a point on the curve. An arbitrary field of vectors tangent to C is then given by

$$t \to v^\alpha(t) = f(x)g(t)\frac{\mathrm{d}x^\alpha}{\mathrm{d}t},$$

where $f(x)$ is an arbitrary function of the coordinates and $g(t)$ is an arbitrary function of the parameter. When the coordinate system is changed from (x) to (x'), we have

$$v^{\alpha'}(t) = f(x'(x))g(t)\frac{\mathrm{d}x^{\alpha'}}{\mathrm{d}t} = f(x)g(t)x^{\alpha'},_\alpha\frac{\mathrm{d}x^\alpha}{\mathrm{d}t},$$

in agreement with (3.2).

3.5 Covariant vectors

The functions $u_\alpha(x)$, $\alpha = 1, \ldots, n$ are said to be coordinates of a covariant vector field when, by a change of coordinates on M_n, they transform by the law

$$u_{\alpha'}(x'(x)) = x^\alpha{}_{,\alpha'} u_\alpha(x). \tag{3.3}$$

By convention, the indices of contravariant vectors are placed as superscripts, whereas those of covariant vectors are placed as subscripts.

An example of a covariant vector is the gradient of a scalar function $\varphi(x)$. For such a function we have

$$\varphi_{,\alpha'}(x'(x)) = x^\alpha{}_{,\alpha'} \varphi_{,\alpha}(x),$$

in agreement with (3.3).

Note that the quantity $v^\alpha u_\alpha$ is a scalar, since we have

$$
\begin{aligned}
v^{\alpha'} u_{\alpha'} &= x^{\alpha'}{}_{,\alpha} v^\alpha x^\beta{}_{,\alpha'} u_\beta = \left(x^\beta{}_{,\alpha'} x^{\alpha'}{}_{,\alpha} \right) v^\alpha u_\beta \\
&= x^\beta{}_{,\alpha} v^\alpha u_\beta = \delta^\beta{}_\alpha v^\alpha u_\beta = v^\alpha u_\alpha.
\end{aligned}
$$

Another example of a scalar field is the directional derivative of a scalar field along a contravariant vector field, $v(\varphi) \overset{\text{def}}{=} v^\alpha \varphi_{,\alpha}$.

3.6 Tensors of second rank

Scalars are sometimes called tensors of rank zero, to emphasise that they have no indices. The contravariant and covariant vectors are collectively called tensors of rank 1. The tensors of rank 2 are objects whose components are labelled by two indices. There are three kinds of them:

1. Doubly contravariant tensors. Their components $T^{\alpha\beta}(x)$ transform under a coordinate transformation $x \to x'$ on M_n as follows:

$$T^{\alpha'\beta'}(x'(x)) = x^{\alpha'}{}_{,\alpha} x^{\beta'}{}_{,\beta} T^{\alpha\beta}(x). \tag{3.4}$$

2. Doubly covariant tensors. These are quantities whose components $T_{\alpha\beta}(x)$ transform by the rule

$$T_{\alpha'\beta'}(x'(x)) = x^\alpha{}_{,\alpha'} x^\beta{}_{,\beta'} T_{\alpha\beta}(x). \tag{3.5}$$

3. Mixed tensors. Their components transform by the rule

$$T_\alpha{}^{\beta'}(x'(x)) = x^\alpha{}_{,\alpha'} x^{\beta'}{}_{,\beta} T_\alpha{}^\beta(x). \tag{3.6}$$

The collection of components of a second-rank tensor is a square matrix that transforms in a prescribed way when coordinates are changed.

An example of a doubly covariant tensor is a matrix of a quadratic form,

$$\Phi(A) = \Phi_{\alpha\beta} A^\alpha A^\beta,$$

where A^α are components of a contravariant vector, and the value of $\Phi(A)$ is a scalar.

An example of a mixed tensor of rank 2 is a matrix of a mapping of one vector space into another,

$$V^\alpha = B^\alpha{}_\beta W^\beta,$$

where V^α and W^β are contravariant vectors in different vector spaces.

We will meet examples of doubly contravariant vectors later in this book. The simplest example is the inverse matrix to a matrix of a quadratic form, but in order to be able to prove this we have to learn about some other objects.

The quantity T^α_α for a mixed tensor (the sum of its diagonal components) is called the **trace** of T and is a scalar. Quantities like $\sum_\alpha T^{\alpha\alpha}$ for a contravariant second-rank tensor and $\sum_\alpha T_{\alpha\alpha}$ for a covariant second-rank tensor are not tensorial objects. Summations over indices standing on the same level occur only exceptionally in differential geometry – for example, when a calculation is done in a chosen coordinate system.

3.7 Tensor densities

A tensor density differs from the corresponding tensor in that, when transformed from one coordinate system to another, it gets multiplied by a certain power of the Jacobian of the transformation. The exponent of the Jacobian is called the **weight** of the density. For example, a scalar density of weight w transforms as follows:

$$\Phi'(x') = \left[\frac{\partial(x')}{\partial(x)}\right]^w \Phi(x), \tag{3.7}$$

a contravariant vector density of weight w transforms by the rule

$$v^{\alpha'}(x') = \left[\frac{\partial(x')}{\partial(x)}\right]^w x^{\alpha'}{}_{,\alpha} v^\alpha(x), \tag{3.8}$$

and so on.

An example of a scalar density is the element of volume in a multidimensional integral. It transforms by the law

$$d_n x = \frac{\partial(x)}{\partial(x')} d_n x', \tag{3.9}$$

so it is a scalar density of weight $+1$.

An arbitrary tensor is by definition a tensor density of weight zero.

3.8 Tensor densities of arbitrary rank

The components of a tensor density of weight w, k times contravariant and l times covariant, transform by the law

$$T^{\alpha'_1 \alpha'_2 \ldots \alpha'_k}_{\beta'_1 \beta'_2 \ldots \beta'_l}(x'(x)) = \left[\frac{\partial(x')}{\partial(x)} \right]^w x^{\alpha'_1}{}_{,\alpha_1} x^{\alpha'_2}{}_{,\alpha_2} \ldots x^{\alpha'_k}{}_{,\alpha_k}$$

$$\times x^{\beta_1}{}_{,\beta'_1} x^{\beta_2}{}_{,\beta'_2} \ldots x^{\beta_l}{}_{,\beta'_l} T^{\alpha_1 \alpha_2 \ldots \alpha_k}_{\beta_1 \beta_2 \ldots \beta_l}(x). \qquad (3.10)$$

Such an object is called a tensor density of type $[w, k, l]$.

For general tensor densities one can carry out an operation that is analogous to finding the trace of a mixed tensor of rank 2. This operation is called **contraction**. It consists in making an upper index equal to a lower index, and summing over all its allowed values. The resulting density is of type $[w, k-1, l-1]$, thus

$$[\text{contracted } T]^{\alpha_1 \ldots \alpha_{i-1} \alpha_{i+1} \ldots \alpha_k}_{\beta_1 \ldots \beta_{j-1} \beta_{j+1} \ldots \beta_l}(x) = T^{\alpha_1 \ldots \alpha_{i-1} \rho \alpha_{i+1} \ldots \alpha_k}_{\beta_1 \ldots \beta_{j-1} \rho \beta_{j+1} \ldots \beta_l}(x) \qquad (3.11)$$

(note that a sum over all values of ρ is implied above). The indices over which the summing is carried out are called 'dummy indices' since they do not show up in the transformation law of the contracted density.

The contraction may be done over several pairs of indices at the same time. Then, one must take care to give different names to each pair of dummy indices, to avoid confusion.

3.9 Algebraic properties of tensor densities

Here is a list of the most basic properties of tensor densities.

1. If $T^{\cdots}_{\cdots} \equiv 0$ in one coordinate system, then $T^{\cdots}_{\cdots} \equiv 0$ in all coordinate systems (this follows easily from the transformation law).
2. A linear combination of two tensor densities of type $[w, k, l]$ is a tensor density of the same type. (Adding tensor densities of different types makes no sense.)
3. The collection of quantities obtained when each component of one tensor density (of type $[w, k, l]$) is multiplied by each component of another tensor density (of type $[w', k', l']$) is called a **tensor product** of the two densities, and is a tensor density of type $[w+w', \ k+k', \ l+l']$. For example, out of u_α and v^α one can form such tensor products as $v^\alpha v^\beta$, $v^\alpha u_\beta$, $u_\alpha u_\beta$, $v^\alpha v^\beta v^\gamma$, $v^\alpha u_\beta v^\gamma$, $v^\alpha u_\alpha v^\beta u_\gamma v^\delta$. The tensor product is denoted by $\otimes$, thus for example $v^\alpha u_\beta = (v \otimes u)^\alpha{}_\beta$.
4. If a tensor density does not change its value when two indices (either both upper or both lower) are interchanged, then it is called **symmetric** with respect to this pair of indices. If it only changes sign, then it is called **antisymmetric** in this pair of indices. The property of being symmetric or antisymmetric with respect to a given pair of indices is preserved under transformations of coordinates.

5. This last property allows us to define the **symmetrisation** and the **antisymmetrisation**. The **symmetric part** of a tensor density with respect to the indices $\{\alpha_1, \ldots, \alpha_k\}$ is the quantity

$$T_{(\alpha_1 \ldots \alpha_k)} \stackrel{\text{def}}{=} \frac{1}{k!} \sum_{\substack{\text{over all permutations} \\ i_1, \ldots, i_k}} T_{\alpha_{i_1} \ldots \alpha_{i_k}}. \tag{3.12}$$

The **antisymmetric part** of a tensor density with respect to the indices $\{\alpha_1, \ldots, \alpha_k\}$ is the quantity

$$T_{[\alpha_1 \ldots \alpha_k]} \stackrel{\text{def}}{=} \frac{1}{k!} \sum_{\substack{\text{over all permutations} \\ i_1, \ldots, i_k}} (\text{sign of the permutation}) T_{\alpha_{i_1} \ldots \alpha_{i_k}}. \tag{3.13}$$

One can carry out the symmetrisation or antisymmetrisation (i.e., respectively, *symmetrise* or *antisymmetrise*) also with respect to the upper indices, and, in each case, over only a subset of all the indices that a tensor density has. For example, a tensor of rank 4 $T_{\alpha\beta\gamma\delta}$ can be symmetrised over only two of its indices

$$T_{(\alpha\beta)\gamma\delta} = \frac{1}{2} \left(T_{\alpha\beta\gamma\delta} + T_{\beta\alpha\gamma\delta} \right),$$

or over three indices

$$T_{(\alpha\beta\gamma)\delta} = \frac{1}{6} \left(T_{\alpha\beta\gamma\delta} + T_{\alpha\gamma\beta\delta} + T_{\beta\alpha\gamma\delta} + T_{\beta\gamma\alpha\delta} + T_{\gamma\alpha\beta\delta} + T_{\gamma\beta\alpha\delta} \right).$$

If we want to emphasise that a certain index is excluded from symmetrisation or antisymmetrisation, then we put it between vertical strokes. For example, the antisymmetrisation of $T_{\alpha\beta\gamma\delta}$ with respect to α and γ only would be denoted $T_{[\alpha|\beta|\gamma]\delta}$.

The symmetrisation and the antisymmetrisation with respect to just two indices are complementary operations, as can easily be verified, thus

$$T_{(\alpha\beta)\gamma\delta\ldots} + T_{[\alpha\beta]\gamma\delta\ldots} \equiv T_{\alpha\beta\gamma\delta\ldots}.$$

However, this is no longer true for symmetrisations/antisymmetrisations over larger numbers of indices. The other parts of $T_{\alpha\beta\gamma\delta\ldots}$ that enter neither $T_{(\alpha\beta\gamma)\delta\ldots}$ nor $T_{[\alpha\beta\gamma]\delta\ldots}$ are obtained when different rules of assigning signs to the different terms in (3.12) and (3.13) are chosen. We will not encounter those operations in this book.

If the antisymmetrisation is done with respect to a larger number of indices than the dimension of the manifold, then the result is a tensor density that identically equals zero. This is because, with n possible different values of each index, there must be at least one value that is repeated in a set of $m > n$ indices. Then, each term in (3.13) has its counterpart that is identical, but enters the sum with an opposite sign.

3.10 Mappings between manifolds

The following notation will be used.

M_n and P_m are two arbitrary differentiable manifolds, $\dim M_n = n$, $\dim P_m = m$; $\{x^\alpha\}, \alpha = 1, \ldots, n$ and $\{y^a\}, a = 1, \ldots, m$ are the coordinate systems on M_n and P_m, respectively. (Recall that several coordinate neighbourhoods may be needed to cover each manifold, but here we consider a single coordinate patch on each one.)

$F: M_n \to P_m$ is an arbitrary mapping of class C^1, represented by the set of functions $y^a = F^a(\{x\})$.

Now consider a function $f: P_m \to \mathbb{R}^1$ defined on P_m. For those points of P_m that are images of some points of M_n, the mapping F and the function f automatically define a function acting on M_n. Let $P_m \ni q = F(p)$, where $p \in M_n$, and let $f(q) = r \in \mathbb{R}^1$. Then $(f \circ F)(p) = r$, and so $f \circ F: M_n \to \mathbb{R}^1$. We can thus say that the mapping F that takes points of M_n to points of P_m defines an associated mapping of functions on P_m to functions on M_n. This associated mapping will be denoted F_0^*. (The zero denotes that this is a mapping of functions, i.e. tensors with zero indices, and the asterisk placed as a superscript denotes that the functions are sent in the opposite direction to points of the manifold M_n. An associated mapping that sends objects backwards with respect to the main mapping is called a **pullback**.) The function associated to f will thus be denoted $F_0^* f$, so

$$F_0^* f: M_n \to \mathbb{R}^1, \qquad (F_0^* f)(p) \stackrel{\text{def}}{=} f(F(p)). \qquad (3.14)$$

Now consider a contravariant continuous vector field v^α on M_n. This field defines a family of curves tangent to the vectors of this field by

$$v^\alpha = \frac{dx^\alpha}{d\tau}. \qquad (3.15)$$

Suppose that an arc of the curve $x^\alpha(\tau)$ is in the domain of the mapping F. Then points of this curve have their images in P_m, $F(x^\alpha(\tau))$, or, in terms of coordinates, $y^a(\tau) = F^a(x^\alpha(\tau))$. Hence, the arc of the image-curve in P_m is automatically parametrised by the same parameter τ. Since the functions $x^\alpha(\tau)$ are of class C^1 (as integrals of continuous functions), and F is of the same class by assumption, we can differentiate the functions $y^a(\tau)$ by τ. The derivatives are components of the field of vectors w^a tangent to the curve $y^a(\tau)$. The mapping F thus defines an associated mapping of vector fields on M_n to vector fields on P_m:

$$w^a(F(x)) = \frac{d}{d\tau}(F^a(x)) = \frac{\partial F^a}{\partial x^\alpha}\frac{dx^\alpha}{d\tau} = \frac{\partial F^a}{\partial x^\alpha} v^\alpha. \qquad (3.16)$$

If $\dim P_m < \dim M_n$, then some curves in M_n will be mapped onto single points of P_m, and then, according to (3.16), the image of the vector field v^α tangent to such a curve will be a zero vector on P_m (because the image of such a curve $y^a(\tau)$, will in fact be independent of τ).

Note the similarity of (3.16) to the transformation law of contravariant vectors; we will come back to it at the end of this section.

We have thus found that the mapping of manifolds $F: M_n \to P_m$ defines an associated mapping of vector fields on M_n to vector fields on P_m. We will denote this associated mapping F_{1*}. (This time it is a mapping of tensors with one index, and it takes objects in the same direction as the main mapping. Such a mapping is sometimes called a **pushforward**.) If v is a vector field on M_n, then $F_{1*}(v)$ is a vector field on P_m. The mapping F_{1*} is a linear mapping between vector spaces, determined via (3.16) by the matrix of derivatives $\| F^a,_\alpha \|$.

Vector fields v^α on M_n and w^a on P_m can be uniquely represented by directional derivatives of functions. Let $g : M_n \to \mathbb{R}^1$ and $f : P_m \to \mathbb{R}^1$, then

$$v^\alpha(x) \longleftrightarrow v(g) \stackrel{\text{def}}{=} v^\alpha \frac{\partial g}{\partial x^\alpha},$$

$$w^a(y) \longleftrightarrow w(f) \stackrel{\text{def}}{=} w^a \frac{\partial f}{\partial y^a}.$$

$$(3.17)$$

Then the mapping F_{1*} can be defined in another, equivalent way,

$$v(F_0^* f) = (F_{1*} v)(f),$$

$$(3.18)$$

where v is a vector field defined on M_n.

A field of covariant vectors ω_a on P_m can be understood as a linear form ω that maps vector fields on P_m to $\mathbb{R}^1$:

$$\omega(w) \stackrel{\text{def}}{=} \omega_a w^a.$$

$$(3.19)$$

If a field of forms ω is defined on P_m, then, for those points of P_m that are images of points of M_n, we can define a field of forms on M_n by

$$(F_1^* \omega)(v) = \omega(F_{1*} v),$$

$$(3.20)$$

where v is a vector field on M_n. (This is an associated mapping of tensors with one index that is again a pullback.) In terms of coordinates, this can be written as follows:

$$(F_1^* \omega)_\alpha v^\alpha = \omega_a F^a{}_{,\alpha} v^\alpha,$$

$$(3.21)$$

or equivalently

$$(F_1^* \omega)_\alpha = \omega_a F^a{}_{,\alpha}.$$

$$(3.22)$$

Note the similarity of (3.22) to the transformation law of covariant vectors.

To sum up: a mapping $F : M_n \to P_m$ between manifolds defines the mapping F_{1*} of vector fields on M_n to vector fields on P_m and the mappings F_0^* and F_1^* of, respectively, functions and forms on P_m to functions and forms on M_n. These definitions can be extended further, in a rather obvious way, to mappings of contravariant tensor fields of arbitrary rank from M_n to P_m and of covariant tensor fields from P_m to M_n. Note that tensor densities can be defined only for nonsingular transformations, while F can be singular. Therefore, no general definition of an associated mapping of tensor densities can be given.

If $n = m$ and F is a diffeomorphism of class C^1, then F^{-1} exists and is also of class C^1. In this case, tensors of arbitrary rank, including mixed tensors, can be transported in both directions between M_n and P_m. If $F : M_n \to P_m$ is not reversible, then mixed tensors cannot be transported in any direction.

Now observe the following. Coordinate transformations are in fact nothing else than mappings of $\mathbb{R}^n$ into itself – because coordinate patches are subsets of $\mathbb{R}^n$ by definition. Hence, if such a mapping occurs within the image of a single map, a mapping of the manifold into itself is associated with it. Consequently, coordinate transformations can be interpreted as mappings of the underlying manifold into itself.

Let us now specialise the considerations of this section to the case when $n = m$, and M_n and P_m are subsets of the same manifold, $M_n \subset Q_n$ and $P_m \subset Q_n$. Then it follows from (3.14), (3.16), (3.22), and from their extensions to tensor fields of higher ranks, that the transformation laws of tensors are consistent with the interpretation of a coordinate transformation as a mapping of the manifold into itself. We shall make use of this analogy in Chapter 8.

3.11 The Levi-Civita symbol

Let $\epsilon_{\alpha_1 \ldots \alpha_n}$ be a tensor density (whose weight is as yet unknown) defined on a differentiable manifold M_n, which is antisymmetric with respect to any pair of its indices. Because of the antisymmetry, each component that has at least a pair of identical indices will be equal to zero. Only those components can be different from zero for which the set $\{\alpha_1, \ldots, \alpha_n\}$ is a permutation of the set $\{1, \ldots, n\}$. For even permutations $\epsilon_{\alpha_1 \ldots \alpha_n} = \epsilon_{1 \ldots n}$; for odd permutations $\epsilon_{\alpha_1 \ldots \alpha_n} = -\epsilon_{1 \ldots n}$. Hence, defining the component $\epsilon_{1 \ldots n}$ suffices to determine the whole tensor density $\epsilon_{\alpha_1 \ldots \alpha_n}$. We thus define

$$\epsilon_{1 \ldots n} = +1. \tag{3.23}$$

The quantity $\epsilon_{\alpha_1 \ldots \alpha_n}$ is called the **Levi-Civita symbol**.

Let $A^\alpha{}_\beta$ be an arbitrary matrix. Let us investigate the quantity

$$[D(A)]_{\beta_1 \ldots \beta_n} \stackrel{\text{def}}{=} \epsilon_{\alpha_1 \ldots \alpha_n} A^{\alpha_1}{}_{\beta_1} \ldots A^{\alpha_n}{}_{\beta_n}. \tag{3.24}$$

Let α_i label the columns and β_j the rows of the matrix. The expression above is a sum of n-tuple products of elements of the matrix $A^\alpha{}_\beta$. In each product, each factor comes from a different column (because, if two indices α_i and α_j are equal, then $\epsilon_{\alpha_1 \ldots \alpha_n} = 0$). The ith factor in each product is always from the same (ith) row, but each time from a different column, and, in the whole sum, runs over all columns. For even permutations $\alpha_1 \ldots \alpha_n$, the product enters the sum with a plus sign, for odd permutations – with a minus sign. The quantity (3.24) is also antisymmetric in all the β_i indices (see Exercise 1). Hence, $[D(A)]_{\beta_1 \ldots \beta_n}$ vanishes if any two indices β_i, β_j are equal (i.e. if the same row of the matrix A appears in the positions (i, j)).

Thus $[D(A)]_{\beta_1 \ldots \beta_n}$ has all the properties of the determinant of A. It will equal $+ \det(A)$ when the permutation $\{1, \ldots, n\} \to \{\beta_1, \ldots, \beta_n\}$ is even, and $- \det(A)$ when the permutation is odd. Hence

$$\epsilon_{\alpha_1 \ldots \alpha_n} A^{\alpha_1}{}_{\beta_1} \ldots A^{\alpha_n}{}_{\beta_n} = \epsilon_{\beta_1 \ldots \beta_n} \det(A). \tag{3.25}$$

Now let us apply this formula to the matrix of derivatives (the Jacobi matrix) $x^\alpha{}_{,\alpha'}$ of the coordinate transformation $\{x\} \to \{x'\}$. We have

$$\epsilon_{\alpha_1' \ldots \alpha_n'} = \frac{\partial(x')}{\partial(x)} x^{\alpha_1}{}_{,\alpha_1'} \ldots x^{\alpha_n}{}_{,\alpha_n'} \epsilon_{\alpha_1 \ldots \alpha_n}. \tag{3.26}$$

This shows that $\epsilon_{\alpha_1 \ldots \alpha_n}$ is a tensor density of type $[1, 0, n]$.

By a similar method one can verify that $\epsilon^{\alpha_1 \ldots \alpha_n}$ is a tensor density of type $[-1, n, 0]$.

3.12 Multidimensional Kronecker deltas

Let us recall the definition of the ordinary Kronecker delta symbol:

$$\delta^{\alpha}{}_{\beta} = \begin{cases} 1 & \text{when } \alpha = \beta \\ 0 & \text{when } \alpha \neq \beta \end{cases} \tag{3.27}$$

(this is the unit matrix). This is a tensor, since, on the one hand

$$\delta'^{\alpha}{}_{\beta} = \delta^{\alpha}{}_{\beta} \tag{3.28}$$

by definition, and on the other hand

$$\delta'^{\alpha'}{}_{\beta'} = x^{\alpha'}{}_{,\beta'} = x^{\alpha'}{}_{,\rho} \, x^{\rho}{}_{,\beta'} = x^{\alpha'}{}_{,\rho} \, x^{\sigma}{}_{,\beta'} \, \delta^{\rho}{}_{\sigma}, \tag{3.29}$$

which is the transformation law of a mixed tensor of rank 2.

A multidimensional Kronecker delta is defined as follows:

$$\delta^{\alpha_1 \dots \alpha_k}_{\beta_1 \dots \beta_k} = \begin{vmatrix} \delta^{\alpha_1}{}_{\beta_1} & \cdots\cdots & \delta^{\alpha_1}{}_{\beta_k} \\ \vdots & & \vdots \\ \vdots & & \vdots \\ \delta^{\alpha_k}{}_{\beta_1} & \cdots\cdots & \delta^{\alpha_k}{}_{\beta_k} \end{vmatrix}. \tag{3.30}$$

From the definition we have at once

$$\delta^{\alpha_1 \dots \alpha_k}_{\beta_1 \dots \beta_k} = \delta^{[\alpha_1 \dots \alpha_k]}_{\beta_1 \dots \beta_k} = \delta^{\alpha_1 \dots \alpha_k}_{[\beta_1 \dots \beta_k]}, \tag{3.31}$$

and it follows that k must not be greater than the dimension of the manifold, n, or else $\delta^{\alpha_1 \dots \alpha_k}_{\beta_1 \dots \beta_k} \equiv 0$. Other than this, k is unrelated to n.

Now let us consider the special case $k = n$. Since an object that is antisymmetric in all n indices has just one independent component, it must be proportional to the Levi-Civita symbol with the same indices. For the upper indices of the multidimensional delta we thus have $\delta^{\alpha_1 \dots \alpha_n}_{\beta_1 \dots \beta_n} = T_{\beta_1 \dots \beta_n} \epsilon^{\alpha_1 \dots \alpha_n}$. But $T_{\beta_1 \dots \beta_n}$ is antisymmetric in $\{\beta_1, \dots, \beta_n\}$, so it must be proportional to $\epsilon_{\beta_1 \dots \beta_n}$, thus $\delta^{\alpha_1 \dots \alpha_n}_{\beta_1 \dots \beta_n} = \lambda \epsilon_{\beta_1 \dots \beta_n} \epsilon^{\alpha_1 \dots \alpha_n}$. To calculate λ it now suffices to substitute in the last formula any sets of indices for which both sides are nonzero. We substitute $\{\alpha_1, \dots, \alpha_n\} = \{\beta_1, \dots, \beta_n\} = \{1, \dots, n\}$, and we see that $\lambda = 1$. Hence, finally

$$\delta^{\alpha_1 \dots \alpha_n}_{\beta_1 \dots \beta_n} = \epsilon_{\beta_1 \dots \beta_n} \epsilon^{\alpha_1 \dots \alpha_n}. \tag{3.32}$$

This, in consequence of the properties of the Levi-Civita symbol, is a tensor (i.e. has the weight zero).

From (3.30) it follows that $\delta^{\alpha_1 \dots \alpha_k}_{\beta_1 \dots \beta_k} \neq 0$ only when $\{\alpha_1, \dots, \alpha_k\}$ are all different and are a permutation of $\{\beta_1 \dots \beta_k\}$. If any pair of upper indices has equal values, or if any pair of lower indices has equal values, then the determinant in (3.30) has two identical rows or two identical columns and is zero. If any index β_i of the set $[\beta_1 \dots \beta_k]$ is different from all the αs in $\{\alpha_1, \dots, \alpha_k\}$, the determinant in (3.30) has only zeros in the ith column and is zero again. If $\delta^{\alpha_1 \dots \alpha_k}_{\beta_1 \dots \beta_k} \neq 0$, then $\delta^{\alpha_1 \dots \alpha_k}_{\beta_1 \dots \beta_k} = +1$ when the lower indices are an even permutation of the upper ones, and $\delta^{\alpha_1 \dots \alpha_k}_{\beta_1 \dots \beta_k} = -1$ when they are an odd permutation.

Using these properties one can verify that

$$\delta^{\alpha_1...\alpha_k\rho}_{\beta_1...\beta_k\rho} = (n-k)\delta^{\alpha_1...\alpha_k}_{\beta_1...\beta_k}. \tag{3.33}$$

This is seen as follows. If $\{\beta_1, ..., \beta_k\}$ are not a permutation of $\{\alpha_1, ..., \alpha_k\}$, or if any index in either of the two sets is repeated, then both sides of (3.33) are zero and the equation holds. When all $\{\alpha_1, ..., \alpha_k'\}$ are different, while $\{\beta_1, ..., \beta_k\}$ are their permutation, each term in the sum on the left-hand side is equal to $\delta^{\alpha_1...\alpha_k}_{\beta_1...\beta_k}$ when $\rho \notin \{\alpha_1, ..., \alpha_k\}$, and is zero when $\rho \in \{\alpha_1, ..., \alpha_k\}$. Consequently, there are $n-k$ values of ρ with which there are nonzero contributions on the left-hand side, and each contribution is equal to the delta on the right. $\square$

The following equations are simple consequences of (3.33):

$$\delta^{\alpha_1...\alpha_{n-1}\rho}_{\beta_1...\beta_{n-1}\rho} = \delta^{\alpha_1...\alpha_{n-1}}_{\beta_1...\beta_{n-1}}, \tag{3.34}$$

$$\delta^{\rho_1...\rho_s}_{\rho_1...\rho_s} = (n-s+1)(n-s+2)...n = \frac{n!}{(n-s)!}, \tag{3.35}$$

$$\delta^{\rho_1...\rho_n}_{\rho_1...\rho_n} = n!, \tag{3.36}$$

$$\delta^{\alpha_1...\alpha_k\rho_{k+1}...\rho_n}_{\beta_1...\beta_k\rho_{k+1}...\rho_n} = (n-k)!\delta^{\alpha_1...\alpha_k}_{\beta_1...\beta_k}. \tag{3.37}$$

3.13 Examples of applications of the Levi-Civita symbol and of the multidimensional Kronecker delta

The Levi-Civita symbols and the multidimensional deltas are useful in calculations with determinants or antisymmetrisations: they allow us to replace tricky reasonings with simple computational rules.

With the help of (3.25), (3.32) and (3.36) we can verify that

$$\det(A^{\cdot}_{\cdot}) = \frac{1}{n!}\delta^{\beta_1...\beta_n}_{\alpha_1...\alpha_n} A^{\alpha_1}{}_{\beta_1} \cdots A^{\alpha_n}{}_{\beta_n}, \tag{3.38}$$

so the determinant of a mixed tensor is a scalar;

$$\det(B_{..}) = \frac{1}{n!}\epsilon^{\alpha_1...\alpha_n}\epsilon^{\beta_1...\beta_n} B_{\alpha_1\beta_1} \cdots B_{\alpha_n\beta_n}, \tag{3.39}$$

so the determinant of a doubly covariant tensor is a scalar density of weight -2;

$$\det(C^{..}) = \frac{1}{n!}\epsilon_{\alpha_1...\alpha_n}\epsilon_{\beta_1...\beta_n} C^{\alpha_1\beta_1} \cdots C^{\alpha_n\beta_n}, \tag{3.40}$$

so the determinant of a doubly contravariant tensor is a scalar density of weight $(+2)$.

One can also verify that the antisymmetrisation with respect to any set of indices can be written as follows:

$$T_{[\alpha_1...\alpha_k]} = \frac{1}{k!}\delta^{\rho_1...\rho_k}_{\alpha_1...\alpha_k} T_{\rho_1...\rho_k}; \tag{3.41}$$

and similarly for upper indices.

3.14 Exercises

1. Verify that the quantity $D(A)$ defined in (3.24) is antisymmetric with respect to the β-indices.
   Hint. Interchange the names of any two αs in the sum and then move the new αs to their old positions by transposing the indices of ϵ and interchanging the As in the product.

2. Prove that the cofactor of the element $A^{\alpha}{}_{\beta}$ in a mixed tensor $\{A^{\alpha}{}_{\beta}\}$ is given by the equation

$$M^{\beta}{}_{\alpha} = \frac{1}{(n-1)!} \delta^{\beta\beta_1\ldots\beta_{n-1}}_{\alpha\alpha_1\ldots\alpha_{n-1}} A^{\alpha_1}{}_{\beta_1} \cdots A^{\alpha_{n-1}}{}_{\beta_{n-1}}. \qquad (3.42)$$

3. Find the formulae for the cofactors of the elements $B_{\alpha\beta}$ and $C^{\alpha\beta}$ in a doubly covariant and a doubly contravariant tensor, respectively. Note that the cofactor has in each case its indices positioned opposite to its corresponding element.

4. Find the formula for the coefficient of λ^i in the characteristic equation for a matrix $M^{\mu}{}_{\nu}$:

$$\det (M^{\mu}{}_{\nu} - \lambda\delta^{\mu}{}_{\nu}) = 0.$$

Prove that all the coefficients of this polynomial are scalars. Note the coefficients of λ^0 and of λ^{n-1} – what functions of the matrix are they?

4

Covariant derivatives

4.1 Differentiation of tensors

Calculate the derivative of a contravariant vector field v^α after it has been transformed from the $\{x\}$-coordinates to the $\{x'\}$-coordinates:

$$v^{\alpha'}{}_{,\beta'} = \left(x^{\alpha'}{}_{,\alpha} v^\alpha\right)_{,\beta'} = x^{\alpha'}{}_{,\alpha\beta} x^\beta{}_{,\beta'} v^\alpha + x^{\alpha'}{}_{,\alpha} x^\beta{}_{,\beta'} v^\alpha{}_{,\beta}. \tag{4.1}$$

This is not a tensor field, in consequence of the term $x^{\alpha'}{}_{,\alpha\beta} x^\beta{}_{,\beta'} v^\alpha$. The same is true for most other tensor fields. There are only a few cases in which the derivatives of tensor fields are themselves tensor fields. One example is the derivative of a scalar field, which is a covariant vector field. Other examples are the following.

1. The derivatives of the Levi-Civita symbols and of all the Kronecker deltas are identically equal to zero, and hence are tensors.
2. If $T_{\alpha_1 \ldots \alpha_k}$ is a tensor field (of weight 0), then $T_{[\alpha_1 \ldots \alpha_k, \alpha_{k+1}]}$ (a generalisation of rotation) is a tensor field, too.
3. If $T^{\alpha_1 \ldots \alpha_k}$ is a tensor density field of weight -1, and is antisymmetric in all the indices, then $T^{\alpha_1 \ldots \alpha_k}{}_{,\alpha_k}$ (a generalisation of the divergence of a vector field) is also a tensor density field of weight -1.

We will verify the third example.

By assumption, when the coordinates are transformed from $\{x\}$ to $\{x'\}$, $T^{\alpha_1 \ldots \alpha_k}$ transforms as follows:

$$T^{\alpha'_1 \ldots \alpha'_k} = \left(\frac{\partial(x')}{\partial(x)}\right)^{-1} x^{\alpha'_1}{}_{,\alpha_1} \ldots x^{\alpha'_k}{}_{,\alpha_k} T^{\alpha_1 \ldots \alpha_k}. \tag{4.2}$$

Let us differentiate this by $x^{\alpha'_k}$ and contract the result by α'_k. In that term, in which the differentiation acts on the determinant $(\partial(x')/\partial(x))^{-1} = \partial(x)/\partial(x')$, we will develop the determinant according to (3.38). We have

$$T^{\alpha'_1 \ldots \alpha'_k}{}_{,\alpha'_k} = \frac{1}{n!} \sum_{i=1}^{n} \delta^{\rho'_1 \ldots \rho'_n}_{\sigma_1 \ldots \sigma_n} x^{\sigma_1}{}_{,\rho'_1} \ldots x^{\sigma_i}{}_{,\rho'_i \alpha'_k} \ldots x^{\sigma_n}{}_{,\rho'_n}$$

$$\times x^{\alpha'_1}{}_{,\alpha_1} \ldots x^{\alpha'_k}{}_{,\alpha_k} T^{\alpha_1 \ldots \alpha_k}$$

26

$$+ \frac{\partial(x)}{\partial(x')} \sum_{i=1}^{k-1} x^{\alpha'_1}{}_{,\alpha_1} \ldots x^{\alpha'_i}{}_{,\alpha_i\alpha'_k} \ldots x^{\alpha'_k}{}_{,\alpha_k} T^{\alpha_1\ldots\alpha_k}$$

$$+ \frac{\partial(x)}{\partial(x')} x^{\alpha'_1}{}_{,\alpha_1} \ldots x^{\alpha'_{k-1}}{}_{,\alpha_{k-1}} x^{\alpha'_k}{}_{,\alpha_k\alpha'_k} T^{\alpha_1\ldots\alpha_k}$$

$$+ \frac{\partial(x)}{\partial(x')} x^{\alpha'_1}{}_{,\alpha_1} \ldots x^{\alpha'_k}{}_{,\alpha_k} T^{\alpha_1\ldots\alpha_k}{}_{,\rho} x^{\rho}{}_{,\alpha'_k}. \tag{4.3}$$

In the first term we recall that δ is antisymmetric with respect to both upper and lower indices, so interchanging simultaneously a pair of upper indices and a pair of lower indices does not change it. Hence, the products in the first term can be ordered so that the factor with the second derivative is contracted with the last indices of δ. After such an ordering we see that all the n components of the sum are identical, so

$$\text{first term} = \frac{1}{(n-1)!} \delta^{\rho'_1\ldots\rho'_n}_{\sigma_1\ldots\sigma_n} x^{\sigma_1}{}_{,\rho'_1} \ldots x^{\sigma_{n-1}}{}_{,\rho'_{n-1}} x^{\sigma_n}{}_{,\rho'_n\alpha'_k}$$

$$\times x^{\alpha'_1}{}_{,\alpha_1} \ldots x^{\alpha'_k}{}_{,\alpha_k} T^{\alpha_1\ldots\alpha_k}. \tag{4.4}$$

In the second term of (4.3), each component of the sum contains $x^{\alpha'_i}{}_{,\alpha_i\alpha'_k} x^{\alpha'_k}{}_{,\alpha_k} = x^{\alpha'_i}{}_{,\alpha_i\alpha_k}$, which is symmetric in (α_i, α_k). However, it is then contracted with respect to both (α_i, α_k) with the $T^{\alpha_1\ldots\alpha_k}$, which is antisymmetric in these two indices. Such a contraction is always identically equal to zero, hence the second term is zero.

In the third term, we note that $x^{\alpha'_k}{}_{,\alpha_k}$ is an element of the inverse matrix to $[x^{\alpha}{}_{,\alpha'}]$. Thus, $x^{\alpha'_k}{}_{,\alpha_k}$ is equal to the cofactor of the element transposed to $\binom{\alpha'_k}{\alpha_k}$ in the matrix $[x^{\alpha}{}_{,\alpha'}]$, divided by the determinant of $[x^{\alpha}{}_{,\alpha'}]$. The element transposed to $\binom{\alpha'_k}{\alpha_k}$ in $[x^{\alpha}{}_{,\alpha'}]$ is $x^{\alpha_k}{}_{,\alpha'_k}$, while $\det[x^{\alpha}{}_{,\alpha'}] = \partial(x)/\partial(x')$. Hence, using (3.42), we have

$$x^{\alpha'_k}{}_{,\alpha_k\alpha'_k} = \left[\left(\frac{\partial(x)}{\partial(x')} \right)^{-1} \frac{1}{(n-1)!} \delta^{\alpha'_k\mu'_1\ldots\mu'_{n-1}}_{\alpha_k\nu_1\ldots\nu_{n-1}} x^{\nu_1}{}_{,\mu'_1} \ldots x^{\nu_{n-1}}{}_{,\mu'_{n-1}} \right]_{,\alpha'_k}. \tag{4.5}$$

We see at once that the differentiation of the factors $x^{\nu_i}{}_{,\mu'_i}$ by $x^{\alpha'_k}$ will give zero contributions because the results of these differentiations, $x^{\nu_i}{}_{,\mu'_i\alpha'_k}$, are symmetric in (μ'_i, α'_k) and will be contracted with the delta which is antisymmetric in the same indices. The only nonzero contribution will thus be from the derivative of the determinant, so

$$(x^{\alpha'_k}{}_{,\alpha_k})_{,\alpha'_k} = - \left(\frac{\partial(x)}{\partial(x')} \right)^{-2} \cdot \frac{1}{(n-1)!} \cdot \delta^{\rho'_1\ldots\rho'_n}_{\sigma_1\ldots\sigma_n} x^{\sigma_1}{}_{,\rho'_1} \ldots x^{\sigma_{n-1}}{}_{,\rho'_{n-1}} x^{\sigma_n}{}_{,\rho'_n\alpha'_k}$$

$$\times \frac{1}{(n-1)!} \delta^{\alpha'_k\mu'_1\ldots\mu'_{n-1}}_{\alpha_k\nu_1\ldots\nu_{n-1}} x^{\nu_1}{}_{,\mu'_1} \ldots x^{\nu_{n-1}}{}_{,\mu'_{n-1}}. \tag{4.6}$$

Note that the last line above is just

$$\left(\frac{\partial(x)}{\partial(x')} x^{\alpha'_k}{}_{,\alpha_k} \right).$$

Using this we see that the third term in (4.3) has the same absolute value as the first one, but is of opposite sign. Hence, the first and third terms cancel each other, and the final result in (4.3) is the last term, which equals

$$T^{\alpha'_1 \ldots \alpha'_k}{}_{,\alpha'_k} = \frac{\partial(x)}{\partial(x')} \, x^{\alpha'_1}{}_{,\alpha_1} \ldots x^{\alpha'_{k-1}}{}_{,\alpha_{k-1}} \, T^{\alpha_1 \ldots \alpha_k}{}_{,\alpha_k}, \tag{4.7}$$

where we have used $x^{\alpha'_k}{}_{,\alpha_k} x^\rho{}_{,\alpha'_k} = x^\rho{}_{,\alpha_k} = \delta^\rho{}_{\alpha_k}$. Hence, $T^{\alpha_1 \ldots \alpha_k}{}_{,\alpha_k}$ is indeed a tensor density field of weight -1.

The statement proved above also holds for the case when T^α has just one index and is a contravariant vector density field. Then, the first and third terms still cancel each other while the second term in (4.3) simply does not exist.

The fact that partial derivatives of tensor fields are not tensor fields themselves is unfortunate because the laws of physics are usually formulated as differential equations. We would like these equations to have the form (a tensor) $= 0$, since this would hold in all coordinate systems. This suggests the following idea: let us define a 'generalised differentiation', which will yield tensor fields when acting on tensor fields, and will coincide with ordinary differentiation when acting on scalars and the Kronecker deltas, for which partial derivatives are tensors. Then, we will replace the partial derivatives with the generalised derivatives in the basic equations. We guess that this generalised differentiation, called **covariant differentiation**, will reduce to ordinary differentiation in certain privileged coordinate systems.

4.2 Axioms of the covariant derivative

We want the covariant differentiation to have all the algebraic properties of an ordinary differentiation, and in addition to yield tensor densities when acting on tensor densities. We will denote the covariant derivative by ∇_α or $|_\alpha$ or $D/\partial x^\alpha$. The symbols $T_i[w, k, l]$ will denote tensor densities whose explicit indices are irrelevant.

Specifically, we want ∇_α to have the following properties:

1. To be distributive with respect to addition:

$$\nabla_\alpha(T_1[w, k, l] + T_2[w, k, l]) = \nabla_\alpha(T_1[w, k, l]) + \nabla_\alpha(T_2[w, k, l]). \tag{4.8}$$

2. To obey the Leibniz rule when acting on a tensor product:

$$\nabla_\alpha(T_1[w_1, k_1, l_1] \otimes T_2[w_2, k_2, l_2]) = (\nabla_\alpha T_1[w_1, k_1, l_1]) \otimes T_2[w_2, k_2, l_2]$$
$$+ (T_1[w_1, k_1, l_1]) \otimes (\nabla_\alpha T_2[w_2, k_2, l_2]). \tag{4.9}$$

3. To reduce to the partial derivative when acting on a scalar field:

$$\nabla_\alpha \Phi = \Phi_{,\alpha}. \tag{4.10}$$

4. To yield zero when acting on the Levi-Civita symbols and Kronecker deltas:

$$\nabla_\alpha \epsilon^{\alpha_1 \ldots \alpha_n} = 0, \qquad \nabla_\alpha \epsilon_{\alpha_1 \ldots \alpha_n} = 0, \qquad \nabla_\alpha \delta^\alpha{}_\beta = 0. \tag{4.11}$$

The last equation implies at once that

$$\nabla_\alpha \delta^{\alpha_1 \dots \alpha_k}_{\beta_1 \dots \beta_k} = 0 \tag{4.12}$$

for any k. It also implies that ∇_α commutes with contraction.

5. When acting on a tensor density field of type $[w, k, l]$, it should produce a tensor density field of type $[w, k, l+1]$, thus

$$\nabla_\alpha (T_1[w, k, l]) = T_2[w, k, l+1].$$

Only the last property does not hold for partial derivatives.

From these postulated properties we will now derive an operational formula for the covariant derivative.

4.3 A field of bases on a manifold and scalar components of tensors

In every tangent space to an n-dimensional manifold M_n we can choose a set of n linearly independent contravariant vectors, $\{e_1{}^\alpha, \dots, e_n{}^\alpha\}$. The indices $a, b, c, \dots = 1, \dots, n$ will label vectors (while Greek indices label coordinate components of tensors). After such a basis is chosen at every $x \in M_n$, we consider the set of n vector *fields*:

$$x \to e_a{}^\alpha(x), \qquad a = 1, \dots, n.$$

The collection of quantities $\{e_a{}^\alpha(x)\}$, $\alpha = 1, \dots, n$, $a = 1, \dots, n$ forms a matrix whose elements are functions on the manifold. Since all the vectors are linearly independent at every x, the matrix is non-singular, so there exists an inverse matrix $e^a{}_\alpha$ that obeys

$$e^a{}_\alpha e_a{}^\beta = \delta^\beta_\alpha. \tag{4.13}$$

A subset of the matrix $\|e^a{}_\alpha\|$ defined by a fixed a is then a covariant vector field. The set of fields corresponding to all values of a forms a **dual** basis to $\{e_a{}^\alpha(x)\}$, $a = 1, \dots, n$. By virtue of the $\{e_a{}^\alpha(x)\}$ being linearly independent, Eq. (4.13) implies the following:

$$e^a{}_\alpha e_b{}^\alpha = \delta^a{}_b. \tag{4.14}$$

For any tensor field (of weight 0) $T^{\alpha_1 \dots \alpha_k}_{\beta_1 \dots \beta_l}$, the collection of quantities

$$T^{a_1 \dots a_k}_{b_1 \dots b_l} := e^{a_1}{}_{\alpha_1} \dots e^{a_k}{}_{\alpha_k} e_{b_1}{}^{\beta_1} \dots e_{b_l}{}^{\beta_l} T^{\alpha_1 \dots \alpha_k}_{\beta_1 \dots \beta_l}, \tag{4.15}$$

labelled by the indices $a_1, \dots, a_k, b_1, \dots, b_l = 1, \dots, n$, is a set of n^{k+l} scalar fields that uniquely represents the set of n^{k+l} coordinate components of the tensor field $T^{\alpha_1 \dots \alpha_k}_{\beta_1 \dots \beta_l}$. This is so because, in consequence of (4.13) and (4.14), an inverse formula to (4.15) exists that allows one to calculate $T^{\alpha_1 \dots \alpha_k}_{\beta_1 \dots \beta_l}$ when $T^{a_1 \dots a_k}_{b_1 \dots b_l}$ are given:

$$T^{\alpha_1 \dots \alpha_k}_{\beta_1 \dots \beta_l} = e_{a_1}{}^{\alpha_1} \dots e_{a_k}{}^{\alpha_k} e^{b_1}{}_{\beta_1} \dots e^{b_l}{}_{\beta_l} T^{a_1 \dots a_k}_{b_1 \dots b_l} \tag{4.16}$$

Let us denote

$$e := \det \|e_a{}^\alpha\| = \frac{1}{n!} \epsilon^{a_1 \dots a_n} \epsilon_{\alpha_1 \dots \alpha_n} e_{a_1}{}^{\alpha_1} \dots e_{a_n}{}^{\alpha_n}. \tag{4.17}$$

Now, $\epsilon_{\alpha_1 \dots \alpha_n}$ is a tensor density of weight $+1$, whereas $\epsilon^{a_1 \dots a_n}$ is a set of scalars because it depends only on the basis in the vector space, and not on the coordinate system. Hence, e is a scalar density of weight $+1$.

The quantity e, together with $\{e_a{}^\alpha\}$ and $\{e^a{}_\alpha\}$, can be used to represent tensor density fields by sets of scalar fields. Let $T^{\alpha_1 \dots \alpha_k}_{\beta_1 \dots \beta_l}$ now be a tensor density field of type $[w, k, l]$; then each element of the set

$$T^{a_1 \dots a_k}_{b_1 \dots b_l} := e^{-w} e^{a_1}{}_{\alpha_1} \dots e^{a_k}{}_{\alpha_k} e_{b_1}{}^{\beta_1} \dots e_{b_l}{}^{\beta_l} T^{\alpha_1 \dots \alpha_k}_{\beta_1 \dots \beta_l} \qquad (4.18)$$

is a scalar field. The set $T^{a_1 \dots a_k}_{b_1 \dots b_l}$ then uniquely defines $T^{\alpha_1 \dots \alpha_k}_{\beta_1 \dots \beta_l}$ via a formula analogous to (4.16), with the factor e^{+w} added. The weight w has to be given as an extra bit of information.

4.4 The affine connection

We now define the set of quantities

$$\Gamma^\alpha{}_{\beta\gamma} = -e_s{}^\alpha \left(\nabla_\gamma - \partial_\gamma \right) e^s{}_\beta. \qquad (4.19)$$

The elements of this set are called the coefficients of **affine connection**. When specified explicitly, they tell us how the covariant derivative acts on the basis vector fields. Later we will consider manifolds in which these coefficients can be calculated from more basic objects (see Chapter 7). For now, we consider manifolds in which the $\Gamma^\alpha{}_{\beta\gamma}$ are just given.

The $\Gamma^\alpha{}_{\beta\gamma}$ do not depend on the choice of basis. Let us assume that $\{e_a{}^\alpha\}$ and $\{e_{a'}{}^\alpha\}$ are two different bases. The vector fields of the second basis can then be decomposed in the first basis

$$e_{a'}{}^\alpha = A^b{}_{a'} e_b{}^\alpha, \qquad (4.20)$$

and then the elements of the transformation matrix

$$A^b{}_{a'} = e^b{}_\alpha e_{a'}{}^\alpha \qquad (4.21)$$

are scalar fields. Hence, $A^b{}_{a'|\alpha} = A^b{}_{a',\alpha}$ and $\left(A^{-1} \right)^{c'}{}_{d|\alpha} = \left(A^{-1} \right)^{c'}{}_{d,\alpha}$. Then, calculating the $\Gamma^\alpha{}_{\beta\gamma}$ in the basis $\{e_{a'}{}^\alpha\}$, we have

$$\left(\Gamma^\alpha{}_{\beta\gamma} \right)_{e'} = -A^r{}_{s'} e_r{}^\alpha \left(\nabla_\gamma - \partial_\gamma \right) \left[\left(A^{-1} \right)^{s'}{}_s e^s{}_\beta \right]$$

$$= -\delta^r{}_s e_r{}^\alpha \left(\nabla_\gamma - \partial_\gamma \right) e^s{}_\beta = \left(\Gamma^\alpha{}_{\beta\gamma} \right)_e. \qquad (4.22)$$

Now let us note that the $\Gamma^\alpha{}_{\beta\gamma}$ *are not tensor fields*. When coordinates are transformed, these coefficients change as follows:

$$\Gamma^{\alpha'}{}_{\beta'\gamma'} = x^{\alpha'}{}_{,\alpha} x^\beta{}_{,\beta'} x^\gamma{}_{,\gamma'} \Gamma^\alpha{}_{\beta\gamma} + x^{\alpha'}{}_{,\rho} x^\rho{}_{,\beta'\gamma'}. \qquad (4.23)$$

However, the antisymmetric part

$$\Omega^\alpha{}_{\beta\gamma} \overset{\text{def}}{=} \Gamma^\alpha{}_{[\beta\gamma]} \qquad (4.24)$$

is a tensor, since $x^\rho{}_{,[\beta'\gamma']} = 0$. It is called the **torsion tensor**.

4.5 The explicit formula for the covariant derivative of tensor density fields

In order to obtain an explicit formula for the covariant derivative, we need to know two other properties of the connection coefficients:

$$(\text{I}) \qquad \Gamma^{\alpha}{}_{\beta\gamma} = e^{s}{}_{\beta}\left(\nabla_{\gamma} - \partial_{\gamma}\right)e_{s}{}^{\alpha}, \qquad (4.25)$$

$$(\text{II}) \qquad \nabla_{\alpha}(e^{w}) = we^{w-1}\,\nabla_{\alpha}e. \qquad (4.26)$$

The verification of (I) is easy. To verify (II), consider the quantity

$$F_{\alpha}(w) \overset{\text{def}}{=} e^{-w}\,\nabla_{\alpha}(e^{w}). \qquad (4.27)$$

Using the postulated properties of the covariant derivative, we obtain

$$F_{\alpha}(w_{1} + w_{2}) = F_{\alpha}(w_{1}) + F_{\alpha}(w_{2}). \qquad (4.28)$$

Every continuous function that has the property $f(w_{1} + w_{2}) = f(w_{1}) + f(w_{2})$ for all real w_{1} and w_{2} also has the property $f(w) = f(1)w$. Hence $e^{-w}\nabla_{\alpha}(e^{w}) = we^{-1}\nabla_{\alpha}e$, which is equivalent to (4.26).

Equation (4.26) holds also for partial derivatives, so

$$e^{-w}\left(\nabla_{\gamma} - \partial_{\gamma}\right)(e^{w}) = we^{-1}\left(\nabla_{\gamma} - \partial_{\gamma}\right)e. \qquad (4.29)$$

Now, using Eqs. (3.38) and (3.42), we obtain

$$\left(\nabla_{\gamma} - \partial_{\gamma}\right)e = ee^{a_{n}}{}_{\rho_{n}}\left(\nabla_{\gamma} - \partial_{\gamma}\right)e_{a_{n}}{}^{\rho_{n}} = e\Gamma^{\rho}{}_{\rho\gamma}. \qquad (4.30)$$

At this point, we can derive the formula for the covariant derivative of an arbitrary tensor density field. Let us convert the tensor density field to the set of scalar fields given by (4.18). Axiom 3 implies then

$$\left(\nabla_{\gamma} - \partial_{\gamma}\right)T^{a_{1}\ldots a_{k}}_{b_{1}\ldots b_{l}} = 0. \qquad (4.31)$$

On the other hand, using (4.18) and axiom 2, we obtain

$$\begin{aligned}
\left(\nabla_{\gamma} - \partial_{\gamma}\right)T^{a_{1}\ldots a_{k}}_{b_{1}\ldots b_{l}} =\ & -we^{-w}\Gamma^{\rho}{}_{\rho\gamma}e^{a_{1}}{}_{\alpha_{1}}\ldots e^{a_{k}}{}_{\alpha_{k}}e_{b_{1}}{}^{\beta_{1}}\ldots e_{b_{l}}{}^{\beta_{l}}T^{\alpha_{1}\ldots\alpha_{k}}_{\beta_{1}\ldots\beta_{l}} \\
& + e^{-w}\sum_{i=1}^{k}e^{a_{1}}{}_{\alpha_{1}}\ldots\left(\nabla_{\gamma} - \partial_{\gamma}\right)e^{a_{i}}{}_{\alpha_{i}}\ldots e^{a_{k}}{}_{\alpha_{k}}e_{b_{1}}{}^{\beta_{1}}\ldots e_{b_{l}}{}^{\beta_{l}}T^{\alpha_{1}\ldots\alpha_{k}}_{\beta_{1}\ldots\beta_{l}} \\
& + e^{-w}\sum_{j=1}^{l}e^{a_{1}}{}_{\alpha_{1}}\ldots e^{a_{k}}{}_{\alpha_{k}}e_{b_{1}}{}^{\beta_{1}}\ldots\left(\nabla_{\gamma} - \partial_{\gamma}\right)e_{b_{j}\beta_{j}}\ldots e_{b_{l}}{}^{\beta_{l}}T^{\alpha_{1}\ldots\alpha_{k}}_{\beta_{1}\ldots\beta_{l}} \\
& + e^{-w}e^{a_{1}}{}_{\alpha_{1}}\ldots e^{a_{k}}{}_{\alpha_{k}}e_{b_{1}}{}^{\beta_{1}}\ldots e_{b_{l}}{}^{\beta_{l}}\left(\nabla_{\gamma} - \partial_{\gamma}\right)T^{\alpha_{1}\ldots\alpha_{k}}_{\beta_{1}\ldots\beta_{l}}. \qquad (4.32)
\end{aligned}$$

Now we convert this back to coordinate components, by contracting it with $e^w e_{a_1}{}^{\alpha_1} \ldots e_{a_k}{}^{\alpha_k} e^{b_1}{}_{\beta_1} \ldots e^{b_l}{}_{\beta_l}$ and using Eqs. (4.19) and (4.25):

$$0 = w\Gamma^\rho{}_{\rho\gamma} T^{\alpha_1 \ldots \alpha_k}_{\beta_1 \ldots \beta_l} - \sum_{i=1}^{k} \Gamma^{\alpha_i}{}_{\rho_i \gamma} T^{\alpha_1 \ldots \rho_i \ldots \alpha_k}_{\beta_1 \ldots \beta_l} + \sum_{j=1}^{l} \Gamma^{\rho_j}{}_{\beta_j \gamma} T^{\alpha_1 \ldots \alpha_k}_{\beta_1 \ldots \rho_j \ldots \beta_l}$$

$$+ \left(\nabla_\gamma - \partial_\gamma \right) T^{\alpha_1 \ldots \alpha_k}_{\beta_1 \ldots \beta_l}. \tag{4.33}$$

From this, finally

$$\nabla_\gamma T^{\alpha_1 \ldots \alpha_k}_{\beta_1 \ldots \beta_l} = \partial_\gamma T^{\alpha_1 \ldots \alpha_k}_{\beta_1 \ldots \beta_l} + w\Gamma^\rho{}_{\rho\gamma} T^{\alpha_1 \ldots \alpha_k}_{\beta_1 \ldots \beta_l} + \sum_{i=1}^{k} \Gamma^{\alpha_i}{}_{\rho_i \gamma} T^{\alpha_1 \ldots \rho_i \ldots \alpha_k}_{\beta_1 \ldots \beta_l}$$

$$- \sum_{j=1}^{l} \Gamma^{\rho_j}{}_{\beta_j \gamma} T^{\alpha_1 \ldots \alpha_k}_{\beta_1 \ldots \rho_j \ldots \beta_l}. \tag{4.34}$$

The following special cases of (4.34) occur frequently:

$$A^\alpha{}_{|\gamma} = A^\alpha{}_{,\gamma} + \Gamma^\alpha{}_{\rho\gamma} A^\rho; \tag{4.35}$$

$$a_{\alpha|\gamma} = a_{\alpha,\gamma} - \Gamma^\rho{}_{\alpha\gamma} a_\rho; \tag{4.36}$$

$$T^{\alpha\beta}{}_{|\gamma} = T^{\alpha\beta}{}_{,\gamma} + \Gamma^\alpha{}_{\rho\gamma} T^{\rho\beta} + \Gamma^\beta{}_{\rho\gamma} T^{\alpha\rho}; \tag{4.37}$$

$$T_{\alpha\beta|\gamma} = T_{\alpha\beta,\gamma} - \Gamma^\rho{}_{\alpha\gamma} T_{\rho\beta} - \Gamma^\rho{}_{\beta\gamma} T_{\alpha\rho}; \tag{4.38}$$

$$T^\alpha{}_{\beta|\gamma} = T^\alpha{}_{\beta,\gamma} + \Gamma^\alpha{}_{\rho\gamma} T^\rho{}_\beta - \Gamma^\rho{}_{\beta\gamma} T^\alpha{}_\rho. \tag{4.39}$$

Note that, unlike the partial derivative, the covariant derivative does not act on single components of tensor densities. It is an operator that acts on the whole tensor density and produces another tensor density.

4.6 Exercises

1. What is the condition for the 'covariant rotation' $T_{[\alpha|\beta]}$ of a covariant vector field T_α to coincide with the ordinary rotation $T_{[\alpha,\beta]}$?

2. Let $g_{\alpha\beta} = g_{(\alpha\beta)}$ be a doubly covariant tensor that is nonsingular, i.e. $\det \| g_{\alpha\beta} \| \neq 0$. Let $g^{\alpha\beta}$ be its inverse matrix, i.e.

$$g^{\alpha\rho} g_{\rho\beta} = \delta^\alpha_\beta.$$

Show that the object defined as

$$\left\{ \begin{matrix} \alpha \\ \beta\gamma \end{matrix} \right\} = \frac{1}{2} g^{\alpha\rho} \left(g_{\beta\rho,\gamma} + g_{\gamma\rho,\beta} - g_{\beta\gamma,\rho} \right)$$

transforms under coordinate transformations by the same law as the coefficients of affine connection. What is the torsion in this case?

3. Verify that (4.28) implies $F_\alpha(w) = w F_\alpha(1)$.

5

Parallel transport and geodesic lines

5.1 Parallel transport

Let a curve C be given in a manifold with affine connection, and let v be a field of vectors along the curve, $C \ni x \rightarrow v^\alpha(x)$. In a Euclidean space, in Cartesian coordinates, the vectors $v(x)$ are parallel when

$$\frac{dv^\alpha}{d\tau} = \frac{\partial v^\alpha}{\partial x^\beta} \frac{dx^\beta}{d\tau} = 0, \tag{5.1}$$

where τ is a parameter along C, while $x^\alpha(\tau)$ are coordinates of a point on C. Then, for two points on C corresponding to τ_1 and τ_2,

$$v^\alpha(\tau_1) \underset{*}{=} v^\alpha(\tau_2) \tag{5.2}$$

(the asterisk means that the equation holds only in some coordinate systems; for example, (5.2) does not hold in polar coordinates on a plane). We will generalise the definition of parallelism so that it does not depend on the coordinates. The following definition suggests itself:

$$\frac{Dv^\alpha}{d\tau} \overset{\text{def}}{=} \left(\nabla_\rho v^\alpha\right) \frac{dx^\rho}{d\tau} = 0, \tag{5.3}$$

which is at the same time the definition of a covariant derivative along a curve. Using (4.35), we obtain from this

$$v^\alpha{}_{,\rho} \frac{dx^\rho}{d\tau} + \Gamma^\alpha_{\sigma\rho} v^\sigma \frac{dx^\rho}{d\tau} = 0. \tag{5.4}$$

This can be written equivalently as

$$\frac{dv^\alpha}{d\tau} = -\Gamma^\alpha_{\sigma\rho} v^\sigma \frac{dx^\rho}{d\tau}. \tag{5.5}$$

Thus, the vector $v^\alpha(\tau_1)$, while being parallelly transported from the point $x^\alpha(\tau_1)$ to the point $x^\alpha(\tau_2)$, changes in the following way:

$$v^\alpha_\parallel(\tau_2) = v^\alpha(\tau_1) - \int_{x(\tau_1)}^{x(\tau_2)} \Gamma^\alpha_{\sigma\rho}(x) v^\sigma(x) dx^\rho. \tag{5.6}$$

33

The expression under the integral may not be a perfect differential. Then, the result of integration will depend on the curve C. Thus, in general, for the parallel transport so defined the result depends on the path of transport, and for parallel transport along a closed curve

$$v_{\parallel}^{\alpha}(\tau_1) \overset{\text{def}}{=} v^{\alpha}(\tau_1) - \oint_C \Gamma_{\sigma\rho}^{\alpha}(\tau)v^{\sigma}(\tau)\frac{dx^{\rho}}{d\tau}\,d\tau \neq v^{\alpha}(\tau_1). \tag{5.7}$$

The conditions under which parallel transport is independent of the path, i.e. under which (5.7) does not occur, will be given in Section 6.3.

The parallel transport of an arbitrary tensor density field $T_{\beta_1...\beta_l}^{\alpha_1...\alpha_k}$ is defined analogously to (5.3):

$$\frac{D}{D\tau}T_{\beta_1...\beta_l}^{\alpha_1...\alpha_k} = T_{\beta_1...\beta_l|\rho}^{\alpha_1...\alpha_k}\frac{dx^{\rho}}{d\tau} = 0. \tag{5.8}$$

5.2 Geodesic lines

A **geodesic line** is a curve G whose tangent vector $v^{\alpha} = dx^{\alpha}/d\tau$, after being parallely transported along it from $x^{\alpha} \in G$ to $x^{\alpha}(\tau) \in G$, is collinear with the tangent vector that is defined at $x(\tau)$. Hence

$$v_{\parallel}^{\alpha}(\tau)\big|_{\tau_0 \to \tau} = \lambda(\tau)v^{\alpha}(\tau), \tag{5.9}$$

or, according to (5.6),

$$v_{\parallel}^{\alpha}(\tau)\big|_{\tau_0 \to \tau} = v^{\alpha}(\tau_0) - \int_G^{\tau}\tau_0\Gamma_{\sigma\rho}^{\alpha}(t)\lambda(t)v^{\sigma}(t)v^{\rho}(t)dt = \lambda(\tau)v^{\alpha}(\tau). \tag{5.10}$$

An example is a straight line in a Euclidean space. In that case, the integral in (5.10) is zero and, *if the length of arc is chosen as the parameter τ, $\lambda(\tau) = 1$.* A geodesic line is a generalisation of the notion of a straight line to any manifold with affine connection.

Let us differentiate (5.10) by τ. Since $v^{\alpha} = dx^{\alpha}/d\tau$, the result is

$$\lambda\left(\frac{d^2x^{\alpha}}{d\tau^2} + \Gamma_{\sigma\rho}^{\alpha}(\tau)\frac{dx^{\sigma}}{d\tau}\frac{dx^{\rho}}{d\tau}\right) = -\frac{d\lambda}{d\tau}\frac{dx^{\alpha}}{d\tau}. \tag{5.11}$$

Now let us change the parameter as follows:

$$\tau \to s(\tau) \overset{\text{def}}{=} \int_{\tau_0}^{\tau}\frac{c}{\lambda(t)}\,dt, \tag{5.12}$$

where c and τ_0 are arbitrary constants. Then, Eq. (5.11) becomes

$$\frac{d^2x^{\alpha}}{ds^2} + \Gamma_{\sigma\rho}^{\alpha}(s)\frac{dx^{\sigma}}{ds}\frac{dx^{\rho}}{ds} = 0, \tag{5.13}$$

or, equivalently,

$$\frac{D}{Ds}\left(\frac{dx^{\alpha}}{ds}\right) = 0. \tag{5.14}$$

The form (5.13) of the **geodesic equation** is privileged in that, in the parametrisation (5.12), the tangent vector transported parally along the geodesic is not only collinear with the locally defined tangent vector, but coincides with it. The parameter s that has this property exists for any $\lambda(\tau)$ such that $\lambda(\tau) \neq 0$ for every τ, and it is called the **affine parameter**. It is defined up to the linear transformations

$$s' = as + b, \qquad a, b = \text{constant}. \tag{5.15}$$

Equation (5.13) allows us to prove the following theorem:

Theorem 5.1 *In a manifold M_n with an affine connection, for every point x and for every vector $\mathbf{v}$ tangent to M_n at x, there exists a geodesic line passing through x that is tangent to $\mathbf{v}$.*

This is so because a solution of a second-order differential equation is uniquely determined by its value at one point and its first derivative at that point.

However, two points of a manifold with affine connection cannot always be connected by one and only one geodesic. A geodesic joining two points may not exist in a non-connected manifold, like a two-sheeted hyperboloid. On the other hand, any two points on the surface of a cylinder (where the geodesic lines are straight lines, circles and screw-lines) can be connected by an infinite number of geodesics. (Just imagine a screw-line that connects two points p and q by the shortest arc, then another one that runs one extra time around the cylinder between p and q, then another one that runs two times around the cylinder, and so on.)

Note that only the symmetric part of the connection gives a nonzero contribution to the geodesic equation.

5.3 Exercises

1. Consider a vector on a Euclidean plane being transported parally along a straight line. Find how its components change when they are given in polar coordinates.
2. Do the same for a vector in a 3-dimensional Euclidean space when its components are given in spherical coordinates. From the result, read out the connection coefficients of the Euclidean space in spherical coordinates.

6

The curvature of a manifold; flat manifolds

6.1 The commutator of second covariant derivatives

The covariant derivative was introduced in order that derivatives of tensor densities are still tensor densities. This advantage has a few inconvenient consequences. One of these we already know: parallel transport defined via covariant differentiation depends on the path. Another one will appear now: the second covariant derivatives do not commute.

For a proper scalar field (with weight $w = 0$) we have

$$\left(\nabla_\delta \nabla_\gamma - \nabla_\gamma \nabla_\delta\right) T = \nabla_\delta \left(T,_\gamma\right) - \nabla_\gamma \left(T,_\delta\right) = -2\Omega^\rho{}_{\gamma\delta} T,_\rho. \tag{6.1}$$

Hence, the second covariant derivatives of a scalar field commute only in torsion-free manifolds.

For a scalar density field of weight w we have

$$\begin{aligned}
\left(\nabla_\delta \nabla_\gamma - \nabla_\gamma \nabla_\delta\right) \check{T} &= \left(\nabla_\gamma \check{T}\right),_\delta + w\Gamma^\rho{}_{\rho\delta} \nabla_\gamma \check{T} - \Gamma^\rho{}_{\gamma\delta} \nabla_\rho \check{T} \\
&\quad - \left(\nabla_\delta \check{T}\right),_\gamma - w\Gamma^\rho{}_{\rho\gamma} \nabla_\delta \check{T} + \Gamma^\rho{}_{\delta\gamma} \nabla_\rho \check{T} \\
&= w\left(\Gamma^\rho{}_{\rho\gamma,\delta} - \Gamma^\rho{}_{\rho\delta,\gamma}\right) \check{T} - 2\Omega^\rho{}_{\gamma\delta} \nabla_\rho \check{T}.
\end{aligned} \tag{6.2}$$

Let us leave this formula without comment for a while and let us note the analogous result for a covariant vector field:

$$\begin{aligned}
\left(\nabla_\delta \nabla_\gamma - \nabla_\gamma \nabla_\delta\right) T_\beta &= -\left(\Gamma^\rho{}_{\beta\gamma,\delta} - \Gamma^\rho{}_{\beta\delta,\gamma}\right) T_\rho \\
&\quad + \left(\Gamma^\rho{}_{\sigma\gamma}\Gamma^\sigma{}_{\beta\delta} - \Gamma^\rho{}_{\sigma\delta}\Gamma^\sigma{}_{\beta\gamma}\right) T_\rho - 2\Omega^\sigma{}_{\gamma\delta} T_{\beta|\sigma}.
\end{aligned} \tag{6.3}$$

Now let us define

$$\begin{aligned}
B^\alpha{}_{\beta\gamma\delta} &\stackrel{\text{def}}{=} -\Gamma^\alpha{}_{\beta\gamma,\delta} + \Gamma^\alpha{}_{\beta\delta,\gamma} + \Gamma^\alpha{}_{\sigma\gamma}\Gamma^\sigma{}_{\beta\delta} - \Gamma^\alpha{}_{\sigma\delta}\Gamma^\sigma{}_{\beta\gamma} \\
&\equiv -2\Gamma^\alpha{}_{\beta[\gamma,\delta]} + 2\Gamma^\alpha{}_{\sigma[\gamma}\Gamma^\sigma{}_{|\beta|\delta]}.
\end{aligned} \tag{6.4}$$

The quantity $B^\alpha{}_{\beta\gamma\delta}$ is called the **curvature tensor**. In order to see that it is indeed a tensor, note that the left-hand side of (6.3) is a tensor by definition. The quantities $\Omega^\sigma{}_{\gamma\delta}$ and $T_{\beta|\sigma}$ are tensors, too, hence the last term on the right-hand side is a tensor. It follows that $B^{\rho}{}_{\beta\gamma\delta}T_{\rho}$ is a tensor. Now, since T_{ρ} is an arbitrary covariant vector, $B^{\rho}{}_{\beta\gamma\delta}$ must be a tensor itself. (This can also be verified using (6.4) and (4.23).)

Now let us observe that

$$B^{\rho}{}_{\rho\gamma\delta} = -\Gamma^{\rho}{}_{\rho\gamma,\delta} + \Gamma^{\rho}{}_{\rho\delta,\gamma}. \tag{6.5}$$

Hence, Eq. (6.2) can be rewritten as follows

$$\left(\nabla_{\delta}\nabla_{\gamma} - \nabla_{\gamma}\nabla_{\delta}\right)\check{T} = -wB^{\rho}{}_{\rho\gamma\delta}\check{T} - 2\Omega^{\rho}{}_{\gamma\delta}\check{T}_{|\rho}. \tag{6.6}$$

Thus, the same curvature tensor determines also the commutator of the covariant derivatives of a scalar density field.

Finally, for the commutator acting on a contravariant vector field

$$\left(\nabla_{\delta}\nabla_{\gamma} - \nabla_{\gamma}\nabla_{\delta}\right)T^{\alpha} = -B^{\alpha}{}_{\rho\gamma\delta}T^{\rho} - 2\Omega^{\sigma}{}_{\gamma\delta}T^{\alpha}{}_{|\sigma}. \tag{6.7}$$

It can be verified that the commutator of covariant derivatives acts on the tensor product of two tensor density fields in the following way:

$$\left(\nabla_{\delta}\nabla_{\gamma} - \nabla_{\gamma}\nabla_{\delta}\right)(T_1 \otimes T_2)$$
$$= \left[\left(\nabla_{\delta}\nabla_{\gamma} - \nabla_{\gamma}\nabla_{\delta}\right)T_1\right] \otimes T_2 + T_1 \otimes \left[\left(\nabla_{\delta}\nabla_{\gamma} - \nabla_{\gamma}\nabla_{\delta}\right)T_2\right] \tag{6.8}$$

(the weights and indices of T_1 and T_2 are irrelevant here). Thus, the operator $(\nabla_{\delta}\nabla_{\gamma} - \nabla_{\gamma}\nabla_{\delta})$ has the property of an ordinary differentiation. Then, an arbitrary tensor density field of type $[w, k, l]$ behaves under covariant differentiation like a tensor product of a single scalar density of weight w, k contravariant vectors and l covariant vectors – see (4.34). Hence, we can guess the formula for the commutator of second covariant derivatives acting on an arbitrary tensor density field:

$$\left(\nabla_{\delta}\nabla_{\gamma} - \nabla_{\gamma}\nabla_{\delta}\right)T^{\alpha_1...\alpha_k}_{\beta_1...\beta_l} = -wB^{\rho}{}_{\rho\gamma\delta}T^{\alpha_1...\alpha_k}_{\beta_1...\beta_l} - \sum_{i=1}^{k}B^{\alpha_i}{}_{\rho_i\gamma\delta}T^{\alpha_1...\rho_i...\alpha_k}_{\beta_1...\beta_l}$$

$$+ \sum_{j=1}^{l}B^{\rho_j}{}_{\beta_j\gamma\delta}T^{\alpha_1...\alpha_k}_{\beta_1...\rho_j...\beta_l} - 2\Omega^{\rho}{}_{\gamma\delta}T^{\alpha_1...\alpha_k}_{\beta_1...\beta_l|\rho}. \tag{6.9}$$

(This can be formally derived by projecting $T^{\alpha_1...\alpha_k}_{\beta_1...\beta_l}$ on the basis fields and calculating the commutator for the projection in two ways.) Equation (6.9) is called the **Ricci formula**. Let us note, from (6.4), that

$$B^{\alpha}{}_{\beta\gamma\delta} = B^{\alpha}{}_{\beta[\gamma\delta]}. \tag{6.10}$$

We will see in Section 7.9 that the name 'curvature tensor' evokes the right associations.

6.2 The commutator of directional covariant derivatives

Let k^α and l^α be two linearly independent contravariant vector fields determined on a manifold with affine connection, M_n. Let us calculate, for an arbitrary tensor density field, the commutator of second directional covariant derivatives along k and l, thus

$$[\nabla_k, \nabla_l] T^{\alpha_1...\alpha_k}_{\beta_1...\beta_l} \stackrel{\text{def}}{=} k^\mu \nabla_\mu \left(l^\rho \nabla_\rho T^{\alpha_1...\alpha_k}_{\beta_1...\beta_l} \right) - l^\mu \nabla_\mu \left(k^\rho \nabla_\rho T^{\alpha_1...\alpha_k}_{\beta_1...\beta_l} \right)$$

$$= \left(k^\mu l^\rho{}_{,\mu} - l^\mu k^\rho{}_{,\mu} \right) T^{\alpha_1...\alpha_k}_{\beta_1...\beta_l|\rho} - 2k^\mu l^\nu \Omega^\rho{}_{\mu\nu} T^{\alpha_1...\alpha_k}_{\beta_1...\beta_l|\rho}$$

$$+ k^\mu l^\rho \left(\nabla_\mu \nabla_\rho - \nabla_\rho \nabla_\mu \right) T^{\alpha_1...\alpha_k}_{\beta_1...\beta_l}. \tag{6.11}$$

The quantity

$$[k, l]^\rho \stackrel{\text{def}}{=} k^\mu l^\rho{}_{,\mu} - l^\mu k^\rho{}_{,\mu} \tag{6.12}$$

is called the **commutator** of k and l. It is a tensor field, since

$$[k, l]^\rho \equiv k^\mu l^\rho{}_{|\mu} - l^\mu k^\rho{}_{|\mu} - 2\Omega^\rho{}_{\mu\nu} k^\mu l^\nu. \tag{6.13}$$

The second term in the last part of Eq. (6.11) cancels the term with torsion that will appear from (6.9). In the end, we obtain

$$[\nabla_k, \nabla_l] T^{\alpha_1...\alpha_k}_{\beta_1...\beta_l} = [k, l]^\rho T^{\alpha_1...\alpha_k}_{\beta_1...\beta_l|\rho} - wB^\rho{}_{\rho\nu\mu} k^\mu l^\nu T^{\alpha_1...\alpha_k}_{\beta_1...\beta_l}$$

$$- \sum_{i=1}^{k} B^{\alpha_i}{}_{\rho_i\nu\mu} k^\mu l^\nu T^{\alpha_1...\rho_i...\alpha_k}_{\beta_1...\beta_l} + \sum_{j=1}^{l} B^{\rho_j}{}_{\beta_j\nu\mu} k^\mu l^\nu T^{\alpha_1...\alpha_k}_{\beta_1...\rho_j...\beta_l}. \tag{6.14}$$

Suppose in addition that $[k, l]^\alpha = 0$. This is a necessary and sufficient condition for the existence of coordinates **adapted** simultaneously to both k and l, call them τ^a and τ^b:

$$k^\alpha = \frac{\partial x^\alpha}{\partial \tau^a}, \qquad l^\alpha = \frac{\partial x^\alpha}{\partial \tau^b}, \tag{6.15}$$

so that, if τ^a and τ^b are chosen as two of the coordinates, then, with $x'^1 = \tau^a$ and $x'^2 = \tau^b$, we have $k'^\alpha = \delta^\alpha_1$, $l'^\alpha = \delta^\alpha_2$. (For the proof see Exercise 2.) For such fields, (6.14) simplifies further to

$$[\nabla_{\tau_a}, \nabla_{\tau_b}] T^{\alpha_1...\alpha_k}_{\beta_1...\beta_l} = -wB^\rho{}_{\rho\nu\mu} \frac{\partial x^\nu}{\partial \tau^b} \frac{\partial x^\mu}{\partial \tau^a} T^{\alpha_1...\alpha_k}_{\beta_1...\beta_l}$$

$$+ \frac{\partial x^\nu}{\partial \tau^b} \frac{\partial x^\mu}{\partial \tau^a} \left(-\sum_{i=1}^{k} B^{\alpha_i}{}_{\rho_i\nu\mu} T^{\alpha_1...\rho_i...\alpha_k}_{\beta_1...\beta_l} + \sum_{j=1}^{l} B^{\rho_j}{}_{\beta_j\nu\mu} T^{\alpha_1...\alpha_k}_{\beta_1...\rho_j...\beta_l} \right). \tag{6.16}$$

6.3 The relation between curvature and parallel transport

By definition, a tensor density field $T^{\alpha_1 \ldots \alpha_k}_{\beta_1 \ldots \beta_l}$ is parallely transported along a curve $x^\alpha(\tau)$ when (5.8) holds. This is a linear homogeneous set of first-order differential equations with respect to the functions $T^{\alpha_1 \ldots \alpha_k}_{\beta_1 \ldots \beta_l}(\tau)$. As is well known, the solutions of such equations are linear functions of the initial conditions. Let the initial values of $T^{\alpha_1 \ldots \alpha_k}_{\beta_1 \ldots \beta_l}(\tau)$ be given at $\tau = 0$. Then the solution of (5.8) can be represented as

$$T^{\alpha_1 \ldots \alpha_k}_{\beta_1 \ldots \beta_l}(\tau) = P^{\alpha_1 \ldots \alpha_k \ \ \overline{\beta_1} \ldots \overline{\beta_l}}_{\overline{\alpha_1} \ldots \overline{\alpha_k} \ \ \beta_1 \ldots \beta_l}(\tau) T^{\overline{\alpha_1} \ldots \overline{\alpha_k}}_{\overline{\beta_1} \ldots \overline{\beta_l}}(0), \tag{6.17}$$

where $P^{\alpha_1 \ldots \alpha_k \ \ \overline{\beta_1} \ldots \overline{\beta_l}}_{\overline{\alpha_1} \ldots \overline{\alpha_k} \ \ \beta_1 \ldots \beta_l}(\tau)$ is a linear operator that maps the initial value $T^{\overline{\alpha_1} \ldots \overline{\alpha_k}}_{\overline{\beta_1} \ldots \overline{\beta_l}}(0)$ into the running value $T^{\alpha_1 \ldots \alpha_k}_{\beta_1 \ldots \beta_l}(\tau)$. It has two sets of indices, those with a bar correspond to the initial point $x^\alpha(0)$, those without a bar correspond to the running point $x^\alpha(\tau)$. We will call it the **propagator of parallel transport**. It depends on the two points and, apart from exceptional cases, on the curve along which the transport occurs. It transforms like a tensor density at the point $x^\alpha(\tau)$, with the indices $^{\alpha_1 \ldots \alpha_k}_{\beta_1 \ldots \beta_l}$, multiplied in the sense of a tensor product by a tensor density at the point $x^\alpha(0)$, with the indices $^{\overline{\alpha_1} \ldots \overline{\alpha_k}}_{\overline{\beta_1} \ldots \overline{\beta_l}}$. Such an object is called a **bi-tensor**. In consequence of (5.8), the parallel propagator must obey

$$\frac{D}{d\tau} P^{\alpha_1 \ldots \alpha_k \ \ \overline{\beta_1} \ldots \overline{\beta_l}}_{\overline{\alpha_1} \ldots \overline{\alpha_k} \ \ \beta_1 \ldots \beta_l}(\tau) = 0, \tag{6.18}$$

and, in consequence of (6.17), it must obey the initial condition

$$P^{\alpha_1 \ldots \alpha_k \ \ \overline{\beta_1} \ldots \overline{\beta_l}}_{\overline{\alpha_1} \ldots \overline{\alpha_k} \ \ \beta_1 \ldots \beta_l}(0) = \delta^{\alpha_1}_{\overline{\alpha_1}} \ldots \delta^{\alpha_k}_{\overline{\alpha_k}} \delta^{\overline{\beta_1}}_{\beta_1} \ldots \delta^{\overline{\beta_l}}_{\beta_l}. \tag{6.19}$$

In Eq. (6.18), the covariant differentiation applies only to the indices without a bar because the initial point does not depend on τ.

We will give a special name to the propagator of parallel transport along a closed curve. Let $0 \leq \tau \leq 1$ be a parameter on a closed curve and let $x^\alpha(0) = x^\alpha(1)$ be its beginning–end point. We denote

$$P^{\alpha_1 \ldots \alpha_k \ \ \overline{\beta_1} \ldots \overline{\beta_l}}_{\overline{\alpha_1} \ldots \overline{\alpha_k} \ \ \beta_1 \ldots \beta_l}(1) = S^{\alpha_1 \ldots \alpha_k \ \ \overline{\beta_1} \ldots \overline{\beta_l}}_{\overline{\alpha_1} \ldots \overline{\alpha_k} \ \ \beta_1 \ldots \beta_l}. \tag{6.20}$$

In the symbol $S^{\cdots}_{\cdots}$ all the indices refer to the point $x^\alpha(0) = x^\alpha(1)$, hence $S^{\cdots}_{\cdots}$ is an ordinary tensor density field of type $[w, k+l, k+l]$.

Now let us define elementary propagators: P_w, to transport scalar densities, $P^\alpha_{\overline{\alpha}}$, to transport contravariant vectors, and $P^{\overline{\beta}}_\beta$, to transport covariant vectors. All of them obey (6.18) and the initial conditions

$$P_w(0) = 1, \qquad P^\alpha_{\overline{\alpha}}(0) = \delta^\alpha_{\overline{\alpha}}, \qquad P^{\overline{\beta}}_\beta(0) = \delta^{\overline{\beta}}_\beta. \tag{6.21}$$

The following equation holds:

$$P^{\alpha_1 \ldots \alpha_k \ \ \overline{\beta_1} \ldots \overline{\beta_l}}_{\overline{\alpha_1} \ldots \overline{\alpha_k} \ \ \beta_1 \ldots \beta_l} = P_w P^{\alpha_1}_{\overline{\alpha_1}} \ldots P^{\alpha_k}_{\overline{\alpha_k}} P^{\overline{\beta_1}}_{\beta_1} \ldots P^{\overline{\beta_l}}_{\beta_l} \tag{6.22}$$

(because both its sides obey the same set of differential equations (6.18) and the same set of initial conditions (6.19)).

Similarly, the general propagator of transport along a closed curve, $S^{\cdots\cdots}_{\cdots\cdots}$, is just the tensor product of elementary propagators

$$S^{\alpha_1\ldots\alpha_k \ \ \overline{\beta}_1\ldots\overline{\beta}_l}_{\overline{\alpha}_1\ldots\overline{\alpha}_k \ \ \beta_1\ldots\beta_l} = S_w S^{\alpha_1}_{\overline{\alpha}_1}\ldots S^{\alpha_k}_{\overline{\alpha}_k} S^{\overline{\beta}_1}_{\beta_1}\ldots S^{\overline{\beta}_l}_{\beta_l}. \tag{6.23}$$

Hence, in order to investigate the properties of general propagators, it suffices to deal with the elementary propagators.

Let us consider parallel transport along a curve $x^\alpha(\tau)$ from the point $x^\alpha(0)$ to an arbitrary point $x^\alpha(\tau)$, of the following scalar field:

$$V_\alpha(\tau)W^\alpha(\tau) \equiv P^{\overline{\alpha}}_\alpha(\tau)P^\alpha_{\overline{\beta}}(\tau)V_{\overline{\alpha}}(0)W^{\overline{\beta}}(0). \tag{6.24}$$

Let us take, for the beginning, the special case when the above scalar is constant along the whole curve. Then

$$V_\alpha(\tau)W^\alpha(\tau) = V_{\overline{\alpha}}(0)W^{\overline{\alpha}}(0) = \delta^{\overline{\alpha}}_{\overline{\beta}}V_{\overline{\alpha}}(0)W^{\overline{\beta}}(0). \tag{6.25}$$

Comparing this with the previous equation, we see that

$$\left(P^{\overline{\alpha}}_\alpha(\tau)P^\alpha_{\overline{\beta}}(\tau) - \delta^{\overline{\alpha}}_{\overline{\beta}}\right)V_{\overline{\alpha}}(0)W^{\overline{\beta}}(0) = 0. \tag{6.26}$$

However, the vectors $V_{\overline{\alpha}}(0)$ and $W^{\overline{\beta}}(0)$ are arbitrary, so

$$P^{\overline{\alpha}}_\rho(\tau)P^\rho_{\overline{\beta}}(\tau) = \delta^{\overline{\alpha}}_{\overline{\beta}}. \tag{6.27}$$

Now let us contract this equation with $P^{\overline{\beta}}_\gamma(\tau)V_{\overline{\alpha}}(0)$. The result is

$$\left(P^\rho_{\overline{\beta}}P^{\overline{\beta}}_\gamma - \delta^\rho_\gamma\right)P^{\overline{\alpha}}_\rho(\tau)V_{\overline{\alpha}}(0) = \left(P^\rho_{\overline{\beta}}P^{\overline{\beta}}_\gamma - \delta^\rho_\gamma\right)V_\rho(\tau) = 0. \tag{6.28}$$

Since $V_\rho(\tau)$ is an arbitrary vector, we have as a consequence

$$P^\alpha_{\overline{\beta}}P^{\overline{\beta}}_\gamma = \delta^\alpha_\gamma. \tag{6.29}$$

Equations (6.27) and (6.29) show that the operators $P^{\overline{\alpha}}_\alpha$ and $P^\beta_{\overline{\beta}}$ are inverse to each other: $P^\beta_{\overline{\beta}}$ is not only the propagator of parallel transport of a contravariant vector from $x^\alpha(0)$ to $x^\alpha(\tau)$, but at the same time also the propagator of parallel transport of a covariant vector from $x^\alpha(\tau)$ to $x^\alpha(0)$ *along the same curve*. A similar duality exists for $P^{\overline{\alpha}}_\alpha$.

In consequence of (6.27) and (6.29), the following also hold:

$$S^\alpha_{\overline{\rho}}S^{\overline{\rho}}_\beta = \delta^\alpha_\beta, \qquad S^{\overline{\gamma}}_\sigma S^\sigma_{\overline{\delta}} = \delta^{\overline{\gamma}}_{\overline{\delta}}. \tag{6.30}$$

Therefore we shall consider only the propagators P_w, S_w, $P^{\overline{\alpha}}_\beta$ and $S^{\overline{\alpha}}_\beta$, because $P^\alpha_{\overline{\beta}}$ and $S^\alpha_{\overline{\beta}}$ are algebraically determined by the first four.

Equation (6.18) for $P_w(\tau)$ becomes

$$\frac{dP_w}{d\tau} + w\Gamma^\rho{}_{\rho\sigma}\frac{dx^\sigma}{d\tau}P_w = 0. \tag{6.31}$$

With the initial condition $P_w(0) = 1$, this has the following solution:

$$P_w(\tau) = \exp\left(-w\int_{0}^{\tau}{}_C\Gamma^\rho{}_{\rho\sigma}(t)\frac{dx^\sigma}{dt}(t)dt\right). \tag{6.32}$$

Hence

$$S_w = \exp\left(-w\oint_C \Gamma^\rho{}_{\rho\sigma}(x)dx^\sigma\right). \tag{6.33}$$

Now let us assume that the loop C can be contracted to a point (this assumption will be made in all that follows). Then we can use the Stokes theorem and express the integral along C through the integral over an arbitrary 2-surface leaf S_C spanned on C:

$$\oint_C \Gamma^\rho{}_{\rho\sigma}(x)dx^\sigma = -\int_{S_C}\Gamma^\rho{}_{\rho[\gamma,\delta]}dx^\gamma \wedge dx^\delta = \frac{1}{2}\int_{S_C} B^\rho{}_{\rho\gamma\delta}\,dx^\gamma \wedge dx^\delta. \tag{6.34}$$

Hence, in (6.33)

$$S_w = \exp\left(-\frac{1}{2}w\int_{S_C} B^\rho{}_{\rho\gamma\delta}\,d_2x^{\gamma\delta}\right), \tag{6.35}$$

where $d_2x^{\gamma\delta}$ denotes the surface element $dx^\gamma \wedge dx^\delta$.

To solve (6.18) for $P^{\bar\beta}_\beta$, we span a 2-surface leaf S_C on the loop C, and then we embed C in a one-parameter family of loops $C(\epsilon)$ so that $C(0) = x^\alpha(0) = x^\alpha(1)$ is the single initial/final point of C, and $C(1) \equiv C$ (see Fig. 6.1). Points on S_C will be labelled by τ and ϵ, in such a way that $x^\alpha(0, \epsilon) = x^\alpha(1, \epsilon)$ and $x^\alpha(\tau_1, 0) = x^\alpha(\tau_2, 0)$ for all τ_1 and τ_2.

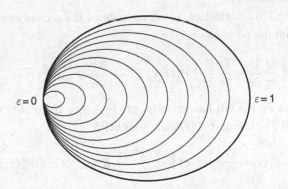

$\varepsilon = 0$ $\varepsilon = 1$

Fig 6.1. Embedding a loop in a one-parameter family of loops. The thicker line is the loop along which we consider the parallel transport. It corresponds to the parameter value $\epsilon = 1$. As ϵ goes down to 0 through all the values in $[0, 1]$, the loop becomes shorter and, in the limit $\epsilon \to 0$, degenerates to the single point where all the loops are tangent. The loops in this figure are ellipses given by the parametric equations $x = \epsilon a[1 - \cos(2\pi t)], y = \epsilon b\sin(2\pi t)$.

Then, the surface element under the integral is

$$dx^\gamma \wedge dx^\delta = \left(\frac{\partial x^\gamma}{\partial \tau} \frac{\partial x^\delta}{\partial \epsilon} - \frac{\partial x^\gamma}{\partial \epsilon} \frac{\partial x^\delta}{\partial \tau} \right) d\tau\, d\epsilon, \tag{6.36}$$

so, for an arbitrary function $F(\tau, \epsilon)$,

$$\int_{S_C} F(\tau, \epsilon) dx^\gamma \wedge dx^\delta = \int_0^1 d\tau \int_0^1 d\epsilon\, F(\tau, \epsilon) \left(\frac{\partial x^\gamma}{\partial \tau} \frac{\partial x^\delta}{\partial \epsilon} - \frac{\partial x^\gamma}{\partial \epsilon} \frac{\partial x^\delta}{\partial \tau} \right). \tag{6.37}$$

Along each loop $C(\epsilon)$ the propagator $P^{\bar\beta}_\beta(\tau, \epsilon)$ is defined by the equation

$$\frac{D}{\partial \tau} P^{\bar\beta}_\beta(\tau, \epsilon) \underset{\epsilon}{\equiv} 0, \tag{6.38}$$

with the initial condition

$$P^{\bar\beta}_\beta(0, \epsilon) \underset{\epsilon}{\equiv} \delta^{\bar\beta}_\beta. \tag{6.39}$$

We also have

$$S^{\bar\beta}_\beta(\epsilon) = P^{\bar\beta}_\beta(1, \epsilon) \tag{6.40}$$

(note that $S^{\bar\beta}_\beta$ in general depends on the curve, i.e. is a function of ϵ).

On the leaf S_C we can consider also the covariant derivatives by the parameter ϵ (i.e. along the curves given by $\tau = $ constant). We have

$$\frac{D}{\partial \epsilon} S^{\bar\beta}_\beta(\epsilon) = \frac{d}{d\epsilon} S^{\bar\beta}_\beta(\epsilon) \tag{6.41}$$

because under a transformation of the parameter ϵ the propagator $S^{\bar\beta}_\beta(\epsilon)$ transforms like a scalar (by substitution only).

Now let us use (6.16) to calculate the commutator of the second covariant derivatives by τ and ϵ acting on the propagator $P^{\bar\beta}_\beta(\tau, \epsilon)$:

$$\left(\frac{D}{\partial \tau} \frac{D}{\partial \epsilon} - \frac{D}{\partial \epsilon} \frac{D}{\partial \tau} \right) P^{\bar\beta}_\beta(\tau, \epsilon) = \frac{\partial x^\gamma}{\partial \tau} \frac{\partial x^\delta}{\partial \epsilon} B^\rho{}_{\beta\gamma\delta} P^{\bar\beta}_\rho(\tau, \epsilon). \tag{6.42}$$

But the second term on the left-hand side is zero by virtue of (6.38). Knowing this, we contract (6.42) with $P^\beta_{\bar\alpha}$. Using (6.38) again, we obtain

$$\frac{D}{\partial \tau} \left(P^\beta_{\bar\alpha}(\tau, \epsilon) \frac{D}{\partial \epsilon} P^{\bar\beta}_\beta(\tau, \epsilon) \right) = \frac{\partial x^\gamma}{\partial \tau} \frac{\partial x^\delta}{\partial \epsilon} B^\rho{}_{\beta\gamma\delta} P^{\bar\beta}_\rho(\tau, \epsilon) P^\beta_{\bar\alpha}(\tau, \epsilon). \tag{6.43}$$

The expression in parentheses on the left-hand side is a scalar with respect to the non-barred indices. For covariant differentiation with respect to τ those quantities that carry only barred indices are scalars (the barred indices refer to the initial points of curves, where $\tau \equiv 0$). Hence, the covariant derivative by τ on the left-hand side of (6.43) reduces to the partial derivative by τ. Then, we integrate the resulting equation by τ from 0 to 1, we make use of (6.39), of the covariant constancy of $\delta_\beta^{\overline{\beta}}$ (axiom 4 of the covariant derivative) and of (6.40). The result is

$$S_{\overline{\alpha}}^\beta(\epsilon)\frac{D}{\partial\epsilon}S_\beta^{\overline{\beta}}(\epsilon) = \oint_0^1 d\tau\,\frac{\partial x^\gamma}{\partial\tau}\frac{\partial x^\delta}{\partial\epsilon}B^\rho{}_{\beta\gamma\delta}P_\rho^{\overline{\beta}}(\tau,\epsilon)P_{\overline{\alpha}}^\beta(\tau,\epsilon). \tag{6.44}$$

Now we contract both sides of the above equation with $S_\alpha^{\overline{\alpha}}$, we make use of (6.41) and integrate the result by ϵ from 0 to 1, obtaining

$$S_\alpha^{\overline{\beta}} - \delta_\alpha^{\overline{\beta}} = \int_0^1 d\epsilon\oint_0^1 d\tau\,\frac{\partial x^\gamma}{\partial\tau}\frac{\partial x^\delta}{\partial\epsilon}B^\rho{}_{\beta\gamma\delta}P_\rho^{\overline{\beta}}(\tau,\epsilon)P_{\overline{\alpha}}^\beta(\tau,\epsilon)S_\alpha^{\overline{\alpha}}(\epsilon)$$

$$= \frac{1}{2}\int_{S_C}B^\rho{}_{\beta\gamma\delta}P_\rho^{\overline{\beta}}(\tau,\epsilon)P_{\overline{\alpha}}^\beta(\tau,\epsilon)S_\alpha^{\overline{\alpha}}(\epsilon)d_2x^{\gamma\delta}. \tag{6.45}$$

Now it is seen that, if $B^\alpha{}_{\beta\gamma\delta} = 0$, then $S_\alpha^{\overline{\beta}} = \delta_\alpha^{\overline{\beta}}$ – the parallel transport of an arbitrary covariant vector along an arbitrary loop that can be contracted to a point reproduces the initial vector. In consequence of (6.30), the same is true for an arbitrary contravariant vector and, in consequence of (6.35), for any scalar density. Then, in consequence of (6.23), with $B^\alpha{}_{\beta\gamma\delta} = 0$, *any* tensor density will return to its initial value when transported around any loop that can be contracted to a point.

The converse theorem also holds: if the parallel transport reproduces the initial tensor density *for every closed loop that can be contracted to a point*, then $B^\alpha{}_{\beta\gamma\delta} = 0$. This is because the integral in (6.45) is then zero for any arbitrary surface leaf S_C. We have thus

Theorem 6.1 *Parallel transport along any closed loop that can be contracted to a point reproduces the initial value of every tensor density if and only if the curvature tensor is zero.*

This is at the same time a necessary and sufficient condition for the parallel transport to be independent of the path.

The manifolds for which $B^\alpha{}_{\beta\gamma\delta} = 0$ are called **flat**.

6.4 Covariantly constant fields of vector bases

Suppose that a set of n vector fields exists on a manifold M_n such that at every point of M_n the vectors defined by these fields are linearly independent, and, moreover, all the fields are covariantly constant: $e^\alpha{}_{a|\beta} \equiv 0$, $a = 1, ..., n$. Then we have

$$0 = e^\alpha{}_{a|\gamma\delta} - e^\alpha{}_{a|\delta\gamma} = B^\alpha{}_{\rho\gamma\delta}e^\rho{}_a. \tag{6.46}$$

Contracting both sides of this with $e^a{}_\beta$ we obtain $B^\alpha{}_{\beta\gamma\delta} = 0$ – the necessary condition for the existence of such a field of bases. We will show that this is a sufficient condition, too. Assume that $B^\alpha{}_{\beta\gamma\delta} = 0$. Then the result of parallel transport does not depend on the path. Hence, with a fixed initial point and fixed initial data at that point, this is a well-defined operation with a unique result at any other point of the manifold. Choose then an arbitrary point $x_0 \in M_n$, and in the tangent space to M_n at x_0 choose an arbitrary basis of vectors $e^\alpha{}_a(x_0)$. Then define bases in spaces tangent to M_n at other points as sets of vectors obtained from $e^\alpha{}_a(x_0)$ by parallel transport. Such a set of vector fields will be covariantly constant on M_n.

The field of bases $e^a{}_\alpha$, dual to $e^\alpha{}_a$, is then also covariantly constant. Choosing such bases to calculate the connection coefficients we obtain

$$\Gamma^\alpha{}_{\beta\gamma} = e_s{}^\alpha e^s{}_{\beta,\gamma} = -e^s{}_\beta e_s{}^\alpha{}_{,\gamma}. \tag{6.47}$$

6.5 A torsion-free flat manifold

If a manifold is not only flat, but also torsion-free, then, for a covariantly constant field of bases, we obtain from (6.47):

$$0 = -e_s{}^\alpha \left(e^s{}_{\beta,\gamma} - e^s{}_{\gamma,\beta} \right). \tag{6.48}$$

Contracting this with $e^a{}_\alpha$ we obtain $e^a{}_{\beta,\gamma} - e^a{}_{\gamma,\beta} \underset{a}{\equiv} 0$. This equation implies that, in a neighbourhood of every point of the manifold, a set of scalar functions $\phi^a(x)$ exists such that $e^a{}_\beta = \phi^a{}_{,\beta}$, $a = 1, \ldots, n$. Since we assumed that the vectors $e^a{}_\alpha$ are linearly independent at every point, i.e. $\det \|e^a{}_\alpha\| \neq 0$, this now implies $\partial(\phi^1, \ldots, \phi^n)/\partial(x^1, \ldots, x^n) \neq 0$. This means that the connection between the functions $\{\phi^a\}_{a=1,\ldots,n}$ and the coordinates $\{x^\alpha\}$ is unique and thus reversible, so the functions $\{\phi^a\}$ can be chosen as new coordinates. Then, let $x^{\alpha'}(x) = \phi^{\alpha'}(x)$. In these coordinates, $e^a{}_{\alpha'} = \phi^a{}_{,\alpha'} = \partial\phi^a/\partial\phi^{\alpha'} = \delta^a{}_{\alpha'}$ and $e_a{}^{\alpha'} = \partial\phi^{\alpha'}/\partial\phi^a = \delta^{\alpha'}{}_a$. Hence, in (6.47) we have $\Gamma^{\alpha'}{}_{\beta'\gamma} = 0$.

In a flat torsion-free manifold one can thus choose such coordinates in which the connection coefficients are zero. In these coordinates, the covariant derivatives reduce to ordinary partial derivatives; we shall call these coordinates Cartesian. A special case (but not the only one) of a flat torsion-free manifold is a Euclidean space (of any dimension).

6.6 Parallel transport in a flat manifold

Since the parallel transport in a flat manifold does not depend on the path, the propagators of parallel transport are functions that depend only on the initial point $\bar{x}$ and on the final point x, but not on the curve that is chosen to join $\bar{x}$ to x. For covariantly constant fields of bases $\{e_A{}^\alpha\}$ and $\{e^A{}_\alpha\}$ we then have

$$P_w = \left[\frac{e(x)}{e(\bar{x})} \right]^w, \qquad P^\alpha{}_{\bar\alpha} = e_s{}^\alpha(x) e^s{}_{\bar\alpha}(\bar{x}), \qquad P^{\bar\beta}{}_\beta = e^s{}_\beta(x) e_s{}^{\bar\beta}(\bar{x}). \tag{6.49}$$

These equations follow because their right-hand sides obey the differential equations for propagators and the corresponding initial conditions.

If the manifold is torsion-free in addition, then $e^A{}_\alpha$ are gradients of scalar functions. Choosing these functions as new coordinates we obtain

$$P_w(x, \bar{x}) = 1, \qquad P^\alpha{}_{\bar\alpha} = \delta^\alpha{}_{\bar\alpha}, \qquad P^{\bar\beta}{}_\beta = \delta_\beta{}^{\bar\beta}. \tag{6.50}$$

Thus, in a flat torsion-free manifold, in Cartesian coordinates, a vector that is transported parallely has the same components at every point and the point to which the vector is attached becomes irrelevant.

6.7 Geodesic deviation

As we will see in Section 12.3, the inertial forces that act on a body moving along a geodesic exactly cancel the gravitational forces. On the other hand, the geodesic lines are the only curves privileged by the geometry of a manifold with connection, hence only they can be used as a privileged reference system. Since a gravitational field cannot be detected on a geodesic, one can only observe neighbouring geodesics from the given one and draw conclusions about the gravitational field on the basis of those observations. The vector field of geodesic deviation is a measure of the position of a particle on a neighbouring geodesic with respect to a particle on a given geodesic.

Let $U \subset M_n$ be an open subset of the manifold M_n with connection. Choose in U a one-parameter family of geodesics labelled by the parameter ϵ. Let $\epsilon = 0$ correspond to that geodesic from which we will be observing the neighbourhood. Every point on any of the geodesics of the family may be identified by the values of the parameter ϵ (whose value identifies the geodesic) and the affine parameter s on that geodesic, $x^{\alpha}|_U = x^{\alpha}(s, \epsilon)$. The **geodesic deviation** is the vector field

$$\delta x^{\alpha}(s) \stackrel{\text{def}}{=} \frac{\partial x^{\alpha}}{\partial \epsilon}\bigg|_{\epsilon=0} \tag{6.51}$$

defined along our geodesic $\epsilon = 0$. We shall find how the deviation changes as we move down this geodesic. Since the set $\{x^{\alpha}(s, \epsilon)\}$ with fixed ϵ and changing s is a geodesic, the following is true:

$$\frac{D}{ds}\frac{\partial x^{\alpha}}{\partial s}\bigg|_{\epsilon=\text{const}} = 0. \tag{6.52}$$

Then, from (5.8),

$$\left(\frac{D}{ds}\frac{D}{d\epsilon} - \frac{D}{d\epsilon}\frac{D}{ds}\right)\frac{\partial x^{\alpha}}{\partial s} = B^{\alpha}{}_{\rho\mu\nu}\frac{\partial x^{\rho}}{\partial s}\frac{\partial x^{\mu}}{\partial s}\frac{\partial x^{\nu}}{\partial \epsilon}. \tag{6.53}$$

But, by virtue of (6.52), the second term on the left-hand side is zero. Now note that

$$\begin{aligned}
\frac{D}{d\epsilon}\frac{\partial x^{\alpha}}{\partial s} &= \frac{\partial^2 x^{\alpha}}{\partial \epsilon \partial s} + \Gamma^{\alpha}{}_{\rho\sigma}\frac{\partial x^{\rho}}{\partial s}\frac{\partial x^{\sigma}}{\partial \epsilon} \\
&= \frac{\partial^2 x^{\mu}}{\partial s \partial \epsilon} + \Gamma^{\alpha}{}_{\sigma\rho}\frac{\partial x^{\sigma}}{\partial \epsilon}\frac{\partial x^{\rho}}{\partial s} + 2\Omega^{\alpha}{}_{\rho\sigma}\frac{\partial x^{\rho}}{\partial s}\frac{\partial x^{\sigma}}{\partial \epsilon} \\
&= \frac{D}{ds}\frac{\partial x^{\alpha}}{\partial \epsilon} + 2\Omega^{\alpha}{}_{\rho\sigma}\frac{\partial x^{\rho}}{\partial s}\frac{\partial x^{\sigma}}{\partial \epsilon}.
\end{aligned} \tag{6.54}$$

Making use of (6.54) in (6.52) we obtain

$$\frac{\mathrm{D}^2}{\mathrm{d}s^2}\frac{\partial x^\alpha}{\partial\epsilon} + 2\frac{\partial x^\rho}{\partial s}\frac{\mathrm{D}}{\mathrm{d}s}\left(\Omega^\alpha{}_{\rho\sigma}\frac{\partial x^\sigma}{\partial\epsilon}\right) = B^\alpha{}_{\rho\mu\nu}\frac{\partial x^\rho}{\partial s}\frac{\partial x^\mu}{\partial s}\frac{\partial x^\nu}{\partial\epsilon}. \tag{6.55}$$

Denoting $k^\alpha = \partial x^\alpha/\partial s$ (the vector field tangent to the geodesic) and δx^α as in (6.51), then substituting $\epsilon = 0$ in the above, we finally obtain

$$\frac{\mathrm{D}^2}{\mathrm{d}s^2}\delta x^\alpha + 2k^\rho\frac{\mathrm{D}}{\mathrm{d}s}\left(\Omega^\alpha{}_{\rho\sigma}\delta x^\sigma\right) - B^\alpha{}_{\rho\mu\nu}k^\rho k^\mu\delta x^\nu = 0. \tag{6.56}$$

This is the **geodesic deviation equation**. It is useful in experimental tests of general relativity, since it allows one (in principle – in practice this is difficult) to calculate the curvature by measuring relative displacements of bodies moving along neighbouring geodesics.

In a flat torsion-free manifold and in the Cartesian coordinates, Eq. (6.56) simplifies to $\mathrm{d}^2(\delta x^\alpha)/\mathrm{d}s^2 = 0$, so its solution is $\delta x^\alpha = A^\alpha s + B^\alpha$, where A^α and B^α are constant vectors. From (6.51) then, the position of the point under observation, $x^\alpha(s, \epsilon_0)$, relative to the corresponding point $x^\alpha(s, 0)$ of the geodesic $\epsilon = 0$ is given by

$$x^\alpha(s, \epsilon_0) = x^\alpha(s, 0) + \epsilon_0\delta x^\alpha. \tag{6.57}$$

Hence, in a flat torsion-free manifold two neighbouring geodesics either diverge or converge with a constant velocity, or remain parallel (when $A^\alpha = 0$). In curved manifolds and in manifolds with torsion, solutions of the geodesic deviation equation are more complicated. (Actually, it is rarely possible to find the solution explicitly. Usually, solutions are found numerically or by perturbations.)

Note that the form of (6.54) does not change when the parameter ϵ is transformed by $\epsilon = \epsilon(\epsilon')$. The derivatives $\mathrm{d}\epsilon'/\mathrm{d}\epsilon$ are constant along geodesics (because ϵ itself is constant), hence the factor $\mathrm{d}\epsilon'/\mathrm{d}\epsilon$ will cancel in (6.56). Changing the parametrisation means merely choosing a different basis in the space of solutions of Eq. (6.56).

6.8 Algebraic and differential identities obeyed by the curvature tensor

The curvature tensor obeys two other important sets of identities, in addition to (6.10). Let us calculate

$$\begin{aligned}
B^\alpha{}_{[\beta\gamma\delta]} &= \frac{1}{3!}\delta^{\rho\sigma\mu}_{\beta\gamma\delta}B^\alpha{}_{\rho\sigma\mu} = \frac{1}{3!}\delta^{\rho\sigma\mu}_{\beta\gamma\delta}\left(-2\Gamma^\alpha{}_{\rho[\sigma,\mu]} + 2\Gamma^\alpha{}_{\nu[\sigma}\Gamma^\nu{}_{|\rho|\mu]}\right) \\
&= \frac{1}{3}\delta^{\rho\sigma\mu}_{\beta\gamma\delta}\left(-\Gamma^\alpha{}_{\rho\sigma,\mu} + \Gamma^\alpha{}_{\nu\sigma}\Gamma^\nu{}_{\rho\mu}\right) = \frac{1}{3}\delta^{\rho\sigma\mu}_{\beta\gamma\delta}\left(-\Omega^\alpha{}_{\rho\sigma,\mu} + \Gamma^\alpha{}_{\nu\sigma}\Omega^\nu{}_{\rho\mu}\right) \\
&= \frac{1}{3}\delta^{\rho\sigma\mu}_{\beta\gamma\delta}\left(-\Omega^\alpha{}_{\rho\sigma|\mu} + \Gamma^\alpha{}_{\nu\mu}\Omega^\nu{}_{\rho\sigma} - \Gamma^\nu{}_{\rho\mu}\Omega^\alpha{}_{\nu\sigma} - \Gamma^\nu{}_{\sigma\mu}\Omega^\alpha{}_{\rho\nu} + \Gamma^\alpha{}_{\nu\sigma}\Omega^\nu{}_{\rho\mu}\right).
\end{aligned} \tag{6.58}$$

(We used the fact that in the contraction over three indices with the antisymmetric $\delta^{\rho\sigma\mu}_{\beta\gamma\delta}$ the antisymmetrisation with respect to $\beta\gamma\delta$ occurs automatically, and we expressed the partial derivative of $\Omega^\alpha{}_{\rho\sigma}$ through the corresponding covariant derivative.) Now let us note that the last term is equal to the second with an opposite sign. Thus

$$B^\alpha{}_{[\beta\gamma\delta]} = \frac{1}{3}\delta^{\rho\sigma\mu}_{\beta\gamma\delta}\left(-\Omega^\alpha{}_{\rho\sigma|\mu} - 2\Omega^\alpha{}_{\nu\sigma}\Omega^\nu{}_{\rho\mu}\right)$$

$$= -2\Omega^\alpha{}_{[\beta\gamma|\delta]} - 4\Omega^\alpha{}_{\nu[\gamma}\Omega^\nu{}_{\beta\delta]}. \tag{6.59}$$

The curvature tensor obeys also the following set of differential identities:

$$B^\alpha{}_{\beta[\gamma\delta|\epsilon]} = -2\Omega^\nu{}_{[\gamma\delta}B^\alpha{}_{|\beta|\epsilon]\nu}, \tag{6.60}$$

called the **Bianchi identities**. For their derivation see Exercise 3.

Recall that the curvature tensor arose by calculating the commutator of covariant derivatives. Commutators of linear operators obey the well-known Jacobi identity. The Bianchi identities arise from the commutators of covariant derivatives in the same way as the Jacobi identity arises from commutators of linear operators.

The Bianchi identities are important for the physical interpretation of relativity. They ensure that equations of motion of material media follow from the field equations and need not be postulated separately.

6.9 Exercises

1. Observe that coordinates can always be adapted to *one* vector field k^α so that, in the new coordinates, $k^\alpha = \delta^\alpha{}_1$.

2. Using the result of the previous exercise show that coordinates can be simultaneously adapted to two vector fields k^α and l^α so that $k'^\alpha = \delta^\alpha_1$ and $l'^\alpha = \delta^\alpha_2$ if and only if the two vector fields commute, $[k, l]^\alpha = 0$. It follows that, if i vector fields all commute, then coordinates can be adapted simultaneously to all of them.

   Hint. To show that this is sufficient, adapt the coordinates to k^α, then find the transformations of coordinates that preserve the property $k'^\alpha = \delta^\alpha_1$, and, using only these transformations, try to adapt the coordinates to l^α.

3. Derive the Bianchi identities (6.60).

   Hint. Use Kronecker 3-deltas to calculate antisymmetrisations, in the same way in which they were used in deriving (6.59).

7

Riemannian geometry

The most detailed sources for the subject of this chapter are Eisenhart (1940), a thorough presentation of differential geometry of curves and 2-dimensional surfaces in a Euclidean space, and Eisenhart (1964), a textbook on Riemannian geometry in n dimensions.

7.1 The metric tensor

Up to here, we have dealt with manifolds on which the only additional object were the affine connection coefficients. That structure allowed us to define the parallel transport and thereby to compare directions of vectors attached to different points. However, we could not calculate distances between points or angles between vectors.

Now we will add a new object that will allow for that – a symmetric second-rank covariant tensor, $g_{\alpha\beta} = g_{(\alpha\beta)}$, called the **metric tensor**. Using it, we can define the **metric form**, also called the **metric**:

$$ds^2 = g_{\alpha\beta}\,dx^\alpha dx^\beta. \tag{7.1}$$

This expression (a scalar) in fact makes sense only under an integral. The length of arc of a curve $x^\alpha(\lambda)$ between the points $x^\alpha(\lambda_0)$ and $x^\alpha(\lambda_1)$ is

$$l_{\lambda_0\lambda_1} = \int_{\lambda_0}^{\lambda_1} ds = \int_{\lambda_0}^{\lambda_1} \left| g_{\alpha\beta}(\lambda)\frac{dx^\alpha}{d\lambda}\frac{dx^\beta}{d\lambda} \right|^{1/2} d\lambda. \tag{7.2}$$

However, (7.1) is a convenient shorthand notation. (We take the absolute value above because the form need not be positive-definite. We will explain later the meaning of $g_{\alpha\beta}\,dx^\alpha dx^\beta \le 0$.)

An example of a manifold with a metric tensor is a Euclidean space (of arbitrary dimension). Its metric tensor in rectangular Cartesian coordinates is the unit matrix, and its metric form is

$$ds^2 = \left(dx^1\right)^2 + \left(dx^2\right)^2 + \cdots + \left(dx^n\right)^2. \tag{7.3}$$

7.2 Riemann spaces

It is logical to require that a vector $k^\alpha(\lambda_0)$ (attached to the point on the curve $x^\alpha(\lambda)$ given by $\lambda = \lambda_0$), when parallely transported along that curve in the affine parametrisation, preserve its length. What condition does such a requirement impose on the metric tensor?

Let $\lambda = s$ be an affine parameter. The length of the vector $k^\alpha(s)$, parallely transported along the curve $x^\alpha(\lambda)$, will not change when

$$0 = \frac{d}{ds}\left(g_{\alpha\beta}k^\alpha k^\beta\right) = \frac{D}{ds}\left(g_{\alpha\beta}k^\alpha k^\beta\right) = \frac{Dg_{\alpha\beta}}{ds}k^\alpha k^\beta \tag{7.4}$$

(see (5.3)). If this equation is to hold for every vector field k^α, then

$$0 = \frac{Dg_{\alpha\beta}}{ds} = \frac{dx^\rho}{ds}g_{\alpha\beta|\rho}. \tag{7.5}$$

Now, this should hold for every curve, which implies in the end

$$g_{\alpha\beta|\rho} = 0, \tag{7.6}$$

or, explicitly,

$$g_{\alpha\beta,\rho} - \Gamma^\sigma{}_{\alpha\rho}g_{\sigma\beta} - \Gamma^\sigma{}_{\beta\rho}g_{\alpha\sigma} = 0. \tag{7.7}$$

Those manifolds, on which a metric tensor obeying (7.6) is defined, and on which in addition

$$\Omega^\alpha{}_{\beta\gamma} = 0 \Longrightarrow \Gamma^\alpha{}_{\beta\gamma} = \Gamma^\alpha{}_{(\beta\gamma)} \tag{7.8}$$

are called **Riemann spaces**. For brevity, we shall use the same name also for those manifolds on which the metric tensor is not positive-definite, although this does not agree with the habit of mathematicians. In mathematics, those manifolds for which $g_{\alpha\beta}$ is not positive-definite are called *pseudo-Riemannian* in general, and sometimes have special names, e.g. the metric used in relativity (both special and general) is called **Lorentzian**. We prefer the name 'Riemann' or 'Riemannian' because the statements we will make apply to the proper Riemannian as well as to pseudo-Riemannian geometries. In order to be perfectly precise, we would thus have to say 'Riemannian and pseudo-Riemannian' every time.

The Riemann spaces are the mathematical basis of relativity. Manifolds with a metric tensor and with torsion are sometimes considered. However, torsion could not so far be connected with any observable effects. Since this exposition is concentrated on the physical aspects of relativity, we shall always keep the assumption (7.8).

7.3 The signature of a metric, degenerate metrics

The expressions $g_{\alpha\beta}\,dx^\alpha dx^\beta$ and $g_{\alpha\beta}k^\alpha k^\beta$ are quadratic forms. A coordinate transformation $\{x\} \to \{x'\}$ changes the form $g_{\alpha\beta}k^\alpha k^\beta$ to $g_{\alpha'\beta'}k^{\alpha'}k^{\beta'}$, where $k^{\alpha'} = x^{\alpha'}{}_{,\alpha}k^\alpha$ and $g_{\alpha'\beta'} = x^\alpha{}_{,\alpha'}x^\beta{}_{,\beta'}g_{\alpha\beta}$. Let X denote the matrix $\|x^{\alpha'}{}_{,\alpha}\|$, let g denote the matrix $\|g_{\alpha\beta}\|$ and let k denote the vector k^α. In the terminology of quadratic forms, the transition $k \to k' = Xk$,

$g \to g' = (X^{-1})^{\mathsf{T}} g X^{-1}$, corresponding to the coordinate transformation $k^{\alpha} \to k^{\alpha'}$, $g_{\alpha\beta} \to g_{\alpha'\beta'}$ is called a transformation of the basis. For matrices of quadratic forms, the Sylvester *theorem on the inertia of forms* holds. It says that, whatever method is used to diagonalise a real symmetric matrix, the numbers of diagonal elements that are positive, zero and negative are always the same. This set of positive, negative and zero values is denoted by the symbol

$$(\underbrace{+, \cdots, +}_{n_1 \text{ times}}, \underbrace{-, \cdots, -}_{n_2 \text{ times}}, \underbrace{0, \cdots, 0}_{n_3 \text{ times}}), \tag{7.9}$$

where n_1 is the number of positive elements, n_2 is the number of negative elements and n_3 is the number of zero elements, $n_1 + n_2 + n_3 = n = \dim V_n$. The symbol (7.9), called the **signature** of the metric tensor, is a basis-independent characteristic property of every matrix, and so is a coordinate-independent characteristic of a point of a Riemann space V_n.

When $n_3 > 0$, the metric is called degenerate. When $n_2 = n_3 = 0$, the metric is positive-definite.

In relativity, a 4-dimensional Riemann space is used, with the signature $(+ - - -)$ or $(- + + +)$ or $(+ + + -)$, depending on the convention (in this text, the signature will always be $(+ - - -)$). Since such a metric is not positive-definite, the Riemann spaces of relativity are not metric spaces. If the distance $d(x, y)$ between the points $x \in V_n$ and $y \in V_n$ is defined by the integral (7.2) along any curve (e.g. a geodesic), then $d(x, y) = 0$ does not imply $x = y$; the integral can vanish also for such pairs of points that do not coincide. The geometrical and physical meaning of its being zero will be explained later.

Coordinate transformations preserve the signature of the metric tensor at each single point. No theorem guarantees that the signature must be the same at different points of the manifold. However, a region with a signature different from $(+ - - -)$ would have no physical interpretation. Usually, the subset on which the signature changes to an unphysical one is, in one sense or another, an 'edge' of the manifold or of the allowed coordinate patch. However, transitions of the type $(+ - - -) \to (- + - -)$ do occur, in which two coordinates interchange their roles and another becomes the privileged one. This happens, for example, at the horizon of a black hole (see Section 14.11).

For a degenerate metric, its determinant is zero and the inverse matrix to $g_{\alpha\beta}$ does not exist. With such a metric, the mapping $k^{\alpha} \to g_{\alpha\beta} k^{\beta}$ is analogous to projecting a vector on a subspace of lower dimension, so no inverse mapping exists. For nondegenerate metrics, the matrix inverse to $g_{\alpha\beta}$ does exist, and is denoted $g^{\alpha\beta}$. It obeys the equation $g^{\alpha\rho} g_{\rho\beta} = \delta^{\alpha}{}_{\beta}$. Thus, in a Riemann space with a nondegenerate metric, for every contravariant vector k^{α} there exists the corresponding covariant vector $g_{\alpha\beta} k^{\beta}$ which is denoted k_{α}, and for every covariant vector m_{α} there exists the corresponding contravariant vector $g^{\alpha\beta} m_{\beta}$ which is denoted m^{α}. The mapping $k^{\alpha} \to k_{\alpha} = g_{\alpha\beta} k^{\beta}$ is called **lowering the index**, the inverse mapping $m_{\alpha} \to m^{\alpha} = g^{\alpha\beta} m_{\beta}$ is called **raising the index**. In such a Riemann space, we can thus consider contravariant and covariant *components* of the same vector. In a general manifold, with no metric defined on it, there is no relation between covariant and contravariant vectors.

7.4 Christoffel symbols

Now we shall solve Eq. (7.7) for $\Gamma^\alpha{}_{\beta\gamma}$ assuming that the metric tensor is nondegenerate. Let us rewrite (7.7) three times, each time with a different permutation of the indices:

$$g_{\alpha\beta,\rho} - \Gamma^\sigma{}_{\alpha\rho}g_{\sigma\beta} - \Gamma^\sigma{}_{\beta\rho}g_{\alpha\sigma} \doteq 0, \tag{7.10}$$

$$g_{\beta\rho,\alpha} - \Gamma^\sigma{}_{\beta\alpha}g_{\sigma\rho} - \Gamma^\sigma{}_{\rho\alpha}g_{\beta\sigma} = 0, \tag{7.11}$$

$$g_{\rho\alpha,\beta} - \Gamma^\sigma{}_{\rho\beta}g_{\sigma\alpha} - \Gamma^\sigma{}_{\alpha\beta}g_{\rho\sigma} = 0. \tag{7.12}$$

Let us add the last two equations, and subtract the first one from the result. Using (7.8) and contracting with $g^{\gamma\rho}$ we obtain

$$\Gamma^\gamma{}_{\alpha\beta} = \frac{1}{2}g^{\gamma\rho}\left(g_{\alpha\rho,\beta} + g_{\beta\rho,\alpha} - g_{\alpha\beta,\rho}\right). \tag{7.13}$$

The affine connection coefficients that are built of the metric tensor in this way are called the **Christoffel symbols** and are denoted $\left\{{}^{\gamma}_{\alpha\beta}\right\}$.

7.5 The curvature of a Riemann space

The curvature tensor of a Riemann space, built of the Christoffel symbols standing in place of the connection coefficients, is called the **Riemann tensor** and denoted $R^\alpha{}_{\beta\gamma\delta}$. Since the Riemann spaces are torsion-free, the identities (6.59) and (6.60) for the Riemann tensor reduce to

$$R^\alpha{}_{[\beta\gamma\delta]} = 0 \tag{7.14}$$

and

$$R^\alpha{}_{\beta[\gamma\delta,\epsilon]} = 0 \tag{7.15}$$

(note that the covariant derivative in a Riemann space is denoted by a semicolon). The Riemann tensor obeys one more set of identities. Applying (6.9) and recalling that a metric is covariantly constant, we obtain

$$0 = g_{\alpha\beta;\gamma\delta} - g_{\alpha\beta;\delta\gamma} = R^\rho{}_{\alpha\gamma\delta}g_{\rho\beta} + R^\rho{}_{\beta\gamma\delta}g_{\alpha\rho}. \tag{7.16}$$

This can be written as

$$R_{\beta\alpha\gamma\delta} + R_{\alpha\beta\gamma\delta} = 0 \Longleftrightarrow R_{\alpha\beta\gamma\delta} = R_{[\alpha\beta]\gamma\delta}. \tag{7.17}$$

From the above, one more identity can be derived. Equation (7.14), in consequence of $R_{\alpha\beta\gamma\delta} = R_{\alpha\beta[\gamma\delta]}$, can be rewritten in the form

$$R_{\alpha\beta\gamma\delta} + R_{\alpha\gamma\delta\beta} + R_{\alpha\delta\beta\gamma} = 0. \tag{7.18}$$

Let us rewrite this equation three more times, each time taking a cyclic permutation of all four indices

$$R_{\beta\gamma\delta\alpha} + R_{\beta\delta\alpha\gamma} + R_{\beta\alpha\gamma\delta} = 0,\tag{7.19}$$

$$R_{\gamma\delta\alpha\beta} + R_{\gamma\alpha\beta\delta} + R_{\gamma\beta\delta\alpha} = 0,\tag{7.20}$$

$$R_{\delta\alpha\beta\gamma} + R_{\delta\beta\gamma\alpha} + R_{\delta\gamma\alpha\beta} = 0.\tag{7.21}$$

Now add the first and fourth equation, and subtract the second and third from the result. Taking into account the antisymmetry in the first pair of indices and in the last pair of indices, we obtain

$$R_{\alpha\beta\gamma\delta} = R_{\gamma\delta\alpha\beta},\tag{7.22}$$

which means that the Riemann tensor is symmetric with respect to the interchange of the first pair of indices with the second.

Since $R_{\alpha\beta\gamma\delta}$ is a tensor of rank 4, it has n^4 components, i.e. 256 in a 4-dimensional space. However, in consequence of the symmetries, only some of the components are algebraically independent. Antisymmetry in (α, β) provides $n(n+1)/2$ equations $R_{(\alpha\beta)\gamma\delta} = 0$ for each set (γ, δ), which leaves us with at most $n^4 - n^2 \cdot n(n+1)/2 = n^2 \cdot n(n-1)/2$ independent components. However, because of the antisymmetry in (γ, δ), we obtain $[n(n+1)/2] \cdot [n(n-1)/2]$ additional equations $R_{\alpha\beta(\gamma\delta)} = 0$, which leaves us with at most $[n(n-1)/2]^2$ independent components. From this, we have to subtract $n\binom{n}{3}$ equations (7.14). This gives finally

$$[n(n-1)/2]^2 - n\binom{n}{3} = \frac{1}{12}n^2\left(n^2 - 1\right)\tag{7.23}$$

independent components, i.e. only 20 when $n = 4$.

7.6 Flat Riemann spaces

If $R^{\alpha}{}_{\beta\gamma\delta} = 0$ in a Riemann space, then the results of Section 6.5 apply – one can choose coordinates so that $\{{}^{\alpha}_{\beta\gamma}\} = 0$. This equation is preserved by all linear coordinate transformations. In such coordinates, as seen from (7.7), the metric has constant components. Then, linear coordinate transformations can be used so that the metric tensor becomes diagonal and all its diagonal elements are either $+1$, -1 or 0, thus

$$ds^2 = \sum_{i=1}^{n}\epsilon_i\left(dx^i\right)^2,$$

where $\epsilon_i = +1, 0$ or -1.

7.7 Subspaces of a Riemann space

Let a subspace S_m of a Riemann space V_n be given by the parametric equations $x^\alpha = f^\alpha (\tau^1, \tau^2, \ldots, \tau^m)$, $\alpha = 1, \ldots, n$. The metric of the space V_n for the pair of points $\{\tau_0^a\} \in S_m$ and $\{\tau_0^a + d\tau^a\} \in S_m$ is

$$g_{\alpha\beta} \, dx^\alpha \, dx^\beta = g_{\alpha\beta} \frac{\partial x^\alpha}{\partial \tau^a} \frac{\partial x^\beta}{\partial \tau^b} \, d\tau^a \, d\tau^b. \qquad (7.24)$$

Hence, $g_{\alpha\beta}(\partial x^\alpha / \partial \tau^a)(\partial x^\beta / \partial \tau^b)$ plays the role of a metric tensor in S_m. We say that the metric of V_n **induces** the metric in S_m by the formula

$$h_{ab} = g_{\alpha\beta} \frac{\partial x^\alpha}{\partial \tau^a} \frac{\partial x^\beta}{\partial \tau^b}. \qquad (7.25)$$

In the 3-dimensional Euclidean space, in rectangular Cartesian coordinates, the length of a curve is

$$l_{\lambda_0, \lambda_1} = \int_{\lambda_0}^{\lambda_1} \left[\left(\frac{dx}{d\lambda} \right)^2 + \left(\frac{dy}{d\lambda} \right)^2 + \left(\frac{dz}{d\lambda} \right)^2 \right]^{1/2} d\lambda. \qquad (7.26)$$

Hence, the metric form here is $ds^2 = dx^2 + dy^2 + dz^2$, and the metric tensor is the unit matrix. Knowing this, we can use (7.25) to find the metric tensor of an arbitrary 2-dimensional surface in the Euclidean space. For example, for the spherical surface of radius a, with the centre at $x = y = z = 0$, the parametric equations in spherical coordinates are:

$$x = a \sin \vartheta \cos \varphi, \qquad y = a \sin \vartheta \sin \varphi, \qquad z = a \cos \vartheta. \qquad (7.27)$$

Hence

$$ds^2 = dx^2 + dy^2 + dz^2 = a^2 \left(d\vartheta^2 + \sin^2 \vartheta \, d\varphi^2 \right), \qquad (7.28)$$

and the metric tensor of the sphere is the matrix

$$\|h_{ab}\| = \begin{pmatrix} a^2 & 0 \\ 0 & a^2 \sin^2 \vartheta \end{pmatrix}. \qquad (7.29)$$

7.8 Flat Riemann spaces that are globally non-Euclidean

A flat space is not necessarily Euclidean. As an example, take a cylinder of radius a. It has parametric equations

$$x = a \cos \varphi, \qquad y = a \sin \varphi, \qquad z = z, \qquad (7.30)$$

i.e. z will be used as the second parameter. Hence

$$dx^2 + dy^2 + dz^2 = a^2 \, d\varphi^2 + dz^2. \qquad (7.31)$$

Thus, the metric tensor of a cylinder has constant coefficients in these coordinates, so the Christoffel symbols will all be zero in the same coordinates, hence, the Riemann tensor will be zero – in these coordinates, and so, being a tensor, also in all other coordinate systems. It follows that the surface of a cylinder is flat in the sense of Riemann geometry. However, it is not Euclidean – because one can travel along a geodesic (one of the circles that are perpendicular to generators) still in the same direction and yet arrive back at the starting point in the end. The path of this journey will be a closed curve that is not continuously contractible to a point. In a Euclidean plane such curves do not exist. Hence, the metric tensor determines the local geometry, but does not determine the global topology. Every 2-dimensional subset of a cylinder that is continuously contractible to a point is isometric to a certain subset of a Euclidean plane.

The surface of an ordinary torus in a Euclidean 3-space has nonzero curvature (see Exercise 1). But a torus can be flat – provided it is embedded in a 4-dimensional space. To see this, let us introduce polar coordinates in the plane $\mathbb{R}^2$:

$$x = r \cos \varphi, \qquad y = r \sin \varphi.$$

In these coordinates, the metric form of the plane becomes:

$$ds_2{}^2 = dx^2 + dy^2 = dr^2 + r^2 d\varphi^2,$$

where, consequently, $\varphi \in [0, 2\pi]$, the point (r, φ) is identical to the point $(r, \varphi + 2\pi)$ and the curves $r = $ constant are circles.

Now take the 4-dimensional Euclidean space with the metric

$$ds_4{}^2 = dx^2 + dy^2 + dz^2 + du^2$$

and introduce the polar coordinates in the planes $(z = $ constant, $u = $ constant$)$ and in the planes $(x = $ constant, $y = $ constant$) : x = r \cos \varphi, y = r \sin \varphi, z = \rho \cos \psi, u = \rho \sin \psi$, where $\varphi \in [0, 2\pi]$, $\psi \in [0, 2\pi]$ and the following points are identical: $(r, \varphi, \rho, \psi) \equiv (r, \varphi, \rho, \psi + 2\pi) \equiv (r, \varphi + 2\pi, \rho, \psi)$. The curves on which (r, ρ, ψ) are constant and the curves on which (r, φ, ρ) are constant are all circles. Now choose the 2-surface given by $r = r_0 = $ constant, $\rho = \rho_0 = $ constant. The metric form of this surface is

$$ds_2{}^2 = r_0{}^2 d\varphi^2 + \rho_0{}^2 d\psi^2.$$

This is a flat surface (because the metric tensor has constant coefficients), but it has all the topological properties of a torus.

7.9 The Riemann curvature versus the normal curvature of a surface

In an n-dimensional Riemann space, the curvature tensor has $n^2(n^2 - 1)/12$ algebraically independent components, which makes 1 when $n = 2$. Hence, the curvature of a 2-dimensional surface is determined by just one function, for example $R = g^{\alpha\beta} R^\rho{}_{\alpha\rho\beta}$. This quantity turns out to be equal to the Gauss curvature, whose description is given below. For the full reasoning see Eisenhart (1940).

We assume that the reader knows the definition of the curvature of a curve. Now consider a 2-dimensional surface S and a point p ∈ S, at which we want to calculate the curvature of S. Draw the straight line L through p that is perpendicular to S, and then consider an arbitrary plane F containing L. The plane intersects the surface along a curve C called the **normal section** of S. Now imagine F being rotated around L, and consider the curvatures at p of the resulting normal sections. It can happen that all these curvatures will be equal – this is the case at every point of a sphere, or when L is a symmetry axis of S. But in general the curvature of C changes when F is rotated, and in the collection of all curvatures there will be a maximal and a minimal value. (If the point p is nonsingular, then all the curvatures will be finite.) The Gauss curvature of the surface S at p is the product of the greatest curvature of C by the smallest curvature. When all curvatures are equal, the curvature of the surface is the square of the curvature of normal sections. Knowing this, one can verify that the Gauss curvature is equal to $g^{\alpha\beta}R^{\rho}{}_{\alpha\rho\beta}$. This shows that the name 'curvature tensor' evokes the right association.

Now it is seen why the curvature of a cylinder is equal to zero: at any point of the cylinder one of the normal sections is a straight line whose curvature is zero, while all the other normal sections have positive curvatures. Hence, the smallest curvature of a normal section is zero. The curvature of a one-sheeted hyperboloid is negative (see Exercise 1). Hence, even though a straight line is among the normal sections of such a hyperboloid, the other normal sections have positive and negative curvatures. Then, the zero curvature is neither maximal nor minimal and does not enter the product contained in the Gauss curvature.

7.10 The geodesic line as the line of extremal distance

If the length of a curve arc is defined on a manifold, then in the collection of all arcs joining two given points we can look for the arc of extremal (i.e. greatest or smallest) length. Thus, we look for a curve on which, with fixed λ_0 and λ_1, the quantity

$$\int_{\lambda_0}^{\lambda_1} \left| g_{\alpha\beta}(x) \frac{dx^{\alpha}}{d\lambda} \frac{dx^{\beta}}{d\lambda} \right|^{1/2} d\lambda \tag{7.32}$$

takes the extremal value. It must obey the Euler–Lagrange equations

$$\frac{d}{d\lambda}\left(\frac{\partial f}{\partial \dot{x}^{\gamma}}\right) - \frac{\partial f}{\partial x^{\gamma}} = 0, \tag{7.33}$$

where f is the integrand in (7.32), and $\dot{x}^{\gamma} \overset{\text{def}}{=} dx^{\gamma}/d\lambda$.

For convenience, let us write

$$\left| g_{\alpha\beta}\dot{x}^{\alpha}\dot{x}^{\beta} \right| = \varepsilon g_{\alpha\beta}\dot{x}^{\alpha}\dot{x}^{\beta}, \tag{7.34}$$

where $\varepsilon = +1$ or $\varepsilon = -1$, as appropriate. Then, from (7.33),

$$\frac{d}{d\lambda}\left(\frac{g_{\gamma\beta}\dot{x}^{\beta}}{\sqrt{\varepsilon g_{\mu\nu}\dot{x}^{\mu}\dot{x}^{\nu}}}\right) - \frac{1}{2}\frac{g_{\alpha\beta,\gamma}\dot{x}^{\alpha}\dot{x}^{\beta}}{\sqrt{\varepsilon g_{\mu\nu}\dot{x}^{\mu}\dot{x}^{\nu}}} = 0. \tag{7.35}$$

The case $g_{\mu\nu}\dot{x}^{\mu}\dot{x}^{\nu} = 0$ requires separate treatment. We shall do this further on, but for now we assume that $g_{\mu\nu}\dot{x}^{\mu}\dot{x}^{\nu} \neq 0$. Let us introduce the new parameter $s(\lambda)$ defined by

$$\frac{ds}{d\lambda} = \sqrt{\varepsilon g_{\mu\nu}\dot{x}^{\mu}\dot{x}^{\nu}}. \tag{7.36}$$

Then we have in (7.35):

$$\frac{ds}{d\lambda}\frac{d}{ds}\left(g_{\gamma\beta}\frac{dx^{\beta}}{ds}\right) - \frac{1}{2}g_{\alpha\beta,\gamma}\frac{ds}{d\lambda}\frac{dx^{\alpha}}{ds}\frac{dx^{\beta}}{ds} = 0. \tag{7.37}$$

Hence

$$g_{\gamma\beta,\alpha}\frac{dx^{\alpha}}{ds}\frac{dx^{\beta}}{ds} + g_{\gamma\beta}\frac{d^{2}x^{\beta}}{ds^{2}} - \frac{1}{2}g_{\alpha\beta,\gamma}\frac{dx^{\alpha}}{ds}\frac{dx^{\beta}}{ds} = 0 \tag{7.38}$$

and, after contracting this with $g^{\alpha\gamma}$,

$$\frac{d^{2}x^{\alpha}}{ds^{2}} + \left\{\begin{matrix}\alpha\\\mu\nu\end{matrix}\right\}\frac{dx^{\mu}}{ds}\frac{dx^{\nu}}{ds} = 0. \tag{7.39}$$

This is the geodesic equation, (5.13). Thus, in Riemann spaces, a geodesic has another characteristic property: it extremises the distance.

Note that the geodesic extremises not only (7.32) but also the functional $\int_{\lambda_0}^{\lambda_1} g_{\mu\nu}\dot{x}^{\mu}\dot{x}^{\nu}d\lambda$. In this case, the assumption $g_{\mu\nu}\dot{x}^{\mu}\dot{x}^{\nu} \neq 0$ is not necessary, and a curve of zero length can also be the extremal.

7.11 Mappings between Riemann spaces

Now let us apply the considerations of Section 3.10 to the Riemann spaces.

If a Riemann space P_m is an image of another manifold M_n (or of a subset thereof) under the mapping $F: M_n \rightarrow P_m$, then the metric tensor of P_m can be pulled back to M_n by the mapping F_2^*. However, M_n can itself be a Riemann space and have its own metric tensor, in general different from the one pulled back. Hence, more than one metric may be defined on the same manifold.

An example of such a situation is a geographic map. It is a projection of a subset of the surface of the Earth into the plane of the sheet of paper. Within the domain of the map, the projection is invertible, i.e. the inverse mapping is defined. This inverse mapping can then be used to pull back the metric of the Earth surface into the plane of the map. Thus, the surface of the page in the atlas has its own Euclidean metric, and also the metric of the surface of the Earth, pulled back by the inverse projection. Navigators who travel over large areas are in fact reading out the metric of a sphere from their maps.

7.12 Conformally related Riemann spaces

Let V_n and U_n be Riemann spaces of the same dimension, and let $\{x^{\alpha}\}$ and $\{y^{a}\}$ be the coordinates in V_n and U_n, respectively, where $\alpha, a = 1, \ldots, n$. Let $F: V_n \rightarrow U_n$ be a diffeomorphism of class C^1, and let $g_{\alpha\beta}$ and h_{ab} be the metric tensors on V_n and U_n. On each of the spaces we then have two metric tensors. For example on U_n there will be its own metric tensor h_{ab}, and the metric pulled back from V_n by

$$\left\{ \left[(F_2)^{-1} \right]^* g \right\}_{ab} (y) = \frac{\partial x^\alpha}{\partial y^a} (y) \frac{\partial x^\beta}{\partial y^b} (y) g_{\alpha\beta} (x(y)). \tag{7.40}$$

If there exists a scalar function $\phi_F : U_n \to \mathbb{R}^1$ such that for any $y \in U_n$

$$\left\{ \left[(F_2)^{-1} \right]^* g \right\}_{ab} (y) = (\phi_F(y))^{-2} h_{ab}(y), \tag{7.41}$$

or, equivalently,

$$\frac{\partial x^\alpha}{\partial y^a} (y) \frac{\partial x^\beta}{\partial y^b} (y) \, g_{\alpha\beta}(x(y)) = (\phi_F(y))^{-2} h_{ab}(y), \tag{7.42}$$

then we call F a **conformal mapping**.[1] The inverse mapping F^{-1} is then conformal, too, and $\phi_{F^{-1}} = 1/\phi_F$. Riemann spaces that can be related by a conformal mapping are called **conformally related**. A Riemann space that is conformally related to a flat Riemann space is called **conformally flat**.

A conformal mapping does not change the angles between vectors. Take two vectors $k^\alpha(x_0)$ and $l^\alpha(x_0)$ of nonzero length at $x_0 \in V_n$, then the angle between them is

$$\cos \alpha^{(V)} = \frac{g_{\alpha\beta} k^\alpha l^\beta}{\sqrt{|g_{\rho\sigma} k^\rho k^\sigma|} \sqrt{|g_{\mu\nu} l^\mu l^\nu|}}. \tag{7.43}$$

Now, using (3.16), the analogue of (3.22) for covariant tensors of rank 2 and (7.42), we obtain for the images of the vectors k and l in the tangent space to U_n at $y_0 = F(x_0)$:

$$h_{rs} (F_{1*}k)^r (F_{1*}k)^s = \phi_F^2 \frac{\partial x^\rho}{\partial y^r} \frac{\partial x^\sigma}{\partial y^s} \frac{\partial y^r}{\partial x^\lambda} \frac{\partial y^s}{\partial x^\tau} g_{\rho\sigma} k^\lambda k^\tau$$

$$= \phi_F^2 \delta^\rho_\lambda \delta^\sigma_\tau g_{\rho\sigma} k^\lambda k^\tau = \phi_F^2 g_{\rho\sigma} k^\rho k^\sigma \tag{7.44}$$

and similarly for the remaining scalar products in (7.43). Hence, the angle $\alpha^{(U)}$ between the images of k and l is determined by

$$\cos \alpha^{(U)} = \frac{h_{ab}(F_1^* k)^a (F_1^* l)^b}{\sqrt{|h_{rs}(F_1^* k)^r (F_1^* k)^s|} \sqrt{|h_{mn}(F_1^* l)^m (F_1^* l)^n|}}$$

$$= \frac{g_{\alpha\beta} k^\alpha l^\beta}{\sqrt{|g_{\rho\sigma} k^\rho k^\sigma|} \sqrt{|g_{\mu\nu} l^\mu l^\nu|}} = \cos \alpha^{(V)}. \tag{7.45}$$

---

[1] The $\phi_F{}^{-2}$ is just a convention that simplifies some of the further formulae; the important point is that the two metrics differ only by a scalar factor.

Equation (7.42) also shows that a conformal mapping maps vectors of zero length on V_n onto vectors of zero length on U_n.

7.13 Conformal curvature

A second-rank tensor can be constructed from the Riemann tensor

$$R_{\alpha\beta} \stackrel{\text{def}}{=} R^{\rho}{}_{\alpha\rho\beta}, \tag{7.46}$$

called the **Ricci tensor**. It is symmetric in $(\alpha\beta)$. In addition, we calculate its trace

$$R \stackrel{\text{def}}{=} g^{\alpha\beta} R_{\alpha\beta} \equiv R^{\alpha}{}_{\alpha}, \tag{7.47}$$

and its trace-free part

$$\overline{R}^{\alpha}{}_{\beta} \stackrel{\text{def}}{=} R^{\alpha}{}_{\beta} - \frac{1}{n}\delta^{\alpha}{}_{\beta}R. \tag{7.48}$$

Using these quantities we can now define a new tensor of rank 4 called the **conformal curvature tensor** or the **Weyl tensor**:

$$C^{\alpha\beta}{}_{\gamma\delta} \stackrel{\text{def}}{=} R^{\alpha\beta}{}_{\gamma\delta} + \frac{1}{n-2}\delta^{\alpha\beta\rho}_{\gamma\delta\sigma}\overline{R}^{\sigma}{}_{\rho} - \frac{1}{n(n-1)}\delta^{\alpha\beta}_{\gamma\delta}R. \tag{7.49}$$

Note: this definition makes sense only for $n > 2$. We shall deal with the case $n = 2$ separately.

The Weyl tensor has all the same symmetries as the Riemann tensor, and in addition all of its traces are zero. In relativity, the Weyl tensor describes that part of the gravitational field that propagates into vacuum and is detectable outside the sources, gravitational waves among other things. It is the same for conformally related manifolds: if $\tilde{g}_{\alpha\beta} = \phi^{-2}g_{\alpha\beta}$, then $\tilde{C}^{\alpha}{}_{\beta\gamma\delta} = C^{\alpha}{}_{\beta\gamma\delta}$ (note the positions of indices; with other positions the two tensors are proportional, but not equal). This fact will be verified in the course of proving the following theorem.

Theorem 7.1 *If in a Riemann space* V_n *the Weyl tensor is zero, then the metric of* V_n *is conformally flat, i.e. there exists a function* ϕ *such that* $g_{\alpha\beta} = \phi^{-2}\eta_{\alpha\beta}$, *where* $R^{\alpha\beta}{}_{\gamma\delta}(\eta) = 0$. *When* $n = 3$, $C^{\alpha\beta}{}_{\gamma\delta} \equiv 0$, *but not every 3-dimensional metric is conformally related to a flat one. The necessary and sufficient condition for a 3-dimensional Riemann space to be conformally flat is the vanishing of the Cotton–York tensor*

$$C^{\alpha\beta} \stackrel{\text{def}}{=} 2\epsilon^{\alpha\gamma\delta}\left(R^{\beta}{}_{\gamma,\delta} - \frac{1}{4}\delta^{\beta}_{\gamma}R_{,\delta}\right), \tag{7.50}$$

where $\epsilon^{\alpha\gamma\delta}$ *is the Levi-Civita symbol. In two dimensions, the Weyl tensor is undetermined, but every 2-dimensional metric is conformally flat.*

Proof: (adapted from Raszewski (1958, pp. 516–521))

Part I: $n > 3$.

We shall show that if $C^{\alpha\beta}{}_{\gamma\delta} = 0$, then the equations determining the function ψ such that $\psi^2 g_{\alpha\beta} = \eta_{\alpha\beta}$ do have a solution. The quantities without a tilde will be those determined by $g_{\alpha\beta}$; those with a tilde are determined by $\tilde{g}_{\alpha\beta} \overset{\text{def}}{=} \psi^{-2} g_{\alpha\beta}$. We introduce the following conventions: all covariant derivatives will be with respect to $g_{\alpha\beta}$, indices of objects without a tilde will be manipulated by $g_{\alpha\beta}$ and indices of objects with a tilde will be manipulated by $\tilde{g}_{\alpha\beta}$.

We first find that

$$\widetilde{\left\{ \begin{matrix} \alpha \\ \beta\gamma \end{matrix} \right\}} = \left\{ \begin{matrix} \alpha \\ \beta\gamma \end{matrix} \right\} - \psi^{-1} \left(\delta^{\alpha}{}_{\beta} \psi,_{\gamma} + \delta^{\alpha}{}_{\gamma} \psi,_{\beta} - g_{\beta\gamma} \psi^{;\alpha} \right). \tag{7.51}$$

(The last term is in fact a new symbol: the derivative of ψ, which is a covariant vector, has its index raised.) Then we find the corresponding relations between the other quantities:

$$\tilde{R}^{\alpha\beta}{}_{\gamma\delta} = \psi^2 \left(R^{\alpha\beta}{}_{\gamma\delta} - \psi^{-1} \delta^{\rho\lambda}_{\gamma\delta} \delta^{\alpha\beta}_{\sigma\rho} \psi;_{\lambda}{}^{;\sigma} - \psi^{-2} \delta^{\alpha\beta}_{\gamma\delta} \psi_{\rho} \psi^{;\rho} \right), \tag{7.52}$$

$$\tilde{R}_{\alpha\beta} = R_{\alpha\beta} + (n-2)\psi^{-1}\psi;_{\alpha\beta} + g_{\alpha\beta}\psi^{-1}\psi;_{\rho}{}^{;\rho} - (n-1)g_{\alpha\beta}\psi^{-2}\psi,_{\rho}\psi^{;\rho}, \tag{7.53}$$

$$\tilde{R} = \psi^2 R + (n-1)\left(2\psi\psi;_{\rho}{}^{;\rho} - n\psi,_{\rho}\psi^{;\rho} \right). \tag{7.54}$$

Using these equations one can now verify that $\tilde{C}^{\alpha}{}_{\beta\gamma\delta} = C^{\alpha}{}_{\beta\gamma\delta}$. Then we require $\tilde{R}^{\alpha}{}_{\beta\gamma\delta} = 0$ (which automatically implies $\tilde{C}^{\alpha}{}_{\beta\gamma\delta} = C^{\alpha}{}_{\beta\gamma\delta} = 0$). By virtue of (7.52) this means

$$R^{\alpha\beta}{}_{\gamma\delta} = \psi^{-1} \delta^{\rho\lambda}_{\gamma\delta} \delta^{\alpha\beta}_{\sigma\rho} \psi;_{\lambda}{}^{;\sigma} + \psi^{-2} \delta^{\alpha\beta}_{\gamma\delta} \psi_{\rho} \psi^{;\rho}. \tag{7.55}$$

This implies that also $\tilde{R}_{\alpha\beta} = 0$ and $\tilde{R} = 0$, so, from (7.54),

$$\psi;_{\rho}{}^{;\rho} = -\frac{1}{2(n-1)}\psi R + \frac{1}{2}n\psi^{-1}\psi,_{\rho}\psi^{;\rho}, \tag{7.56}$$

and, from (7.53) with use of (7.56):

$$\psi;_{\alpha\beta} = -\frac{1}{n-2}\psi R_{\alpha\beta} + \frac{1}{2(n-1)(n-2)}g_{\alpha\beta}\psi R + \frac{1}{2}g_{\alpha\beta}\psi^{-1}\psi,_{\rho}\psi^{;\rho}. \tag{7.57}$$

This equation can be rewritten in the form $\psi,_{\alpha\beta} = $ [the appropriate expression]. Hence, if such a ψ exists, then the integrability conditions $\psi,_{\alpha\beta\gamma} - \psi,_{\alpha\gamma\beta} = 0$ should be fulfilled by the right-hand sides. They are equivalent to $\psi;_{\alpha\beta\gamma} - \psi;_{\alpha\gamma\beta} = R^{\rho}{}_{\alpha\beta\gamma}\psi,_{\rho}$. In this, we must substitute for $R^{\rho}{}_{\alpha\beta\gamma}$ from (7.55) and for all second covariant derivatives of ψ from (7.57). After a long calculation, the following result emerges:

$$-R_{\alpha\beta;\gamma} + R_{\alpha\gamma,\beta} + \frac{1}{2(n-1)}\left(g_{\alpha\beta}R,_{\gamma} - g_{\alpha\gamma}R,_{\beta} \right) = 0. \tag{7.58}$$

The Bianchi identities $R^\alpha{}_{\beta[\gamma\delta;\epsilon]} = 0$, with (7.49) and $C^{\alpha\beta}{}_{\gamma\delta} = 0$ substituted into them, when contracted over (α, γ), become

$$(n-3)(-R_{\alpha\beta;\gamma} + R_{\alpha\gamma,\beta}) + g_{\alpha\beta}R^\rho{}_{\gamma,\rho} - g_{\alpha\gamma}R^\rho{}_{\beta;\rho}$$
$$+ \frac{1}{n-1}(g_{\alpha\gamma}R_{,\beta} - g_{\alpha\beta}R_{,\gamma}) = 0. \tag{7.59}$$

This, when contracted over (α, β), becomes:

$$0 = 3g^{\alpha\beta}R^\rho{}_{\alpha[\rho\beta;\gamma]} = R^\rho{}_{\gamma,\rho} - \frac{1}{2}R_{,\gamma}. \tag{7.60}$$

With $n > 3$, Eq. (7.58) follows from (7.59) and (7.60), and so is fulfilled in consequence of the Bianchi identities. This means that (7.57) is then integrable, so a function ψ exists such that $g_{\alpha\beta} = \psi^2 \eta_{\alpha\beta}$, where $R^\alpha{}_{\beta\gamma\delta}(\eta) = 0$. $\square$

Part II: $n = 3$.
When $n = 3$, Eq. (7.59) follows from (7.60) and does not determine $(-R_{\alpha\beta;\gamma} + R_{\alpha\gamma,\beta})$, so it cannot be equivalent to (7.58). In this case, (7.58) is an additional condition that must be obeyed by the curvature in order that the metric can be conformally flat. This is equivalent to $C^{\alpha\beta} = 0$, where $C^{\alpha\beta}$ is the Cotton–York tensor of (7.50). However, with $n = 3$, the Weyl tensor is identically zero, as can be verified by substituting consecutively all the values of all indices in (7.49). $\square$

Part III: $n = 2$.
With $n = 2$, Eq. (7.49) does not apply since the second term becomes undetermined. However, then

$$R^{\alpha\beta}{}_{\gamma\delta} = \frac{1}{2}\delta^{\alpha\beta}{}_{\gamma\delta}R, \tag{7.61}$$

because any tensor antisymmetric in $(\alpha\beta)$ in two dimensions must be proportional to $\epsilon^{\alpha\beta}$, and similarly for the lower indices. Equations (7.52)–(7.54) still apply, but (7.53) and (7.54) follow from (7.52). Consequently, the only limitation on ψ is in this case the equation $\tilde{R} = 0$, i.e.

$$\psi;{}_\rho{}^{;\rho} = -\frac{1}{2}\psi R + \psi^{-1}\psi,{}_\rho\psi^{;\rho}.$$

This can be equivalently rewritten as

$$(\ln \psi);{}_\rho{}^{;\rho} = \frac{1}{2}R,$$

which is a linear inhomogeneous equation of the type of the Poisson equation (for positive-definite signature) or of the wave equation (for indefinite signature). Hence, in every case it will have a solution, so every 2-dimensional metric is conformally flat. $\square$

7.14 Timelike, null and spacelike intervals in a 4-dimensional spacetime

The physically important 4-dimensional Riemann spaces are those with the signature $(+ - - -)$, as already mentioned. They are called **spacetimes**. Consider the following equation in a spacetime:

$$ds^2 = g_{\alpha\beta}\, dx^\alpha dx^\beta = 0. \tag{7.62}$$

Let us choose a point $p_0 \in V_4$. We can choose coordinates so that at p_0

$$g_{\alpha\beta}\,(p_0) = \begin{bmatrix} +1 & 0 & 0 & 0 \\ 0 & -1 & 0 & 0 \\ 0 & 0 & -1 & 0 \\ 0 & 0 & 0 & -1 \end{bmatrix},$$

i.e. the metric becomes identical with the Minkowski metric of special relativity. (Note: it becomes Minkowskian only at p_0, and it remains approximately Minkowskian in a small neighbourhood of p_0.) In these coordinates, called **locally Cartesian**, Eq. (7.62) taken at p_0 becomes

$$ds^2 = \left(dx^0\right)^2 - \left(dx^1\right)^2 - \left(dx^2\right)^2 - \left(dx^3\right)^2 = 0. \tag{7.63}$$

In a 3-dimensional Euclidean space, the equation $(z - z_0)^2 - (x - x_0)^2 - (y - y_0)^2 = 0$ describes a cone with the vertex at (x_0, y_0, z_0) and the axis parallel to the z-axis. By analogy, the hypersurface in spacetime determined by (7.62) is called a **light cone** (see Fig. 7.1). The coordinate x^0 is called the **time coordinate**, the remaining ones are

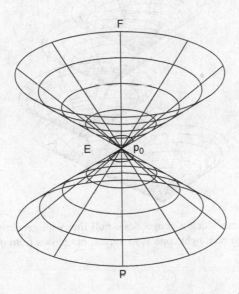

Fig. 7.1. Equation (7.62) determines the light cone that divides the neighbourhood of p_0 into the future F of p_0, the past P of p_0 and 'elsewhere' E. A light cone looks so simple only in the flat spacetime in which the coordinates with the property (7.63) can be introduced globally.

space coordinates. The light cone divides the neighbourhood of p_0 into three disjoint regions. Each point lying on the light cone can be connected to p_0 by a geodesic arc of zero length. All curves of zero length, geodesic or not, are called **null curves**, so the points on the cone are said to be in a **null relation** to p_0. The tangent vector to a null curve at any point has zero length; such vectors are called **null vectors**. Each point in the regions F and P inside the cone can be connected to p_0 by a curve $x^\alpha(\lambda)$ on which $g_{\alpha\beta}\frac{dx^\alpha}{d\lambda}\frac{dx^\beta}{d\lambda} > 0$ everywhere. These points are said to be in a **timelike relation** with p_0, and the vectors of positive length are called **timelike vectors**. Finally, each point in the region E outside the light cone can be connected to p_0 by a curve on which $g_{\alpha\beta}\frac{dx^\alpha}{d\lambda}\frac{dx^\beta}{d\lambda} < 0$ everywhere. These points are said to be in a **spacelike relation** to p_0, and vectors with negative length are called **spacelike vectors**. Justifications of these names come from special relativity. The region F inside the cone, in which the x^0-coordinates of all the points are greater than the x^0-coordinate of p_0, is called the **future** of p_0. The region P is called the **past** of p_0. Finally, the region E does not have a name and is colloquially called 'elsewhere' with respect to p_0.[1]

Note that not every curve lying on the light cone (7.62) has zero length. Only the generators of the cone, which are geodesics, or, more precisely, **null geodesics**, have this property. Other curves on the cone, for example spirals winding on its surface towards p_0, are spacelike curves whose tangent vectors have negative $g_{\alpha\beta}\frac{dx^\alpha}{d\lambda}\frac{dx^\beta}{d\lambda}$. Curves that are null but not geodesic are at each point q tangent to the light cone of q, but veer from one cone to another. Figure 7.2 shows a broken null line whose straight segments are null geodesic

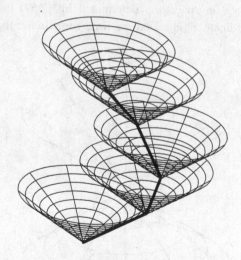

Fig. 7.2. Future light cones along a nongeodesic null line with geodesic segments (the thicker line). The line is tangent to each light cone at its vertex, but passes from one cone to another as it proceeds.

---

[1] The whole of Minkowski spacetime can be divided into F, P, E and the light cone relative to every p_0. In curved spacetimes, the light cones can be complicated hypersurfaces that neither are axially symmetric nor have straight generators, and can have self-intersections. The latter necessarily happens in a sufficiently large neighbourhood of a black hole (see Section 14.11). This is why the division of spacetime by a light cone is well defined only in a finite neighbourhood of each p_0.

arcs, but at the corners it goes from one cone to another. A general nongeodesic null line can be imagined as a limit of a sequence of such broken null lines as the extent of each geodesic segment (as measured, for example, by the range of the affine parameter) goes to zero. Such null lines through p_0 enter the region F, or reach p_0 from within P.

Similarly, not every curve arc in the regions F and P will be timelike. These regions contain also null nongeodesic curves, as described above, and spacelike curves. The characteristic property of the region F∪P is that for each of its points a timelike curve joining it to p_0 *exists*. Such curves do not exist on the light cone and in E. Curves on the light cone that reach p_0 are either null or spacelike, and every curve arc in E that reaches p_0 will have at least a segment that is spacelike.

It should now be clear that a light cone exists at every point of a spacetime, and the analysis above applies to every point. Since ds^2 is a scalar, Eq. (7.62) is covariant, so the light cone is a geometric object; it does not depend on the choice of coordinates.

Finally, let us emphasise again that the equation of a light cone can be reduced to the form (7.63) only at each point separately. Light cones in curved spacetimes do not look as simple as in the Minkowski space – in general they are not axially symmetric, and their spacelike diameters do not uniformly increase with the growth of x^0. They can have self-intersections (caustics) and the structure of a multi-sheeted surface.

7.15 Embeddings of Riemann spaces in Riemann spaces of higher dimension

Let W_m be a subspace of the Riemann space V_n, and let the metric tensor of V_n be $g_{\alpha\beta}$. Let W_m be defined by the parametric equations $x^\alpha = f^\alpha(\tau^1, \ldots, \tau^m)$. Then, as we observed in Section 7.7, W_m is itself a Riemann space of m dimensions, with the metric tensor (7.25).

Now we ask a question reciprocal to the one answered in Section 7.7. When can a given Riemann space V_n be a subspace of another Riemann space? We do not consider *null subspaces* (i.e. subspaces whose normal vectors have zero length) because in them the determinant of the metric is zero, and they require a separate treatment. Also, we will investigate only the question of *local embeddings*, i.e. whether an open subset of V_n can be embedded in another Riemann space. Global embeddings pose additional problems that we will not discuss. If V_n is a subspace of U_N of dimension $N > n$, then a set of functions

$$Y^A = f^A(x^1, \ldots, x^n), \qquad A = 1, \ldots, N, \tag{7.64}$$

should exist such that

$$g_{\alpha\beta} = G_{AB} Y^A{}_{,\alpha} Y^B{}_{,\beta}, \tag{7.65}$$

where G_{AB} is the metric tensor of U_N. Equations (7.64) are parametric equations of our V_n as a subspace of U_N. Note that Y^A and G_{AB} are scalars with respect to coordinate transformations in V_n.

The question of whether a given V_n can be embedded in a U_N can be answered with the help of the reasoning presented below (see Eisenhart (1964) for a more detailed exposition). Let $X_{\widehat{A}}{}^{A}$ be a set of vector fields on U_N (where $\widehat{A} = n+1, \ldots, N$ labels vectors and $A = 1, \ldots, N$ labels their components) that are orthogonal to V_n and orthogonal to each other, with each field being of unit length:

$$G_{AB}X_{\widehat{A}}{}^{A}X_{\widehat{A}}{}^{B} = \varepsilon_{\widehat{A}} = \pm 1 \qquad \text{(no sum over } \widehat{A}\text{)}, \tag{7.66}$$

$$G_{AB}X_{\widehat{A}}{}^{A}X_{\widehat{B}}{}^{B} = 0 \qquad \text{for } \widehat{A} \neq \widehat{B}. \tag{7.67}$$

Since $Y^A{}_{,\alpha}$ are tangent to V_n and $X_{\widehat{A}}{}^{A}$ are orthogonal to V_n, we have

$$G_{AB}Y^A{}_{,\alpha}X_{\widehat{A}}{}^{B} = 0 \tag{7.68}$$

for all α and $\widehat{A}$. We differentiate (7.65) covariantly by x^γ. Since Y^A and G_{AB} are scalars in V_n, we obtain

$$\frac{\partial G_{AB}}{\partial Y^C}Y^A{}_{,\alpha}Y^B{}_{,\beta}Y^C{}_{,\gamma} + G_{AB}\left(Y^A{}_{;\alpha\gamma}Y^B{}_{,\beta} + Y^A{}_{,\alpha}Y^B{}_{;\beta\gamma}\right) = 0. \tag{7.69}$$

Rewriting this equation with indices and signs permuted as $-\{\alpha\beta\gamma\} + \{\alpha\gamma\beta\} + \{\beta\gamma\alpha\}$ and adding all three equations we obtain

$$G_{AB}Y^B{}_{,\gamma}\left(Y^A{}_{;\alpha\beta} + \left\{\begin{matrix} A \\ MN \end{matrix}\right\}_G Y^M{}_{,\alpha}Y^N{}_{,\beta}\right) = 0. \tag{7.70}$$

For a fixed γ, $Y^B{}_{,\gamma}$ is the collection of components of a vector in U_N tangent to V_n and, since γ runs through all n values, the collection $\{Y^B{}_{,\gamma}\}_{\gamma=1,\ldots,n}$ is a basis of the tangent space to V_n at a fixed point $\{x^\alpha\} \in V_n$. Equation (7.70) means that the object in parentheses is orthogonal to all the n tangent vectors to V_n, and so must be spanned on the $(N-n)$ vectors $X_{\widehat{A}}{}^{A}$ that are orthogonal to V_n:

$$Y^A{}_{;\alpha\beta} + \left\{\begin{matrix} A \\ MN \end{matrix}\right\}_G Y^M{}_{,\alpha}Y^N{}_{,\beta} = \sum_{\widehat{S}=n+1}^{N} \varepsilon_{\widehat{S}}\Omega_{(\widehat{S})\alpha\beta}X_{\widehat{S}}{}^{A}, \tag{7.71}$$

where $\Omega_{(\widehat{S})\alpha\beta}$ are the coefficients to be determined below. For a fixed $\widehat{S}$, $\Omega_{(\widehat{S})\alpha\beta}$ is a tensor in V_n, symmetric in $(\alpha\beta)$, and a scalar in U_N; the index $\widehat{S}$ labels different Ωs. Using (7.65)–(7.68), we find:

$$\Omega_{(\widehat{B})\alpha\beta} = G_{AB}Y^A{}_{;\alpha\beta}X_{\widehat{B}}{}^{B} + G_{AB}\left\{\begin{matrix} A \\ MN \end{matrix}\right\}_G Y^M{}_{,\alpha}Y^N{}_{,\beta}X_{\widehat{B}}{}^{B}. \tag{7.72}$$

In the case $N = n+1$, the single quantity $\Omega_{\alpha\beta}$ defined by (7.72) is called the **second fundamental form** of the subspace V_n. Note that

$$Y^A{}_{;\alpha\beta} + \left\{\begin{matrix} A \\ MN \end{matrix}\right\}_G Y^M{}_{,\alpha}Y^N{}_{,\beta} \equiv \left(Y^A{}_{,\alpha}\right)_{|N}Y^N{}_{,\beta}$$

is the directional covariant derivative of $Y^A{}_{,\alpha}$ along $Y^N{}_{,\beta}$ (the subscript $|N$ denotes a covariant derivative in U_N). Consequently, this quantity measures the rate of change of the αth tangent vector $Y^A,_\alpha$ as we move along the βth tangent vector field $Y^N,_\beta$. Then, $\Omega_{(\widehat{B})\alpha\beta}$ is the projection of this rate of change on the $\widehat{B}$th normal vector field to V_n. For this reason, the $\Omega_{(\widehat{B})\alpha\beta}$ are sometimes called the **extrinsic curvatures** of the subspace V_n embedded in U_N. They allow us to 'view' the geometry of V_n from an enveloping space and to see a difference between Riemann spaces that have the same intrinsic Riemann geometry. For example, a plane and a cylinder in a Euclidean 3-space that have identical intrinsic geometries have different second fundamental forms.

In order to know whether a given V_n can be embedded in a given U_N, we have to find out whether the functions Y^A obeying (7.65) exist. They must obey (7.71), which determine the second covariant derivatives of $Y^A,_\alpha$ in V_n. These equations will be solvable if the integrability condition – the Ricci formula (6.9) – is fulfilled:

$$Y^A;_{\alpha\beta\gamma} - Y^A;_{\alpha\gamma\beta} = R^\rho{}_{\alpha\beta\gamma}(g)Y^A,_\rho, \tag{7.73}$$

where $R^\rho{}_{\alpha\beta\gamma}(g)$ is the Riemann tensor of V_n. In calculating the third covariant derivatives of Y^A from (7.71), we will encounter the first derivatives of $X_{\widehat{S}}{}^A$, so we need to know more about them.

Differentiating (7.68) covariantly by x^β, eliminating $G_{AB}Y^A;_{\alpha\beta}X_{\widehat{B}}{}^B$ with use of (7.72) and making use of covariant constancy of G_{AB}:

$$G_{AB,\beta} \equiv G_{AB,C}Y^C,_\beta = \left(\left\{\begin{matrix} R \\ AC \end{matrix}\right\}_G G_{RB} + \left\{\begin{matrix} R \\ BC \end{matrix}\right\}_G G_{AR}\right)Y^C,_\beta \tag{7.74}$$

we obtain another expression for $\Omega_{(\widehat{B})\alpha\beta}$, which is equivalent to (7.72):

$$\Omega_{(\widehat{A})\alpha\beta} = -G_{AB}Y^A,_\alpha X_{\widehat{A}}{}^B,_\beta - G_{AR}\left\{\begin{matrix} R \\ BC \end{matrix}\right\}_G Y^A,_\alpha Y^C,_\beta X_{\widehat{A}}{}^B. \tag{7.75}$$

This can be equivalently written as

$$-\Omega_{(\widehat{A})\alpha\beta} = G_{AB}Y^A,_\alpha \left(X_{\widehat{A}}{}^B{}_{|C}\right)Y^C,_\beta, \tag{7.76}$$

which is a covariant derivative of $X_{\widehat{A}}{}^B$ in U_N projected onto vectors tangent to V_n. This provides another interpretation of the second fundamental forms: $\left(X_{\widehat{A}}{}^B{}_{|C}\right)Y^C,_\beta$ is the rate of change of the $\widehat{A}$th normal vector to V_n as we move along the βth tangent field $Y^C,_\beta$, and $\Omega_{(\widehat{A})\alpha\beta}$ is the projection of that rate of change on the αth tangent vector field $Y^A,_\alpha$. Equation (7.76) clearly shows that $\Omega_{(\widehat{A})\alpha\beta}$ are scalars in U_N.

Now we define the following set of vector fields on V_n:

$$\mu_{|\widehat{RS}|\beta} \overset{\text{def}}{=} G_{AB}X_{\widehat{R}}{}^A \left(X_{\widehat{S}}{}^B{}_{|C}\right)Y^C,_\beta, \tag{7.77}$$

where $\widehat{R}, \widehat{S} = n+1, \ldots, N$. Applying the identity

$$G_{AB}X_{\widehat{R}}{}^A X_{\widehat{S}}{}^B,_\beta = \left(G_{AB}X_{\widehat{R}}{}^A X_{\widehat{S}}{}^B\right),_\beta - \left(G_{AB}X_{\widehat{R}}{}^A\right),_\beta X_{\widehat{S}}{}^B,$$

then making use of (7.66), (7.67) and (7.74) we note that the $\mu_{\widehat{[RS]}\beta}$ are antisymmetric in $[\widehat{RS}]$:

$$\mu_{\widehat{[RS]}\beta} = -\mu_{\widehat{[SR]}\beta},\tag{7.78}$$

so they vanish identically when $N = n+1$.

The derivatives $X_{\widehat{S}}{}^{B}{}_{,\beta}$ are not tensors in U_N. Still, they are objects with one contravariant index in U_N, and so in every fixed coordinate system they can be decomposed in the vector basis $\{Y^{B}{}_{,\beta}, X_{\widehat{S}}{}^{B}\}$; only the coefficients of the decomposition will not be scalars:

$$X_{\widehat{S}}{}^{B}{}_{,\beta} = A_{\widehat{(S)}\beta}{}^{\gamma} Y^{B}{}_{,\gamma} + \sum_{\widehat{P}=n+1}^{N} B_{\widehat{(PS)}\beta} X_{\widehat{P}}{}^{B},\tag{7.79}$$

where the coefficients $A_{\widehat{(S)}\beta}{}^{\gamma}$ and $B_{\widehat{(PS)}\beta}$ are determined when (7.79) is substituted in (7.75) and (7.77). Making use of (7.65)–(7.68) and contracting the result of substitution in (7.75) with $g^{\alpha\gamma}$, we obtain

$$A_{\widehat{(A)}\beta}{}^{\gamma} = -g^{\alpha\gamma}\Omega_{\widehat{(A)}\alpha\beta} - g^{\alpha\gamma}G_{AR}\left\{\begin{matrix} R \\ BC \end{matrix}\right\}_{G} Y^{A}{}_{,\alpha} Y^{C}{}_{,\beta} X_{\widehat{A}}{}^{B},\tag{7.80}$$

$$B_{\widehat{(RS)}\beta} = \varepsilon_{\widehat{R}}\mu_{\widehat{[RS]}\beta} - \varepsilon_{\widehat{R}}G_{BR}\left\{\begin{matrix} R \\ MN \end{matrix}\right\}_{G} Y^{M}{}_{,\beta} X_{\widehat{S}}{}^{N} X_{\widehat{R}}{}^{B}.\tag{7.81}$$

Note now that the set $\{Y^{A}{}_{,\alpha}, X_{\widehat{B}}{}^{A}\}$, $\alpha = 1, \ldots, n; \widehat{B} = n+1, \ldots, N$ is a field of vector bases on U_N, of exactly the kind we used in Section 4.3. In agreement with our considerations there, the metric tensor G_{AB} in U_N can be represented through its scalar components in these bases:

$$\widehat{G}_{\alpha\beta} = Y^{A}{}_{,\alpha} Y^{B}{}_{,\beta} G_{AB} = g_{\alpha\beta}, \qquad \widehat{G}_{\widehat{A}\beta} = X_{\widehat{A}}{}^{A} Y^{B}{}_{,\beta} G_{AB},$$

$$\widehat{G}_{\widehat{A}\widehat{B}} = X_{\widehat{A}}{}^{A} X_{\widehat{B}}{}^{B} G_{AB}, \qquad \widehat{A}, \widehat{B} = n+1, \ldots, N.\tag{7.82}$$

Since $X_{\widehat{A}}{}^{A}$ are orthogonal to all $Y^{A}{}_{,\alpha}$ and to other $X_{\widehat{B}}{}^{A}$, we have:

$$\widehat{G}_{\widehat{A}\beta} = 0, \qquad \widehat{G}_{\widehat{A}\widehat{B}} = \varepsilon_{\widehat{A}}\delta_{\widehat{A}\widehat{B}}.\tag{7.83}$$

Therefore, the inverse metric $\widehat{G}^{AB}$ has the same block-diagonal form

$$\widehat{G}^{\alpha\beta} = g^{\alpha\beta}, \qquad \widehat{G}^{\widehat{A}\beta} = 0, \qquad \widehat{G}^{\widehat{A}\widehat{B}} = \varepsilon_{\widehat{A}}\delta^{\widehat{A}\widehat{B}}.\tag{7.84}$$

The same coefficients $\{Y^{A}{}_{,\alpha}, X_{\widehat{B}}{}^{A}\}$ can then be used to represent the inverse metric G^{AB} via the $\widehat{G}^{\alpha\beta}$, $\widehat{G}^{\widehat{A}\beta}$ and $\widehat{G}^{\widehat{A}\widehat{B}}$. Adapting Eq. (4.16) to our present notation, we have

$$G^{AB} = Y^{A}{}_{,\alpha} Y^{B}{}_{,\beta} g^{\alpha\beta} + \sum_{\widehat{P}=n+1}^{N} \varepsilon_{\widehat{P}} X_{\widehat{P}}{}^{A} X_{\widehat{P}}{}^{B}.\tag{7.85}$$

Using this to eliminate $Y^A,_\alpha Y^B,_\gamma g^{\alpha\gamma}$, we find from (7.80):

$$A_{\widehat{(S)}\beta}{}^\gamma Y^B,_\gamma = -g^{\alpha\gamma}\Omega_{\widehat{(A)}\alpha\beta}Y^B,_\gamma - \begin{Bmatrix} B \\ SC \end{Bmatrix}_G Y^C,_\beta X_{\widehat{A}}{}^S$$

$$+G_{AR}\begin{Bmatrix} R \\ SC \end{Bmatrix}_G Y^C,_\beta X_{\widehat{A}}{}^S \sum_{\widehat{P}=n+1}^N \varepsilon_{\widehat{P}} X_{\widehat{P}}{}^A X_{\widehat{P}}{}^B. \tag{7.86}$$

Using this and (7.81) in (7.79), we see that two of the sums over $\widehat{P}$ cancel out and the result is

$$X_{\widehat{S}}{}^B,_\beta = -g^{\mu\gamma}\Omega_{\widehat{(S)}\mu\beta}Y^B,_\gamma - \begin{Bmatrix} B \\ SC \end{Bmatrix}_G Y^C,_\beta X_{\widehat{S}}{}^S + \sum_{\widehat{P}=n+1}^N \varepsilon_{\widehat{P}}\mu_{[\widehat{PS}]\beta}X_{\widehat{P}}{}^B. \tag{7.87}$$

Now we can employ the integrability condition (7.73). Substituting for $Y^A;_{\alpha\beta}$ from (7.71), then using (7.71) and (7.87) to eliminate the second derivatives of Y^A and the derivatives of $X_{\widehat{S}}{}^A$, we obtain

$$R^A{}_{MNP}(G)Y^M,_\alpha Y^N,_\beta Y^P,_\gamma$$

$$+ \sum_{\widehat{S}=n+1}^N \varepsilon_{\widehat{S}} X_{\widehat{S}}{}^A \left(\Omega_{\widehat{(S)}\alpha\beta;\gamma} - \Omega_{\widehat{(S)}\alpha\gamma;\beta} \right)$$

$$+ \sum_{\widehat{S}=n+1}^N \varepsilon_{\widehat{S}} Y^A,_\rho g^{\rho\mu} \left(\Omega_{\widehat{(S)}\alpha\gamma}\Omega_{\widehat{(S)}\mu\beta} - \Omega_{\widehat{(S)}\alpha\beta}\Omega_{\widehat{(S)}\mu\gamma} \right)$$

$$+ \sum_{\widehat{S}=n+1}^N \sum_{\widehat{P}=n+1}^N \varepsilon_{\widehat{S}}\varepsilon_{\widehat{P}} X_{\widehat{P}}{}^A \left(\Omega_{\widehat{(S)}\alpha\beta}\mu_{[\widehat{PS}]\gamma} - \Omega_{\widehat{(S)}\alpha\gamma}\mu_{[\widehat{PS}]\beta} \right)$$

$$- R^\rho{}_{\alpha\beta\gamma}(g)Y^A,_\rho = 0, \tag{7.88}$$

where $R^A{}_{MNP}(G)$ is the Riemann tensor of U_N calculated at points of V_n. Since $\{Y^A,_\alpha, X_{\widehat{B}}{}^A\}$ are a basis of the tangent space, Eqs. (7.88) are equivalent to the collection of projections of (7.88) on $\{Y^A,_\alpha\}$ and $\{X_{\widehat{B}}{}^A\}$. Contracting (7.88) with $G_{AQ}Y^Q,_\delta$ and with $G_{AQ}X_{\widehat{T}}{}^Q$ and using (7.65)–(7.68), we obtain, respectively

$$R_{\delta\alpha\beta\gamma}(g) = R_{QMNP}(G)Y^Q,_\delta Y^M,_\alpha Y^N,_\beta Y^P,_\gamma$$

$$+ \sum_{\widehat{S}=n+1}^N \varepsilon_{\widehat{S}} \left(\Omega_{\widehat{(S)}\alpha\gamma}\Omega_{\widehat{(S)}\delta\beta} - \Omega_{\widehat{(S)}\alpha\beta}\Omega_{\widehat{(S)}\delta\gamma} \right), \tag{7.89}$$

$$\Omega_{\widehat{(T)}\alpha\beta;\gamma} - \Omega_{\widehat{(T)}\alpha\gamma;\beta} = -R_{QMNP}(G)X_{\widehat{T}}{}^Q Y^M,_\alpha Y^N,_\beta Y^P,_\gamma$$

$$- \sum_{\widehat{S}=n+1}^N \varepsilon_{\widehat{S}} \left(\Omega_{\widehat{(S)}\alpha\beta}\mu_{[\widehat{TS}]\gamma} - \Omega_{\widehat{(S)}\alpha\gamma}\mu_{[\widehat{TS}]\beta} \right). \tag{7.90}$$

Equations (7.89)–(7.90) are called the **Gauss–Codazzi equations**. When $N = n+1$, this is the full set of necessary and sufficient conditions for V_n to be embeddable in U_N. If $N > n+1$, then (7.89)–(7.90) must be supplemented with the integrability conditions of (7.87), $X_{\widehat{S}}{}^B{}_{;\beta\gamma} - X_{\widehat{S}}{}^B{}_{;\gamma\beta} = 0$. Using (7.87), (7.90) and eliminating second derivatives of Y^A by (7.71), we find these conditions to be

$$\sum_{\widehat{P}=n+1}^{N} \varepsilon_{\widehat{P}} X_{\widehat{P}}{}^B \left(\mu_{[\widehat{PS}]\beta;\gamma} - \mu_{[\widehat{PS}]\gamma;\beta} \right)$$

$$+ \sum_{\widehat{P}=n+1}^{N} \sum_{\widehat{R}=n+1}^{N} \varepsilon_{\widehat{P}}\varepsilon_{\widehat{R}} X_{\widehat{R}}{}^B \left(\mu_{[\widehat{PS}]\beta}\mu_{[\widehat{RP}]\gamma} - \mu_{[\widehat{PS}]\gamma}\mu_{[\widehat{RP}]\beta} \right)$$

$$+ g^{\mu\nu} \sum_{\widehat{P}=n+1}^{N} \varepsilon_{\widehat{P}} X_{\widehat{P}}{}^B \left(\Omega_{(\widehat{S})\mu\gamma}\Omega_{(\widehat{S})\nu\beta} - \Omega_{(\widehat{S})\mu\beta}\Omega_{(\widehat{S})\nu\gamma} \right)$$

$$+ R^B{}_{ACD}(G) Y^C{}_{,\beta} Y^D{}_{,\gamma} X_{\widehat{S}}{}^A$$

$$+ g^{\mu\nu} Y^B{}_{,\nu} R_{QMNP}(G) X_{\widehat{T}}{}^Q Y^M{}_{,\mu} Y^N{}_{,\beta} Y^P{}_{,\gamma} = 0. \tag{7.91}$$

As before, this is equivalent to the set of projections on $\{Y^A{}_{,\alpha}\}$ and $\{X_{\widehat{B}}{}^A\}$. However, the projection on $\{Y^A{}_{,\alpha}\}$ is zero, so the other one fully represents (7.91). Contracting (7.91) with $G_{BQ} X_{\widehat{T}}{}^Q$ we obtain

$$\mu_{[\widehat{TS}]\beta;\gamma} - \mu_{[\widehat{TS}]\gamma;\beta} + \sum_{\widehat{P}=n+1}^{N} \varepsilon_{\widehat{P}} \left(\mu_{[\widehat{PS}]\beta}\mu_{[\widehat{TP}]\gamma} - \mu_{[\widehat{PS}]\gamma}\mu_{[\widehat{TP}]\beta} \right)$$

$$+ g^{\mu\nu} \left(\Omega_{(\widehat{T})\mu\gamma}\Omega_{(\widehat{T})\nu\beta} - \Omega_{(\widehat{T})\mu\beta}\Omega_{(\widehat{T})\nu\gamma} \right)$$

$$+ R_{QACD}(G) X_{\widehat{T}}{}^Q Y^C{}_{,\beta} Y^D{}_{,\gamma} X_{\widehat{S}}{}^A = 0. \tag{7.92}$$

In relativity, Eqs. (7.89), (7.90) and (7.92) appear almost always in the special case $N = n+1$ (actually, most often with $N = 4$ and $n = 3$, i.e. for hypersurfaces in spacetime). In that case, they simplify. Equations (7.92) are fulfilled identically (because, as already stated, $\mu_{[\widehat{RS}]\beta} = 0$ in this case, and the indices with a hat run through just one value $N = n+1$, while all terms in (7.92) are antisymmetric in $[\widehat{TS}]$). The Gauss–Codazzi equations then become

$$R_{\delta\alpha\beta\gamma}(g) = R_{QMNP}(G) Y^Q{}_{,\delta} Y^M{}_{,\alpha} Y^N{}_{,\beta} Y^P{}_{,\gamma}$$

$$+ \varepsilon \left(\Omega_{\alpha\gamma}\Omega_{\delta\beta} - \Omega_{\alpha\beta}\Omega_{\delta\gamma} \right), \tag{7.93}$$

$$\Omega_{\alpha\beta;\gamma} - \Omega_{\alpha\gamma;\beta} = -R_{QMNP}(G) X^Q Y^M{}_{,\alpha} Y^N{}_{,\beta} Y^P{}_{,\gamma}, \tag{7.94}$$

where X^Q is the single normal vector to V_n and $\varepsilon = G_{AB} X^A X^B$.

The expression (7.76) may be simplified further when the coordinates in U_{n+1} are adapted to V_n as follows. Through every point of V_n we run a curve $C(p)$ in U_{n+1} orthogonal to V_n and choose the arc length s on the curves as the Y^{n+1} coordinate in U_{n+1} in such a way that $Y^{n+1} = A =$ constant on V_n. The equations $A \neq Y^{n+1} =$ constant then define other hypersurfaces in U_{n+1}. The $\{Y^1, \ldots, Y^n\}$ coordinates in U_{n+1} are chosen so that in V_n they coincide with the intrinsic coordinates of V_n, $Y^\alpha = x^\alpha$, $\alpha = 1, \ldots, n$ on V_n. In such coordinates $G_{(n+1)\alpha}(V_n) = 0$ and $G_{\alpha\beta}(V_n) = g_{\alpha\beta}$. In these coordinates, (7.76) may be written as

$$\Omega_{\alpha\beta} = -X_{\alpha;\beta}. \tag{7.95}$$

In some textbooks, Eq. (7.95) is used as the definition of the second fundamental form of a hypersurface. Although correct in principle, it is rather misleading, since it has the appearance of a fully covariant definition, which it is not. It holds only in the adapted coordinates, and the semicolon in (7.95) denotes not the covariant derivative in V_n, but the V_n components of the covariant derivative in U_N.

Another context in which the Gauss–Codazzi equations sometimes appear in relativity is the problem of embedding a given spacetime V_n in a *flat* Riemann space of higher dimension. Then, Eqs. (7.89), (7.90) and (7.92) should be fulfilled with $R_{ABCD} = 0$. With flat G_{AB}, Eqs. (7.65) are a set of $n(n+1)/2$ differential equations for N unknown functions Y^A. In fact, we do not know what signature G_{AB} should have, so the signs in the canonical form of G_{AB} are additional, discrete unknowns. A simple accounting suggests that $N = n(n+1)/2$, then the set should have a solution (for $n = 4$, $N = 10$). However, this does not take into account various subtle possibilities. For example, if G_{AB} is positive-definite while $g_{\alpha\beta}$ is not, the set (7.65) will be unsolvable with any N.

This problem of embedding in flat Riemann spaces has not been solved in general, and $N \leq n(n+1)/2$ is only a plausible hint. However, for various special cases the embeddings were demonstrated by explicit calculations and often the dimension of U_N is considerably smaller than $n(n+1)/2$. For the smallest N for which an embedding of a given Riemann space in a flat Riemann space is possible, the number $(N - n)$ is called the **class** of the Riemann space. For example, all conformally flat Riemann spaces can be embedded in flat spaces of dimension $(n+2)$ (Plebański, 1967) and thus are of class 2; the 4-dimensional manifold corresponding to a spherically symmetric gravitational field in vacuum can be embedded in a 6-dimensional flat space (see Section 14.10), i.e. is also of class 2. Several other 4-dimensional Riemann spaces can be embedded in a 5-dimensional flat space (Stephani, 1967b).

In the adapted coordinates, Eq. (7.75) reduces to another useful, although non-covariant, form. In these coordinates, X^B has only the $(n+1)$st component, thus $X^\alpha = 0$. Since $Y^\alpha = x^\alpha$ on V_n, the first term in (7.75) becomes $-G_{\alpha(n+1)}X^{n+1},_\beta = 0$ (because $G_{\alpha(n+1)} = 0$). The second term becomes $g_{\alpha\rho}\left\{\begin{matrix}\rho\\(n+1)\beta\end{matrix}\right\}_G X^{n+1}$. It is easy to verify that $\left\{\begin{matrix}\rho\\(n+1)\beta\end{matrix}\right\}_G = -\frac{1}{2}g^{\rho\sigma}g_{\beta\sigma,(n+1)}$, thus

$$\Omega_{\alpha\beta} = -\frac{1}{2}g_{\alpha\beta,(n+1)}X^{n+1}. \tag{7.96}$$

Thus, in the adapted coordinates, $\Omega_{\alpha\beta}$ is proportional to the directional derivative of the metric in V_n along the normal vector.

7.16 The Petrov classification

One unsolved problem of general relativity is that of detecting invariant differences between metrics. Two metrics that seem to be different might be representations of the same metric in two coordinate systems. Finding a transformation between them or proving that no such transformation exists requires solving a complicated set of partial differential equations, for which no general methods of investigation are known. Therefore, each invariant criterion that allows us to detect coordinate-independent differences between metrics is very useful.

One of such criteria is the Petrov classification of algebraic types of the Weyl tensor. It was first introduced by Petrov (1954). Later, a few other methods of introducing the same classification were presented by Pirani (1957) and Debever (1959), and the simplest one, based on spinorial techniques, was proposed by Penrose (1960) (see descriptions of all methods by Stephani *et al.* (2003); the Penrose method is presented in our Chapter 11). We shall use here the method that can be introduced in the briefest way (Ehlers and Kundt, 1962; Barnes, 1984), but it has the disadvantage that it is not simple in practical application. **Note:** this classification applies only in four dimensions and for the Lorentzian signature $(+ - - -)$. Analogous classifications may exist for higher dimensions and for other signatures, but they have not been considered in the literature.

Let u^α be an arbitrary timelike vector field of unit length, $g_{\mu\nu}u^\mu u^\nu = 1$. We define the following two tensors:

$$E_{\alpha\gamma} \overset{\text{def}}{=} C_{\alpha\beta\gamma\delta}u^\beta u^\delta, \tag{7.97}$$

$$H_{\alpha\gamma} \overset{\text{def}}{=} \frac{1}{2}\sqrt{-g}\epsilon_{\alpha\beta\mu\nu}C^{\mu\nu}{}_{\gamma\delta}u^\beta u^\delta, \tag{7.98}$$

called, respectively, the **electric part** and the **magnetic part** of the Weyl tensor; g is the determinant of the metric tensor, necessarily negative because of the signature. (These colloquial names have nothing to do with physical interpretation; they refer to the algebraic analogy between (7.97)–(7.98) and the decomposition of the tensor of electromagnetic field into the electric field and the magnetic field – see Chapter 12.) Both these tensors are symmetric, although the symmetry of $H_{\alpha\gamma}$ is not self-evident (it follows from the traces of $C^{\mu\nu}{}_{\gamma\delta}$ being zero and from the properties of $\epsilon_{\alpha\beta\mu\nu}$). Also, both these tensors represent the Weyl tensor uniquely, since inverse formulae exist (see Exercise 9). We use these tensors to form a new complex tensor,

$$Q_{\alpha\gamma} \overset{\text{def}}{=} E_{\alpha\gamma} + iH_{\alpha\gamma} = Q_{\gamma\alpha}. \tag{7.99}$$

Since $E_{\alpha\gamma}u^\gamma = H_{\alpha\gamma}u^\gamma \equiv 0$, we have $Q_{\alpha\gamma}u^\gamma \equiv 0$, so $Q_{\alpha\gamma}$ is in fact a tensor in a 3-dimensional space. Let us investigate the minimal equation for the matrix $Q = \|Q_{\alpha\gamma}\|$. Since

$$g^{\alpha\gamma}Q_{\alpha\gamma} \equiv 0, \tag{7.100}$$

the sum of eigenvalues of Q must be zero. The following possibilities then arise (I denotes the unit matrix):

The minimal equation	Petrov type
$(Q - \lambda_1 I)(Q - \lambda_2 I)(Q - \lambda_3 I) = 0$	I
$(Q - \lambda I)^2 (Q + 2\lambda I) = 0$	II
$(Q - \lambda I)(Q + 2\lambda I) = 0$	D
$Q^3 = 0$	III
$Q^2 = 0$	N
$Q = 0$	0

The relations between different Petrov types are shown in Fig. 7.3.

The Petrov classification is important because it is coordinate-independent. Two metrics whose Weyl tensors are of different Petrov types cannot be different coordinate representations of the same metric (but one metric *can* be a limiting case of the other). If the Petrov type is the same for two metrics, then the question is still undecided, and other criteria must be used. In general, this **equivalence problem** is not algorithmic, but attempts to solve it 'in practice' are under way, with some success already (Stephani *et al.*, 2003, Chapter 9; Lake, www.grdb.org).

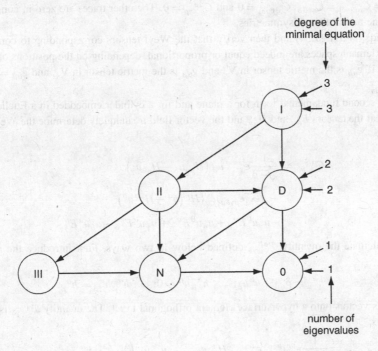

Fig. 7.3. The Petrov classification. Arrows show possible specialisations.

As seen from the formulae above, in order to determine the Petrov type by this particular method, one must choose a timelike vector field u^α. But it can be verified that the Petrov type thus established does not depend on u^α (see Exercise 10).

7.17 Exercises

1. Find the metric tensors of the following surfaces: (*i*) a cone; (*ii*) a paraboloid of revolution; (*iii*) one- and two-sheeted hyperboloids of revolution; (*iv*) a torus; (*v*) a cylinder. Calculate the curvature of a cone, a cylinder, a one-sheeted hyperboloid and a torus.

2. Show that the geodesics on a 2-dimensional sphere are great circles.
   Hint. First derive the equation of a great circle in spherical coordinates, then solve the geodesic equation in the same coordinates.

3. Solve the geodesic equation on the surface of a cylinder. What curves are the geodesics?

4. Show that the Mercator mapping used in cartography is a conformal mapping of a sphere onto a cylinder. This mapping is obtained when a cylinder is wrapped around the globe so that it is tangent to the globe all along the equator. The images of the points on Earth's surface are obtained by drawing straight lines through the centre of the globe. The image of the point where the straight line intersects the globe's surface is the point where the same line intersects the cylinder.

5. Show that a conformal image of a geodesic line of zero length in V_n is also a geodesic line of zero length in U_n.

6. Verify that the Weyl tensor indeed has the properties mentioned after Eq. (7.49), i.e. $C_{\alpha\beta\gamma\delta} = C_{[\alpha\beta]\gamma\delta} = C_{\alpha\beta[\gamma\delta]} = C_{\gamma\delta\alpha\beta}$, $C_{\alpha[\beta\gamma\delta]} = 0$ and $C^{\alpha\rho}{}_{\gamma\rho} = 0$. The other traces are zero in consequence of this one and of the antisymmetries.

7. Verify Eqs. (7.51)–(7.54) and then verify that the Weyl tensors corresponding to conformally related Riemann spaces are indeed equal or proportional, depending on the positions of indices. Namely, if $\widetilde{g}_{\alpha\beta}$ is the metric tensor in $\widetilde{V}_n$ and $g_{\alpha\beta}$ is the metric tensor in V_n, and $\widetilde{g}_{\alpha\beta} = \phi^{-2}g_{\alpha\beta}$, then $\widetilde{C}^\alpha{}_{\beta\gamma\delta} = C^\alpha{}_{\beta\gamma\delta}$.

8. Find the second fundamental form for a plane and for a cylinder embedded in a Euclidean E^3.

9. Verify that the tensors $E_{\alpha\gamma}$ and $H_{\alpha\gamma}$ and the vector field u^α uniquely determine the Weyl tensor $C^{\alpha\beta}{}_{\gamma\delta}$ by

$$C^{\alpha\beta}{}_{\gamma\delta} = \frac{1}{\sqrt{-g}}\epsilon^{\rho\alpha\beta\sigma}u_\sigma\left(H_{\rho\gamma}u_\delta - H_{\rho\delta}u_\gamma\right)$$

$$+\sqrt{-g}\,\epsilon_{\rho\gamma\delta\sigma}u^\sigma\left(H^{\rho\beta}u^\alpha - H^{\rho\alpha}u^\beta\right)$$

$$-u_\gamma u^\beta E^\alpha{}_\delta + u_\delta u^\beta E^\alpha{}_\gamma + u_\gamma u^\alpha E^\beta{}_\delta - u_\delta u^\alpha E^\beta{}_\gamma. \qquad (7.101)$$

Hint. Calculate the quantity $\mathcal{P}^{\alpha\beta}{}_{\gamma\delta}$ defined below in two ways. First introduce the auxiliary operator

$$h^\alpha{}_\mu = \delta^\alpha{}_\mu - u^\alpha u_\mu; \qquad h^\alpha{}_\mu u^\mu = 0; \qquad h^\alpha{}_\mu h^\mu{}_\beta = h^\alpha{}_\beta. \qquad (7.102)$$

It projects vectors onto a hypersurface element orthogonal to u^α. The quantity $\mathcal{P}^{\alpha\beta}{}_{\gamma\delta}$ is defined as follows:

$$\mathcal{P}^{\alpha\beta}{}_{\gamma\delta} \overset{\text{def}}{=} \frac{1}{\sqrt{-g}}\epsilon^{\rho\mu_1\mu_2\mu_3}\epsilon^{\sigma\nu_1\nu_2\nu_3}H_{\rho\sigma}\epsilon_{\lambda_1\lambda_2\nu_2\nu_3}u_{\mu_1}u_{\nu_1}h^{\lambda_1}{}_\gamma h^{\lambda_2}{}_\delta h^\alpha{}_{\mu_2}h^\beta{}_{\mu_3}. \qquad (7.103)$$

The two ways to calculate it are these:

(i) Substitute $\epsilon^{\sigma \nu_1 \nu_2 \nu_3} \epsilon_{\lambda_1 \lambda_2 \nu_2 \nu_3} = 2 \delta^{\sigma \nu_1}_{\lambda_1 \lambda_2}$, and write out all h-operators as in (7.102). You will find that $\mathcal{P}^{\alpha \beta}{}_{\gamma \delta} = 0$.

(ii) Substitute for $H_{\rho \sigma}$ from (7.101), and use the following identity:

$$C^{\tau_2 \tau_3}{}_{\gamma \tau_4} = \frac{1}{2} \delta^{\epsilon_1 \epsilon_2}_{\gamma \tau_4} C^{\tau_2 \tau_3}{}_{\epsilon_1 \epsilon_2} = \frac{1}{4} \epsilon^{\epsilon_1 \epsilon_2 \epsilon_3 \epsilon_4} \epsilon_{\gamma \tau_4 \epsilon_3 \epsilon_4} C^{\tau_2 \tau_3}{}_{\epsilon_1 \epsilon_2}. \tag{7.104}$$

Then, using $\epsilon^{\rho \alpha \beta \sigma} \epsilon_{\rho \tau_1 \tau_2 \tau_3} = \delta^{\alpha \beta \sigma}_{\tau_1 \tau_2 \tau_3}$, express $\mathcal{P}^{\alpha \beta}{}_{\gamma \delta}$ through $C^{\alpha \beta}{}_{\gamma \delta}$ and the (single and double) projections of $C^{\alpha \beta}{}_{\gamma \delta}$ on u^δ. Equation (7.101) results with the help of the following two auxiliary formulae:

$$C^{\rho \sigma}{}_{\gamma \delta} u^\delta = \frac{1}{\sqrt{-g}} \epsilon^{\alpha \rho \sigma \lambda} u_\lambda H_{\alpha \gamma} - u^\rho E^\sigma{}_\gamma + u^\sigma E^\rho{}_\gamma, \tag{7.105}$$

$$\epsilon_{\alpha \beta \gamma \delta} = g^{-1} g_{\alpha \mu} g_{\beta \nu} g_{\gamma \rho} g_{\delta \sigma} \epsilon^{\mu \nu \rho \sigma}. \tag{7.106}$$

10. Show that the Petrov type determined by the method of Section 7.16 does not depend on the choice of the vector field u^α.

Hint. Consider a Lorentz transformation that changes u^α into another timelike vector w^α and verify what happens to Q and its minimal equation in consequence of this. The matrix of a Lorentz transformation, $u^\alpha = L^\alpha{}_\beta w^\beta$, has the property $g_{\alpha \beta} L^\alpha{}_\gamma L^\beta{}_\delta = g_{\gamma \delta}$.

8

Symmetries of Riemann spaces, invariance of tensors

8.1 Symmetry transformations

We noted in Section 3.10 that a coordinate transformation on a manifold may be interpreted as a mapping of the manifold into itself. Now we shall interpret the associated transformations of the tensor fields.

Let $F: M_n \to M_n$ be an isomorphism between two subsets of the manifold M_n, $p \in M_n$ and $F(p) = p' \in M_n$. Then, the mappings associated with F carry tensors from p to p'. Thereby, a tensor T that was attached to p before the transformation becomes T' attached to p'. Now consider a subset $U \subset M_n$ and its image $F(U) \subset M_n$. Suppose that F is an element of a continuous group of mappings $\{F_t\}$, with $\{F_{t_0}\}$ being the identity map. If $F = F_{t_1}$ and $|t_1 - t_0|$ is sufficiently small, then $F(U) \cap U \neq \emptyset$. So let $p, p' \in F(U) \cap U$. Then p is an image of another point q, $p = F(q)$, and the tensors that were attached to q before the transformation were sent into p (see Fig. 8.1). Hence, we have two tensors attached to each point p: $T(p)$ that was there before the transformation and $T'(p)$ that was sent to p from q by the transformation. The latter can be calculated from the old $T(q)$ by (3.10). Consequently, we can compare $T'(p)$ with $T(p)$.

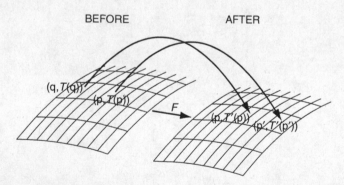

Fig. 8.1. A mapping F of the manifold M_n into itself takes the point q to p, and the associated mapping transforms the tensors $T(p)$ and $T(q)$ into $T'(p')$ and $T'(p)$, respectively. Thus, after the transformation we have two tensors at the same point: $T(p)$ that was there before, and $T'(p)$ that was brought to p by the mapping. If $T'(p) = T(p)$ for all $p \in M_n$, then the tensor field T is invariant under F.

If it so happens that, under the mapping F, $T'(p) = T(p)$ for all points of the manifold, then we call the tensor field T **invariant under the action of** F, and we call F an **invariance transformation** of T. If M_n is a Riemann space, and the metric tensor of M_n is invariant under F, then the mapping F is called a **symmetry** or an **isometry** of M_n.

There is no general theory of all kinds of invariances; for example, no theory covers discrete invariances (like reflections or the groups known in crystallography). However, there exists a theory of those mappings that constitute continuous groups, and it will be the subject of this chapter.

8.2 The Killing equations

Let Γ be a one-parameter family of mappings of a manifold M_n into itself such that to every value of the parameter t from the range $[t_1, t_2] \stackrel{\text{def}}{=} B \subset \mathbb{R}^1$ there corresponds a mapping $f_t : M_n \to M_n$:

$$\mathbb{R}^1 \supset B \stackrel{\text{def}}{=} [t_1, t_2] \ni t \to x'^\alpha = f^\alpha(t, \{x\}), \tag{8.1}$$

where f_t is the collection of all the functions f^α at a given t and Γ is the collection of all the mappings f_t for every value of $t \in B$. Let us also assume that for $t = t_0$ where $t_1 \le t_0 \le t_2$ the mapping f_{t_0} is an identity:

$$f^\alpha(t_0, \{x\}) = x^\alpha. \tag{8.2}$$

Example: let $B = [0, 2\pi]$, $M_n = \mathbb{R}^3$, and let f_t be the rotation of $\mathbb{R}^3$ around a fixed axis A by the angle t. Γ is then the collection of rotations of $\mathbb{R}^3$ around A by all angles in the range $0 \le t < 2\pi$ and $t_0 = 0$.

Now let $p \in M_n$, and apply to p the mappings f_t corresponding to all $t \in B$. The collection of all images of p will then be an arc of a curve in M_n passing through $p = f_{t_0}(p)$, and each $p \in M_n$ may be used to generate such an arc. The arc is called the **orbit of p under the action of** Γ, and p is called the initial point of the orbit (although in 'practical' instances the orbits are closed or infinite curves with no endpoints).

We assume that (1) the functions $f^\alpha(t, \{x\})$ are of class C^2 with respect to t; (2) each f_t is invertible; and (3) its inverse (denoted f_t^{-1}) is also of class C^2. Assumption (1) implies that along each orbit a field of tangent vectors exists and is continuously differentiable. Assumption (2) implies that the mappings of the family Γ and their inverses form a group G. The group multiplication is the superposition of the mappings:

$$\left(f_{t_2} \circ f_{t_1}\right)(x) = f_{t_2}\left(f_{t_1}(x)\right), \tag{8.3}$$

where f_{t_1} is represented by $f^\alpha(t_1, x)$ and f_{t_2} is represented by $f^\alpha\left(t_2, f_{t_1}(x)\right)$. Assumption (3) guarantees that the orbits generated by G will have a continuously differentiable field of tangent vectors. For each mapping f_t we may then write (from Taylor's formula):

$$x'^\alpha = x^\alpha + \left.\frac{\partial f^\alpha}{\partial t}\right|_{t=t_0} (t - t_0) + O\left(\epsilon^2\right), \tag{8.4}$$

where $x'^\alpha = f^\alpha(t, \{x\})$, $x^\alpha = f^\alpha(t_0, \{x\})$, $\epsilon \overset{\text{def}}{=} t - t_0$ and $O(\epsilon^2)$ has the property

$$\lim_{\epsilon \to 0} \frac{O(\epsilon^2)}{\epsilon} = 0. \tag{8.5}$$

Note that we are not making any approximation here; Eq. (8.4) is exact, but the form of $O(\epsilon^2)$ will in the end turn out to be irrelevant. All such irrelevant terms will be denoted by the same symbol $O(\epsilon^2)$ even though they may not be identical to each other.

The quantities

$$k^\alpha \overset{\text{def}}{=} \frac{\partial f^\alpha}{\partial t}\bigg|_{t=t_0} \tag{8.6}$$

are components of the vector field tangent to the orbits at their initial points and are called the generators of the group G.

Suppose that a tensor $T_{\alpha\beta}$ is invariant under all the transformations in Γ:

$$T'_{\alpha\beta}(\mathrm{p}) = T_{\alpha\beta}(\mathrm{p}) \qquad \text{for all } \mathrm{p} \in \mathrm{M}_n \text{ and } t \in \mathrm{B}. \tag{8.7}$$

What analytic condition must $T_{\alpha\beta}$ fulfil? Again from Taylor's formula:

$$T'_{\alpha\beta}(\mathrm{p}') = T'_{\alpha\beta}(\mathrm{p}) + \epsilon T'_{\alpha\beta,\mu}(\mathrm{p})k^\mu + O(\epsilon^2), \tag{8.8}$$

where p' has the coordinates $x'^\alpha = f^\alpha(t, \{x\})$ and p has the coordinates $x^\alpha = f^\alpha(t_0, \{x\})$. From (8.4) and (8.6) we have

$$\frac{\partial x^\mu}{\partial x'^\alpha} = \frac{\partial}{\partial x'^\alpha}\left[x'^\mu - \epsilon k^\mu - O(\epsilon^2)\right] = \delta^\mu{}_\alpha - \epsilon k^\mu{}_{,\rho}\frac{\partial x^\rho}{\partial x'^\alpha} - O\left(\epsilon^2\right)$$

$$= \delta^\mu{}_\alpha - \epsilon k^\mu{}_{,\rho}\left(\delta^\rho{}_\alpha - \epsilon k^\rho{}_{,\sigma}\frac{\partial x^\sigma}{\partial x'^\alpha} - O(\epsilon^2)\right) - O(\epsilon^2)$$

$$= \delta^\mu{}_\alpha - \epsilon k^\mu{}_{,\alpha} + O(\epsilon^2). \tag{8.9}$$

(We do not differentiate ϵ because we have advanced along the orbits by the fixed parameter distance $\epsilon = t - t_0$ which is coordinate-independent.) Hence, from (8.9) and the transformation law for $T_{\alpha\beta}$

$$T'_{\alpha\beta}(\mathrm{p}') = \frac{\partial x^\mu}{\partial x'^\alpha}\frac{\partial x^\nu}{\partial x'^\beta}T_{\mu\nu}(\mathrm{p})$$

$$= T_{\alpha\beta}(\mathrm{p}) - \epsilon k^\mu{}_{,\alpha}T_{\mu\beta}(\mathrm{p}) - \epsilon k^\nu{}_{,\beta}T_{\alpha\nu}(\mathrm{p}) + O(\epsilon^2). \tag{8.10}$$

Comparing (8.8) with (8.10) and using (8.7), we have

$$\epsilon\left(k^\mu T_{\alpha\beta,\mu} + k^\mu{}_{,\alpha}T_{\mu\beta} + k^\nu{}_{,\beta}T_{\alpha\nu}\right) + O(\epsilon^2) = 0. \tag{8.11}$$

We now divide (8.11) by ϵ, let $\epsilon \to 0$ and recall (8.5). The result is

$$k^\mu T_{\alpha\beta,\mu} + k^\mu{}_{,\alpha}T_{\mu\beta} + k^\mu{}_{,\beta}T_{\alpha\mu} = 0. \tag{8.12}$$

These are the **Killing equations**, and their solutions $k^\mu(\{x\})$ are called **Killing vector fields**. Every field of tangent vectors to orbits of invariance transformations of $T_{\alpha\beta}$ must fulfil (8.12), and every solution of (8.12) generates an invariance group of $T_{\alpha\beta}$. How to find the invariance transformations given k^α and vice versa will be shown in the next section.

Equations (8.12) can be rewritten in an equivalent form that shows explicitly their covariance:

$$k^\mu T_{\alpha\beta;\mu} + k^\mu{}_{;\alpha} T_{\mu\beta} + k^\mu{}_{;\beta} T_{\alpha\mu} = 0. \tag{8.13}$$

If $T_{\alpha\beta} = g_{\alpha\beta}$ (the metric tensor), then, in view of $g_{\alpha\beta;\gamma} = 0$, Eq. (8.13) may be rewritten as

$$k_{\alpha;\beta} + k_{\beta;\alpha} \equiv 2k_{(\alpha;\beta)} = 0. \tag{8.14}$$

In this form, the Killing equations are most easy to remember, but less convenient to work with (and apply only to the metric tensor).

Note that Eq. (8.12) applies only to a doubly covariant tensor field, and only in this case are the generators of invariances called Killing vector fields. We shall deal with invariances of other tensor fields in Section 8.5 and thereafter.

8.3 The connection between generators and the invariance transformations

If $x'^\alpha = f^\alpha(t, \{x\})$ is a family of invariance transformations of a certain tensor field, then the corresponding generator is given by (8.6), where $t = t_0$ corresponds to the identity transformation.

Finding the family of invariances given k^α is less straightforward. Any orbit $B \ni t \to x'^\alpha(t) = f^\alpha(t, \{x\})$ is at its every point tangent to $k^\alpha(p')$ (where $\{x'\}$ are the coordinates of p'). Hence the orbits must obey

$$\frac{dy^\alpha}{dt} = k^\alpha(\{y\}) \tag{8.15}$$

with the initial conditions

$$y^\alpha\big|_{t=t_0} = x^\alpha. \tag{8.16}$$

Equations (8.15)–(8.16) are to be understood as follows. A solution to (8.15) will be a family of curves $y^\alpha = f^\alpha(t, C_1, \ldots, C_n)$, labelled by n parameters $(C_1, \ldots, C_n)$. The condition (8.16) allows one to express the constants C_α in terms of the coordinates x^α of the initial points of the curves. In this way, we obtain the set of functions

$$y^\alpha = f^\alpha(t, \{x\}) \tag{8.17}$$

that satisfies Eqs. (8.15) and the initial conditions (8.16).

8.4 Finding the Killing vector fields

The Killing equations are applied to two kinds of problems:

1. Finding the metric tensor of a Riemann space whose symmetries are assumed – then they are equations determining $g_{\alpha\beta}$, with k^α given.
2. Finding the symmetries of a Riemann space whose metric tensor is given – then they are equations determining k^α with $g_{\alpha\beta}$ given.

An example of the first application will be shown in Section 8.9. The second application requires additional explanation. The Killing equations are linear and homogeneous in k^α, which means that if k^α and l^α are Killing fields, then so is $(Ak^\alpha + Bl^\alpha)$, where A and B are arbitrary constants. A general solution of the Killing equations should thus be a linear combination of basis solutions.

Does there exist a finite basis in the space of solutions of the Killing equations? The answer is: yes, but only for the proper Killing equations, i.e. for the generators of invariances of the metric tensor. The proof given below (borrowed from Stephani (1990)) does not work if K_α generates an invariance group of a tensor field other than the metric tensor (and examples of infinite bases are known (Krasiński, 1983)).

For a field K_α generating symmetries of M_n we have from (8.14)

$$K_{\alpha;\beta} = -K_{\beta;\alpha}, \tag{8.18}$$

and from the Ricci identity

$$K_{\alpha;\beta\gamma} - K_{\alpha;\gamma\beta} = R^\rho{}_{\alpha\beta\gamma} K_\rho. \tag{8.19}$$

Because $R^\rho{}_{[\alpha\beta\gamma]} \equiv 0$, we have from the above

$$\left(K_{\alpha;\beta} - K_{\beta;\alpha}\right)_{;\gamma} + \left(K_{\gamma;\alpha} - K_{\alpha;\gamma}\right)_{;\beta} + \left(K_{\beta;\gamma} - K_{\gamma;\beta}\right)_{;\alpha} = 0. \tag{8.20}$$

Using now (8.18), the above reduces to

$$K_{\alpha;\beta\gamma} + K_{\gamma;\alpha\beta} + K_{\beta;\gamma\alpha} = 0, \tag{8.21}$$

and, again from (8.18), this yields

$$K_{\gamma;\alpha\beta} = K_{\beta;\alpha\gamma} - K_{\beta;\gamma\alpha} = R^\rho{}_{\beta\alpha\gamma} K_\rho. \tag{8.22}$$

Thus, in a given Riemannian manifold (where $g_{\alpha\beta}$ and, consequently, $R^\alpha{}_{\beta\gamma}$ are given as functions of $\{x\}$ on open neighbourhoods of any nonsingular point), Eq. (8.22) allows us to calculate $K_{\gamma;\alpha\beta}(p_0)$ **algebraically** if $K_\gamma(p_0)$ is given. If $K_{\gamma;\delta}(p_0)$ is given as well, then from the derivative of (8.22) we can **algebraically** calculate $K_{\gamma;\alpha\beta\epsilon}(p_0)$. By differentiating (8.22) consecutively, we can then calculate all covariant derivatives of K_γ at p_0. Further, having all these derivatives (and hence, equivalently, all partial derivatives of K_γ at p_0), we can calculate $K_\gamma(p)$ where $p \in M_n$ lies in such a neighbourhood of p_0 in which the Taylor series for $K_\alpha(p)$ is convergent. However, after each differentiation a new derivative of the Riemann tensor appears, so, in order that the series is convergent, $R^\rho{}_{\alpha\beta\gamma}$ must be analytic in that neighbourhood.[1] From this argument we see that $K_\gamma(p_0)$ and $K_{\gamma;\delta}(p_0)$ at any chosen $p_0 \in M$ are the data which are needed to determine $K_\gamma(p)$ uniquely (if $R^\rho{}_{\alpha\beta\gamma}(p_0)$ are not sufficiently differentiable, then simply another initial point is needed, not more data). But $K_{\gamma;\delta}$ obey (8.18), so $K_{\gamma;\delta}(p_0)$ are $\frac{1}{2}n(n-1)$ constants, and $K_\gamma(p_0)$ are n constants in an n-dimensional manifold. Thus, the Taylor series for $K_\gamma(p)$ will contain at most $\frac{1}{2}n(n+1)$ arbitrary constants multiplying various functions of $\{x\}$. The multipliers of those constants will be the basis solutions, and hence their number cannot exceed $\frac{1}{2}n(n+1)$. $\square$

The prescription for finding the basis of the Killing vector fields for the metric tensor is therefore the following:

1. Solve the Killing equations. The general solution will then depend on $N \le \frac{1}{2}n(n+1)$ arbitrary constants, $k^\mu = K^\mu(A_1, \ldots, A_N, \{x\})$.
2. Calculate

$$k^\mu_{(i)} \overset{\text{def}}{=} \frac{\partial K^\mu}{\partial A_i}, \qquad i = 1, \ldots, N \tag{8.23}$$

– the basis. Each $k^\mu_{(i)}$ generates a one-parameter subgroup of symmetries discussed in Section 8.2.

A possible confusion has to be explained here. For brevity, we say 'Killing vectors', but in truth these are *vector fields*, whose components are functions. Hence, the number of linearly independent Killing vector *fields* can be larger than the dimension of the manifold. For example, in a flat Riemann space the number of linearly independent Killing vector fields is equal to the maximal one, $\frac{1}{2}n(n+1)$.

For tensor fields other than the metric tensor a finite basis may not exist, i.e. the general solution of the invariance equations will contain arbitrary functions rather than arbitrary constants. This is the case e.g. for the invariance group of $R^{\alpha\beta}{}_{\gamma\delta}$ (note the positions of indices!) in a space of constant curvature: any arbitrary vector field l has the property that $R'^{\alpha\beta}{}_{\gamma\delta}(\{x\}) = R^{\alpha\beta}{}_{\gamma\delta}(\{x\})$ for transformations generated by l (see Krasiński (1983) for less trivial examples). Thus, any coordinate transformation $x'^\alpha = f^\alpha(\{x\})$ is an invariance transformation of that Riemann tensor.

8.5 Invariance of other tensor fields

We investigated the conditions of invariance of the metric tensor in more detail because they are the most important and are most frequently met. Sometimes, though, we need to know the invariance transformations of other tensor fields. Repeating the reasoning (8.7)–(8.12) for the field of contravariant vectors, that is, assuming the condition $V'^\alpha(x) = V^\alpha(x)$, we would obtain the following equation:

$$k^\rho V^\alpha{}_{,\rho} - V^\rho k^\alpha{}_{,\rho} = 0, \tag{8.24}$$

where k^α is the generator of the transformation group.

Invariance conditions for other tensor fields are given in the exercises.

---

[1] For more on the existence of symmetries see Section 8.11. In truth, the curvature tensor does not have to be analytic.

8.6 The Lie derivative

If we trace the procedure that led to the Killing equations (8.12), and also the procedures leading to the other equations listed in Section 8.5 and in the exercises, then we will notice that the invariance equations are obtained in the following steps:

1. Take the tensor field T (arbitrary indices) to be $T(t_0)$, where t_0 is the value of the orbit parameter corresponding to the identity transformation.
2. Using the transformation law for T under coordinate changes, calculate $T(t)$ – the value of T transported to another point along the group orbit. The calculation is done up to terms linear in $(t - t_0)$ (the remaining terms are not neglected, but left in implicit form).
3. Calculate the quantity

$$-\lim_{t \to t_0} \frac{T(t) - T(t_0)}{t - t_0} \overset{\text{def}}{=} \underset{k}{\pounds} T, \qquad (8.25)$$

where $k^\alpha = \mathrm{d}x'^\alpha / \mathrm{d}t|_{t=t_0}$, and equate the result to zero.

The quantity $\left(- \underset{k}{\pounds} T\right)$ defined in (8.25) is thus the derivative of the tensor field T by the parameter of the transformation. As seen from (8.25), that derivative measures the speed of changes of the field T transported along the orbits tangent to k^α with respect to the values of T defined at the consecutive points. If $\underset{k}{\pounds} T = 0$, then the field transported along the orbit everywhere coincides with the tensor T defined before the transformation. The quantity $\underset{k}{\pounds} T$ is called the **Lie derivative** of the tensor field T along the vector field k and can be calculated also if T is not invariant under the action of Γ.[1] We have thus:

$$\left(\underset{k}{\pounds} T = 0\right) \Longleftrightarrow (T'(P) \equiv T(P)). \qquad (8.26)$$

The Lie derivative has all the algebraic properties of ordinary differentiation: it is linear with respect to addition, gives zero when acting on a constant, and when acting on a tensor product it obeys:

$$\underset{k}{\pounds} (T_1 \otimes T_2) = \left(\underset{k}{\pounds} T_1\right) \otimes T_2 + T_1 \otimes \left(\underset{k}{\pounds} T_2\right). \qquad (8.27)$$

These properties allow us to derive the formula for the components of the Lie derivative of any tensor field $T^{\alpha_1 \ldots \alpha_k}_{\beta_1 \ldots \beta_l}$ along any vector field k^α using the results of Section 8.5 and of Exercises 7 and 8:

$$\underset{k}{\pounds} T^{\alpha_1 \ldots \alpha_k}_{\beta_1 \ldots \beta_l} = k^\rho T^{\alpha_1 \ldots \alpha_k}_{\beta_1 \ldots \beta_l, \rho} - \sum_{i=1}^{k} k^{\alpha_i}{}_{, \rho_i} T^{\alpha_1 \ldots \rho_i \ldots \alpha_k}_{\beta_1 \ldots \beta_l} + \sum_{j=1}^{k} k^{\rho_j}{}_{, \beta_j} T^{\alpha_1 \ldots \alpha_k}_{\beta_1 \ldots \rho_j \ldots \beta_l}, \qquad (8.28)$$

---

[1] The notion of the Lie derivative was introduced by Ślebodziński (1931), but the name was proposed by van Dantzig and made popular by Schouten (Schouten and van Kampen, 1934; Schouten and Struik, 1935).

where sums over i and j extend over all positions of ρ_i and ρ_j, respectively, in the series of indices. Thus, the Lie derivative acts similarly to a directional covariant derivative along k, the factors $(-k^\mu,_\nu)$ playing the role of the Christoffel symbols projected onto k,

$-k^\mu,_\nu \longleftrightarrow \left\{{\mu \atop \nu\rho}\right\}k^\rho$. With the help of (8.28) one can now verify that for any tensor field T and any vector fields k and l we have

$$\underset{[k,l]}{\pounds}\, T \equiv [\underset{k}{\pounds}, \underset{l}{\pounds}]T, \tag{8.29}$$

so $[k, l]$ always generates an invariance of T if k and l do.

8.7 The algebra of Killing vector fields

Since for the Killing fields generating symmetries of M_n a finite basis exists, we conclude from the last statement of the previous section that there exist such constants $C^l{}_{ij}$ that, for the basis fields,

$$[\underset{(i)}{k}, \underset{(j)}{k^\alpha}] = C^l{}_{ij}\underset{(l)}{k^\alpha} \quad \text{(sum over } l\text{)}. \tag{8.30}$$

The constants $C^l{}_{ij}$ are called **structure constants** of the symmetry group. We see thus that the set of Killing fields for a given manifold M_n is a Lie algebra. For generators of invariances of other tensor fields, the coefficients $C^l{}_{ij}$ will not necessarily be constant.

8.8 Surface-forming vector fields

Let us consider two linearly independent vector fields, k and l, defined on M_n. They define two families of curves that are everywhere tangent to these fields, by the equations $k^\alpha = \mathrm{d}x^\alpha/\mathrm{d}\lambda$ and $l^\alpha = \mathrm{d}x^\alpha/\mathrm{d}\lambda$, where λ is a parameter on each curve. Take a single curve C of the family defined by the field l, and then consider all the curves defined by the field k that intersect C (see Fig. 8.2). They form a surface S out of a single curve tangent to l and of curves tangent to k. It is clear that other vectors of the field l attached to points of this surface need not be tangent to it. However, if they are tangent to every such S, then the vector fields k and l are called **surface-forming**.

What is the condition for two vector fields to be surface-forming? Consider the family of curves $x^\alpha(\lambda)$ tangent to the vectors of the field k, lying in the surface S. They define a family F of mappings of the surface S into itself (the image of a point $S \ni p = x^\alpha(\lambda_0)$ is the point $p' = x^\alpha(\lambda_0 + \Delta\lambda)$, where $\Delta\lambda$ is the same for all $p \in S$). The mapping F_{1*} associated to F maps then vectors tangent to S onto other vectors tangent to S. Hence, starting with the vectors l that are tangent to our initial curve C, we can construct the field $(F_{1*}l)$ of vectors tangent to S. The vectors of the field k attached to points of S are tangent to S, too – because S was constructed in this way. Hence, the vectors of the field

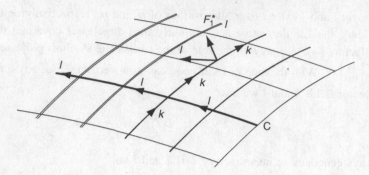

Fig. 8.2.   We take a single curve C of the curves tangent to the field l and consider all curves tangent to the field k that intersect C. In this way, we form a surface S, to which other vectors of the field l may or may need not be tangent. If they are, then the vector fields k and l are called surface-forming.

l that are attached to points of S will be everywhere tangent to S if they are everywhere spanned on the vectors k and $(F_{1*}l)$, thus

$$l^\alpha = \widetilde{a}k^\alpha + \widetilde{b}\,(F_{1*}l)^\alpha, \tag{8.31}$$

where $\widetilde{a}(x)$ and $\widetilde{b}(x)$ are arbitrary scalar functions and $\widetilde{b}(x) \neq 0$ (with $\widetilde{b}(x) = 0$, the fields k and l would be linearly dependent, contrary to our assumption). This condition may be rewritten as follows:

$$l^\alpha - (F_{1*}l)^\alpha = \frac{\widetilde{a}}{\widetilde{b}}k^\alpha + \frac{\widetilde{b}-1}{\widetilde{b}}l^\alpha \overset{\text{def}}{=} ak^\alpha + bl^\alpha. \tag{8.32}$$

But the rate of change of $[l^\alpha - (F_{1*}l)^\alpha]$ along the curves tangent to k is, by definition, the Lie derivative $\left(\underset{k}{\pounds}\,l\right)^\alpha$. Finally then, the necessary and sufficient condition for the vector fields k and l to be surface-forming is

$$[k, l]^\alpha \equiv \left(\underset{k}{\pounds}\,l\right)^\alpha = ak^\alpha + bl^\alpha. \tag{8.33}$$

8.9 Spherically symmetric 4-dimensional Riemann spaces

We call a Riemann space **spherically symmetric** when the group of rotations around a point, O(3), is its isometry group. Its metric tensor must thus obey the Killing equations for each of the three generators of the group O(3). We shall first derive the formulae for these generators.

The orbits of O(3) are 2-dimensional spheres. Each sphere can be embedded in a 3-dimensional Euclidean space E^3. Its equation is then

$$x^2 + y^2 + z^2 = R^2, \tag{8.34}$$

where R is the radius of the sphere, and x, y, z are Cartesian coordinates in E^3. The rotation around the centre of the sphere by the angle α in the plane (x^i, x^j) is then described by the transformation

$$x'^i = x^i \cos \alpha + x^j \sin \alpha, \qquad x'^i = -x^i \sin \alpha + x^j \cos \alpha,$$

$$x'^k = x^k \qquad \text{for } i \neq k \neq j. \tag{8.35}$$

The angle α is here the group parameter. By using Eq. (8.6) we then find that the corresponding Killing vector is

$$k^\mu_{[i,j]} = x^j \delta^\mu{}_i - x^i \delta^\mu{}_j. \tag{8.36}$$

An arbitrary transformation of the sphere into itself can be described as a composition of three consecutive rotations around different axes. Hence, a basis of the space of Killing vectors will be three generators corresponding to rotations around three different axes.

It is often convenient to represent the Killing vectors by the corresponding operators of directional derivatives, also called generators:

$$J_{(i)} \overset{\text{def}}{=} k^\mu_{(i)} \frac{\partial}{\partial x^\mu}. \tag{8.37}$$

We will choose as our basis of Killing fields the generators of rotations around the three axes of the rectangular Cartesian coordinate system:

$$J_{[xy]} = x \frac{\partial}{\partial y} - y \frac{\partial}{\partial x}, \qquad J_{[yz]} = y \frac{\partial}{\partial z} - z \frac{\partial}{\partial y}, \qquad J_{[xz]} = x \frac{\partial}{\partial z} - z \frac{\partial}{\partial x}. \tag{8.38}$$

Now let us transform the generators to the spherical coordinates

$$x = r \sin \vartheta \cos \varphi, \qquad y = r \sin \vartheta \sin \varphi, \qquad z = r \cos \vartheta. \tag{8.39}$$

In these coordinates the generators become, up to sign,

$$J_{[xy]} = \frac{\partial}{\partial \varphi}, \qquad J_{[yz]} = \sin \varphi \frac{\partial}{\partial \vartheta} + \cos \varphi \cot \vartheta \frac{\partial}{\partial \varphi},$$

$$J_{[xz]} = \cos \varphi \frac{\partial}{\partial \vartheta} - \sin \varphi \cot \vartheta \frac{\partial}{\partial \varphi}. \tag{8.40}$$

Since the coordinates ϑ and φ are defined inside the spheres, we can use them as coordinates in the whole Riemann space. Let us denote the two remaining coordinates t and r. We shall now solve the Killing equations for the metric tensor $g_{\alpha\beta}(t, r, \vartheta, \varphi)$, where $x^0 = t$, $x^1 = r$, $x^2 = \vartheta$, $x^3 = \varphi$, with the Killing vectors given by (8.40), thus

$$k^\alpha_{(1)} = \delta^\alpha{}_3, \qquad k^\alpha_{(2)} = \sin \varphi \, \delta^\alpha{}_2 + \cos \varphi \cot \vartheta \, \delta^\alpha{}_3,$$

$$k^\alpha_{(3)} = \cos \varphi \delta^\alpha{}_2 - \sin \varphi \cot \vartheta \delta^\alpha{}_3. \tag{8.41}$$

Note that in assuming the form (8.41) of the Killing vectors for the whole Riemann space we have tacitly made one more assumption. Namely, we assumed that the (ϑ, φ) coordinates on different spheres are correlated in such a way that the rotation of the whole space is described by the same formula as the rotation of a single sphere. One can easily find examples of coordinate systems that do not obey this condition. For example, suppose that we choose the (r, φ) coordinates in the E^2 plane so that the curves $r = $ constant are non-concentric circles, and the azimuthal angle is measured on each circle independently, beginning from a certain reference direction chosen to be $\varphi = 0$. In such coordinates, the rotation of the plane by the angle α

will not be described by $\varphi' = \varphi + \alpha$; the φ' will in general be a nonlinear function of r and φ, which will reduce to $(\varphi + \alpha)$ only on that circle whose centre coincides with the centre of rotation of the plane (and such a circle will not necessarily exist).

As another example of an untypical coordinate system, we may choose the 'generalised spherical coordinates' in the Euclidean space E^3, in which the surfaces $r = $ constant will be concentric spheres, but the poles of the spherical coordinates on different spheres will not lie on a straight line. A rotation by the angle α around an axis will be described by $\varphi' = \varphi + \alpha$ only on that sphere whose pole lies on the axis. On other spheres the rotation will be given by a complicated function of r, ϑ and φ, and the generators will not have the form (8.40).

We should thus expect that the Killing equations for (8.41) will give us not only a limitation on the geometrical properties of the space, but also a limitation on the coordinate systems resulting from the assumed form of generators (8.40). We shall come back to this question – see the paragraph containing (8.53).

The Killing equations for the vector $k^\alpha_{(1)}$ reduce to $g_{\alpha\beta,3} = 0$, that is, the whole metric tensor is independent of $x^3 = \varphi$. For $k^\alpha_{(2)}$ and $k^\alpha_{(3)}$ the Killing equations reduce to

$$\sin\varphi\frac{\partial}{\partial\vartheta}g_{\alpha\beta} + (\sin\varphi)_{,\alpha}\,g_{2\beta} + (\sin\varphi)_{,\beta}\,g_{\alpha2}$$
$$+ (\cos\varphi\cot\vartheta)_{,\alpha}\,g_{3\beta} + (\cos\varphi\cot\vartheta)_{,\beta}\,g_{\alpha3} = 0, \qquad (8.42)$$

$$\cos\varphi\frac{\partial}{\partial\vartheta}g_{\alpha\beta} + (\cos\varphi)_{,\alpha}\,g_{2\beta} + (\cos\varphi)_{,\beta}\,g_{\alpha2}$$
$$- (\sin\varphi\cot\vartheta)_{,\alpha}\,g_{3\beta} - (\sin\varphi\cot\vartheta)_{,\beta}\,g_{\alpha3} = 0. \qquad (8.43)$$

In order to simplify further calculations, we will replace these equations by two combinations thereof. Multiply (8.42) by $\cos\varphi$, (8.43) by $\sin\varphi$ and subtract one result from the other. Using the two identities

$$\sin\varphi(\sin\varphi)_{,\alpha} + \cos\varphi(\cos\varphi)_{,\alpha} \equiv 0,$$
$$\cos\varphi(\sin\varphi)_{,\alpha} - \sin\varphi(\cos\varphi)_{,\alpha} \equiv \varphi_{,\alpha} \qquad (8.44)$$

we obtain

$$\varphi_{,\alpha}\,g_{2\beta} + \varphi_{,\beta}\,g_{\alpha2} + (\cot\vartheta)_{,\alpha}\,g_{3\beta} + (\cot\vartheta)_{,\beta}\,g_{\alpha3} = 0. \qquad (8.45)$$

Now multiply (8.42) by $\sin\varphi$, multiply (8.43) by $\cos\varphi$, and add the results. Using (8.44) again, we obtain

$$\frac{\partial}{\partial\vartheta}g_{\alpha\beta} - \varphi_{,\alpha}\cot\vartheta g_{3\beta} - \varphi_{,\beta}\cot\vartheta g_{\alpha3} = 0. \qquad (8.46)$$

Equation (8.45) is algebraic. Taking consecutively the various values of the indices α and β, we obtain from it the following results:

- For $2 \neq \alpha \neq 3$, $2 \neq \beta \neq 3$ the equation is fulfilled identically.
- For $2 \neq \alpha \neq 3$, $\beta = 2$:

$$g_{\alpha3} = 0, \qquad \alpha = 0, 1. \qquad (8.47)$$

- For $2 \neq \alpha \neq 3$, $\beta = 3$:

$$g_{\alpha2} = 0, \qquad \alpha = 0, 1. \qquad (8.48)$$

- For $\alpha = 2, \beta = 2$:

$$g_{23} + g_{32} \equiv 2g_{23} = 0. \tag{8.49}$$

- For $\alpha = 2, \beta = 3$:

$$g_{33} = g_{22} \sin^2 \vartheta. \tag{8.50}$$

- For $\alpha = \beta = 3$ Eq. (8.50) follows once more.

Now we take the same cases for Eq. (8.46) and obtain the following:

- For $2 \neq \alpha \neq 3, 2 \neq \beta \neq 3 : (\partial/\partial\vartheta)g_{\alpha\beta} = 0, \alpha, \beta = 0, 1$.
- For $2 \neq \alpha \neq 3, \beta = 2$ an identity as a consequence of (8.48).
- For $2 \neq \alpha \neq 3, \beta = 3$ an identity as a consequence of (8.47).
- For $\alpha = 2, \beta = 2 : (\partial/\partial\vartheta)g_{22} = 0$.
- For $\alpha = 2, \beta = 3$ an identity as a consequence of (8.49).
- For $\alpha = \beta = 3$ an identity as a consequence of (8.50).

Hence, finally, the solution of (8.45)–(8.46) is

$$ds^2 = \alpha(t, r)dt^2 + 2\beta(t, r)dt\,dr + \gamma(t, r)dr^2 + \delta(t, r)\left(d\vartheta^2 + \sin^2\vartheta\,d\varphi^2\right). \tag{8.51}$$

This is the general 4-dimensional spherically symmetric metric form. Note that, in consequence of the 2-spheres being subspaces of the Riemann space (they are the orbits of the symmetry group), we have obtained a limitation on the signature: the signs of $d\vartheta^2$ and of $d\varphi^2$ must be the same. This is an illustration of the remark made in Section 7.15 on embedding the Riemann spaces in spaces of higher dimension: an inconsistency between the signatures may render the embedding impossible.

Note that we have not assumed anything about the subspaces of the variables (t, r). Hence, arbitrary nonsingular coordinate transformations can be carried out within those subspaces:

$$t = f(t', r'), \qquad r = g(t', r'), \tag{8.52}$$

where f and g are arbitrary functions subject to the condition that $\partial(t, r)/\partial(t', r') \neq 0$. After such a transformation the function $\delta(t, r)$ will preserve its value, while α, β and γ will change to combinations of α, β and γ, but the combinations will still depend only on t' and r'.

Now we can illustrate the remarks made after (8.41). We can carry out a coordinate transformation on (8.51) that does not obey (8.52), for example

$$r = r' + h(\vartheta, \varphi), \tag{8.53}$$

where $h(\vartheta, \varphi)$ is an arbitrary function. The result will be

$$\begin{aligned}
ds^2 = {} & \alpha(t, r)dt^2 + 2\beta(t, r)dt\,dr' + 2\beta(t, r)h_{,\vartheta}\,dt\,d\vartheta + 2\beta(t, r)h_{,\varphi}\,dt\,d\varphi \\
& + \gamma(t, r)dr'^2 + 2\gamma(t, r)h_{,\vartheta}\,dr'\,d\vartheta + 2\gamma(t, r)h_{,\varphi}\,dr'\,d\varphi \\
& + \left(\delta(t, r) + \gamma(t, r)h_{,\vartheta}^2\right)d\vartheta^2 + 2\gamma(t, r)h_{,\vartheta}\,h_{,\varphi}\,d\vartheta\,d\varphi \\
& + \left(\delta(t, r)\sin^2\vartheta + \gamma(t, r)h_{,\varphi}^2\right)d\varphi^2
\end{aligned} \tag{8.54}$$

(note that both r and r' are present here; r' is one of the coordinates, while r is the function defined by (8.53)). The metric (8.54) is still spherically symmetric because it resulted from (8.51) by a coordinate transformation, and the existence of symmetries does not depend on the coordinates. Nevertheless, Eq. (8.54) is not of the form (8.51) because after the transformation (8.53) the spherical coordinates on different spheres ($t =$ constant, $r =$ constant) are not correlated in the simple way implicit in (8.41).

The orbits of the group O(3) are the subspaces $\{t = $ constant, $r = $ constant$\}$ whose metric form, in the coordinates of (8.51), is $ds_2{}^2 = \delta(d\vartheta^2 + \sin^2\vartheta\,d\varphi^2)$ (in these subspaces δ is constant). The centre of symmetry is where $\delta(t, r) = 0$. However, there is no guarantee that such a point exists within the manifold. For example, if δ is constant throughout the Riemann space (which is a property invariant under (8.52)), then the centre of symmetry does not exist. This is an analogy to a cylinder or a one-sheeted hyperboloid: these surfaces are rotationally symmetric, but no point on the surface is the centre of rotation.

If the functions α, β, γ and δ in (8.51) are independent of r, then there exists a fourth Killing field $k^\alpha_{(4)} = \delta^\alpha{}_1$. The symmetry group of such spacetimes is called the **Kantowski–Sachs symmetry** (Kantowski and Sachs, 1966). In these spacetimes those hypersurfaces $t =$ constant for which $\delta(t) \neq 0$ have no centre of symmetry. We will come back to them in Chapter 10.

8.10 * Conformal Killing fields and their finite basis

If, after a coordinate transformation, the new expression for the metric tensor is conformally equivalent to the old one, then such a transformation is called a **conformal symmetry** of the Riemann space (and of the metric). The generators of these transformations are called **conformal Killing vectors** and obey the equations

$$k^\rho g_{\alpha\beta,\rho} + k^\rho{}_{,\alpha} g_{\rho\beta} + k^\rho{}_{,\beta} g_{\alpha\rho} \equiv k_{\alpha;\beta} + k_{\beta;\alpha} = \lambda g_{\alpha\beta}, \tag{8.55}$$

where λ is a scalar function.

By a similar method to that in Section 8.4 one can show that for the Riemann spaces of dimension $n > 2$ a finite basis of generators of conformal symmetries always exists; i.e. that the dimension of the group of conformal symmetries is finite if $n > 2$. (We shall deal with the case $n = 2$ at the end of this section – then, no finite basis exists since all 2-dimensional Riemann spaces are conformally equivalent.)

Contracting both sides of (8.55) with $g^{\alpha\beta}$ we obtain

$$\lambda = \frac{2}{n} k^\mu{}_{;\mu}. \tag{8.56}$$

From the Ricci identities we obtain (8.19) and (8.20) again, but this time instead of (8.18) we have

$$-k_{\beta;\alpha} = k_{\alpha;\beta} - \lambda g_{\alpha\beta}. \tag{8.57}$$

Hence, using (8.55) and $g_{\alpha\beta;\gamma} = 0$ in (8.20), we obtain

$$k_{\alpha;\beta\gamma} + k_{\gamma;\alpha\beta} + k_{\beta;\gamma\alpha} = \frac{1}{2}\left(\lambda,_\gamma g_{\alpha\beta} + \lambda,_\beta g_{\alpha\gamma} + \lambda,_\alpha g_{\beta\gamma}\right). \tag{8.58}$$

From (8.57) and (8.58):

$$k_{\gamma;\alpha\beta} = k_{\beta;\alpha\gamma} - k_{\beta;\gamma\alpha} + \frac{1}{2}\left(-\lambda_{,\gamma}g_{\alpha\beta} + \lambda_{,\beta}g_{\alpha\gamma} + \lambda_{,\alpha}g_{\beta\gamma}\right). \tag{8.59}$$

To the first two terms we now apply the Ricci formula, and we obtain

$$k_{\gamma;\alpha\beta} = R^{\rho}{}_{\beta\alpha\gamma}k_{\rho} + \frac{1}{2}\left(-\lambda_{,\gamma}g_{\alpha\beta} + \lambda_{,\beta}g_{\alpha\gamma} + \lambda_{,\alpha}g_{\beta\gamma}\right). \tag{8.60}$$

The second derivatives of the conformal Killing fields are thus determined by the fields k_{α} themselves and by the gradient of λ.

The first derivatives of the generators are connected by Eqs. (8.55) that determine the symmetric part of the matrix $k_{\alpha;\beta}$. Hence, in order to know the first derivatives of k_{α}, we have to know the antisymmetric part of $k_{\alpha;\beta}$, i.e. $\frac{1}{2}n(n-1)$ function values, plus the value of λ. Together with k_{α} and $\lambda_{,\alpha}$ this makes

$$\frac{1}{2}n(n-1) + n + n + 1 = \frac{1}{2}(n+1)(n+2) \tag{8.61}$$

constants that must be known in order to calculate $k_{\gamma;\alpha\beta}$.

Now take the covariant derivative of (8.60) with the index δ:

$$k_{\gamma;\alpha\beta\delta} = R^{\rho}{}_{\beta\alpha\gamma;\delta}k_{\rho} + R^{\rho}{}_{\beta\alpha\gamma}k_{\rho;\delta} + \frac{1}{2}\left(-\lambda_{;\gamma\delta}g_{\alpha\beta} + \lambda_{;\beta\delta}g_{\alpha\gamma} + \lambda_{;\alpha\delta}g_{\beta\gamma}\right). \tag{8.62}$$

From this, we subtract the corresponding expression with the indices $(\beta\delta)$ interchanged, and we apply the Ricci formula on the left-hand side:

$$R^{\rho}{}_{\gamma\beta\delta}k_{\rho;\alpha} + R^{\rho}{}_{\alpha\beta\delta}k_{\gamma;\rho} = R^{\rho}{}_{\beta\alpha\gamma;\delta}k_{\rho} - R^{\rho}{}_{\delta\alpha\gamma;\beta}k_{\rho} + R^{\rho}{}_{\beta\alpha\gamma}k_{\rho;\delta} - R^{\rho}{}_{\delta\alpha\gamma}k_{\rho;\beta}$$
$$+ \frac{1}{2}\left(-\lambda_{;\gamma\delta}g_{\alpha\beta} + \lambda_{;\gamma\beta}g_{\alpha\delta} + \lambda_{;\alpha\delta}g_{\beta\gamma} - \lambda_{;\alpha\beta}g_{\delta\gamma}\right). \tag{8.63}$$

If (8.63) is solvable with respect to $\lambda_{;\beta\delta}$, then the second derivatives of λ will be determined by k_{α} and $k_{\alpha;\beta}$. In that case, to calculate the third derivatives of k_{α} from (8.62) we need to know only those constants that were counted in (8.61).

In order to find $\lambda_{;\gamma\delta}$ from (8.63), we contract it with $g^{\alpha\beta}$:

$$R^{\rho}{}_{\gamma}{}^{\alpha}{}_{\delta}k_{\rho;\alpha} - R^{\rho}{}_{\delta}k_{\gamma;\rho} = -R^{\rho}{}_{\gamma;\delta}k_{\rho} - R^{\rho}{}_{\delta}{}^{\alpha}{}_{\gamma;\alpha}k_{\rho} - R^{\rho}{}_{\gamma}k_{\rho;\delta} - R^{\rho}{}_{\delta}{}^{\alpha}{}_{\gamma}k_{\rho;\alpha}$$
$$+ \frac{1}{2}\left[-(n-2)\lambda_{;\gamma\delta} - g_{\gamma\delta}\lambda^{;\alpha}{}_{;\alpha}\right]. \tag{8.64}$$

The first term on the left and the last term containing curvature on the right can be combined as follows:

$$R^{\rho}{}_{\gamma}{}^{\alpha}{}_{\delta}k_{\rho;\alpha} + R^{\rho}{}_{\delta}{}^{\alpha}{}_{\gamma}k_{\rho;\alpha} = R^{\rho}{}_{\gamma}{}^{\alpha}{}_{\delta}(k_{\rho;\alpha} + k_{\alpha;\rho}) = R^{\rho}{}_{\gamma}{}^{\alpha}{}_{\delta}\lambda g_{\alpha\rho} = \lambda R_{\gamma\delta}. \tag{8.65}$$

Now using (8.65), we rewrite (8.64) as follows:

$$(n-2)\lambda_{;\gamma\delta} + g_{\gamma\delta}\lambda^{;\mu}{}_{;\mu} = -2\lambda R_{\gamma\delta} + R^{\rho}{}_{\delta}k_{\gamma;\rho} - 2R^{\rho}{}_{\gamma;\delta}k_{\rho}$$
$$- 2R^{\rho}{}_{\delta}{}^{\alpha}{}_{\gamma;\alpha}k_{\rho} - 2R^{\rho}{}_{\gamma}k_{\rho;\delta}. \tag{8.66}$$

We contract this with $g^{\gamma\delta}$ and obtain

$$(n-1)\lambda^{;\mu}{}_{;\mu} = -\lambda R - 2R^{\rho\gamma}{}_{;\gamma}k_\rho = -\lambda R - R_{,\mu}k^\mu \tag{8.67}$$

(in the second step we used the contracted Bianchi identities $\left(R^{\mu\nu} - \frac{1}{2}g^{\mu\nu}R\right)_{;\nu} \equiv 0$). From here we substitute for $\lambda^{;\mu}{}_{;\mu}$ in (8.66) and obtain

$$(n-2)\lambda_{;\gamma\delta} = \frac{1}{n-1}g_{\gamma\delta}\left(\lambda R + R_{,\mu}k^\mu\right) - 2\lambda R_{\gamma\delta} + 2R^\rho{}_\delta k_{\gamma;\rho}$$
$$-2R^\rho{}_{\gamma;\delta}k_\rho - 2R^\rho{}_\delta{}^\alpha{}_{\gamma;\alpha}k_\rho - 2R^\rho{}_\gamma k_{\rho;\delta}. \tag{8.68}$$

(It can be verified, with use of (7.15) and (8.55), that the right-hand side of (8.68) is symmetric in γ and δ.) Hence, for $n > 2$, the second derivatives of λ are determined by $k_\alpha, k_{\alpha;\beta}$ and λ itself. Thus, $\lambda, \lambda_{,\alpha}, k_\alpha$ and $k_{(\alpha;\beta)}$ determine $k_{\alpha;\beta\gamma}, k_{\alpha;\beta\gamma\delta}$ and $\lambda_{;\gamma\delta}$, and, consequently, determine all higher derivatives of k_α and λ. Then, the functions $k_\alpha(x)$ and $\lambda(x)$ are determined in that neighbourhood of the point investigated within which these functions are analytic.[1] Consequently, with $n > 2$, the conformal symmetry transformation can have at most $\frac{1}{2}(n+1)(n+2)$ parameters, i.e. a finite basis of generators exists. Note that $\frac{1}{2}(n+1)(n+2)$ is at the same time the maximal dimension of the isometry group of an $(n+1)$-dimensional manifold.

It remains to verify the subcase $n = 2$. Then, from (7.61) we obtain:

$$R^{\alpha\beta}{}_{\gamma\delta} = \frac{1}{2}R\delta^{\alpha\beta}_{\gamma\delta}, \qquad R^\alpha{}_\beta = \frac{1}{2}R\delta^\alpha_\beta,$$

$$R^\rho{}_\delta{}^\alpha{}_\gamma = g_{\delta\mu}g^{\alpha\sigma}R^{\rho\mu}{}_{\sigma\gamma} = \frac{1}{2}R\left(g_{\delta\gamma}g^{\alpha\rho} - \delta^\alpha_\delta\delta^\rho_\gamma\right). \tag{8.69}$$

After substituting these in (8.68) we obtain the identity $0 \equiv 0$. Hence, for 2-dimensional surfaces, the only limitation that follows from (8.66) is (8.67). But (8.66) is just one consequence of (8.63), which was the full integrability condition of (8.62). Hence, we must now reconsider (8.63).

In two dimensions, every object that is antisymmetric in two indices must be proportional to the Levi-Civita symbol. Since the last parenthesis in (8.63) is antisymmetric in both $[\beta\delta]$ and $[\alpha\gamma]$, the following must hold:

$$-\lambda_{;\gamma\delta}g_{\alpha\beta} + \lambda_{;\beta\gamma}g_{\alpha\delta} + \lambda_{;\alpha\delta}g_{\beta\gamma} - \lambda_{;\alpha\beta}g_{\delta\gamma} = \psi\epsilon_{\alpha\gamma}\epsilon_{\beta\delta}. \tag{8.70}$$

Contracting this with $\epsilon^{\alpha\gamma}\epsilon^{\beta\delta}$ and using $\epsilon^{\beta\delta} = gg^{\beta\nu}g^{\delta\mu}\epsilon_{\mu\nu}$ we obtain

$$\psi = -g\lambda^{;\mu}{}_{;\mu}. \tag{8.71}$$

Now substituting (8.70), (8.71) and (8.69) in (8.63) gives:

$$g_{\gamma\sigma}R\delta^{\rho\sigma}_{\beta\delta}k_{\rho;\alpha} + g_{\alpha\sigma}R\delta^{\rho\sigma}_{\beta\delta}k_{\gamma;\rho} - g_{\beta\sigma}R_{,\delta}k_\rho\delta^{\rho\sigma}_{\alpha\gamma} + g_{\delta\sigma}R_{,\beta}k_\rho\delta^{\rho\sigma}_{\alpha\gamma}$$

$$-g_{\beta\sigma}R\delta^{\rho\sigma}_{\alpha\gamma}k_{\rho;\delta} + g_{\delta\sigma}R\delta^{\rho\sigma}_{\alpha\gamma}k_{\rho;\beta} = -g\lambda^{;\mu}{}_\mu\epsilon_{\alpha\gamma}\epsilon_{\beta\delta}. \tag{8.72}$$

---

[1] The argument presented in the next section applies to conformal symmetries as well, so this limitation on the number of conformal symmetries applies also for non-analytic functions.

The left-hand side above is antisymmetric in $[\beta\delta]$. Using (8.55) one can verify that it is also antisymmetric in $[\alpha\gamma]$. Hence, (8.72) will hold if the contractions of both sides of it with $\epsilon^{\alpha\gamma}\epsilon^{\beta\delta}$ are equal. Doing this contraction and using again $\epsilon^{\beta\delta} = gg^{\beta\nu}g^{\delta\mu}\epsilon_{\mu\nu}$ we obtain

$$R_{,\rho}k^{\rho} + Rk^{\mu}{}_{;\mu} = -\lambda^{;\mu}{}_{;\mu}. \tag{8.73}$$

Comparing this with (8.56) in the case $n = 2$ we see that (8.73) is equivalent to (8.67).

Consequently, the only integrability condition of (8.62) is (8.67). This equation is of the type of d'Alembert equation that puts a limit only on $(\lambda^{;1}{}_{;1} + \lambda^{;2}{}_{;2})$, leaving $\lambda_{;11}$ and $\lambda_{;12}$ undetermined. Hence, when differentiated, Eq. (8.62) will produce still higher derivatives of λ. Let us also note the form of (8.62) in the case $n = 2$:

$$(k_{\gamma;\alpha\beta})_{;\delta} = \frac{1}{2}g_{\beta\sigma}\delta^{\rho\sigma}_{\gamma\delta}(Rk_{\rho})_{;\delta} + \frac{1}{2}(-\lambda_{,\gamma}g_{\alpha\beta} + \lambda_{,\beta}g_{\alpha\gamma} + \lambda_{,\alpha}g_{\beta\gamma})_{;\delta}. \tag{8.74}$$

Both sides of this are covariant derivatives of some expressions, and subsequent differentiations of them will have their integrability conditions identically fulfilled. In order to determine λ and k_{α} we will thus need an infinite number of constants, i.e. no finite basis of the conformal Killing fields exists in two dimensions. This is a consequence of the fact that every 2-dimensional metric is conformally flat, so there exists an infinite family of transformations preserving the explicitly conformally flat representation of the metric.

8.11 * The maximal dimension of an invariance group

In Section 8.4 we proved that the dimension of the symmetry group of an n-dimensional Riemann space cannot be larger than $\frac{1}{2}n(n+1)$. However, in the proof we assumed that the Riemann tensor was analytic in the neighbourhood considered. We will argue now that the result applies generally. It will follow from the sequence of theorems presented below.

Theorem 8.1

Assumptions:

(1) *The metric tensor $g_{\alpha\beta}$ of the Riemann space V_n depends on a number of constant parameters $u \overset{\text{def}}{=} \{u_1, \ldots, u_k\}$.*

(2) *The metric $h_{\alpha\beta}$ results from $g_{\alpha\beta}$ when some of the parameters go to certain limits, thus*

$$u_0 \overset{\text{def}}{=} \{u_{10}, \ldots, u_{i0}, u_{i+1}, \ldots, u_k\}, \qquad h_{\alpha\beta} = \lim_{u \to u_0} g_{\alpha\beta}, \tag{8.75}$$

where $i \leq k$.

(3) *The limit is non-singular, i.e. no vector field defined on V_n becomes identically zero in the limit.*

Thesis:

The dimension of the symmetry group of $h_{\alpha\beta}$ is larger than or equal to the dimension of the symmetry group of $g_{\alpha\beta}$.

Proof:

Let $k_{(1)}, \ldots, k_{(p)}$ be the generators of symmetries of $g_{\alpha\beta}$, and consider the equations $\underset{k_{(m)}}{\pounds} g_{\alpha\beta} = 0$, $m = 1, \ldots, p$. Since they are fulfilled at all values of the parameters $\{u_1, \ldots, u_k\}$, they must still hold in the limit. Since the limit is nonsingular, none of the generators will become zero in the limit. Hence, $h_{\alpha\beta}$ must inherit all the symmetries of $g_{\alpha\beta}$. $\square$

Comments:

(a) Limits of metric tensors are coordinate-dependent, but the theorem holds for every nonsingular limit.

(b) New symmetries may appear in the limit if there exists a vector field w^α on V_n such that $\underset{w}{\pounds} g_{\alpha\beta} \neq 0$ but $\lim_{u \to u_0} \underset{w}{\pounds} g_{\alpha\beta} = 0$.

(c) The structure of the symmetry group H of $h_{\alpha\beta}$ may be different from the structure of the symmetry group G of $g_{\alpha\beta}$, even if $\dim H = \dim G$. This may happen because some of the structure constants of G may become zero or take other privileged values in the limit.

(d) The theorem holds for invariance groups of any tensor fields, not just for isometries. Thus, for example, the group of invariance transformations of the Riemann tensor (called **collineations**) may only increase or preserve its dimension in a limit.

(e) The theorem holds for conformal symmetries, too.

Theorem 8.2 *The flat Riemann space is contained as a limit in every nonsingular region of every curved Riemann space.*

Proof:

For any Riemann space, the flat space of the corresponding signature is its tangent space. Thus, since in a nonsingular region the tangent space exists at every point, we may imagine the flat limit as follows: we decrease the curvature to zero (this may require introducing some free parameters by a coordinate transformation), and in the process open subsets of the tangent spaces become isometric with open subsets of the manifold itself. $\square$

Comments:

(a) This theorem is in fact one of the fundamental postulates of relativity. In the physical language it reads as follows: special relativity is the zero-curvature limit of general relativity, i.e. every curved spacetime has the Minkowski spacetime as a limit.

(b) The construction will fail at singular points, where the curvature is infinite (such as the vertex of a cone). At those points, no tangent space exists.

(c) In general, we do not expect the tangent spaces at different points to coincide in the limit. For example, a cylinder is flat, but its tangent planes do not coincide.

Theorem 8.3 *No Riemann space can have a symmetry group of higher dimension than the flat Riemann space, i.e. of more dimensions than $\frac{1}{2}n(n+1)$.*

Proof:

The proof follows from Theorems 8.1 and 8.2 and from the result of Exercise 10. □

Comment:

There exist Riemann spaces whose symmetry groups have exactly $\frac{1}{2}n(n+1)$ parameters, and which are not locally isometric to the flat space. They are the **spaces of constant curvature**, with the Riemann tensor $R^{\alpha\beta}{}_{\gamma\delta} = R\delta^{\alpha\beta}_{\gamma\delta}$. Their symmetry groups are different (not isomorphic) in each of the three cases $R > 0$, $R = 0$ and $R < 0$. In relativity, they are the de Sitter spacetimes, see Exercise 12 and Section 14.4.

8.12 Exercises

1. Find the coordinate transformation corresponding to the field of generators $k^\alpha = \delta^\alpha{}_{\alpha_0}$, where α_0 is the label of one of the coordinates.
2. Find the coordinate transformation corresponding to the field of generators $k^\mu = x^i\delta^\mu{}_j - x^j\delta^\mu{}_i$, where i and j are the labels of fixed coordinates. Show that, when (x^i, x^j) are Cartesian coordinates, the transformation found here is the rotation in the plane of (x^i, x^j).
3. Solve the Killing equations for $k^\alpha = \delta^\alpha{}_{\alpha_0}$.
4. Show that if the parameter λ of the integral lines of a generator $k^\alpha = dx^\alpha/d\lambda$ is chosen as a coordinate in the Riemann space, then a tensor invariant under the transformations generated by k^α is simply independent of λ.
5. Show that if there exist at least two linearly independent fields k^α and l^α generating invariances of a tensor field $T_{\alpha\beta}$, then the corresponding orbit parameters λ and τ can be chosen as coordinates on M_n, if and only if $[k, l]^\alpha = k^\rho l^\alpha{}_{,\rho} - l^\rho k^\alpha{}_{,\rho} = 0$. In that case $\partial T_{\alpha\beta}/\partial\lambda = \partial T_{\alpha\beta}/\partial\tau = 0$.
   Note: This is just a different wording of Exercise 2 in Chapter 6.
6. Prove that if k^α and l^α are Killing fields in a certain Riemann space, then so is their commutator $[k, l]^\alpha$.
7. Show that the condition of invariance of a scalar field ϕ with respect to the transformation group generated by the vector field k^α is

$$k^\alpha\phi_{,\alpha} = 0. \tag{8.76}$$

8. Show that the condition of invariance of a covariant vector field ω_α with respect to the transformation group generated by k^α is

$$k^\rho\omega_{\alpha,\rho} + k^\rho{}_{,\alpha}\omega_\rho = 0. \tag{8.77}$$

9. Prove that the conformal Killing vectors obey (8.55) and that λ is related to the conformal factor Φ in $g'_{\alpha\beta}(x') = \Phi(x')g_{\alpha\beta}(x')$ by $\lambda = k^\alpha\Phi_{,\alpha}$.
   Hint. Take first the conformal condition $g'_{\alpha\beta}(x') = \Phi(x')g_{\alpha\beta}(x')$ and note that x'^α are functions of the group parameters t_i. Then take the limit of this equation as $t_i \to t_{0i}$, where $x'^\alpha(t_{0i}, x) = x^\alpha$, i.e. is an identity transformation. It follows that Φ must obey $\Phi(x')|_{x'=x} = 1$.
10. Find and interpret all the Killing fields for the Minkowski spacetime in the Cartesian coordinates, in which $ds^2 = dt^2 - dx^2 - dy^2 - dz^2$. Find the corresponding isometries. Identify the isometries that should be known to you from a course on special relativity: the special Lorentz transformations along the x-, y- and z-directions, and the rotations in the planes $\{x, y\}$, $\{y, z\}$

and $\{x, z\}$. Verify that they are indeed isometries (i.e. substitute these transformations into the metric form and see what happens). Calculate all the structure constants of the full group.

11. Find all the conformal Killing fields for the Minkowski spacetime in the Cartesian coordinates, in which $ds^2 = dt^2 - dx^2 - dy^2 - dz^2$. Show that those generators that do not correspond to isometries generate the following mappings of the Minkowski spacetime into itself:

(a) Dilatation:

$$x'^{\alpha} = x^{\alpha}/C, \tag{8.78}$$

where C is the group parameter.

(b) The so-called acceleration transformations (for which Plebański proposed the name 'Haantjes transformations' (Haantjes, 1937, 1940))[1]:

$$x'^{\alpha} = \frac{x^{\alpha} + C^{\alpha} x_{\rho} x^{\rho}}{1 + 2C_{\sigma} x^{\sigma} + C_{\nu} C^{\nu} x_{\sigma} x^{\sigma}}, \tag{8.79}$$

where $C^{\mu}, \mu = 0, 1, 2, 3$, are group parameters, and the indices are raised and lowered by the Minkowski metric. Verify that this is an Abelian group, that the composition of two Haantjes transformations, with parameters C^{μ} and D^{μ}, is a Haantjes transformation with parameters $(C^{\mu} + D^{\mu})$, and that the transformation inverse to (8.79) is a Haantjes transformation with parameters $\widetilde{C}^{\mu} = -C^{\mu}$. Verify that (8.79) is a composition of the following three transformations:

- the inversion in the pseudosphere of radius L with its centre at $x^{\mu} = 0$:

$$x'^{\mu} = \frac{L^2}{x_{\rho} x^{\rho}} x^{\mu}; \tag{8.80}$$

- translation by the vector C^{μ}:

$$x''^{\mu} = x'^{\mu} + C^{\mu}; \tag{8.81}$$

- the inversion in the pseudosphere of radius L with its centre at $x''^{\mu} = 0$:

$$x'''^{\mu} = \frac{L^2}{x''_{\rho} x''^{\rho}} x''^{\mu}. \tag{8.82}$$

The transformations (8.80) and (8.82) are discrete conformal symmetries of the Minkowski spacetime in Cartesian coordinates, and (8.81) is a conformal symmetry of the Minkowski spacetime in the coordinates x'^{μ}.

Verify that the Haantjes transformations are conformal symmetries also for a flat space of any dimension and with any signature. Equations (8.79)–(8.82) apply then with the indices running through n values.

Hint. Verify first that (8.79) has the following properties:

$$x_{\mu} x^{\mu} = x'_{\mu} x'^{\mu}/T, \qquad dx_{\mu} dx^{\mu} = dx'_{\mu} dx'^{\mu}/T^2, \tag{8.83}$$

---

[1] The papers by Haantjes (1937) contain only special cases of (8.79), corresponding to (i) $C_{\mu} C^{\mu} = 0$ (the 1937 paper) and (ii) only one of the parameters C^{μ} being nonzero (the 1940 paper). The oldest source known to us in which (8.79) is given in full generality is the text by Plebański (1967).

where

$$T \overset{\text{def}}{=} 1 + 2C_\sigma x^\sigma + C_\nu C^\nu x_\sigma x^\sigma. \tag{8.84}$$

12. One of the coordinate representations of 4-dimensional spaces of constant curvature with the signature $(+ - - -)$ (the de Sitter (1917) spacetimes) is

$$ds^2 = \left(1 - \Lambda r^2\right) dt^2 - \frac{1}{1 - \Lambda r^2} dr^2 - r^2 \left(d\vartheta^2 + \sin^2 \vartheta \, d\varphi^2\right), \tag{8.85}$$

where Λ is an arbitrary constant. Find all the Killing fields for this metric, in each of the cases $\Lambda > 0$, $\Lambda < 0$ and $\Lambda = 0$ (the last case is just the Minkowski spacetime in spherical coordinates). Find the structure constants of these groups. Take the limit $\Lambda \to 0$ of the first and second cases and see what happens with the structure constants.

Hint. You may prefer to find the Killing fields for the case $\Lambda = 0$ by transforming the results of Exercise 10 to spherical coordinates.

9

Methods to calculate the curvature quickly – Cartan forms and algebraic computer programs

9.1 The basis of differential forms

Let us recall the conclusion of Section 4.3: if a field of bases of contravariant (covariant) vectors is given on a manifold, then this field determines the dual field of bases of covariant (contravariant) vectors via (4.13) or (4.14), and then both bases can be used to uniquely represent arbitrary tensor densities by sets of scalars.

In Chapter 4 we assumed the bases to be given, but they could be arbitrary. We are thus allowed to choose such a basis that the set of scalars representing a given tensor field is particularly simple. For example, the scalars representing the metric tensor

$$\eta_{ij} = e_i{}^\alpha e_j{}^\beta g_{\alpha\beta} \tag{9.1}$$

(where i and j label different vectors) can all be constant. This is possible for any $g_{\alpha\beta}$: the transition from $g_{\alpha\beta}$ to η_{ij} is equivalent, in the language of linear algebra, to a transformation of the matrix of a quadratic form induced by a change of basis of the underlying vector space. Equation (9.1) does not determine the basis uniquely, but up to the transformations that preserve η_{ij}. For example, when $\dim V_n = 3$ and $\|\eta_{ij}\| = \mathrm{diag}(1, 1, 1)$, the $e_i{}^\alpha$ are determined up to the orthogonal transformations O(3), and when $\dim V_n = 4$ and $\|\eta_{ij}\| = \mathrm{diag}(+1, -1, -1, -1)$, the $e_i{}^\alpha$ are determined up to the Lorentz transformations L(1,3).

A given basis $e_i{}^\alpha$ and a given scalar image of the metric η_{ij} uniquely determine the metric tensor by a formula inverse to (9.1):

$$g_{\alpha\beta} = e^i{}_\alpha e^j{}_\beta \eta_{ij}. \tag{9.2}$$

In a 4-dimensional manifold, the basis $e_i{}^\alpha$ is called a **tetrad of vector fields**, and the scalar image of the metric η_{ij} is called a **tetrad metric**.

As already shown in Section 3.2, every coordinate system naturally defines a field of vector bases; they are the vectors orthogonal to the hypersurfaces $f^i = $ constant, defined by

$$e_{(i)}{}^\alpha f^{(i)}{}_{,\alpha} = 1, \qquad e_i{}^\alpha f^j{}_{,\alpha} = 0, \qquad \text{for } j \neq i \tag{9.3}$$

(with no summation over i in the first equation), where $\{f^j\}, j = 1, \ldots, n$ are the coordinates. However, not every field of bases defines a coordinate system because vector fields are in general not orthogonal to families of hypersurfaces. For general fields, Eqs. (9.3) will not be integrable.

Covariant fields $e^i{}_\alpha$ can be uniquely represented by differential forms

$$e^i = e^i{}_\alpha \, dx^\alpha. \tag{9.4}$$

Multiplying both sides of this equation by $e_i{}^\beta$ and using (4.13), we obtain

$$dx^\beta = e_i{}^\beta e^i. \tag{9.5}$$

From (9.2) and (9.4) we see that the metric can be represented in the basis $\{e^i\}$ in the following way:

$$ds^2 = g_{\alpha\beta} \, dx^\alpha \, dx^\beta = \eta_{ij} e^i e^j. \tag{9.6}$$

9.2 The connection forms

The quantities

$$\Gamma^i{}_{jk} \overset{\text{def}}{=} -e^i{}_{\rho;\sigma} e_j{}^\rho e_k{}^\sigma \tag{9.7}$$

are called the **Ricci rotation coefficients**. Comparing this with Eq. (4.19) for the Christoffel symbols we find

$$\Gamma^i{}_{jk} = e^i{}_\alpha e_j{}^\beta e_k{}^\gamma \left\{ \begin{matrix} \alpha \\ \beta\gamma \end{matrix} \right\} - e^i{}_{\beta,\gamma} e_j{}^\beta e_k{}^\gamma, \tag{9.8}$$

$$\left\{ \begin{matrix} \alpha \\ \beta\gamma \end{matrix} \right\} = e_i{}^\alpha e^j{}_\beta e^k{}_\gamma \Gamma^i{}_{jk} + e_i{}^\alpha e^i{}_{\beta,\gamma}. \tag{9.9}$$

The basis vectors and the Ricci rotation coefficients are thus a unique representation of the Christoffel symbols.

Now let us calculate the exterior derivative of the forms e^i:

$$de^i = -e^i{}_{[\beta,\gamma]} \, dx^\beta \wedge dx^\gamma = -e^i{}_{[\beta;\gamma]} e_j{}^\beta e_k{}^\gamma e^j \wedge e^k = \Gamma^i{}_{[jk]} e^j \wedge e^k \tag{9.10}$$

(along the way, we have used the antisymmetry of the exterior products and the symmetry of the Christoffel symbols in their subscripts that allowed us to replace $e^i{}_{[\beta,\gamma]}$ with $e^i{}_{[\beta;\gamma]}$). Hence, calculating all de^i and then decomposing the result in the basis $e^j \wedge e^k$ we find the $\Gamma^i{}_{[jk]}$. Now we will show that $\Gamma_{ijk} \overset{\text{def}}{=} \eta_{is} \Gamma^s{}_{jk}$ are antisymmetric in (ij) by virtue of their definition. Note that $\Gamma^i{}_{jk}$ can be equivalently written as

$$\Gamma^i{}_{jk} = e^i{}_\rho e_j{}^\rho{}_{;\sigma} e_k{}^\sigma. \tag{9.11}$$

Note also that the following is true:

$$e^i{}_\alpha = \eta^{is} g_{\alpha\rho} e_s{}^\rho, \qquad e_i{}^\alpha = \eta_{is} g^{\alpha\rho} e^s{}_\rho. \tag{9.12}$$

To see this, multiply both sides of the first equation by $e_j{}^\alpha$, to obtain $e^i{}_\alpha e_j{}^\alpha = \delta^i{}_j$, which means that the $e^i{}_\alpha$ defined by (9.12) are identical with the inverse matrices to $e_i{}^\alpha$, i.e. with the $e^i{}_\alpha$ as originally defined by (4.13) and (4.14). Our notation is thus self-consistent: the $e^i{}_\alpha$ result from the $e_i{}^\alpha$ by raising the scalar index with η^{ij} and lowering the tensor index with $g_{\alpha\beta}$ (where η^{ij} is the inverse matrix to η_{ij}). The second equation in (9.12) is verified in a similar way.

Now let us calculate the directional derivatives of (9.1) along $e_k{}^\gamma$:

$$
\begin{aligned}
0 = \left(\eta_{ij}\right)_{,\gamma} e_k{}^\gamma &= \left(e_i{}^\alpha e_j{}^\beta g_{\alpha\beta}\right)_{;\gamma} e_k{}^\gamma \\
&= e_i{}^\alpha{}_{;\gamma} e_j{}^\beta g_{\alpha\beta} e_k{}^\gamma + e_i{}^\alpha e_j{}^\beta{}_{;\gamma} g_{\alpha\beta} e_k{}^\gamma \\
&= e_i{}^\alpha{}_{;\gamma} g^{\beta\rho} \eta_{js} e^s{}_\rho g_{\alpha\beta} e_k{}^\gamma + g^{\alpha\rho} \eta_{is} e^s{}_\rho e_j{}^\beta{}_{;\gamma} g_{\alpha\beta} e_k{}^\gamma \\
&= \eta_{js} e_i{}^\rho{}_{;\gamma} e^s{}_\rho e_k{}^\gamma + \eta_{is} e_j{}^\rho{}_{;\gamma} e^s{}_\rho e_k{}^\gamma = \Gamma_{jik} + \Gamma_{ijk}.
\end{aligned}
\tag{9.13}
$$

(Along the way we used: η_{ij} being constant, the $e_i{}^\alpha e_j{}^\beta g_{\alpha\beta}$ being scalars, $g_{\alpha\beta}$ being covariantly constant, Eqs. (9.12), the equation $g^{\alpha\rho} g_{\beta\rho} = \delta^\alpha{}_\beta$ and (9.11).) The final result is as announced:

$$
\Gamma_{ijk} = -\Gamma_{jik}.
\tag{9.14}
$$

Using this property, one may easily verify that

$$
\Gamma_{ijk} = \Gamma_{i[jk]} - \Gamma_{j[ik]} - \Gamma_{k[ij]}.
\tag{9.15}
$$

Thus, having found $\Gamma^i{}_{[jk]}$ from (9.10), we can find the full $\Gamma^i{}_{jk} = \eta^{is}\Gamma_{sjk}$ from the above. We now use the Ricci rotation coefficients to form differential 1-forms called **connection forms**:

$$
\Gamma^i{}_j = \Gamma^i{}_{jk} e^k.
\tag{9.16}
$$

The tetrad image of the covariant derivative of any tensor field can be expressed fully in terms of the Ricci rotation coefficients in place of the Christoffel symbols and directional derivatives in place of partial derivatives. Thus, for example, for the Riemann tensor

$$
\begin{aligned}
e^a{}_\alpha e_b{}^\beta e_c{}^\gamma e_d{}^\delta e_e{}^\varepsilon R^\alpha{}_{\beta\gamma\delta;\varepsilon} &\equiv e_e{}^\rho R^a{}_{bcd,\rho} \\
&\quad + \Gamma^a{}_{se} R^s{}_{bcd} - \Gamma^s{}_{be} R^a{}_{scd} - \Gamma^s{}_{ce} R^a{}_{bsd} - \Gamma^s{}_{de} R^a{}_{bcs}.
\end{aligned}
\tag{9.17}
$$

9.3 The Riemann tensor

The expression $d\Gamma^i{}_j + \Gamma^i{}_s \wedge \Gamma^s{}_j$ is a 2-form that can be decomposed in the basis $e^k \wedge e^l$:

$$
d\Gamma^i{}_j + \Gamma^i{}_s \wedge \Gamma^s{}_j = \frac{1}{2} R^i{}_{jkl} e^k \wedge e^l.
\tag{9.18}
$$

This is a definition of the coefficients $R^i{}_{jkl}$, but they turn out to be the scalar image of the Riemann tensor

$$
R^i{}_{jkl} = e^i{}_\alpha e_j{}^\beta e_k{}^\gamma e_l{}^\delta R^\alpha{}_{\beta\gamma\delta}.
\tag{9.19}
$$

This is verified as follows:

$$d\Gamma^i{}_j + \Gamma^i{}_s \wedge \Gamma^s{}_j = d\left(\Gamma^i{}_{jk}e^k\right) + \Gamma^i{}_{sk}\Gamma^s{}_{jl}e^k \wedge e^l$$
$$= \left(-\Gamma^i{}_{j[k,|\alpha|}e_l]^\alpha + \Gamma^i{}_{js}\Gamma^s{}_{[kl]} + \Gamma^i{}_{s[k}\Gamma^s{}_{|j|l]}\right)e^k \wedge e^l, \qquad (9.20)$$

where we have used (9.16), (9.10), the antisymmetry of the exterior product and (9.5). Hence, from (9.18) and (9.20) we obtain:

$$\frac{1}{2}R^i{}_{jkl} = -\Gamma^i{}_{j[k,|\alpha|}e_l]^\alpha + \Gamma^i{}_{js}\Gamma^s{}_{[kl]} + \Gamma^i{}_{s[k}\Gamma^s{}_{|j|l]}. \qquad (9.21)$$

Now, using (9.8) and (6.4), we obtain (9.19).

It is now easy to verify that

$$R_{jl} \overset{\text{def}}{=} R^i{}_{jil} \equiv e_j{}^\alpha e_l{}^\beta R_{\alpha\beta}, \qquad (9.22)$$

i.e. that the R_{jl} defined above is the scalar image of the Ricci tensor $R_{\alpha\beta}$ given by (7.46). It is also easy to see that the scalar curvature is

$$R = \eta^{ij}R_{ij} \equiv g^{\alpha\beta}R_{\alpha\beta}. \qquad (9.23)$$

The differential forms allow us to carry out many calculations faster and easier than the traditional tensor calculus. Indeed, some modern courses of relativity avoid tensor calculus altogether, using the Cartan forms as a primary device. However, that approach, although more efficient from the point of view of calculations, makes it more difficult to explain the historical roots of relativity.

As examples of the great simplifications to which the Cartan forms lead, let us note that the identity $R^\alpha{}_{[\beta\gamma\delta]} \equiv 0$ is an immediate consequence of (9.10). Since the antisymmetrisation in (9.10) is carried out automatically (it was written out explicitly only for better clarity), we can make use of (9.16) and write (9.10) in the form

$$de^i = -\Gamma^i{}_j \wedge e^j. \qquad (9.24)$$

Substituting (9.24) and (9.18) in the identity $d^2e^i \equiv 0$ we obtain $\frac{1}{2}R^i{}_{jkl}e^j \wedge e^k \wedge e^l \equiv 0$, which is equivalent to $R^i{}_{[jkl]} \equiv 0$. This identity is thus the integrability condition for (9.24). Similarly, the Bianchi identities (7.15) are obtained as the integrability condition of (9.18).

In relativity, with its 4-dimensional Riemann spaces of signature $(+ - -\,-)$, the most frequently used tetrads are: the **orthonormal** one (in which $\|\eta_{ij}\| = \mathrm{diag}(+1, -1, -1, -1)$), the **null tetrad** (in which $\eta_{01} = \eta_{10} = 1$, $\eta_{22} = \eta_{33} = -1$, all other components of η are zero) and the **double-null tetrad** (in which $\eta_{01} = \eta_{10} = -\eta_{23} = -\eta_{32} = 1$, all other components of η are zero). See Sections 16.5 and 16.6 for an application of the double-null tetrad; in it the forms e^3 and e^4 are complex and conjugate to each other.

9.4 Using computers to calculate the curvature

Calculating the curvature tensor from a given metric tensor is tedious and time-consuming, and seemingly innocent errors made along the way cause great chaos in the final results. In order to obtain a reliable result, every step of the calculation must be carefully verified. A relatively simple calculation typically takes several hours; in complicated cases it can extend to months. At the same time, this is routine work that does not really require intelligence, but just careful application of a set of rules. (Intelligence is required only to understand the rules.) Consequently, it has all the typical features of a task for a computer.

This fact was noticed and made use of long ago. Over the last 40 years, several computer systems have been created that can calculate the Riemann tensor and the associated quantities from a given metric. (We mean here symbolic calculations, in which the computer just transforms mathematical expressions, without expecting them to have numerical values.) Some of those systems are parts of big general-purpose algebraic systems (like Maple or Mathematica), some are specialised programs written in generally accessible programming languages. The language of choice for computer algebra is Lisp. Among the specialised programs that are still available are Sheep, created in Stockholm, and Ortocartan, written by one of the authors of this text (Krasiński, 2001b). Information on them is best obtained from the Internet (the expression to look for is 'computer algebra').

The modern computer algebra programs are fairly easy to use, and the reduction in time and effort required to do the calculation is dramatic. Instead of doing routine calculations for weeks, one can have the result in less than a minute. Of course, this does not include the time required to write the data, and usually several runs are needed before the result becomes acceptable (for example, additional simplifications might be needed). Still, the gain is obvious, and these calculations are done 'by hand' essentially only by students for educational purposes. In research work, the computers have taken over the field completely.

9.5 Exercises

1. Verify Eq. (9.17).
2. Calculate the exterior derivative of (9.18) and show that it is equivalent to the Bianchi identities (7.15).
3. Take a simple metric, for example the de Sitter metric of Eq. (8.85), and calculate the Riemann tensor first by tensor calculus, then by using the Cartan forms in the orthonormal tetrad. Compare the time and effort required in the two cases. Verify Eq. (9.19).

10

The spatially homogeneous Bianchi type spacetimes

10.1 The Bianchi classification of 3-dimensional Lie algebras

As already mentioned, generators of symmetries of a manifold form a Lie algebra. For reasons to be explained further, the case when this algebra is 3-dimensional is important for relativity. The aim of the Bianchi classification is to sort out those 3-dimensional algebras which are inequivalent in the following sense.

A basis of generators is not defined uniquely. If $k_{(i)}$ is a basis, then

$$\widetilde{k}^{\alpha}_{(i')} \stackrel{\text{def}}{=} A_{i'}{}^{j} \, k^{\alpha}_{(j)} \qquad (\text{sum over } j) \tag{10.1}$$

is also a basis, provided that the matrix $A_{i'}{}^{j}$ (of constant elements) is non-singular. Such a change of basis induces the following change of the structure constants:

$$\widetilde{C}^{l'}_{i'j'} = \left(A^{-1}\right)^{l'}{}_{l} A_{i'}{}^{i} A_{j'}{}^{j} C^{l}_{ij}. \tag{10.2}$$

Thus, two sets of structure constants that are related by (10.2) correspond to two bases that span isomorphic algebras. How can one recognise whether any matrix A obeying (10.2) exists for two given sets of C^{l}_{ij}? The answer is provided by the Bianchi classification. Note that no reference will be made in this section to the way in which the corresponding group acts on the manifold.

The method of presentation used here was invented by Schücking in the 1950s, but introduced only during a seminar talk. It diffused into public knowledge via notes taken by Kundt that were published only recently (Kundt, 2003); see Krasiński et al. (2003) for a description of the story. The earliest papers that used this approach and that introduced it into the literature were by Estabrook, Wahlquist and Behr (1968), Ellis and MacCallum (1969) and by MacCallum and Ellis (1970). The classification was originally introduced by Bianchi (1898), but his method requires a long presentation and is no longer in use.[1] Bianchi sorted out the different types by the properties of the derived algebras.

---

[1] Bianchi's classification was considered from the point of view of the theory of Lie algebras. Its importance for relativity was recognised much later. The first paper that explicitly introduced the Bianchi types into relativity was that by Taub (1951), but the inspiration is said to have come from Gödel (1949). See Jantzen (2001) for more on this story.

From (8.30) it follows that $C^l{}_{ij} = -C^l{}_{ji}$, so for an N-dimensional group the number of structure constants cannot exceed $N \cdot \frac{1}{2}N(N-1)$. Through the Jacobi identity

$$[[\underset{(i)}{k}, \underset{(j)}{k}], \underset{(l)}{k}] + [[\underset{(j)}{k}, \underset{(l)}{k}], \underset{(i)}{k}] + [[\underset{(l)}{k}, \underset{(i)}{k}], \underset{(j)}{k}] = 0, \tag{10.3}$$

the definition (8.30) implies a further limitation on $C^l{}_{ij}$:

$$C^m{}_{ij}C^n{}_{ml} + C^m{}_{jl}C^n{}_{mi} + C^m{}_{li}C^n{}_{mj} = 0. \tag{10.4}$$

We shall take (10.4) into account later.

For $N = 3$, $\frac{1}{2}N^2(N-1) = 9$ and equals the number of elements of a general 3×3 matrix. Thus for a 3-dimensional algebra all structure constants can be put into a single 3×3 matrix, let us call it H^{ab}. The one-to-one correspondence between $C^l{}_{ij}$ and the elements of H^{ab} is

$$H^{ab} = \frac{1}{2}\epsilon^{akl}C^b{}_{kl} \Longleftrightarrow C^i{}_{jk} = \epsilon_{sjk}H^{si}, \tag{10.5}$$

where ϵ^{akl} and ϵ_{sjk} are the Levi-Civita symbols. The matrix H^{ab} can now be split into its symmetric part

$$n^{ab} \overset{\text{def}}{=} H^{(ab)} = \frac{1}{2}\left(H^{ab} + H^{ba}\right) \tag{10.6}$$

and its antisymmetric part $H^{[ab]}$. But in a 3-dimensional space there is a one-to-one correspondence between antisymmetric matrices and vectors, so $H^{[ab]}$ can be represented by the vector a defined by

$$a_i = -\frac{1}{2}\epsilon_{ijk}H^{[jk]} \Longleftrightarrow H^{[ij]} = -\epsilon^{ijm}a_m. \tag{10.7}$$

From (10.5)–(10.7) we have

$$C^l{}_{jk} = \epsilon_{sjk}n^{sl} - \delta^{lm}_{jk}a_m. \tag{10.8}$$

The identity (10.4) for $N = 3$ may be written as $C^m{}_{ij}C^n{}_{ml}\epsilon^{ijl} = 0$, and with use of (10.8) this implies that

$$n^{is}a_s = 0, \tag{10.9}$$

i.e. either $a_i = 0$ or else n^{is} has at least one zero eigenvalue.

From (10.1) and (10.2) it is seen that n^{ab} will transform, under the change of basis (10.1), like a twice contravariant tensor density

$$\tilde{n}^{i'j'} = (\det A)^{-1}\left(A^{-1}\right)^{i'}{}_i\left(A^{-1}\right)^{j'}{}_j n^{ij}, \tag{10.10}$$

while a_i will transform like a covariant vector, $\tilde{a}_{i'} = A_{i'}{}^i a_i$. The transformations (10.10) can be used to diagonalise n^{ij}. Let us then assume that the basis $\underset{(i)}{k}$ was chosen appropriately and that n^{ij} is of the form

$$n^{ij} = \begin{pmatrix} n_1 & 0 & 0 \\ 0 & n_2 & 0 \\ 0 & 0 & n_3 \end{pmatrix}. \tag{10.11}$$

This does not yet fix the basis uniquely; transformations that permute (n_1, n_2, n_3) are still allowed. Therefore, if any of the n_i is zero, it can always be moved by transformations of the basis to the upper left corner in (10.11), i.e. we can always assume that

$$n_1 = 0 \quad \text{if} \quad a_i \neq 0. \tag{10.12}$$

If $n_1 = 0$ is the single zero eigenvalue, then in view of (10.9) the vector a will assume, after such a choice of basis, the form

$$a_i = [a, 0, 0]. \tag{10.13}$$

If $n_1 = 0$ is a multiple eigenvalue, then (10.12) still allows us to rotate a within the eigenspace of the $n_1 = 0$ eigenvalue. We can then rotate a so that it will assume the form (10.13). Equation (10.13) covers the subcase $a_i = 0$. In such a basis, (10.9) becomes

$$an_1 = 0. \tag{10.14}$$

Using all the information about $C^l{}_{ij}$, the commutators become:

$$\left[\underset{(1)}{k}, \underset{(2)}{k}\right] = a \underset{(2)}{k} + n_3 \underset{(3)}{k}, \quad \left[\underset{(2)}{k}, \underset{(3)}{k}\right] = n_1 \underset{(1)}{k}, \tag{10.15}$$

$$\left[\underset{(3)}{k}, \underset{(1)}{k}\right] = n_2 \underset{(2)}{k} - a \underset{(3)}{k}. \tag{10.16}$$

This form of the commutation relations was obtained using rotations of the basis vector fields $\underset{(i)}{k}$. From now on, no further rotations are allowed, but we may still scale $\underset{(i)}{k}$ without changing their directions, $\underset{(i)}{k} = C_i \underset{(i)}{k'}$ (no sum). After such a scaling, (10.15)–(10.16) change to

$$\left[\underset{(1)}{k}, \underset{(2)}{k}\right] = \frac{a}{C_1} \underset{(2)}{k} + \frac{C_3}{C_1 C_2} n_3 \underset{(3)}{k}, \quad \left[\underset{(2)}{k}, \underset{(3)}{k}\right] = \frac{C_1}{C_2 C_3} n_1 \underset{(1)}{k}, \tag{10.17}$$

$$\left[\underset{(3)}{k}, \underset{(1)}{k}\right] = \frac{C_2}{C_1 C_3} n_2 \underset{(2)}{k} - \frac{a}{C_1} \underset{(3)}{k} \tag{10.18}$$

(primes were dropped for clarity). We now want to use these scalings to simplify a, n_1, n_2 and n_3. Using C_1, C_2 and C_3 we can scale those of the parameters (a, n_1, n_2, n_3) that are nonzero. A nonzero value can never be made zero by scaling. Hence, the preliminary classification is as shown in Table 10.1, where S stands for 'something' (different from zero). However, not all the entries in Table 10.1 are inequivalent. Permutations of the basis vectors that do not violate (10.14) are still allowed. Clearly, any permutation of (n_1, n_2, n_3) is allowed when $a = 0$, and with $a \neq 0$ we are still allowed to permute n_2 and n_3. Hence, only the cases indicated in the last line of the table have a chance to be inequivalent.

We will see that the classification of the algebras into inequivalent types does not match the columns of Table 10.1 – the table will serve merely to make the presentation orderly. The labels for the types were introduced by Bianchi (1898), who, as already mentioned, used a different method. By tradition, his numbering is still in use, although it does not seem natural in the derivation presented below.

Table 10.1 *A preliminary Bianchi classification*

a	0	0	0	0	0	0	0	0	S	S	S	S
n_1	0	0	0	0	S	S	S	S	0	0	0	0
n_2	0	0	S	S	0	0	S	S	0	0	S	S
n_3	0	S	0	S	0	S	0	S	0	S	0	S
Case number	1	/	/	/	2	/	3	4	5	6	/	7

Let us consider the consecutive columns of Table 10.1.

(1) $a = n_1 = n_2 = n_3 = 0$.
This is Bianchi type I where all commutators are zero.

(2) $a = n_2 = n_3 = 0$, $n_1 \neq 0$.
Taking $C_1 = C_2 C_3 / n_1$ we obtain $n_1' = 1$. This is Bianchi type II.

(3) $a = n_3 = 0$, $n_1 \neq 0 \neq n_2$.
Taking $C_1 = C_2 C_3 / n_1$ we obtain $n_1' = 1$. However, as seen from (10.17)–(10.18), we have then $n_2' = n_1 n_2 / C_3^2$ and we will not be able to change the sign of n_2' by choice of C_3. Therefore we must consider two cases:

(3a) $n_1 n_2 > 0$.
Then we take $C_3 = (n_1 n_2)^{1/2}$ and obtain $n_2' = 1$. This is a subcase of Bianchi's type VII. Bianchi himself called it type VII_1; today it is called type VII_0.

(3b) $n_1 n_2 < 0$.
Then we take $C_3 = (-n_1 n_2)^{1/2}$ and obtain $n_2' = -1$. This is a subcase of Bianchi's type VI, called today VI_0. Bianchi noted that this case requires a separate treatment, but the final result fits well within the general type VI, so he did not give it any special name.

(4) $a = 0$, $n_1 n_2 n_3 \neq 0$.
We take $C_1 = C_2 C_3 / n_1$ and obtain $n_1' = 1$. But then, as before, $n_2' = n_1 n_2 / C_3^2$, and the two signs of $n_1 n_2$ have to be considered separately.

(4a) $n_1 n_2 > 0$.
We take $C_3 = (n_1 n_2)^{1/2}$ and obtain $n_2' = 1$. However, $n_3' = n_1 n_3 / C_2^2$, and two further subcases arise:

(4a$_1$) $n_1 n_3 > 0$.
Then we take $C_2 = (n_1 n_3)^{1/2}$ and obtain $n_3' = 1$. Thus finally $n_1' = n_2' = n_3' = 1$, $a = 0$. This is Bianchi type IX.

(4a$_2$) $n_1 n_3 < 0$.
Then we take $C_2 = (-n_1 n_3)^{1/2}$ and obtain $n_3' = -1$. Hence finally $n_1' = n_2' = 1 = -n_3'$. This is Bianchi type VIII.

(4b) $n_1 n_2 < 0$.
We then take $C_3 = (-n_1 n_2)^{1/2}$ and obtain $n_2' = -1$. But then we must again consider the same two subcases as before for C_2:

(4b$_1$) $n_1 n_3 > 0$.

Then we take $C_2 = (n_1 n_3)^{1/2}$ and obtain $n'_3 = 1$. Through the basis change (still allowed!)

$$\left(\underset{(3)}{k}, \underset{(2)}{k}\right) = \left(\underset{(2)}{\tilde{k}}, -\underset{(3)}{\tilde{k}}\right)$$

we then obtain the same parameter values as in case (4a$_2$).

(4b$_2$) $n_1 n_3 < 0$.

Then we take $C_2 = (-n_1 n_3)^{1/2}$ and obtain $n'_3 = -1$. Again, through the basis change

$$\left(\underset{(1)}{k}, \underset{(3)}{k}, \underset{(2)}{k}\right) = \left(-\underset{(3)}{\tilde{k}}, -\underset{(2)}{\tilde{k}}, -\underset{(1)}{\tilde{k}}\right),$$

we arrive back at case (4a$_2$).

(5) $a \neq 0$, $n_1 = n_2 = n_3 = 0$.

Taking $C_1 = a$ we obtain $a' = 1$. This is Bianchi type V.

(6) $a \neq 0 \neq n_3$, $n_1 = n_2 = 0$.

We take $a = C_1$, $C_3 = aC_2/n_3$ and obtain $a' = 1 = n'_3$. This is Bianchi type IV.

(7) $a n_2 n_3 \neq 0$, $n_1 = 0$.

Taking $C_2 = C_1 C_3/n_2$ we obtain $n'_2 = 1$. We lose again the possibility of changing the sign of $n'_3 = n_2 n_3/C_1{}^2$ and must consider two cases:

(7a) $n_2 n_3 > 0$.

Then we take $C_1 = (n_2 n_3)^{1/2}$ and obtain $n'_3 = 1$. Since, however, we fix C_1 in this way, we have no possibility left to scale a. Hence, with $n'_3 = 1$, the value of a remains arbitrary. Then the algebras corresponding to $n_2 = n_3 = 1$ but different values of a are *not* equivalent. Bianchi called this case VII$_2$, but, as opposed to previous types, this is a one-parameter family of inequivalent types. With $a = 0$, an algebra equivalent to the one obtained in (3a) results. This is seen if, with $a = 0$, we carry out the basis change

$$\left(\underset{(1)}{k}, \underset{(3)}{k}\right) = \left(\underset{(3)}{\tilde{k}}, -\underset{(1)}{\tilde{k}}\right).$$

The general type VII, with $a \neq 0$, is today denoted VII$_h$.

(7b) $n_2 n_3 < 0$.

Then we take $C_1 = (-n_2 n_3)^{1/2}$ and obtain $n'_3 1 = -1$. Just as before, we are left then with no possibility of changing a. Again, this is a one-parameter family of inequivalent types. Bianchi denoted it type VI; today it is denoted VI$_h$. With $a = 0$, the algebra obtained in (3b) results, i.e. type VI$_0$. To see this, the same basis change as in (7a) is necessary.

Table 10.2 *The Bianchi classification*

a	0	0	0	0	0	0	1	1	a	a	1
n_1	0	1	1	1	1	1	0	0	0	0	0
n_2	0	0	1	-1	1	1	0	0	1	1	1
n_3	0	0	0	0	1	-1	0	1	1	-1	-1
Bianchi type	I	II	VII_0	VI_0	IX	VIII	V	IV	VII_h	VI_h	III

In Bianchi's scheme the subcase of type VI corresponding to $a = 1$ emerged as a separate type which he called type III. In it, $a = n_2 = 1 = -n_3$, $n_1 = 0$.

The final classification is shown in Table 10.2.

10.2 The dimension of the group versus the dimension of the orbit

In Section 8.2 we defined the orbits of one-parameter families of mappings. At this point it should be clear that the notion of an orbit can be defined also for a family (or group) of mappings that has several parameters; such orbits can be multidimensional curved spaces.

Let us recall that the Bianchi classification was done without any reference to the way in which the generators act on the manifold M_n. Two situations are possible:

1. The three generators (which are by definition linearly independent as *vector fields*) may also be linearly independent as *vectors* at each single point of M_n. This happens e.g. for generators of translations in $\mathbb{R}^3$.
2. The three generators may be linearly dependent at each single point of M_n. This happens e.g. for the generators of the group O(3), which is a 3-dimensional group, but has 2-dimensional orbits.

A 3-dimensional isometry group cannot have 1-dimensional orbits (from the Killing equations, if the orbits are 1-dimensional, then any two generators will be proportional to each other with a constant factor).

If the orbits are 2-dimensional, then their curvature is characterised by one scalar that must be invariant under the action of the group, i.e. constant. The constant curvature may be positive (the orbits are then 2-dimensional spheres), zero (the orbits are then 2-planes) or negative (such a surface has, in polar coordinates, the metric $ds^2 = (1 + a^2 r^2)^{-1} dr^2 + r^2 d\phi^2$, but cannot be embedded in the $\mathbb{R}^3$ with a positive-definite Euclidean metric). Then, the three algebras of generators are of Bianchi types IX, VII_0 and VIII, respectively.

If a 3-dimensional group has 3-dimensional orbits, then the scalar curvature of the orbits must also be constant. However, in three dimensions the scalar curvature does not fully characterise the curvature of space; there exists also the Ricci tensor. Therefore more geometries are possible; they will be briefly discussed in Section 10.6.

10.3 Action of a group on a manifold

If a 3-dimensional group of mappings of M_n into itself has 3-dimensional orbits, then several situations are possible, e.g. the following:

1. The mappings may or may not be symmetries of M_n (they could be, for example, conformal symmetries of M_n).
2. The orbits may be timelike, spacelike or null hypersurfaces in M_n.

In each situation one can consider the various Bianchi types. Examples of solutions of Einstein's equations are known for which the orbits are timelike (e.g. Krasiński (1974) for Bianchi type I and Krasiński (1998 and 2001a) for more general types; in the 1998 and 2001a papers there are examples of null orbits). A systematic investigation of all the spacetimes with 3-dimensional timelike orbits of the Bianchi groups was done by Harness (1982). Examples are also known in which 3-dimensional groups act as groups of symmetries on certain preferred 3-dimensional submanifolds of M_n, but are not symmetries of the whole M_n (these are the so-called spacetimes with intrinsic symmetries (Collins, 1979); for examples see Krasiński (1981) and Wolf (1985)). A brief discussion of general properties of spacetimes for which a 3-dimensional symmetry group has 3-dimensional null orbits is given in Stephani *et al.* (2003, Section 24.2). A systematic investigation was done of the case when the group has 3-dimensional spacelike orbits and the group transformations are conformal symmetries of the manifold with the conformal factor ϕ in $\underset{k}{£}g_{\alpha\beta} = \phi g_{\alpha\beta}$ being constant (these are the so-called **self-similar** spacetimes (Eardley 1974a)).

Most effort went into investigating the case when the 3-dimensional orbits are spacelike and the group in question is a group of symmetries of M_n (but examples are known where the orbits are spacelike in one part of M_n and timelike elsewhere; see Collins and Wainwright (1983) and Collins and Ellis (1979)). Before we consider this case in more detail, a few definitions must be given.

10.4 Groups acting transitively, homogeneous spaces

Let Orb(p, G) denote the orbit of the point $p \in M_n$ under the action of the group G. We say that G **acts transitively** on a manifold S when, for every $q \in S$, Orb(q, G) = S (in further considerations S will be a submanifold of M_n). *Examples*: The group O(3) acts transitively on the surface of a sphere, the group of arbitrary translations in $\mathbb{R}^n$ acts transitively on $\mathbb{R}^n$. The space S on which G acts transitively is called **homogeneous with respect to G**.

If, for every $q \in S$ (assumed homogeneous with respect to G) there exists a subgroup $H \subset G$ such that Hq = q, then G is said to act **multiply transitively** on S. A group acting transitively, but not multiply, acts **simply transitively**. *Examples*: The group of arbitrary translations in $\mathbb{R}^n$ acts simply transitively on $\mathbb{R}^n$, the group O(3) acts multiply transitively on a 2-sphere since each point p of the sphere remains unchanged by rotations around the axis which passes through p.

A 4-dimensional manifold M_n with the metric tensor g of signature $(+---)$ is called a **spatially homogeneous spacetime of Bianchi type** if it has the following properties:

1. It has a 3-dimensional symmetry group G.
2. The orbits of the group are spacelike hypersurfaces in M_n on which G acts simply transitively.

A Bianchi-type spacetime may have a larger symmetry group; the definition requires then that the full symmetry group has a 3-dimensional subgroup that acts simply transitively on spacelike hypersurfaces.

10.5 Invariant vector fields

As noted in Section 8.5, the vector field X is invariant under the group of transformations generated by the vector field k if

$$\underset{k}{\pounds} X \equiv [k, X] = 0. \tag{10.19}$$

Now let X be a vector field on S, let $\underset{(i)}{k}, i = 1, \ldots, m$ be a set of generators of invariances of X such that, at every point $p \in S$, $\left\{ \underset{(i)}{k}(p) \right\}_{i=1}^{m}$ is a basis in the tangent space to S at p (i.e. the orbits of the group generated by $\underset{(i)}{k}$ are m-dimensional and so is S). In that case the matrix K where $K_i{}^\alpha = \underset{(i)}{k}{}^\alpha(p)$ is non-singular at every $p \in S$, and so defines the inverse matrix κ. Let us write (10.19) for all the fields $\underset{(i)}{k}$:

$$\underset{(i)}{k}{}^\rho X^\alpha{}_{,\rho} - X^\rho \underset{(i)}{k}{}^\alpha{}_{,\rho} = 0 \tag{10.20}$$

and multiply this set of equations by the matrix κ. The result will be

$$X^\alpha{}_{,\beta} - \kappa^i_\beta \underset{(i)}{k}{}^\alpha{}_{,\rho} X^\rho = 0 \qquad \text{(sum over } i\text{)}. \tag{10.21}$$

This formula is similar to the condition of covariant constancy of a vector field, where the quantity

$$G^\alpha{}_{\beta\gamma} \overset{\text{def}}{=} - \kappa^i_\beta \underset{(i)}{k}{}^\alpha{}_{,\gamma} \qquad \text{(sum over } i\text{)} \tag{10.22}$$

plays the role of the affine connection. Let us follow this analogy and calculate the curvature tensor defined by $G^\alpha{}_{\beta\gamma}$. It turns out to be equal to zero. Therefore Eq. (10.21) defines a transport of the vector field X along the vector fields $\underset{(i)}{k}$ that is formally analogous to the parallel transport on S and has vanishing curvature. Consequently, the transport is path-independent. Thus, if $X(p)$ is defined at any single point $p \in S$, then (10.21) will uniquely define X at any other point of S.

We can then define a vector *basis* $X_{(i)}(p)$ at $p \in S, i = 1, \ldots, m$, and use Eq. (10.21) (called the **Lie transport**) to transplant the basis to all other point of S. The vector fields $X_{(i)}$ on S thus obtained will be automatically invariant with respect to the transformations generated by the fields $k_{(i)}$. We have thus proven:

Theorem 10.1 *If a set of vector fields $k_{(i)}$ on S exists such that $k_{(i)}(p)$ form a basis of the tangent space to S at each $p \in S$, then there exists a set of vector fields $X_{(i)}$ on S that are invariant under the group of transformations generated by $k_{(i)}$ and also form a basis at each $p \in S$.*

Now let us assume in addition that all the fields $k_{(i)}$ are Killing fields, and let us calculate the Lie derivatives along $k_{(i)}$ of the quantities

$$g_{(j)(k)} \overset{def}{=} X_{(j)}^\alpha X_{(k)}^\beta g_{\alpha\beta}, \tag{10.23}$$

where $g_{\alpha\beta}$ is the metric tensor on S. Since $X_{(i)}$ are invariant, we have $\pounds_{k_{(i)}} X_{(j)} = 0$, and since $k_{(i)}$ are now Killing fields, we have $\pounds_{k_{(i)}} g_{\alpha\beta} = 0$, and so $\pounds_{k_{(i)}} g_{(j)(k)} = 0$. But $g_{(j)(k)}$ are scalars, so from (8.76) it follows that

$$0 = \pounds_{k_{(i)}} g_{(j)(k)} = k_{(i)}^\rho g_{(j)(k),\rho} \tag{10.24}$$

and, since $k_{(i)}$ form a basis, this means that $g_{(j)(k)}$ are constants.

If $X_{(i)}$ and $X_{(j)}$ are invariant, then so is $[X_{(i)}, X_{(j)}]$. Since $X_{(i)}(p)$ form a basis at every $p \in S$, it follows that $[X_{(i)}, X_{(j)}]$ can be decomposed in this basis,

$$[X_{(i)}, X_{(j)}] = D^l{}_{ij} X_{(l)}, \tag{10.25}$$

where the scalar coefficients $D^l{}_{ij}$ could be expected to depend on the point p. However, calculating the Lie derivative of (10.25) along $k_{(i)}$, using the invariance of $[X_{(i)}, X_{(j)}]$ and of $X_{(l)}$, and using the fact that $X_{(l)}$ form a basis at each $p \in S$, we conclude that $D^l{}_{ij}$ are constants.

The basis $X_{(i)}$ was defined by choosing $X_{(i)}(p_0)$ arbitrarily at a $p_0 \in S$ and transporting it off p_0 by (10.21). Let us assume then that at p_0 we have $X_{(i)}^\alpha(p_0) = k_{(i)}^\alpha(p_0)$. In this case it can be proven that

$$D^l{}_{ij} = -C^l{}_{ij}, \tag{10.26}$$

where $C^l{}_{ij}$ are the structure constants defined by the commutators of $k_{(i)}$. A hint for the proof: since both $k_{(i)}$ and $X_{(i)}$ are bases at every $p \in S$ and $k_{(i)} = X_{(i)}$ at p_0, there exists a matrix $M(\{x\})$ (point-dependent!) such that $X^\alpha_{(i)} = M_i{}^j k^\alpha_{(j)}$ (sum over j) and $M_i{}^j(p_0) = \delta_i{}^j$. Knowing this, play with the commutators at p_0. (Without the condition $M_i{}^j(p_0) = \delta_i{}^j$ the constants $D^l{}_{ij}$ would be linear combinations of the $C^l{}_{ij}$.)

10.6 The metrics of the Bianchi-type spacetimes

Let $k_{(i)}$ be the three Killing fields required by the definition, and let x^I be the coordinates in the homogeneous hypersurfaces ($i, I = 1, 2, 3$).

The homogeneous hypersurfaces are tangent to $k_{(i)}$ (by definition) and uniquely define the vector field m orthogonal to them,

$$m^\alpha = |g|^{-1/2} \epsilon^{\alpha\beta\gamma\delta} k_\beta k_\gamma k_\delta,$$
$$\quad\quad\quad\quad {}_{(1)\,(2)\,(3)}$$
(10.27)

where $g = \det||g_{\alpha\beta}||$. Let us choose the parameter on the integral lines of m as the t-coordinate in M_n: $m^\alpha = \partial x^\alpha/\partial t$. Since m is orthogonal to all $k_{(i)}$, in such coordinates we have

$$g_{0I} = 0.$$
(10.28)

From the (00) component of the Killing equations, using (10.28) and $k^0_{(i)} = 0$, we further obtain $k^I_{(i)} g_{00,I} = 0$, i.e. $g_{00} = g_{00}(t)$. By the next transformation, $t' = \int \sqrt{g_{00}(t)}\,dt$, we obtain (dropping the prime)

$$ds^2 = dt^2 - g_{IJ}\,dx^I\,dx^J.$$
(10.29)

From the $(0, i)$, $i = 1, 2, 3$, components of the Killing equations we then conclude that $k^I_{(i),0} = 0$ and that $k_{(i)}$ are Killing fields also for the 3-metric g_{IJ}. Since for the homogeneous hypersurfaces with the 3-metric g_{IJ} the fields $k_{(i)}$ form a basis at each point, we can apply the results of Section 10.5 to each single hypersurface. Here, Eq. (10.24) will read $k^R_{(i)} g_{(j)(k),R} = 0$, but $(\partial/\partial t)g_{(j)(k)}$ is not determined. Consequently, the $g_{(j)(k)}$ will be functions of t. Let us denote by $\omega^{(i)}{}_I$ the matrix inverse to $X^I_{(i)}$ (where $X_{(i)}$ are the vector fields invariant with respect to the transformations generated by $k_{(i)}$); then, from (10.23):

$$g_{IJ} = g_{(j)(k)}(t)\omega^{(j)}{}_I \omega^{(k)}{}_J$$
(10.30)

and

$$ds^2 = dt^2 - g_{(j)(k)}(t)\omega^{(j)}{}_I \omega^{(k)}{}_J \, dx^I \, dx^J.$$
(10.31)

In order to find $\omega^{(i)}{}_I$ one has to construct the Killing fields and their invariant fields for each Bianchi type separately. The resulting formulae can be found e.g. in Stephani *et al.* (2003, Chapters 13 and 14, in particular Table 13.4 and p. 209).

10.7 The isotropic Bianchi-type (Robertson–Walker) spacetimes

In astrophysics, such Bianchi-type spacetimes are important that are not only spatially homogeneous but also spherically symmetric (**isotropic**). We shall deal with their (astro)physical implications in Chapter 17. Here, we will only derive the appropriate metric form.

The form (8.51) of the metric is preserved by the transformations $t = f(t', r')$, $r = g(t', r')$ which can be used to simplify (8.51) further. In order to transform (8.51) into the form (10.31), we can choose f and g so that the new $\beta = 0$. In the new form, only those transformations will be permissible which preserve the hypersurfaces $t = $ constant. They are $t = f(t')$, $r = g(r')$. In agreement with (10.29), α should then depend only on t and be transformable to 1. So, finally, a spherically symmetric metric can possibly be homogeneous in the Bianchi sense only if it can be put in the form

$$ds^2 = dt^2 + \gamma(t, r)dr^2 + \delta(t, r)\left(d\vartheta^2 + \sin^2 \vartheta \, d\phi^2\right). \tag{10.32}$$

With such a metric one should now solve the Killing equations. We will present the solution and give hints on how to derive it later in this section.

Some properties of the solution can be guessed in advance. The spacetime with the metric (10.32) has O(3) as a symmetry group. Since O(3) has 2-dimensional orbits, it cannot be the homogeneity group H existing in the Bianchi-type spacetimes. Hence, the full symmetry group must contain O(3) and H as two subgroups. O(3) cannot be a subgroup of H because it acts multiply transitively. O(3) and H cannot have any common 2-dimensional subgroup because O(3) has no 2-dimensional subgroups at all. All 1-dimensional subgroups of O(3) are rotations around an axis, which act multiply transitively and so cannot be subgroups of H. Thus, O(3) and H cannot have any common subgroup apart from the identity transformation. Hence, the full symmetry group will have *at least* six parameters, three of them connected with O(3) and three with H. On the other hand, we showed in Section 8.4 that an n-dimensional manifold can have a symmetry group of *at most* $\frac{1}{2}n(n+1)$ parameters, i.e. at most six when $n = 3$. Consequently, the spatially homogeneous and isotropic spacetimes have symmetry groups with exactly six parameters.

We shall now indicate how to solve the Killing equations for the metric (10.32). The calculation is laborious, but uses only routine mathematics. We recall (see Section 10.6) that in a general Bianchi-type spacetime the Killing fields have no time-component ($k^0 = 0$), while the components (00) and (0, i), $i = 1, 2, 3$, of the Killing equations have already been solved with the result $k^I{}_{,t} = 0$, $I = 1, 2, 3$. (We could proceed without using this information, but then among the solutions there would be those with 4-dimensional orbits of the symmetry group – the Minkowski and de Sitter metrics – that are just subcases of the Bianchi spacetimes.)

The remaining Killing equations are

$$k^1 \gamma_{,r} + 2k^1{}_{,r}\, \gamma = 0, \tag{10.33}$$

$$k^2{}_{,r}\, \delta + k^1{}_{,\vartheta}\, \gamma = 0, \tag{10.34}$$

$$k^3{}_{,r}\,\delta\sin^2\vartheta + k^1{}_{,\varphi}\,\gamma = 0, \tag{10.35}$$

$$k^1\delta_{,r} + 2k^2{}_{,\vartheta}\,\delta = 0, \tag{10.36}$$

$$k^3{}_{,\vartheta}\sin^2\vartheta + k^2{}_{,\varphi} = 0, \tag{10.37}$$

$$k^1\delta_{,r} + 2k^2\cot\vartheta\,\delta + 2k^3{}_{,\varphi}\,\delta = 0. \tag{10.38}$$

We seek solutions with the physical signature $(+---)$, so $\gamma < 0$ and $\delta < 0$. We also demand that $k^1 \neq 0$, for otherwise the orbits of the symmetry group would come out 2-dimensional, contrary to a basic property of the Bianchi-type spacetimes. The set (10.33)–(10.38) is overdetermined, so, whichever equation we solve first, the solution will be further limited by the remaining equations. Limitations are thereby imposed not only on the Killing fields, but also on the metric components γ and δ. Along the way, some of the equations lead to alternatives of the type $(ab = 0 \Longrightarrow a = 0 \text{ or } b = 0)$, and this is where the special cases appear.

One of the special solutions that emerge is the metric[1]

$$ds^2 = dt^2 - R^2(t)\,dr^2 - S^2(t)\left(d\vartheta^2 + \sin^2\vartheta\,d\varphi^2\right), \tag{10.39}$$

whose generators of symmetries are

$$\underset{(1)}{J} = \frac{\partial}{\partial r}, \qquad \underset{(2)}{J} = \cos\varphi\,\frac{\partial}{\partial\vartheta} - \sin\varphi\cot\vartheta\,\frac{\partial}{\partial\varphi},$$

$$\underset{(3)}{J} = \sin\varphi\,\frac{\partial}{\partial\vartheta} + \cos\varphi\cot\vartheta\,\frac{\partial}{\partial\varphi}, \qquad \underset{(4)}{J} = \frac{\partial}{\partial\varphi}. \tag{10.40}$$

The generators $\underset{(2)}{J}$, $\underset{(3)}{J}$ and $\underset{(4)}{J}$ generate the O(3) group that was assumed from the beginning. The four-parameter group generated by all of (10.40) has 3-dimensional orbits, and has no three-parameter simply transitive subgroup. This is because O(3) alone has 2-dimensional orbits and so cannot be simply transitive, and has no 2-dimensional subgroups that could be combined with the transformations generated by $\underset{(1)}{J}$ into a three-parameter group. Hence, (10.39) does not belong among the Bianchi-type spacetimes. It is a metric of the Kantowski–Sachs (1966) class.[2]

The generic solution of (10.33)–(10.38) is

$$\underset{(1)}{J} = -V\cos\vartheta\,\frac{\partial}{\partial r} + W\sin\vartheta\,\frac{\partial}{\partial\vartheta},$$

$$\underset{(2)}{J} = V\sin\vartheta\cos\varphi\,\frac{\partial}{\partial r} + W\cos\vartheta\cos\varphi\,\frac{\partial}{\partial\vartheta} - W\frac{\sin\varphi}{\sin\vartheta}\,\frac{\partial}{\partial\varphi},$$

---

[1] A transformation of the r-coordinate is required to achieve the form (10.39).

[2] Metrics with the Kantowski–Sachs symmetry have a longer history than most people suspect. A *generalisation* of such a metric first appeared in a paper by Datt (1938), but the author instantly dismissed it as unphysical. Some physical properties of the metrics (10.39) were investigated by Kompaneets and Chernov (1964). The symmetry was noted and investigated by Kantowski (1965), and became a classical piece of knowledge after the paper by Kantowski and Sachs (1966). The geometric properties of the Datt metric were first investigated by Ruban (1968, 1969). See the reprints of the Datt (1938), Kantowski (1965) and Ruban papers for more on this story.

$$J_{(3)} = V \sin \vartheta \sin \varphi \, \frac{\partial}{\partial r} + W \cos \vartheta \sin \varphi \, \frac{\partial}{\partial \vartheta} + W \frac{\cos \varphi}{\sin \vartheta} \frac{\partial}{\partial \varphi},$$

$$J_{(4)} = \cos \varphi \, \frac{\partial}{\partial \vartheta} - \cot \vartheta \sin \varphi \, \frac{\partial}{\partial \varphi},$$

$$J_{(5)} = \sin \varphi \, \frac{\partial}{\partial \vartheta} + \cot \vartheta \cos \varphi \, \frac{\partial}{\partial \varphi}, \qquad J_{(6)} = -\frac{\partial}{\partial \varphi}, \tag{10.41}$$

where

$$W(r) \overset{\text{def}}{=} \frac{1}{r} - \frac{1}{4}kr, \qquad V(r) \overset{\text{def}}{=} 1 + \frac{1}{4}kr^2. \tag{10.42}$$

The metric with these symmetries is

$$ds^2 = dt^2 - \frac{R^2(t)}{\left(1 + \frac{1}{4}kr^2\right)^2} \left[dr^2 + r^2 \left(d\vartheta^2 + \sin^2 \vartheta \, d\phi^2\right)\right], \tag{10.43}$$

where $R(t)$ is an arbitrary function and k is an arbitrary constant.

Special cases of this metric form, corresponding to $k > 0$ and $k < 0$, were first derived, by a rather loose argument, by Friedmann (1922), who thus became, unknowingly, the father of modern cosmology. (He died long before he could witness his success.) A mathematically rigorous derivation, by methods different from ours, and with all signs of k included, was given independently by Robertson (1929, 1933) and Walker (1935). The metric form (10.43) is thus frequently called the **Robertson–Walker metric**. Other names attached to it in various combinations are Friedmann and Lemaître, but those refer to various special cases of (10.43); we shall come back to this point in Chapter 17.

The last three generators in (10.41) are easily recognised as those of O(3). We shall use the abbreviation $[i, j] \overset{\text{def}}{=} [J_{(i)}, J_{(j)}]$. The commutators are

$$[1, 2] = k \, J_{(4)}$$

$$[1, 3] = k \, J_{(5)} \qquad [2, 3] = k \, J_{(6)}$$

$$[1, 4] = -J_{(2)} \qquad [2, 4] = J_{(1)} \qquad [3, 4] = 0$$

$$[1, 5] = -J_{(3)} \qquad [2, 5] = 0 \qquad [3, 5] = J_{(1)} \tag{10.44}$$

$$[1, 6] = 0 \qquad [2, 6] = -J_{(3)} \qquad [3, 6] = J_{(2)}$$

$$[4, 5] = J_{(6)} \qquad [4, 6] = -J_{(5)} \qquad [5, 6] = J_{(4)}.$$

The last three commutators are those of the algebra of O(3). In (10.44) such relations should be found that correspond to the Bianchi algebras. However, the Bianchi classification introduced standard bases, and (10.41) may contain transformed generators. Indeed, some of the Bianchi bases are linear combinations of (10.41).

In the standard Bianchi bases, the nonzero structure constants were scaled to $+1$ or -1 whenever possible. Thus, for comparing (10.44) with (10.15)–(10.16) and Table 10.2, we have to scale out k in (10.44); the scaling is

$$\left(\underset{(1)}{J}, \underset{(2)}{J}, \underset{(3)}{J} \right) = \sqrt{|k|} \left(\underset{(1)}{\tilde{J}}, \underset{(2)}{\tilde{J}}, \underset{(3)}{\tilde{J}} \right),$$

with other generators unchanged. The result is as if $k = -1$ when $k < 0$ and $k = +1$ when $k > 0$. The formulae below show the scaled generators, with tildes omitted.

For $k > 0$, the Bianchi sub-basis is of type IX:

$$\underset{(1)}{L} = \frac{1}{2} \left(\underset{(1)}{J} + \underset{(6)}{J} \right), \qquad \underset{(2)}{L} = \frac{1}{2} \left(\underset{(2)}{J} - \underset{(5)}{J} \right), \qquad \underset{(3)}{L} = \frac{1}{2} \left(\underset{(3)}{J} + \underset{(4)}{J} \right). \tag{10.45}$$

The new generators are determined up to arbitrary orthogonal transformations because the matrix n^{ij} has a triple eigenvalue.

For $k < 0$, two Bianchi algebras are found in (10.41), one of type V:

$$\underset{(1)}{L} = -\underset{(1)}{J}, \qquad \underset{(2)}{L} = \underset{(2)}{J} + \underset{(4)}{J}, \qquad \underset{(3)}{L} = \underset{(3)}{J} + \underset{(5)}{J}, \tag{10.46}$$

and one of type VII$_h$:

$$\underset{(1)}{l} = -a\underset{(1)}{J} + \underset{(6)}{J}, \qquad \underset{(2)}{l} = \underset{(2)}{J} + \underset{(4)}{J}, \qquad \underset{(3)}{l} = \underset{(3)}{J} + \underset{(5)}{J}. \tag{10.47}$$

For $k = 0$, two standard Bianchi bases are contained in (10.41): $\left\{ \underset{(1)}{J}, \underset{(2)}{J}, \underset{(3)}{J} \right\}$ of Bianchi type I and $\left\{ \underset{(1)}{J}, \underset{(2)}{J}, \underset{(4)}{J} \right\}$ of Bianchi type VII$_0$.

It is easy to verify that the examples of bases shown are indeed the standard Bianchi bases of the types indicated. It is, however, more difficult to prove that only these Bianchi algebras are contained in (10.41). This fact was apparently first discovered by Grishchuk (1967). The relation between the symmetries of (10.33) and the possible Bianchi groups is discussed in more detail in Ellis and MacCallum (1969).

10.8 Exercises

1. Verify that the curvature defined by (10.22) is indeed zero.
2. Verify Eq. (10.26).
3. Since the invariant fields are defined by $[\underset{(i)}{k}, X] = 0$ and the $\underset{(i)}{k}$ do not depend on t, we can solve this set assuming that X are independent of t. However, a general solution does depend on time. Show that the time-dependent fields are $X^I(t) = B_j{}^k(t) X^I_{(k)}(t_0)$, where t_0 is an initial value and $B_j{}^k(t)$ is a nonsingular 3×3 matrix. Hence, the time dependence can be hidden in $\tilde{g}_{(j)(k)}(t) = g_{(r)(s)}(t) B^{-1}{}^r{}_j B^{-1}{}^s{}_k$, and we can use $X^I(t_0)$ as the new invariant fields.
   Hint. Show first that if X is an invariant field, then so is dX/dt, and thus can be decomposed in the basis of the invariant fields.

11

* The Petrov classification by the spinor method

11.1 What is a spinor?

Spinors are tensor densities in a 2-dimensional vector space over the body of complex numbers. A covariant 1-spinor ψ_A is a covariant vector that has two complex components. When the basis in the vector space is transformed by $e^{A'} = L^{A'}{}_A e^A$, $A, A' = 1, 2$, the ψ_A transforms by

$$\psi_{A'} = (L^{-1})_{A'}{}^A \psi_A. \tag{11.1}$$

In the body of complex numbers, we have to consider the complex conjugation, which reduces to identity when applied to real numbers. Consequently, a new classification of objects, in addition to that into covariant and contravariant, has to be introduced: some spinors transform <u>linearly</u>, some transform **antilinearly** under a change of basis. Antilinearly means that $\overline{(L^{-1})_{A'}{}^A}$ appears in the transformation law instead of $(L^{-1})_{A'}{}^A$; the overbar denotes complex conjugation. Indices of objects that transform antilinearly are marked by an overdot, thus

$$\overline{\psi_A} = \psi_{\dot{A}}. \tag{11.2}$$

The same spinor may have all four kinds of indices, for example $\chi^{\dot{A}B}{}_{\dot{C}D}$.

The spinor methods are used here as an auxiliary tool for just one task. However, spinors are a powerful tool that often makes complicated problems miraculously simple. A useful brief introduction to spinors in relativity is the paper by Penrose (1960); extended treatments can be found in Penrose and Rindler (1984) and Plebański (1974).

There is no metric in the space of spinors; indices are raised and lowered by the Levi-Civita symbols $\epsilon^{\dot{A}\dot{B}}$, $\epsilon_{\dot{A}\dot{B}}$, ϵ^{AB} and ϵ_{AB}. Since ϵ^{AB} is antisymmetric, a convention on manipulating indices has to be fixed. We adopt the convention that subscripts go before superscripts, and the dummy indices must be made adjacent (Plebański, 1974), so

$$\psi^A = \psi_S \epsilon^{SA}, \qquad \psi_A = \epsilon_{AS} \psi^S, \tag{11.3}$$

and similarly for dotted indices. Because of the antisymmetry and of the convention, the following are true:

$$\psi_A \phi^A = -\psi^A \phi_A, \qquad \psi^A \psi_A \equiv 0, \qquad \forall \, \psi_A. \tag{11.4}$$

11.2 Translating spinors to tensors and vice versa

Spinors will be used here to represent tensors in a 4-dimensional spacetime of signature $(+---)$. We introduce four Hermitean 2×2 **Pauli matrices** $g^{\alpha \dot{A}B}$ that are a basis of the space of 2×2 complex matrices. With respect to the index α they transform like components of a contravariant vector; with respect to the other two indices they transform as spinors with one contravariant antilinear index and one contravariant linear index. We will find below that we must in fact require that the Pauli matrices are spinor *densities*. The fact that they are Hermitean means that

$$\overline{g^{\alpha \dot{A}B}} \equiv g^{\alpha A \dot{B}} = g^{\alpha \dot{B}A}. \tag{11.5}$$

The dotted indices label rows, the non-dotted ones label columns of the matrix. The spinor image of a covariant vector v_α is defined as follows:

$$v^{\dot{A}B} = v_\alpha g^{\alpha \dot{A}B}. \tag{11.6}$$

The $v^{\dot{A}B}$ is a Hermitean spinor (density! – see below), and a scalar with respect to coordinate transformations on the manifold. Since the $g^{\alpha \dot{A}B}$ are a basis in the space of complex 2×2 matrices, the coefficients of decomposition of $v^{\dot{A}B}$ in that basis are uniquely determined, so there must be an inverse linear mapping from $v^{\dot{A}B}$ to v_α; we denote it

$$v_\alpha = \frac{1}{2} g_{\alpha \dot{A}B} v^{\dot{A}B}. \tag{11.7}$$

Since this must hold for arbitrary vectors v_α and spinors $v^{\dot{A}B}$, the Pauli matrices and their reciprocal matrices $g_{\alpha \dot{A}B}$ must obey

$$\frac{1}{2} g_{\alpha \dot{A}B} g^{\beta \dot{A}B} = \delta^\beta{}_\alpha, \qquad \frac{1}{2} g_{\alpha \dot{C}D} g^{\alpha \dot{A}B} = \delta^{\dot{A}}{}_{\dot{C}} \delta^B{}_D. \tag{11.8}$$

This notation is consistent: the reciprocal Pauli matrices are obtained from the proper Pauli matrices by lowering the tensor index with a metric and by lowering the spinor indices with the Levi-Civita symbols.

The following objects may be formed out of the Pauli matrices:

$$\Lambda^{\alpha\beta} = \frac{1}{2} \epsilon_{AB} \epsilon_{\dot{R}\dot{S}} g^{\alpha \dot{R}A} g^{\beta \dot{S}B}; \tag{11.9}$$

$$S^{\alpha\beta AB} = \frac{1}{2} \epsilon_{\dot{R}\dot{S}} \left(g^{\alpha \dot{R}A} g^{\beta \dot{S}B} - g^{\beta \dot{R}A} g^{\alpha \dot{S}B} \right). \tag{11.10}$$

The first object coincides with the metric tensor. We would like the metric to be a scalar with respect to spinor transformations. Because the Levi-Civita symbols are not spinors, but densities of weight $w = 1$ and $\dot{w} = 1$, the Pauli matrices must be taken to be spinor densities of weights $w = \dot{w} = -1/2$. This is a convention that defines how the Pauli matrices will be transformed under a change of basis.

The object $S^{\alpha\beta AB}$ is called a **spin-tensor**. It is a proper spinor (not a density), and it has the following properties:

$$S^{\alpha\beta AB} = S^{[\alpha\beta]AB} = S^{\alpha\beta(AB)}. \tag{11.11}$$

It can be used to transform tensors antisymmetric in two indices into spinors (symmetric in two indices). Two other properties are:

$$S^{\alpha\beta AB} = \frac{i}{2\sqrt{-g}} \epsilon^{\alpha\beta\cdots}{}_{\gamma\delta} S^{\gamma\delta AB},$$

$$S^{\alpha\beta\dot{A}\dot{B}} = -\frac{i}{2\sqrt{-g}} \epsilon^{\alpha\beta\cdots}{}_{\gamma\delta} S^{\gamma\delta\dot{A}\dot{B}},$$

(11.12)

where $g = \det\|g_{\alpha\beta}\|$. (The dots in $\epsilon^{\cdots}$ denote the places from which the indices were lowered. Since $\epsilon^{\cdots}$ and $\epsilon_{\cdots}$ are tensor densities of different weights, the covariant one is not equal to the contravariant one with lowered indices.) The first property means that the spin tensor is **self-dual**, the second property means that the complex-conjugate spin-tensor is **anti-self-dual**. A further property follows:

$$S^{\alpha\beta AB} S_{\alpha\beta\dot{C}\dot{D}} = 0.$$

(11.13)

The spinor image of a null vector has a special property. Let k^α be a null vector, $k^\alpha k_\alpha = 0$. From (11.8) this implies that the spinor image has then the property $k_{\dot{A}B} k^{\dot{A}B} = 0$. But because of the Levi-Civita symbols used to manipulate indices, $k_{\dot{A}B} k^{\dot{A}B} = 2\det\|k^{\dot{A}B}\|$. A Hermitean matrix whose determinant is zero has the property $k_{\dot{A}B} = k_{\dot{A}} k_B$, i.e. there exists a spinor with one index k_A that obeys this. Hence, the spinor image of a null vector is a spinor with one index. (Note: the k_A is determined only up to the phase, i.e. k_A and $e^{i\phi} k_A$, where ϕ is a real function, determine the same null vector field k^α.)

The Pauli matrices could be defined as just any basis of Hermitean 2×2 matrices. By tradition, for the flat Minkowski manifold, the Pauli matrices are defined as follows:

$$g^{i\dot{A}B} = \left[\begin{pmatrix} 1 & 0 \\ 0 & 1 \end{pmatrix}, \begin{pmatrix} 0 & 1 \\ 1 & 0 \end{pmatrix}, \begin{pmatrix} 0 & -i \\ i & 0 \end{pmatrix}, \begin{pmatrix} 1 & 0 \\ 0 & -1 \end{pmatrix} \right].$$

(11.14)

The Minkowski metric η_{ij} is contained as the background in every curved spacetime; one can always find a set of vectors $e^i{}_\alpha$, $i, \alpha = 0, 1, 2, 3$, such that $g_{\alpha\beta} e_i{}^\alpha e_j{}^\beta = \eta_{ij}$ (see Chapter 9). Thus, vectors from the Minkowski space can always be transformed into the corresponding vector fields in a curved spacetime by the mapping defined by the matrices $e_i{}^\alpha$. The image of the vector v^i of the Minkowski spacetime in the curved spacetime with metric $g_{\alpha\beta}$ is $v^\alpha = e_i{}^\alpha v^i$. Consequently, the Pauli matrices for a curved spacetime with the metric $g_{\alpha\beta}$ are defined by

$$g^{\alpha\dot{A}B} = e_i{}^\alpha g^{i\dot{A}B}.$$

(11.15)

Example. For the metric (the Schwarzschild metric)

$$ds^2 = \left(1 - \frac{2m}{r} \right) dt^2 - \frac{1}{1 - 2m/r} dr^2 - r^2 \left(d\vartheta^2 + \sin^2\vartheta \, d\varphi^2 \right)$$

the basis of *orthonormal* contravariant vectors that maps this metric onto the Minkowski metric is (indices number vectors, not their components)

$$e^0 = \left(1 \Big/ \sqrt{1 - \tfrac{2m}{r}}, 0, 0, 0\right), \qquad e^1 = \left(0, \sqrt{1 - \tfrac{2m}{r}}, 0, 0\right),$$

$$e^2 = (0, 0, 1/r, 0), \qquad e^3 = (0, 0, 0, 1/(r\sin\vartheta)).$$

11.3 The spinor image of the Weyl tensor

The spinor image of the Weyl tensor is defined as follows:

$$C_{ABCD} = \frac{1}{64} S^{\alpha\beta}{}_{AB} S^{\gamma\delta}{}_{CD} C_{\alpha\beta\gamma\delta}. \tag{11.16}$$

Because of the properties of $S^{\alpha\beta}{}_{AB}$ and $C_{\alpha\beta\gamma\delta}$, the following is true:

$$S^{\alpha\beta}{}_{AB} S^{\gamma\delta}{}_{\dot C\dot D} C_{\alpha\beta\gamma\delta} \equiv 0. \tag{11.17}$$

Now one of the miraculous properties of spinors shows up: all the complicated symmetries and antisymmetries of the Weyl tensor $C_{\alpha\beta\gamma\delta}$ translate into one simple property of C_{ABCD}: it is symmetric in all its indices,

$$C_{ABCD} = C_{(ABCD)}. \tag{11.18}$$

Because of (11.17), the inverse mapping to (11.16) is quite simple:

$$C_{\alpha\beta\gamma\delta} = S_{\alpha\beta}{}^{AB} S_{\gamma\delta}{}^{CD} C_{ABCD} + S_{\alpha\beta}{}^{\dot A\dot B} S_{\gamma\delta}{}^{\dot C\dot D} C_{\dot A\dot B\dot C\dot D}. \tag{11.19}$$

11.4 The Petrov classification in the spinor representation

Consider the following 4-linear form:

$$\Omega(\zeta) = C_{ABCD}\zeta^A \zeta^B \zeta^C \zeta^D, \tag{11.20}$$

where $\zeta^A = (\zeta^1, \zeta^2)$ is an arbitrary spinor. Assume that $\zeta^1 \neq 0$ and let $z = \zeta^2/\zeta^1$. Then $\Omega(\zeta) = (\zeta^1)^4 P_4(z)$, where $P_4(z)$ is a polynomial of 4th degree. It has four complex roots, and so can be factorised:

$$\Omega(\zeta) = (\zeta^1)^4 a\,(z_1 - z)\,(z_2 - z)\,(z_3 - z)\,(z_4 - z)$$
$$= (\alpha_A \zeta^A)\,(\beta_A \zeta^A)\,(\gamma_A \zeta^A)\,(\delta_A \zeta^A), \tag{11.21}$$

where $\{z_1, z_2, z_3, z_4\}$ are the roots of P_4, and the spinors $\{\alpha_A, \beta_A, \gamma_A, \delta_A\}$ are defined by the above equation, for example $\alpha_A = (az_1\zeta_1, -a\zeta_2)$, $\beta_A = (z_2\zeta_1, -\zeta_2)$, etc. Hence, finally:

$$C_{ABCD}\zeta^A \zeta^B \zeta^C \zeta^D = \alpha_A\beta_B\gamma_C\delta_D\zeta^A \zeta^B \zeta^C \zeta^D = \alpha_{(A}\beta_B\gamma_C\delta_{D)}\zeta^A \zeta^B \zeta^C \zeta^D. \tag{11.22}$$

Since the spinor ζ^A was arbitrary, this means that

$$C_{ABCD} = \alpha_{(A}\beta_B\gamma_C\delta_{D)} \tag{11.23}$$

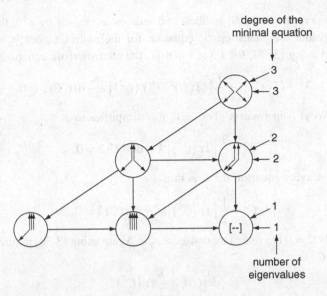

Fig. 11.1.   The Petrov classification by the spinor method. Arrows show possible specialisations.

(thanks to the total symmetry of C_{ABCD}, no bit of information was lost in considering the form (11.20)).

The quantities $\{\alpha_A, \beta_B, \gamma_C, \delta_D\}$ are called the **principal spinors** of the Weyl tensor, also **Debever spinors**. In a 2-dimensional space, only two of these spinors can be linearly independent, but we can ask how many pairs of linearly independent spinors can be chosen from this set. There are six different cases: (I) no two Debever spinors are collinear; (II) exactly two Debever spinors are collinear; (III) three Debever spinors are collinear; (D) two Debever spinors are collinear, the two others are collinear, too, but these pairs are distinct; (N) all four Debever spinors are collinear; (0) the Weyl tensor is identically zero. We have denoted the different cases by the same symbols as the Petrov types because these *are* exactly the Petrov types. This approach was first introduced by Penrose (1960). The resulting diagram can be drawn in a more illustrative way than in Section 7.16; see Fig. 11.1. It remains to prove that this is indeed the same classification. First, however, we shall determine some other algebraic properties of the Weyl spinor.

11.5 The Weyl spinor represented as a 3 × 3 complex matrix

Now let us interpret the Weyl spinor C_{ABCD} as a 3 × 3 complex matrix:

$$C \overset{\text{def}}{=} \left\| C^{AB}{}_{CD} \right\|, \tag{11.24}$$

whose elements will be labelled by the 'superindices' (AB) and (CD). Each 'superindex' takes three values: $(AB) = \{(11), (12), (22)\}$. In the set of such matrices, the unit matrix has components

$$\mathbb{I} = \left\| \delta^{(A}{}_C \delta^{B)}{}_D \right\|. \tag{11.25}$$

Since C_{ABCD} is symmetric in all indices, and indices are raised by ϵ^{AB}, the matrix C is trace-free. Consider the characteristic equation for the matrix C, $\det ||C - \lambda \mathbb{I}|| = 0$. As can be verified using (3.38), for a 3×3 matrix, the characteristic equation is

$$\lambda^3 + (\mathrm{Tr}\, C)\lambda^2 + \frac{1}{2}\left[(\mathrm{Tr}\, C)^2 - \mathrm{Tr}\left(C^2\right)\right]\lambda - \det(C)\,\mathbb{I} = 0. \tag{11.26}$$

Since for the Weyl 'supermatrix' $\mathrm{Tr}\, C = 0$, this simplifies to

$$\lambda^3 - \frac{1}{2}\left[\mathrm{Tr}\left(C^2\right)\right]\lambda - \det(C)\,\mathbb{I} = 0. \tag{11.27}$$

The Hamilton–Cayley equation for C is thus

$$C^3 - \frac{1}{2}\left[\mathrm{Tr}\left(C^2\right)\right]C - \det(C)\,\mathbb{I} = 0. \tag{11.28}$$

The property $\mathrm{Tr}\, C = 0$ has one more consequence: Again using (3.38) in three dimensions, we obtain for C

$$\det(C) = \frac{1}{3}\mathrm{Tr}\left(C^3\right). \tag{11.29}$$

In the equations below we will use the following shorthand notation:

$$(\beta\delta) \stackrel{\text{def}}{=} \beta_A \delta^A \equiv -\delta_A \beta^A = -(\delta\beta). \tag{11.30}$$

Using Eq. (11.23) and the above notation, we find for $\mathrm{Tr}\left(C^2\right)$:

$$\mathrm{Tr}\left(C^2\right) = \frac{1}{24}\left\{[(\alpha\beta)(\gamma\delta) + (\alpha\gamma)(\beta\delta) + (\alpha\delta)(\beta\gamma)]^2 \right.$$
$$\left. - 4(\alpha\beta)(\beta\gamma)(\gamma\delta)(\alpha\delta)\right\}. \tag{11.31}$$

The general formula for the determinant is too unwieldy, but we will not need it in the most general case. When two Debever spinors, say α^A and β^A, are collinear, $\alpha^A = \mu\beta^A$, the determinant is

$$\det(C) = \frac{1}{3}\mathrm{Tr}\left(C^3\right) \equiv \frac{1}{3}C^{AB}{}_{CD}C^{CD}{}_{EF}C^{EF}{}_{AB} = -\frac{1}{108}\mu^3(\beta\gamma)^3(\beta\delta)^3. \tag{11.32}$$

In the same special case, the formula for $\mathrm{Tr}\left(C^2\right)$ simplifies to

$$\mathrm{Tr}\left(C^2\right) = \frac{1}{6}\mu^2(\beta\gamma)^2(\beta\delta)^2. \tag{11.33}$$

Knowing all this, we can now verify that the following is true.

- In general, Eq. (11.27) has three different roots, so Eq. (11.28) is the minimal equation, of degree 3.
- When two of the Debever spinors are collinear, say $\alpha^A = \mu\beta^A$, Eqs. (11.32) and (11.33) apply, and (11.28) becomes

$$\left(C - \frac{a}{6}\mathbb{I}\right)^2\left(C + \frac{a}{3}\mathbb{I}\right) = 0, \qquad a \stackrel{\text{def}}{=} \mu(\beta\gamma)(\beta\delta). \tag{11.34}$$

With no further limitations on the Debever spinors, this is still the minimal equation, but there are only two distinct eigenvalues here.

- When $\alpha^A = \mu\beta^A$ and $\gamma^A = \nu\beta^A$ (three Debever spinors are collinear), we have $\mathrm{Tr}\,(C^2) = \det(C) = 0$, and (11.28) becomes

$$C^3 = 0. \tag{11.35}$$

With no further limitations on the Debever spinors, this is the minimal equation, and the only eigenvalue is 0, triply degenerate.

- When $\alpha^A = \mu\beta^A$ and $\gamma^A = \rho\delta^A$ (i.e. the Debever spinors (α^A, β^A) are collinear and (γ^A, δ^A) are collinear, but these two pairs are distinct), Eqs. (11.32) and (11.33) simplify further to

$$\det(C) = -\frac{1}{108}\mu^3\rho^3(\beta\delta)^6, \qquad \mathrm{Tr}\,(C^2) = \frac{1}{6}\mu^2\rho^2(\beta\delta)^4. \tag{11.36}$$

Now Eq. (11.28) becomes

$$\left(C - \frac{b}{6}\mathbb{I}\right)^2 \left(C + \frac{b}{3}\mathbb{I}\right) = 0, \qquad b \overset{\text{def}}{=} \mu\rho(\beta\delta)^2, \tag{11.37}$$

but the minimal equation here is of degree 2:

$$\left(C - \frac{b}{6}\mathbb{I}\right)\left(C + \frac{b}{3}\mathbb{I}\right) = 0. \tag{11.38}$$

In verifying this, Eq. (11.25) is helpful and so is the following observation: in two dimensions, every spinor with two indices that is antisymmetric must be proportional to the corresponding Levi-Civita symbol, e.g. $\beta_A\delta_B - \beta_B\delta_A = x\epsilon_{AB}$. Contracting this with ϵ^{AB} we find $x = -(\beta\delta)$. In consequence, we also have $\beta^A\delta_B - \beta_B\delta^A = -(\beta\delta)\delta^A{}_B$.

- When all four Debever spinors are collinear, thus $\alpha^A = \mu\delta^A$, $\gamma^A = \rho\delta^A$ and $\beta^A = \sigma\delta^A$, Eq. (11.35) still holds, but the minimal equation is obtained from (11.38) and it is $C^2 = 0$.
- When $C = 0$, this is the minimal equation.

In this way, we have verified all the information in Figs. 7.3 and 11.1.

11.6 The equivalence of the Penrose classes to the Petrov classes

It remains to verify that the algebraic types found by the Penrose (1960) method in this chapter coincide with those found by the Ehlers–Kundt (1962) method used in Section 7.16. For this purpose, recall the formula

$$\epsilon_{\alpha\beta\mu\nu} = g^{-1}g_{\alpha\rho}g_{\beta\sigma}g_{\mu\lambda}g_{\nu\tau}\epsilon^{\rho\sigma\lambda\tau}. \tag{11.39}$$

Using this, together with (11.19), (11.12) and (7.97)–(7.99), we find

$$Q_{\alpha\gamma} = E_{\alpha\gamma} + iH_{\alpha\gamma}$$

$$= S_{\alpha\beta}{}^{AB} u^\beta S_{\gamma\delta}{}^{CD} u^\delta C_{ABCD} + S_{\alpha\beta}{}^{\dot{A}\dot{B}} u^\beta S_{\gamma\delta}{}^{\dot{C}\dot{D}} u^\delta C_{\dot{A}\dot{B}\dot{C}\dot{D}}$$

$$- \frac{i}{2\sqrt{-g}} \epsilon_{\alpha\beta}{}^{\mu\nu} S_{\mu\nu AB} S_{\gamma\delta}{}^{CD} u^\beta u^\delta C_{ABCD}$$

$$- \frac{i}{2\sqrt{-g}} \epsilon_{\alpha\beta}{}^{\mu\nu} S_{\mu\nu \dot{A}\dot{B}} S_{\gamma\delta}{}^{\dot{C}\dot{D}} u^\beta u^\delta C_{\dot{A}\dot{B}\dot{C}\dot{D}}$$

$$= 2 \left(u^\beta S_{\alpha\beta}{}^{\dot{A}\dot{B}} \right) \left(u^\delta S_{\gamma\delta}{}^{\dot{C}\dot{D}} \right) C_{\dot{A}\dot{B}\dot{C}\dot{D}}. \tag{11.40}$$

The quantity $\left(u^\beta S_{\alpha\beta}{}^{\dot{A}\dot{B}} \right)$ is in effect a 3×3 matrix. It has one symmetric 'superindex' $(\dot{A}\dot{B})$ that takes three values, and it obeys (because of antisymmetry in $(\alpha\beta)$) the identity $u^\alpha \left(u^\beta S_{\alpha\beta}{}^{\dot{A}\dot{B}} \right) \equiv 0$; hence it operates in the 3-dimensional hypersurface element orthogonal to u^α in the space of tensors. Consequently, the tensor index α takes values in a 3-dimensional space. Thus, $\left(u^\beta S_{\alpha\beta}{}^{\dot{A}\dot{B}} \right)$ provides an invertible linear mapping from the complex vector space V_1 of vectors labelled by (AB) to the complex vector space V_2 of vectors labelled by the index α. Equation (11.40) is the corresponding bilinear mapping of matrices over V_1 to matrices over V_2. Such mappings preserve the eigenvalues and other invariants of matrices. Hence, the classification by the Penrose method is equivalent to that by the Ehlers–Kundt method. $\square$

11.7 The Petrov classification by the Debever method

The spinor image of the Weyl tensor has the following properties:

- In the completely general case, each of the Debever spinors obeys

$$\alpha^A \alpha^B \alpha^C \alpha^D C_{ABCD} \equiv 0, \tag{11.41}$$

$$\alpha_{[A} C_{B]CD[E} \alpha_{F]} \alpha^C \alpha^D \equiv 0. \tag{11.42}$$

Both equations are easy to verify using (11.23).
- For type II, the nondegenerate Debever spinors obey (11.41)–(11.42), while the preferred (double) Debever spinor α_A obeys

$$\alpha^A \alpha^B \alpha^C C_{ABCD} \equiv 0, \tag{11.43}$$

$$\alpha_{[A} C_{B]CDE} \alpha^C \alpha^D \equiv 0. \tag{11.44}$$

- For type D, each of the (double) Debever spinors α_A and γ_A obeys (11.43) and (11.44).
- For type III, the nondegenerate Debever spinor obeys (11.41)–(11.42), while the preferred (triple) Debever spinor α_A obeys

$$\alpha^A \alpha^B C_{ABCD} \equiv 0. \tag{11.45}$$

- For type N, the only existing (quadruple) Debever spinor α_A obeys

$$\alpha^A C_{ABCD} \equiv 0. \tag{11.46}$$

In order to express these properties in the tensor language, we must observe that the spin-tensor has the following property:

$$S_{\alpha\beta AB} S^{\gamma\delta AB} + S_{\alpha\beta\dot A\dot B} S^{\gamma\delta\dot A\dot B} = 4\delta^{\gamma\delta}_{\alpha\beta}. \tag{11.47}$$

This can be verified using the Hermitean property of the Pauli matrices. Thus, using the spin-tensor, we may uniquely represent any antisymmetric second-rank tensor $T_{\alpha\beta}$ by its (symmetric) spinor image

$$T_{AB} = S_{\alpha\beta AB} T^{\alpha\beta}; \qquad T^{\alpha\beta} = \frac{1}{8}\left(S^{\alpha\beta AB} T_{AB} + S^{\alpha\beta\dot A\dot B} T_{\dot A\dot B}\right). \tag{11.48}$$

In particular, $T^{\alpha\beta} \equiv 0$ if and only if $T_{AB} \equiv 0$.

Knowing this, let

$$k_\mu \stackrel{\text{def}}{=} \frac{1}{2} g_{\mu\dot A B} \alpha^{\dot A} \alpha^B \tag{11.49}$$

be the null vector defined by the Debever spinor and calculate

$$k_{[\alpha} C_{\beta]\mu\nu[\gamma} k_{\delta]} k^\mu k^\nu. \tag{11.50}$$

The calculation is long and tedious, and the result is that this quantity is always zero. The null vector k^μ defined by (11.49) is called a **Debever vector**. In verifying that (11.50) is zero, one must calculate

$$S^{\alpha\beta AB} S^{\gamma\delta CD} k_\alpha C_{\beta\mu\nu\gamma} k_\delta k^\mu k^\nu, \tag{11.51}$$

express the Weyl tensor $C_{\beta\mu\nu\gamma}$ through its spinor image by (11.19), use the complete symmetry of C_{ABCD}, and use the following:

$$S_{\beta\mu EF} k^\mu = -\frac{1}{2} g_{\beta\dot P E} \alpha^{\dot P} \alpha_F - \frac{1}{2} g_{\beta\dot P F} \alpha^{\dot P} \alpha_E; \tag{11.52}$$

$$S^{\alpha\beta AB} g_{\beta\dot P E} = -\epsilon_{\dot P\dot S}\left(\delta^B_{\ E} g^{\alpha\dot S A} + \delta^A_{\ E} g^{\alpha\dot S B}\right). \tag{11.53}$$

In the end, the first part of (11.19) will give zero contribution because it will create factors like $\alpha^{\dot P}\alpha_{\dot P} \equiv 0$, while the second part will be proportional to $C^{\dot E\dot F\dot G\dot H}\alpha_{\dot E}\alpha_{\dot F}\alpha_{\dot G}\alpha_{\dot H}$, which is zero by virtue of (11.41).

The quantity (11.50) vanishes even in the most general case (Petrov type I). With more special Petrov types, the vanishing quantity becomes progressively simpler. Specifically:

- For Petrov type II, each of the three nondegenerate Debever vectors obeys

$$k_{[\alpha} C_{\beta]\mu\nu[\gamma} k_{\delta]} k^\mu k^\nu = 0, \tag{11.54}$$

but the Debever vector that is the tensor image of the double Debever spinor obeys

$$k_{[\alpha} C_{\beta]\mu\nu\gamma} k^\mu k^\nu = 0, \tag{11.55}$$

by virtue of (11.43) and its complex conjugate.

- For Petrov type D, both Debever vectors obey (11.55).
- For Petrov type III, the tensor image of the nondegenerate Debever spinor obeys (11.54), while the tensor image of the triple Debever spinor obeys

$$C_{\beta\mu\nu\gamma} k^\mu k^\nu = 0, \tag{11.56}$$

by virtue of (11.45) and its complex conjugate.

- For Petrov type N, the only existing Debever vector obeys

$$C_{\beta\mu\nu\gamma} k^\mu = 0, \tag{11.57}$$

by virtue of (11.45) and its complex conjugate.

The method of verification is similar in each case. For (11.55), we calculate $S^{\alpha\beta AB} k_\alpha C_{\beta\mu\nu\gamma} k^\mu k^\nu = 0$, whereas for (11.56) and (11.57) we only express $C_{\beta\mu\nu\gamma}$, k^μ and k^ν through their spinor images.

This approach to the Petrov classification was introduced by Debever (1959, 1964).

11.8 Exercises

1. Show that the reciprocal Pauli matrices $g_{\alpha\dot A B}$ result from the proper Pauli matrices $g^{\alpha\dot A B}$ by lowering the tensor index with the metric and lowering the spinor indices with $\epsilon_{\dot A \dot B}$ and ϵ_{AB}.
2. Prove that the $\Lambda_{\alpha\beta}$ defined by (11.9) is equal to the metric tensor.
3. Verify that the spin-tensor is symmetric in its spinorial indices.
4. Verify Eqs. (11.12) and (11.13).
5. Prove that a Hermitean 2-spinor $k_{\dot A B}$ whose determinant is zero does indeed define a spinor with one index by $k_{\dot A B} = k_{\dot A} k_B$.
6. Verify Eq. (11.17).
7. Verify Eq. (11.18).
8. Verify that (11.19) is the inverse transformation to (11.16).
9. Prove that (11.22) implies (11.23).
   Hint. Since the spinor ζ^A in (11.22) is arbitrary, assume that $\zeta^A = (1, z)$ and consider the polynomial in z defined by (11.22).
10. Verify Eq. (11.29).
11. Verify Eq. (11.31).
12. Verify Eq. (11.38).
13. Verify Eqs. (11.41) and (11.42).
14. Verify Eqs. (11.47) and (11.48).
15. Verify Eqs. (11.52)–(11.54).

Part II

The theory of gravitation

12

The Einstein equations and the sources of a gravitational field

12.1 Why Riemannian geometry?

As argued in Section 1.4, gravitational forces can be simulated by inertial forces in accelerated motion. Special relativity describes relations between objects in uniform motion with respect to inertial frames, while gravitational interactions are neglected. The metric of the Minkowski spacetime in an inertial reference frame has constant coefficients. If we transform that metric to an accelerated frame, its components will become functions. Hence, a gravitational field should have the same effect: in a gravitational field the metric should also have non-constant components. Unlike in the Minkowski spacetime, in a gravitational field it should not be possible to make the metric components constant by a coordinate transformation. This was, in great abbreviation, the basic observation that led Einstein (1916) to general relativity.

This idea had to be supplemented with equations that would generalise the Newtonian laws of gravitation, and would relate the metric form to the gravitational field. The derivation of these equations, together with several related matters, will be presented in this chapter.

12.2 Local inertial frames

Let us recall the conclusion of Chapter 1: the Universe is permeated by gravitational fields that cannot be screened. Their intensity can be decreased by going away from the sources, but one can never decrease that intensity below the minimum determined by the local mean density of matter in the Universe. For this reason, no body in the Universe moves freely in the sense of Newton's mechanics, and consequently inertial frames can be realised only approximately, with a limited precision. Moreover, there exists no natural standard of a straight line, so the departures of real motions from rectilinearity cannot be measured.

However, let us recall that for a body falling freely in a gravitational field the inertial force caused by the acceleration balances the gravitational force. Assume for the beginning that the gravitational field is homogeneous. Then, two bodies falling freely in it will have the same acceleration all the time, so their relative acceleration will be zero and, relative to each other, they will either be at rest or move with a constant velocity. It is known from very precise experiments that the gravitational mass is for all material bodies proportional to the inertial mass, and the factor of proportionality is a universal constant (taken to be just 1). Hence, all bodies will experience the same acceleration in a given gravitational field. Consequently, the frame of reference connected with a body falling freely in a homogeneous gravitational field is inertial.

Homogeneous gravitational fields do not exist in Nature. However, along the trajectory of a body falling freely in any real gravitational field, the gravitational force and the inertial force cancel each other at every point. Hence, if the gravitational field is continuous (which is the case in all practical instances), then, in a *sufficiently small neighbourhood* of the falling body, the inertial forces will be arbitrarily small. Given the precision ε attainable in measurements, a falling body B will, at every point of its trajectory, determine a sphere of radius δ inside which the inertial force will be smaller than ε, i.e. unmeasurable. Inside that sphere the reference system defined by the body B will thus be 'practically' inertial. It is called the **local inertial frame** of the body B. (In fact, local inertial frames are defined not by individual bodies, but by their trajectories that do not depend on mass.) It is called 'local' because it differs from the universal inertial frame postulated in Newton's theory: there is an infinity of local inertial frames. It is easy to see that two different local inertial frames will in general move with acceleration relative to each other. For example, consider two local inertial frames connected with bodies falling freely towards the Earth from opposite directions. Thus, at a large distance from a freely falling body its local inertial frame ceases to be inertial.

12.3 Trajectories of free motion in Einstein's theory

Let us recall one more conclusion of Chapter 1: since no standard for a straight line exists in Nature, it will be simpler to assume that the geometry of our space is non-Euclidean, and in that geometry the trajectories of material bodies are free-motion trajectories. What was called 'gravitational field' in Newton's theory will be a consequence of the non-Euclidean geometry in which the free motions take place. By the argument of Section 12.1, the Riemann geometries should be appropriate. They do contain the Euclidean geometry as a special case and, because of the coordinate-independent formalism they use, are consistent with the postulate of equivalence of all reference systems.

The generalisation of the notion of a straight line that exists in a Riemann space is the geodesic line. If in a Riemann space the curvature goes to zero, the geometry goes over into the Euclidean geometry and the geodesic lines go over into straight lines. Hence, the geodesic line is a natural candidate for the trajectory of a free motion. In addition, the geodesic lines have the following property:

Theorem 12.1 *For a timelike geodesic G, the coordinates in a Riemann space can be chosen so that the Christoffel symbols vanish along G.*

This is an additional argument that the geodesics should be taken as the trajectories of free fall. The vanishing of the Christoffel symbols means that along G the local gravitational field will be approximately zero. (It is not exactly zero because the derivatives of the Christoffel symbols will not vanish, and so the curvature will be nonzero.)

Proof: (Latin indices will label vectors; the lower case ones will run through the values 0, 1, 2, 3, whereas the upper case ones will run through the values 1, 2, 3. Tensor indices running through the values 1, 2, 3 will also be denoted by latin letters. Wherever confusion might arise, vector indices will have a hat over them.)

At a point $p_0 \in G$, we choose such a basis in the tangent space that $e_{\hat{0}}{}^\alpha(p_0)$ is tangent to G, and the other vectors, $e_A{}^\alpha(p_0)$, $A = 1, 2, 3$ are orthogonal to $e_{\hat{0}}{}^\alpha(p_0)$, i.e. $g_{\alpha\beta} e_{\hat{0}}{}^\alpha e_A{}^\alpha \big|_{p_0} = 0$. Then we define the bases $e_i{}^\alpha(p)$ in a neighbourhood of G as follows:

1. For $p \in G$, we transport $e_i{}^\alpha(p_0)$ parallely from p_0 to p along G.
2. For $p \notin G$ we draw through p a geodesic G'_p that intersects G orthogonally (see Fig. 12.1). Let $p' \in G$ be the point of intersection of G'_p and G. Then we transport the basis $e_i{}^\alpha(p')$ (already defined in point 1) to p along G'_p.

This procedure works provided that there are no singular points on G'_p. It gives a unique result provided that p is not too distant from G, otherwise the geodesics orthogonal to G might intersect each other.

The following equations hold on G:

$$e_i{}^\alpha{}_{;\beta}\big|_G = 0, \qquad e^i{}_{\alpha;\beta}\big|_G = 0. \tag{12.1}$$

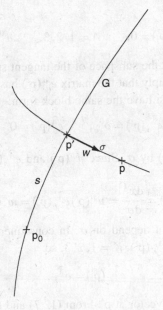

Fig. 12.1. Construction of the Fermi coordinates in which the Christoffel symbols vanish along a given timelike geodesic G. More explanation in the text.

The first equation follows because (i) $e_i{}^\alpha{}_{;\mu} e_{\hat{0}}{}^\mu\big|_G = 0$ in consequence of the basis $e_i{}^\alpha$ being transported parallely along G and $e_{\hat{0}}{}^\mu$ being tangent to G and (ii) $e_i{}^\alpha{}_{;\mu} e_S{}^\mu\big|_G = 0 (S = 1, 2, 3)$ in consequence of $e_S{}^\mu$ at G being tangent to one of the geodesics used to transplant the bases parallely to other points. The second of (12.1) follows from $e_i{}^\alpha e^i{}_\beta = \delta^\alpha{}_\beta$.

Now, using (12.1) in (4.19), we obtain

$$\left\{ \begin{matrix} \alpha \\ \beta\gamma \end{matrix} \right\} \Bigg|_G = e_s{}^\alpha e^s{}_{\beta,\gamma}\big|_G . \tag{12.2}$$

Since the Christoffel symbols are symmetric, this implies at once that

$$e^i{}_{\beta,\gamma}\big|_G = e^i{}_{\gamma,\beta}\big|_G . \tag{12.3}$$

We have defined the basis (up to rotations of $e_A{}^\alpha(p_0)$, $A = 1, 2, 3$), but we have not so far defined the coordinates. Let w^α be the tangent vector to G'_p of unit length (see Fig. 12.1), and let σ be the length of the segment of G'_p between p and G ($\sigma = 0$ for points on G). Let s be the length of the segment of G between p_0 and p'. For the point p we define the time coordinate $x^0 = s$ and the space coordinates

$$x^A = \sigma w^\rho(p') e^{\hat{A}}{}_\rho(p'), \qquad A = 1, 2, 3. \tag{12.4}$$

In the coordinates (s, x^A) thus defined, called **Fermi coordinates**,

$$e_{\hat{0}}{}^\alpha(p') = \frac{\partial x^\alpha}{\partial s}\bigg|_G \underset{*}{=} \delta^\alpha{}_0, \tag{12.5}$$

since $e_{\hat{0}}{}^\alpha$ is tangent to G, and

$$e_A{}^0(p') \underset{*}{=} 0, \qquad A = 1, 2, 3, \qquad w^0 \underset{*}{=} 0 \tag{12.6}$$

because $e_A{}^\alpha(p')$ and w^α lie in the subspace of the tangent space in which $x^0 =$ constant. Equations (12.5) and (12.6) imply that the matrix $e_i{}^\alpha(p')$ has a block form. Consequently, the inverse matrix $e^i{}_\alpha(p')$ must have the same block form:

$$e^{\hat{0}}{}_\alpha(p') \underset{*}{=} \delta^0{}_\alpha, \qquad e^{\hat{A}}{}_0(p') \underset{*}{=} 0. \tag{12.7}$$

Now let us differentiate (12.4) by σ. Since $w^\rho(p')$ and $e^{\hat{A}}{}_\rho(p')$ do not depend on σ, we have

$$w^A(\sigma) \overset{\text{def}}{=} \frac{dx^A}{d\sigma} = w^\rho(p') e^{\hat{A}}{}_\rho(p') = w^A(p'), \tag{12.8}$$

which means that w^A does not depend on σ. In consequence of (12.6) and (12.7) this implies that $w^A(p') \underset{*}{=} w^K(p') e^{\hat{A}}{}_K(p')$, $K = 1, 2, 3$, so

$$e^{\hat{A}}{}_K(p') \underset{*}{=} \delta^{\hat{A}}{}_K, \tag{12.9}$$

since $w^K(p')$ is an arbitrary 3-vector at p'. From (12.7) and (12.9) we then obtain

$$e^i{}_\alpha(p') \underset{*}{=} \delta^i{}_\alpha, \qquad e_i{}^\alpha(p') \underset{*}{=} \delta^\alpha{}_i. \tag{12.10}$$

This implies that $e_i{}^\alpha(p')$ is constant along G, so

$$e^i{}_{\alpha,\beta} e^\beta_{\hat{0}}\big|_G = \frac{\partial}{\partial s} e^i{}_\alpha(p') \underset{*}{=} e^i{}_{\alpha,0}\big|_G = 0, \qquad e^i{}_{0,\alpha}\big|_G = 0. \tag{12.11}$$

(The last equation follows from the previous one by virtue of (12.3).)

Now note that

$$\frac{\mathrm{d}}{\mathrm{d}\sigma}\left(w^\alpha e^i{}_\alpha\right) \equiv \frac{\mathrm{D}}{\mathrm{d}\sigma}\left(w^\alpha e^i{}_\alpha\right) = 0, \tag{12.12}$$

because w^α is tangent to a geodesic on which σ is the affine parameter, while $e^i{}_\alpha$ is by definition parallely transported along that geodesic. Moreover, (12.8) implies that $\mathrm{d}w^\alpha/\mathrm{d}\sigma = 0$, hence

$$0 = \frac{\mathrm{d}}{\mathrm{d}\sigma}\left(w^\alpha e^i{}_\alpha\right) = w^\alpha \frac{\mathrm{d}e^i{}_\alpha}{\mathrm{d}\sigma} = w^\alpha w^\beta e^i{}_{\alpha,\beta} = w^A w^B e^i{}_{A,B}, \tag{12.13}$$

which means that $e^i{}_{(A,B)}\big|_G = 0$, because along G w^A is an arbitrary vector orthogonal to G. Then, from (12.3),

$$e^i{}_{A,B}\big|_G = 0. \tag{12.14}$$

Equations (12.11) and (12.14) imply that $e^i{}_{\alpha,\beta}\big|_G = 0$, so, from (12.2),

$$\left\{\begin{matrix} \alpha \\ \beta\gamma \end{matrix}\right\}\bigg|_G \underset{*}{=} 0. \; \square \tag{12.15}$$

The physical meaning of (12.15) is that, in a sufficiently small neighbourhood of each $p' \in G$, other geodesics emanating from p' are, up to first-order terms, approximated by straight lines. This neighbourhood is thus a local inertial frame. This is one more suggestion that geodesics should be the trajectories of free motion in relativity.

12.4 Special relativity versus gravitation theory

We have so far been dealing with the postulate that general relativity should reduce to the Newtonian kinematics of free motion in the limit of vanishing gravitational field. However, in between these extremes there is special relativity that describes the kinematics of free particles in the absence of gravitation, but takes into account velocities comparable to the velocity of light. In the previous sections we have thus been discussing a two-stage limiting transition: with the gravitational field to zero and with the velocity of light to infinity. When we switch off the gravitational field, but put no limits on the velocities, the geometric theory of gravitation should reproduce special relativity.

The spacetime of special relativity is a 4-dimensional flat Riemann space of signature $(+---)$. The Riemann space of general relativity should thus be of the same signature. This is because a change of signature means either a discontinuity in some metric components or at least one component of the metric passing through zero value, while we expect that the 'switching off' of the gravitational field can be done in a continuous way and does not lead through any singularities.

12.5 The Newtonian limit of relativity

We assumed that with the gravitational field switched off, general relativity should reproduce special relativity and, in addition, with the velocity of light becoming infinite it should reproduce the Newtonian kinematics of free motion. A logical consequence of these two postulates is the requirement that when the velocity of light is made infinite while the gravitational field is still there, general relativity should reproduce the Newtonian theory of gravitation. Hence, the field equations of relativity should, in the limit $c \to \infty$, reproduce the Poisson equation

$$\Delta\phi = 4\pi G\rho, \tag{12.16}$$

where ϕ is the gravitational potential and ρ is the density distribution of the matter generating the gravitational field.

12.6 Sources of the gravitational field

In the theory of gravitation that we are now constructing, the gravitational field should manifest its presence as non-flatness of the metric. Consequently, the metric tensor should be the device to describe gravitation. The metric tensor in four dimensions has in the most general case 10 components. Hence, we will need 10 equations to determine it.

The description of the sources in the equations should thus be correspondingly elaborate. The source in the equations generalising (12.16) should be a quantity that is a generalisation of mass-density. According to special relativity, the mass of a body depends on its energy. Hence, the energy of motion of a continuous medium should contribute to the gravitational field and so should the internal energy, e.g. pressure in a fluid and stresses in a solid.

In special relativity, the physical state of a continuous medium is described by the energy-momentum tensor $T_{\alpha\beta}$. In any chosen coordinate system, its component T_{00} is equal to the energy-density ρc^2, the components $T_{0I}, I = 1, 2, 3$, form the 3-dimensional vector of energy stream through a unit area of surface orthogonal to the direction of flow, and the components T_{ij} form the stress tensor. The tensor $T_{\alpha\beta}$ is symmetric, and so has in general 10 independent components. Consequently, it is a natural candidate for the source in the field equations of gravitation.

The description of motion of matter, i.e. the distribution of velocities inside matter, is part of the energy-momentum tensor. Consequently, the equations of motion are differential relations between various components of the energy-momentum tensor. In the Cartesian coordinates, the equations of motion are $T^{\alpha\beta}{}_{,\beta} = 0$. In any other coordinates then, these equations take the form

$$T^{\alpha\beta}{}_{;\beta} = 0. \tag{12.17}$$

The field equations determining the metric tensor in which the energy-momentum tensor should be the source must be consistent with (12.17).

12.7 The Einstein equations

Since the gravitational field should be a consequence of the geometry being non-flat, it should be connected with the curvature tensor which contains second derivatives of the metric. If the curvature is contained in the left-hand side of the field equations, then these equations will be of second order in the metric. This suggests that the components of the metric should be analogues of the Newtonian gravitational potential. In that case, the trajectories of motion (the geodesics) would be determined by the first derivatives of the potential, just like in Newton's theory.

The Riemann tensor is of rank 4, while the energy-momentum tensor is of rank 2. Hence, if the Riemann tensor were equal to some quantity constructed from the energy-momentum tensor, then the source of the gravitational field would be a quadratic function of the matter density, and this would make the transition to Newton's theory complicated. We should thus rather equate a certain quantity constructed from the Riemann tensor to the energy-momentum tensor. We already know one good candidate, the Ricci tensor (7.46). It is symmetric and linear in the second derivatives of the metric. The field equations might thus read $R_{\alpha\beta} = \kappa T_{\alpha\beta}$, where κ is a constant coefficient. However, such equations are not consistent with (12.17) because in general $R^{\alpha\beta}{}_{;\beta} \neq 0$. But the Ricci tensor obeys an identity similar to (12.17), which is a consequence of the Bianchi identities (7.15). They may be written as

$$R^{\alpha}{}_{\beta\gamma\delta;\epsilon} + R^{\alpha}{}_{\beta\delta\epsilon;\gamma} + R^{\alpha}{}_{\beta\epsilon\gamma;\delta} = 0. \tag{12.18}$$

Contracting this equation with $\delta^{\alpha}{}_{\gamma} g^{\beta\delta}$ we obtain

$$G^{\alpha\beta}{}_{;\beta} = 0, \tag{12.19}$$

where

$$G^{\alpha\beta} \stackrel{\text{def}}{=} R^{\alpha\beta} - \frac{1}{2} g^{\alpha\beta} R \tag{12.20}$$

is called the **Einstein tensor**. It is symmetric, linear in second derivatives of the metric and obeys (12.19), which is identical in form to (12.17). Hence, $G^{\alpha\beta}$ is a better candidate for the left-hand side of the field equations, which should thus read

$$R_{\alpha\beta} - \frac{1}{2} g^{\alpha\beta} R = \kappa T_{\alpha\beta}. \tag{12.21}$$

These equations are called the **Einstein equations**. The coefficient κ will be determined in Section 12.11 from the condition of correspondence of (12.21) to the Poisson equation (12.16).

In vacuum, the Einstein equations take the form $G^{\alpha\beta} = 0$, which is equivalent to $R^{\alpha\beta} = 0$. This equation, together with (7.48)–(7.49), shows that outside the sources the Riemann tensor coincides with the Weyl tensor. Thus, the Weyl tensor represents that part of the gravitational field that propagates into vacuum. The Ricci tensor represents that part of curvature that is algebraically determined by matter and vanishes in vacuum.

It may seem unbelievable that such important and nontrivial equations should be derived almost without calculations and without experimental hints, by a speculation that was in places non-unique. However, this is how it was; Einstein actually guessed his equations by a reasoning described above in great abbreviation (it had taken him about 10 years to arrive at them). Therefore, it is not easy to present a convincing, brief and logical derivation of the Einstein equations. The most convincing route to them is Einstein's own. It was described in more detail by Mehra (1974) and also in Einstein's papers (Einstein *et al.*, 1923).

Because of the non-uniqueness of the reasoning leading to the Einstein equations, the Einstein theory is not the only geometric theory of gravitation that can be built upon the Newton theory and special relativity. Almost every one of the intuitive assumptions made on the way can be modified so that another set of equations will result. Some of those alternative theories will be briefly presented in Section 12.16. However, Einstein's theory has successfully passed all experimental tests, whereas the other theories were either proved wrong or else found to require such small modifications of Einstein's relativity that it does not really make sense to use them.

12.8 Hilbert's derivation of the Einstein equations

The derivation of Einstein's equations given in the previous section was based on the original reasoning of Einstein himself. However, David Hilbert had been working on deriving these equations simultaneously with Einstein by a different method. There was some exchange of information between them, but, taking things formally, Hilbert was the first to publish the correct result (see more details in Mehra (1974)).[1]

Hilbert proposed that all theories in mathematics and physics should be derived by deduction from sets of axioms. This programme contributed greatly to clarifying the logical structure of physical theories. However, the physical justification of some of the postulates is still lacking.

For the gravitation theory, Hilbert postulated to use a variational principle. The reasoning leading to the postulated integrand, just like Einstein's own reasoning, contains a few questionable and mysterious assumptions, which are justified only intuitively. Most of the other geometric theories of gravitation were derived from variational principles, which confirms that this method is not unique, either.

Hilbert proposed the following axioms:

I. The field equations of gravitation should follow from a variational principle. The independent variables in the action integral should be the components of the metric tensor.
II. The action functional should be a scalar.
III. The Euler–Lagrange equations (i.e. the field equations) that will follow should be differential equations of second order in $g_{\mu\nu}$.

---

[1] Hilbert presented a general subclass of Eq. (12.21), corresponding to $T_{\alpha\beta} = 0$ and to $T_{\alpha\beta}$ being the energy-momentum tensor of the electromagnetic field, at the meeting of the Royal Academy of Sciences at Göttingen on 20 November 1915. Einstein presented the final correct form of Eq. (12.21) at the meeting of the Prussian Academy of Sciences in Berlin on 25 November 1915. This game with dates is only a historical curiosity. Einstein was the unquestionable spiritual father of relativity, and Hilbert himself made that point repeatedly. Einstein had worked on relativity since about 1907 and had published several papers explaining the basic ideas and preliminary results. Hilbert joined in about 1913 and was influenced by Einstein's ideas from the beginning.

From this point on, further reasoning leads to Einstein's equations almost uniquely, but the postulates themselves leave several questions open. For example, some unanswered questions regarding axiom I are the following:

1. What is the geometrical and physical interpretation of the action functional?
2. What is the geometrical and physical meaning of the extremum found from the Euler–Lagrange equations?
3. Why should the field equations of gravitation be equivalent to an extremum of some abstract quantity?

Axiom II is justified by the observation that if an integral should be a tensorial object, then the integrand must be a scalar. Integrals of other tensorial objects are not tensors. Deriving covariant equations from a non-covariant action integral would thus be inconsistent with the underlying philosophy of the theory.

Axiom III makes use of the correspondence principle in the simplest way. Since the Poisson equation, which should follow from the geometric equations as a limit, is of second order, and most other physical theories are based on second-order equations, the field equations of gravitation should be of second order, too. However, this is not a unique argument either: theories of gravitation with field equations of fourth order do exist. In the limiting transition to Newton's theory, the terms with derivatives of order higher than 2 disappear.

The conclusions of Hilbert's axioms are the following:

I. In a 4-dimensional spacetime, the action integral should be $\int_{V_4} \mathcal{H} \, \mathrm{d}_4 x$, where V_4 is a certain 4-volume in the spacetime, on whose boundary ∂V_4 the variations $\delta g_{\mu\nu}$ vanish. The function $\mathcal{H}$ depends on the metric tensor $g_{\mu\nu}$ and its derivatives $g_{\mu\nu,\lambda}$, $g_{\mu\nu,\lambda\sigma}$, etc.

II. If $\int_{V_4} \mathcal{H} \, \mathrm{d}_4 x$ should be a scalar, then $\mathcal{H}$ should be a scalar density of weight -1 because the volume element $\mathrm{d}_4 x$ is a scalar density of weight $+1$. The simplest scalar density of weight -1 is $\sqrt{-g}$. Thus $\mathcal{H} = \sqrt{-g} H$, where H is a proper scalar.

III. The order of the Euler–Lagrange equations is twice the order of the highest derivative in the action integral. It would thus be best if H were a function only of $g_{\mu\nu}$ and $g_{\mu\nu,\lambda}$. However, one cannot form a non-trivial scalar by combining algebraically $g_{\mu\nu}$ and $g_{\mu\nu,\lambda} = \begin{Bmatrix} \rho \\ \mu\lambda \end{Bmatrix} g_{\nu\rho} + \begin{Bmatrix} \rho \\ \nu\lambda \end{Bmatrix} g_{\mu\rho}$, since the Christoffel symbols are not tensors.

The second derivatives of $g_{\mu\nu}$, if present in H, will not give a contribution to the Euler–Lagrange equations if they can be collected into a divergence of an expression that vanishes on the boundary of the integration volume. This will be possible if $g_{\mu\nu,\rho\sigma}$ enter H linearly, i.e. are contracted only with terms that do not contain $g_{\mu\nu,\rho\sigma}$. Expressions linear in the Riemann tensor are of this form.

The only expressions that can be built from the Riemann tensor and are linear in $g_{\mu\nu,\rho\sigma}$ are $R^{\rho}{}_{\beta\rho\delta} = R_{\beta\delta} = -R^{\rho}{}_{\beta\delta\rho}$ and $g^{\alpha\beta} R_{\alpha\beta} = R$. Hence $H = g^{\alpha\beta} R_{\alpha\beta} = R$ and

$$W_g = \int_{V_4} \sqrt{-g} R \, \mathrm{d}_4 x. \tag{12.22}$$

More exactly, this should be the action integral for the left-hand side of the field equations. The full action integral should be

$$W = W_g + \int_{V_4} L \, d_4 x,$$ (12.23)

where $L = 0$ in vacuum.

Before we calculate $\delta W_g / \delta g_{\alpha\beta}$, let us note two auxiliary equations:

$$\delta g = \frac{1}{3!} \epsilon^{\alpha_1 \dots \alpha_4} \epsilon^{\beta_1 \dots \beta_4} g_{\alpha_1 \beta_1} g_{\alpha_2 \beta_2} g_{\alpha_3 \beta_3} \delta g_{\alpha_4 \beta_4}$$

$$= \frac{1}{3!} g \epsilon^{\alpha_1 \dots \alpha_4} g^{\beta_1 \rho_1} \dots g^{\beta_4 \rho_4} \epsilon_{\rho_1 \dots \rho_4} g_{\alpha_1 \beta_1} g_{\alpha_2 \beta_2} g_{\alpha_3 \beta_3} \delta g_{\alpha_4 \beta_4}$$

$$= \frac{1}{3!} g \delta^{\alpha_1 \dots \alpha_4}_{\rho_1 \dots \rho_4} \delta^{\rho_1}{}_{\alpha_1} \delta^{\rho_2}{}_{\alpha_2} \delta^{\rho_3}{}_{\alpha_3} g^{\beta_4 \rho_4} \delta g_{\alpha_4 \beta_4} = g g^{\alpha\beta} \delta g_{\alpha\beta},$$ (12.24)

$$\delta g^{\alpha\beta} = -g^{\alpha\rho} g^{\beta\sigma} \delta g_{\rho\sigma}.$$ (12.25)

In (12.24) we used (3.39), its derivative, (7.106), (3.32) and (3.37). Equation (12.25) follows by differentiating the equation $g^{\alpha\rho} g_{\beta\rho} = \delta^\alpha{}_\beta$.

Note that the variations of the Christoffel symbols are differences between Christoffel symbols calculated from different metric tensors,

$$\delta \begin{Bmatrix} \alpha \\ \beta\gamma \end{Bmatrix} = \begin{Bmatrix} \alpha \\ \beta\gamma \end{Bmatrix} [g_{..} + \delta g_{..}] - \begin{Bmatrix} \alpha \\ \beta\gamma \end{Bmatrix} [g].$$ (12.26)

Hence, $\delta \begin{Bmatrix} \alpha \\ \beta\gamma \end{Bmatrix}$ is a tensor because the non-tensorial terms in (4.23) will cancel out in the combination (12.26). Consequently, it makes sense to calculate the covariant derivative of $\delta \begin{Bmatrix} \alpha \\ \beta\gamma \end{Bmatrix}$. Writing it out and comparing it with $\delta R_{\alpha\beta}$, which is calculated from the definition of $R_{\alpha\beta}$, we find that

$$\delta R_{\alpha\beta} = - \left(\delta \begin{Bmatrix} \rho \\ \alpha\rho \end{Bmatrix} \right)_{;\beta} + \left(\delta \begin{Bmatrix} \rho \\ \alpha\beta \end{Bmatrix} \right)_{;\rho}.$$ (12.27)

Hence

$$\delta W_g = \int_{V_4} d_4 x \left\{ - \frac{1}{2\sqrt{-g}} g g^{\alpha\beta} \delta g_{\alpha\beta} R - \sqrt{-g} g^{\alpha\rho} g^{\beta\sigma} R_{\alpha\beta} \delta g_{\rho\sigma} \right.$$

$$\left. + \sqrt{-g} g^{\alpha\beta} \left[- \left(\delta \begin{Bmatrix} \rho \\ \alpha\rho \end{Bmatrix} \right)_{;\beta} + \left(\delta \begin{Bmatrix} \rho \\ \alpha\beta \end{Bmatrix} \right)_{;\rho} \right] \right\}.$$ (12.28)

The term in square brackets can be written as

$$-\left(\sqrt{-g}g^{\alpha\beta}\delta\left\{{\rho\atop\alpha\rho}\right\}\right)_{;\beta}+\left(\sqrt{-g}g^{\alpha\beta}\delta\left\{{\rho\atop\alpha\beta}\right\}\right)_{;\rho} \qquad (12.29)$$

because $\sqrt{-g}$ and $g^{\alpha\beta}$ are covariantly constant. The expressions in parentheses are vector densities of weight -1. Thus, as shown in Section 4.1, their covariant divergences equal their ordinary divergences:

$$\sqrt{-g}g^{\alpha\beta}\left[-\left(\delta\left\{{\rho\atop\alpha\rho}\right\}\right)_{;\beta}+\left(\delta\left\{{\rho\atop\alpha\beta}\right\}\right)_{;\rho}\right]$$

$$=\left(-\sqrt{-g}g^{\alpha\beta}\delta\left\{{\rho\atop\alpha\rho}\right\}+\sqrt{-g}g^{\alpha\rho}\delta\left\{{\beta\atop\alpha\rho}\right\}\right)_{,\beta} \qquad (12.30)$$

(in the second term we interchanged ρ and β). Now we see that (12.30) is an ordinary divergence. Hence, from the Stokes theorem,

$$\int_{V_4}\left(-\sqrt{-g}g^{\alpha\beta}\delta\left\{{\rho\atop\alpha\rho}\right\}+\sqrt{-g}g^{\alpha\rho}\delta\left\{{\beta\atop\alpha\rho}\right\}\right)_{,\beta}d_4x$$

$$=\int_{\partial V_4}\left(-\sqrt{-g}g^{\alpha\beta}\delta\left\{{\rho\atop\alpha\rho}\right\}+\sqrt{-g}g^{\alpha\rho}\delta\left\{{\beta\atop\alpha\rho}\right\}\right)n_\beta d_3x, \qquad (12.31)$$

where n_β is the normal vector to the boundary ∂V_4 of the region V_4, while d_3x is the volume element in ∂V_4. From the assumption $\delta g_{\alpha\beta}(\partial V_4)=0$ it now follows that the integral of (12.29) is zero. Hence, in (12.28):

$$\delta W_g=\int_{V_4}d_4x\left(\frac{1}{2}\sqrt{-g}g^{\alpha\beta}R-\sqrt{-g}R^{\alpha\beta}\right)\delta g_{\alpha\beta}=0. \qquad (12.32)$$

Thus, if we denote $\delta L/\delta g_{\alpha\beta}=-\kappa\sqrt{-g}T^{\alpha\beta}$, then the equation $\delta W=0$ implies (12.21) by virtue of (12.23) and (12.32).

Hilbert's method does not say what $T^{\alpha\beta}$ should be; he managed to specify it only for the electromagnetic field in vacuum.

The Hilbert variational principle works safely only in deriving the Einstein equations in full generality. With a less-than-general metric, for example with symmetries, the variational principle can lead to a false result. This is known to happen for a large subset of the Bianchi-type models, because of their spatial homogeneity. When all fiduciary metrics are spatially homogeneous, the variations are spatially homogeneous, too. Hence, they cannot be assumed to vanish on the boundary – they vanish either everywhere or nowhere. In consequence, the boundary terms cannot be neglected. See MacCallum (1979) for more details.

12.9 The Palatini variational principle

Einstein's equations can be derived from a still more general variational principle, called the Palatini principle.[1] In this approach, the metric tensor and the connection coefficients are treated as two independent sets of variables. The connection is assumed only to be symmetric, $\Gamma^{\alpha}{}_{\beta\gamma} = \Gamma^{\alpha}{}_{(\beta\gamma)}$. The vanishing of coefficients of variations of $\Gamma^{\alpha}{}_{\beta\gamma}$ implies then Eq. (12.21), while the vanishing of coefficients of variations of the metric implies $\Gamma^{\alpha}{}_{\beta\gamma} = \left\{ \begin{matrix} \alpha \\ \beta\gamma \end{matrix} \right\}$, i.e. the covariant constancy of $g_{\alpha\beta}$.

12.10 The asymptotically Cartesian coordinates and the asymptotically flat spacetime

In the next section we will investigate the limiting transitions from Einstein's theory to Newton's theory and to special relativity. This section is a comment on the interpretation of those operations.

A flat space (spacetime) exists as a background in both of the limiting theories. When we go away from the sources of the gravitational field, the field becomes ever weaker and disappears in the limit of infinite distance. In that limit, the metric tensor of general relativity should tend to the flat metric of special relativity. This is only a 'thought experiment' because, as mentioned earlier, in the real Universe the intensity of the gravitational field can never be smaller than the value determined by the average matter density in the Universe.

The spacetime whose geometry becomes flat in the limit of large distances from the sources of the gravitational field is called **asymptotically flat**. Only in such spacetimes can we consider the Newtonian limit of general relativity. The coordinates that, in the same limit, go over into the Cartesian coordinates of special relativity are called **asymptotically Cartesian**.

12.11 The Newtonian limit of Einstein's equations

From the considerations of Section 12.5 it follows that in the limit $c \to \infty$ general relativity should reduce to Newton's theory of gravitation. Hence, in the same limit, the

---

[1] This is how this approach is commonly called, but Ferraris, Francaviglia and Reina (1982) argued that it was not Palatini (1919) who invented it, but Einstein (1925).

equation of a geodesic should reduce to the Newtonian equation of motion of a particle in a gravitational field,

$$m\frac{dv_I}{dt} = -m\phi_{,I},$$ (12.33)

where v_I is the velocity of the particle and ϕ is the gravitational potential. We will deal with the field equations later in this section.

Newton's equations of motion follow from the variational principle

$$\delta \int_{t_1}^{t_2} L \, dt = 0, \qquad L \stackrel{\text{def}}{=} \frac{1}{2}m\delta_{IJ}v^I v^J - m\phi,$$ (12.34)

while the equations of a geodesic follow from

$$\delta \int_{t_1}^{t_2} \mathcal{L} \, dt = 0, \qquad \mathcal{L} \stackrel{\text{def}}{=} \frac{ds}{dt}.$$ (12.35)

Investigating the limit on the equations of motion themselves would be difficult. It will be simpler to do it on the Lagrangians.

The Euler–Lagrange equations do not change when the Lagrange function is multiplied by a constant or when a constant is added to it. Hence, the condition of correspondence for the Lagrangians is

$$\lim_{c \to \infty} (C_1\mathcal{L} + C_2) = L,$$ (12.36)

where C_1 and C_2 are constants, as yet unknown. The Newtonian Lagrangian has the dimension of energy. The relativistic Lagrangian is

$$\mathcal{L} = \frac{ds}{dt} = c\sqrt{g_{00} + 2g_{0I}\frac{v^I}{c} + g_{IJ}\frac{v^I v^J}{c^2}},$$ (12.37)

where $I, J = 1, 2, 3$ and $v^I = dx^I/dt$ is the Newtonian velocity of the particle. Since the expression under the square root is dimensionless, the dimension of $\mathcal{L}$ is c. Hence, in order that the dimensions of both sides in (12.36) are the same, we must have $C_1 = \alpha mc$, where α is a dimensionless coefficient.

The Newtonian Lagrangian contains the kinetic energy of the particle. In special relativity, the kinetic energy is a part of the total energy $mc^2/\sqrt{1 - (v/c)^2}$. Hence, the constant C_2 in (12.36) must compensate for the rest energy contained in $C_1\mathcal{L}$ so that $(C_1\mathcal{L} + C_2)$ contains only the kinetic energy. We do not know yet with what sign the rest energy will be contained in $C_1\mathcal{L}$, so we look for C_2 in the form $C_2 = \beta mc^2$, $\beta = \pm 1$, and the sign of β will follow later. Finally, (12.36) becomes

$$\lim_{c \to \infty} mc^2 \left(\alpha\sqrt{g_{00} + 2g_{0I}\frac{v^I}{c} + g_{IJ}\frac{v^I v^J}{c^2}} + \beta \right) = \frac{1}{2}\delta_{IJ}mv^I v^J - m\phi.$$ (12.38)

Developing the square root by the Taylor formula up to terms of second degree in the velocity we obtain

$$\sqrt{g_{00} + 2g_{0I}\frac{v^I}{t} + g_{IJ}\frac{v^I v^J}{c^2}} = \sqrt{g_{00}} + \frac{g_{0I}}{\sqrt{g_{00}}}\frac{v^I}{c}$$
$$+ \frac{1}{2}\left(\frac{g_{IJ}}{\sqrt{g_{00}}} - \frac{g_{0I}g_{0J}}{g_{00}^{3/2}}\right)\frac{v^I v^J}{c^2} + O\left(\frac{v^3}{c^3}\right). \tag{12.39}$$

Since there are no terms linear in v^I/c on the right-hand side in (12.38), $cg_{0I}/\sqrt{g_{00}} \xrightarrow[c\to\infty]{} 0$ must hold, i.e.

$$\frac{g_{0I}}{\sqrt{g_{00}}} = O\left(\frac{1}{c^2}\right). \tag{12.40}$$

Then $(-g_{0I}g_{0J}/g_{00}^{3/2})$ in (12.39) becomes a correction of order $1/c^4$ to $g_{IJ}/\sqrt{g_{00}}$. Hence, using (12.39) and (12.40), we obtain in (12.38)

$$\lim_{c\to\infty}\left[c^2\left(\alpha\sqrt{g_{00}} + \beta\right) + \alpha\frac{\tilde{g}_{IJ}v^I v^J}{2\sqrt{g_{00}}} + O\left(\frac{1}{c}\right)\right] = \frac{1}{2}\delta_{IJ}v^I v^J - \phi, \tag{12.41}$$

where $\tilde{g}_{IJ} = g_{IJ} - O\left(1/c^4\right)$. The equation above should be an identity in v^I. Equating the terms of the same order in v^I we obtain

$$c^2\left(\alpha\sqrt{g_{00}} + \beta\right) = -\phi + O\left(\frac{1}{c}\right), \tag{12.42}$$

$$\frac{\alpha}{\sqrt{g_{00}}}\tilde{g}_{IJ} = \delta_{IJ} + O\left(\frac{1}{c}\right). \tag{12.43}$$

From (12.42) we have

$$\sqrt{g_{00}} = \frac{1}{\alpha}\left[-\beta - \frac{\phi}{c^2} + O\left(\frac{1}{c^3}\right)\right]. \tag{12.44}$$

We have not yet made use of the condition that, in the limit of vanishing gravitation, the general relativistic Lagrangian (12.35) should go over into the special relativistic Lagrangian for geodesic motion. The term $O\left(1/c^3\right)$ is of higher order than ϕ/c^2, so, if $\phi/c^2 \xrightarrow[c\to\infty]{} 0$, then $O\left(1/c^3\right) \xrightarrow[c\to\infty]{} 0$ all the more, so

$$\frac{1}{\alpha}\left[-\frac{\phi}{c^2} + O\left(\frac{1}{c^3}\right)\right] \xrightarrow[\phi\to 0]{} 0. \tag{12.45}$$

Moreover, for $\phi = 0$ we should obtain the Minkowski metric, hence $-\beta/\alpha \xrightarrow[\phi\to 0]{} 1$. But α and β are constants, so the above means that $-\beta/\alpha = 1$. In the signature $(+---)$ it is seen from (12.43) that $\alpha < 0$, while $\beta = \pm 1$ by definition, so the above implies $\alpha = -1, \beta = +1$. The final result is thus, from (12.44), (12.40) and (12.43):

$$g_{00} = 1 + \frac{2\phi}{c^2} + O\left(\frac{1}{c^3}\right), \qquad g_{0I} = O\left(\frac{1}{c^2}\right), \qquad g_{IJ} = -\delta_{IJ} + O\left(\frac{1}{c}\right). \tag{12.46}$$

Equations (12.46) follow from the conditions that the general relativistic equations of motion go over into the Newtonian equations of motion when $c \to \infty$ and into the special relativistic equations of motion when $\phi \to 0$. Now we shall investigate the limit $c \to \infty$ for the field equations.

From the equation $g_{\alpha\rho}g^{\beta\rho} = \delta^\beta{}_\alpha$, taking the cases $\{\alpha = 0, \beta = I \neq 0\}$, $\{\alpha = \beta = 0\}$ and $\{\alpha = I \neq 0, \beta = J \neq 0\}$, we conclude that

$$g^{00} = 1 - \frac{2\phi}{c^2} + O\left(\frac{1}{c^3}\right), \qquad g^{0I} = O\left(\frac{1}{c^2}\right), \qquad g^{IJ} = -\delta^{IJ} + O\left(\frac{1}{c}\right). \qquad (12.47)$$

Using these in the formulae for the Christoffel symbols we obtain

$$\left\{ \begin{matrix} 0 \\ 00 \end{matrix} \right\} = O\left(\frac{1}{c^3}\right), \qquad (12.48)$$

$$\left\{ \begin{matrix} 0 \\ 0I \end{matrix} \right\} = \frac{\phi_{,I}}{c^2} + O\left(\frac{1}{c^3}\right) = \left\{ \begin{matrix} I \\ 00 \end{matrix} \right\}, \qquad (12.49)$$

$$\left\{ \begin{matrix} 0 \\ IJ \end{matrix} \right\} = O\left(\frac{1}{c^2}\right) = \left\{ \begin{matrix} I \\ 0J \end{matrix} \right\}, \qquad (12.50)$$

$$\left\{ \begin{matrix} I \\ JK \end{matrix} \right\} = O\left(\frac{1}{c}\right). \qquad (12.51)$$

Hence, further

$$R_{00} = \frac{\phi_{,II}}{c^2} + O\left(\frac{1}{c^3}\right) \qquad \text{(sum over } I), \qquad (12.52)$$

$$R_{0I} = O\left(\frac{1}{c^2}\right), \qquad R_{IJ} = O\left(\frac{1}{c}\right). \qquad (12.53)$$

The Einstein equations (12.21) can be written in the equivalent form

$$R_{\alpha\beta} = \kappa\left(T_{\alpha\beta} - \frac{1}{2}g_{\alpha\beta}T\right), \qquad T \overset{\text{def}}{=} g^{\alpha\beta}T_{\alpha\beta}. \qquad (12.54)$$

How should the right-hand side of (12.54) behave in the Newtonian limit? Since T_{00} is the energy-density, it contains a contribution from the rest energy. All other contributions to T_{00} must be by at least one order in c smaller than that, so we expect that $T_{00} = \rho c^2 + O(c)$, where ρ is the mass-density. The components T_{0I} form the vector of energy stream. In the limit $c \to \infty$, motion of matter should have no influence on the gravitational field it generates.[1] Consequently, T_{0I} should be by at least one order in c smaller than T_{00} in order that they become negligible in the limit $c \to \infty$, thus $T_{0I} = O(c)$. The components

---

[1] If the mass distribution does not change with time, then the motion of matter is not detectable via its gravitational field. For example, the exterior gravitational field of a mass in axisymmetric rotation in Newtonian theory is not distinguishable from the field generated by a static mass with the same mass distribution.

T_{IJ} describe the stress energy-density. Compared to the rest energy, any other kind of energy is by two orders in c smaller, so $T_{IJ} = O(1)$. Hence, finally,

$$T = g^{\alpha\beta} T_{\alpha\beta} = \rho c^2 + O(c). \tag{12.55}$$

From this

$$R_{00} = \kappa \left(\frac{1}{2} \rho c^2 + O(c) \right), \tag{12.56}$$

$$R_{0I} = \kappa O(c), \qquad R_{IJ} = \kappa O(c^2). \tag{12.57}$$

From (12.56) and (12.52) we have

$$\frac{\phi_{,II}}{c^2} + O\left(\frac{1}{c^3} \right) = \kappa \left(\frac{1}{2} \rho c^2 + O(c) \right). \tag{12.58}$$

In the limit $c \to \infty$ this should be equivalent to the Poisson equation (12.16), so

$$\kappa = \frac{8\pi G}{c^4}. \tag{12.59}$$

The remaining equations in the set, (12.57), impose limitations on the terms $O(1/c^2)$ in g_{0I} and $O(1/c)$ in g_{IJ}. These can be read out if we want to consider the Einstein theory as a small perturbation imposed on the Newton theory in a weak gravitational field. There are various approaches to this application of general relativity, the most elaborate of them is the Parametrised Post-Newtonian (PPN) formalism (Misner, Thorne and Wheeler, 1973, Will, 1981). We will briefly describe one such approach (the weak-field approximation) in Section 12.18. Formally, in the limit $c \to \infty$, Eqs. (12.57) are fulfilled identically.

In the equation $R_{00} = \kappa \left(T_{00} - \frac{1}{2} g_{00} T \right)$ we had to include terms of order $1/c^2$ because the coordinate was $x^0 = ct$, so every differentiation by x^0 introduced the factor $1/c$. In order to liberate R_{00} from the factor $1/c^2$ thus introduced, we had to multiply (12.56) by c^2. To achieve the same in the first equation of (12.57) we would have to multiply it by c, and then we would obtain the identity $0 = 0$ in the limit $c \to \infty$.

12.12 Examples of sources in the Einstein equations: perfect fluid and dust

A fluid whose pressure obeys the Pascal law, while the transport of energy occurs only by means of mass flow is called a **perfect fluid**. It does not conduct heat or electric current, and its viscosity is zero. From this definition, we will now deduce the form of its energy-momentum tensor.

If u^α is the velocity field of an arbitrary continuous medium, while s is the proper time on the lines of flow of that medium, then

$$ds^2 = g_{\alpha\beta} \, dx^\alpha \, dx^\beta, \qquad u^\alpha = \frac{dx^\alpha}{ds}. \tag{12.60}$$

These imply

$$g_{\alpha\beta}u^{\alpha}u^{\beta} = 1. \tag{12.61}$$

This holds *for every continuous velocity field* and for any medium.

Now let us choose, for a while, the coordinates adapted to u^{α}, in which $u^{\alpha} \underset{*}{=} \delta^{\alpha}{}_{0}$, i.e. in which the time-coordinate x^0 is the affine parameter on the flow lines of matter, and $dx^{I}(s)/ds \underset{*}{=} 0$, $I = 1, 2, 3$, where $x^{I}(s)$ are spatial coordinates of the particles of the fluid. In these coordinates, $x^{I} = $ constant, which means that the particles of the fluid do not move with respect to the timelike hypersurfaces of the coordinate system. Such coordinates are called **comoving**, and they are frequently used in relativistic hydrodynamics. Their construction can be visualised as follows. We choose an arbitrary hypersurface S that intersects all flow lines of the fluid, and an arbitrary coordinate system within S. To each particle of the fluid we then assign the spatial coordinate x^{I} of the point where it crossed S. To assign the time-coordinate to a point p in spacetime we take the flow line C_p that passes through p. The time-coordinate of p is the proper time that elapsed between the event of C_p intersecting S and the event of C_p passing through p.

By definition, $T_{00}(p)$ is, in any coordinate system, the energy density measured at a given point p. In the comoving coordinates, the only energy that a particle can have is its inner energy $\epsilon = $ (rest energy) + (energy of thermal motion of its particles) + (chemical energy). Hence, $T_{00} \underset{*}{=} \epsilon$ and, at the same time, $T_{00} \underset{*}{=} T_{\alpha\beta}u^{\alpha}u^{\beta}$, so

$$T_{\alpha\beta}u^{\alpha}u^{\beta} = \epsilon. \tag{12.62}$$

This is now an equality of two scalars, so it holds in any coordinates.

By definition, $T^{I}{}_{0}$, $I = 1, 2, 3$, is the vector of the energy stream. But in the comoving coordinates, by the definition of a perfect fluid, there are no energy flows, so consequently $T^{I}{}_{0} \underset{*}{=} T^{I}{}_{\beta}u^{\beta} \underset{*}{=} 0$. This implies that $T^{\alpha}{}_{\beta}u^{\beta} \underset{*}{=} \lambda u^{\alpha}$, where λ is an unknown coefficient. From (12.61) and (12.62) we find that $\lambda = \epsilon$, so

$$T^{\alpha}{}_{\beta}u^{\beta} = \epsilon u^{\alpha}, \tag{12.63}$$

which is again a tensor equation.

Now choose a point q in the spacetime and an arbitrary vector $v^{\alpha}(q)$ at q that is orthogonal to $u^{\alpha}(q)$:

$$u^{\alpha}(q)v_{\alpha}(q) = 0. \tag{12.64}$$

The vector $v^{\alpha}(q)$ points from q towards a neighbouring particle. Since Eq. (12.64) says that the projection of the velocity on $v^{\alpha}(q)$ is zero, it follows that the particle to which $v^{\alpha}(q)$ points does not move relative to q. The collection of all vectors having the property (12.64) thus determines a 3-dimensional volume element comoving with the particle that was at q. Consequently, the Pascal law must apply in this volume element: the pressure p exerted on the surface element σ in the fluid creates the force $f = p\sigma$ in the direction n^{I}, $I = 1, 2, 3$, orthogonal to σ. The pressure p and the force f do not depend on the direction of n^{I}, i.e. on the orientation of σ within the fluid. Let $-T^{I}{}_{J}$ denote the Newtonian (3-dimensional) stress tensor. (The minus sign is a consequence of the signature $(+---)$; in this signature the spatial part of the energy-momentum tensor $T^\alpha{}_\beta$ is not the stress tensor $\tau^I{}_J$ itself, but $-\tau^I{}_J$.) By the definition of the stress tensor, the following must hold:

$$-T^I{}_J \sigma n^J = f n^I \equiv p \sigma n^I, \tag{12.65}$$

which implies

$$T^I{}_J n^J = -p n^I. \tag{12.66}$$

The vector n^I was an arbitrary vector within the volume element of the fluid comoving with q, i.e. an arbitrary vector in the 3-dimensional subspace orthogonal to $u^\alpha(q)$. Equation (12.66) shows that every such vector is an eigenvector of the matrix $T^I{}_J$ connected with the eigenvalue $(-p)$, which implies

$$T^I{}_J = -p \delta^I{}_J. \tag{12.67}$$

An arbitrary timelike vector field u^α in general is not orthogonal to a family of hypersurfaces, so the vectors n^I in (12.65)–(12.67) are in general neither tangent nor orthogonal to hypersurfaces, and so cannot define a coordinate system. However, at every point q of the manifold, in the tangent space we can choose a subspace orthogonal to u^α, and choose in that subspace the orthonormal basis $e_{\widehat{I}}{}^\alpha$, $I = 1, 2, 3$. In that basis, Eq. (12.67) will be fulfilled, where

$$T_{\widehat{I}\widehat{J}} = e_{\widehat{I}}{}^\alpha e_{\widehat{J}}{}^\beta T_{\alpha\beta}. \tag{12.68}$$

(In other words, we choose coordinates that are adapted to the basis vectors $e_{\widehat{I}}{}^\alpha$ at one point q only, which is always possible.) Now, from (12.62)–(12.63), (12.67) and (12.68) we can deduce the general formula for the energy-momentum tensor of a perfect fluid. Let us choose an orthonormal tetrad $e_i{}^\alpha$, $i = 0, 1, 2, 3$, in spacetime such that $e_{\widehat{0}}{}^\alpha = u^\alpha$, while each $e_{\widehat{I}}{}^\alpha$, $I = 1, 2, 3$, obeys (12.66). Then

$$g_{\alpha\beta} e_i{}^\alpha e_j{}^\beta = \eta_{ij} = \mathrm{diag}(+1, -1, -1, -1) \tag{12.69}$$

and from (12.62)–(12.63), (12.68) and (12.69) we have

$$T_{\widehat{0}\widehat{0}} = T_{\alpha\beta} u^\alpha u^\beta = \epsilon, \tag{12.70}$$

$$T_{\widehat{0}\widehat{A}} = T_{\alpha\beta} u^\alpha e_{\widehat{A}}{}^\beta = \epsilon u_\beta e_{\widehat{A}}{}^\beta = 0, \qquad A = 1, 2, 3, \tag{12.71}$$

$$T_{\widehat{A}\widehat{B}} = -T^{\widehat{A}}{}_{\widehat{B}} = p \delta^{\widehat{A}}{}_{\widehat{B}} = -p \eta_{\widehat{A}\widehat{B}}. \tag{12.72}$$

These are the components of the scalar image of $T_{\alpha\beta}$. Applying the inverse projection, (4.16), we obtain

$$\begin{aligned}
T_{\alpha\beta} &= e^i{}_\alpha e^j{}_\beta T_{ij} = u_\alpha u_\beta T_{\widehat{0}\widehat{0}} + e^{\widehat{A}}{}_\alpha e^{\widehat{B}}{}_\beta T_{\widehat{A}\widehat{B}} \\
&= \epsilon u_\alpha u_\beta - p \eta_{\widehat{A}\widehat{B}} e^{\widehat{A}}{}_\alpha e^{\widehat{B}}{}_\beta - p u_\alpha u_\beta + p u_\alpha u_\beta \\
&= (\epsilon + p) u_\alpha u_\beta - p g_{\alpha\beta}. \tag{12.73}
\end{aligned}$$

A perfect fluid whose pressure is identically zero is called **dust**. It follows that for dust

$$T_{\alpha\beta} = (\epsilon + p)u_\alpha u_\beta. \tag{12.74}$$

12.13 Equations of motion of a perfect fluid

The equations of motion of the sources of a gravitational field, $T^{\alpha\beta}{}_{;\beta} = 0$, for a general perfect fluid are equivalent to

$$(\epsilon + p),_\beta u^\alpha u^\beta + (\epsilon + p)u^\alpha{}_{;\beta} u^\beta + (\epsilon + p)u^\alpha u^\beta{}_{;\beta} - p,_\beta g^{\alpha\beta} = 0. \tag{12.75}$$

The identity (12.61) implies

$$u^\alpha u_{\alpha;\beta} = 0. \tag{12.76}$$

Contracting (12.75) with u_α and using (12.61) and (12.76) we obtain

$$\epsilon,_\beta u^\beta + (\epsilon + p)u^\beta{}_{;\beta} = 0. \tag{12.77}$$

This is the energy conservation equation which says that the volume work $-pu^\beta{}_{;\beta}$ generates the energy stream ϵu^β. Now using (12.77) in (12.75), we obtain

$$(\epsilon + p)u^\alpha{}_{;\beta} u^\beta - p,_\beta g^{\alpha\beta} + p,_\beta u^\beta u^\alpha = 0. \tag{12.78}$$

These are the general relativistic equations of motion of a perfect fluid. In the Newtonian limit $(c \to \infty)$ and in asymptotically Cartesian coordinates, they go over into the Euler equations of motion $\rho\, d\mathbf{v}/dt = -\text{grad}\,p$.

Equations (12.77) and (12.78) simplify for dust, when $p = 0$. Equation (12.77) then becomes

$$(\epsilon u^\beta){}_{;\beta} = 0, \tag{12.79}$$

which is the relativistic equation of continuity (mass conservation) that in the Newtonian limit goes over into $\partial\rho/\partial t + \text{div}(\rho\mathbf{v}) = 0$. Equation (12.78) becomes $u^\alpha{}_{;\beta} u^\beta = 0$. This means that the covariant derivative of the vector field tangent to the flow lines along these lines is zero. This fulfils the requirements of the definition of a geodesic in affine parametrisation, see Section 5.2. Consequently, *dust moves along geodesics*, and *the proper time of the particles of dust is an affine parameter on these geodesics*.

In fact, the necessary and sufficient condition for a geodesic motion of a perfect fluid is somewhat weaker than $p = 0$. As seen from (12.78), it is $p,_\beta (g^{\alpha\beta} - u^\alpha u^\beta) = 0$. We have already met the operator $h^{\alpha\beta} = g^{\alpha\beta} - u^\alpha u^\beta$ once; see Eq. (7.102). It projects tensors on the hypersurface elements orthogonal to the vector field u^α. The equation $h^{\alpha\beta} p,_\alpha = 0$ means that the gradient of pressure is collinear with the velocity field, i.e. there are no spatial gradients of pressure in the comoving coordinates.

12.14 The cosmological constant

When general relativity was created, it was clear that its predictions would significantly differ from those of Newton's theory in two situations:

(1) in strong gravitational fields, when the extent of the regions where local inertial frames exist becomes small; and
(2) in large sub-volumes of the Universe, where even small departures of the real geometry from the Euclidean geometry cumulate over large distances and become visible during observations of distant objects.

Consequently, when Einstein thought about physical applications of his theory, one of the first things he tried was to construct a model of the Universe. At that time everybody was *sure* that the geometry and matter distribution in the Universe were time- and space-independent. Calculations showed, however, that these assumptions are self-contradictory – such a model of the Universe did not exist in relativity.

At that moment, Einstein was on the verge of making another great discovery. The year was 1916, and the first observations proving the expansion of the Universe were published only in 1927. Yet this time Einstein's belief in prejudice turned out to be stronger than his tendency to think boldly. While searching for a reason of the failure of his first attempt, Einstein did not question the *assumption* that the Universe is static, but turned his suspicion against his equations (12.21). He soon found a gap in his reasoning: the left-hand side of the field equations that has all the required properties need not necessarily be the Einstein tensor $R_{\mu\nu} - \frac{1}{2}g_{\mu\nu}R$. One can add to it any symmetric tensor $H_{\mu\nu}$ whose covariant divergence vanishes and that does not depend on the derivatives of the metric. Such a correction will not increase the order of the field equations, while the equation $\left(R^{\mu\nu} - \frac{1}{2}g^{\mu\nu}R + H^{\mu\nu}\right)_{;\nu} = 0$ will still hold. The simplest tensor of this property is $\Lambda g^{\mu\nu}$, where Λ should be a universal constant. The modified Einstein equations

$$R^{\mu\nu} - \frac{1}{2}g^{\mu\nu}R + \Lambda g^{\mu\nu} = \kappa T^{\mu\nu} \tag{12.80}$$

in the limit $c \to \infty$ go over into the modified Poisson equation

$$\Delta\phi - \Lambda c^2 = 4\pi G\rho. \tag{12.81}$$

Hence, the constant Λ, called the **cosmological constant**, describes an effect that is absent from the ordinary Newton theory: a universal attraction (for $\Lambda > 0$) or repulsion (for $\Lambda < 0$) of matter particles.

The modified equations (12.80) allowed for the existence of a static, homogeneous and isotropic solution of the form

$$ds^2 = c^2\,dt^2 - R^2\,d\chi^2 - R^2\sin^2\chi\left(d\vartheta^2 + \sin^2\vartheta\,d\varphi^2\right), \tag{12.82}$$

where

$$R = \frac{1}{\sqrt{-\Lambda}} = \frac{c}{2\sqrt{\pi G\rho}}, \tag{12.83}$$

ρ being the average mass density in the Universe, constant by assumption. The spacetime corresponding to the metric (12.82) is, by tradition, still called the **Einstein Universe**, although it is no longer considered a model of the real Universe. Since $\Lambda < 0$ here, the 'cosmological repulsion' balances the gravitational attraction, which allows the system to be static (but unstable, as we will see in Section 17.6).

This brief history of the introduction of the cosmological constant shows that in fact it appeared in consequence of an error. Had Einstein not insisted on obtaining a static model of the Universe, he would have had a chance to predict that the Universe should expand or collapse, 11 years before the expansion was observed. When he realised later how close he was to making that prediction, he said that the introduction of the constant was 'the biggest blunder of his life' (Misner, Thorne and Wheeler, 1973, pp. 410–411).[1]

Today, the cosmological constant is more popular among particle physicists than among specialists in relativity, and the observations of brightness of distant supernovae are even said to indicate that Λ is strictly negative; see Section 17.9. Its absolute value must be very small (less than 10^{-50} cm^{-2} (Misner, Thorne and Wheeler, 1973, pp. 410–411)), so it can play a role only in the evolution of the Universe. In the Solar System, it has no observable influence on the motion of planets. A great number of solutions of the modified Einstein equations (12.80) is known (Stephani *et al.*, 2003, section 24.2), both inside matter distributions and in vacuum. Examples will be given later in this text (see Section 14.4).

12.15 An example of an exact solution of Einstein's equations: a Bianchi type I spacetime with dust source

The Einstein equations had for many years been rumoured to be very difficult, almost too difficult to find any exact solution. This opinion lingers until today and is even repeated in some publications, although it ceased to have any basis decades ago. No formula for a general solution of Einstein's equations has been found that would generalise $\phi(\mathbf{x}) = -\int d_3 x' G\rho(\mathbf{x}')/|\mathbf{x} - \mathbf{x}'|$, which is the formal solution of the Poisson equation. The reason of the difficulty is the nonlinearity of the Einstein equations. In general relativity, the superposition of two solutions is not their sum and the law of composition of solutions is not known. Consequently, finding a solution for a simple situation does not really help in looking for more general solutions, except that it increases the basis of experience and knowledge. Nevertheless, for special cases (like high symmetry, special properties of the source or special properties of the Weyl tensor), hundreds if not thousands of solutions are known, and any attempt to list and compare them is a major undertaking

---

[1] That story has a complicated continuation. Hubble, who is credited with the discovery of the expansion of the Universe (Hubble, 1929), had not believed, until the end of his life, that the Universe is actually expanding. He insisted, even in his last paper that appeared in print after his death (Hubble, 1953), that expressing the observed redshifts in spectra of galaxies through their equivalent velocities of recession is merely a convenient mathematical device (Krasiński and Ellis, 1999). Then, Milne and McCrea (1934) (see Section 17.5) showed that expansion of the Universe could and should have been predicted on the basis of Newton's theory of gravitation in the eighteenth century since all the mathematical knowledge necessary for that purpose had existed already at that time. The prediction was not made because nobody tried – just because everybody was sure that the Universe was static.

(Stephani *et al.*, 2003; Krasiński, 1997). In order to prove to the reader that the task of finding a solution is not at all hopeless and to demonstrate a few characteristic methods of calculation, we shall present a derivation of a certain exact solution.

We will assume that the spacetime is spatially homogeneous, with the symmetry group of Bianchi type I, and that the source in the equations is dust. The cosmological constant will be assumed zero. We thus begin with the metric form (10.29). For a Bianchi type I algebra all generators commute to zero. Hence (see Exercise 5 in Chapter 8) we can choose the spatial coordinates (x, y, z) so that the generators become

$$k^\alpha_{(I)} = \delta^\alpha{}_I, \qquad x^1 = x, \quad x^2 = y, \quad x^3 = z \tag{12.84}$$

(which means that x^I is a parameter on the curve tangent to $k^\alpha{}_{(I)}$ and obeys $k^\alpha{}_{(I)} = \partial x^\alpha / \partial x^I$). For the generators (12.84), the Killing equations are

$$g_{IJ,K} = 0, \qquad I, J, K = 1, 2, 3, \tag{12.85}$$

i.e. the components g_{IJ} depend only on t. Knowing this, we find the following formulae for the Christoffel symbols, the Riemann tensor and the Ricci tensor (the components not listed are zero):

$$\left\{ \begin{matrix} 0 \\ IJ \end{matrix} \right\} = \frac{1}{2} g_{IJ,t}, \qquad \left\{ \begin{matrix} I \\ 0J \end{matrix} \right\} = \frac{1}{2} g^{IS} g_{JS,t}; \tag{12.86}$$

$$R^0{}_{IOJ} = \frac{1}{2} g_{IJ,tt} - \frac{1}{4} g^{RS} g_{IR,t} g_{JS,t}, \tag{12.87}$$

$$R^I{}_{JKL} = \frac{1}{4} g^{IR} g_{KR,t} g_{JL,t} - \frac{1}{4} g^{IR} g_{LR,t} g_{JK,t}; \tag{12.88}$$

$$R_{00} = -\frac{1}{2} g^{RS} g_{RS,tt} + \frac{1}{4} g^{LM} g^{RS} g_{LR,t} g_{MS,t}, \tag{12.89}$$

$$R_{JL} = \frac{1}{2} g_{JL,tt} - \frac{1}{4} g^{RS} g_{JR,t} g_{LS,t} + \frac{1}{4} g^{KR} g_{KR,t} g_{JL,t}. \tag{12.90}$$

The identity $R_{0I} \equiv 0$ implies $u_0 u_I = 0$. Since u_0 cannot vanish (then u_α would be a spacelike vector, i.e. the dust would move with a superluminal velocity), we must have

$$u_I = 0 = u^I \Longrightarrow u_0 = u^0 = 1,$$
$$T_{00} = \epsilon, T_{\alpha I} = 0, \qquad I = 1, 2, 3, \qquad \alpha = 0, 1, 2, 3, \tag{12.91}$$

where ϵ is the energy-density of the dust.

In the following, we will use the auxiliary formulae that follow immediately from (12.24) and (12.25):

$$g^{RS} g_{RS,t} = g_{,t}/g, \tag{12.92}$$

$$g^{MS}{}_{,t} = -g^{LM} g^{RS} g_{LR,t}, \tag{12.93}$$

where $g = \det \| g_{\alpha\beta} \| = -\det \| g_{KL} \|$.

From (12.79) we have $\left(\sqrt{-g}\epsilon u^{\alpha}\right),_{\alpha}=0$, which in our coordinates becomes $\left(\sqrt{-g}\epsilon\right),_{t}=0$. Since the components of the metric tensor depend only on t, also g and ϵ depend only on t, so

$$\kappa\epsilon\sqrt{-g}=M=\text{constant} \tag{12.94}$$

From (12.90) and (12.91), using (12.92) and (12.93), we obtain

$$R=R_{00}-g^{JL}R_{JL}=-g^{RS}g_{RS,tt}-\frac{3}{4}g^{RS},_{t}g_{RS,t}-\frac{1}{4}\left(g,_{t}/g\right)^{2}. \tag{12.95}$$

From $R_{00}-\frac{1}{2}g_{00}R=\kappa\epsilon$, using (12.90) and (12.93)–(12.95), we now obtain

$$\frac{1}{8}g^{RS},_{t}g_{RS,t}+\frac{1}{8}\left(g,_{t}/g\right)^{2}=M/\sqrt{-g}. \tag{12.96}$$

This equation will be later used to eliminate $g^{RS},_{t}g_{RS,t}$.

The Einstein equations imply that for dust

$$R=g^{\alpha\beta}R_{\alpha\beta}=-\kappa\epsilon. \tag{12.97}$$

We substitute (12.95) and (12.94) in the above and obtain

$$g^{RS}g_{RS,tt}+\frac{3}{4}g^{RS},_{t}g_{RS,t}+\frac{1}{4}\left(g,_{t}/g\right)^{2}$$

$$\equiv\left(g^{RS}g_{RS,t}\right),_{t}-\frac{1}{4}g^{RS},_{t}g_{RS,t}+\frac{1}{4}\left(g,_{t}/g\right)^{2}=M/\sqrt{-g}. \tag{12.98}$$

Now we use (12.96) to eliminate $g^{RS},_{t}g_{RS,t}$, and we obtain

$$\left(g^{RS}g_{RS,t}\right),_{t}+\frac{1}{2}\left(g,_{t}/g\right)^{2}=3M/\sqrt{-g}. \tag{12.99}$$

Using (12.92) we obtain $\left(\sqrt{-g}\right),_{tt}=3M/2$, which is integrated to give

$$\sqrt{-g}=\frac{3}{4}Mt^{2}+At+B, \tag{12.100}$$

where A and B are arbitrary constants. We still have to integrate the equations $G_{JL}=0$. Using (12.97), (12.94), (12.92) and (12.93) they are reduced to

$$0=-G^{K}{}_{L}=g^{KJ}R_{JL}-\frac{1}{2}\delta^{K}{}_{L}M/\sqrt{-g}$$

$$=\frac{1}{2}g^{KS}g_{LS,tt}+\frac{1}{2}g^{KS},_{t}g_{LS,t}+\frac{1}{4}\frac{g,_{t}}{g}g^{KS}g_{LS,t}-\frac{1}{2}\delta^{K}{}_{L}M/\sqrt{-g}. \tag{12.101}$$

Hence

$$\sqrt{-g}\left(g^{KS}g_{LS,t}\right),_{t}-\frac{1}{2}\left(g,_{t}/\sqrt{-g}\right)g^{KS}g_{LS,t}-M\delta^{K}{}_{L}=0, \tag{12.102}$$

which is integrated with the result

$$\sqrt{-g}g^{KS}g_{LS,t}=Mt\delta^{K}{}_{L}+A^{K}{}_{L}=0, \tag{12.103}$$

where $A^K{}_L$ is a matrix of constant elements. From this, contracting with g_{KM}, we obtain

$$g_{LM,t} = \frac{Mt}{\sqrt{-g}} g_{LM} + \frac{A^K{}_L g_{KM}}{\sqrt{-g}}. \tag{12.104}$$

This shows at once that $A_{ML} = A_{LM}$. Furthermore, note that for any fixed instant $t = t_0$ the matrix $g_{KM}(t_0)$ can be transformed into the unit matrix by transformations of the form $x'^I = b^I{}_J x^J$, where the matrix $b^I{}_J$ has constant coefficients. Since A_{KL} is symmetric, we can then use the orthogonal transformations preserving the simple form of $g'_{KM}(t_0)$ to diagonalise A_{KL}, too. Assume that we have done so and have chosen the coordinates in which $g_{KM}(t_0)$ and A_{KL} have these simple forms. Then, with use of (12.100), (12.104) becomes

$$g_{LM,t} = \frac{Mt + A^L{}_L}{\frac{3}{4}Mt^2 + At + B} g_{LM} \qquad \text{(no sum over } L\text{)}, \tag{12.105}$$

and the solution of this is

$$g_{LM}(t) = g_{LM}(t_0) \exp\left(\int_{t_0}^{t} \frac{M\tau + A^L{}_L}{\frac{3}{4}M\tau^2 + A\tau + B} \, d\tau \right). \tag{12.106}$$

Thus, if $g_{LM}(t_0)$ is diagonal, then so is $g_{LM}(t)$ at all times.

(But it is not always possible to diagonalise the metric tensor in a Bianchi-type space-time. In our case the diagonal form followed from the field equations.)

Equation (12.106) is not the final result yet because the components g_{LM} still have to obey (12.100) and (12.96). The constants $A^L{}_L$ are by tradition parametrised as follows:

$$A^L{}_L = 2p_L A \qquad \text{(no sum over } L\text{)} \tag{12.107}$$

and then, by virtue of (12.100) and (12.96), the new parameters p_L must obey

$$p_1 + p_2 + p_3 = 1,$$
$$\frac{1}{2} A^2 \left[1 - (p_1^2 + p_2^2 + p_3^2) \right] = MB. \tag{12.108}$$

In the case $M = 0$ the metric obtained above becomes a vacuum solution:

$$ds^2 = dt^2 - t^{2p_1} \, dx^2 - t^{2p_2} \, dy^2 - t^{2p_3} \, dz^2, \tag{12.109}$$

in which the constants p_1, p_2 and p_3 obey

$$p_1 + p_2 + p_3 = p_1^2 + p_2^2 + p_3^2 = 1. \tag{12.110}$$

This solution was found by Kasner (1921). It was one of the earliest exact solutions of Einstein's equations to be published in the literature.

12.16 * Other gravitation theories

As mentioned at the end of Section 12.7, there are several gravitation theories that are alternatives to Einstein's. In this section we will present a brief overview of a few of them. They are not necessarily those that ever had the greatest chance to replace general relativity as *the* theory of gravitation – the examples are meant to illustrate the various possibilities used by different authors. Except for Rosen's theory, all of them are generalisations of general relativity and contain it as a limit. A much more extended overview of alternative gravitation theories, and a comparison of them with experimental tests, can be found in the book by Will (1981). The Kaluza–Klein theory is one of the generalisations, but its presentation must be preceded by a description of the Maxwell theory in curved spacetime, therefore we postpone it to the next chapter.

12.16.1 The Brans–Dicke theory

The theory was first published in 1961 (Brans and Dicke, 1961), and the best source for studying it is the book by Dicke (1964). The field equations of the Brans–Dicke (BD) theory are

$$R_{\alpha\beta} - \frac{1}{2}g_{\alpha\beta}R = \frac{8\pi}{\phi}T_{\alpha\beta} + \frac{\omega}{\phi^2}\left(\phi_{,\alpha}\,\phi_{,\beta} - \frac{1}{2}g_{\alpha\beta}\phi_{,\sigma}\,\phi^{,\sigma}\right)$$

$$+ \frac{1}{\phi}\left(\phi_{;\alpha\beta} - g_{\alpha\beta}g^{\mu\nu}\phi_{;\mu\nu}\right), \tag{12.111}$$

$$g^{\mu\nu}\phi_{;\mu\nu} - \frac{1}{2\phi}\phi_{,\mu}\,\phi^{,\mu} + \frac{\phi}{2\omega}R = 0, \qquad \omega = \text{constant}. \tag{12.112}$$

This is a generalisation of Einstein's theory in which the gravitational constant is replaced by the scalar field ϕ. In this theory, the gravitational 'constant' c^4/ϕ thus varies in time and space, and its changes are induced by the scalar curvature R via Eq. (12.112). The scalar curvature in turn depends on the distribution of matter via (12.111). In this way, the distribution of matter influences the intensity of gravitation, which is a partial realisation of the Mach principle.

General relativity (GR) follows from the BD theory as the limit $\phi = c^4/G = \text{constant}$, $\omega \to \infty$. Observable modifications of GR would be predicted if $\omega \le 6$; in particular, part of the anomalous orbital motion of Mercury would then be caused by the oblateness of the Sun. However, detailed observations imply that $\omega \ge 23$. With that value, the predictions of the BD theory are observationally indistinguishable from the predictions of Einstein's theory. For this reason, the BD theory is currently out of favour with theoreticians. However, this may be a temporary situation. This theory is conceptually the most developed of all competing theories of gravitation, with its logical structure and interpretation well clarified, and may still have a role to play in the future.

12.16.2 The Bergmann–Wagoner theory

Original references for this theory are Bergmann (1968) and Wagoner (1970). Its field equations are

$$R_{\mu\nu} - \frac{1}{2}g_{\mu\nu}R - \lambda(\phi)g_{\mu\nu} = \frac{8\pi}{\phi}T_{\mu\nu} + \frac{\omega(\phi)}{\phi^2}\left(\phi_{,\mu}\,\phi_{,\nu} - \frac{1}{2}g_{\mu\nu}\phi_{,\sigma}\,\phi^{,\sigma}\right)$$

$$+ \frac{1}{\phi}\left(\phi_{;\mu\nu} - g_{\mu\nu}g^{\rho\sigma}\phi_{;\rho\sigma}\right), \tag{12.113}$$

$$g^{\mu\nu}\phi_{;\mu\nu} + \frac{1}{2}\phi_{,\mu}\,\phi^{,\mu}\frac{\mathrm{d}}{\mathrm{d}\phi}\ln[\omega(\phi)/\phi] + \frac{\phi}{2\omega(\phi)}\left(R + 2\frac{\mathrm{d}}{\mathrm{d}\phi}(\phi\lambda(\phi))\right) = 0, \tag{12.114}$$

where $\lambda(\phi)$ and $\omega(\phi)$ are arbitrary functions. This is a generalisation of the Brans–Dicke theory, which results as the limit $\omega = \mathrm{constant}$, $\lambda = 0$. It is presented here just as a formal curiosity because it is far from being well understood. Its main weakness is the existence of two arbitrary functions that are not determined by any field equations. The function $\lambda(\phi)$ becomes the cosmological constant when $\phi = \mathrm{constant}$.

12.16.3 The conformally invariant Canuto theory

The original reference for this theory is Canuto *et al.* (1977). Its field equations are

$$R_{\mu\nu} - \frac{1}{2}g_{\mu\nu}R + \frac{2\beta_{;\mu\nu}}{\beta} - \frac{4\beta_{,\mu}\beta_{,\nu}}{\beta^2} - g_{\mu\nu}\left(\frac{2\beta^{;\lambda}_{\;\;;\lambda}}{\beta} - \frac{\beta^{,\lambda}\beta_{,\lambda}}{\beta^2}\right)$$

$$= \frac{8\pi}{c^4}G(\beta)T_{\mu\nu}(\beta) + \Lambda g_{\mu\nu}. \tag{12.115}$$

This theory was designed to be such a generalisation of general relativity that is invariant under arbitrary conformal transformations $g_{\mu\nu} \to \beta^2 g_{\mu\nu}$. The left-hand side of (12.115) is indeed invariant under such a transformation; the invariance of the right-hand side is a postulate of the theory. The Einstein theory follows as the limit $\beta = 1$. The changes in the energy-momentum tensor induced by the transformation $g_{\mu\nu} \to \beta^2 g_{\mu\nu}$ should be compensated by $G(\beta)$, which should be determined for every case. It is assumed that $G(\beta)T_{\mu\nu}(\beta) = G(1)T_{\mu\nu}(1)$, the latter being the right-hand side of the Einstein equations. The arbitrary function β is interpreted as the local change of scale.

The physical interpretation of this theory is not quite clear, and the arbitrary function β makes it difficult, just as in the Bergmann–Wagoner theory, to calculate observable effects. The Canuto theory has never really caught on with the physics community and is not currently considered a viable alternative to Einstein's theory.

12.16.4 The Einstein–Cartan theory

In this theory (Trautman, 1972) the connection coefficients are not symmetric. The Ricci tensor is built of the full, nonsymmetric connection coefficients and obeys equations formally identical to the Einstein equations (12.21), but $R_{\mu\nu}$ and $T_{\mu\nu}$ are not necessarily symmetric. The torsion tensor $Q^\rho{}_{\mu\nu}$ obeys a separate set of equations:

$$Q^\rho{}_{\mu\nu} + \delta^\rho{}_\mu Q^\sigma{}_{\nu\sigma} - \delta^\rho{}_\nu Q^\sigma{}_{\mu\sigma} = \kappa s^\rho{}_{\mu\nu}, \tag{12.116}$$

where $s^\rho{}_{\mu\nu}$ is the tensor describing the stream of energy connected with the spin of matter, and it vanishes in vacuum. When $s^\rho{}_{\mu\nu} = 0$, also $Q^\rho{}_{\mu\nu} = 0$. Thus, in this theory torsion does not propagate out into vacuum; it may be nonzero only inside matter. With $s^\rho{}_{\mu\nu} = 0$, the Einstein–Cartan (EC) theory reduces to the Einstein theory. In vacuum, it is not distinguishable from the Einstein theory.

The advantage of the EC theory is that it allows a singularity-free Universe model. (General relativity predicts that every model of the Universe must have a singularity in the past or in the future – see Section 15.3.) However, it has not been accepted as a replacement of relativity, mainly because of the impossibility of distinguishing it experimentally from GR – since all experiments testing GR are carried out in vacuum.

12.16.5 The bi-metric Rosen theory

It is not a generalisation, but an alternative to Einstein's theory (Rosen, 1973). There are two metric tensors in it: $g_{\alpha\beta}$, of the same interpretation as in general relativity, and the flat tensor $\gamma_{\alpha\beta}$, for which the curvature tensor $B^\alpha{}_{\beta\gamma\delta}(\gamma) \equiv 0$. Its field equations are

$$N_{\mu\nu} - \frac{1}{2} g_{\mu\nu} N = \kappa T_{\mu\nu}, \tag{12.117}$$

where

$$N_{\mu\nu} = \frac{1}{2} \gamma^{\rho\sigma} g_{\mu\nu|\rho\sigma} - \frac{1}{2} \gamma^{\rho\sigma} g^{\lambda\tau} g_{\mu\lambda|\rho} g_{\nu\tau|\sigma}, \tag{12.118}$$

and the vertical stroke denotes a covariant derivative with respect to $\gamma_{\mu\nu}$. The author's argument in favour of this theory was its greater simplicity compared to Einstein's theory. Its experimental predictions are not much different from those of general relativity. When the gravitational field vanishes, $g_{\mu\nu} = \gamma_{\mu\nu}$. However, a majority of relativists consider the flat background metric $\gamma_{\mu\nu}$ to be an artificial and unnatural element that spoils the elegance of the relativistic approach (the opinion of the author himself was exactly the opposite). Many papers had been published on this theory, but in the end it has not established itself as a viable alternative to general relativity.

12.17 Matching solutions of Einstein's equations

Just as in electrodynamics, in gravitation theory we sometimes have to match solutions of Einstein's equations obtained separately for different spacetime regions. Most often, we want to determine whether a given vacuum solution can be interpreted as the exterior gravitational field to a material body for whose interior we have found another solution.

Matching results in the arbitrary constants in the vacuum region being determined in terms of the parameters of the interior metric.

Sometimes we are interested in describing a surface matter distribution on the boundary between two regions, but we shall exclude this case here. Generally, we assume that there are no singularities of the type of Dirac δ function in components of the curvature tensor and that the hypersurface across which we match the two metrics is non-null. (Together with surface matter distributions we thus exclude shock waves in which the discontinuity in the Riemann tensor is matched to vacuum solutions on both sides. Null matching hypersurfaces pose additional problems.)

With these assumptions, the components of the Riemann tensor can at worst be discontinuous across the matching hypersurface Σ. We have to allow discontinuities because, for example, the mass-density on the surface of a perfect fluid body, which equals the component $T^0{}_0$ of the energy-momentum tensor in comoving coordinates, is nonzero at the surface, but zero at adjacent vacuum points.

In order to discuss these conditions in detail, it is most convenient to use the coordinates adapted to Σ, the same that were introduced at the end of Section 7.15. This time, $N = 4$, $n = 3$, $V_n = \Sigma$, and the spacetime metric in the adapted coordinates is

$$ds^2 = g_{IJ}\,dx^I\,dx^J + \varepsilon N^2\left(dx^4\right)^2,\tag{12.119}$$

where the x^4 coordinate may be timelike (then $\varepsilon = +1$) or spacelike ($\varepsilon = -1$). The normal vector to the boundary $x^4 = A = \text{constant}$ is $X^\alpha = (0, 0, 0, 1/N)$. For $x^4 > A$ we have one metric $g^+_{\alpha\beta}$ (e.g. vacuum); for $x^4 < A$ we have another $g^-_{\alpha\beta}$ (e.g. the interior of a material body). Since Σ has to be the same, whichever of the two 4-metrics is used to describe it, the components g_{IJ} have to be continuous across Σ. Since on Σ they are functions of the x^I-coordinates, the same functions no matter which 4-metric is used to calculate the g_{IJ}, the continuity of all the derivatives of g_{IJ} along Σ, i.e. of the derivatives by x^K, is automatically guaranteed, and we need only take extra care about $g_{IJ,4}$.

So far, we have made sure that the boundary hypersurface Σ has the same intrinsic geometry in both metrics. However, if the spacetime regions on the opposite sides of Σ are to be parts of the same manifold, then Σ must have the same extrinsic geometry with respect to both metrics, i.e. be embeddable in both in the same way. Without that, we might end up trying to identify a cylinder with a plane, for example. Hence, the second fundamental form of Σ must be the same, whichever of the two metrics is used to calculate it.

Thus, finally, the conditions that two metrics, $g^+_{\alpha\beta}$ and $g^-_{\alpha\beta}$, can be considered to describe two parts of the same manifold are as follows:

A hypersurface Σ must exist such that, in coordinates adapted to Σ as in (12.119), the metrics induced on Σ are the same ($g^+_{\alpha\beta}(\Sigma) = g^-_{\alpha\beta}(\Sigma)$), and the second fundamental form of Σ must be the same, whichever metric, $g^+_{\alpha\beta}$ or $g^-_{\alpha\beta}$, is used to calculate it.

Now we will investigate the consequences of these conditions for the Riemann and Einstein tensors of the spacetime and for the Riemann tensor of Σ. We will use the Gauss–Codazzi equations (7.93)–(7.94), but we will have to adapt the notation. Now U_{n+1}

is the spacetime (here denoted V_4) and the V_n of Section 7.15 is the boundary hypersurface Σ. Thus, the capital Latin indices of Section 7.15 now become the Greek spacetime indices, and the Greek indices of Section 7.15 now become the capital Latin indices for Σ. The covariant derivative in the V_n of Section 7.15, denoted by a semicolon, now becomes the covariant derivative in Σ, denoted by the vertical stroke $|$. With $X^\alpha = (0, 0, 0, 1/N)$ and Y^α coinciding with x^I in Σ for $\alpha, I = 1, 2, 3$, Eqs. (7.93)–(7.94) become

$$R_{DABC}(\Sigma) = R_{DABC}(V_4) + \varepsilon\,(\Omega_{AC}\Omega_{DB} - \Omega_{AB}\Omega_{DC}), \tag{12.120}$$

$$\Omega_{IJ|K} - \Omega_{IK|J} = -R_{4IJK}(V_4)/N. \tag{12.121}$$

Since $R_{DABC}(\Sigma)$ and Ω_{AB} are continuous across Σ, $R_{DABC}(V_4)$, i.e. the Σ-components of the 4-dimensional Riemann tensor, will be continuous across Σ, too. The covariant derivatives of Ω_{AB} in (12.121) are taken within Σ, so they are also continuous. This means that $R_{4IJK}(V_4)/N$ has to be continuous across Σ, but N and $R_{4IJK}(V_4)$ individually need not be – they may have a discontinuity that cancels out in the quotient.

For completeness, we have now to consider the components $R_{4I4J}(V_4)$ that are not determined by the Gauss–Codazzi equations. They are

$$R_{4I4J} = \varepsilon N^2 R^4{}_{I4J}$$

$$= \varepsilon N^2 \left[-\frac{1}{4}g^{44}{}_{,J}\,g_{44,I} - \frac{1}{2}g^{44}g_{44,IJ} - \frac{1}{4}g^{44}{}_{,4}\,g_{IJ,4} - \frac{1}{2}g^{44}g_{IJ,44} \right.$$

$$\left. + \frac{1}{2}g^{44}g_{44,R}\left\{ \begin{matrix} r \\ IJ \end{matrix} \right\}_{(\Sigma)} + \frac{1}{4}g^{44}g^{RS}g_{RJ,4}g_{IS,4} \right]$$

$$= \varepsilon N \left(-N_{|IJ} + \varepsilon N g^{RS}\Omega_{RJ}\Omega_{IS} + \varepsilon N\Omega_{IJ,S}X^S \right). \tag{12.122}$$

These components contain second derivatives of g_{IJ} by x^4, so they can be discontinuous across Σ.

Now we find for the components of the Einstein tensor:

$$G^4{}_4 = -\frac{1}{2}g^{IJ}R^S{}_{ISJ}, \qquad G_{4I} = R^S{}_{4SI},$$

$$G_{IJ} = R^S{}_{ISJ} + R^4{}_{I4J} - \frac{1}{2}g_{IJ}g^{KL}R^S{}_{KSL} - g_{IJ}g^{KL}R^4{}_{K4L}. \tag{12.123}$$

This shows that $G^4{}_4$ is continuous across Σ, while G_{4I} and G_{IJ} can be discontinuous.

Note what this means for matching a perfect fluid solution to vacuum across a spacelike hypersurface such that the velocity has no component in the x^4 direction. Then $G^4{}_4 = \kappa p$, and since $G^4{}_4 = 0$ in vacuum and has to be continuous across Σ, this means that $p = 0$ must hold on Σ, which agrees with the expectation based on physics.

12.18 The weak-field approximation to general relativity

Although considerable progress has been made in the search for exact solutions of Einstein's equations (Stephani *et al.*, 2003), real physical or astrophysical situations are often too complicated to be captured by an exact solution. In those cases, one must resort to approximate calculations, for which several methods have been developed. We shall introduce one such method here; it is useful also for interpreting exact solutions because it allows one to recognise the physical meaning of parameters in the metric. The presentation is based on Stephani (1990), where more details can be found.

Suppose that the metric includes the Minkowski limit $\eta_{\mu\nu}$, write it as

$$g_{\mu\nu} = \eta_{\mu\nu} + h_{\mu\nu}, \tag{12.124}$$

and choose coordinates in which the $\eta_{\mu\nu}$ has the Lorentzian form $\eta_{\mu\nu} = \text{diag}(+1, -1, -1, -1)$. Then assume that $h_{\mu\nu}$ is a small correction to $\eta_{\mu\nu}$, thus $|h_{\mu\nu}| \ll 1$ for all indices and $|h_{\mu\nu}h_{\rho\sigma}| \ll |h_{\alpha\beta}|$ for all (μ, ν, ρ, σ) and all $h_{\alpha\beta} \neq 0$. Assume that also the first and second derivatives of $h_{\mu\nu}$ are small, i.e. $|h_{\mu\nu,\alpha}| \ll 1$, $|h_{\mu\nu,\alpha\beta}| \ll 1$. Under these assumptions, all terms in the Christoffel symbols and in the curvature tensor that are nonlinear in $h_{\mu\nu}$ can be neglected.

This set of assumptions is called the **weak-field approximation** to the Einstein theory, and the resulting scheme is sometimes called the 'linearised theory of gravitation'. Within this scheme, it is assumed that the energy-momentum tensor $T_{\mu\nu}$ has its special-relativistic form and obeys the equations of motion in flat space

$$T^{\mu\nu}{}_{,\nu} = 0. \tag{12.125}$$

Since the metric $h_{\mu\nu}$ does not appear in (12.125) (there are no covariant derivatives in it), the gravitational field has no influence on the motion of the sources in this approximation. This is a strong qualitative difference between the exact and the approximate theory. If, however, this assumption of no **back-reaction** holds to a satisfactory degree of precision in a given situation, then the gravitational field generated by a given source can be calculated (approximately) in a much simpler way than in the full Einstein theory.

We adopt the convention that all indices are raised and lowered by means of the flat metric $\eta_{\mu\nu}$. Since the determinant of $g_{\mu\nu}$ has a form similar to (12.124) ($g = -1 +$ (small perturbation)), it follows that also $g^{\mu\nu}$ has the property $g^{\mu\nu} = \eta^{\mu\nu} + f^{\mu\nu}$, $|f^{\mu\nu}| \ll 1$, and that

$$g^{\mu\nu} = \eta^{\mu\nu} - h^{\mu\nu} \equiv \eta^{\mu\nu} - \eta^{\mu\rho}\eta^{\nu\sigma}h_{\rho\sigma},$$

$$\left\{ \begin{matrix} \alpha \\ \beta\gamma \end{matrix} \right\} = \frac{1}{2}\eta^{\alpha\rho}\left(h_{\beta\rho,\gamma} + h_{\gamma\rho,\beta} - h_{\beta\gamma,\rho}\right), \tag{12.126}$$

$$R^{\alpha}{}_{\beta\gamma\delta} = \frac{1}{2}\eta^{\alpha\rho}\left(h_{\rho\delta,\beta\gamma} - h_{\rho\gamma,\beta\delta} + h_{\beta\gamma,\rho\delta} - h_{\beta\delta,\rho\gamma}\right).$$

The linearised Riemann tensor is seen to have all the required symmetries in the indices. The linearised Einstein equations are

$$-\frac{1}{2}\left(h_{\alpha\beta}{}^{,\rho}{}_{,\rho}+h^{\rho}{}_{\rho,\alpha\beta}-h_{\alpha}{}^{\rho}{}_{,\beta\rho}-h_{\beta}{}^{\rho}{}_{,\alpha\rho}\right)+\frac{1}{2}\eta_{\alpha\beta}\left(h^{\rho}{}_{\rho}{}^{,\sigma}{}_{,\sigma}-h^{\rho\sigma}{}_{,\rho\sigma}\right)=\kappa T_{\alpha\beta}.$$

(12.127)

The field equations become simpler when we transform the functions in them by analogy with electrodynamics. We define $\widetilde{h}_{\alpha\beta}$ by

$$\widetilde{h}_{\alpha\beta}\overset{\text{def}}{=}h_{\alpha\beta}-\frac{1}{2}\eta_{\alpha\beta}h^{\rho}{}_{\rho}$$

$$\Longrightarrow\widetilde{h}^{\rho}{}_{\rho}=-h^{\rho}{}_{\rho},\qquad h_{\alpha\beta}=\widetilde{h}_{\alpha\beta}-\frac{1}{2}\eta_{\alpha\beta}\widetilde{h}^{\rho}{}_{\rho},$$

(12.128)

$$g^{\alpha\beta}=\eta^{\alpha\beta}-\eta^{\alpha\rho}\eta^{\beta\sigma}\widetilde{h}_{\rho\sigma}+\frac{1}{2}\eta^{\alpha\beta}\widetilde{h}^{\rho}{}_{\rho},$$

(12.129)

and we obtain in (12.127)

$$\widetilde{h}_{\alpha\beta}{}^{,\rho}{}_{,\rho}+\widetilde{h}_{\alpha}{}^{\rho}{}_{,\beta\rho}+\widetilde{h}_{\beta}{}^{\rho}{}_{,\alpha\rho}-\eta_{\alpha\beta}\widetilde{h}^{\rho\sigma}{}_{,\rho\sigma}=2\kappa T_{\alpha\beta}.$$

(12.130)

We have not yet specified the coordinates. We carry out the transformation

$$\widetilde{x}^{\alpha}=x^{\alpha}+b^{\alpha}(x),$$

(12.131)

where, to preserve the property $|h_{\alpha\beta}|\ll 1$, the functions b^{α} are assumed to be of the same order as $h_{\alpha\beta}$. This means that $|b^{\alpha}b^{\beta}|\ll|b^{\gamma}|$, $|b^{\alpha}h_{\beta\gamma}|\ll|h_{\delta\epsilon}|$, $|b^{\alpha}b^{\beta}|\ll|h^{\gamma\delta}|$, and similarly for the derivatives. We will later impose a condition on b^{α} that will further simplify the equations. In deriving the formulae below, it must be remembered that it is the full metric $g_{\alpha\beta}$ which undergoes a linearised tensor transformation, not $\widetilde{h}_{\alpha\beta}$ itself. The transformation law for $\widetilde{h}_{\alpha\beta}$ follows by linearising the equations $g_{\alpha'\beta'}=x^{\rho}{}_{,\alpha'}x^{\sigma}{}_{,\beta'}g_{\rho\sigma}$. The flat background metric preserves its Lorentzian form in consequence of the assumed approximations.

Assuming that after the transformation $\widetilde{\widetilde{g}}^{\alpha\beta}$ is represented in terms of $\widetilde{\widetilde{h}}^{\alpha\beta}$ still by the rule (12.129), we obtain

$$\widetilde{h}^{\mu}{}_{\mu}=\widetilde{\widetilde{h}}^{\mu}{}_{\mu}-2b^{\mu}{}_{,\mu},\qquad\widetilde{h}^{\alpha\beta}=\widetilde{\widetilde{h}}^{\alpha\beta}+b^{\alpha,\beta}+b^{\beta,\alpha}-\eta^{\alpha\beta}b^{\mu}{}_{,\mu}.$$

(12.132)

Calculating $\partial/\partial x^{\alpha}$ of the second equation in (12.132) we find

$$\widetilde{h}^{\alpha\beta}{}_{,\alpha}=\widetilde{\widetilde{h}}^{\alpha\beta}{}_{,\alpha}+b^{\beta,\mu}{}_{,\mu}.$$

(12.133)

This simplifies if we impose the following condition on b^{β}:

$$b^{\beta,\mu}{}_{,\mu}=\widetilde{h}^{\alpha\beta}{}_{,\alpha},$$

(12.134)

whereupon

$$\widetilde{\widetilde{h}}^{\alpha\beta}{}_{,\alpha}=0.$$

(12.135)

However, since (12.129) held with $\tilde{h}^{\alpha\beta}$ replaced by $\widetilde{\tilde{h}}^{\alpha\beta}$, so does (12.130), and consequently (12.135) implies that

$$-2\kappa T_{\alpha\beta} = \eta^{\mu\nu}\widetilde{\tilde{h}}_{\alpha\beta,\mu\nu} \equiv \Box\,\widetilde{\tilde{h}}_{\alpha\beta}. \tag{12.136}$$

This is the inhomogeneous wave equation, for which a general formal solution is known.

For brevity, we omitted several instructive analogies of the linearised Einstein equations to Maxwell's equations in electrodynamics. In fact, the whole reasoning leading to (12.136) was inspired by those analogies; they are pointed out in more detail in Stephani (1990).

The interesting property of (12.136) is that for each pair of indices $(\alpha\beta)$ the equation is independent of the others. Thus, a given component of the metric is influenced by just one component of the energy-momentum tensor. This simplifies the interpretation of parameters in the metric when the weak-field approximation can be applied.

We shall now decompose the energy-momentum tensor into its physical components. We will be omitting the second tilde over symbols from now on. A formal solution of (12.136) in terms of retarded potentials is the same as in electrodynamics,

$$\tilde{h}_{\alpha\beta}(t, \mathbf{r}) = -\frac{\kappa}{2\pi} \int_V \frac{1}{|\mathbf{r}-\mathbf{r}'|} T_{\alpha\beta}(\mathbf{r}', t-|\mathbf{r}-\mathbf{r}'|/c)\mathrm{d}_3 r', \tag{12.137}$$

where $\mathbf{r}$ are the spatial coordinates of the observer's location, $\mathbf{r}'$ are the spatial coordinates of a field point and the volume of integration is the set on which $T_{\alpha\beta} \neq 0$.

Assume now that the volume V is finite and that the observer, located at $\mathbf{r}$, is far from V, thus $|\mathbf{r}| \gg |\mathbf{r}'|$ for all points inside V. Then $|\mathbf{r}-\mathbf{r}'|$ and $1/|\mathbf{r}-\mathbf{r}'|$ can be expanded in power series with respect to $\mathbf{r}'$:

$$|\mathbf{r}-\mathbf{r}'| = \sqrt{\mathbf{r}^2 - 2\mathbf{r}\cdot\mathbf{r}' + \mathbf{r}'^2}$$

$$= r - \frac{x^I x'^I}{r} - \frac{x^I x^J}{2r^3}\left(x'^I x'^J - r'^2\delta^{IJ}\right) + \cdots, \tag{12.138}$$

$$1/|\mathbf{r}-\mathbf{r}'| = \frac{1}{r} + \frac{x^I x'^I}{r^3} + \frac{x^I x^J}{2r^5}\left(3x'^I x'^J - r'^2\delta^{IJ}\right) + \cdots, \tag{12.139}$$

where $I, J, K = 1, 2, 3$. The higher-order terms will not appear explicitly.

Then, expand $T_{\alpha\beta}$ with respect to $(t - |\mathbf{r}-\mathbf{r}'|/c)$ around $(t - |\mathbf{r}|/c)$. It is justifiable to truncate the expansion because $t - |\mathbf{r}-\mathbf{r}'|/c - (t - |\mathbf{r}|/c) \equiv (|\mathbf{r}| - |\mathbf{r}-\mathbf{r}'|)/c$ is small compared with $|\mathbf{r}|/c$, see Fig. 12.2. Thus

$$T_{\alpha\beta}(\mathbf{r}', t - |\mathbf{r}-\mathbf{r}'|/c)$$

$$= T_{\alpha\beta}(\mathbf{r}', t - |\mathbf{r}|/c) + \frac{1}{c}\dot{T}_{\alpha\beta}(\mathbf{r}', t - |\mathbf{r}|/c)(r - |\mathbf{r}-\mathbf{r}'|)$$

$$+ \frac{1}{2c^2}\ddot{T}_{\alpha\beta}(\mathbf{r}', t - |\mathbf{r}|/c)(r - |\mathbf{r}-\mathbf{r}'|)^2 + \cdots. \tag{12.140}$$

Now we substitute (12.138)–(12.140) in (12.137) and truncate the series at terms of order r^{-5}, remembering that $|\mathbf{r}'| \ll |\mathbf{r}|$. We also assume that there are no fast motions inside

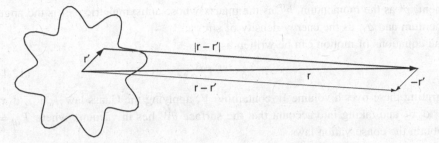

Fig. 12.2. Vectors and distances involved in calculating the linearised metric of a finite body in the weak-field approximation.

the body, so $(dT_{\alpha\beta}/dt)/c$ is of order no larger than $1/r$, so the truncation will be done at the *combined* order r^{-3} (i.e. at r^{-3} at $T_{\alpha\beta}$, r^{-2} at $\dot{T}_{\alpha\beta}$ and r^{-1} at $\ddot{T}_{\alpha\beta}$). The result is

$$\frac{1}{|\mathbf{r}-\mathbf{r}'|}T_{\alpha\beta}(\mathbf{r}', t-|\mathbf{r}-\mathbf{r}'|/c)$$

$$= T_{\alpha\beta}\left(\frac{1}{r}+\frac{x^I x'^I}{r^3}+\frac{x^I x^J}{2r^5}\left(3x'^I x'^J - r'^2\delta^{IJ}\right)\right)$$

$$+\frac{1}{c}\dot{T}_{\alpha\beta}\left(\frac{x^I x'^I}{r^2}+\frac{x^I x^J}{2r^4}\left(3x'^I x'^J - r'^2\delta^{IJ}\right)\right)+\frac{1}{2c^2}\ddot{T}_{\alpha\beta}\frac{x^I x^J}{r^3}x'^I x'^J \qquad (12.141)$$

(the arguments of $T_{\alpha\beta}$ on the right-hand side are $\mathbf{r}'$ and $(t-r/c)$).

Now, in calculating the integral in (12.137) it must be remembered that the coordinate x^0 is ct and so, if there are no fast motions in the source (which we assume), then T^{IJ} is smaller by one order in v/c than T^{0J}, and this in turn is smaller by one order than T^{00}. We are working within the first-order post-Newtonian approximation (i.e. in the linearised Einstein theory), and the lowest non-special-relativistic correction to T^{IJ} is of the order $v/c \approx r'/r$. Accordingly, we have to calculate T^{0I} and T^{00} to orders $(v/c)^2 \approx (r'/r)^2$ and $(v/c)^3 \approx (r'/r)^3$, respectively. The following quantities will appear in the calculations (the arguments of $T_{\alpha\beta}$ are everywhere the same as in (12.141)):

$$M \stackrel{\text{def}}{=} \int_V T_{00}\, d_3x', \qquad d^I \stackrel{\text{def}}{=} \int_V T_{00}x'^I\, d_3x',$$

$$d^{IJ} \stackrel{\text{def}}{=} \int_V T_{00}x'^I x'^J\, d_3x', \qquad p^I \stackrel{\text{def}}{=} \int_V T^{0I}\, d_3x',$$

$$b^{IJ} \stackrel{\text{def}}{=} -\int_V T^{0I}x'^J\, d_3x', \qquad a_{IJ} \stackrel{\text{def}}{=} \int_V T_{IJ}\, d_3x' \qquad (12.142)$$

(all integrals are over the volume V of the body). From the physical interpretation of the components of $T_{\alpha\beta}$ we recognise (up to the choice of units, i.e. up to a constant factor) M as the mass of the source, d^I as the mass dipole moment, d^{IJ} as the mass quadrupole moment, p^I as the momentum, b^{IJ} as the matrix whose antisymmetric part is the angular momentum and a_{IJ} as the energy-density of stress.

The equations of motion can be written as

$$T^{\alpha 0}{}_{,0} = -T^{\alpha S}{}_{,S}.$$ (12.143)

Integrating these over a volume W containing V, applying the Gauss law $\int_W v^I{}_{,I}\, d_3 x' = \int_{\partial W} v^I\, d_2 x'^I$ and taking into account that the surface ∂W lies in vacuum, where $T_{\alpha\beta} = 0$, we obtain the conservation laws

$$M = \text{constant} \qquad \text{for } \alpha = 0,$$
$$p^I = \text{constant} \qquad \text{for } \alpha = I.$$ (12.144)

Applying the same rule to the integral defining d^I we obtain

$$\frac{1}{c}\frac{dd^I}{dt} = \int_W T^{00}{}_{,0}\, x'^I\, d_3 x' = -\int_W T^{0S}{}_{,S}\, x'^I\, d_3 x'$$
$$= -\int_W \left(T^{0S} x'^I\right)_{,S}\, d_3 x' + \int_W T^{0I}\, d_3 x'$$
$$= -\int_{\partial W} T^{0S} x'^I\, d_2 x'^S + p^I \equiv p^I.$$ (12.145)

The quantity $B^{IJ} \overset{\text{def}}{=} T^{0I} x'^J - T^{0J} x'^I$ is the density of angular momentum. Applying again the same rule to it we obtain

$$B^{IJ} \overset{\text{def}}{=} T^{0I} x'^J - T^{0J} x'^I = \text{constant}.$$ (12.146)

The momentum p^I can be set to zero by transforming the coordinates to the rest frame of the body. This is a Lorentz transformation that does not change the property (12.124) and will change only the numerical values of the other quantities in (12.142), keeping constant those that were constant. With $p^I = 0$, d^I becomes constant and can be set to zero by moving the origin of coordinates to the centre of mass of the body.

The equation $T^{00}{}_{,0} + T^{0S}{}_{,S} = 0$ implies that

$$\left(T^{00} x'^I x'^J\right)_{,0} + \left(T^{0S} x'^I x'^J\right)_{,S} = T^{0I} x'^J + T^{0J} x'^I.$$ (12.147)

Differentiating this by x^0 and applying (12.143) again, we obtain

$$\left(T^{00} x'^I x'^J\right)_{,00} = \left(T^{RS} x'^I x'^J\right)_{,RS} - 2\left(T^{IS} x'^J + T^{JS} x'^I\right)_{,S} + 2T^{IJ}.$$ (12.148)

By integrating (12.147) and (12.148) over the volume W and applying the Gauss law to eliminate the divergence terms we obtain, respectively,

$$-\frac{1}{c}\dot{d}^{IJ} = b^{IJ} + b^{JI}, \qquad \frac{1}{2c^2}\ddot{d}^{IJ} = a^{IJ}.$$ (12.149)

Thus, b^{IJ} are determined by the angular momentum and the time-derivative of the quadrupole moment, while the stresses are determined by the second time-derivative of the quadrupole moment. Thus, finally, the quantities M, B^{IJ} and d^{IJ} fully characterise the body, to the desired order of approximation.

Substituting (12.141) in (12.137), employing the definitions (12.142) and the relations (12.149), we obtain for the metric

$$
\begin{aligned}
-\frac{2\pi}{\kappa}\tilde{h}_{00} &= \frac{M}{r} + \frac{x^I x^J}{2r^5}\left(3d^{IJ} - d^{SS}\delta^{IJ}\right) \\
&\quad + \frac{x^I x^J}{2cr^4}\left(3\dot{d}^{IJ} - \dot{d}^{SS}\delta^{IJ}\right) + \frac{x^I x^J}{2c^2 r^3}\ddot{d}^{IJ}, \\
-\frac{2\pi}{\kappa}\tilde{h}_{0I} &= \frac{B_{IJ}x^J}{2r^3} - \frac{\dot{d}_{IJ}x^J}{2cr^3} - \frac{\ddot{d}_{IJ}x^J}{2c^2 r^2}, \\
-\frac{2\pi}{\kappa}\tilde{h}_{IJ} &= \frac{\ddot{d}_{IJ}}{2c^2 r}.
\end{aligned}
\tag{12.150}
$$

Finally, we wish to find the $h_{\alpha\beta}$ defined in (12.124). They are calculated from (12.128), and then simplified by a coordinate transformation of the form (12.131), with b^α given as follows (note that now we do not require these b^α to obey (12.134) and that they are expressed in the coordinates of (12.150), i.e. in the xs before the transformation):

$$
\begin{aligned}
-\frac{2\pi}{\kappa}b_0 &= \frac{x^I x^J}{8r^4}\left(3d_{IJ} - \delta_{IJ}d_{SS}\right) + \frac{x^I x^J}{8cr^3}\left(\dot{d}_{IJ} + \delta_{IJ}\dot{d}_{SS}\right), \\
-\frac{2\pi}{\kappa}b_I &= -\frac{3d_{IJ}x^J}{4r^3} - \frac{\dot{d}_{IJ}x^J}{2cr^2} + \frac{3d_{JK}x^J x^K x^I}{4r^5} \\
&\quad + \frac{\left(\dot{d}_{KL} + \delta_{KL}\dot{d}_{SS}\right)x^K x^L x^I}{8cr^4}.
\end{aligned}
\tag{12.151}
$$

These were designed so as to cancel the time-derivatives in h_{00} and h_{0I} to the desired order. The transformed $h_{\alpha\beta}$ are

$$
h'_{\alpha\beta} = h_{\alpha\beta} - b_{\alpha,\beta} - b_{\beta,\alpha},
\tag{12.152}
$$

and the final result for the linearised metric, up to the desired order (combined order in v/c and $r'/r \approx v/c$), is[1]

$$
g_{00} = 1 - \frac{2m}{r} - \frac{x^I x^J}{2r^5}\left(3D^{IJ} - D^{SS}\delta^{IJ}\right) + O(1/r^4),
\tag{12.153}
$$

$$
g_{0I} = -\frac{2\epsilon_{IJK}x^J P^K}{r^3} + O(1/r^3),
\tag{12.154}
$$

$$
g_{IJ} = \eta_{IJ}\left(1 + \frac{2m}{r} + O(1/r^2)\right),
\tag{12.155}
$$

---

[1] Stephani's formula for g_{IJ} in Stephani (1990) seems to be incorrect.

where ϵ_{IJK} is the Levi-Civita symbol and the new constants are defined in terms of the old ones as follows:

$$m \overset{\text{def}}{=} \frac{\kappa M}{8\pi}, \qquad P^I \overset{\text{def}}{=} \frac{\kappa}{8\pi} \epsilon^{IJK} B^{JK}, \qquad D^{IJ} \overset{\text{def}}{=} \frac{\kappa}{8\pi} d^{IJ}. \tag{12.156}$$

For m and D^{IJ} these are just new units, while P^I is the **vector of angular momentum**.

The presence of P^I in the metric marks an important difference between general relativity and the Newtonian theory of gravitation. Let us take the Minkowski metric in the Lorentzian coordinates, $\mathrm{d}s^2 = -\delta_{IJ}\,\mathrm{d}x^I\,\mathrm{d}x^J + \mathrm{d}t^2$, and let us transform it to the rotating coordinates, in which $x^I = x'^I + \epsilon^{IKL}\omega^K x'^L t$. Assuming that the angular velocity of rotation ω^K is small and neglecting the terms proportional to ω^2, we see[1] that g_{0I} is of the same form as in (12.154) (except that the contribution from rotation in (12.154) decreases quickly with distance, because of the r^3 in the denominator). The coordinates of (12.153)–(12.155) become inertial at infinity ($r \to \infty$), which means that the local inertial coordinate system is rotating with respect to the inertial system of a distant observer. Thus, a rotating body in relativity is **dragging the inertial frames** with it; this phenomenon is also known as the **Lense–Thirring (1918) effect**. An experiment to measure this effect in the gravitational field of the rotating Earth, known under the working name 'Gravity Probe B', was launched recently (Gravity Probe B mission update, 2004). In contrast, in the Newtonian gravitation there is no direct influence of rotation of the source on the exterior field. There is only indirect influence because the centrifugal force caused by rotation flattens the source, and the changed distribution of mass causes changes in the exterior field (by producing a quadrupole moment).

12.19 Exercises

1. Verify that the Newtonian limit of Eqs. (12.78) are the Euler equations of motion of a perfect fluid $\rho\,\mathrm{d}\mathbf{v}/\mathrm{d}t = -\mathrm{grad}\,p$.
   Hint. Use the asymptotically Cartesian coordinates in which the metric has the form (12.46)–(12.47), and observe that in these coordinates the 4-velocity u^α has the form $u^0 = 1$, $u^I = v^I/c$, $I = 1, 2, 3$.

2. Verify that Eq. (12.79) becomes the continuity equation $\partial\rho/\partial t + \mathrm{div}(\rho\mathbf{v}) = 0$ in the Newtonian limit.

3. Verify that Eqs. (12.82)–(12.83) do indeed represent a solution of the modified Einstein equations (12.80). The energy-momentum tensor is that of a perfect fluid, the velocity field is $u^\alpha \overset{*}{=} \delta^\alpha{}_0/c$. Do the calculation both by the tensor method and by using the Cartan forms. Find the pressure.

---

[1] Strictly speaking, this approximation is incorrect because the square term in ω is multiplied by time, so the error of approximation is cumulative.

13

The Maxwell and Einstein–Maxwell equations and the Kaluza–Klein theory

13.1 The Lorentz-covariant description of electromagnetic field

The separation of the electromagnetic field into electric and magnetic fields is not covariant with the Lorentz transformation; it depends on the motion of the observer. In relativity electromagnetic field is described by the antisymmetric **tensor of electromagnetic field** $F_{\mu\nu}$. In any fixed coordinate system, the electric field is

$$F^{0I} = E^I, \qquad I = 1, 2, 3 \tag{13.1}$$

and the magnetic field is

$$H^I = \frac{1}{2}\epsilon^{IJK} F_{JK}, \qquad I, J, K = 1, 2, 3. \tag{13.2}$$

The formula inverse to (13.2) is $F_{IJ} = \epsilon_{IJK} H^K$. If E^I and H^I are transformed by the Lorentz transformation, then $F_{\mu\nu}$ transforms like a tensor. The change in $F_{\mu\nu}$ implied by a coordinate transformation is interpreted as the relation between the electric and magnetic fields measured by two observers.

13.2 The covariant form of the Maxwell equations

In special relativity, the Maxwell equations in vacuum are written as

$$F^{\mu\nu}{}_{,\nu} = \frac{4\pi}{c} j^\mu, \tag{13.3}$$

$$F_{[\mu\nu,\lambda]} = 0, \tag{13.4}$$

where j^μ is the 4-vector of current. Equation (13.3) is equivalent to the set $\{\text{div } \mathbf{E} = 4\pi\sigma, \ \text{rot } \mathbf{H} = (4\pi/c)\mathbf{j} + (1/c)\partial \mathbf{E}/\partial t\}$ and Eq. (13.4) to the set $\{\text{div } \mathbf{H} = 0, \ \text{rot } \mathbf{E} + (1/c)\partial \mathbf{H}/\partial t = 0\}$, where σ is the density of electric charge and $\mathbf{j}$ is the ordinary current vector. The covariance of the Maxwell equations with the Lorentz transformation is automatically guaranteed by $F_{\mu\nu}$ being a tensor. We also have

$$j^0 = c\sigma, \qquad \{j^I\} = \mathbf{j}. \tag{13.5}$$

In general relativity, the partial derivatives in (13.3)–(13.4) are replaced with covariant derivatives, thus (13.3) becomes

$$F^{\mu\nu};_{\nu} = \frac{4\pi}{c} j^{\mu}. \tag{13.6}$$

Equation (13.4) need not be changed because the Christoffel symbols cancel out in the combination $F_{[\mu\nu;\lambda]}$.

One can define the differential form $F = F_{\mu\nu} \, dx^{\mu} \wedge dx^{\nu}$, and then (13.4) becomes $dF = 0$. This is the condition of existence of a 1-form A such that $F = dA$, where $A = A_{\mu} \, dx^{\mu}$. The consequence (and solution) of (13.4) is thus the existence of the 4-potential A_{μ}. The continuity equation $j^{\mu};_{\mu} = 0$ is an identical consequence of (13.6).

13.3 The energy-momentum tensor of an electromagnetic field

The electromagnetic field $(\mathbf{E}, \mathbf{H})$ has the energy-density

$$u = \frac{1}{8\pi} \left(\mathbf{E}^2 + \mathbf{H}^2 \right), \tag{13.7}$$

its energy stream is

$$\mathbf{s} = \frac{1}{4\pi} \left(\mathbf{E} \times \mathbf{H} \right) \tag{13.8}$$

and its stress tensor is

$$T_{IJ} = \frac{1}{4\pi} \left(\mathbf{E}_I \mathbf{E}_J + \mathbf{H}_I \mathbf{H}_J - \frac{1}{2} \left(\mathbf{E}^2 + \mathbf{H}^2 \right) \delta_{IJ} \right). \tag{13.9}$$

Just as we have done with the perfect fluid in Section 12.12, we can now compose these quantities into a 4-dimensional tensor of energy-momentum of electromagnetic field:

$$T^{\alpha\beta} = \frac{1}{4\pi} \left(F^{\alpha\mu} F_{\mu}{}^{\beta} + \frac{1}{4} g^{\alpha\beta} F_{\mu\nu} F^{\mu\nu} \right). \tag{13.10}$$

It has the property $T^{\rho}{}_{\rho} \equiv 0$. This can be written in the equivalent form

$$T^{\alpha\beta} = \frac{1}{8\pi} \left(F^{\alpha\mu} F_{\mu}{}^{\beta} + {}^{*}F^{\alpha\mu*} F_{\mu}{}^{\beta} \right), \tag{13.11}$$

where

$${}^{*}F_{\mu\nu} \overset{\text{def}}{=} \frac{1}{2} \sqrt{-g} \, \epsilon_{\mu\nu\rho\sigma} F^{\rho\sigma} \tag{13.12}$$

is called the **dual** tensor to $F^{\mu\nu}$. In this form, it is easy to verify that the energy-momentum tensor does not change when the electromagnetic field is transformed by the following operation:

$$\widetilde{F}_{\mu\nu} = F_{\mu\nu} \cos \delta + {}^{*}F_{\mu\nu} \sin \delta, \tag{13.13}$$

where δ is an arbitrary constant. This operation is called the **duality rotation**. Equation (13.13) implies that

$$^*\widetilde{F}_{\mu\nu} = -F_{\mu\nu}\sin\delta + {}^*F_{\mu\nu}\cos\delta. \tag{13.14}$$

Consequently, although a given electromagnetic tensor uniquely defines the electromagnetic energy-momentum tensor (13.10), the converse is not true. Given $T^{\alpha\beta}$, the $F^{\mu\nu}$ is defined only up to duality rotations.

From (13.12) it is seen that the duality operation interchanges the electric and magnetic fields: the electric field of $F_{\mu\nu}$ becomes the magnetic field of $^*F_{\mu\nu}$ and vice versa. The new electric and magnetic fields defined by $\widetilde{F}_{\mu\nu}$ are linear combinations of the old fields defined by $F_{\mu\nu}$.

The second set of Maxwell's equations, (13.4), is equivalent to

$$^*F^{\mu\nu}{}_{;\nu} = 0. \tag{13.15}$$

This means that no magnetic monopoles (and currents) exist. The 'duality-rotated' $^*\widetilde{F}_{\mu\nu}$ will not always have this property. But if $^*\widetilde{F}^{\mu\nu}{}_{;\nu} = m^\mu \neq 0$, then the inverse duality rotation (Eq. (13.13) with δ replaced by $-\delta$) may sometimes be used to define such an $F^{\mu\nu}$ for which (13.15) holds. The condition for this to be possible is that the 'magnetic current' $^*\widetilde{F}^{\mu\nu}{}_{;\nu}$ and the electric current $F^{\mu\nu}{}_{;\nu}$ are linearly dependent.

13.4 The Einstein–Maxwell equations

The energy-momentum tensor of electromagnetic field is a source of a gravitational field on equal footing with the energy-momentum tensor of matter. In the presence of electromagnetic field, the field equations of gravitation are

$$R_{\alpha\beta} - \frac{1}{2}g_{\alpha\beta}R = \kappa \overset{\text{mat}}{T}_{\alpha\beta} + \overset{\text{em}}{T}_{\alpha\beta}, \tag{13.16}$$

where $\overset{\text{mat}}{T}_{\alpha\beta}$ and $\overset{\text{em}}{T}_{\alpha\beta}$ are the energy-momentum tensors of matter and of electromagnetic field, (13.10), respectively. The electromagnetic field must obey the Maxwell equations (13.4) and (13.6). The set $\{(13.4), (13.6), (13.10), (13.16)\}$ is called the **Einstein–Maxwell equations**.

The contracted Bianchi identities $\left(R^{\alpha\beta} - \frac{1}{2}g^{\alpha\beta}R\right)_{;\beta} = 0$ imply the equations of motion $T^{\alpha\beta}{}_{;\beta} = 0$ for the *whole* energy-momentum tensor $T_{\alpha\beta}$. In the absence of matter $(\overset{\text{mat}}{T}_{\alpha\beta} = 0)$ and currents $(j^\mu = 0)$, the Maxwell equations $\{(13.4), (13.6)\}$ do imply $\overset{\text{em}}{T}^{\alpha\beta}{}_{;\beta} = 0$. But the converse is not true: the Maxwell equations do not follow from the equations of motion and have to be postulated independently.

13.5 * The variational principle for the Einstein–Maxwell equations

The Maxwell equations follow from the variational principle $\delta \overset{\text{em}}{W} = 0$, where

$$\overset{\text{em}}{W} \overset{\text{def}}{=} \int_{\mathcal{V}} \mathrm{d}_4 x \sqrt{-g} \left(\frac{1}{8\pi} F_{\alpha\beta} F^{\alpha\beta} + \frac{2}{c} j_\alpha A^\alpha \right), \qquad F_{\alpha\beta} \overset{\text{def}}{=} A_{\alpha,\beta} - A_{\beta,\alpha}, \qquad (13.17)$$

i.e. the sourceless Maxwell equations (13.4) are pre-assumed and built into the Lagrangian. The independent variables are the components of the 4-potential A_α that obey the boundary condition $A_\alpha|_{\partial\mathcal{V}} = 0$.

Then, the full set of Einstein–Maxwell equations follows from the variational principle with the action integral

$$W_g + \overset{\text{em}}{W} + \overset{\text{em}}{W} = \int_{\mathcal{V}} \mathrm{d}_4 x \sqrt{-g} \left(R + \frac{1}{8\pi} F_{\alpha\beta} F^{\alpha\beta} + \frac{2}{c} j_\alpha A^\alpha \right) + \overset{\text{mat}}{W}. \qquad (13.18)$$

The independent variables are $g_{\alpha\beta}$, A_α and matter variables. In vacuum, $\overset{\text{mat}}{W} = 0$ and $j_\alpha = 0$.

13.6 * The Kaluza–Klein theory

In the 1920s, the only interactions known were electromagnetism and gravitation. Finding a geometric theory that would unite them was thus an attractive goal. The first such attempt was undertaken by Kaluza (1921). Several improvements on this theory were made in later years, but complete success has still not been achieved, as will be seen below. The other basic reference is Klein (1926). The exposition given in this section is based on Appelquist and Chodos (1983).

The basic idea of Kaluza was the assumption that our world has five dimensions, that the electromagnetic 4-potential forms the components of the metric tensor along the fifth dimension and that the Einstein–Maxwell equations follow from vacuum Einstein equations in five dimensions.

In the formulae below, the capital Latin indices $A, B, C, \ldots$ will run through the values 0, 1, 2, 3, 4, and the Greek indices will run through the values 0, 1, 2, 3. We make the following assumptions (Appelquist and Chodos, 1983).

1. The 5-dimensional metric tensor is

$$\|G_{AB}\| = \left(\begin{array}{c|c} g_{\mu\nu} + \phi A_\mu A_\nu & \phi A_\mu \\ \hline \phi A_\nu & \phi \end{array} \right), \qquad (13.19)$$

where ϕ is a scalar field, A_μ is the electromagnetic 4-potential and $g_{\mu\nu}$ are components of the 4-dimensional metric tensor.

2. The vector field $k^\mu = \delta^\mu_4$ tangent to the 5-dimensional manifold along the fifth dimension is a Killing field of the 5-dimensional metric tensor, i.e. all components of G_{AB} are independent of x^4.
3. The fifth dimension is not observable since the 5-space is closed in the x^4-direction, and the circumference along it is very small.

One of the disadvantages of the theory is visible already here: its formulation is not covariant with respect to 5-dimensional coordinate transformations. The metric (13.19) has the determinant

$$G = \det \|G_{AB}\| = \phi g, \tag{13.20}$$

where $g = \det \|g_{\mu\nu}\|$. Hence

$$\|G^{AB}\| = \left(\begin{array}{c|c} g^{\mu\nu} & -A^\mu \\ \hline -A^\nu & \phi^{-1} + A_\mu A^\mu \end{array} \right). \tag{13.21}$$

For the Christoffel symbols and all subsequent quantities we will indicate the dimension of the space to which they refer by a subscript in parentheses. The 5-dimensional Christoffel symbols are

$$\left\{ \begin{array}{c} \alpha \\ \beta\gamma \end{array} \right\}_{(5)} = \left\{ \begin{array}{c} \alpha \\ \beta\gamma \end{array} \right\}_{(4)} + \frac{1}{2}\phi \left(A_\beta F^\alpha{}_\gamma + A_\gamma F^\alpha{}_\beta \right) - \frac{1}{2}g^{\alpha\rho}\phi_{,\rho}A_\beta A_\gamma,$$

$$\left\{ \begin{array}{c} \alpha \\ \beta 4 \end{array} \right\}_{(5)} = \phi F^\alpha{}_\beta - \frac{1}{2}g^{\alpha\rho}\phi_{,\rho}A_\beta,$$

$$\left\{ \begin{array}{c} \alpha \\ 44 \end{array} \right\}_{(5)} = -\frac{1}{2}g^{\alpha\rho}\phi_{,\rho},$$

$$\left\{ \begin{array}{c} 4 \\ \beta\gamma \end{array} \right\}_{(5)} = -\frac{1}{2}\phi A^\rho A_\beta F_{\rho\gamma} - \frac{1}{2}\phi A^\rho A_\gamma F_{\rho\beta} + \frac{1}{2}\phi_{,\rho}A^\rho A_\beta A_\gamma \tag{13.22}$$

$$+ \frac{1}{2\phi}\left(\phi_{,\gamma}A_\beta + \phi_{,\beta}A_\gamma\right) + \frac{1}{2}\left(A_{\beta;\gamma} + A_{\gamma;\beta}\right),$$

$$\left\{ \begin{array}{c} 4 \\ \beta 4 \end{array} \right\}_{(5)} = -\frac{1}{2}\phi A^\rho F_{\rho\beta} + \frac{1}{2}A^\rho \phi_{,\rho}A_\beta + \frac{1}{2\phi}\phi_{,\beta},$$

$$\left\{ \begin{array}{c} 4 \\ 44 \end{array} \right\}_{(5)} = \frac{1}{2}A^\rho \phi_{,\rho}.$$

The components of the Ricci tensor are now found to be

$$
R_{(5)\beta\delta} = R_{(4)\beta\delta} - \frac{1}{2}\phi F^{\rho}{}_{\beta}F_{\rho\delta} + \frac{3}{4}A_{\beta}\phi_{,\rho}F^{\rho}{}_{\delta} + \frac{3}{4}A_{\delta}\phi_{,\rho}F^{\rho}{}_{\beta}
$$
$$
+ \frac{1}{2}\phi A_{\beta}F^{\rho}{}_{\delta;\rho} + \frac{1}{2}\phi A_{\delta}F^{\rho}{}_{\beta;\rho} - \frac{1}{2}\phi^{;\rho}{}_{;\rho}A_{\beta}A_{\delta}
$$
$$
+ \frac{1}{4}\phi^2 A_{\beta}A_{\delta}F^{\sigma\rho}F_{\sigma\rho} + \frac{1}{4\phi}A_{\beta}A_{\delta}g^{\sigma\rho}\phi_{,\sigma}\phi_{,\rho}
$$
$$
+ \frac{1}{4\phi^2}\phi_{,\beta}\phi_{,\delta} - \frac{1}{2\phi}\phi_{;\beta\delta}, \tag{13.23}
$$

$$
R_{(5)4\delta} = \frac{1}{2}\phi F^{\rho}{}_{\delta;\rho} - \frac{1}{2}\phi^{;\rho}{}_{;\rho}A_{\delta} + \frac{3}{4}\phi_{,\rho}F^{\rho}{}_{\delta} + \frac{1}{4}\phi^2 A_{\delta}F^{\sigma\rho}F_{\sigma\rho}
$$
$$
+ \frac{1}{4\phi}A_{\delta}g^{\sigma\rho}\phi_{,\sigma}\phi_{,\rho},
$$

$$
R_{(5)44} = -\frac{1}{2}\phi^{;\rho}{}_{;\rho} + \frac{1}{4}\phi^2 F^{\sigma\rho}F_{\sigma\rho} + \frac{1}{4\phi}g^{\sigma\rho}\phi_{,\sigma}\phi_{,\rho},
$$

and the 5-dimensional scalar curvature is

$$
R_{(5)} = R_{(4)} - \phi^{-1}\phi^{;\rho}{}_{;\rho} + \frac{1}{4}\phi F^{\sigma\rho}F_{\sigma\rho} + \frac{1}{2\phi^2}g^{\sigma\rho}\phi_{,\sigma}\phi_{,\rho}. \tag{13.24}
$$

The scalar field ϕ is an element of unknown interpretation; no object has been observed in Nature that would be described by it. This was initially seen as a weakness of the Kaluza–Klein theory, before the scalar field became one of the favourite subjects in particle physics.

With $\phi = -1/(2\pi) = \text{constant}$, the 5-dimensional vacuum Einstein equations $R_{(5)AB} = 0$ reproduce the 4-dimensional Einstein equations with electromagnetic field as a source and the Maxwell equations $F^{\rho}{}_{\delta;\rho} = 0$, but impose the additional condition $F^{\sigma\rho}F_{\sigma\rho} = 0$, which means that $\mathbf{E}^2 = \mathbf{H}^2$. Thus $R_{(5)AB} = 0$ do not reproduce the Einstein–Maxwell equations in their full generality, and this is another weakness of the Kaluza–Klein theory that remains uncorrected to this day.

However, the full Einstein–Maxwell equations may be recovered by the following trick. Take the variational principle for the 5-dimensional vacuum Einstein equations, $\delta W_{(5)} = 0$, where

$$
W_{(5)} = \int_{\mathcal{V}_{(5)}} d_5 x \sqrt{-G}\, R_{(5)}. \tag{13.25}
$$

Since the circumference of the 5-dimensional space along the fifth dimension, ℓ, is finite and G_{AB} does not depend on x^4, one can integrate with respect to x^4 in (13.25). Using (13.20) and (13.24), we obtain

$$W_{(5)} = \ell \int_{\mathcal{V}_{(4)}} d_4 x \sqrt{-g} \, \phi \left(R_{(4)} - \phi^{-1} \phi^{;\rho}{}_{;\rho} + \frac{1}{4} \phi F^{\sigma\rho} F_{\sigma\rho} \right.$$

$$\left. + \frac{1}{2\phi^2} g^{\sigma\rho} \phi_{,\sigma} \, \phi_{,\rho} \right). \qquad (13.26)$$

With $\phi = -1/(2\pi) = \text{constant}$, $j_\alpha = 0$ and $\overset{\text{mat}}{W} = 0$, Eq. (13.26) reduces to (13.15). In this crooked sense, the Kaluza–Klein theory is the unified theory of gravitation and electromagnetism.[1]

The Kaluza–Klein theory became a prototype for other attempts to unify physical theories. The common idea is to take a multidimensional metric and interpret its components in higher dimensions as physical fields in spacetime. Although none of these theories passed experimental verification, the idea still remains popular and inspires new attempts.

13.7 Exercises

1. Deduce (13.10) from (13.7)–(13.9) by the method of Section 12.12.
2. Show that the duality operation (13.13) has the property

$$^*\left(^*F_{\mu\nu}\right) = -F_{\mu\nu}. \qquad (13.27)$$

Derive (13.14) from (13.13). Verify that the duality rotation does not change $T_{\alpha\beta}$.

3. Verify that in the absence of matter ($\overset{\text{mat}}{T}_{\alpha\beta} = 0$) the Maxwell equations {(13.4), (13.6)} and the equations of motion $\overset{\text{em}}{T}{}^{\alpha\beta}{}_{;\beta} = 0$ imply $F^{\mu\nu} j_\mu = 0$. Use the Maxwell equations to prove that if $j^\mu \neq 0$, then this is possible only with rotating currents for which $j_{[\mu,\nu]} \neq 0$.

---

[1] As we showed in Section 12.8, a variational principle with the Lagrangian (13.25) leads to the full Einstein equations (12.21) with $T_{\alpha\beta} = 0$. This result is independent of the dimension of the manifold. The fact that taking the symmetry $g_{AB,4} = 0$ into account and transforming (13.25) to (13.26) leads to a *different* set of field equations is an illustration of the remark made at the end of Section 12.8.

14
Spherically symmetric gravitational fields of isolated objects

In this chapter we will be dealing with spherically symmetric gravitational fields generated by isolated objects, such as stars and black holes.[1] Just as in Newton's theory, the spherically symmetric case is a foundation for the description of many phenomena taking place in real, more complicated objects.

14.1 The curvature coordinates

In Section 8.9 we showed that the metric form of a spherically symmetric spacetime, in a suitably limited class of coordinate systems, has the form (8.51). The coordinates ϑ and φ are spherical coordinates whose values fix a point on a 2-sphere; t and r are arbitrary coordinates whose values are constant on the spheres which are the orbits of the symmetry group. The general form of (8.51) and the formulae for the Killing fields of the O(3) group (8.41) do not change under transformations that are composites of the following two classes:

1. The isometries generated by the Killing fields via Eqs. (8.15)–(8.16). These transformations preserve not just the form (8.51) but also the values of the individual components of the metric – the functions α, β, γ and δ among them. An example of such a transformation is a rotation by the angle α_0 around the axis $\{\vartheta = \pi/2, \varphi = 0\}$ (this axis passes through the centre of the sphere and through a point on the equator). In the (ϑ, φ)-coordinates, such a rotation is given by

$$\vartheta = \arctan\left(\frac{\sqrt{N}}{\cos\alpha_0 \cos\vartheta' - \sin\alpha_0 \sin\vartheta' \sin\varphi'}\right),$$

$$N \overset{\text{def}}{=} \sin^2\vartheta' \cos^2\varphi' + (\cos\alpha_0 \sin\vartheta' \sin\varphi' + \sin\alpha_0 \cos\vartheta')^2,$$

$$\varphi = \arctan\left(\cos\alpha_0 \tan\varphi' + \sin\alpha_0 \frac{\cot\vartheta'}{\cos\varphi'}\right). \tag{14.1}$$

---

[1] Spherically symmetric solutions of Einstein's equations are also applied in cosmology, and these applications will be discussed separately in Chapter 18.

2. The transformations (8.52) that preserve just the general form (8.51), but mix the functions α, β, γ and δ, and are of the form

$$t = F(t', r'), \qquad r = G(t', r'), \tag{14.2}$$

where F and G are arbitrary functions of two variables subject to $\partial(F, G)/\partial(t', r') \neq 0$. After such a transformation, α, β and γ change as follows:

$$\tilde{\alpha} = \alpha F^2{}_{,t'} + 2\beta F{}_{,t'} G{}_{,t'} + \gamma G^2{}_{,t'}, \tag{14.3}$$

$$\tilde{\beta} = \alpha F{}_{,t'} F{}_{,r'} + \beta (F{}_{,t'} G{}_{,r'} + F{}_{,r'} G{}_{,t'}) + \gamma G{}_{,t'} G{}_{,r'}, \tag{14.4}$$

$$\tilde{\gamma} = \alpha F^2{}_{,r'} + 2\beta F{}_{,r'} G{}_{,r'} + \gamma G^2{}_{,r'}, \tag{14.5}$$

while δ transforms like a scalar, $\tilde{\delta}(t', r') = \delta(F, G)$.

Thus, choosing the functions F and G appropriately, one can manipulate α, β and γ, but not δ. Because of the signature $(+ - - -)$ assumed throughout, the determinant of the metric must be negative,

$$\alpha\gamma - \beta^2 < 0. \tag{14.6}$$

Thanks to this inequality, the equation $\tilde{\alpha} = 0$ can always be solved for $F{}_{,t'}$, and the solution is

$$F{}_{,t'} = \frac{1}{\alpha} \left(-\beta \pm \sqrt{\beta^2 - \alpha\gamma} \right) G{}_{,t'}. \tag{14.7}$$

We can thus take an arbitrary G and choose F to obey (14.7). Consequently, we lose no generality when we begin with coordinates in which $\alpha = 0$ already. Beginning with such coordinates, we see from (14.4) that we can carry out such a transformation, after which $\tilde{\beta} = 0$ (but after this transformation, the new $\tilde{\alpha}$ will be nonzero). Achieving $\tilde{\beta} = 0$ while $\alpha = 0$ amounts to solving the equation

$$\beta G{}_{,r'} F{}_{,t'} + \beta G{}_{,t'} F{}_{,r'} + \gamma G{}_{,t'} G{}_{,r'} = 0. \tag{14.8}$$

Thus, G can be chosen arbitrarily, and then (14.8) becomes a linear inhomogeneous partial differential equation for F, for which standard methods of solving exist.

We achieved $\beta = 0$ in two steps, and in both G was arbitrary. Hence, we can now choose G so as to achieve some further simplification.

This is where most textbooks make a mistake – they choose G so that $\tilde{\delta} = -r'^2$. But δ is a scalar under the allowed transformations. Hence, if $\delta = $ constant before the transformation, then $\tilde{\delta} = \delta = $ constant and no condition can be imposed on it. Thus, the further simplification of the form (8.51) must be considered more carefully.

Suppose that we wish to transform the coordinates of (8.51) so that

$$\tilde{\beta} = 0, \qquad \tilde{\delta} = -r'^2. \tag{14.9}$$

When is this possible? From the second of (14.9) we have

$$r' = \sqrt{-\delta(F(t',r'),G(t',r'))}.$$ (14.10)

Differentiating this by t' and by r' we obtain

$$\delta_{,F} F_{,t'} + \delta_{,G} G_{,t'} = 0,$$ (14.11)

$$\delta_{,F} F_{,r'} + \delta_{,G} G_{,r'} = -2\sqrt{-\delta} = -2r'.$$ (14.12)

While solving this set for $(G_{,t'}, G_{,r'})$, two cases must be considered:

- Ia. $\delta_{,G} \neq 0$. Then, substituting for $G_{,t'}$ from (14.11) in (14.3), we obtain

$$\tilde{\alpha} = \frac{F_{,t'}{}^2}{\delta_{,G}{}^2} \left(\alpha \delta_{,G}{}^2 - 2\beta \delta_{,F} \delta_{,G} + \gamma \delta_{,F}{}^2 \right).$$ (14.13)

We want t' to be the time coordinate, so $\tilde{\alpha} > 0$, which means that

$$\alpha \delta_{,G}{}^2 - 2\beta \delta_{,F} \delta_{,G} + \gamma \delta_{,F}{}^2 > 0.$$ (14.14)

This is easily verified to be equivalent to

$$g^{\alpha\beta} \delta_{,\alpha} \delta_{,\beta} < 0.$$ (14.15)

Hence, if $\delta_{,G} \neq 0$, then the condition $\tilde{\delta} = -r'^2$ can be fulfilled only if the gradient of δ is a spacelike vector.

- Ib. $\delta_{,G} = 0$. Then, from (14.11), either $\delta_{,F} = 0$, which means that $\delta = $ constant (this case will be discussed separately below), or $\delta_{,F} \neq 0$ and $F_{,t'} = 0$. In this second case, (14.3) implies that

$$\tilde{\alpha} = \gamma G_{,t'}{}^2.$$ (14.16)

Hence, $\tilde{\alpha} > 0$ only if $\gamma > 0$. With $\delta_{,G} = 0$, $\gamma > 0$ implies (14.15). (Equation (14.16) shows that in this case it is r that is the time coordinate before the transformation.)

Hence, in either case the following lemma applies:

Lemma 14.1 (Case I) *The condition $\tilde{\delta} = -r'^2$ can be fulfilled only if the gradient of δ is a spacelike vector.*

However, there are three other cases in which the set (14.11)–(14.12) cannot be solved at all. We leave it as an exercise for the reader to prove the remaining lemmas.

Lemma 14.2 (Case II) *If the gradient of δ is a timelike vector, then one can choose the coordinates of (8.51) so that*

$$\tilde{\beta} = 0, \qquad \tilde{\delta} = -t'^2.$$ (14.17)

Lemma 14.3 (Case III) *If the gradient of δ is a nontrivial null vector, so that $g^{\alpha\beta} \delta_{,\alpha} \delta_{,\beta} = 0$ but $\delta_{,\alpha} \neq 0$, then one can choose the coordinates of (8.51) so that $\tilde{\delta} = -r'^2$, but this automatically implies $\tilde{\alpha} = 0$. Hence, it is not possible to simultaneously achieve $\tilde{\beta} = 0$ because of Eq. (14.6), but it is possible to have in addition $\tilde{\gamma} = 0$.*

(**Case IV**) When δ is constant, no condition at all can be imposed on it. The spacetime is then a Cartesian product of a sphere of radius $l = \sqrt{-\delta}$ and a 2-dimensional surface with the metric $\alpha(t, r)\mathrm{d}t^2 + 2\beta(t, r)\mathrm{d}t\,\mathrm{d}r + \gamma(t, r)\mathrm{d}r^2$.

The existence of regions of spacetimes with $\delta_{,\alpha}$ being either spacelike or timelike was first noted by Novikov (1962b), who called them *R-regions* and *T-regions*, respectively.

Case II is ignored by most textbooks, even by those that subsequently describe the extension of the Schwarzschild manifold (see Section 14.9). Cases III and IV are ignored almost universally, with the notable exception of Stephani *et al.* (2003). The situation of these cases versus the Einstein equations is this:

For the vacuum equations with zero cosmological constant, the cases $g^{\alpha\beta}\delta_{,\alpha}\delta_{,\beta} < 0$ and $g^{\alpha\beta}\delta_{,\alpha}\delta_{,\beta} > 0$ are complementary to each other; they cover different parts of the same manifold, as we will see in Section 14.9.

However, cases III and IV require separate treatment. Case III has been investigated by Foyster and McIntosh (1972). It leads to a contradictory set of equations in pure vacuum, and the solution of the vacuum Einstein–Maxwell equations with this geometry is given in Stephani *et al.* (2003, Eq. (15.18)). Case IV leads to a contradiction when $T_{\alpha\beta} = 0 = \Lambda$, but a few solutions of Einstein's equations for more general sources can be found in the literature. The vacuum solution with $\Lambda \neq 0$ was found by Nariai (1950), and discussed at some length by Krasiński and Plebański (1980). The solutions of the Einstein–Maxwell equations with this geometry and with the source being the vacuum electrostatic field were found by Bertotti (1959) and Robinson (1959); they are compared with the Nariai solution in Krasiński (1999).

In this chapter, we will deal only with the case $g^{\alpha\beta}\delta_{,\alpha}\delta_{,\beta} < 0$. We then choose the coordinates so as to achieve (14.9). In order to make subsequent calculations easier, we denote $\widetilde{\alpha} = \mathrm{e}^{2\nu(t,r)}$, $\widetilde{\gamma} = -\mathrm{e}^{2\mu(t,r)}$, and then

$$\mathrm{d}s^2 = \mathrm{e}^{2\nu(t,r)}\,\mathrm{d}t^2 - \mathrm{e}^{2\mu(t,r)}\,\mathrm{d}r^2 - r^2\big(\mathrm{d}\vartheta^2 + \sin^2\vartheta\,\mathrm{d}\varphi^2\big). \tag{14.18}$$

This form of the metric still allows a subgroup of the transformations (14.2). Namely, we have $\delta = -r^2$ before the transformation and $\widetilde{\delta} = -r'^2$ after the transformation. Since δ transforms like a scalar, this means that $r = r'$ and, consequently, $G_{,t'} = 0$ and $G_{,r'} = 1$ in Eqs. (14.3)–(14.5). We also require that $\beta = 0$ before the transformation and $\widetilde{\beta} = 0$ after the transformation, so from (14.4)

$$\alpha F_{,t'} F_{,r'} = 0. \tag{14.19}$$

But $\alpha \neq 0$ necessarily, or else the metric would be singular. Also $F_{,t'} \neq 0$ because, with $G_{,t'} = 0$ already required, the Jacobian of the transformation would be zero. Hence, the consequence of (14.19) is $F_{,r'} = 0$, i.e. the transformations still allowed are $t = F(t')$, where F is an arbitrary function of one variable.

The coordinates of (14.18) are called **curvature coordinates** because in them the radial coordinate r is connected with the curvature $\mathcal{R}$ of the symmetry orbits in the same way as in the flat space, $\mathcal{R} = 1/r^2$.

14.2 Symmetry inheritance

Solutions of nonvacuum Einstein equations give rise to the so-called **symmetry inheritance** problem, which is the following: to what extent do the sources in the Einstein equations, coded in the energy-momentum tensor, inherit the symmetries of the metric of the spacetime?

For a perfect fluid, the matter-density, pressure and velocity field are uniquely determined by the metric tensor. The energy-momentum tensor $T_{\alpha\beta}$ is simply a function of the metric, the matter-density and pressure are the eigenvalues of $T_{\alpha\beta}$ and the velocity field is the unique timelike eigenvector of $T_{\alpha\beta}$. Hence, if the metric tensor does not change under some transformation, then neither does $T_{\alpha\beta}$, so also the matter-density, pressure and velocity field remain unchanged.

However, if the source in the Einstein equations is an electromagnetic field, then, even though the electromagnetic energy-momentum tensor (13.10) does inherit all the symmetries of the metric, the tensor of electromagnetic field $F^{\mu\nu}$ does not necessarily do the same. The equation $\underset{k}{\pounds} T_{\alpha\beta} = 0$ puts some limitation on $F^{\mu\nu}$, but does not imply $\underset{k}{\pounds} F^{\mu\nu} = 0$, and examples of $F^{\mu\nu}$ not being invariant under the isometries of the metric are known (Wainwright and Yaremovicz, 1976; Li and Liang, 1985). However, if $T_{\alpha\beta}$ is invariant under the whole group O(3), then so is $F^{\mu\nu}$, as proven by Wainwright and Yaremovicz (1976). We will accept this result without re-deriving it, and will make use of it in Section 14.4.

14.3 Spherically symmetric electromagnetic field in vacuum

In Section 8.9 we solved the equations $\underset{\substack{k\\(i)}}{\pounds} g_{\alpha\beta} = 0$ for the generators of the group O(3), given by (8.41). This result can be used in solving the equations $\underset{\substack{k\\(i)}}{\pounds} F_{\alpha\beta} = 0$ for the same generators. There is one point to be observed: since the metric tensor is symmetric, Eq. (8.49) that had the form $g_{23} + g_{32} = 0$ implied $g_{23} = 0$. However, the electromagnetic field tensor is antisymmetric, so (8.49) is identically fulfilled for it, i.e. the component F_{23} does not have to be zero. Also, because of antisymmetry we have immediately $F_{00} = F_{11} = F_{22} = F_{33} = 0$. Knowing this, and proceeding exactly like in Section 8.9, we conclude that only the components F_{01} and F_{23} of a spherically symmetric electromagnetic field can be nonzero, and then they must be of the form

$$F_{01} = f_{01}(t, r), \qquad F_{23} = f_{23}(t, r)\sin\vartheta, \tag{14.20}$$

where f_{01} and f_{23} are arbitrary functions of two variables.

Substituting Eqs. (14.20) in the Maxwell equations in Minkowski spacetime we find that the component F_{23} plays the role of the exterior field of a magnetic monopole. In agreement with one of the postulates of classical electrodynamics ('no magnetic monopoles exist') we would thus tend to assume $F_{23} = 0$. However, this is an additional assumption that does not follow from spherical symmetry, but from experiments – Maxwell's equations do allow such a solution. We will keep F_{23} nonzero for a while, and then we will find out that in vacuum the magnetic monopole may be eliminated by the duality rotation (13.13).

The second set of Maxwell equations, $F_{[\mu\nu,\lambda]} = 0$, does not impose any limitation on F_{01}, but for F_{23} it implies

$$f_{23} = \sqrt{8\pi}q = \text{constant} \tag{14.21}$$

(the coefficient $\sqrt{8\pi}$ was introduced in order to simplify later calculations). The first set of Maxwell's equations in vacuum, $F^{\mu\nu}{}_{;\nu} = 0$, may be written in the form $\left(\sqrt{-g}F^{\mu\nu}\right)_{,\nu} = 0$; then they become

$$\left(r^2 e^{\mu+\nu} F^{01}\right)_{,t} = 0 = \left(r^2 e^{\mu+\nu} F^{01}\right)_{,r}, \tag{14.22}$$

while $F^{23} = \sqrt{8\pi}q/(r^4 \sin\vartheta)$ obeys $\left(\sqrt{-g}F^{\mu\nu}\right)_{,\nu} = 0$ identically. The solution of (14.22) is

$$F^{01} = \sqrt{8\pi}e e^{-\mu-\nu}/r^2, \tag{14.23}$$

where e is another arbitrary constant (and the coefficient $\sqrt{8\pi}$ was again introduced for later convenience). Now it may be verified that the duality rotation (13.13) with the parameter $\delta = -\arctan(q/e)$ leads to the new field $\widetilde{F}^{\mu\nu}$ in which $\widetilde{F}_{23} = 0$, thus $\widetilde{q} = 0$ and $\widetilde{e} = \sqrt{e^2 + q^2}$, while the electromagnetic energy-momentum tensor does not change.[1] Hence, with no loss of generality we may now assume $q = 0 = F_{23}$.

14.4 The Schwarzschild and Reissner–Nordström solutions

For calculating the energy-momentum tensor of the electromagnetic field given by (14.23), the following formulae are useful:

$$F_{0\rho}F^\rho{}_0 = 8\pi e^2 e^{2\nu}/r^4, \qquad F_{1\rho}F^\rho{}_1 = -8\pi e^2 e^{2\mu}/r^4,$$
$$F_{\mu\nu}F^{\mu\nu} = -16\pi e^2/r^4; \tag{14.24}$$

the components of $F_{\alpha\rho}F^\rho{}_\beta$ that are not listed are identically zero. For the metric (14.18), the most convenient tetrad of differential forms is the orthonormal one

$$e^0 = e^\nu \, dt, \qquad e^1 = e^\mu \, dr, \qquad e^2 = r \, d\vartheta, \qquad e^3 = r \sin\vartheta \, d\varphi. \tag{14.25}$$

By whatever method (preferably, by using an algebraic computer program) we find now that the tetrad components of the Ricci tensor are

$$R_{\widehat{00}} = e^{-2\mu}\nu'' - e^{-2\mu}\nu'\mu' + e^{-2\mu}\nu'^2 + 2e^{-2\mu}\nu'/r$$
$$\qquad - e^{-2\nu}\ddot{\mu} + e^{-2\nu}\dot{\nu}\dot{\mu} - e^{-2\nu}\dot{\mu}^2, \tag{14.26}$$

$$R_{\widehat{01}} = 2e^{-\nu-\mu}\dot{\mu}/r, \tag{14.27}$$

---

[1] Many solutions of the Einstein–Maxwell equations include the electric charge and magnetic charge on equal footing, through the combination $e^2 + q^2$. The generalisation with respect to pure electric charge is thus rather illusory, since the magnetic charge may then be generated/removed/changed by duality rotations and does not influence the geometry of spacetime in an independent way.

$$R_{\widehat{11}} = -e^{-2\mu}\nu'' + e^{-2\mu}\nu'\mu' - e^{-2\mu}\nu'^2 + 2e^{-2\mu}\mu'/r$$
$$+ e^{-2\nu}\ddot{\mu} - e^{-2\nu}\dot{\nu}\dot{\mu} + e^{-2\nu}\dot{\mu}^2, \tag{14.28}$$

$$R_{\widehat{22}} = R_{\widehat{33}} = e^{-2\mu}(\mu' - \nu')/r + (1 - e^{-2\mu})/r^2, \tag{14.29}$$

where $\dot{\mu} = \partial\mu/\partial t$ and $\mu' = \partial\mu/\partial r$. The nonzero tetrad components of $F_{\alpha\rho}F^\rho{}_\beta$ are now found from (14.24) and (14.25):

$$F_{\widehat{0s}}F^{\widehat{s}}{}_{\widehat{0}} = -F_{\widehat{1s}}F^{\widehat{s}}{}_{\widehat{1}} = 8\pi e^2/r^4. \tag{14.30}$$

We will find the solutions of the Einstein–Maxwell equations with the cosmological constant and with the energy-momentum tensor (13.10), so

$$R_{\alpha\beta} - \frac{1}{2}g_{\alpha\beta}R + \Lambda g_{\alpha\beta} = \frac{1}{4\pi}\left(F_{\alpha\rho}F^\rho{}_\beta + \frac{1}{4}g_{\alpha\beta}F_{\mu\nu}F^{\mu\nu}\right), \tag{14.31}$$

but first we will transform them to a more convenient form. The trace of (14.31) implies $R = 4\Lambda$. Substituting this in (14.31) and transforming the result to tetrad components we obtain

$$R_{ij} = \Lambda g_{ij} + \frac{1}{4\pi}\left(F_{is}F^s{}_j + \frac{1}{4}g_{ij}F_{\mu\nu}F^{\mu\nu}\right). \tag{14.32}$$

Since $\|g_{ij}\| = \mathrm{diag}(+1, -1, -1, -1)$, Eqs. (14.30) and (14.32) imply

$$R_{\widehat{00}} + R_{\widehat{11}} = 0. \tag{14.33}$$

Substituting (14.26) and (14.28) in this we obtain $2e^{-2\mu}(\mu' + \nu')/r = 0$, which solves as

$$\nu = -\mu + f(t), \tag{14.34}$$

where $f(t)$ is an arbitrary function. Then, from $R_{\widehat{01}} = 0$ we obtain $\mu = \mu(r)$ (so far an arbitrary function). Now recall that the metric (14.18) does not change its form under the transformations $t' = g(t)$, where $g(t)$ is an arbitrary function. Consequently, we transform t in (14.18) by

$$t' = \int e^{f(t)}\,dt, \tag{14.35}$$

where $f(t)$ is the function that appeared in (14.34). The result of (14.35) is as if $f(t) \equiv 0$. Hence, with no loss of generality, $\nu = -\mu$, and the whole metric tensor becomes independent of the time coordinate.[1] Moreover, in (14.23) we obtain $F^{01} = \sqrt{8\pi}e/r^2$, which means that the electromagnetic tensor is also independent of time.

---

[1] The statement thus obtained is sometimes called the **Birkhoff theorem**: the spherically symmetric gravitational field in vacuum is static. (Here we have in fact derived a generalisation – our source is a vacuum electromagnetic field.) However, calling it a theorem is rather inappropriate. It follows from straightforward calculations and finds no other application. Moreover, it is, in this formulation, false: it holds only for those spacetime regions where the gradient of δ is spacelike.

From the equation $R_{\bar{2}\bar{2}} = -\Lambda - [1/(16\pi)]F_{\mu\nu}F^{\mu\nu}$ we now obtain, with the help of (14.29) and (14.33):

$$1 - e^{-2\mu} + 2re^{-2\mu}\mu' = -\Lambda r^2 + e^2/r^2, \tag{14.36}$$

which is easily integrated with the result

$$e^{-2\mu} = 1 + \frac{C}{r} + \frac{1}{3}\Lambda r^2 + \frac{e^2}{r^2} = e^{2\nu}, \tag{14.37}$$

where C is an arbitrary constant. The components $i = j = 0$ and $i = j = 1$ of (14.32) now turn out to be fulfilled identically.

For the strictly vacuum field ($\Lambda = 0 = e$) we obtain from here

$$g_{00} = e^{2\nu} = 1 + C/r, \tag{14.38}$$

while the consideration of the Newtonian limit indicated that for weak gravitational fields (12.46) should hold. In Newton's theory, for a spherically symmetric gravitational field in vacuum we have

$$\phi = -GM/r, \tag{14.39}$$

where G is the gravitational constant and M is the mass of the source of the field. Equations (12.46) and (14.39) are consistent with (14.38) when $C = -2GM/(c^2) \overset{\text{def}}{=} -2m$.

Finally then, we found the following solution:

$$ds^2 = e^{2\nu}\,dt^2 - e^{-2\nu}\,dr^2 - r^2\left(d\vartheta^2 + \sin^2\vartheta\,d\varphi^2\right), \tag{14.40}$$

where

$$e^{2\nu} = 1 - \frac{2m}{r} + \frac{1}{3}\Lambda r^2 + \frac{e^2}{r^2}. \tag{14.41}$$

Special cases of this solution were derived by different authors during the first few years after the theory of general relativity had been published in 1915 and today they are known under different names. The case $\Lambda = e = 0$ was found by Schwarzschild (1916a) and was historically the first exact solution of the Einstein equations ever found.[1] A large part of relativistic astrophysics and several experimental tests of general relativity are based on it. The subcase $\Lambda = 0$ was found by Reissner (1916) and Nordström (1918). The subcase $m = e = 0$ was found by de Sitter (1916). The subcase $e = 0$ was found by Kottler (1918). In full generality, the metric (14.40)–(14.41) appeared as a subcase of still more general solutions found by Cahen and Defrise (1968) and by Kinnersley (1969).

We will mostly deal with the Schwarzschild solution for which

$$e^{2\nu} = 1 - \frac{2m}{r}. \tag{14.42}$$

---

[1] A year later, Droste (1917) presented another, mathematically more elegant, derivation of the same solution and discussed its geometrical properties. His paper was surprisingly insightful and ahead of its time. Consequently, it would be quite appropriate to refer to this metric as the Schwarzschild–Droste solution (Rothman, 2002). An English translation of the Schwarzschild paper was published in *Gen. Rel. Grav.* **35**, 951 (2003), but the editorial note to it makes incorrect claims about its interpretation.

Note that for $r \to 2m$ this metric apparently has a singularity, since $g_{00} \to 0$ and $g_{11} \to -\infty$. However, if we substitute in (14.40) and (14.42) a value $r < 2m$, then the metric is still nonsingular, but $g_{00} < 0$ and $g_{11} > 0$, i.e. the r-coordinate becomes time and the t-coordinate becomes a measure of the radial distance. The region $r < 2m$ is the complementary part to $r > 2m$ that was mentioned after Lemma 14.3. In that region $g^{\alpha\beta} \delta_{,\alpha} \delta_{,\beta} > 0$, and the 'Birkhoff theorem' does not hold.[1] We will discuss the geometrical and physical interpretation of the spurious singularity at $r = 2m$ in Section 14.11.

14.5 Orbits of planets in the gravitational field of the Sun

In general relativity, just as in Newton's theory, it is assumed that the planets in their motion on the orbits behave like point masses, while their own gravitational fields are weak and can be neglected in comparison with the gravitational field of the Sun. These two assumptions in fact contradict each other. The gravitational field of a point mass is singular at the mass' position, and so stronger than any exterior field. In Newton's theory this difficulty is solved by the observation that the centre of mass of any extended body follows the same trajectory as a point body with no self-gravitation would follow. At the centre of mass, the body's own gravitational field vanishes.

In relativity, so far there is not even a generally accepted definition of the centre of mass, although work on this problem is being done. Thus, when we consider the orbits of point bodies in relativity, we are in fact extending the theory into the domain in which it has not been worked out yet. Nevertheless, these results agree with observational tests.

We assume that the gravitational field of the Sun is spherically symmetric, that the space around the Sun does not contain electromagnetic fields and that the cosmological constant is zero. This means that the spacetime will be described by the metric form (14.40)–(14.42). The orbits of planets should be timelike geodesics in this spacetime, and hence should be solutions of (5.14), where the derivative D/ds is calculated along dx^α/ds and $x^\alpha(s)$ is the equation of the geodesic. Contracting (5.14) with $g_{\alpha\beta} dx^\beta/ds$ and using (5.14) in the result, we obtain

$$0 = \frac{D}{ds}\left(g_{\alpha\beta}\frac{dx^\alpha}{ds}\frac{dx^\beta}{ds}\right) = \frac{d}{ds}\left(g_{\alpha\beta}\frac{dx^\alpha}{ds}\frac{dx^\beta}{ds}\right). \tag{14.43}$$

Hence, the quantity $g_{\alpha\beta}(dx^\alpha/ds)(dx^\beta/ds)$ is constant along every geodesic. This means in particular that its sign cannot change during motion; a geodesic is thus timelike, null or spacelike along its whole length.

An orbit of a planet is timelike, so $g_{\alpha\beta}(dx^\alpha/ds)(dx^\beta/ds) > 0$. The affine parameter s is determined up to inhomogeneous linear transformations, so for timelike geodesics it can be scaled so that

$$g_{\alpha\beta}\frac{dx^\alpha}{ds}\frac{dx^\beta}{ds} = 1. \tag{14.44}$$

---

[1] This region is an example of a spacetime with the Kantowski–Sachs symmetry; see the footnote after (10.40). In fact, it is the unique vacuum solution of Einstein's equations with this symmetry.

The affine parameter is then $s = c\tau$, where c is the velocity of light and τ is the proper time of the particle that follows the geodesic. In the calculation of the orbit we wish to use the physical units, so in calculating the Christoffel symbols our time coordinate will be ct. With this, the Christoffel symbols for the metric (14.40)–(14.42) are:

$$\begin{Bmatrix} 0 \\ 01 \end{Bmatrix} = \frac{m}{r^2} \frac{1}{1-2m/r} = -\begin{Bmatrix} 1 \\ 11 \end{Bmatrix},$$

$$\begin{Bmatrix} 1 \\ 00 \end{Bmatrix} = \frac{mc^2}{r^2} - \frac{2m^2c^2}{r^3},$$

$$\begin{Bmatrix} 1 \\ 22 \end{Bmatrix} = 2m - r = \begin{Bmatrix} 1 \\ 33 \end{Bmatrix} \Big/ \sin^2 \vartheta, \qquad (14.45)$$

$$\begin{Bmatrix} 2 \\ 12 \end{Bmatrix} = \frac{1}{r} = \begin{Bmatrix} 3 \\ 13 \end{Bmatrix},$$

$$\begin{Bmatrix} 2 \\ 33 \end{Bmatrix} = -\cos \vartheta \sin \vartheta, \qquad \begin{Bmatrix} 3 \\ 23 \end{Bmatrix} = \cot \vartheta.$$

Hence, the equations of a geodesic are these:

$$c\frac{d^2 t}{ds^2} + \frac{2m}{r^2} \frac{1}{1-2m/r} c\frac{dt}{ds}\frac{dr}{ds} = 0, \qquad (14.46)$$

$$\frac{d^2 r}{ds^2} + \left(\frac{mc^2}{r^2} - \frac{2m^2c^2}{r^3} \right)\left(\frac{dt}{ds} \right)^2 - \frac{m}{r^2}\frac{1}{1-2m/r}\left(\frac{dr}{ds} \right)^2$$

$$+ (2m-r)\left(\frac{d\vartheta}{ds} \right)^2 + (2m-r)\sin^2 \vartheta \left(\frac{d\varphi}{ds} \right)^2 = 0, \qquad (14.47)$$

$$\frac{d^2 \vartheta}{ds^2} + \frac{2}{r}\frac{d\vartheta}{ds}\frac{dr}{ds} - \cos \vartheta \sin \vartheta \left(\frac{d\varphi}{ds} \right)^2 = 0, \qquad (14.48)$$

$$\frac{d^2 \varphi}{ds^2} + \frac{2}{r}\frac{dr}{ds}\frac{d\varphi}{ds} + 2\cot \vartheta \frac{d\vartheta}{ds}\frac{d\varphi}{ds} = 0. \qquad (14.49)$$

The last equation may be written as $(d/ds)(r^2 \sin^2 \vartheta \, (d\varphi/ds)) = 0$, which is easily integrated with the result

$$\frac{d\varphi}{ds} = \frac{J_0}{r^2 \sin^2 \vartheta}, \qquad (14.50)$$

where J_0 is an arbitrary constant. Substituting this in (14.48) we obtain

$$\frac{d^2 \vartheta}{ds^2} + \frac{2}{r}\frac{dr}{ds}\frac{d\vartheta}{ds} - \frac{J_0^2 \cos \vartheta}{r^4 \sin^3 \vartheta} = 0. \qquad (14.51)$$

A first integral of this equation is

$$r^4 \left(\frac{d\vartheta}{ds} \right)^2 + J_0{}^2 \cot^2 \vartheta = A = \text{constant}. \tag{14.52}$$

We still have the freedom to rotate the axes of the spherical coordinates – the metric is invariant under such rotations, and we have not applied any of them so far. Suppose that at the initial instant on the orbit $\vartheta(s_0) = \vartheta_0$. Rotate the axis of the spherical coordinates so that at s_0 the planet is on the equator, i.e. $\vartheta_0 = \pi/2$. Then we still have the freedom to rotate the spherical coordinates around the single axis passing through $x^\alpha(s_0)$. Apply this rotation by such an angle that the initial velocity of the planet will lie in the plane of the equator, i.e. $d\vartheta/ds|_{s=s_0} = 0$. Substituting the initial conditions $\vartheta_0 = \pi/2$ and $d\vartheta/ds|_{s=s_0} = 0$ in (14.52) we obtain $A = 0$. But A is a sum of two non-negative terms, hence both of them must be zero at all times, which means that

$$\vartheta = \pi/2 \tag{14.53}$$

on the whole orbit. Substituting (14.53) in (14.50) we obtain

$$\frac{d\varphi}{ds} = \frac{J_0}{r^2}. \tag{14.54}$$

A first integral of (14.46) is

$$\left(1 - \frac{2m}{r} \right) c \frac{dt}{ds} = E, \tag{14.55}$$

where E is an arbitrary constant. In this way we have made use of (14.48), (14.46) and (14.49). Only (14.47) still remains, but we can replace it by the first integral (14.44). Using (14.53)–(14.55) it becomes

$$\frac{E^2}{1 - 2m/r} - \frac{1}{1 - 2m/r} \left(\frac{dr}{ds} \right)^2 - \frac{J_0{}^2}{r^2} = 1, \tag{14.56}$$

which can be written as[1]

$$\frac{dr}{ds} = \left[E^2 - \left(1 + \frac{J_0{}^2}{r^2} \right) \left(1 - \frac{2m}{r} \right) \right]^{1/2}. \tag{14.57}$$

Now we have three first-order equations to solve, (14.54), (14.55) and (14.57). In most situations we are interested only in the shape of the orbit, which is given by (14.54) and (14.57). The third equation gives information on the dependence of coordinates of the planet on time.

---

[1] We omit the minus sign in front of the right-hand side in (14.57) because the resulting solutions describe motions along the same orbits in the opposite direction. In a spherically symmetric spacetime, the reversal of the sense of orbital motion is equivalent to a coordinate transformation (reflection). This is not so when the source of the gravitational field is rotating, where orbital motion can be **direct**, i.e. with angular velocity parallel to that of the source, or **retrograde**, i.e. with the two angular velocities antiparallel. We shall meet this problem in the Kerr spacetime; see Section 20.7.

If $J_0 = 0$, then, from (14.54), $\varphi = \varphi_0 = $ constant, which means that the 'planet' falls directly towards the centre of the Sun. We will omit this uninteresting case, and assume $J_0 \neq 0$. Then (14.54) and (14.57) can be replaced by just one equation:

$$\left(\frac{J_0}{r^2}\frac{dr}{d\varphi}\right)^2 = E^2 - \left(1 + \frac{J_0^2}{r^2}\right)\left(1 - \frac{2m}{r}\right). \tag{14.58}$$

Now we introduce the new variable $\sigma = 1/r$ and obtain in (14.58)

$$\left(J_0\frac{d\sigma}{d\varphi}\right)^2 = E^2 - \left(1 + J_0^2\sigma^2\right)\left(1 - 2m\sigma\right). \tag{14.59}$$

An exact solution of this equation would lead to elliptic integrals. We prefer to solve it approximately, up to first order in the small parameter defined below. For the perturbative calculation, it is more convenient to go one step back and replace (14.59) by a second-order equation. We differentiate (14.59) by φ and obtain

$$\frac{d\sigma}{d\varphi}\cdot 2J_0^2\frac{d^2\sigma}{d\varphi^2} = \frac{d\sigma}{d\varphi}\left[-2J_0^2\sigma + 2m\left(1 + 3J_0^2\sigma^2\right)\right]. \tag{14.60}$$

One special solution of this is $d\sigma/d\varphi = 0$, i.e. motion on a circular orbit with $r = $ constant. When $d\sigma/d\varphi \neq 0$, (14.60) becomes

$$\frac{d^2\sigma}{d\varphi^2} + \sigma = \frac{m}{J_0^2} + 3m\sigma^2. \tag{14.61}$$

This equation differs from the corresponding Newtonian equation by the last term that vanishes in the Newtonian limit $c \to \infty$. In order that the first term on the right assumes the Newtonian form $GM\mu^2/J^2$ in this limit (where M is the mass of the source, μ is the mass of the planet on the orbit and J is the orbital angular momentum of the planet), we must assume $J_0 = J/(mc)$. To simplify the notation, we denote

$$\frac{mc^2\mu^2}{J^2} = \frac{GM\mu^2}{J^2} \overset{\text{def}}{=} \frac{1}{p}, \qquad 3m \equiv \frac{3GM}{c^2} \overset{\text{def}}{=} \alpha. \tag{14.62}$$

In the Newtonian limit, α becomes zero, while p is the **parameter** of the orbit: the distance from the Sun to the planet at that instant at which the Sun–planet direction is perpendicular to the Sun–perihelion direction. In this notation, Eq. (14.61) becomes

$$\frac{d^2\sigma}{d\varphi^2} + \sigma = \frac{1}{p} + \alpha\sigma^2. \tag{14.63}$$

We will solve this equation approximately, up to terms linear in α, taking α as the small parameter. More precisely, since α has the dimension of distance, the small parameter will be the dimensionless number $\alpha\sigma$. For the Sun, $\alpha \approx 4.5\,\text{km}$, while the radius of the Sun is approximately $7 \times 10^5\,\text{km}$, so the planetary orbits of necessity have still larger radii, and $\alpha\sigma < 0.64 \times 10^{-5}$. Since σ and $1/p$ are comparable, $\alpha\sigma^2 \ll \sigma$ and $\alpha\sigma^2 \ll 1/p$, and the last term in (14.63) is small compared to all other terms. However, for strongly concentrated objects such as neutron stars or black holes, the orbits may come much closer to $r = 0$, and the approximation used here becomes invalid. In those cases, one has to resort either to numerical calculations or to calculations with elliptic functions.[1]

In the zeroth order of approximation, the solution of (14.63) is the Newtonian orbit

$$\sigma_0 = \frac{1}{p} + \frac{\varepsilon}{p} \cos(\varphi - \varphi_0),$$ (14.64)

where ε is the **eccentricity** of the elliptic orbit, which is also a small quantity for all solar planets (for Pluto, $\varepsilon = 0.2444$, for Mercury $\varepsilon = 0.205\,628$; for other planets it is still smaller). In the first approximation, we assume σ to be of the form $\sigma_{[1]} = \sigma_0 + \sigma_1$, where σ_1 is of the order of α, i.e. $\sigma_1 = 0$ when $\alpha = 0$. We substitute this σ in (14.63), neglect terms of order α^2, and obtain

$$\frac{d^2\sigma_1}{d\varphi^2} + \sigma_1 = \frac{\alpha}{p^2}\left[1 + 2\varepsilon\cos(\varphi - \varphi_0) + \varepsilon^2\cos^2(\varphi - \varphi_0)\right].$$ (14.65)

Since we assumed $\sigma_1 = 0$ when $\alpha = 0$, we must take $\sigma_{1j} = 0$ as the solution of the homogeneous part of (14.65). The problem is thus to find a special solution of the full inhomogeneous equation. This can be done by textbook methods; the solution is

$$\sigma_1 = \frac{\alpha}{p^2}\left(1 + \frac{\varepsilon^2}{3}\right) + \frac{\alpha\varepsilon}{p^2}(\varphi - \varphi_0)\sin(\varphi - \varphi_0) + \frac{\alpha\varepsilon^2}{3p^2}\sin^2(\varphi - \varphi_0).$$ (14.66)

However, the solution $\sigma_0 + \sigma_1$ implied by this σ_1 is seen to be in violent disagreement both with Newton's theory and with observations. At $\varphi = n\pi$, $n = 1, 2, 3, \ldots$, the curve given by (14.66) passes through always the same two points, but in the sectors where $\sin(\varphi - \varphi_0) > 0$ σ_1 increases systematically with every revolution, with no upper limit, i.e. $r \to 0$ as $\varphi \to \infty$. In the sectors where $\sin(\varphi - \varphi_0) < 0$ σ_1 decreases systematically with every revolution, with no lower limit, and will at a certain instant cause that $\sigma_{[1]} = 0$, i.e. $r \to \infty$. The term responsible for this unusual behaviour is $(\alpha\varepsilon/p^2)(\varphi - \varphi_0)\sin(\varphi - \varphi_0)$, which becomes arbitrarily large as φ increases. On the other hand, analysing the function on the right-hand side of (14.59) by the methods of classical mechanics one can prove that with $E < 1$, $J_0 > 2\sqrt{3}GM/c^2$ and the initial σ being sufficiently small the solution $\sigma(\varphi)$ must be bounded (see Exercise 4). We conclude that the trouble-causing term must be the first term of a power series representing a bounded function and that in the first order solution to (14.63) we must include that function in full. Going to higher orders of approximation to identify the unknown function is a laborious way of solving the problem, since, with every added order, many new terms appear. We want the solution to reduce to (14.66) only to first order, and any bounded function with this property will do. After some guesswork, the function

$$\frac{\varepsilon}{p}\cos\left[\left(1 - \frac{\alpha}{p}\right)(\varphi - \varphi_0)\right]$$ (14.67)

---

[1] In real astrophysical situations, orbits are perturbed in so many ways that numerical treatment becomes necessary anyway. Examples: gravitation of other planets, rotation of the star or black hole, lack of spherical symmetry caused by rotation, loss of orbital angular momentum because of friction with interplanetary matter.

is found to be the right one. This function is bounded for all values of α, but replacing it by the 'approximation' linear in α caused the nonsensical behaviour of (14.66). This approximation holds only for a limited range of values of φ and becomes divergent when φ is allowed to increase without bounds. We will come back to this at the end of this section.

Hence, the correct approximation to the solution of (14.63) up to terms linear in α is:

$$\sigma = \frac{1}{p} + \frac{\alpha}{p^2}\left(1 + \frac{\varepsilon^2}{3}\right) + \frac{\varepsilon}{p}\cos\left[\left(1 - \frac{\alpha}{p}\right)(\varphi - \varphi_0)\right] + \frac{\alpha\varepsilon^2}{3p^2}\sin^2(\varphi - \varphi_0). \qquad (14.68)$$

Since $\varepsilon\alpha/p \ll 1$, the third term on the right dominates over the last one, which can thus be neglected. After the planet has made one full revolution, $\varphi \to \varphi + 2\pi$, σ does not return to its initial value. Instead, it returns (approximately) to its initial value after the planet has revolved by $2\pi/(1 - \alpha/p) > 2\pi$. Consequently, the aphelion and the perihelion move while the planet goes around its orbit, in the same direction in which the planet revolves. Two consecutive perihelia are separated by the angle $\Delta\varphi = 2\pi/(1 - \alpha/p) - 2\pi = 2\pi\alpha/p + O((\alpha/p)^2)$, and the formula for this perihelion shift can be written in three equivalent ways:

$$\Delta\varphi \approx \frac{2\pi\alpha}{p} = \frac{6\pi GM}{c^2 p} = \frac{6\pi GM}{c^2 b(1 + \varepsilon)} = \frac{6\pi GM}{c^2 a(1 - \varepsilon^2)}, \qquad (14.69)$$

where b is the minimal distance of the planet from the Sun and a is the semimajor axis of the orbit. The shape of the orbit is shown in Fig. 1.1 (where the value of $\Delta\varphi$ is greatly exaggerated).

Note that $\Delta\varphi$ is larger the smaller b is. Hence, Mercury offers the best conditions for measuring this effect. Since $\Delta\varphi$ is very small for a single revolution, but cumulates with time, the most convenient unit to measure $\Delta\varphi$ in astronomical practice is not radians per revolution, as in (14.69), but arc seconds per century. In these units, we have

$$\Delta\Phi = \frac{C}{T} \cdot \frac{360 \cdot 60 \cdot 60}{2\pi} \cdot \frac{6\pi GM}{c^2 a(1 - \varepsilon^2)}, \qquad (14.70)$$

where $C = 100$ years, T is the orbital period of Mercury and the second factor converts radians to arc seconds. From the tables (Allen, 1973) we find $G = 6.670 \times 10^{-8}$ cm^3 g^{-1} s^{-2} for the gravitational constant, $M = 1.989 \times 10^{33}$ g for the mass of the Sun, $c^2 = 8.987\,554 \times 10^{20}$ cm^2 s^{-2} for the square of the velocity of light, $a = 57.9 \times 10^{11}$ cm for the semimajor axis of Mercury's orbit, $\varepsilon = 0.205\,628$ for the eccentricity of Mercury's orbit and $T = 0.240\,85$ terrestrial years for Mercury's orbital period. Substituting all these values in (14.70) we obtain[1] $\Delta\Phi = 43.03$ arc seconds per century.

Equation (14.69) is the famous result (the **relativistic perihelion shift**) that predicted the orbital motion of Mercury in agreement with astronomical observations. As described

---

[1] Most pocket calculators will probably show a slightly different result, because of inconsistent roundoff errors. The value quoted here as the official result has been obtained by carefully tuning the precision of all factors.

in Chapter 1, when this correction is added to the perturbations of Mercury's orbit by other planets, the sum is in good agreement with the observed value.

The observational tests of this effect are done by registering the positions of planets at different instants, reconstructing their orbits in space and determining the positions of perihelia. As mentioned in Chapter 1, the perihelion shift of Mercury had been observed already in the first half of the nineteenth century, and its Newtonian origins were identified by LeVerrier in 1859 (Dicke, 1964). He was also the person who noted the discrepancy between observational results and the prediction of Newton's theory. It has to be mentioned that in addition to perturbations of Mercury's orbit by other planets there is one more component in the observed effect: the astronomical observations are carried out from the Earth, so their raw results are expressed in the geocentric reference system. This gives the largest component of the perihelion shift. According to modern observations (Will, 1981; Lang, 1974), the full observed perihelion shift of Mercury is $5599.74 \pm 0.41''$ per century, of which approximately $5000''$ per century is caused by the geocentric reference system, approximately $280''$ is caused by the gravitational perturbations from Venus, approximately $150''$ by perturbations from Jupiter and approximately $100''$ by perturbations from the remaining planets (Will, 1981). The residue is $43.11 \pm 0.45''$ per century (Lang, 1974), which agrees very well with the value calculated from relativity.

For other planets, the relativistic perihelion shift is no larger than a few arc seconds per century (Will, 1981). Moreover, their orbits have small eccentricities, so the positions of their perihelia are known with smaller precision and they are less suitable for testing relativity.

Now a few comments on the procedure we used to obtain the approximate orbit equation:

1. We were able to recognise that (14.66) is an incorrect solution because we knew from elsewhere what properties the orbit should have. However, perturbative methods are often applied in relativistic astrophysics in situations where nothing is known about the expected result. The experience with Mercury's orbit should be a warning that one must be careful about drawing conclusions from approximate results.

2. While going to higher orders of approximation, an avalanche of new terms appears. This is typical of all perturbative schemes. Thus, the hope that something useful can be inferred about the exact result by considering consecutive approximations may be illusory in most instances. There are some results, for example, in the third post-Newtonian order of the relativistic two-body problem, but the formulae are so unwieldy that they can be handled only by algebraic computer programs.

3. We emphasise again that the results (14.69)–(14.70) were obtained under the assumption that the radius of the orbit, at every point, is much larger than the gravitational radius of the central body, GM/c^2. They *do not* apply to those orbits around neutron stars and black holes that come close to the central body.

14.6 Deflection of light rays in the Schwarzschild field

We will now investigate the orbit of a light ray in the spherically symmetric gravitational field. Hence, this time we have to solve the equations of null geodesics in the Schwarzschild geometry.

The consideration in the previous section that led to the conclusion that coordinates may be chosen so that $\vartheta = \pi/2$ on the whole orbit remains valid also for null orbits. The only thing that changes is the value of the first integral (14.44); this time

$$g_{\alpha\beta} \frac{dx^\alpha}{ds} \frac{dx^\beta}{ds} = 0. \tag{14.71}$$

Moreover, now s is not time, but just a parameter on the curve.

The partly integrated equations of a geodesic are (14.54), (14.56) and

$$E^2 - \left(\frac{dr}{ds}\right)^2 - \frac{J_0^2}{r^2}\left(1 - \frac{2m}{r}\right) = 0. \tag{14.72}$$

It can be seen that, with $J_0 = 0$, radial motion is possible here also. With $J_0 \neq 0$, we obtain from (14.55) and (14.72):

$$\frac{dr}{d\varphi} = \frac{r^2}{J_0}\sqrt{E^2 - \frac{J_0^2}{r^2}\left(1 - \frac{2m}{r}\right)}. \tag{14.73}$$

(We take only the $+$ sign for the square root for the same reason as before.) We will handle this equation by a method similar to the one used in the previous section. Introducing $\sigma = 1/r$, we obtain

$$\left(J_0 \frac{d\sigma}{d\varphi}\right)^2 = E^2 - J_0^2\sigma^2\left(1 - 2m\sigma\right). \tag{14.74}$$

From here, differentiating by φ,

$$\frac{d\sigma}{d\varphi} \cdot \frac{d^2\sigma}{d\varphi^2} = \frac{d\sigma}{d\varphi}\left(-\sigma + 3m\sigma^2\right). \tag{14.75}$$

It is seen that again we have two cases here: either $d\sigma/d\varphi = 0$, which means motion on a circular orbit, or $d\sigma/d\varphi \neq 0$, in which case

$$\frac{d^2\sigma}{d\varphi^2} + \sigma = \alpha\sigma^2, \tag{14.76}$$

where $\alpha = 3m$, as before. We solve this equation by the same method as (14.63). In the zeroth (Newtonian) approximation, the solution is

$$\frac{1}{r} = \sigma_0 = \frac{1}{R}\cos(\varphi - \varphi_0), \tag{14.77}$$

where $R = $ constant. In the Newtonian limit, (r, ϑ, φ) are the spherical coordinates in the Euclidean space. Then (14.77) is an equation of a straight line passing at the distance R from the point $r = 0$.

In the first approximation we look for a solution in the form $\sigma_{[1]} = \sigma_0 + \sigma_1$, where $\sigma_1/\sigma_0 \propto \alpha$. The function σ_1 must then obey

$$\frac{d^2\sigma_1}{d\varphi^2} + \sigma_1 = \frac{\alpha}{R^2} \cos^2(\varphi - \varphi_0). \tag{14.78}$$

A particular integral of (14.78) is

$$\sigma_1 = \frac{\alpha}{3R^2} \left[1 + \sin^2(\varphi - \varphi_0)\right]. \tag{14.79}$$

The full solution of (14.78) up to terms of order α is thus

$$\frac{1}{r} = \frac{1}{R} \cos(\varphi - \varphi_0) + \frac{\alpha}{3R^2} \left[1 + \sin^2(\varphi - \varphi_0)\right]. \tag{14.80}$$

This may be solved for φ with the result

$$\varphi_{\pm} = \varphi_0 \pm \arccos\left[\frac{3R}{2\alpha}\left(1 - \sqrt{1 + \frac{8\alpha^2}{9R^2} - \frac{4\alpha}{3r}}\right)\right]. \tag{14.81}$$

This orbit has two asymptotes whose directions are calculated from here in the limit $r \to \infty$. The angle between them is (see Fig. 14.1)

$$\Delta\varphi = \lim_{r \to \infty}(\varphi_+ - \varphi_-) - \pi = 2\arccos\left[\frac{3R}{2\alpha}\left(1 - \sqrt{1 + \frac{8\alpha^2}{9R^2}}\right)\right] - \pi. \tag{14.82}$$

Equation (14.80) is an approximate solution to (14.76), so it will apply only with a limited precision (up to terms linear in α). Hence, we do not have to consider Eq. (14.82) in its exact form – it suffices to take it with the same precision, and then it becomes

$$\Delta\varphi = \frac{4\alpha}{3R} = \frac{4GM}{c^2R}. \tag{14.83}$$

This is another famous result. Before we confront it with observations, we will first rectify a misunderstanding that still lingers in the literature.

Some authors wish to calculate the angle of gravitational deflection of a light ray in a quick and easy way, and for that purpose they use a combination of special relativity and Newton's theory. They assume that a photon of energy E has the mass E/c^2 that interacts with the central body by the ordinary law of gravitation. Consequently, the orbit of a light ray is just an orbit of a particle of mass E/c^2. We will show below that the result obtained in this way is exactly half of the correct (verified observationally!) Eq. (14.83).

The Newtonian orbit in polar coordinates is given by (14.77). From there

$$\varphi_{\pm} = \varphi_0 \pm \arccos\left[\frac{1}{\varepsilon}\left(\frac{R}{r} - 1\right)\right]. \tag{14.84}$$

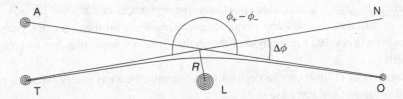

Fig. 14.1. Gravitational deflection of light rays in a spherically symmetric gravitational field. T is the true position of the source of light, A is the apparent position seen by the observer O in consequence of the deflection. According to Newton's theory, the light ray emitted from T in the direction TN would move along the straight line TN. In reality, and according to relativity, it follows the curve TO. The straight line TN is the asymptote of the initial part of the ray's path; AO is the asymptote of the final part. L is the deflecting body, R is the distance between the centre of the body L and the Newtonian path of the light ray, $\phi_+ - \phi_-$ is the angle between the asymptotes, calculated from (14.81) as $\phi_+ - \phi_- = \lim_{r\to\infty}(\varphi_+ - \varphi_-)$, and $\Delta\phi$ is the angle of deflection. The origin of the polar coordinates is at the centre of the deflecting body L; ϕ increases in the clockwise direction.

The deflection angle is thus

$$\Delta\varphi = \lim_{r\to\infty}(\varphi_+ - \varphi_-) - \pi = 2\arccos\left(-\frac{1}{\varepsilon}\right) - \pi = 2\arcsin\left(\frac{1}{\varepsilon}\right). \tag{14.85}$$

For a Newtonian orbit the eccentricity ε is

$$\varepsilon = \sqrt{1 + \frac{2E_0 J^2}{G^2 M^2 \mu^3}} \tag{14.86}$$

($\varepsilon > 1$ because the orbit is hyperbolic), where E_0 is the total energy of the particle on the orbit, J is its angular momentum and μ is its mass. In our case the 'particle' is a photon moving at the velocity c, so

$$E_0 = \frac{1}{2}mc^2, \qquad J = mcR. \tag{14.87}$$

Hence, in (14.86):

$$\varepsilon = \sqrt{1 + \frac{R^2 c^4}{G^2 M^2}} \gg 1. \tag{14.88}$$

Up to linear terms in $GM/(c^2 R)$ we thus have from (14.85)

$$\Delta\varphi = \frac{2GM}{c^2 R}, \tag{14.89}$$

i.e. half of the prediction of general relativity. Note that we arranged for the discrepancy to be small: had we taken the special relativistic equation $E = mc^2$ instead of (14.87), the final result would have been $\Delta\varphi = \sqrt{2}GM/(c^2 R)$, differing from (14.83) by still more.

The lesson to be drawn is this: simplified intuitive arguments constructed by picking out elements of different theories and doing simple arithmetics on them are not guaranteed to lead to correct results. Each physical theory has its inner logic that has to be applied consistently.

The same result (14.89) can be derived from the pure Newton theory, without mixing in special relativity; see Exercise 6.

14.7 Measuring the deflection of light rays

It can be seen from (14.83) that $\Delta\varphi$ is inversely proportional to R, thus the angle of deflection is greater for those rays that approach the central star to within a smaller distance. For a light ray grazing the surface of the Sun, $R = 6.9599 \times 10^{10}$ cm. Taking this number, and the other quantities in (14.83) as in Section 14.5, we obtain $\Delta\varphi = 1.75''$. This is the maximum value available in observations.[1] For stars other than the Sun the necessary parameters M and R cannot be measured with sufficient precision, and the deflection angles would be much too small to be measured (see later in this section). Planets cause unmeasurably small deflection angles.

We discussed the deflection of light rays, but of course the result applies to all kinds of radiation and other objects that move on null geodesics, e.g. X rays, γ rays, microwaves and neutrinos, provided the latter do indeed have zero mass. However, during the first decades of the twentieth century the whole of observational astronomy relied on optical observations; other kinds of radiation from space had been either impossible to detect with the technology of those times, or were not yet known. Consequently, in the first attempt to measure the deflection of light in the gravitational field of the Sun optical observations were applied.

The attempt was undertaken by Eddington in 1919 (Dyson, Eddington and Davidson, 1920). Observing the light rays that graze the Sun's surface is possible only during a total eclipse. The idea was this:

1. Find two stars that will be visible at the very edge of the Sun during the eclipse, best of all on opposite ends of the Sun's diameter.
2. Take a photograph of these stars during the eclipse, with the darkened Sun visible between them.
3. Take a photograph of the same two stars several months later, when the Sun is safely away from them in the sky.
4. Measure the differences between the positions of the stars in the two photographs.

---

[1] Today, gravitational deflection of light is observed also for distant galaxies, in the **gravitational lenses**. In those configurations, light rays deflected on opposite sides of the deflecting mass intersect on the Earth. However, the gravitational lenses are not useful for testing relativity because the parameters of the deflecting objects – their mass distributions and radii – are not known with sufficient precision. Also, in order to calculate the deflection angle in a gravitational lens correctly, the equation of a null geodesic should be solved in a different geometry – a model of the Universe with an inhomogeneous distribution of matter. No such results are available as yet.

The geometry of this measurement is shown in Fig. 14.2. With the Sun being away, the observer O sees the two stars at their true positions T_1 and T_2, at angular separation α_1. With the Sun between them, the stars are seen at the apparent positions A_1 and A_2, at angular separation $\alpha_2 > \alpha_1$. The angle of deflection is $\Delta\varphi = \alpha_2 - \alpha_1$.

With the technology of 1919, this measurement posed severe technical difficulties. The expected effect was so small that the mechanical deformations of photographic plates during the months when they had to be stored could seriously disturb the result. Total eclipses of the Sun typically occur in the tropical zone – in the oceans or in jungles and deserts, far away from well-equipped observatories. Eddington's expedition consisted of two groups, carrying out the observations in Sobral in Brazil and on Principe Island in the Guinea Bay off Africa.

Reaching such a destination on time is a problem in itself. In addition, one has to carry out precise measurements under field conditions. During an eclipse the temperature of the atmosphere drops rapidly, which causes some turbulence and the cooling of the telescope. The precision achievable under such conditions was limited, but, nevertheless, the effect predicted by relativity was definitely confirmed. The deflection angle measured at Sobral was $1.98 \pm 0.16''$, that on Principe Island was $1.61 \pm 0.40''$ (Will, 1981, p. 5). Unlike the anomalous perihelion shift of Mercury that had been known for many years before relativity was created, the deflection of light rays was *predicted* by relativity. Apart from the old and long-forgotten attempts by Cavendish and Soldner (Will, 1988; Soldner, 1804; Schneider, Ehlers and Falco, 1992), nobody had expected gravitation to influence the propagation of light. Eddington's positive result catapulted relativity and Einstein personally to the fame and public fascination that they still have today.

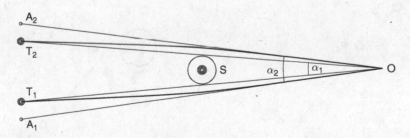

Fig. 14.2. Measuring the deflection of light rays by the Sun by Eddington's method. When the Sun is away, the observer O sees the stars at their true positions T_1 and T_2, at angular separation α_1. Then, the light from the stars reaches the observer along the straight lines T_1O and T_2O. When the Sun S is seen between the stars, their light follows the curved arcs T_1O and T_2O, and the observer sees the images of the stars at the apparent positions A_1 and A_2, at angular separation $\alpha_2 > \alpha_1$. The straight lines A_1O and A_2O are tangent to the arcs T_1O and T_2O at the point O. The prediction of relativity is $\Delta\varphi = \alpha_2 - \alpha_1 = 1.75''$.

Results obtained later from measurements by the same method gave results between 90% and 150% of the relativity value 1.75″, with formal errors estimated to be 6% to 25%, and unknown systematic errors.

Today this measurement is made by a different method. The electromagnetic waves used in the observation are microwaves, and the observing devices are radio telescopes. Three radio sources, denoted by the symbols $0119 + 11, 0116 + 08$ and $0111 + 02$, lie almost exactly on a straight line, inclined at a large angle to the Sun's yearly trajectory through the sky. (The symbols of the radio sources are their coordinates on the sky, namely the right ascension in hours and minutes and the declination in degrees.) Each year, for about 3 weeks in March/April, the Sun passes in front of the middle radio source, see Fig. 14.3. The other two radio sources are then so far from the Sun's disc on the sky that their radiation is not measurably deflected. Thus, the real position of the middle radio source relative to the two others is measured when the Sun is elsewhere in the sky and compared with the observed position while the Sun's edge approaches the line of sight. The Sun is a weak radio source and its own microwave radiation does not disturb the measurement; the object $0116 + 08$ can be observed right up to the moment when it vanishes behind the Sun. The only complication is that the plasma in the solar corona also deflects the microwaves. However, the plasma deflection depends, in a known way, on the wavelength of the radiation, while the gravitational deflection is the same for all wavelengths. Thus, carrying out the measurement at two wavelengths suffices to disentangle the two effects.

This measurement was carried out for the first time in 1974 by Fomalont and Sramek (1975) at the National Radio Astronomy Observatory in Green Bank (West Virginia). With this method, no expeditions to remote sites are necessary – the whole observation is done under laboratory conditions and so is more precise. It has become customary to quote the observational results in the form of the ratio of the measured quantity to the value

Fig. 14.3. Measuring the deflection of microwaves by the Sun. When the Sun is away, the three radio sources P_1, P_2 and P_3 are seen nearly on a straight line (left). The true position of the middle source P_2 can then be measured relative to the other two. With the disc of the Sun approaching (right), the image of P_2 moves away from the Sun. Its true position (small circle) can then be calculated from the positions of P_1 and P_2 (which do not change measurably) and compared with the observed apparent position to calculate the angle of deflection. The shift of the image of P_2 is exaggerated for clarity.

predicted by relativity. The value 1 means perfect agreement. In the radio-astronomical observation, the deflection was measured to be 1.007 ± 0.009 (standard error), so the value 1 is within the error bar.

14.8 Gravitational lenses

A **gravitational lens** is a body deflecting light rays that is situated relative to the observer in such a way that the deflected rays intersect at the observer's position. The theory and observations of gravitational lenses has become a science in itself (Schneider, Ehlers and Falco, 1992), and we will touch on this subject here only very briefly.

A gravitational lens formed by a single star is shown in Fig. 14.4. S is the source of light, L is the deflecting star (the lens) and O is the observer; R is the radius of the star, d_S and d_O are the distances from the lens to the source of light and to the observer, respectively; $\Delta\varphi_S$ and $\Delta\varphi_O$ are the angles filled by the radius R at the position of the source and of the observer, respectively; $\Delta\varphi$ is the deflection angle of the ray that grazes the surface of the lens. Neglecting the curvature of space, we obtain, approximately for small angles,

$$R = d_S\,\Delta\varphi_S = d_O\,\Delta\varphi_O. \tag{14.90}$$

On the other hand, using (14.83) and Fig. 14.4 we have

$$\Delta\varphi = \Delta\varphi_S + \Delta\varphi_O = \frac{4GM}{c^2 R}. \tag{14.91}$$

From (14.90) and (14.91) we obtain the 'equation of a gravitational lens':

$$\frac{1}{d_S} + \frac{1}{d_O} = \frac{4GM}{c^2 R^2}. \tag{14.92}$$

Unlike optical lenses, gravitational lenses do not focus the light rays: the rays flying farther from the optical axis are deflected by smaller angles than are those flying closer to the axis. Hence, it is not possible to 'view' anything through a gravitational lens as if it were a magnifying glass – the image is very distorted. Nevertheless, there is

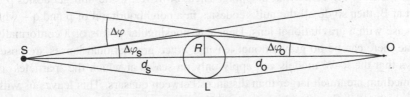

Fig. 14.4. A spherical star as a gravitational lens. In the approximation used here, the radius of the star, R, equals the distance from the star's centre to the point where the two straight lines intersect. Everything else is elementary geometry – see the text.

some intensification of light: the rays that would have dispersed in the absence of a lens intersect again. Gravitational lenses can thus increase the range of optical observations, and are actually used for this purpose (Schneider, Ehlers and Falco, 1992).

Would it be possible for an observer on the Earth to use the Sun as a gravitational lens? Equation (14.92) allows us to answer this question decisively: no. The first rays (those that graze the surface of the Sun) intersect at the distance d_O that is smaller the greater d_S is. Hence, the minimal d_O is calculated from (14.92) in the limit $d_S \to \infty$; it is $d_{min} = (cR)^2/(4GM) = 8.2 \times 10^{10}$ km, while the radius of the Earth's orbit is $1.495\,978\,92 \times 10^8$ km. (The distance d_{min} is more than 13 times the radius of the orbit of Pluto, i.e. far beyond the Solar System.) Likewise, there is no chance to observe the 'lensing' by other stars by measuring the deflection angles. Even for the closest star, which is 4.5 light years away from the Earth, assuming it has the same radius as the Sun, the angle $\Delta \varphi_O$ would be $3.4 \times 10^{-3''}$ – too little to be measured.[1]

By extending the formula (14.83) to intergalactic distances (which is not really correct) we can conclude that galaxies do have a chance to be gravitational lenses, and indeed many such lenses have been observed (Schneider, Ehlers and Falco, 1992). Substituting in (14.92) the mass of our Galaxy, $M = 1.4 \times 10^{11} M_\odot$ and its smallest diameter $R = 5$ kpc, with 1 kpc $= 3.0857 \times 10^{21}$ cm, we obtain $d_{min} = 9.33 \times 10^2$ Mpc. According to the Hubble formula the luminosity distance is $D_L = zc/H_0$, where $H_0 \approx 50$ km s^{-1} Mpc^{-1} is the Hubble constant, the distance d_{min} corresponds to the redshift $z \approx 0.15$. Taking the largest diameter of our Galaxy, $R = 30$ kpc, we get $z = 5.6$. The quasars are in the middle of this range. For a galaxy of diameter 30 kpc at the distance 3×10^4 Mpc the angle $\Delta \varphi_O$ is $0.2''$, which is measureable. Thus, it should be not surprising that most of the observed gravitational lenses are quasars.

However, (14.83) and (14.92) do not apply to quasars. The first of them applies (approximately) only in the gravitational field of a single star; the second is an even cruder approximation in the same geometry. The distance from the Earth to quasars is large on the cosmological scale. For calculating the deflection of light over such distances one should consider null geodesics in a model of the Universe. The models usually used in astronomy are the Robertson–Walker (R–W) metrics derived in Section 10.7. But even these are not general enough because they are conformally flat – the relations among null geodesics in the R–W spacetimes are the same as in the Minkowski spacetime. In particular, different null geodesics issued from point A can intersect at B $\neq$ A only if B is a singular point of the conformal mapping. Then, however, if the null geodesics p and q intersect at B, then so do all the null geodesics in a neighbourhood of p and q – which is not the case with a gravitational lens. Hence, at nonsingular points of a conformally flat spacetime there can be no gravitational lensing. Since gravitational lenses are observed, it follows that the R–W models can apply only on scales at which the 'particles' of the cosmic medium are much larger than distances between quasars. This leaves us with just a couple of 'particles' with which to build the whole observed region of the Universe.

---

[1] Such 'microlensing' has successfully been observed by measuring the changes in intensity of light from more distant stars when they are eclipsed by the lenses.

Models that should apply on any finer scale must be conformally nonflat, and hence cannot belong to the R–W family.

In astronomical practice, gravitational lenses are described by a sort of geometric optics based on the Newtonian description of propagation of light (Schneider, Ehlers and Falco, 1992). This description is not a theory in the sense that the word 'theory' has in physics – rather, it is a collection of heuristic rules. Nevertheless, it gives testable results that are in approximate agreement with observations.

14.9 The spurious singularity of the Schwarzschild solution at $r = 2m$

The Schwarzschild solution given by (14.40) and (14.42) has a singularity at $r = 2m = 2GM/c^2$: as $r \to 2m$, the component g_{00} of the metric goes to zero, while g_{11} becomes infinite. However, for the determinant of the metric, $g = -r^4 \sin^4 \vartheta$, the value $r = 2m$ is not in any way special. Also, the tetrad components of the Riemann tensor, R_{ijkl}, in the tetrad defined by (14.25), are all regular at $r = 2m$: they are all of the form $\alpha m/r^3$, where $\alpha = 0, \pm 1, \pm 2$. These facts suggest that the singularity at $r = 2m$ of the solution given by (14.40) and (14.42) is spurious – created by the coordinates used there, and not by the geometry of the manifold.[1] To see that this is possible, let us recall the transformation law of tensors:

$$g_{\alpha'\beta'} = \frac{\partial x^\alpha}{\partial y^{\alpha'}} \frac{\partial x^\beta}{\partial y^{\beta'}} g_{\alpha\beta}. \tag{14.93}$$

It is seen that if $g_{\alpha\beta}$ is regular in the old coordinates at $x^\alpha = x_0^\alpha$, but some of the functions $\partial x^\alpha/\partial y^{\alpha'}$ have a singularity at $x^\alpha = x_0^\alpha$, then some components of $g_{\alpha'\beta'}$ will be singular at $y^\alpha = y^\alpha(x_0)$. Such a singularity clearly can be removed by the inverse transformation $y^\alpha \to x^\alpha$.

There is no general criterion that would allow one to distinguish a 'true' singularity of the Riemannian geometry from a singularity introduced artificially together with the coordinate system.[2] However, if a coordinate transformation that removes the singularity is found, then this is proof that the singularity was spurious. For the Schwarzschild solution several such transformations were discovered. The most general of them is the one found (independently and almost simultaneously) by Kruskal (1960) and by G. Szekeres (1960).[3] This transformation leads to coordinates that reveal the global structure of the Schwarzschild manifold.

---

[1] But the singularity at $r = 0$ is real, as will be seen from the following.

[2] If scalars connected with the geometry of the manifold, such as tetrad components of the curvature, become infinite at certain points, then this is an indication that the singularity is genuine. Even so, one must be careful and verify whether these singularities occur within the range of the map. For example, the singularity of the Schwarzschild solution at $r = 0$ occurs beyond the allowed range of the curvature coordinates, and we can deal properly with that singularity only after introducing coordinates that cover its neighbourhood.

[3] The Kruskal–Szekeres transformation was a crowning of a whole series of less general results. Probably the oldest proof that the singularity at $r = 2m$ is spurious was given by Lemaître (1933a), who noticed that in coordinates connected with freely falling observers the surface $r = 2m$ is no obstacle to their motion. (In fact, this interpretation of the Lemaître coordinates was provided much later by Novikov (1964b).) Other transformations removing the spurious singularity at $r = 2m$ were found by Raychaudhuri (1953) and by Finkelstein (1958).

Let us write the Schwarzschild solution in the form[1]

$$ds^2 = \left(1 - \frac{2m}{r}\right)\left[dt^2 - \frac{dr^2}{(1-2m/r)^2}\right] - r^2\left(d\vartheta^2 + \sin^2\vartheta\,d\varphi^2\right) \tag{14.94}$$

and transform the coordinates as follows:

$$\widetilde{p} = t + \xi, \qquad \widetilde{q} = t - \xi, \tag{14.95}$$

where

$$\xi \overset{\text{def}}{=} \int \frac{dr}{1 - 2m/r} = r + 2m\ln\left(\frac{r}{2m} - 1\right). \tag{14.96}$$

Then the metric becomes

$$ds^2 = \left(1 - \frac{2m}{r}\right)d\widetilde{p}\,d\widetilde{q} - r^2\left(d\vartheta^2 + \sin^2\vartheta\,d\varphi^2\right), \tag{14.97}$$

where $r = r(\xi)$ is the function inverse to the $\xi(r)$ of (14.96). Since $\xi_{,r} > 0$ for $r > 2m$, the transformation (14.96) is invertible in the region in which the Schwarzschild solution was originally defined. The function (14.96) can be formally extended to the values $r < 2m$, by writing $\xi = r + 2m\ln|r/(2m) - 1|$; Fig. 14.5 shows a graph of the extension. The inverse functions exist in both domains, $r > 2m$ and $r < 2m$, but there exists no single inverse function in the domain $0 < r < \infty$.

Now let us carry out the next coordinate transformation:

$$(p, q) = \left(e^{\widetilde{p}/a}, e^{\widetilde{q}/a}\right) \Longleftrightarrow (\widetilde{p}, \widetilde{q}) = a(\ln p, \ln q), \tag{14.98}$$

where a is a constant whose value will be chosen later. In these coordinates we have

$$ds^2 = -\left(1 - \frac{2m}{r}\right)\frac{a^2}{pq}\,dp\,dq - r^2\left(d\vartheta^2 + \sin^2\vartheta\,d\varphi^2\right), \tag{14.99}$$

where this time $r = r(p, q)$; the transformation $(\widetilde{p}, \widetilde{q}) \longrightarrow (p, q)$ is invertible for all positive values of p and q. Now let us substitute (14.98) and (14.95) in (14.99). The result is:

$$ds^2 = -\left(1 - \frac{2m}{r}\right)a^2 e^{-2r/a}\left(\frac{r}{2m} - 1\right)^{-4m/a}dp\,dq$$
$$- r^2\left(d\vartheta^2 + \sin^2\vartheta\,d\varphi^2\right). \tag{14.100}$$

---

[1] Even though the set $r = 2m$ is the boundary of the region covered by the curvature coordinates, Eq. (14.94) still makes sense when $0 < r < 2m$. In that region, t becomes a spacelike coordinate and r becomes time. The resulting metric is the unique vacuum Kantowski–Sachs (K–S) spacetime (Kantowski and Sachs, 1966) (see also Sections 8.9, 10.7 and 19.4 for brief descriptions of the K–S metrics). This is the case that is commonly overlooked in textbooks and that was mentioned in Section 14.1, Eq. (14.17).

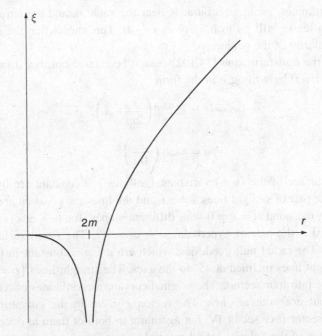

Fig. 14.5. A graph of the function $\xi(r) = r + 2m \ln |r/(2m) - 1|$.

If we choose $a = 4m$, then the factor that causes the singularity will cancel out and the result will be

$$ds^2 = -\frac{32m^3}{r} e^{-r/(2m)} \, dp \, dq - r^2 \left(d\vartheta^2 + \sin^2 \vartheta \, d\varphi^2\right). \tag{14.101}$$

The hypersurfaces of constant p and of constant q are null, which is not always convenient. In order to go back to ordinary timelike–spacelike coordinates, we now introduce the further new coordinates v and u by:

$$v = \frac{1}{2}(p - q) \equiv e^{r/(4m)} \sqrt{\frac{r}{2m} - 1} \, \sinh\left(\frac{t}{4m}\right),$$

$$u = \frac{1}{2}(p + q) \equiv e^{r/(4m)} \sqrt{\frac{r}{2m} - 1} \, \cosh\left(\frac{t}{4m}\right). \tag{14.102}$$

In the coordinates $(v, u, \vartheta, \varphi)$, the metric (14.101) becomes

$$ds^2 = \frac{32m^3}{r} e^{-r/(2m)} \left(dv^2 - du^2\right) - r^2 \left(d\vartheta^2 + \sin^2 \vartheta \, d\varphi^2\right). \tag{14.103}$$

By following all the intermediate transformations, it is seen that the transformation $(t, r) \rightarrow (v, u)$ is single-valued and nonsingular for all $r \geq 2m$, so it is invertible. At $r = 2m$, the transformation is also well defined, but $\partial v/\partial r$ and $\partial u/\partial r$ go to infinity as $r \rightarrow 2m$. This is the reason why (14.94) has a singularity at $r = 2m$.

In these coordinates, r can go arbitrarily near the value 0, and the tetrad components of the Riemann tensor still go to infinity as $r \to 0$. This shows that the set defined by $r = 0$ is a singularity of the geometry.

For $r < 2m$, the transformation (14.102) cannot be carried out, but it can be formally extended to all $r > 0$ by writing it in the form

$$u^2 - v^2 = e^{r/(2m)} \left(\frac{r}{2m} - 1 \right), \tag{14.104}$$

$$v/u = \tanh \left(\frac{t}{4m} \right). \tag{14.105}$$

Hence, in the surface of the (v, u) variables, the lines $r = $ constant are hyperbolae, the set $r = 2m$ is the pair of straight lines $u = \pm v$ and the lines $t = $ constant are straight lines passing through the point $u = v = 0$ with different slopes; for $t \to \pm\infty$, $v/u = \pm 1$. The singularity $r = 0$ is the pair of hyperbolae $u^2 - v^2 = -1$. The (u, v) surface thus looks as in Fig. 14.6. The radial null geodesics, which are $u \pm v = $ constant in (14.94), are in this figure straight lines inclined at 45° to the axes. The straight lines $\{r = 2m, t = \pm\infty\}$ divide the figure into four sectors. The original curvature coordinates cover only sector I or sector III, but not both at once. The region $r < 2m$ in the curvature coordinates corresponds to sector II or sector IV, but again not to both of them at once. We will use this representation of the Schwarzschild manifold again later.

Figure 14.6 is called the **Kruskal diagram**. It not only provides new useful coordinates for the Schwarzschild solution but is at the same time the **maximal extension** of the original Schwarzschild spacetime. The incompleteness of the Schwarzschild spacetime as represented in the curvature coordinates was recognised when it turned out that there exist timelike and null geodesics that escape the range of the Schwarzschild map without hitting any singularity. The Kruskal–Szekeres representation of the Schwarzschild solution does not have this defect – any timelike or null (or spacelike, too) geodesic can either be continued to infinite values of the affine parameter, or hits the singularity $r = 0$. Note that, contrary to what the original Schwarzschild representation would have us expect, the radial null geodesics reaching the observer in sector I come from a different part of the manifold than the one to which the observer can send his light signals, and the singularity set consists of two separate parts. An observer in sector I can never get an answer to a signal that he would send through the hypersurface $r = 2m$, because any such answer will hit the future singularity. However, observers in sector IV do have such a possibility: after moving into sector II, they can get an answer from sectors I and III to the signals they had sent in sector IV, but they will have no chance to reply.

A similar consideration was applied to the Reissner–Nordström solution by Graves and Brill (1960), and the result was even more strange (see Section 14.15). When the electric charge obeys $e^2 < m^2$, there are two hypersurfaces analogous to Schwarzschild's $r = 2m$, and the maximally extended manifold consists of an infinite number of copies of a set somewhat similar to the one in Fig. 14.6. The same strange behaviour is found in the Kerr solution that generalises Schwarzschild's for rotation of the source; we will deal with it in Chapter 20.

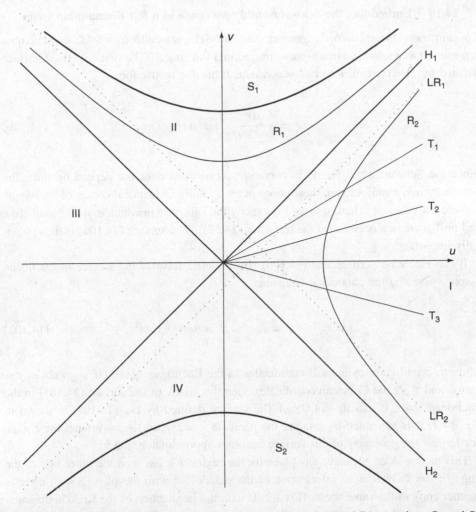

Fig. 14.6. The Kruskal diagram of the maximally extended Schwarzschild spacetime. S_1 and S_2 are the singularities at $r = 0$; the regions above and below them are not parts of the spacetime. The straight lines H_1 and H_2 are the event horizons; they divide the plane into four sectors denoted I,...,IV. The curvature coordinates cover only sector I. Curvature coordinates applied inside the horizons cover sectors II and IV, although one does not see then that the two sectors do not coincide. The dotted lines LR_1 and LR_2 are paths of radial light rays; LR_1 passes out from the past singularity S_2 and from under the horizon and escapes to infinity, LR_2 falls from infinity and is trapped inside the horizon, eventually hitting the future singularity S_1. The hyperbolae R_1 and R_2 are lines of constant r, one inside the horizon (with $r < 2m$) and the other outside (with $r > 2m$). The straight lines T_1, T_2 and T_3 are lines of constant t. Note that, on the horizons, $r = 2m$ and $|t| = \infty$, with $t = +\infty$ on H_1 and $t = -\infty$ on H_2. This misled some early researchers into believing that the surface $r = 2m$ can never be reached by a material object. But the value of t is not the physical time; t is just a badly behaving parameter used to measure time.

14.10 * Embedding the Schwarzschild spacetime in a flat Riemannian space

To gain more insight into the geometry of the Schwarzschild manifold, we will now represent it as a subspace of a higher-dimensional flat space. Take first only the 2-surface defined by $t = $ constant, $\vartheta = \pi/2 = $ constant. It has the metric form

$$ds^2 = \frac{dr^2}{1 - 2m/r} + r^2 \, d\varphi^2. \tag{14.106}$$

Since the Schwarzschild metric in curvature coordinates does not depend on time, the Schwarzschild manifold is in these coordinates a family of identical copies of the hyper-surface $t = $ constant. Then, $\vartheta = \pi/2$ is exactly the subspace in which motion on timelike and null geodesics occurs (see the text after (14.52)). The surface (14.106) is thus physically important.

It can easily be verified that (14.106) is the metric form of the surface in Euclidean 3-space given by the parametric equations

$$r = r, \qquad \varphi = \varphi, \qquad z^2 = 8mr - 16m^2, \tag{14.107}$$

where r, φ and z are cylindrical coordinates in the Euclidean space. (If $x = r \cos \varphi$, $y = r \sin \varphi$ and z are the Cartesian coordinates, then the metric of the surface (14.107) in the Euclidean space is exactly (14.106).) The surface defined by Eq. (14.107) is shown in Fig. 14.7. It is generated by rotating the parabola $r = z^2/(8m) + 2m$ around the z-axis. For large r the geometry of this surface becomes approximately flat.[1]

This picture does not leave any place for the region $r < 2m$ – on the other side of the ring $r = 2m$ (which is an intersection of the paraboloid with the plane $z = 0$) there is another copy of the same sheet. This agrees with the implication of the Kruskal diagram, and once more shows that the curvature coordinates do not cover the whole manifold. The surface shown in Fig. 14.7 coincides with the surface $\{v = 0, \vartheta = \pi/2\}$ in the Kruskal–Szekeres coordinates because $v = 0$ at $t = 0$. From Fig. 14.6 it is seen that the smallest value of r on the surface $\{v = $ constant, $\vartheta = \pi/2\}$, which is attained at $u = 0$, is $r = 2m$. Also from Fig. 14.6 it is seen that on other surfaces with $\{v = $ constant, $\vartheta = \pi/2\}$, where $0 < v < 1$, the smallest value of r (the radius of the smallest ring in the throat) is smaller than $2m$, until it becomes zero at $v = \pm 1$ (see Eq. (14.104)). For $|v| = $ constant > 1, the bridge connecting the two sheets disappears; each of those surfaces is composed of two disjoint subsets.

---

[1] Nonetheless, it is not correct to draw this surface as if it were actually becoming a plane far away from the 'throat' at $z = 0$. It is a simple exercise to verify that no asymptotic plane exists for this surface.

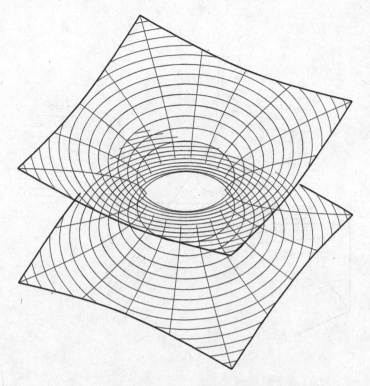

Fig. 14.7.   The surface $\{t = \text{constant}, \vartheta = \pi/2\}$ in the Schwarzschild spacetime. The z-axis runs vertically through the middle of the 'throat', the r-coordinate is the distance from that axis.

The full 4-dimensional Schwarzschild spacetime can be embedded in a flat 6-dimensional Riemannian space. This was first shown by Fronsdal (1959). The metric form of the 6-dimensional space we will use is

$$ds_6{}^2 = dZ_1{}^2 - dZ_2{}^2 - dZ_3{}^2 - dZ_4{}^2 - dZ_5{}^2 - dZ_6{}^2, \tag{14.108}$$

where

$$Z_1 = 4m\sqrt{1 - \frac{2m}{r}}\,\sinh\left(\frac{t}{4m}\right),$$

$$Z_2 = 4m\sqrt{1 - \frac{2m}{r}}\,\cosh\left(\frac{t}{4m}\right),$$

$$Z_3 = \pm \int \left[\frac{2m}{r} + \left(\frac{2m}{r}\right)^2 + \left(\frac{2m}{r}\right)^3\right]^{1/2} dr \tag{14.109}$$

$$(Z_3 > 0 \text{ for } r > 2m \text{ and } Z_3 = 0 \text{ for } r = 2m),$$

$$Z_4 = r\sin\vartheta\cos\varphi, \qquad Z_5 = r\sin\vartheta\sin\varphi, \qquad Z_6 = r\cos\vartheta.$$

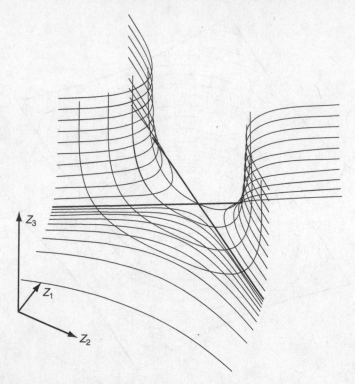

Fig. 14.8. Embedding of the Schwarzschild spacetime in six dimensions projected on the space of the variables (Z_1, Z_2, Z_3). The thick straight lines are the horizons $r = 2m$; they cross at the saddle where $Z_1 = Z_2 = Z_3 = 0$. r increases from bottom to top, at $r \to 0$ all Z_i go to $\pm\infty$ (except when along a curve $Z_2 = $ constant). Along the $Z_2 = 0$ curve, $|\lim_{r \to 0} dZ_3/dZ_1| = 1$. The curves running horizontally are the $Z_3 = $ constant curves; the U-shaped curves are given by $Z_1 = $ constant, with $Z_1 = 0$ running through the saddle. The surface is mirror-symmetric with respect to the $Z_1 = 0$ plane, but the back side below the saddle is not shown, for the sake of clarity.

Figures 14.8 and 14.9 show the projections of the 6-dimensional space of variables $(Z_1, \ldots, Z_6)$ on the 3-dimensional subspaces of the variables (Z_1, Z_2, Z_3) (Fig. 14.8) and (Z_3, Z_4, Z_5) (Fig. 14.9).[1] In order to draw the first projection, one must observe that the surfaces $r = $ constant are surfaces $Z_3 = $ constant, and that

$$Z_2{}^2 - Z_1{}^2 = 16m^2 \left(1 - \frac{2m}{r}\right), \tag{14.110}$$

so $Z_2{}^2 - Z_1{}^2 = $ constant when $Z_3 = $ constant. The surface of the Schwarzschild variables (t, r) is the saddle-shaped surface in Fig. 14.8. The pair of thicker straight lines is the

---

[1] Reconstructing a 3-dimensional structure in a 3-dimensional Euclidean space from projections on different 2-planes is what engineers routinely do. Are you good enough to be an engineer in a 6-dimensional space?

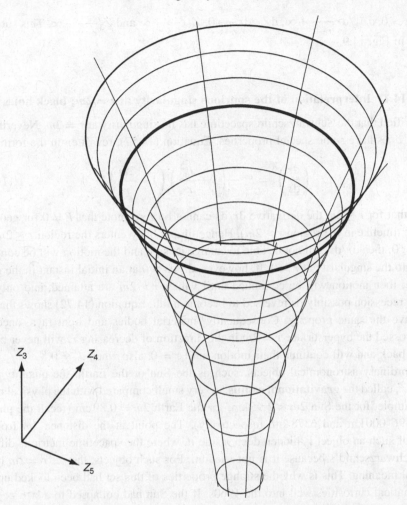

Fig. 14.9. Embedding of the Schwarzschild spacetime in six dimensions projected onto the space of the variables (Z_3, Z_4, Z_5). r increases from bottom to top, with $r \to 0$ at the tip of the funnel (where $Z_3 \to -\infty$, $Z_4 \to 0$ and $Z_5 \to 0$). The thick circle is the horizon $r = 2m$.

image of the set $r = 2m$. Further properties of this surface that are helpful in reading the diagram are

$$\left| \frac{dZ_3}{dZ_2} \right| \underset{\substack{r \to \infty \\ Z_1 = \text{const}}}{\longrightarrow} +\infty, \qquad \left| \frac{dZ_3}{dZ_1} \right| \underset{\substack{r \to 0 \\ Z_2 = 0}}{\longrightarrow} 1,$$

$$|Z_1| \underset{\substack{r \to 0 \\ Z_2 = \text{const}}}{\longrightarrow} +\infty, \qquad Z_3 \underset{r \to 0}{\longrightarrow} -\infty.$$

The surface of the Schwarzschild variables (r, φ) in the subspace (Z_3, Z_4, Z_5) can be drawn after investigating the properties of dZ_3/dr, where $r^2 = Z_4{}^2 + Z_5{}^2$. It is seen that

$dZ_3/dr > 0$, $dZ_3/dr \xrightarrow[r \to 0]{} +\infty$, $dZ_3/dr \xrightarrow[r \to \infty]{} 0$, $Z_3 \xrightarrow[r \to \infty]{} +\infty$ and $Z_3 \xrightarrow[r \to 0]{} -\infty$. This surface is shown in Fig. 14.9.

14.11 Interpretation of the spurious singularity at $r = 2m$; black holes

We verified that the Schwarzschild spacetime has no singularity at $r = 2m$. Nevertheless, this set does have some special properties. Equation (14.57), rewritten in the form

$$\left(\frac{dr}{ds}\right)^2 = E^2 - \left(1 + \frac{J_0^2}{r^2}\right)\left(1 - \frac{2m}{r}\right), \qquad (14.111)$$

shows that for $r \leq 2m$ the derivative dr/ds cannot be zero (note that $E \neq 0$ for geodesics that are timelike in the region $r > 2m$). Hence, if any body enters the region $r \leq 2m$ with $dr/ds < 0$, then dr/ds will not be able to change its sign and the motion will be continued right into the singularity at $r = 0$. If, however, $dr/ds > 0$ at an initial instant in the region $r \leq 2m$, then motion will be continued until values $r > 2m$ are attained, and only then can the recession possibly be reversed to become a fall. Equation (14.72) shows that light rays have the same property. Consequently, material bodies and light rays, once they have crossed the hypersurface $r = 2m$ in the direction of decreasing r, will never be able to turn back and will continue their motion until $r = 0$, also when $J_0 \neq 0$.

For ordinary astronomical objects, such as the Sun or the Earth, the quantity $2m = 2GM/c^2$, called the **gravitational radius**, is very small compared with the physical radius. For example, for the Sun $2m \approx 2.95$ km, for the Earth $2m = 0.89$ cm (recall the physical radii: 696 000 km and 6378 km, respectively). The point at the distance $2m$ from the centre of such an object is hidden deep inside it, where the spacetime metric is different from Schwarzschild's because it is not vacuum. For such objects, the set $r = 2m$ has no physical meaning. This is why the strange properties of this set had been looked upon as mathematical curiosities well into the 1960s. If the Sun had collapsed to a size $r_0 \leq 2m$, then the mean density of matter in the Sun would go up to $\rho_0 \geq 1.85 \times 10^{16}$ g cm^{-3}, which is 100 times more than the density inside an atomic nucleus. Note, however, that $2m$ is proportional to the mass M of the object, while the physical radius r is approximately proportional to $M^{1/3}$ (because $M = (4/3)\pi\bar{\rho}r^3$, where $\bar{\rho}$ is the mean mass-density in the object). Hence, when M is sufficiently large, $2m$ increases much faster than r with increasing M. For every finite density such a mass exists, at which the whole object will be inside the radius $r = 2m$. For $\rho = 1$ g cm^{-3} (the density of water) this mass is equal to 4.8×10^{41} g $\approx 3 \times 10^8$ solar masses ≈ 0.002 of the mass of our Galaxy. The corresponding radius is 4.7 astronomical units, which is less than the radius of the orbit of Jupiter.

Those objects whose physical radii are smaller than $2m$ are called **black holes** because everything that falls into them can never come back out. Note, however, that the theory predicts also the inverse behaviour: objects of radius smaller than $2m$ may emit matter as long as there is any supply inside. These (so far only hypothetical) objects are called

white holes. They can be imagined as the initial cosmological singularity (the **Big Bang**, see Sections 17.4 and 18.3) still going off at an isolated location.

Astronomers have several candidates for black holes among observed objects, and the number of such objects is systematically increasing. It is commonly believed today that every galaxy contains a very massive black hole at its centre. Signatures of such candidate black holes are unusually high energy output in the form of X rays (caused by matter spiralling into the black hole and radiating out energy in consequence of collisions) or unusually high mass concentrated in a small volume (detected through the high orbital velocities of stars in the galaxy). See Frolov and Novikov (1998) for an exhaustive description of the theory and observations of black holes. We will come back to the subject of black holes in Section 18.9, where we will discuss the formation of spherically symmetric black holes in the Universe, and in Chapter 20, where we will briefly discuss rotating black holes.

So far, the only sure identification of a white hole is the whole Universe. As we will see in Chapters 16 and 17, in the currently accepted cosmological models the evolution of the Universe begins with an explosion called the Big Bang. In the most commonly used models, those of Robertson–Walker class, the Big Bang is a single event in the spacetime. The process of expansion away from the Big Bang is the time-reverse of the collapse to a singularity, i.e. it is precisely what a white hole should do. In fact, as we will see in Chapter 17, the R–W models are exceptional in almost every possible respect. In more general models (see Chapters 18 and 19), the Big Bang is not a single event, but a process extended in time. In such models, 'lagging cores' of expansion may exist that would be visible to distant observers as localised white holes. They were once proposed as the explanation of the energy source in quasars (Novikov, 1964a; Neeman and Tauber, 1967), but this explanation was later abandoned in favour of black holes with orbiting discs of matter.

In Newton's theory of gravitation, the escape velocity from the surface of a spherical object of mass M and radius r is $v^2 = 2GM/r$. Note then that $r_g = 2GM/c^2$ is the radius of the object of mass M from whose surface the escape velocity equals the velocity of light. This observation was first made by Laplace in the eighteenth century (Laplace, 1795). Hence, the notion of a black hole had in fact already been introduced then, although the name was coined only in the 1960s.[1]

The Kruskal diagram allows us to follow an object falling into a black hole. The lines $u \pm v = $ constant are intersections of the Kruskal diagram with light cones. Any timelike line (not necessarily geodesic) in that diagram must thus have its tangent inclined to the v-axis at an angle smaller than 45° at every point. Imagine an object proceeding towards the black hole and emitting light signals at regular time intervals. Suppose that they are picked up by an observer resting at $r = r_0 \gg 2m$. When the emitter approaches the surface $r = 2m$, the observer receives the signals at increasing intervals. These intervals tend to

---

[1] However, it is risky to take the analogy between a black hole and Laplace's 'dark star' literally. The surface of the black hole, $r = 2m$, is absolutely impenetrable from inside out. In Newton's theory, no object can escape from the 'dark star' *to infinity*, but the surface $r = 2GM/c^2$ is freely traversable in both directions.

infinity as $r_{\text{sender}} \to 2m$. The signal sent at $r = 2m$ will stay in the set $r = 2m$ for ever. For this reason, this set is called the **event horizon**. All signals sent at $r < 2m$ will hit the singularity at $r = 0$. Thus, from the point of view of the distant observer, the process of falling into a black hole lasts infinitely long. The same is true for matter of an object that collapses to form a black hole. Hence, one should not imagine a black hole as a 'finished' object existing somewhere out there. A distant observer only has a chance to see a black hole at the stage of formation, as an object whose light is becoming darker and redder until it disappears from sight. Objects falling into a black hole will disappear from sight *before* they hit the event horizon.[1]

For an observer who decided to fall into a black hole, the proper time needed to reach the event horizon is finite. This is seen from (14.111), which, for radial motion towards the centre (with $J_0 = 0$), becomes

$$\frac{\mathrm{d}r}{\mathrm{d}s} = -\left(E^2 - 1 + \frac{2m}{r} \right)^{1/2}, \tag{14.112}$$

and then the time of flight from $r = r_0$ to $r = 2m$ is

$$s_{\text{h}} = -\int_{r_0}^{2m} \left(E^2 - 1 - \frac{2m}{r} \right)^{-1/2} \mathrm{d}r. \tag{14.113}$$

This integral has a well-defined finite value for every $r_0 < \infty$.

14.12 The Schwarzschild solution in other coordinate systems

The curvature coordinates were the oldest introduced for the Schwarzschild solution, and in a way the most natural. However, for investigating some advanced topics, in particular the relation of the Schwarzschild solution to other solutions of Einstein's equations, other coordinates are sometimes more useful. Two coordinate systems are used most often.

The first are the **isotropic coordinates**, in which the subspace $t = \text{constant}$ is explicitly conformally flat and in which a large number of nonvacuum generalisations of the Schwarzschild solution was found (Krasiński, 1997). Take the Schwarzschild solution in curvature coordinates, (14.40) and (14.42), and transform the r-coordinate as follows:

$$r = r' \left(1 + \frac{m}{2r'} \right)^2. \tag{14.114}$$

The resulting metric is:

$$\mathrm{d}s^2 = \frac{[1 - m/(2r')]^2}{[1 + m/(2r')]^2} \, \mathrm{d}t^2 - \left(1 + \frac{m}{2r'} \right)^4 \left[\mathrm{d}r'^2 + r'^2 \left(\mathrm{d}\vartheta^2 + \sin^2 \vartheta \, \mathrm{d}\varphi^2 \right) \right]. \tag{14.115}$$

---

[1] Because of the high symmetry of the Schwarzschild solution, the event horizon in it coincides with two other entities that are in general distinct. One of them is the *apparent horizon*. The future apparent horizon is a hypersurface in spacetime within which all null geodesics can proceed only towards the singularity, never away from it. As we will see in Chapter 18, in nonstatic spacetimes the apparent horizon and the event horizon in general do not coincide. Then, in gravitational fields of rotating bodies, the *infinite redshift hypersurface* (IRH) does not in general coincide with the event horizon, as we will see in Chapter 20. From within this hypersurface, light signals arrive at infinity being infinitely redshifted, yet the IRH is freely traversable both ways for material particles and light rays.

The spurious singularity now appears at $r' = m/2$.

The other important coordinates are sometimes called **geodesic coordinates**, **Lemaître coordinates** or **Novikov coordinates**. They are comoving coordinates of a congruence of radially freely falling observers. The Schwarzschild metric in them is:

$$ds^2 = dt^2 - \frac{R_{,r}^2 \, dr^2}{1 + 2E(r)} - R^2(t, r) \left(d\vartheta^2 + \sin^2 \vartheta \, d\varphi^2 \right), \tag{14.116}$$

where the function $R(t, r)$ obeys the equation

$$R_{,t}^2 = \frac{2m}{R} + 2E(r), \tag{14.117}$$

$E(r)$ being an arbitrary function. A subcase of this coordinate system, corresponding to $E = 0$, was first introduced by Lemaître (1933a).[1] The remaining cases were introduced by Novikov (1964b), who provided the physical interpretation of all cases. Equation (14.117) is the equation of radial free fall in the spherically symmetric gravitational field generated by the mass m, with $E(r)$ being the kinetic energy of the observer at infinity (and the total conserved energy of the motion). An observer at fixed coordinate r (note that r is in fact arbitrary in this metric; transformations of the form $r = f(r')$ do not change it) proceeds with time to other values of R, i.e. is either receding from or approaching the centre of symmetry, in accordance with the equation of free fall. When $E > 0$, the observers can recede to infinity and still have nonzero kinetic energy there. With $E = 0$, they can still recede to infinity, but their kinetic energy decreases to zero. With $E < 0$, the observers can fly away from the centre of symmetry only out to a finite distance, and then fall back. For each sign of E, Eq. (14.117) may be explicitly solved for $t(R)$, but for $E \neq 0$ the solutions cannot be inverted to define an explicit elementary function $R(t, r)$.

With $E \geq 0$, the metric (14.116) has no singularity anywhere apart from $R = 0$. As mentioned before, this is how Lemaître first noticed that the singularity of the Schwarzschild solution at $r = 2m$ is only an artefact of the coordinates used.

The Schwarzschild solution in the form (14.116)–(14.117) emerges as the vacuum limit of the **Lemaître–Tolman** cosmological model that will be discussed in Chapter 18. Equation (14.117) frequently appears in cosmology – it is the evolution equation that governs the Lemaître–Tolman model and the Friedmann model that is a spatially homogeneous subcase of the former. It is in fact simpler to obtain this form by assuming geodesic coordinates (in which $g_{00} = 1$, $g_{01} = 0$, other components are those appropriate for a spherically symmetric metric, (8.51)) and solving the Einstein equations.

14.13 The equation of hydrostatic equilibrium

We will now investigate the Einstein equations *inside* a spherically symmetric body of perfect fluid, with the additional assumption that the matter inside the body is at rest in

---

[1] Lemaître's form was somewhat unreadable because he parametrised his metric in such a way that the limit of zero cosmological constant could not be directly taken.

the coordinates of Section 14.1. This assumption means that the velocity field has only the t-component:

$$u^\alpha = e^{-\nu}\delta^\alpha{}_0, \qquad u_\alpha = e^\nu\delta^0{}_\alpha, \tag{14.118}$$

where $\nu(t, r)$ is an unknown function, while the pressure and density of the perfect fluid are constant along the lines of flow, i.e. depend only on r. Choosing an orthonormal tetrad in which $e_0{}^\alpha = u^\alpha$ (and consequently the tetrad components of u^α are $u^i = \delta^i{}_0$) and using (14.26)–(14.29), we obtain the following set of equations:[1]

$$G_{00} = e^{-2\mu}\left(\frac{2}{r}\mu' - \frac{1}{r^2}\right) + \frac{1}{r^2} = \frac{8\pi G}{c^4}\epsilon, \tag{14.119}$$

$$G_{01} = \frac{2}{r}e^{-\mu-\nu}\dot\mu = 0, \tag{14.120}$$

$$G_{11} = e^{-2\mu}\left(\frac{2}{r}\nu' + \frac{1}{r^2}\right) - \frac{1}{r^2} = \frac{8\pi G}{c^4}p, \tag{14.121}$$

$$G_{22} = G_{33} = e^{-2\mu}\left(\nu'' - \mu'\nu' + \nu'^2 + \frac{\nu' - \mu'}{r}\right)$$

$$-e^{-2\nu}\left(\ddot\mu - \dot\mu\dot\nu + \dot\mu^2\right) = \frac{8\pi G}{c^4}p. \tag{14.122}$$

Equation (14.120) implies that $\dot\mu = 0$, like in the Schwarzschild solution. Since $\dot p = \dot\epsilon = 0$, differentiating (14.121) by t we obtain $\dot\nu' = 0$, i.e. $\nu = f(t) + g(r)$. Then, the coordinate transformation $t' = \int e^{f}(t)dt$ gives $f = 0$ in the new coordinates, and $\nu = \nu(r)$.

Equation (14.119) can now be formally integrated. Assuming that $\epsilon(0)$ is finite, we obtain

$$e^{-2\mu} = 1 - \frac{8\pi G}{c^4 r}\int_0^r \epsilon(r')r'^2\,dr'. \tag{14.123}$$

We denote

$$M(r) \overset{\text{def}}{=} \frac{4\pi}{c^2}\int_0^r \epsilon(r')r'^2\,dr'. \tag{14.124}$$

This quantity has the dimension of mass and plays in (14.18) the same role as the mass parameter played in the Schwarzschild solution: if $\epsilon(r) \equiv 0$ for $r \geq R$ (i.e. if the surface of the matter distribution is at $r = R$), then $M(R)$ is equal to the mass parameter of the Schwarzschild metric. Consequently, we interpret $M(r)$ as the mass within the sphere of radius r. Note that this mass is smaller than the sum of rest masses of all the particles inside the body. The density of rest mass is $\rho(r) = \epsilon(r)/c^2$, so the sum of rest masses within the sphere of radius r is

$$M_{\text{rest}} = \int_{\text{Vol}_3(r)} \rho(r')\sqrt{-g}\,d_3x = \frac{4\pi}{c^2}\int_0^r \epsilon(r')r'^2\,e^{\mu(r')}\,dr'. \tag{14.125}$$

---

[1] By choosing the metric in the form (14.18) we left aside the cases when the $\delta_{,\alpha}$ in (8.51) is a timelike or null vector, or is constant. Those cases do not contain static sources for the Schwarzschild metric, but they cannot be neglected when considering nonstatic sources.

Since $e^\mu > 1$ (from (14.123)), we have $M_{rest} > M(r)$. This is the large-scale (astrophysical) analogue of the 'mass defect' known from nuclear/elementary particle physics: the mass of a bound object is smaller than the sum of masses of its components. The mass defect multiplied by c^2 is equal to the energy that would have to be supplied in order to break the object up into separate particles.

Using the notation (14.124), we obtain in (14.123)

$$e^{-2\mu} = 1 - \frac{2GM(r)}{c^2 r}. \tag{14.126}$$

The equations of motion $T^{\alpha\beta}{}_{;\beta} = 0$ must hold in consequence of the field equations, and in the present case they will be first integrals of the field equations. For $\alpha = 0, 2, 3$ they are fulfilled identically, while $T^{1\beta}{}_{;\beta} = 0$ implies $e^{-2\mu}[p' + \nu'(\epsilon + p)] = 0$, which means that

$$\nu' = -\frac{p'}{\epsilon + p}. \tag{14.127}$$

This can be integrated after an equation of state is assumed (necessarily of the form $\epsilon = \epsilon(p)$).

Using (14.126) and (14.127), Eq. (14.121) can now be written as

$$\frac{dp}{dr} = -\frac{G\left(\rho + p/c^2\right)\left[M(r) + 4\pi r^3 p/c^2\right]}{r^2\left[1 - 2GM(r)/(c^2 r)\right]}. \tag{14.128}$$

Equation (14.122) is now fulfilled by virtue of those already solved.

Equation (14.128) is called the **equation of hydrostatic equilibrium**. In the Newtonian limit $c \to \infty$ it becomes

$$\frac{dp}{dr} = -\frac{G\rho M(r)}{r^2}. \tag{14.129}$$

Comparing (14.129) with (14.128) we see that in relativity the pressure increases the gravitational attraction because it appears in the equilibrium equation as a positive contribution to the mass and mass density. Given the same mass, density and pressure, the *gradient* of pressure, opposing the gravitational attraction, is greater in (14.128) than in (14.129). The greater pressure gradient implies faster growth of pressure towards the centre of the object, and this increased pressure then requires a still greater gradient. One can thus imagine a situation in which the equilibrium maintained over some time by a certain process (such as energy production in a star) is perturbed, and then the growth of pressure will lead to a loss of stability: no pressure gradient will be able to keep the object static again. A collapse to a size smaller than the gravitational radius $r_g = 2GM/c^2$ will then occur, and a black hole will be created. The Newtonian equilibrium equation (14.129) does not allow such a situation: the gradient of pressure needed to maintain equilibrium is determined by the mass density at the distance r from the centre of the object and does not depend on the value of pressure, so the solution exists for every density and mass.

14.14 The 'interior Schwarzschild solution'

We are interested here in providing a material source for the Schwarzschild metric. Equation (14.127) can be integrated after an equation of state, or a distribution of mass density inside the object, has been defined. Various equations of state and mass distributions intended to imitate the real conditions inside various stellar objects are discussed in relativistic astrophysics. Typically, such general equations of state lead to complicated forms of (14.127) that can be integrated only numerically. We will deal here with an example that is unrealistic and of academic interest only, but it illustrates in a simple way the problems encountered while matching matter solutions with vacuum solutions. We assume that $\epsilon = $ constant, a constant mass density inside the object. From (14.122)–(14.124) we then obtain:

$$e^{-2\mu} = 1 - \frac{8\pi G}{3c^4} \epsilon r^2 \overset{\text{def}}{=} 1 - Dr^2, \qquad (14.130)$$

$$M(r) = \frac{4\pi G}{3c^2} \epsilon r^3 = \frac{c^2}{2G} Dr^3, \qquad (14.131)$$

where we used the abbreviation $D \overset{\text{def}}{=} 8\pi G\epsilon/(3c^4) = $ constant > 0. Subtracting (14.121) from (14.122) and using (14.130) we obtain

$$\left(1 - Dr^2\right)\left(\nu'' + \nu'^2\right) - Dr\nu' - \frac{1}{r}\left(1 - Dr^2\right)\nu' = 0, \qquad (14.132)$$

which is integrated with the result $e^\nu = C - B\sqrt{1 - Dr^2}$, where B and C are constants. The pressure can now be calculated from (14.121):

$$p = \frac{1}{3}\epsilon \frac{3B\sqrt{1 - Dr^2} - C}{C - B\sqrt{1 - Dr^2}}. \qquad (14.133)$$

This pressure obeys (14.128). Hence, we have obtained a complete solution of Eqs. (14.119)–(14.122):

$$ds^2 = \left(C - B\sqrt{1 - Dr^2}\right)^2 dt^2 - \frac{dr^2}{1 - Dr^2} - r^2\left(d\vartheta^2 + \sin^2\vartheta\, d\varphi^2\right). \qquad (14.134)$$

This is called the **interior Schwarzschild solution** (Schwarzschild, 1916b). Its characteristic geometric property is conformal flatness (which is not evident in the coordinates of (14.134) – but its Weyl tensor is zero).

Now let us verify the matching conditions between (14.134) and the vacuum Schwarzschild solution, (14.40) and (14.42), at the hypersurface Σ given by $r = R = $ constant. The coordinate systems on both sides of Σ are already adapted, so we can use the formalism of Section 12.17. Here, $x^4 = r$. The continuity of the metric components g_{IJ}, $I \neq 4 \neq J$, requires

$$\left(C - B\sqrt{1 - DR^2}\right)^2 = 1 - \frac{2GM}{c^2 R}, \qquad (14.135)$$

and then the continuity of the derivatives of the metric of Σ in directions tangent to Σ is guaranteed, as stated in Section 12.17. Following the recipe of Section 12.17, we require the continuity of $g_{IJ,r}/N$ at $r = R$, where $N = \sqrt{1 - 2GM/(c^2R)}$ in the exterior metric and $N = \sqrt{1 - DR^2}$ in the interior metric. This imposes two more equations:

$$\sqrt{1 - DR^2} = \sqrt{1 - \frac{2GM}{c^2R}}, \tag{14.136}$$

$$2DBR\left(C - B\sqrt{1 - DR^2}\right) = \frac{2GM}{c^2R^2}\sqrt{1 - \frac{2GM}{c^2R}}. \tag{14.137}$$

(The first one results from $g_{22,r}/N$ being equal on both sides of Σ, which implies N being the same on both sides.) The solution of Eqs. (14.135)–(14.137) is

$$D = \frac{2GM}{c^2R^3}, \qquad B = \frac{1}{2}, \qquad C = \frac{3}{2}\sqrt{1 - \frac{2GM}{c^2R}}. \tag{14.138}$$

This guarantees that $p(R) = 0$; see (14.133).

14.15 * The maximal analytic extension of the Reissner–Nordström solution

The Reissner–Nordström (R–N) solution, given by (14.40)–(14.41) with $\Lambda = 0$, can have its spurious singularities at those points where $e^{2\nu} = 0$. There are three cases to consider separately:

- 1. When $m^2 - e^2 < 0$, $e^{2\nu}$ does not vanish at any value of r, and no spurious singularity exists. This case has no Schwarzschild limit.
- 2. When $m^2 - e^2 > 0$, $e^{2\nu}$ vanishes at two different values of r,

$$r_- = m - \sqrt{m^2 - e^2}, \qquad r_+ = m + \sqrt{m^2 - e^2}. \tag{14.139}$$

These are spurious singularities – the tetrad components of the Riemann tensor have well-defined values at those points. In the Schwarzschild limit $e \to 0$, the inner spurious singularity at $r = r_-$ collapses onto the genuine singularity at $r = 0$, while the outer one goes over into the event horizon at $r = 2m$.
- 3. When $m^2 - e^2 = 0$, $e^{2\nu}$ vanishes at just one value of r, $r = m$. This case has no Schwarzschild limit, either.

Similarly, as was done with the Schwarzschild solution in Section 14.9, the spurious singularities of the R–N solution can be removed by a coordinate transformation. Following Graves and Brill (1960), we will show how to remove a spurious singularity in a more general static metric,

$$ds^2 = \phi dt^2 - \frac{1}{\phi}dr^2 - r^2\left(d\vartheta^2 + \sin^2\vartheta\, d\varphi^2\right), \tag{14.140}$$

where $\phi(r)$ is any function. Like in the Kruskal method of Section 14.9, we begin by introducing the coordinates $u(t, r)$ and $v(t, r)$ in which

$$ds^2 = f^2(u, v)\left(dv^2 - du^2\right) - r^2(u, v)\left(d\vartheta^2 + \sin^2\vartheta\, d\varphi^2\right). \tag{14.141}$$

The functions f, u and v must obey

$$f^2\left(v_{,t}^{2} - u_{,t}^{2}\right) = \phi(r), \qquad f^2\left(v_{,r}^{2} - u_{,r}^{2}\right) = -\frac{1}{\phi(r)},$$

$$u_{,r}u_{,t} - v_{,r}v_{,t} = 0. \tag{14.142}$$

The last equation says that $v_{,t}/u_{,t} = u_{,r}/v_{,r}$. Dividing the first equation by the second and using this, we obtain $u_{,t}^{2}/v_{,r}^{2} = \phi^2(r)$. From here and from the last equation in (14.142) we then obtain the set

$$u_{,t} = \phi(r)v_{,r}, \qquad v_{,t} = \phi(r)u_{,r}. \tag{14.143}$$

This is easily solved if we introduce the new variable $r^*(r)$ defined by $\mathrm{d}r^*/\mathrm{d}r = 1/\phi$. The general solution is

$$u = h(r^* + t) + g(r^* - t), \qquad v = h(r^* + t) - g(r^* - t), \tag{14.144}$$

where h and g are arbitrary functions. A prime will denote derivatives of h and g by their arguments. Using (14.144), we find from (14.142)

$$f^2 = \frac{\phi(r)}{4h'(r^* + t)g'(r^* - t)}. \tag{14.145}$$

Any singularity or zero of $\phi(r)$ must now be cancelled by the product in the denominator, and the resulting f must be time-independent (success is not guaranteed, but such choices of $g'h'$ were proven to exist for the Schwarzschild metric and for the R–N metric). The product $h'(r^* + t)g'(r^* - t)$ will be independent of t only if

$$h = A\mathrm{e}^{\gamma(r^* + t)} + C, \qquad g = B\mathrm{e}^{\gamma(r^* - t)} + D, \tag{14.146}$$

where A, B, C, D and γ are arbitrary constants; we shall take $A = B$ and $C = D = 0$. The formula for f^2 then becomes

$$f^2 = \frac{\phi(r)}{4A^2\gamma^2\mathrm{e}^{2\gamma r^*}}. \tag{14.147}$$

Now, the constant γ must be chosen so that any zero or singularity in ϕ is cancelled. Substituting (14.147) in (14.144) we obtain the formulae for the transformation $(t, r) \to (v, u)$:

$$\begin{aligned} u &= A\mathrm{e}^{\gamma r^*}\left(\mathrm{e}^{\gamma t} + \mathrm{e}^{-\gamma t}\right) \equiv 2A\mathrm{e}^{\gamma r^*}\cosh(\gamma t), \\ v &= A\mathrm{e}^{\gamma r^*}\left(\mathrm{e}^{\gamma t} - \mathrm{e}^{-\gamma t}\right) \equiv 2A\mathrm{e}^{\gamma r^*}\sinh(\gamma t). \end{aligned} \tag{14.148}$$

In general, the inverse transformation can be given only implicitly because the formula for $r^*(r)$ cannot be inverted in elementary functions. The inverse transformation is

$$F(r) \overset{\mathrm{def}}{=} 4A^2\mathrm{e}^{2\gamma r^*} = u^2 - v^2, \qquad t = \frac{1}{\gamma}\mathrm{artanh}(v/u), \tag{14.149}$$

where artanh is the inverse function to tanh. Thus, in the (v, u) coordinates, lines of constant t are straight lines through the origin, $v/u = $ constant, and lines of constant r are the hyperbolae $u^2 - v^2 = $ constant.

For the R–N metric in the case $e^2 < m^2$ we have

$$\phi = 1 - \frac{2m}{r} + \frac{e^2}{r^2} = \frac{(r - r_+)(r - r_-)}{r^2}, \tag{14.150}$$

and the expression for $r^*(r)$ is

$$r^* = r + \frac{r_+^2}{r_+ - r_-} \ln|r - r_+| - \frac{r_-^2}{r_+ - r_-} \ln|r - r_-|. \tag{14.151}$$

With this ϕ, we have in (14.147)

$$f^2 = \frac{(r - r_+)^{1 - 2\gamma r_+^2/(r_+ - r_-)}(r - r_-)^{1 + 2\gamma r_+^2/(r_+ - r_-)}}{4A^2 \gamma^2 r^2 e^{2\gamma r}}. \tag{14.152}$$

As can be seen, γ can be chosen so that *one* of the spurious singularities is cancelled, but not both at once. Let the index $i = 1$ refer to r_+ and $i = 2$ to r_-. To cancel the r_i singularity γ must be

$$\gamma_i = \frac{r_i - r_j}{2r_i^2}, \tag{14.153}$$

where $i \neq j$.

If we choose γ so as to cancel the spurious singularity at $r = r_+$, then, in the (v, u) coordinates, we can proceed from large r towards smaller r across the set $r = r_+$. In order to continue further, across $r = r_-$, we then have to go back to the original (t, r) coordinates and transform them to such (v, u) that cancel the second spurious singularity. In the new coordinates, we can continue across $r = r_-$. In order to visualise the extended manifold, we first have to locate the spurious singularities and the true singularity in the (v, u) plane.

As can be seen from (14.151) and (14.153), with $\gamma = \gamma_i$, the function $F(r)$ in (14.149) vanishes at at $r = r_i$, so this set, in the coordinate patch that makes it nonsingular, has the equation $u = \pm v$. Note that, just as in the Schwarzschild case, this spurious singularity consists of two lines in the (v, u) plane. At the true singularity $r = 0$, $F(r)$ is a positive constant. Hence, in the (v, u) coordinates, the equation of the singular set is $u^2 - v^2 = $ constant > 0, i.e. it is a pair of hyperbolae that intersect the horizontal u-axis. (Note: unlike in the Schwarzschild spacetime, the hyperbolae here are *timelike*, i.e. they are rotated by 90° with respect to those of the Kruskal diagram.)

Before we draw the picture, we will employ the method known as the **Penrose transformation**. The (v, u) plane is infinite, and in order to visualise its various properties it is often useful to map it into a finite patch of the plane in such a way that null geodesics go over into null geodesics. This kind of mapping is called the Penrose transformation, and the picture of the spacetime in the new coordinates is called the **conformal diagram**.

The radial null geodesics in the metric (14.141) are $du/dv = \pm 1$. Therefore we first go over to the null coordinates

$$p = u + v, \qquad q = u - v, \tag{14.154}$$

and then employ the following Penrose transformation:

$$P = \tanh p, \qquad Q = \tanh q. \tag{14.155}$$

In the (Q, P) coordinates, the equation of the spurious singularity at $u = v$ is $Q = 0$, and that of the one at $u = -v$ is $P = 0$. The (v, u) plane now fits in the square $\{P, Q\} = \{[-1, 1] \times [-1, 1]\}$, and the **null infinities** $p = \pm\infty$, $q = \pm\infty$ are mapped into the sets $P = \pm 1$, $Q = \pm 1$. We can introduce the usual time–space coordinates in this square by

$$U = (P + Q)/2, \qquad V = (P - Q)/2. \tag{14.156}$$

The spurious singularity that was removed still has the equation $U = \pm V$, while the null infinities (infinities of p and q) become the four straight line segments $P = U + V = \pm 1$, $Q = U - V = \pm 1$. The image of the true singularity at $r = 0$ and the lines $r = $ constant are still hyperbolae, given by equations of the form $\alpha^2 \left[(U + 1/\alpha)^2 - V^2 \right] = 1$, where $\alpha = $ constant; see Exercise 13.

Note that the equation of the true singularity, $u^2 - v^2 = 4A^2 e^{2\gamma r^*(0)} \stackrel{\text{def}}{=} \mathcal{A} = $ constant, in the (p, q) coordinates has the form $pq = \mathcal{A}$, so the singular set contains the points $\{p = \pm\infty, q = 0\}$ and $\{p = 0, q = \pm\infty\}$, whose (P, Q) coordinates are $\{P = \pm 1, Q = 0\}$ and $\{P = 0, Q = \pm 1\}$. These four points thus lie also in the images of null infinities and in the spurious singularities $P = 0$ or $Q = 0$. Thus, the spurious singularities $r = r_+$, the true singularities and the images of infinities do have common points, as will be seen in the picture.

Just as in the Schwarzschild spacetime, when we proceed from a point A in the $r > r_+$ region back in time along a $p = $ constant null geodesic and cross the spurious singularity $r = r_+$ at $q = 0$, we land in a different region of spacetime from that which would be reached by proceeding from A to the future along a $q = $ constant null geodesic. By extending these two kinds of null geodesics, we recover the analogues of sectors I, II and IV of Fig. 14.6. By sending null geodesics back in time from sector II and to the future from sector IV, we can also recover the analogue of sector III. Now, when we are in one of the $r < r_+$ regions, we change back to the (t, r) coordinates, transform them so as to cancel the $r = r_-$ spurious singularity, and then carry out the Penrose transformation once more. We draw the new conformal diagrams in such a way that their images of the $r = r_+$ spurious singularities coincide with the $r = r_+$ singularities of the original Penrose diagram (this in fact involves some deformation). The $r = r_-$ singularities are now again straight lines and, as before, we find that their conformal images do have common points with the endpoints of the true singularities. In this way, by patching together conformal diagrams of different parts of the original manifold, we arrive at the manifold shown in the rectangle in the centre of Fig. 14.10.

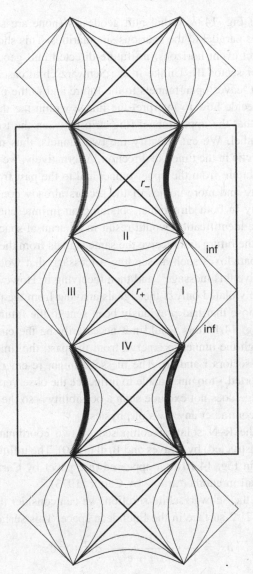

Fig. 14.10.   The conformal diagram of the maximally extended Reissner–Nordström spacetime. Explanation is given in the text.

The thin straight segments in Fig. 14.10 (two of them are marked 'inf') are the conformal images of the null infinities, where $r \to \infty$. Timelike and spacelike infinities are the endpoints of the null infinity segments. The thin hyperbolae segments are the $r =$ constant lines; they are timelike for $r > r_+$ and $r < r_-$ and spacelike for $r_- < r < r_+$. The thick straight segments are the spurious singularities at $r = r_+$ and $r = r_-$. The hatched hyperbolae segments are the true singularities at $r = 0$. Roman numbers label sectors analogous to those of Fig. 14.6. Radial null geodesics (none are shown in Fig. 14.10) would be straight lines parallel to the spurious singularities. This shows that the spurious singularities are in fact event horizons: no future-directed null geodesic can cross from sector II to sector I or sector III. Unlike in the Schwarzschild case, the true singularity here is timelike and it leaves open tunnels to the future and to the past.

Now we have to decide how to interpret the image within the thick-line rectangle in Fig. 14.10. We note that the upper tunnel (the strip between the true singularities) is a copy of the lower tunnel. We can identify the two tunnels, thus making the extended manifold finite and cyclic in the timelike direction. Alternatively, we can continue to send null geodesics to the future from the upper tunnel and to the past from the lower tunnel, thus constructing more and more copies of the sectors already constructed, and extend the picture indefinitely in both directions, obtaining an infinite chain. It has been said (Carter, 1973) that the identification would result in an acausal spacetime, in which one could send signals to the future and receive the same signals from the past. This, the story goes, would lead to paradoxes such as sending a message that would say 'do not send any signal if you receive this message', and then receiving it. However, it must be added that causality could be violated only if the signals arriving from the past were intelligible and recognisable as those that had previously been sent to the future. With a spacetime such as the one in Fig. 14.10 this would not necessarily be the case: while a message would continue through the tunnel to emerge from the past, the time would continue to flow for observers in sectors I and III. The message might re-emerge after a very long time and strongly distorted – too unreadable to influence the observers' decisions. Current experimental knowledge does not exclude such a possibility – so the identification of the two tunnels does not contradict any laws of physics.

The extension of the R–N solution composed of two coordinate patches was first contemplated (and constructed) by Graves and Brill (1960). The infinite mosaic of conformal diagrams shown in Fig. 14.10 first appeared in a paper by Carter (1966a) and was described in more detail in another article by Carter (1973).

Just like we did for the Schwarzschild solution, we can consider the embedding of the $\{t = \text{constant}, \vartheta = \pi/2\}$ 2-surface in the Euclidean space. This surface has the metric

$$ds^2 = \frac{1}{1 - 2m/r + e^2/r^2} \, dr^2 + r^2 \, d\varphi^2, \qquad (14.157)$$

and, if this is going to be the metric of the surface $z(r)$ in the Euclidean space with $ds^2 = dz^2 + dr^2 + r^2 \, d\varphi^2$, then

$$z(r) = \int_{r_+}^{r} \sqrt{\frac{1}{1 - 2m'/r + e^2/r'^2} - 1} \, dr' \equiv \int_{r_+}^{r} \sqrt{\frac{2m'r' - e^2}{r'^2 - 2mr' + e^2}} \, dr'. \qquad (14.158)$$

This integral reduces to (14.107) when $e = 0$. Figure 14.11 shows the comparison of the surface (14.157) with the surface shown in Fig. 14.7. It is seen that the presence of charge makes the throat longer and thinner.

Like in the Schwarzschild spacetime, the surface $\{t = \text{constant}, \vartheta = \pi/2\}$ coincides with the surface $v = 0$ in those (u, v) coordinates that cancel the $r = r_+$ singularity, i.e.

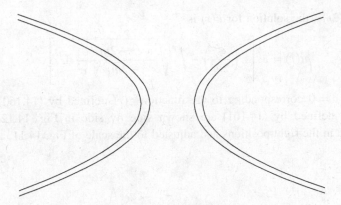

Fig. 14.11. A comparison of the 'throat' in the Schwarzschild spacetime at $v = 0$ (inner curves) and in the R–N spacetime with $e^2 < m^2$ at $r = r_+$ (outer curves). The curves are cross-sections of surfaces like the one in Fig. 14.7 with a vertical plane. The mass parameters for both surfaces are the same. The presence of charge makes the throat longer and thinner; with the charge being close to the limiting value $\pm m$, it can be very much longer. In fact, it becomes infinitely long when $e^2 \to m^2$; see the comment after Eq. (14.162).

with horizontal sections of the manifold shown in Fig. 14.10 that go through the crossing of the $r = r_+$ horizons. Unlike in the Schwarzschild spacetime, when we go to higher values of v, the throat never shrinks to a point and the two sheets of the surface never separate. Imagine the $v = $ constant plane being moved upwards from the position at the crossing of the $r = r_+$ horizons. The minimal value of r in the surface first decreases from r_+ to r_- and then increases to r_+ again. Thus, as some authors like to say, the flux of the electric field through the throat prevents it from collapsing to a point; instead the throat pulsates periodically between the radii r_- and r_+.

The surface $v = $ constant changes its geometry completely as it enters the tunnel between the true singularities – then it does not recede to infinite distances as in Fig. 14.11, but remains finite in extent. Then, with $r \le r_-$, embedding the surface with the metric (14.157) in a Euclidean space by the same method as before requires solving the equation

$$z_{,r}^2 = \frac{2mr - e^2}{r^2 - 2mr + e^2}.$$ (14.159)

A solution will exist only if $e^2/(2m) \le r \le r_-$, because for $r < e^2/(2m)$ the right-hand side of (14.159) is negative. The other part of the $\{t = \text{constant}, \vartheta = \pi/2\}$ surface can be embedded in a flat 3-space with the indefinite metric $(dr^2 + r^2 \, d\varphi^2 - du^2)$. The solution of (14.159) with $e^2/(2m) \le r \le r_-$ is

$$z(r) = \pm \int_{r_-}^{r} \sqrt{\frac{2mr' - e^2}{r'^2 - 2mr' + e^2}} \, dr'.$$ (14.160)

With $r < e^2/(2m)$, the solution for $u(r)$ is

$$u(r) = \pm \left[z\left(\frac{e^2}{2m}\right) - \int_0^r \sqrt{\frac{e^2 - 2mr'}{r'^2 - 2mr' + e^2}}\, dr' \right]. \tag{14.161}$$

The surfaces $v = 0$ corresponding to the function $z(r)$ defined by (14.160) and to the function $u(r)$ defined by (14.161) are shown side by side in Fig. 14.12. The same graphs, placed in the right positions and adjusted to the scale of Fig. 14.11, are shown in Fig. 14.13.

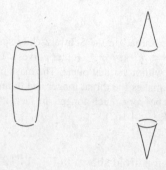

Fig. 14.12. Embeddings of the $v = 0$ surface that passes through $r = r_-$. The figure on the left corresponds to the function $z(r)$ defined by (14.160) and an embedding in an ordinary Euclidean 3-space; the figure on the right corresponds to the function $u(r)$ defined by (14.161) and an embedding in a flat 3-space with the indefinite metric of signature $(+ + -)$. The 'equator' of the left surface corresponds to $r = r_-$; the upper and lower ends of the left surface correspond to $r = e^2/(2m)$, where the embedding in the Euclidean space breaks down. The tips of the right figure correspond to the singularity at $r = 0$; the gap between the two parts of the figure has its edges at $r = e^2/(2m)$, where the embedding in this space also breaks down.

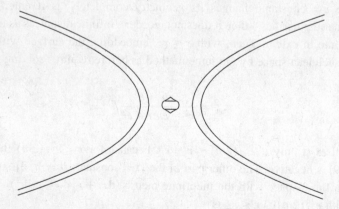

Fig. 14.13. Cross-sections of the surfaces of Fig. 14.12 transformed to the scale of Fig. 14.11 and placed in the correct positions with respect to the $r = r_+$ surface. The horizontal lines separate the part embedded in a Euclidean space (between them) from the part embedded in the space with indefinite metric.

For the extreme case $e^2 = m^2$ the metric can be written as

$$ds^2 = \left(\frac{r-m}{r}\right)^2 dt^2 - \left(\frac{r}{r-m}\right)^2 dr^2 - r^2 \left(d\vartheta^2 + \sin^2\vartheta\, d\varphi^2\right). \tag{14.162}$$

Note that this case is indeed essentially different from the case $e^2 < m^2$. In the latter, the invariant radial distance from any point with coordinate r to the nearest spurious singularity (for $r > r_+$ the distance is equal to $\ell(r) = \int_{r_+}^{r} \left(r'/\sqrt{r'^2 - 2mr' + e^2}\right) dr'$) is finite. With $e^2 = m^2$, the invariant distance from any point $r \neq m$ to the spurious singularity $r = m$ is infinite, as may be verified from the above.

In order to find the extension, we consider the surface $\{\vartheta = \text{constant}, \varphi = \text{constant}\}$, and write the metric as

$$ds^2 = \left(\frac{r-m}{r}\right)^2 \left[dt - \left(\frac{r}{r-m}\right)^2 dr\right]\left[dt + \left(\frac{r}{r-m}\right)^2 dr\right]. \tag{14.163}$$

We introduce the null coordinates p and q by

$$p = t + \zeta, \qquad q = t - \zeta,$$

$$\zeta \stackrel{\text{def}}{=} \int \frac{r^2}{(r-m)^2}\, dr = r - m - \frac{m^2}{r-m} + 2m \ln|r - m|. \tag{14.164}$$

The metric then becomes

$$ds^2 = \left[\left(\frac{r-m}{r}\right)^2 dp\, dq - r^2\left(d\vartheta^2 + \sin^2\vartheta\, d\varphi^2\right)\right]_{r=r(p,q)}. \tag{14.165}$$

The sets $\{r = m, \vartheta = \text{constant}, \varphi = \text{constant}\}$ are seen to be composed of radial null geodesics, so in the (p, q) coordinates they are represented by lines parallel to $p = \text{constant}$ or $q = \text{constant}$.

The metric is regular for all real values of p and q, and the spurious singularity at $r = m$ lies at $p = -\infty$ and $q = +\infty$ when approached from $r > m$, and at $p = +\infty$ and $q = -\infty$ when approached from $r < m$, but these two regions are covered by two different infinite coordinate patches. The infinity $r = \infty$ is at $p = +\infty$ and $q = -\infty$ in the first patch. The singularity $r = 0$ lies on the straight line $p - q = 4m \ln m$ in the second patch. We bring the infinite values of p and q to finite distances by the transformation

$$p = \tan P, \qquad q = \tan Q. \tag{14.166}$$

In this way, the images of $r = m$ in the two patches can be laid side by side; they are now at $P = -\pi/2$ and $Q = \pi/2$ in the $r > m$ patch, and at $P = \pi/2$ and $Q = -\pi/2$ in the $r < m$ patch. The infinity is at $P = \pi/2$ and $Q = -\pi/2$ in the $r > m$ patch. The singularity is at $\tan P - \tan Q = 4m \ln m$ in the $r < m$ patch. Note that, just as in the case $e^2 < m^2$, the image of the singularity includes points at which simultaneously $P = Q = \pi/2$ and those where $P = Q = -\pi/2$, i.e. the singularity has common points with the images of the spurious singularity. However, this time the singularity will have no common points with the infinities. Putting all those bits of information together, we obtain the infinite chain of conformal diagrams shown in Fig. 14.14.

Fig. 14.14.   Conformal diagrams for the maximal extension of the extreme R–N metric, with $e^2 = m^2$. The thin straight segments are the images of the null infinities, where $r \to \infty$. The thin hyperbolae segments are $r =$ constant lines; this time they are timelike except the $r = m$ lines that are null. The thick straight segments are the spurious singularities at $r = m$; they are again event horizons. The hatched curve segments are the true singularities at $r = 0$; they are timelike, too. Just as in the $e^2 < m^2$ case, we can choose to identify the square at the bottom with the next one up.

The extension of this case and Fig. 14.14 were first presented by Carter (1966a, 1973).

Just like we did for the Schwarzschild metric and for the R–N solution with $e^2 < m^2$, also in the present case we can embed the surface $\{t = $ constant, $\vartheta = \pi/2\}$ in a flat 3-dimensional space. As in the previous case, the embedding is different on each side of the spurious singularity $r = m$, and again the region with $r < m$ cannot all be embedded in a Euclidean space. From $\mathrm{d}(z(r))^2 + \mathrm{d}r^2 + r^2\,\mathrm{d}\varphi^2 = r^2\,\mathrm{d}r^2/(r-m)^2 + r^2\,\mathrm{d}\varphi^2$ we obtain

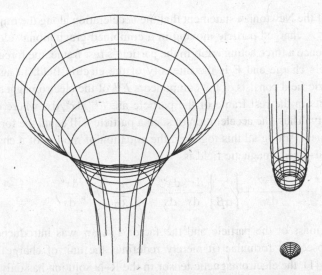

Fig. 14.15. Embeddings of the $\{t = \text{constant}, \vartheta = \pi/2\}$ surface of the extreme $(e^2 = m^2)$ R–N metric. Left: the embedding of the part $r > m$. The funnel is infinitely long downwards, and it approaches asymptotically from the outside of the cylinder of radius m as $r \to m$. It goes infinitely far outwards and upwards. Right: the embedding of the part $0 \leq r < m$ in a flat 3-space. The upper surface goes infinitely far upwards and approaches asymptotically from the inside of the cylinder of the same radius m as $r \to m$. The surface is cut off at $r = m/2$ (where $z = 0$) because the remaining part cannot be embedded in a Euclidean space. The lower cone is the embedding of the part $r \leq m/2$ in the pseudo-Euclidean 3-space with the metric $-\mathrm{d}(u(r))^2 + \mathrm{d}r^2 + r^2 \, \mathrm{d}\varphi^2$. It is cut off at $r = m/2$, and its vertex touches the singularity at $r = 0$.

for the regions $r > m$ and $m/2 \leq r < m$:

$$z(r) = \sqrt{2m} \left(2\sqrt{r - m/2} + \sqrt{\frac{m}{2}} \ln \left| \frac{\sqrt{r - m/2} - \sqrt{m/2}}{\sqrt{r - m/2} + \sqrt{m/2}} \right| \right). \qquad (14.167)$$

This tends to $-\infty$ at $r \to m$, to $+\infty$ at $r \to \infty$ and to 0 at $r \to m/2$. For $r \leq m/2$, we can embed the surface $\{t = \text{constant}, \vartheta = \pi/2\}$ in a flat 3-space with the indefinite metric $-\mathrm{d}(u(r))^2 + \mathrm{d}r^2 + r^2 \, \mathrm{d}\varphi^2$; the equation of embedding is

$$u(r) = -2\sqrt{2m} \left(\sqrt{m/2 - r} - \sqrt{m/2} \arctan \sqrt{1 - 2r/m} \right). \qquad (14.168)$$

This tends to 0 at $r \to m/2$ and to the finite value $-m\pi/2$ at $r = 0$.

The embeddings are shown in Fig. 14.15.

14.16 * Motion of particles in the Reissner–Nordström spacetime with $e^2 < m^2$

This section is mostly borrowed from Graves and Brill (1960).

A free particle in a gravitational field moves on a geodesic $x^\gamma = x^\gamma(s)$, where s is an affine parameter along the trajectory. The geodesic equation $\mathrm{D}^2 x^\gamma / \mathrm{d}s^2 = 0$ is a generalisation of the Newtonian statement that the acceleration along the trajectory of free motion is zero.[1] A charged particle moving in a combined gravitational–electromagnetic field will experience a force acting on it. In the particle's rest frame, the force is $q\mathbf{E}$, where q is the particle's charge and $\mathbf{E}$ is the intensity of the electric field. In agreement with (13.1), the electric field consists of the components F^{0l} of the electromagnetic tensor, and these components in the rest frame of the particle are $F^{\mu}{}_{\nu}\,dx^{\nu}/ds$, where dx^{ν}/ds is the particle's velocity. Thus, the acceleration of such a particle will equal the force divided by the particle's mass. Putting all this together, the equation of motion of a charged particle in a gravitational–electromagnetic field is

$$\frac{d^2x^{\gamma}}{ds^2} + \left\{ \begin{matrix} \gamma \\ \alpha\beta \end{matrix} \right\} \frac{dx^{\alpha}}{ds}\frac{dx^{\beta}}{ds} = \frac{q}{\sqrt{8\pi\mu}}F^{\gamma}{}_{\nu}\frac{dx^{\nu}}{ds}, \tag{14.169}$$

where μ is the mass of the particle and the factor $1/\sqrt{8\pi}$ was introduced in order to simplify the subsequent formulae (it merely redefines the unit of charge). By (14.23), (14.40) and (14.41), the electromagnetic tensor in the R–N solution has only two nonzero components, $F^{01} = -F^{10} = \sqrt{8\pi}e/r^2$. Knowing this, we find that two of the equations in the set (14.169) (those corresponding to $\mu = 2$ and $\mu = 3$) coincide with (14.48) and (14.49), so Eqs. (14.53)–(14.54) remain in force. Using this result, the remaining two equations are

$$\frac{d^2t}{ds^2} + \frac{1}{\phi}\phi_{,r}\frac{dt}{ds}\frac{dr}{ds} = -\frac{qe}{\mu r^2\phi}\frac{dr}{ds}, \tag{14.170}$$

$$\frac{d^2r}{ds^2} + \frac{1}{2}\phi_{,r}\left[\phi\left(\frac{dt}{ds}\right)^2 - \frac{1}{\phi}\left(\frac{dr}{ds}\right)^2\right] - r\phi\left(\frac{d\varphi}{ds}\right)^2 = -\frac{qe\phi}{\mu r^2}\frac{dt}{ds}, \tag{14.171}$$

where we denoted

$$\phi = 1 - \frac{2m}{r} + \frac{e^2}{r^2}. \tag{14.172}$$

Multiplying (14.170) by $2\phi\,dt/ds$, (14.171) by $-2(dr/ds)/\phi$ and adding the results, we obtain an equation that is easily integrated with the result

$$\phi\left(\frac{dt}{ds}\right)^2 - \frac{1}{\phi}\left(\frac{dr}{ds}\right)^2 - \frac{J_0{}^2}{r^2} = \varepsilon, \tag{14.173}$$

where ε is a constant of integration. This is the same integral that exists for geodesics; it says that the tangent vector to the trajectory has a constant length. By adjusting the affine parameter, the constant can be made equal to 1. We shall keep the symbol ε, however, because later we will use (14.173) for null geodesics, for which $\varepsilon = 0$. Equation (14.170) can be integrated with the result

$$\phi\frac{dt}{ds} = \frac{qe}{\mu r} + \Gamma, \tag{14.174}$$

---

[1] See Section 15.2 for the definition of acceleration in relativity.

where Γ is a constant of integration. Using this to eliminate dt/ds from (14.173), we obtain

$$\frac{1}{\phi}\left(\frac{qe}{\mu r}+\Gamma\right)^2 - \frac{1}{\phi}\left(\frac{dr}{ds}\right)^2 = \varepsilon + \frac{J_0^2}{r^2}. \qquad (14.175)$$

Since $\varepsilon = +1$, this shows that the turning points of the trajectory ($dr/ds = 0$) can exist only outside the outer horizon ($r > r_+$) and inside the inner horizon ($r < r_-$), where $\phi > 0$. We now rewrite (14.175) in yet another form:

$$\left(\frac{dr}{ds}\right)^2 = \left(\frac{qe}{\mu r}+\Gamma\right)^2 - \left(\varepsilon + \frac{J_0^2}{r^2}\right)\left(1 - \frac{2m}{r} + \frac{e^2}{r^2}\right). \qquad (14.176)$$

The motion can take place only in those regions where the right-hand side of (14.176) is positive. With $J_0 \neq 0$, however, there is the term $-J_0^2 e^2/r^4$ that will always dominate over the other terms when $r \to 0$, and will render the right-hand side negative. Thus, the neighbourhood of $r = 0$ is inaccessible for motion, which means that no charged particle can ever hit the central singularity. This conclusion still holds with $J_0 = 0$, provided $q^2 < \mu^2$ – then the dominating term is $(q^2/\mu^2 - 1)e^2/r^2$. This means that even a radially moving charged particle will be repelled by the singularity, provided its charge is small enough compared to the mass, irrespective of the signs of the charges of the particle (q) and of the R–N black hole (e). Strangely enough, the conclusion continues to hold for neutral particles for which $q = 0$, even with $J_0 = 0$. Then the dominating terms are $-J_0^2 e^2/r^4$ and $-e^2/r^2$, respectively. Thus, a charge on matter creates effective antigravitation – repulsion that, in spite of its electromagnetic origin, acts also on electrically neutral particles. We will come back to this in Section 19.3 and 19.3.3, where we will discuss the gravitational and electric fields inside a charged dust sphere.

Equation (14.176) can be used to calculate the time it takes to reach the horizon $r = m$ along a timelike radial geodesic in the special case $e^2 = m^2$. In spite of the infinite distance to this horizon within a $t = $ constant space, the time along a timelike geodesic turns out to be finite. The same is true for the value of the affine parameter along a null geodesic; see Exercise 14 for both results. Thus the manifold of the extreme R–N metric in the original R–N coordinates is indeed incomplete.

Equation (14.176) implies that the inner turning point of a radial trajectory of a charged or uncharged particle (the value of r at which, with $J_0 = 0$, $dr/ds = 0$ and the motion reverses from fall to escape) is always within the inner horizon, i.e. at $r < r_-$; see Exercise 15.

14.17 Exercises

1. Prove that (14.1) is indeed a rotation around the axis $\{\vartheta = \pi/2, \varphi = 0\}$ (e.g. find the corresponding Killing field and transform it to Cartesian coordinates). Verify that the components of the metric (8.51) do not change after the transformation (14.1).
2. Prove Lemmas 14.2 and 14.3.

3. Prove that the duality rotation (13.13) with the parameter $\delta = -\arctan(q/e)$ does indeed transform the electromagnetic field (14.23), (14.21) and (14.20), with the metric (14.18), into a new field for which the magnetic charge $\tilde{q} = 0$ and the new electric charge is $\tilde{e} = \sqrt{e^2 + q^2}$.

4. Prove that, with $E < 1$, $J_0 > 2\sqrt{3}GM/c^2$ and the initial σ being sufficiently small, the solution $\sigma(\varphi)$ of (14.59) must be bounded.

   Hint. Assume (14.59) to be the equation of energy conservation on a Newtonian orbit in the potential $[-(1 + J_0^2/r^2)(1 - 2m/r)]$. Consider the possible shapes of this potential depending on J_0. Consider the orbital motion in various ranges of the total energy E.

5. For a light ray on a circular orbit $d\sigma/d\varphi \equiv 0$ in (14.74), so the radius of this orbit must obey $E^2r^3 - J_0^2(r - 2m) = 0$. This equation has at most two positive roots. Most of these orbits are unstable: any perturbation will cause the light ray to go around an orbit on which r oscillates between a maximal and a minimal value. Find the condition that J_0, m and E must obey for circular photon orbits to exist. Show that only one of them is stable, with radius $r = 3m$.

   Hint. Use the same method as in the previous exercise. At the radius of the stable circular orbit the function $E^2 - J_0^2\sigma^2(1 - 2m\sigma)$ has a zero and a local minimum.

6. Prove that the same (incorrect) result (14.89) for the deflection of light rays can be obtained by purely Newtonian methods.

   Hint. Consider a hyperbolic orbit in polar coordinates, Eq. (14.77), where the geometric parameters of the orbit, p and ϵ, are connected with the physical parameters by Eq. (14.86) and $p = J^2/(GM\mu^2)$. Calculate the deflection angle as in (14.85). Then replace the geometric parameters by the physical ones. Note that $J = mRv_R$, where R is the smallest distance between the particle and the central star, and v_R is the velocity at the point of smallest distance. Calculate v_R by equating the energy at that point to the energy at infinity, $\frac{1}{2}\mu v_\infty^2$. Note that the value of the deflection angle does not depend on the mass μ. Assume that the 'particle' is a photon and that $v_\infty = c$. Finally, calculate the result up to linear terms in GM/c^2.

   Remark. This is how the angle of deflection of a light ray was calculated by Cavendish in the eighteenth century, in unpublished notes (Will, 1988), and by von Soldner in 1804 (Soldner, 1804; Schneider, Ehlers and Falco, 1992). Einstein himself found at first this incorrect result, before he formulated his field equations (12.21).

7. Calculate the tetrad components R_{ijkl} of the Riemann tensor for the Schwarzschild solution (14.40)–(14.42) and verify that they are all regular at $r = 2m$.

8. Verify that the integral in (14.113) is finite for every $r_0 < \infty$.

9. Assume geodesic coordinates in (8.51), so that $ds^2 = dt^2 - S^2(t, r)dr^2 - R^2(t, r)(d\vartheta^2 + \sin^2\vartheta\, d\varphi^2)$ (make sure first that the transformations (8.52) do really allow such a choice!). Then solve the vacuum Einstein equations $G_{\mu\nu} = 0$ for this metric and verify that (14.116)–(14.117) is the general solution that results when $R_{,r} \neq 0$.

10. Prove by direct coordinate transformation that (14.116)–(14.117) is indeed a representation of the Schwarzschild solution (14.40) and (14.42).

    Hint. Take the Schwarzschild solution (14.40) and (14.42) and transform the coordinates by $t = f(\tau, u)$, $r = R(\tau, u)$, where f and R are functions as yet unknown. Then demand that in the new coordinates $g_{\tau\tau} = 1$ and $g_{\tau u} = 0$. Solve the resulting equations algebraically for the derivatives of f. Impose the integrability condition $f_{,\tau u} = f_{,u\tau}$. (The τ and u are the (t, r) coordinates of Eqs. (14.116)–(14.117).)

11. Verify that (14.122) is fulfilled by virtue of (14.119)–(14.121).

    Hint. Differentiate (14.121) by r, then manipulate (14.119)–(14.122) and (14.127) until you reproduce (14.122).

12. Verify that the Weyl tensor of (14.134) is zero.
13. The lines $r =$ constant in the subspace $\{\vartheta =$ constant, $\varphi =$ constant$\}$ of the R–N metric are the hyperbolae $u^2 - v^2 =$ constant, as seen from (14.149). Follow the chain of transformations (14.154)–(14.156) to show that in the (U, V) coordinates these lines are still hyperbolae given by

$$\left(\frac{1-C}{1+C}\right)^2 \left[\left(U + \frac{1+C}{1-C}\right)^2 - V^2\right] = 1, \qquad C \stackrel{\text{def}}{=} e^{2pq} = e^{2(u^2 - v^2)}.$$

14. Show that the proper time s needed to reach the horizon $r = m$ of the extreme R–N metric from any point along a radial timelike geodesic is finite. Show that, for a radial null geodesic starting at a finite r, the value of the affine parameter s at $r = m$ is also finite.
    Hint. The curve defined by (14.176) becomes geodesic when $q = 0$.
15. Prove that the inner turning point of a radial trajectory of a charged or uncharged particle is always within the inner horizon if $q^2/\mu^2 < 1$.
    Hint. The right-hand side of (14.176) with $J_0 = 0$ (call it $W(r)$) vanishes at $r = r_1$ and $r = r_2 > r_1$. The inner turning point is at r_1 when $r_1 > 0$ and at r_2 when $r_1 < 0$. One of the two values must be positive because otherwise there would be no turning point at $r > 0$ and the singularity at $r = 0$ would be accessible to particles. Assume that the value of r at the turning point in question is greater than r_+ and verify that it leads to a contradiction in both possible cases, $\Gamma^2 > 1$ and $\Gamma^2 < 1$.

15

Relativistic hydrodynamics and thermodynamics

15.1 Motion of a continuous medium in Newtonian mechanics

Let x_i, $i = 1, 2, 3$, be the rectangular Cartesian coordinates in the Euclidean 3-space. We assume that one line of flow of a fluid passes through every point of a certain region in space. Let the velocity field of the fluid, $v_i(t, x_j)$, be differentiable at every point x_i and at every instant t. Then

$$\frac{dx_j(t)}{dt} = v_j(t, y_i)\big|_{x_j = y_j}. \tag{15.1}$$

The $x_j(t)$ on the left are coordinates of the flowing fluid particle, while the y_i on the right are the coordinates of a point of space.

Let us follow the motion of the particle P, which at the instant t occupies the position $\{x_i\}$ and moves with the velocity $v_j(t, x_i)$, and the motion of an adjacent particle Q that, *at the same instant t*, occupies the position $\{x_i + \delta x_i\}$ and moves with the velocity $v_j(t, x_i + \delta x_i)$. Up to terms linear in δx_i, the velocity of the particle Q with respect to P is

$$(v_{QP})_j(t) \equiv v_j(t, x_i + \delta x_i) - v_j(t, x_i) = v_{j,k}(t, x_i)\delta x_k + O(\delta x^2) \tag{15.2}$$

(sums over all repeated indices are implied in all the equations in this section), where $O(\delta x^2)$ denotes terms of order 2 and higher in δx_k. Hence, at the instant $(t + \Delta t)$ the position of the particle Q relative to P will be given by the vector

$$\delta x_j' = \delta x_j + (v_{QP})_j \Delta t + O(\Delta t^2) = \delta x_j + v_{j,k}\delta x_k \Delta t + O\left(\Delta t^2, (\delta x)^2\right). \tag{15.3}$$

Thus the matrix $v_{j,k}$ determines the relative velocity of two neighbouring particles of the fluid. If we use only Cartesian coordinates, then, under transformations between such coordinates, the matrix $v_{j,k}$ transforms as a tensor. It can be decomposed into three parts, each of which transforms independently of the others:

$$v_{j,k} = \sigma_{jk} + \omega_{jk} + \frac{1}{3}\delta_{jk}\theta, \tag{15.4}$$

where

$$\theta = v_{j,j}, \tag{15.5}$$

$$\sigma_{jk} = v_{(j,k)} - \frac{1}{3}\delta_{jk}\theta, \tag{15.6}$$

222

$$\omega_{jk} = v_{[j,k]}. \tag{15.7}$$

This decomposition can be done for every tensor of rank 2. However, for the tensor $v_{j,k}$, each part has a physical interpretation. The easiest way to read it out is to consider three types of motion, such that in each of them just one of the three parts is nonzero.

- I. Let $\sigma_{jk} = \omega_{jk} = 0 \neq \theta$. Then, from (15.2) and (15.3):

$$\delta x'_j = \left(1 + \frac{1}{3}\theta \Delta t\right)\delta x_j + O\left(\Delta t^2, \delta x^2\right). \tag{15.8}$$

The new vector that connects P and Q at the instant $(t + \Delta t)$ has in this case the same direction as the old vector from P to Q, but a different length, and $\theta = 3(\mathrm{d}/\mathrm{d}t)\ln|\delta\mathbf{x}|$. The particles P and Q either recede from each other (when $\theta > 0$) or approach each other (when $\theta < 0$) along the straight line PQ. Such motion is called **isotropic expansion**, and the quantity θ is called the **scalar of expansion**.

- II. Let $\theta = 0$ and $\sigma_{jk} = 0 \neq \omega_{jk}$. Then, from (15.3) and (15.4):

$$\delta x'_i = (\delta_{ik} + \omega_{ik}\Delta t)\,\delta x_k + O\left(\Delta t^2, \delta x^2\right). \tag{15.9}$$

Let us calculate the length of the vector $\delta x'_i$. In consequence of antisymmetry of ω_{ik}, we obtain

$$\delta\ell' = (\delta x'_i \delta x'_i)^{1/2} = \left[\delta x_k \delta x_k + O(\Delta t^2, \delta x^3)\right]^{1/2}$$
$$= (\delta x_k \delta x_k)^{1/2} + O(\Delta t^2, \delta x^3) = \delta\ell + O(\Delta t^2, \delta x^3). \tag{15.10}$$

Hence, the derivative along v_i of the length $\delta\ell$ is zero. Let us see what happens with the direction of δx_i:

$$(\delta x'_i - \delta x_i)\,\delta x_i = \omega_{ik}\,\delta x_i\,\delta x_k\,\Delta t + O(\Delta t^2, \delta x^3) = O(\Delta t^2, \delta x^3). \tag{15.11}$$

Hence, the rate of change of δx_i projected orthogonally on δx_i is zero. The properties (15.10) and (15.11) are characteristic for rotational motion. Thus, in this type of motion, Q revolves around P.

Let us calculate the vector of angular velocity of this motion, ω. We have, from the definition of angular velocity:

$$\mathbf{v}_{QP} = \omega \times \delta\mathbf{x} \implies (v_{QP})_i = \epsilon_{ikl}\omega_k\,\delta x_l. \tag{15.12}$$

At the same time, from (15.2) in the present case we have $(v_{QP})_i = \omega_{il}\,\delta x_l\,S$. These two equations must hold for every δx_l, hence

$$\omega_{il} = -\epsilon_{ilk}\omega_k. \tag{15.13}$$

Inverting this equation and using (15.7) we obtain

$$\omega_j = \frac{1}{2}\epsilon_{jil}\omega_{li} = \frac{1}{2}\epsilon_{jil}v_{l,i} \implies \omega = \frac{1}{2}\,\mathrm{rot}\ \mathbf{v}. \tag{15.14}$$

In consequence of (15.13) and (15.14), ω_{ik} is called the **rotation tensor**. Note that the quantity $\omega^2 \stackrel{\text{def}}{=} \omega_i \omega_i = \omega_{kl} \omega_{kl}/2$ is always non-negative and vanishes only when $\omega_{kl} = 0$. Hence, also the **scalar of rotation** ω can be used to differentiate between rotational and irrotational motion.

- III. Let $\theta = 0$ and $\omega_{ij} = 0 \neq \sigma_{ij}$. Consider three particles Q_1, Q_2 and Q_3 neighbouring P and occupying, at the instant t, the positions relative to P given by the vectors $\delta\mathbf{x}$, $\delta\mathbf{y}$ and $\delta\mathbf{z}$ attached to P. The volume of the parallelepiped spanned on the vectors $\delta\mathbf{x}$, $\delta\mathbf{y}$ and $\delta\mathbf{z}$ at the instant t is equal to

$$\delta V = \delta\mathbf{x} \cdot (\delta\mathbf{y} \times \delta\mathbf{z}) = \epsilon_{ijk}\, \delta x_i\, \delta y_j\, \delta z_k. \tag{15.15}$$

At the instant $(t + \Delta t)$, the corresponding volume will be equal to

$$\delta V' = \delta V + D\Delta t + O(\Delta t^2, \delta x^4), \tag{15.16}$$

where

$$D \stackrel{\text{def}}{=} (\epsilon_{lmk}\sigma_{kn} + \epsilon_{ljn}\sigma_{jm} + \epsilon_{imn}\sigma_{il})\delta x_l\, \delta y_m\, \delta z_n. \tag{15.17}$$

Using $\sigma_{ij} = \sigma_{ji}$ and $\sigma_{ii} = 0$ one can now verify that $D = 0$, i.e. that $\delta V' = \delta V + O(\Delta t^2, \delta x^4)$. Hence, in this type of motion, the rate of change of volume of the parallelepiped spanned on the vectors $\delta\mathbf{x}$, $\delta\mathbf{y}$ and $\delta\mathbf{z}$ is zero. However, the shape of the parallelepiped is changing because the vector $\sigma_{ik}\delta x_k$ has in general a different direction and different length from δx_i.[1]

This kind of motion is called **shearing motion**, and the tensor σ_{ij} is called the **shear tensor**. Similarly to rotation, the shear tensor vanishes if and only if the scalar σ defined by $\sigma^2 = \sigma_{ij}\sigma_{ij}/2$, called simply **shear**, is zero.

15.2 Motion of a continuous medium in relativistic mechanics

In relativity, the motion of a fluid is described in a similar way to that in Newtonian mechanics. We assume that one line of flow of a fluid passes through every point $\{x^\alpha\}$ of a certain region in spacetime, and that the velocity field of the fluid, $u^\alpha(x)$, tangent to the flow lines, is differentiable in this whole region. Then $dx^\alpha/ds = u^\alpha(x^\beta)$, and, just as in (15.1), the $x^\alpha(s)$ on the left-hand side are coordinates of the flowing fluid element, while the x^β on the right-hand side are the coordinates of that point of spacetime in which the $x^\alpha(s)$ equals x^α. The parameter s is the proper time on the worldlines of the fluid, hence

$$u_\alpha(x)u^\alpha(x) = 1. \tag{15.18}$$

Note now that the tensor

$$h^\alpha{}_\beta \stackrel{\text{def}}{=} \delta^\alpha{}_\beta - u^\alpha u_\beta \tag{15.19}$$

---

[1] If $\delta\mathbf{x}$, $\delta\mathbf{y}$ and $\delta\mathbf{z}$ are collinear with the eigenvectors of the matrix σ, then their directions do not change during the motion, but, in consequence of $\sigma_{ii} \equiv \text{Tr}(\sigma) = 0$, the sum of changes of their lengths must be zero. Hence, as long as $\sigma \neq 0$, if two of the vectors become longer, the third one must become shorter. Consequently, the shape of the parallelepiped will be changed also in this case.

projects vectors on the hypersurface orthogonal to u^α at a given point x. This is because we have $h^\alpha{}_\beta u^\beta = 0$ and, for any arbitrary vector B^α, $u_\alpha \cdot (h^\alpha{}_\beta B^\beta) = 0$. The quantity $B^\alpha_\perp \overset{\text{def}}{=} h^\alpha{}_\beta B^\beta$ is the component of the vector B^α along the direction perpendicular to u_α. Note also that

$$g_{\alpha\beta} B^\alpha_\perp B^\beta_\perp = h_{\alpha\beta} B^\alpha_\perp B^\beta_\perp \qquad (15.20)$$

for any arbitrary vector B^α. Thus, the tensor $h_{\alpha\beta}$ plays the role of the metric tensor on the hypersurfaces orthogonal to u^α.[1]

Let us choose the proper time s on the flow lines as the time coordinate x^0 in spacetime. Then

$$u'^\alpha \underset{*}{=} \delta^\alpha{}_0. \qquad (15.21)$$

Let us next choose three spatial coordinates $x^I, I = 1, 2, 3$, so that their parametric lines are contained in the hypersurfaces $S_P(s)$ orthogonal to one fixed worldline P. In such coordinates $g_{0I}|_P \underset{*}{=} 0$, so, for an arbitrary vector field B^α orthogonal to $u^\alpha_P(s)$, we have $0 = (g_{\alpha\beta} u^\alpha B^\beta)_P \underset{*}{=} g_{00}|_P B^0_P$, i.e. $B^0_P \underset{*}{=} 0$ (since $g_{00} \neq 0$ in consequence of (15.21) and (15.18)). Consider the curves tangent to the vector field B^α, given by $B^\alpha = dx^\alpha/d\lambda$. In our chosen coordinate system we have $0 = B^0_P = dx^0/d\lambda|_P$, i.e. at points of the curve P these lines are tangent to the hypersurface $x^0 = $ constant. Hence, the hypersurfaces $x^0 = $ constant and the hypersurfaces orthogonal to P are tangent to each other along P. Consequently, the hypersurface $S_P(s_0)$ is called the **hypersurface of events simultaneous with $P(s_0)$** or the **hypersurface of constant time** $s = s_0$ for the observer P.

Let the particle moving along the curve P occupy at the instant s the point P_0 in spacetime. Let Q be an adjacent worldline, and let δx^α be a vector joining P_0 to an arbitrary point on Q. The event Q_0 simultaneous with P_0 is then at the position relative to P_0 given by the vector $\delta_\perp x^\alpha = h^\alpha{}_\beta(P_0)\delta x^\beta$. The velocity of the fluid at P_0 is $u^\alpha(x^\beta)$, and the velocity at Q_0 is $u^\alpha(x^\beta + \delta_\perp x^\beta)$. After the time Δs, the particle that at s occupied the point P_0 will be at P_1 of coordinates $x'^\alpha = x^\alpha + u^\alpha \Delta s$. Where will then be the particle that occupied the point Q_0 at s? Its position relative to P_1 will be determined by the vector $\delta_\perp x^\alpha + v^\alpha \Delta s$, attached to P_1, where v^α is determined by $u^\alpha(x^\beta + \delta_\perp x^\beta)$. However, the vector v^α cannot be equal to $u^\alpha(x^\beta + \delta_\perp x^\beta)$, because the latter is attached to the point of coordinates $x^\beta + \delta_\perp x^\beta$, while $\delta_\perp x^\alpha$ is attached to P_0 of coordinates x^β. In order to add the vectors, we must first transport one of them parallely to the point of attachment of the other. Thus v^α must be the vector $u^\alpha(x^\beta + \delta_\perp x^\beta)$ parallely transported from $(x^\beta + \delta_\perp x^\beta)$ to x^β.

Let us apply Eq. (5.6) to our present case. The $v^\alpha_\parallel(\tau_2)$ of (5.6) is our $u^\alpha(x^\beta + \delta_\perp x^\beta)$, and the $v^\alpha(\tau_1)$ of (5.6) is our v^α. Hence we have

$$u^\alpha(x^\beta + \delta_\perp x^\beta) = v^\alpha - \int_{x^\beta}^{x^\beta + \delta_\perp x^\beta} \Gamma^\alpha{}_{\sigma\rho}(x) u^\sigma(x) dx^\rho.$$

---

[1] Each of the hypersurfaces meant here is orthogonal to a single flow line of the fluid. The family of hypersurfaces is in general different for every flow line. For vector fields, the property of being orthogonal to a family of hypersurfaces is rather special, and does not hold in general. As an exercise, readers may wish to verify that such a family of hypersurfaces exists if and only if the rotation tensor, defined later in this section, is zero.

We now apply the mean value theorem to the integral, and also develop the vector on the left-hand side by the Taylor formula up to terms linear in $\delta_\perp x^\beta$. The result is

$$u^\alpha(x^\beta) + u^\alpha,_\rho(x^\beta)\delta_\perp x^\beta + O(\delta_\perp x^2) = v^\alpha - \Gamma^\alpha{}_{\sigma\rho}(\bar{x})u^\sigma(\bar{x})\delta_\perp x^\rho.$$

The point of coordinates $\bar{x}$ is the intermediate point between x^β and $(x^\beta + \delta_\perp x^\beta)$. When we replace it by x^β, the difference will be of the order of $(\delta_\perp x^\beta)$, and, since the whole expression is multiplied by $\delta_\perp x^\beta$, the difference in the equation will be of order $O(\delta_\perp x^2)$. Transferring the expression containing the connection coefficients to the left-hand side of the equation we then obtain

$$v^\alpha = u^\alpha(x^\beta) + u^\alpha,_\rho(x^\beta)\delta_\perp x^\beta + \Gamma^\alpha{}_{\sigma\rho}(x)u^\sigma(x)\delta_\perp x^\rho + O(\delta_\perp x^2).$$

This is equivalent to

$$v^\alpha = u^\alpha(x^\beta) + u^\alpha;_\rho(x^\beta)\delta_\perp x^\beta + O(\delta_\perp x^2). \tag{15.22}$$

Hence, the particle that occupied the point Q_0 at the instant s will at the later instant $(s + \Delta s)$ occupy the point of coordinates

$$x''^\alpha = x^\alpha + \delta_\perp x^\alpha + u^\alpha(x^\beta)\Delta s + u^\alpha;_\rho(x^\beta)\delta_\perp x^\beta \Delta s + O(\delta_\perp x^2, \Delta s^2). \tag{15.23}$$

Hence, the new position of the particle Q relative to the particle P will be given by the vector

$$\delta_\perp x'^\alpha = x''^\alpha - x'^\alpha = \delta_\perp x^\alpha + u^\alpha;_\rho \delta_\perp x^\rho \Delta s + O(\delta_\perp x^2, \Delta s^2). \tag{15.24}$$

Thus, the matrix $u^\alpha;_\rho$ determines the rate of change of the position of the particle Q with respect to the particle P. However, not the whole matrix $u^\alpha;_\rho$ gives a nontrivial contribution to (15.24). We have

$$u^\alpha;_\rho \equiv u^\alpha;_\sigma \delta^\sigma{}_\rho \equiv u^\alpha;_\sigma(h^\sigma{}_\rho + u^\sigma u_\rho). \tag{15.25}$$

Only the first term in (15.25) gives a nonzero contribution in (15.24). Hence, finally

$$\delta_\perp x'^\alpha = \delta_\perp x^\alpha + u^\alpha;_\sigma h^\sigma{}_\rho \delta_\perp x^\rho \Delta s + O(\delta_\perp x^2, \Delta s^2). \tag{15.26}$$

The following identities hold:

$$u^\alpha;_\sigma \equiv h^\alpha{}_\mu u^\mu;_\sigma, \qquad u^\alpha;_\sigma u_\alpha \equiv 0. \tag{15.27}$$

They show that the matrix $u^\alpha;_\sigma h^\sigma{}_\rho$ is an operator acting in the 3-dimensional hypersurface orthogonal to u^α. Comparing (15.27) with (15.3) we see that the quantity $u^\alpha;_\sigma h^\sigma{}_\beta$ plays in relativistic hydrodynamics the same role as the quantity $v_{j,k}$ played in Newtonian hydrodynamics. This time, $u^\alpha;_\sigma h^\sigma{}_\beta$ is a genuine tensor, and we can decompose it into the three independent parts in the same way:

$$u_{\alpha;\sigma} h^\sigma{}_\beta = \sigma_{\alpha\beta} + \omega_{\alpha\beta} + \frac{1}{3}\theta h_{\alpha\beta}, \tag{15.28}$$

where

$$\theta = u^{\alpha}{}_{;\sigma} h^{\sigma}{}_{\alpha} \equiv u^{\alpha}{}_{;\alpha}, \tag{15.29}$$

$$\omega_{\alpha\beta} = u_{[\alpha;|\sigma|]} h^{\sigma}{}_{\beta]} \equiv u_{[\alpha;\beta]} - \dot{u}_{[\alpha} u_{\beta]}, \tag{15.30}$$

$$\dot{u}_{\alpha} = u_{\alpha;\beta} u^{\beta}, \tag{15.31}$$

$$\sigma_{\alpha\beta} = u_{(\alpha;|\sigma|} h^{\sigma}{}_{\beta)} - \frac{1}{3}\theta h_{\alpha\beta} \equiv u_{(\alpha;\beta)} - \dot{u}_{(\alpha} u_{\beta)} - \frac{1}{3}\theta h_{\alpha\beta}. \tag{15.32}$$

These quantities are called, respectively, the **scalar of expansion**, the **rotation tensor**, the **acceleration vector** and the **shear tensor**. The vector $\dot{u}^{\alpha}$ is called acceleration because $\dot{u}^{\alpha} = 0$ is the necessary and sufficient condition for the field u^{α} to be tangent to geodesic lines, i.e. for the fluid to move under the influence of gravitation only, which means free motion in the language of relativity.

The tensor of rotation can be uniquely represented by the vector field

$$w^{\alpha} \overset{\text{def}}{=} \frac{1}{2\sqrt{-g}} \epsilon^{\alpha\beta\gamma\delta} u_{\beta} \omega_{\gamma\delta}, \tag{15.33}$$

and the rotational motion may be characterised by the scalar ω:

$$\omega^{2} \overset{\text{def}}{=} -w_{\alpha} w^{\alpha} = \frac{1}{2} \omega_{\alpha\beta} \omega^{\alpha\beta} \geq 0, \tag{15.34}$$

called the **rotation scalar**. Similarly, the shearing motion may be characterised by the **shear scalar** σ defined by

$$\sigma^{2} \overset{\text{def}}{=} \frac{1}{2} \sigma_{\alpha\beta} \sigma^{\alpha\beta} \geq 0. \tag{15.35}$$

The relativistic definitions of the expansion, rotation, shear and acceleration were introduced by Ehlers (1961).

It can easily be verified that the quantities given by (15.29), (15.30) and (15.32) are, in a suitably chosen coordinate system, proportional to their Newtonian counterparts given by (15.5)–(15.7). Namely, at a fixed point p_0 of the spacetime we can choose a coordinate system in which $\left\{ \begin{matrix} \alpha \\ \beta\gamma \end{matrix} \right\} (p_0) = 0$, i.e. $D/\partial x^{\alpha}|_{p_0} = \partial/\partial x^{\alpha}$. Then, after carrying out all the operations given by (15.29)–(15.32), we substitute $v/c = 0$. Marking the relativistic quantities with the subscript R and their Newtonian counterparts with the subscript N, we obtain

$$\theta_{\text{R}} = \theta_{\text{N}}, \qquad c\omega_{ij} = \omega_{ij}, \qquad c\sigma_{ij} = \sigma_{ij};$$
$$ {}_{\text{R}} {}_{\text{N}} {}_{\text{R}} {}_{\text{N}}$$

$$\omega_{0i} \underset{{}_{\text{R}} \; v/c \to 0}{\longrightarrow} 0, \qquad \sigma_{0i} \underset{{}_{\text{R}} \; v/c \to 0}{\longrightarrow} 0. \tag{15.36}$$

15.3 The equations of evolution of θ, $\sigma_{\alpha\beta}$, $\omega_{\alpha\beta}$ and $\dot{u}^\alpha$; the Raychaudhuri equation

The equations derived in this section will be simply consistency conditions imposed by the Einstein equations on the hydrodynamical scalars, vectors and tensors introduced in the previous section and on the curvature of spacetime. Although they may thus seem to be of secondary meaning, they are surprisingly powerful in their applications. In particular, they show that one cannot make any arbitrary assumptions about the properties of the kinematical tensors – these properties are interrelated and some of them have important consequences.[1]

From Eq. (6.3), applied to the velocity field of a fluid, we have

$$u_{\gamma,\delta\sigma} - u_{\gamma,\sigma\delta} = -R_{\gamma\rho\delta\sigma}u^\rho. \tag{15.37}$$

Let us contract both sides of this equation with $u^\sigma h^\gamma{}_\alpha h^\delta{}_\beta$. In the second term on the left we then transfer the derivative with the index δ from $u_{\gamma,\sigma}$ to u^σ, while on the right we will use the antisymmetries of the Riemann tensor to eliminate some terms. The result is

$$h^\gamma{}_\alpha h^\delta{}_\beta \left(u_{\gamma,\delta}\right)^{\cdot} - h^\gamma{}_\alpha h^\delta{}_\beta \dot{u}_{\gamma,\delta} + h^\gamma{}_\alpha h^\delta{}_\beta u^\sigma{}_{;\delta}\, u_{\gamma,\sigma} = -R_{\alpha\rho\beta\sigma}u^\rho u^\sigma, \tag{15.38}$$

where the overdot denotes the directional covariant derivative along the velocity field, $\overset{\text{def}}{=} u^\mu \nabla_\mu$. Now we contract (15.38) with $g^{\alpha\beta}$ and obtain

$$h^{\gamma\delta} \left(u_{\gamma,\delta}\right)^{\cdot} - h^{\gamma\delta}\dot{u}_{\gamma,\delta} + h^{\gamma\delta} u^\sigma{}_{;\delta}\, u_{\gamma,\sigma} + R_{\rho\sigma}u^\rho u^\sigma = 0. \tag{15.39}$$

Up to this point, these were general equations of differential geometry. Now we use the Einstein equations with a perfect fluid source to replace

$$R_{\alpha\beta} = \kappa(T_{\alpha\beta} - \tfrac{1}{2}g_{\alpha\beta}T) = \kappa[(\epsilon+p)u_\alpha u_\beta + \tfrac{1}{2}(p-\epsilon)g_{\alpha\beta}], \tag{15.40}$$

and then we apply (15.28) to the term $h^{\gamma\delta}u^\sigma{}_{;\delta}$. In the two remaining terms of (15.39) we substitute (15.19) and we transfer the differentiation from the derivatives of u_γ to u^γ. The result is

$$\begin{aligned}
(u^\gamma{}_{;\gamma})^{\cdot} - (u^\gamma u{}_{;\gamma\delta})u^\delta &+ \dot{u}^\gamma u^\delta u{}_{;\gamma\delta} - \dot{u}^\gamma{}_{;\gamma} + (u^\gamma \dot{u}_\gamma){}_{;\delta}\, u^\delta \\[4pt]
&- u^\gamma{}_{;\delta}u^\delta \dot{u}_\gamma + u_{\gamma,\sigma}(\sigma^{\sigma\gamma} + \omega^{\sigma\gamma} + \tfrac{1}{3}\theta h^{\sigma\gamma}) + \tfrac{1}{2}\kappa(\epsilon+3p) = 0.
\end{aligned} \tag{15.41}$$

Now we use (15.31) and (15.27); the latter implies $\dot{u}^\alpha u_\alpha = 0$. We also use the definitions of θ, $\sigma^{\alpha\beta}$ and $\omega^{\alpha\beta}$ and the equations

$$\sigma_{\alpha\beta}u^\beta = \omega_{\alpha\beta}u^\beta = 0 \tag{15.42}$$

---

[1] In the paper in which these equations were first derived (Ellis, 1971), and in probably all papers in which they were applied, the signature $(-+++)$ was used, as opposed to $(+---)$ used here. This is why the equations of this section will differ from those in other sources – but they are equivalent.

that follow from (15.30), (15.32) and (15.27). We obtain then in (15.41):

$$0 = \dot{\theta} + \frac{1}{3}\theta^2 - \dot{u}^\gamma;_\gamma + \sigma^{\sigma\gamma}(\sigma_{\sigma\gamma} + \dot{u}_{(\sigma}u_{\gamma)} + \frac{1}{3}\theta h_{\sigma\gamma})$$

$$+ \omega^{\sigma\gamma}(\omega_{\gamma\sigma} + \dot{u}_{[\gamma}u_{\sigma]}) + \frac{1}{2}\kappa(\epsilon + 3p)$$

$$= \dot{\theta} + \frac{1}{3}\theta^2 - \dot{u}^\gamma;_\gamma + 2(\sigma^2 - \omega^2) + \frac{1}{2}\kappa(\epsilon + 3p). \qquad (15.43)$$

This equation, in the form quoted above, was derived by Ehlers (1961) and is called the **Raychaudhuri equation** (the name seems to have been introduced by Ellis (1971)). The idea, and a subcase of (15.43), corresponding to dust ($p = 0$), without the definitions of the shear and rotation, were first introduced by Raychaudhuri (1955).

Taking the antisymmetric part of Eq. (15.38), then its symmetric part and using in the second one the Raychaudhuri equation to eliminate $\dot{\theta}$, we obtain two other equations:

- the vorticity propagation equation

$$h^\gamma{}_\alpha h^\delta{}_\beta \dot{\omega}_{\gamma\delta} - h^\gamma{}_\alpha h^\delta{}_\beta \dot{u}_{[\gamma;\delta]} + 2\sigma_{\delta[\alpha}\omega^\delta{}_{\beta]} + \frac{2}{3}\theta\omega_{\alpha\beta} = 0; \qquad (15.44)$$

- and the shear propagation equation

$$h^\gamma{}_\alpha h^\delta{}_\beta \dot{\sigma}_{\gamma\delta} - h^\gamma{}_\alpha h^\delta{}_\beta \dot{u}_{(\gamma;\delta)} + \dot{u}_\alpha \dot{u}_\beta + \omega_{\alpha\gamma}\omega^\gamma{}_\beta + \sigma_{\alpha\gamma}\sigma^\gamma{}_\beta$$

$$+ \frac{2}{3}\theta\sigma_{\alpha\beta} + \frac{1}{3}h_{\alpha\beta}\left[2\left(\omega^2 - \sigma^2\right) + \dot{u}^\gamma;_\gamma\right] + E_{\alpha\beta} = 0, \qquad (15.45)$$

where the quantity $E_{\alpha\beta}$ is the 'electric part' of the Weyl tensor, which was defined in (7.97).

In addition, the following three other equations hold:

$$\omega_{[\alpha\beta;\gamma]} + \dot{u}_{[\alpha;\gamma}u_{\beta]} + \dot{u}_{[\alpha}\omega_{\beta\gamma]} = 0, \qquad (15.46)$$

$$h^\alpha{}_\beta(\omega^{\beta\gamma};_\gamma - \sigma^{\beta\gamma};_\gamma + \frac{2}{3}\theta^{;\beta}) - (\omega^\alpha{}_\beta + \sigma^\alpha{}_\beta)\dot{u}^\beta = 0, \qquad (15.47)$$

$$2\dot{u}_{(\alpha}w_{\beta)} - \sqrt{-g}h^\gamma{}_\alpha h^\delta{}_\beta \left(\omega_{(\gamma}{}^{\mu;\nu} + \sigma_{(\gamma}{}^{\mu;\nu}\right)\epsilon_{\delta)\rho\mu\nu}u^\rho = H_{\alpha\beta}, \qquad (15.48)$$

where $H_{\alpha\beta}$ is the 'magnetic part' of the Weyl tensor, defined in (7.98). Equation (15.46) is obtained by antisymmetrising (15.37) in the indices γ, δ and σ and using $-R_{[\gamma|\rho|\delta\sigma]}u^\rho = u^\rho R_{\rho[\gamma\delta\sigma]} = 0$. Equation (15.47) is obtained by contracting (15.37) with $g^{\gamma\delta}h^{\alpha\sigma}$ and using (15.40). In order to obtain (15.48), one has to rewrite (15.37) in the form

$$-u_\delta{}^{;\mu\nu} + u_\delta{}^{;\nu\mu} = R^{\mu\nu}{}_{\delta\sigma}u^\sigma, \qquad (15.49)$$

then act on both sides of (15.49) with the operator $\frac{1}{2}\sqrt{-g}\epsilon_{\gamma\rho\mu\nu}u^\rho h^\gamma{}_\alpha h^\delta{}_\beta$ and then symmetrise the result with respect to α and β. Equations (15.43)–(15.48) are algebraically independent components of Eq. (15.37).

A consequence of Eqs. (15.43)–(15.48) is that assumptions made about the kinematical quantities can lead to restrictive results. For example, assume that $\dot{u}^\alpha = \sigma = \omega = 0$. Then (15.44) and (15.46) are fulfilled identically, (15.47) says that the expansion scalar may change only along the flow lines, while (15.45) and (15.48) imply that the Weyl tensor is zero. The family of perfect fluid solutions of the Einstein equations for which the Weyl tensor vanishes was found by Stephani (1967a) (see also Stephani *et al.* (2003)). They have the properties $\sigma = \omega = 0$, but in general $\dot{u}^\alpha \neq 0$. In the limit $\dot{u}^\alpha = 0$ they reduce to the Robertson–Walker metrics of Section 10.7 – which is a rather strong simplification. The full set of solutions of Einstein's equations with a perfect fluid source for which $\sigma = \omega = 0$ was found by Barnes (1973).

15.4 Singularities and singularity theorems

An important conclusion follows from the Raychaudhuri equation. Let us define the function $\ell(x^\mu)$ through the equation

$$\frac{1}{\ell}\frac{d\ell}{ds} = \frac{1}{3}\theta, \tag{15.50}$$

where $d/ds \overset{\text{def}}{=} u^\rho \partial / \partial x^\rho$. This function is a generalisation of the scale factor $R(t)$ of the Robertson–Walker models; see Section 10.7. It can be seen from (15.24), (15.19), (15.28) and (15.31) that, with $\omega = \sigma = 0 = \dot{u}^\alpha$, the distance between the simultaneous positions of two particles obeys (15.50). In the general case $\ell(x)$ has no direct physical interpretation and is just a convenient representation of the expansion scalar. We will assume that $\omega = 0 = \dot{u}^\alpha$; then the Raychaudhuri equation becomes

$$3\frac{\ddot{\ell}}{\ell} + 2\sigma^2 + \frac{1}{2}\kappa(\epsilon + 3p) = 0. \tag{15.51}$$

Since $\epsilon + 3p > 0$ for all kinds of matter known from laboratory, (15.51) shows that $d^2\ell/ds^2 < 0$. This means that the function $\ell(s)$ is concave in all its range – if at any point p of the curve $\ell(s)$ we draw a straight line tangent to $\ell(s)$, then the *whole* curve $\ell(s)$ will lie below that straight line. There are two possibilities. If at present $(s = s_0)(d\ell/ds)(s_0) > 0$ (i.e. the fluid expands, curve I in Fig. 15.1), then at a certain instant s_P *in the past*, $s_1 < s_P < s_0$, ℓ *was* zero. If, however, $(d\ell/ds)(s_0) < 0$ at present (i.e. the fluid contracts, curve II in Fig. 15.1), then at a certain instant s_F *in the future*, $s_2 > s_F > s_0$, ℓ *will be* zero. Consequently, in every matter model in which $\dot{u}^\alpha = 0 = \omega$ there exists such an instant, in the past or in the future, at which $\ell \to 0$, which implies, via (15.51), that $\epsilon + 3p \to \infty$ or $\sigma \to \infty$. Hence, every such portion of matter must have a singularity either in its future or in its past.

Note that we have proven here the existence of the singularity using the Einstein equations (15.40), without invoking any specific solution.

With the help of a similar analysis, based on various less restrictive assumptions, Penrose, Hawking and Ellis proved several **singularity theorems** that imply that also

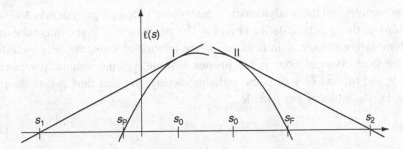

Fig. 15.1. A function that is everywhere concave must go to zero either in the past (at $s_P > s_1$) or in the future (at $s_F < s_2$).

quite general fluid configurations must contain singularities. A summary and overview of these theorems is presented in the book by Hawking and Ellis (1973) that caused a certain revision in the understanding of the relativity theory. The theorems were said to show that general relativity cannot be the ultimate theory of space and time. In order to avoid the singularities, one would have to resort to a more general theory that would be capable of describing the quantum effects taking place at great densities of matter.

However, the singularity theorems are not as general as it was initially claimed. Several interesting solutions of the Einstein equations that *do not* contain any singularities have been found by Senovilla and coworkers (for an extended review see Senovilla (1998)). They have not, so far, been shown to describe any actual astrophysical situation, but their very existence proves that singularities are an inevitable part not of relativity as such, but of the collection of models of matter defined by the assumptions of the singularity theorems.

15.5 Relativistic thermodynamics

The considerations of this section are applied in relativistic astrophysics to interiors of stars or to the Universe as a whole (see Section 16.1). For simplicity, we shall assume that the medium is a one-component perfect fluid. More general media, like viscous, heat-conducting or anisotropic fluids are considered in the literature, but they require more advanced thermodynamics for their description. The equations of motion of a perfect fluid are (12.17), with $T^{\alpha\beta}$ given by (12.73).

Let n denote the particle number density. We shall consider only such processes in which particles are neither created nor annihilated, so the total number of particles contained in a volume at time t_2 will be either the same as at any $t_1 < t_2$, or equal to the sum of the number of particles at t_1 and the number of those that entered/left the volume between t_1 and t_2. In addition to the equations of motion (12.17) we thus postulate the equation of continuity for n:

$$(nu^{\alpha})_{;\alpha} = 0. \qquad (15.52)$$

In phenomenological thermodynamics, if the volume V of a given system is determined, the **enthalpy** of the system is defined by $H = U + pV$, where U is the internal energy of the medium. In cosmology or in considering the interior of stars, the only well-defined volume for local considerations is the **proper volume**, i.e. the volume per particle of the fluid $V_p = 1/n$, and $(\epsilon + p)$ is the enthalpy density. We can thus define the **proper enthalpy**, i.e. the enthalpy per particle,

$$\mathcal{H} = (\epsilon + p)/n. \tag{15.53}$$

Now we postulate that, for a single 'particle' of the fluid, the non-relativistic phenomenological thermodynamics still holds (in the next chapter we will see that in cosmology that 'particle' will be a galaxy cluster or a still larger object). The classical enthalpy obeys the **Gibbs identity**:

$$dH = V\,dp + T\,dS. \tag{15.54}$$

Since ϵ and p are provided by the Einstein equations, and n and $V = 1/n$ have a clear physical interpretation, one may treat H and $\mathcal{H}$ as given. Classical thermodynamics says that at most two state functions are sufficient for a full thermodynamical description of a one-component substance; the other functions can be calculated from the equation of state. Hence, in keeping with classical thermodynamics, one would say that at most two of the three functions p, V and $\mathcal{H}$ are independent. In that case, the differential 1-form $(d\mathcal{H} - V\,dp)$ is a form in only two variables, and thus must have an integrating factor. Denoting this factor $1/T$, we conclude that the form $(1/T)(d\mathcal{H} - V\,dp)$ is a perfect differential of a function S, so

$$d\mathcal{H} = dp/n + T\,dS. \tag{15.55}$$

In this way, we have apparently defined the temperature T and the entropy S. The integrating factor $1/T$ and the function S are not determined uniquely, but, still following the rules of classical phenomenological thermodynamics, one may conclude that T is determined up to linear transformations, i.e. up to the choice of scale (Werle, 1957). Given T, S is determined up to an additive constant.

However, there is a problem with this reasoning. It remained unnoticed for a long time because the solutions of Einstein's equations used in astrophysics are almost exclusively of high symmetry: they are spherically symmetric, or stationary and axisymmetric, or homogeneous of Bianchi type (Robertson–Walker spacetimes being just a subcase of the latter). In the first two cases all the metric components, and thus all the thermodynamical quantities, depend only on two variables, so (15.55) may indeed be considered a definition of T and S. In the third case, all thermodynamical quantities depend on just one variable (the comoving time), so an even simpler equation of state of the type $\epsilon = \epsilon(p)$ is imposed on the matter by the assumed symmetry. However, if the metric has a 1-dimensional symmetry group, or no symmetry at all, then the functions ϵ, p and n depend on three or four variables, respectively, and then the existence of an integrating factor for the differential form $(d\mathcal{H} - V\,dp)$ is an additional postulate, not a certainty.

Spacetimes in which the form $(d\mathcal{H} - V\,dp)$ does have an integrating factor, for which thus temperature and entropy can be defined by the reasoning presented above, are said to admit a **thermodynamical scheme**. The fact that the thermodynamical scheme might not exist in some spacetimes was first noted by Bona and Coll (1985, 1988) and Coll and Ferrando (1989). They also showed that the Stephani Universe (Stephani, 1967a), which has in general no symmetry, acquires a 3-dimensional symmetry group acting on 2-dimensional orbits when the thermodynamical scheme is imposed (Bona and Coll, 1988). The problem of existence of the thermodynamical scheme was further discussed by Quevedo and Sussman (1995) and applied to all currently known cosmological solutions with no symmetry by Krasiński, Quevedo and Sussman (1997). Apart from the Stephani Universe, the problem of existence of the thermodynamical scheme shows up only in the two classes of metrics found by Szafron (1977) (see our Section 19.5) – all other solutions known at present are vacuum, have dust source or have a high symmetry. It turns out that in the $\beta' \neq 0$ family of the Szafron metrics the thermodynamical scheme imposes a 3-dimensional symmetry group. For the $\beta' = 0$ family of Szafron metrics, a metric with no symmetry survives the imposition of the scheme, but it is a limited subcase of the general class.

The conclusion from the results briefly reported above is that, in considering spacetimes of no symmetry, one must allow for a more complicated thermodynamical scheme than a single-component perfect fluid.

Having said this, we will now discuss the thermodynamics of single-component perfect fluids in some more detail.

Using (15.52)–(15.54) one can now write (12.17) as

$$0 = nu^\beta(\mathcal{H}u_\alpha)_{;\beta} - p_{,\alpha} = n\left[u^\beta\,(\mathcal{H}u_\alpha)_{;\beta} - \mathcal{H}_{,\alpha} + TS_{,\alpha}\right]. \tag{15.56}$$

By virtue of $u_\alpha u^\alpha = 1$ and $u^\beta u_{\beta;\alpha} = 0$ this becomes

$$0 = u^\beta\left[(\mathcal{H}u_\alpha)_{;\beta} - (\mathcal{H}u_\beta)_{;\alpha}\right] + TS_{,\alpha} = u^\beta\left[(\mathcal{H}u_\alpha)_\beta - (\mathcal{H}u_\beta)_\alpha\right] + TS_{,\alpha}. \tag{15.57}$$

From here we see easily that $S_{,\alpha} u^\alpha = 0$, i.e. that the entropy is constant along the flow lines. This is a general conclusion that follows from the definition of a one-component perfect fluid. A more special kind of motion is often considered, in which

$$S_{,\alpha} = 0, \tag{15.58}$$

i.e. the entropy is constant in the whole volume under consideration. Such motion is called **isentropic**. Then, from (15.53) and (15.55):

$$\frac{dp}{n} = d\left(\frac{\epsilon + p}{n}\right) = \frac{d\epsilon + dp}{n} - \frac{(\epsilon + p)dn}{n^2}. \tag{15.59}$$

Hence $d\epsilon = (\epsilon + p)dn/n$, which means that $\epsilon = \epsilon(n)$, and consequently $p = p(n)$ and $\epsilon = \epsilon(p)$. This type of equation of state is called **barotropic**. It is almost exclusively used in cosmology (where it is necessitated by the high symmetry of the Robertson–Walker models). However, it is an additional simplifying assumption when used in less symmetric models.

A reverse theorem also holds: for a perfect fluid, if $\epsilon = \epsilon(p)$ and $(nu^\beta)_{;\beta} = 0$, then either $p_{,\alpha} u^\alpha = 0$ or $S_{,\alpha} = 0$.

Proof. For a one-component perfect fluid, we have an equation of state $F(\epsilon, p, n) = 0$. If $\epsilon = \epsilon(p)$, then the equation of state implies immediately that $n = n(p)$, so $\epsilon = \epsilon(n)$. Then, $u_\alpha T^{\alpha\beta}_{;\beta} = 0$, together with (15.55), implies that

$$\epsilon_{,\beta} u^\beta + (\epsilon + p)u^\beta_{;\beta} = 0 \tag{15.60}$$

and the Gibbs identity (15.54) becomes

$$\frac{d\epsilon}{n} - \frac{(\epsilon + p)dn}{n^2} = T \, dS, \tag{15.61}$$

which can be written as

$$\epsilon_{,\alpha} - (\epsilon + p)n_{,\alpha}/n = nTS_{,\alpha}. \tag{15.62}$$

Substituting $u^\beta_{;\beta} = -n_{,\alpha} u^\alpha/n$ (from $(nu^\beta)_{;\beta} = 0$) in (15.60) we obtain $[d\epsilon/dn - (\epsilon + p)/n]n_{,\beta} u^\beta = 0$. Hence, either $n_{,\beta} u^\beta = 0$ (the particle number density does not change along the flow lines), or $d\epsilon/dn = (\epsilon + p)/n$, so $\epsilon_{,\alpha} = (\epsilon + p)n_{,\alpha}/n$. Substitution of the last equation in (15.62) gives $S_{,\alpha} = 0$ □.

15.6 Exercises

1. Show that the quantity D defined in (15.17) is indeed zero.
   Hint. Decompose the vectors δx, δy and δz in the basis of the eigenvectors of the matrix σ. Note that $\text{Tr}(\sigma) = 0$.
2. Verify Eqs. (15.36).

16

Relativistic cosmology I: general geometry

16.1 A continuous medium as a model of the Universe

When describing the Universe as a whole, one assumes that it is filled with a continuous medium (fluid or gas), whose state can be described by physical fields (scalar fields such as mass density and pressure, vector fields such as the velocity of flow, or tensor fields, e.g. an electromagnetic field). This is a rather crude approximation, since our real Universe has a 'granular' structure. Its basic units are stars, and the relevant information from the point of view of observational cosmology is, for example, the number of stars in a given volume rather than the average mass density in that volume. The less-than-perfect adequacy of the fluid approximation is also demonstrated by the fact that the view on which objects should be considered the 'elementary cells' of the cosmic fluid has been changing with time. In the times of Hubble (1920s and 1930s), these were the galaxies. In later times, when galaxy clusters and proper motions of galaxies in clusters were observed, the galaxy clusters took over. In still later years, it was found that galaxies and galaxy clusters tend to occupy edges of large volumes of space that are almost empty inside (called voids). According to current beliefs, the elementary units of the Universe should be groups of voids. These changes in the definition of the elementary unit of the Universe were, characteristically, adopted in order to save the assumption of homogeneity and isotropy of the Universe 'in the large'.

This assumption deserves a separate comment. The astronomical observations provide reliable quantitative information about a relatively small neighbourhood of the Solar System. With increasing distance from the Earth, the precision of this information is quickly degraded. If we want to describe the Universe as a whole, we have to extrapolate the results of local observations to large volumes, and then test the conclusions from the extrapolation. The extrapolations, however, always contain a large amount of arbitrariness. Hence, if observations tell us that a given extrapolation leads to a correct prediction, this does not mean that the extrapolation was the only one possible.

The most fundamental extrapolation is contained in the so-called **cosmological principle**. It stems from the ideas of Copernicus, and so is sometimes called the *Copernican principle*. Copernicus was the first astronomer who noted that the Earth is not at the centre of the Universe, but occupies a relatively unimportant position in the Solar System. Afterwards, the Earth has been 'degraded' a few more times when it was established that even the Sun is not at the centre of the Universe, but is one of a great number of stars that are similar to each other, and that our Galaxy is also one of many, not the greatest one and not placed at any important position. The cosmological principle is a summary of this line of thinking. In its weaker form, it says: we (the inhabitants of the Earth) occupy a position in the Universe that is not in any way preferred. Quite often, however, the cosmological principle is expressed in the extreme form: all positions in the Universe are exactly equivalent; the geometrical and physical properties of the Universe do not depend on the point from which the Universe is observed. No matter whether one believes in the cosmological principle, and in which version of it, it must be remembered that this principle is not a summary of observational results, but an *assumption*, upon which the theory of the structure of the Universe is built. This assumption was a good working hypothesis when theoreticians constructed the first-ever models of the Universe in the 1920s (Friedmann, 1922; Lemaître, 1927) because at that time there were no observational data to contradict it. Today, the cosmological principle still has no direct observational verification,[1] while models not obeying this principle and generalising the Friedmann–Lemaître models are known; see Chapter 18. The fact that virtually the whole of observational cosmology is based on the Friedmann–Lemaître models is a consequence of inertia in thinking and of emotional attachment to the, mathematically elegant and appealing, doctrine of exact equivalence of all positions in the Universe. However, natural sciences, physics and astronomy among them, are said to use the criterion of consistency of theory with observations/experiments. At the very least, in order to verify the cosmological principle, alternatives to it have to be considered and compared with observations. Working always from within the same theory, we make it more difficult to verify its basic assumptions.

The cosmological principle is translated into assumptions about the geometry of space-time in the following way. The observations show that the space is *approximately* isotropic around us.[2] According to the cosmological principle, the space should thus be isotropic around every other point. A space that is isotropic around every point is homogeneous. This argument points to the Robertson–Walker (R–W) spacetimes of Section 10.7.

Before we come to the R–W models, however, we shall discuss the fluid model of the Universe in the background of a general geometry. We assume that each point in the Universe can be assigned energy density, pressure and the 4-vector of velocity of the fluid particle that passes through the point. We also assume that the matter of the Universe treated in this way obeys the equations of hydrodynamics known from laboratory – which is another bold assumption.

---

[1] Because of difficulties in determining the distances to other galaxies. The isotropy of the cosmic microwave background (CMB) radiation imposes only weak limitations on the anisotropy of matter distribution; see Chapter 18.
[2] At present, the main argument supporting this statement is the isotropy of the microwave background radiation, but the statement had been made long before the CMB radiation was discovered.

16.2 Optical observations in the Universe – part I

16.2.1 The geometric optics approximation

The approach presented below is based on the papers by Ellis (1971, 1973) and Kristian and Sachs (1966).

We will apply the geometric optics approximation to the Maxwell equations. We assume that the electromagnetic field propagates into vacuum, with no charges or currents, and that it is weak enough not to influence the geometry of the spacetime (i.e. it is a *test field* in a given geometry). Then, the Maxwell equations are (13.4) and (13.6) with $j^\mu = 0$. We will seek solutions of this set in the form of waves:

$$F_{\alpha\beta} = G_{\alpha\beta} \sin(S + \phi_0), \tag{16.1}$$

where the amplitude $G_{\alpha\beta}$ is assumed to vary slowly compared to the phase S; ϕ_0 is a constant. Thus, we assume that each differentiation of $G_{\alpha\beta}$ introduces a factor ε, where ε is a small parameter. If $G_{\alpha\beta}$ can be developed in a power series with respect to ε, then we can write

$$G_{\alpha\beta} = B^0_{\alpha\beta} + \sum_{i=1}^{\infty} \varepsilon^i B^i_{\alpha\beta}. \tag{16.2}$$

We now assume that the Maxwell equations are fulfilled at each order in ε, and substitute (16.2) in them. Denoting[1]

$$S,_\alpha = k_\alpha, \tag{16.3}$$

we obtain at the zeroth order:

$$(B^0)^{\alpha\beta} k_\beta = 0, \qquad k_{[\alpha} B^0_{\beta\gamma]} = 0 \tag{16.4}$$

and at the first order

$$(B^0)^{\alpha\beta};_\beta \sin(S + \phi) = -\varepsilon (B^1)^{\alpha\beta} k_\beta \cos(S + \phi), \tag{16.5}$$

$$B^0_{[\alpha\beta;\gamma]} \sin(S + \phi) = -\varepsilon B^1_{[\alpha\beta} k_{\gamma]} \cos(S + \phi). \tag{16.6}$$

(Since the $(B^i)^{\alpha\beta}, i = 0, 1, \ldots$, are antisymmetric in $[\alpha\beta]$, the covariant derivative in (16.5) can be written as $(1/\sqrt{-g}) \left[\sqrt{-g} (B^0)^{\alpha\beta} \right],_\beta$, so this expression is small by virtue of our assumptions. The same applies to (16.6) – in this combination, the Christoffel symbols cancel out and the covariant derivative reduces to a partial derivative.) The last two equations show that the first-order terms act as sources in the Maxwell equations for the zeroth order terms. Thus, comparing this scheme with the Maxwell equations in empty space, we see that the electromagnetic wave does not in fact propagate into vacuum – the higher order terms act as a medium with currents and charges, on which the zeroth order wave may be dispersed. Similarly, the first order terms will be influenced by second order terms, and so on. This is the influence of curvature on the propagation

---

[1] With Eq. (16.3) fulfilled, the *rotation* of the vector field k_α, defined later in this section, is zero. Thus, the geometric optics approach turns out to be more general in this respect: the wave description does not allow rotating congruences of rays.

of the electromagnetic wave. (In a flat space in Cartesian coordinates, a constant $B^0_{\alpha\beta}$, with all $B^i_{\alpha\beta} \equiv 0$ for $i \geq 1$ is a solution of (16.5)–(16.6).) Assuming that the scheme is self-consistent (no formal proof of this assumption seems to be available) and that the 'tail' terms remain small, we will now consider the consequences of (16.4). The second equation of (16.4) can be written as

$$k_\alpha B^0_{\beta\gamma} + k_\beta B^0_{\gamma\alpha} + k_\gamma B^0_{\alpha\beta} = 0. \tag{16.7}$$

Contracting this with k^α and making use of (16.4) we immediately obtain

$$k^\alpha k_\alpha = 0, \tag{16.8}$$

i.e. the wave vector of the ray is a null vector. From this it follows that

$$k_{\alpha;\beta}k^\alpha = 0. \tag{16.9}$$

But, since $k_\alpha = S_{,\alpha}$, we have $k_{\alpha;\beta} = k_{\beta;\alpha}$, and then from (16.9)

$$k_{\beta;\alpha}k^\alpha = 0, \tag{16.10}$$

which means that k^α is geodesic, and its parametrisation is affine.

Further, contracting (16.7) with $(B^0)^{\beta\gamma}$ and using (16.4) we obtain

$$B^0_{\alpha\beta}(B^0)^{\alpha\beta} = 0. \tag{16.11}$$

Contracting (16.7) with $(B^0)^{\delta\gamma}$ and using (16.4) once more, we obtain

$$k_\alpha B^0_{\beta\gamma}(B^0)^{\delta\gamma} - k_\beta B^0_{\alpha\gamma}(B^0)^{\delta\gamma} = 0. \tag{16.12}$$

This equation has the form $k_\alpha V_\beta - k_\beta V_\alpha = 0$ (except that V carries also an upper index δ in both places). Such an equation means that the vectors V_α and k_α are proportional (collinear), thus

$$B^0_{\alpha\gamma}(B^0)^{\delta\gamma} = U^\delta k_\alpha, \tag{16.13}$$

where U^δ is the (vectorial) proportionality factor. With the index δ lowered, the left-hand side of (16.13) is symmetric in $(\alpha\delta)$, so $U_\delta k_\alpha - U_\alpha k_\delta = 0$, and by the same argument $U_\alpha = \mu k_\alpha$. Thus, finally

$$B^0_{\alpha\gamma}(B^0)_\delta{}^\gamma = \mu k_\alpha k_\delta. \tag{16.14}$$

From (16.11) and (16.14) we see that the electromagnetic energy-momentum tensor (13.10) of the field (16.1)–(16.2) is

$$T_{\alpha\beta} = \frac{1}{4\pi}\left(\mu k_\alpha k_\beta + O(\varepsilon)\right), \tag{16.15}$$

which is, up to terms linear in ε, a perfect-fluid-type medium with the '4-velocity' k_α. With k_α being null, the velocity of flow equals the velocity of light, which means that (16.15) corresponds to a stream of photons. From $T^{\alpha\beta}{}_{;\beta} = 0$, and making use of (16.10), we obtain

$$(\mu k^\alpha)_{;\alpha} = 0, \tag{16.16}$$

which means that the stream of photons is conserved.

16.2.2 The redshift

An observer moving with the 4-velocity u^α will measure the rate of change of phase of the light wave $v_p = S_{,\alpha} u^\alpha = k_\alpha u^\alpha$. Within a short time-interval Δs, the phase will thus change by $\Delta S = k_\alpha u^\alpha \Delta s$. For another observer, moving with the velocity u_1^α and measuring the change of phase at another spacetime point, where $k^\alpha = k_1^\alpha$, the same change of phase ΔS will in general take a different time-interval, Δs_1: $\Delta S = k_{1\alpha} u_1^\alpha \Delta s_1$. Hence:

$$\frac{\Delta s_1}{\Delta s_2} = \frac{(k_\alpha u^\alpha)_2}{(k_\alpha u^\alpha)_1}. \tag{16.17}$$

If the electromagnetic wave is periodic, then the change of phase is connected with the frequency v by $\Delta S = 2\pi v \Delta s$ (this formula applies for an arbitrary time-interval Δs). Hence, for the same change of phase measured by two different observers we have $v_1 \Delta s_1 = v_2 \Delta s_2$, so

$$\frac{v_2}{v_1} = \frac{\Delta s_1}{\Delta s_2} = \frac{(k_\alpha u^\alpha)_2}{(k_\alpha u^\alpha)_1}. \tag{16.18}$$

This is the formula for the cosmological **redshift**, derived without invoking any definite cosmological model. It can be written in a more familiar form. Let the subscripts e and o denote quantities calculated at the point of emission of the light ray and at the point of detection, respectively. Then, from the definition of the redshift:

$$z = \frac{\lambda_o - \lambda_e}{\lambda_e} = \frac{\lambda_o}{\lambda_e} - 1. \tag{16.19}$$

But $\lambda_o/\lambda_e = v_e/v_o$ (because the locally measured velocity of light $c = \lambda v$ is constant), hence

$$1 + z = \frac{v_e}{v_o} = \frac{(k_\alpha u^\alpha)_e}{(k_\alpha u^\alpha)_o}. \tag{16.20}$$

The light ray with the wave vector k^α is received by the observer from the direction determined by the unit spacelike vector n^α:

$$n_\alpha n^\alpha = -1, \tag{16.21}$$

which is collinear with the projection of k^α on the hypersurface of constant time. Hence, if the 4-velocity of the observer is u^α, then

$$n^\alpha = \rho \left(\delta^\alpha_{\ \beta} - u^\alpha u_\beta \right) k^\beta. \tag{16.22}$$

Substituting (16.22) in (16.21) we obtain $\rho^2 = \left(k_\rho u^\rho \right)^{-2}$; hence

$$n^\alpha = -\frac{1}{k_\rho u^\rho} k^\alpha + u^\alpha \qquad (\Longrightarrow n^\alpha u_\alpha = 0). \tag{16.23}$$

We have chosen $\rho = -\left(k_\rho u^\rho \right)^{-1}$ because n^α denotes the direction *towards* the source of light, opposite to the direction of the wave vector.

The considerations up to this point applied up to an arbitrary distance; the only approximation involved was that connected with introducing the geometric optics. However, in order to apply (16.20) to results of observations, one has to integrate the equations of a null geodesic, which is almost always difficult. In that case, an approximate version of (16.20) may be used, which applies for sources of light with small redshifts, $z \ll 1$. Then (16.19) becomes $z = \mathrm{d}\lambda/\lambda$, and in (16.20) we have

$$z = \frac{(k_\alpha u^\alpha)_e - (k_\alpha u^\alpha)_0}{(k_\alpha u^\alpha)_0} = \frac{\mathrm{d}\,(k_\alpha u^\alpha)}{(k_\alpha u^\alpha)_0}. \tag{16.24}$$

The change $\mathrm{d}(\cdot)$ in (16.24) should be calculated along the light ray connecting the light-source with the observer. Denoting the affine parameter on the ray by v, we have

$$\mathrm{d}\,(k_\alpha u^\alpha) = \mathrm{D}\,(k_\alpha u^\alpha) = \left(k_{\alpha;\beta}k^\beta u^\alpha\right)_0 \mathrm{d}v + \left(k_\alpha u^\alpha{}_{;\beta}k^\beta\right)_0 \mathrm{d}v. \tag{16.25}$$

The first term vanishes in consequence of (16.10), and in the second term we substitute the decomposition (15.28) and use the antisymmetry of $\omega_{\alpha\beta}$. The result is

$$\mathrm{d}\,(k_\alpha u^\alpha) = \left[\left(\sigma_{\alpha\beta}k^\alpha k^\beta\right)_0 - \frac{1}{3}\theta(k_\alpha u^\alpha)_0{}^2 + (k_\alpha \dot{u}^\alpha)_0\,(k_\alpha u^\alpha)_0\right]\mathrm{d}v. \tag{16.26}$$

Using (16.23) and (15.42), we obtain

$$\mathrm{d}\,(k_\alpha u^\alpha) = \left(k_\rho u^\rho\right)_0{}^2 \left(\sigma_{\alpha\beta}n^\alpha n^\beta - \frac{1}{3}\theta - n_\alpha \dot{u}^\alpha\right)_0 \mathrm{d}v. \tag{16.27}$$

Equations (16.23) and (16.8) imply

$$k^\rho n_\rho = k^\rho u_\rho, \tag{16.28}$$

and $[-\left(k^\rho n_\rho\right)_0]\mathrm{d}v$ is the distance in the rest space of the observer travelled by the front of the light wave corresponding to the change $\mathrm{d}v$ in the affine parameter, i.e. it is the distance between the light-source and the observer, which we shall denote by $\delta\ell$. In consequence of this, and using (16.27) in (16.24), we obtain:

$$z = \left(-\sigma_{\alpha\beta}n^\alpha n^\beta + \frac{1}{3}\theta + n_\alpha \dot{u}^\alpha\right)_0 \delta\ell. \tag{16.29}$$

This shows that for $z \ll 1$ rotation has no influence on z. It is seen that $\sigma_{\alpha\beta}$ and $\dot{u}^\alpha$ introduce anisotropy in z, whereas with $\sigma_{\alpha\beta} = 0 = \dot{u}^\alpha$ the redshift should be isotropic. Equation (16.29) is still correct in every cosmological model, but only for light-sources that are near to the observer. Its advantage is that all the quantities contained in it can always be calculated; one does not need to integrate any differential equations.

16.3 The optical tensors

We shall now apply a reasoning similar to that in Section 15.2 to families of null curves. The tensor of projection onto a locally orthogonal space now has to be defined in a different way, because a hypersurface orthogonal to a null vector k contains k. Therefore, given k^α, we first define a second null vector ℓ^α that is tangent to the same light cone and obeys:

$$\ell^\alpha k_\alpha = 1, \qquad \ell^\alpha \ell_\alpha = 0. \tag{16.30}$$

Note that ℓ^α is not defined uniquely. If m^α is an arbitrary spacelike vector of unit length ($m^\alpha m_\alpha = -1$) orthogonal both to k^α and to ℓ^α, then $\ell'^\alpha = \ell^\alpha + \frac{1}{2}b^2 k^\alpha + bm^\alpha$ obeys (16.30) as well, where b is an arbitrary parameter. We will come back to this in Section 16.5.

Then we define the projection tensor on the surface (this time 2-dimensional) that is orthogonal to both ℓ^α and k^α:

$$p_{\alpha\beta} = g_{\alpha\beta} - \ell_\alpha k_\beta - k_\alpha \ell_\beta \implies p_{\alpha\beta} k^\beta = p_{\alpha\beta} \ell^\beta = 0. \tag{16.31}$$

This surface does not include ℓ^α or k^α – because a vector that is orthogonal to two linearly independent null vectors must be spacelike; it is an exercise in special relativity to prove it.

Now assume that k^α is a null vector *field*. It is usually assumed at this point that the field k^α is geodesic and affinely parametrised so that $k^\nu k_{\mu;\nu} = 0$. We will not make this assumption; instead we define

$$\dot{k}^\mu \stackrel{\text{def}}{=} k^\mu{}_{;\nu} k^\nu, \tag{16.32}$$

which we shall call the **acceleration** of a light ray. The specialisation to the geodesic affinely parametrised case will then follow immediately by $\dot{k}^\mu = 0$. We will need the more general formulae in Section 16.5.

Being null, the field k^μ obeys

$$k^\mu k_\mu = 0 = k^\mu k_{\mu;\alpha} \implies k^\mu \dot{k}_\mu = 0. \tag{16.33}$$

We define

$$A_{\alpha\beta} \stackrel{\text{def}}{=} k_{\rho;\sigma} p^\rho{}_\alpha p^\sigma{}_\beta. \tag{16.34}$$

Then the following holds:

$$k_{\alpha;\beta} = A_{\alpha\beta} + a_\alpha k_\beta + k_\alpha b_\beta + \dot{k}_\alpha \ell_\beta, \tag{16.35}$$

where

$$a_\alpha \stackrel{\text{def}}{=} \ell^\rho k_{\alpha;\rho} - \frac{1}{2} k_{\rho;\sigma} \ell^\rho \ell^\sigma k_\alpha,$$

$$b_\alpha \stackrel{\text{def}}{=} \ell^\rho k_{\rho;\alpha} - \frac{1}{2} k_{\rho;\sigma} \ell^\rho \ell^\sigma k_\alpha - \dot{k}^\rho \ell_\rho \ell_\alpha. \tag{16.36}$$

It follows that $a_\alpha k^\alpha = b_\alpha k^\alpha = 0$. Now we apply to $A_{\alpha\beta}$ the decomposition into the trace, the trace-free symmetric part and the antisymmetric part, just like we did in Section 15.2 with the covariant derivative of a unit timelike field:

$$A_{\alpha\beta} = \omega_{\alpha\beta} + \sigma_{\alpha\beta} + p_{\alpha\beta}\theta \tag{16.37}$$

(for some reason, tradition requires one to write the last term without the coefficient 1/2 that would be natural here), where

$$\omega_{\alpha\beta} \stackrel{\text{def}}{=} A_{[\alpha\beta]} \tag{16.38}$$

is called the **rotation** of the family of null curves,

$$\theta \stackrel{\text{def}}{=} \frac{1}{2} g^{\alpha\beta} A_{\alpha\beta} \tag{16.39}$$

is called the **expansion** of the family, and

$$\sigma_{\alpha\beta} \stackrel{\text{def}}{=} A_{(\alpha\beta)} - P_{\alpha\beta}\theta \tag{16.40}$$

is called the **shear** of the family. The geometric interpretation of rotation, expansion and shear is similar to that in hydrodynamics; this time the respective changes apply to images of an object projected by the family of light rays on 2-surfaces orthogonal to the family. Like in hydrodynamics, rotation and shear do or do not vanish simultaneously with the scalars defined below.[1] Also, the following equations are useful:

$$p^{\alpha}{}_{\rho}p^{\rho}{}_{\beta} = p^{\alpha}{}_{\beta}, \qquad g^{\alpha\beta}p_{\alpha\beta} = p^{\alpha\beta}p_{\alpha\beta} = 2,$$

$$p^{\alpha\beta}\sigma_{\alpha\beta} = g^{\alpha\beta}\sigma_{\alpha\beta} = 0, \tag{16.41}$$

$$k^{\beta}\omega_{\alpha\beta} = k^{\beta}\sigma_{\alpha\beta} = \ell^{\beta}\omega_{\alpha\beta} = \ell^{\beta}\sigma_{\alpha\beta} = 0.$$

The scalars of rotation, expansion and shear are then

$$\omega^2 \stackrel{\text{def}}{=} \frac{1}{2}\omega_{\alpha\beta}\omega^{\alpha\beta} = \frac{1}{2}k_{[\alpha;\beta]}\left(k^{\alpha;\beta} - 2\dot{k}^{\alpha}\ell^{\beta}\right) - \frac{1}{4}\left(\dot{k}_{\rho}\ell^{\rho}\right)^2, \tag{16.42}$$

$$\theta = \frac{1}{2}k^{\mu}{}_{;\mu} - \frac{1}{2}\dot{k}_{\rho}\ell^{\rho}, \tag{16.43}$$

$$\sigma^2 \stackrel{\text{def}}{=} \frac{1}{2}\sigma_{\alpha\beta}\sigma^{\alpha\beta}$$

$$= \frac{1}{2}k_{(\alpha;\beta)}\left(k^{\alpha;\beta} - 2\dot{k}^{\alpha}\ell^{\beta}\right) + \frac{1}{4}\left(\dot{k}_{\rho}\ell^{\rho}\right)^2 - \theta^2. \tag{16.44}$$

As can be seen, for affinely parametrised geodesics ($\dot{k}^{\mu} = 0$) these quantities actually depend only on k^{α}, not on the auxiliary field ℓ^{α}.

16.4 The apparent horizon

We have already defined the event horizon for the Schwarzschild metric in Section 14.11. There is one more variety of horizon that is important for studying dynamical black holes, i.e. black holes that keep swallowing up new matter and increase their masses. The notion of the event horizon has the disadvantage that it can be defined only when we know the

---

[1] We recall that, in the wave description of light, rotation is necessarily zero (see footnote cited above (16.3)).

whole future evolution of the spacetime. Hence, it is practically useless in observational cosmology. It is more realistic to use only such notions as can be identified by local observations of short duration. The apparent horizon is such a notion.

In a flat spacetime, a flash of light sent from both sides of a closed surface has the property that the light rays are converging inside the surface and diverging outside. The convergence/divergence is measured by the scalar of expansion for the family of rays, $\theta = \frac{1}{2} k^{\mu}{}_{;\mu}$, defined in the previous section. However, in the vicinity of a singularity the *outward-directed* bundle of rays is convergent. This is most easy to see in the Kruskal diagram (Fig. 14.6): under the horizon H_1 it is impossible to send any bundle of rays in the direction of increasing r because the light rays intersect lines of still-decreasing values of r (recall that each point in the Kruskal diagram represents a sphere $\{t = \text{constant}, r = \text{constant}\}$). Hence, the area of the light front is decreasing for both bundles. A closed surface from which it is impossible to send a diverging bundle of light rays is called a **closed trapped surface**. Then, an **apparent horizon** is the outer envelope of the region in which closed trapped surfaces exist.

There are two kinds of trapped surfaces and two kinds of apparent horizons, as seen in Fig. 14.6: the **future-trapped surfaces** and **future apparent horizons** in region II, and the **past-trapped surfaces** and **past apparent horizons** in region IV. For a past-trapped surface, the light rays converge towards the past. A more physical way to formulate this definition is to say that, for a past-trapped surface S_P, both *ingoing* bundles of rays (i.e. those that simultaneously reach S_P ,both from inside and from outside) are necessarily *diverging*, whereas for a future-trapped surface S_F, both *outgoing* bundles of rays are necessarily converging (where 'outgoing' means starting their journey simultaneously at S_F, both inward and outward).

In the Schwarzschild spacetime, the apparent horizons coincide with the event horizons, but in nonstatic spacetimes these two horizons are different. Examples will be given in Section 18.8.

16.5 * The double-null tetrad

We will set up a field of null vector bases over the spacetime that will include the k^{α} and ℓ^{α} introduced in Section 16.3. The other vectors in the basis, m^{α} and $\overline{m}^{\alpha}$, will be complex conjugate to each other, orthogonal both to k^{α} and to ℓ^{α}, and will obey relations similar to (16.30):

$$g_{\alpha\beta} m^{\alpha} m^{\beta} = g_{\alpha\beta} \overline{m}^{\alpha} \overline{m}^{\beta} = 0, \qquad g_{\alpha\beta} m^{\alpha} \overline{m}^{\beta} = -1,$$

$$g_{\alpha\beta} m^{\alpha} k^{\beta} = g_{\alpha\beta} m^{\alpha} \ell^{\beta} = 0, \qquad g_{\alpha\beta} \overline{m}^{\alpha} k^{\beta} = g_{\alpha\beta} \overline{m}^{\alpha} \ell^{\beta} = 0. \tag{16.45}$$

Thus we have[1]

$$e^{\hat{0}}{}_{\alpha} = k_{\alpha}, \qquad e^{\hat{1}}{}_{\alpha} = \ell_{\alpha}, \qquad e^{\hat{2}}{}_{\alpha} = m_{\alpha}, \qquad e^{\hat{3}}{}_{\alpha} = \overline{m}_{\alpha}, \tag{16.46}$$

---

[1] Hats mark tetrad indices. Where no confusion may arise, they will be omitted.

and, from (9.1), the tetrad metric is

$$[\eta_{ij}] = \begin{bmatrix} 0 & 1 & 0 & 0 \\ 1 & 0 & 0 & 0 \\ 0 & 0 & 0 & -1 \\ 0 & 0 & -1 & 0 \end{bmatrix}. \tag{16.47}$$

The upper and lower tetrad indices are related to each other as follows:

$$v_{\hat{0}} = v^{\hat{1}}, \qquad v_{\hat{1}} = v^{\hat{0}}, \qquad v_{\hat{2}} = -v^{\hat{3}}, \qquad v_{\hat{3}} = -v^{\hat{2}} = \overline{v_{\hat{2}}}. \tag{16.48}$$

The tetrad components of the basis vectors are thus

$$k^i = \delta^i{}_{\hat{1}}, \qquad \ell^i = \delta^i{}_{\hat{0}}, \qquad m^i = -\delta^i{}_{\hat{3}}, \qquad \overline{m}^i = -\delta^i{}_{\hat{2}},$$

$$k_i = \delta_i{}^{\hat{0}}, \qquad \ell_i = \delta_i{}^{\hat{1}}, \qquad m_i = \delta_i{}^{\hat{2}}, \qquad \overline{m}_i = \delta_i{}^{\hat{3}}. \tag{16.49}$$

Some of the Ricci rotation coefficients in this basis, defined by (9.7), vanish in consequence of (9.14), thus

$$\Gamma^0{}_{1a} \equiv \Gamma_{11a} = 0 = \Gamma_{00a} \equiv \Gamma^1{}_{0a}, $$

$$\Gamma^2{}_{3a} \equiv -\Gamma_{33a} = 0 = -\Gamma_{22a} \equiv \Gamma^3{}_{2a}. \tag{16.50}$$

Some other Ricci rotation coefficients have physical interpretation. For example, the nonzero tetrad components of the shear tensor (16.40) are:

$$\sigma_{\hat{2}\hat{2}} = k_{\alpha;\beta} m^\alpha m^\beta = -\Gamma^1{}_{22} \equiv -\Gamma_{022} \equiv \Gamma_{202} \equiv -\Gamma^3{}_{02},$$

$$\sigma_{\hat{3}\hat{3}} = k_{\alpha;\beta} \overline{m}^\alpha \overline{m}^\beta = -\Gamma^1{}_{33} \equiv -\Gamma_{033} \equiv \Gamma_{303} \equiv -\Gamma^2{}_{03}. \tag{16.51}$$

In order to calculate the expansion, we do the following operations

$$k^\mu{}_{;\mu} = g_{\mu\nu} k^{\mu;\nu} = \eta_{ab} e^a{}_\mu e^b{}_\nu k^{\mu;\nu},$$

and similarly for the term $\dot{k}_\rho \ell^\rho$, and then we explicitly run through all the values of a and b. Most of the terms are zero or cancel out, and what remains is

$$\theta = -\frac{1}{2} k^{\mu;\nu} \left(m_\mu \overline{m}_\nu + \overline{m}_\mu m_\nu \right) = \frac{1}{2} \left(\Gamma^1{}_{23} + \Gamma^1{}_{32} \right) \equiv \frac{1}{2} \left(\Gamma^1{}_{23} + \overline{\Gamma}^1{}_{23} \right). \tag{16.52}$$

For the tetrad components of the rotation tensor we find

$$\omega_{23} = -\omega_{32} = -\frac{1}{2} \left(\Gamma^1{}_{23} - \Gamma^1{}_{32} \right) = \overline{\omega}_{32} = -\overline{\omega}_{23}, \tag{16.53}$$

all other components being zero. Thus, the only nonvanishing component of the rotation tensor is pure imaginary, and, moreover, as follows from (16.42) and (16.45), it is connected to the scalar of rotation ω by $\omega_{23}{}^2 = -\omega^2$; thus $\omega_{23} = -i\omega$. As one can see then, the expansion and the rotation are, respectively, the real and the imaginary part of the same complex quantity

$$\theta + i\omega = \Gamma^1{}_{23} \equiv \Gamma_{023} \equiv -\Gamma_{203}. \tag{16.54}$$

Finally, there are several identities among the Ricci rotation coefficients that are not self-evident; examples were given in (16.51) and (16.54).

The double-null tetrad is a basis of a self-contained approach to relativity called the **Newman–Penrose formalism**. Although shuffling indices in it is sometimes confusing (because of the scalar metric being nondiagonal), this tetrad has proven useful and powerful in finding exact solutions of Einstein's equations. For more on this formalism, see Stephani *et al.* (2003) and references cited therein.

The various identities obeyed by the Riemann and Weyl tensors were deduced under the assumption that these tensors arise from commutators of second derivatives of tensors. However, if the basic objects in the theory are the Ricci rotation coefficients, like in the Newman–Penrose formalism, then the curvature tensors are present in first-order equations, and not all the 'identities' will automatically be fulfilled (in fact, only $R_{ijkl} = -R_{ijlk} = -R_{jikl}$ and $C^i{}_{jil} = 0$). In the next two sections, and in Section 20.2, we will see some of those other 'identities' being imposed as equations to fulfil.

The tetrad $(k, \ell, m, \overline{m})$ is not uniquely defined. One usually begins with a given family of null curves, so the *direction* of k^α is fixed, but the field k^α may be rescaled by an arbitrary factor

$$k'^\alpha = Ak^\alpha, \tag{16.55}$$

where A is an arbitrary real function. This change corresponds to changing the parametrisation of the curves tangent to k^α. The vectors m^α and $\overline{m}^\alpha$ can be rotated in their plane by an arbitrary angle ϕ; their scalar products with k^α do not change when a multiple of k^α is added to any of them. Thus $(m^\alpha, \overline{m}^\alpha)$ are defined up to the transformations

$$m'^\alpha = e^{i\phi} m^\alpha + Bk^\alpha, \tag{16.56}$$

where ϕ is a real function and B is a complex function. Finally, ℓ^α may be changed by a fixed multiple of k^α and a fixed multiple of a fixed vector in the $(m^\alpha, \overline{m}^\alpha)$ plane, thus

$$\ell'^\alpha = \frac{1}{A} \left(\ell^\alpha + \overline{B} e^{i\phi} m^\alpha + Be^{-i\phi} \overline{m}^\alpha + B\overline{B} k^\alpha \right). \tag{16.57}$$

As an example of usefulness of this tetrad, let us note how simply it expresses the criterion for the Weyl tensor being algebraically special, i.e. obeying (11.55). Projecting that equation on the double-null tetrad we obtain the equivalent equation

$$C_{200d} = C_{300d} = 0. \tag{16.58}$$

16.6 * The Goldberg–Sachs theorem

There is a connection between properties of the optical tensors of Section 16.3 and the Petrov type of the Weyl tensor. We found a similar connection for the hydrodynamical tensors in Section 15.3, where $\sigma_{\alpha\beta} = \omega_{\alpha\beta} = 0 = \dot{u}_\alpha$ forced the metric to be conformally flat.

In order to prove an analogous, but weaker, limitation imposed by the properties of the optical tensors, we must first list the full set of properties of an algebraically special Weyl tensor in the tetrad $(k, \ell, m, \overline{m})$. With (16.58) assumed, the equations $C^s{}_{0s0} = 0$ imply

$$-C_{0101} = C_{0213} + C_{0312} = -C_{2323},$$

$$C_{0112} = C_{1223}, \qquad C_{0113} = -C_{1323}, \tag{16.59}$$

$$C_{0212} = C_{0223} = C_{0313} = C_{0323} = C_{1213} = 0.$$

The theorem we are going to prove consists of three parts that will be listed as separate theorems.

Theorem 16.1

Assumptions:

1. The Weyl tensor is algebraically special, so that (11.55) and, equivalently, (16.58) are fulfilled.
2. The vacuum Einstein equations are fulfilled, $R_{\alpha\beta} = 0$.

Thesis:

The degenerate Debever vector field k^α in (11.55) is geodesic and shearfree, i.e.

$$\dot{k}^\alpha = \phi k^\alpha, \qquad \sigma_{\alpha\beta} = 0 \Longrightarrow \Gamma_{202} = \Gamma_{303} = 0. \tag{16.60}$$

Proof:

Take the Bianchi identities $R_{\alpha\beta[\gamma\delta;\epsilon]} = 0$, contract them with $g^{\beta\gamma}$ and use the assumption $R_{\alpha\beta} = 0$. The result is equivalent to $C^\rho{}_{\alpha\beta\gamma,\rho} = 0$. Now use (9.17) contracted with δ^a_e:

$$e_a{}^\mu C^a{}_{bcd,\mu} + \Gamma^s{}_{rs} C^r{}_{bcd} - \Gamma^s{}_{br} C^r{}_{scd} - \Gamma^s{}_{cr} C^r{}_{bsd} - \Gamma^s{}_{dr} C^r{}_{bcs} = 0. \tag{16.61}$$

Take the component $(b, c, d) = (0, 0, 2)$ of this equation and use (16.58)–(16.59) together with (16.50). The result is

$$\Gamma_{200} (C_{0213} + C_{0123} - C_{0101}) = 0. \tag{16.62}$$

Then take the component $(0, 0, 3)$ and make similar simplifications. The result is[1]

$$\Gamma_{300} (C_{0312} - C_{0123} - C_{0101}) = 0. \tag{16.63}$$

Thus, either $\Gamma_{200} = 0 \Longrightarrow \Gamma_{300} = 0$, or else the expressions in parentheses must vanish. Consider first the second case. Then:

$$C_{0123} + C_{0213} - C_{0101} = 0,$$

$$-C_{0123} + C_{0312} - C_{0101} = 0, \tag{16.64}$$

---

[1] Note that (16.63) is the complex conjugate of (16.62) – which was to be expected, since the interchanges $2 \leftrightarrow 3$ always correspond to complex conjugation in this tetrad. Thus, one could write down (16.63) on the basis of (16.62) without repeating the calculations. This is one of the advantages of the double-null tetrad.

and also the first of (16.59) and $C_{a[bcd]} = 0$ must be obeyed:

$$C_{0213} + C_{0312} + C_{0101} = 0,$$
$$C_{0123} - C_{0213} + C_{0312} = 0.$$

(16.65)

The solution of the set (16.64)–(16.65) is

$$C_{0123} = C_{0213} = C_{0312} = C_{0101} = C_{2323} = 0.$$

(16.66)

Now take the components $(0, 0, 1)$ and $(0, 2, 3)$ of (16.61), with (16.66) assumed. They are

$$\Gamma_{300} C_{0112} + \Gamma_{200} C_{0113} = 0 = \Gamma_{300} C_{1223} + \Gamma_{200} C_{1323}.$$

(16.67)

Taking into account the middle part of (16.59), we see that with $\Gamma_{200} \neq 0 \neq \Gamma_{300}$ the solution of (16.67) is

$$C_{0112} = C_{0113} = C_{1223} = C_{1323} = 0.$$

(16.68)

Take the components $(0, 1, 2)$ and $(0, 1, 3)$ of (16.61), with (16.66) and (16.68) assumed. They say that $\Gamma_{300} C_{1212} = 0 = \Gamma_{200} C_{1313}$, i.e. $C_{1212} = C_{1313} = 0$. Together with (16.58), (16.66) and (16.68) this means $C_{abcd} = 0$. With $R_{ab} = 0$, this is the Minkowski spacetime, in which a congruence of shearfree null geodesics does exist, as can be verified by calculation in the Cartesian coordinates. Thus, the theorem is then true in a modified form: a shearfree null geodesic congruence exists.

We return to (16.62)–(16.63) and take the other solution

$$\Gamma_{200} = \Gamma_{300} = 0.$$

(16.69)

This means, from the definition of the Ricci rotation coefficients,

$$0 = \Gamma^3{}_{00} = m_{\rho;\sigma} k^\rho k^\sigma = -k_{\rho;\sigma} m^\rho k^\sigma = -m^\rho \dot{k}_\rho,$$
$$0 = \Gamma^2{}_{00} = \overline{m}_{\rho;\sigma} k^\rho k^\sigma = -k_{\rho;\sigma} \overline{m}^\rho k^\sigma = -\overline{m}^\rho \dot{k}_\rho.$$

(16.70)

Thus $\dot{k}^\alpha$ is orthogonal both to m^α and to $\overline{m}^\alpha$ and must lie in the plane spanned by k^α and ℓ^α. Since $\dot{k}^\alpha k_\alpha = 0$ and $\ell^\alpha k_\alpha = 1$, $\dot{k}^\alpha$ must be proportional to k^α, thus the first of (16.60) is proved. We can use the transformations (16.55)–(16.57) to rescale k^α so that $\dot{k}^\alpha = 0$.

Now take (16.61) with the sets of indices $(2, 0, 2)$ and $(3, 0, 3)$, and use (16.69). The result is

$$\Gamma_{022} (C_{0123} + C_{0213} - C_{2323}) = 0 = \Gamma_{033} (-C_{0123} + C_{0312} - C_{2323})$$

(16.71)

(again, the second equation is just a complex conjugate of the first one). Suppose that $\Gamma_{022} \neq 0 \iff \Gamma_{033} \neq 0$. Since $C_{2323} = C_{0101}$, again we obtain a set of equations ((16.71) together with the first of (16.59) and with the cyclic identity – the second of (16.65)) whose solution is (16.66).

The components $(1, 0, 2)$ and $(1, 0, 3)$ of (16.61) now say that $\Gamma_{202} C_{0113} = \Gamma_{303} C_{0112} = 0$. With the assumed $\Gamma_{202} \neq 0 \neq \Gamma_{303}$, and with (16.59), this implies (16.68) again. Then, the components $(1, 0, 1)$ and $(1, 2, 3)$ of (16.61) become $\Gamma_{303}C_{1212} + \Gamma_{202}C_{1313} = 0 = -\Gamma_{303}C_{1212} + \Gamma_{202}C_{1313}$. With $\Gamma_{202} \neq 0 \neq \Gamma_{303}$, the solution of this is $C_{1212} = C_{1313} = 0$, i.e. again $C_{abcd} = 0$. Thus (16.71) implies $\Gamma_{022} = \Gamma_{033} = 0$, which, as seen from (16.51), means that the shear of the k^α congruence is zero. $\square$

Before formulating the next theorem, we must first write out a few equations. Take the Ricci identity for the null field k^α:

$$k_{\alpha;\beta\gamma} - k_{\alpha;\gamma\beta} = R_{\rho\alpha\beta\gamma}k^\rho. \tag{16.72}$$

We will project this on different combinations of the tetrad vectors.

The following equations will be useful:

$$g^{\alpha\beta}k_{\alpha;\beta\gamma} \equiv \left(k^\beta{}_{;\beta}\right)_{;\gamma} = 2\theta_{,\gamma} + \left(\dot{k}_\rho \ell^\rho\right)_{,\gamma}, \tag{16.73}$$

$$k^\gamma g^{\alpha\beta}k_{\alpha;\gamma\beta} \equiv k^\gamma k^\beta{}_{;\gamma\beta} = \dot{k}^\beta{}_{;\beta} - k^{\gamma;\beta}k_{\beta;\gamma}, \tag{16.74}$$

$$k^{\gamma;\beta}k_{\beta;\gamma} = 2\left(\sigma^2 - \omega^2 + \theta^2\right) + b_\beta \dot{k}^\beta + \left(\ell_\beta \dot{k}^\beta\right)^2, \tag{16.75}$$

$$k^\gamma m^\alpha \overline{m}^\beta k_{\alpha;\beta\gamma} = k^\gamma \left(m^\alpha \overline{m}^\beta k_{\alpha;\beta}\right)_{;\gamma} - k^\gamma m^\alpha{}_{;\gamma} \overline{m}^\beta k_{\alpha;\beta} - k^\gamma m^\alpha \overline{m}^\beta{}_{;\gamma} k_{\alpha;\beta}, \tag{16.76}$$

$$k^\gamma m^\alpha \overline{m}^\beta k_{\alpha;\gamma\beta} = \overline{m}^\beta \left(k^\gamma m^\alpha k_{\alpha;\gamma}\right)_{;\beta} - \overline{m}^\beta k^\gamma{}_{;\beta} m^\alpha k_{\alpha;\gamma} - \overline{m}^\beta m^\alpha{}_{;\beta} \dot{k}_\alpha, \tag{16.77}$$

$$m^\alpha{}_{;\gamma} k_{\alpha;\beta} = g^{\alpha\mu} m_{\mu;\gamma} m_{\alpha;\beta} = \eta^{ab} e_a{}^\alpha e_b{}^\beta m_{\mu;\gamma} m_{\alpha;\beta}. \tag{16.78}$$

Equations (16.73) and (16.75) follow from (16.43) and (16.35)–(16.37). In simplifying the right-hand sides of (16.76) and (16.77) we will use (9.7), and in simplifying the right-hand side of (16.78) we will write out explicitly the terms corresponding to different values of a and b.

Contract (16.72) with $k^\gamma g^{\alpha\beta}$; the result is the equation of evolution of θ, analogous to the Raychaudhuri equation (15.43):

$$k^\gamma \theta_{,\gamma} + \sigma^2 - \omega^2 + \theta^2 + \frac{1}{2}A_{(1)} = -\frac{1}{2}R_{\rho\gamma}k^\rho k^\gamma, \tag{16.79}$$

where $A_{(1)}$ is a collection of terms containing acceleration:

$$A_{(1)} \overset{\text{def}}{=} b_\beta \dot{k}^\beta + \left(\ell_\beta \dot{k}^\beta\right)^2 - \dot{k}^\beta{}_{;\beta} + k^\gamma \left(\ell_\beta \dot{k}^\beta\right)_{,\gamma}, \tag{16.80}$$

and vanishes when $\dot{k}^\alpha = 0$ (the index in parentheses is just a label, it does not refer to the coordinates or to the tetrad vectors).

Now contract (16.72) with $k^\gamma \left(m^\alpha \overline{m}^\beta - m^\beta \overline{m}^\alpha\right)$ and use the fact that $R_{\rho[\alpha\beta]\gamma}k^\rho k^\gamma \equiv 0$. The result is

$$k^\gamma \left(\Gamma^1{}_{[23]}\right)_{,\gamma} + \frac{1}{2}\left[\left(\Gamma^1{}_{23}\right)^2 - \left(\Gamma^1{}_{32}\right)^2\right] - \frac{1}{2}A_{(2)} = 0, \tag{16.81}$$

where $A_{(2)}$ is another collection of terms containing acceleration:

$$A_{(2)} \stackrel{\text{def}}{=} -\overline{m}^\beta \left(m^\alpha \dot{k}_\alpha \right)_{,\beta} + m^\beta \left(\overline{m}^\alpha \dot{k}_\alpha \right)_{,\beta} + \overline{m}^\beta m^\alpha_{;\beta} \dot{k}_\alpha$$
$$-m^\beta \overline{m}^\alpha_{;\beta} \dot{k}_\alpha - m_\rho \dot{k}^\rho \left(2\Gamma^1_{13} + \Gamma^2_{10} - \Gamma^1_{31} \right)$$
$$+\overline{m}_\rho \dot{k}^\rho \left(2\Gamma^1_{12} + \Gamma^3_{10} - \Gamma^1_{21} \right), \tag{16.82}$$

and vanishes when $\dot{k}^\alpha = 0$ (in accord with the rest of (16.81), $A_{(2)}$ is imaginary). In consequence of (16.53), Eq. (16.81) can be written as

$$ik^\gamma \omega_{,\gamma} + 2i\theta\omega - \frac{1}{2} A_{(2)} = 0. \tag{16.83}$$

Equations (16.79) and (16.83) can be written as one complex equation:

$$k^\gamma Z_{,\gamma} + Z^2 + \sigma^2 + \frac{1}{2} A_{(1)} - \frac{1}{2} A_{(2)} = -\frac{1}{2} R_{\rho\gamma} k^\rho k^\gamma \equiv -\frac{1}{2} R_{00}, \tag{16.84}$$

where

$$Z \stackrel{\text{def}}{=} \theta + i\omega \equiv \Gamma^1_{23}. \tag{16.85}$$

Next, contract (16.72) with $k^\gamma m^\alpha m^\beta$ and use (16.76)–(16.78). The result is

$$k^\gamma \Gamma^1_{22,\gamma} + 2\Gamma^1_{22} \left(\theta + \Gamma^3_{30} \right) + A_{(3)} = R_{0202}, \tag{16.86}$$

where $A_{(3)}$ is yet another collection of terms with acceleration:

$$A_{(3)} \stackrel{\text{def}}{=} m^\beta \left(m^\alpha \dot{k}_\alpha \right)_{,\beta} - m^\beta m^\alpha_{;\beta} \dot{k}_\alpha + m_\rho \dot{k}^\rho \left(2\Gamma^1_{12} + \Gamma^1_{21} \right). \tag{16.87}$$

The result of contracting (16.72) with $k^\gamma \overline{m}^\alpha \overline{m}^\beta$ can be written at once, as the complex conjugate of (16.86):

$$k^\gamma \Gamma^1_{33,\gamma} + 2\Gamma^1_{33} \left(\theta + \Gamma^2_{20} \right) + \overline{A_{(3)}} = R_{0303}. \tag{16.88}$$

Finally, contract (16.72) with $k^\gamma m^\alpha \overline{m}^\beta$ and use (16.76)–(16.78) to obtain

$$k^\gamma \Gamma^1_{23,\gamma} + \left(\Gamma^1_{23} \right)^2 + \Gamma^1_{22} \Gamma^1_{33} + A_{(4)} = R_{0203}, \tag{16.89}$$

where

$$A_{(4)} \stackrel{\text{def}}{=} -\overline{m}^\beta m^\alpha_{;\beta} \dot{k}_\alpha + \overline{m}^\beta \left(m^\alpha \dot{k}_\alpha \right)_{,\beta} + m_\rho \dot{k}^\rho \left(2\Gamma^1_{13} + \Gamma^2_{10} \right) + \overline{m}_\rho \dot{k}^\rho \Gamma^1_{21}. \tag{16.90}$$

Since $\Gamma^1_{23} = \theta + i\omega = Z$, we can use (16.84) to eliminate $k^\gamma \Gamma^1_{23,\gamma}$ from the above and obtain

$$\frac{1}{2} \left(A_{(2)} - A_{(1)} \right) + A_{(4)} = R_{0203} + \frac{1}{2} R_{00}. \tag{16.91}$$

At this point, we can formulate one more theorem:

Theorem 16.2

Assumptions:
1. *There exists a shearfree geodesic null vector field k^α on a spacetime.*
2. *The vacuum Einstein equations are obeyed, $R_{\alpha\beta} = 0$.*
3. *The Weyl tensor is nonzero.*

Thesis:
The field k^α is collinear with one of the Debever vector fields.

Proof:
In the double-null tetrad, the condition for the field k^α to be collinear with a Debever vector field, i.e. to obey (11.54), is

$$C_{0202} = C_{0303} = C_{0203} = 0. \tag{16.92}$$

Choose the parametrisation on the curves tangent to k^α so that $\dot{k}^\alpha = 0$. Then, with $\dot{k}^\alpha = 0 = \sigma_{ab}$, in consequence of $R_{\alpha\beta} = 0$ and of (16.51), equations (16.86), (16.88) and (16.91) imply (16.92). $\square$

Now contract (16.72) with $m^\alpha \ell^\beta k^\gamma$ to obtain

$$C_{0210} = - k^\gamma \Gamma^1_{21,\gamma} - \Gamma^1_{21} \left(\Gamma^3_{30} + \Gamma^1_{23} \right) + \Gamma^0_{20} \Gamma^1_{23}$$
$$+ \Gamma^1_{22} \left(\Gamma^0_{30} - \Gamma^1_{31} \right) + A_{(5)}, \tag{16.93}$$

where $A_{(5)}$ is another collection of acceleration terms:

$$A_{(5)} \overset{\text{def}}{=} - 2m_\rho \dot{k}^\rho \Gamma^1_{11} - \ell^\beta \left(m_\alpha \dot{k}^\alpha \right)_{,\beta} + \ell^\beta m^\alpha{}_{;\beta} \dot{k}_\alpha - \ell_\rho \dot{k}^\rho \Gamma^1_{21}. \tag{16.94}$$

In order to prove the next theorem, we need the following lemma:

Lemma 16.1 *If $\dot{k}^\alpha = 0 = \sigma$ and $\theta + i\omega \neq 0$, then there exists such a transformation of the basis vectors (16.55)–(16.57) after which $\Gamma^1_{21} = 0$. The direction of the new ℓ^α is uniquely determined.*

Proof:
Apply the transformation (16.55)–(16.57) to Γ^1_{21}; the result is

$$\tilde{\Gamma}^1_{21} = e^{i\phi} \Gamma^1_{21} + B\Gamma^1_{23} + e^{2i\phi} \overline{B} \Gamma^1_{22} + e^{i\phi} B\overline{B} \Gamma^3_{00}. \tag{16.95}$$

In view of (16.51), (16.70) and (16.54) it follows that $\tilde{\Gamma}^1_{21} = 0$ when the assumptions are fulfilled and $B = -e^{i\phi} \Gamma^1_{21}/\Gamma^1_{23}$. The property $\Gamma^1_{21} = 0$ is then preserved by the transformations (16.57) with $B = 0$, so the direction of the new ℓ^α is indeed unique. $\square$

Comments:

1. The result $\tilde{\Gamma}^1_{21} = 0$ could be achieved also with $\Gamma^1_{23} = 0$, provided that $\Gamma^1_{22}\Gamma^3_{00} \neq 0$. However, in the following we will be interested precisely in the case $\Gamma^1_{22} = \Gamma^3_{00} = 0$.

2. With $\dot{k}^\alpha = 0 \,(\Longrightarrow \Gamma^2_{00} = 0 = \Gamma^3_{00})$ the transformation (16.55)–(16.57) does not change Γ^1_{23} and does not violate the equations $\Gamma^1_{22} = \Gamma^1_{33} = 0$.

We can now prove the last part of the main theorem:

Theorem 16.3 *If there exists a shearfree geodesic null vector field k^α on a spacetime and the vacuum Einstein equations are obeyed, $R_{\alpha\beta} = 0$, then the Weyl tensor is algebraically special (i.e. it obeys (11.55)).*

Proof:

In view of Theorem 16.2 and of Eq. (16.58), we need only show that $C_{0102} = 0$ in the double-null tetrad. As Lemma 16.1 demonstrates, the case $\theta + i\omega = 0$ has to be considered separately. Thus we split the proof into two parts.

Part I: $\theta + i\omega = 0$

Contract (16.72) with $k^\gamma \left(m^\alpha \ell^\beta - m^\beta \ell^\alpha\right)$; the result simplifies to

$$k^\gamma \left(\Gamma^1_{12} - \Gamma^1_{21}\right)_{,\gamma} + \Gamma^3_{30}\left(\Gamma^1_{12} - \Gamma^1_{21}\right) + \Gamma^1_{23}\left(\Gamma^0_{20} - \Gamma^1_{21}\right)$$
$$+ \Gamma^1_{32}\left(\Gamma^1_{12} - \Gamma^0_{20}\right) + \Gamma^1_{22}\left(\Gamma^1_{13} - \Gamma^1_{31}\right) + A_{(6)} = 0, \qquad (16.96)$$

where $A_{(6)}$ contains the acceleration terms:

$$A_{(6)} \overset{\text{def}}{=} -m_\rho \dot{k}^\rho \Gamma^1_{11} + \ell_\rho \dot{k}^\rho \left(2\Gamma^1_{12} - \Gamma^1_{21} + \Gamma^3_{10}\right) - \ell^\beta \left(m_\alpha \dot{k}^\alpha\right)_{,\beta}$$
$$+ m^\beta \left(\ell_\alpha \dot{k}^\alpha\right)_{,\beta} + \ell^\beta m^\alpha_{\,;\beta} \dot{k}_\alpha - m^\beta \ell^\alpha_{\,;\beta} \dot{k}_\alpha. \qquad (16.97)$$

Calculate $-R_{02} = R_{0102} + R_{0223} \equiv R_{\rho\alpha\beta\gamma}\left(k^\rho \ell^\alpha k^\beta m^\gamma + k^\rho m^\alpha m^\beta \overline{m}^\gamma\right)$. Since it was assumed that $R_{\alpha\beta} = 0$, it follows that

$$k^\gamma \Gamma^1_{12,\gamma} = -\Gamma^1_{12}\left(\Gamma^1_{32} + \Gamma^1_{23} + \Gamma^3_{30}\right) + \Gamma^1_{21}\left(\Gamma^1_{32} - \Gamma^1_{23}\right)$$
$$+ \Gamma^1_{32}\Gamma^0_{20} + \Gamma^1_{22}\left(\Gamma^0_{30} + 2\Gamma^3_{33}\right) - m^\gamma \Gamma^1_{23,\gamma} + \overline{m}^\gamma \Gamma^1_{22,\gamma} + A_{(7)}, \quad (16.98)$$

where

$$A_{(7)} \overset{\text{def}}{=} m^\gamma \left(\ell_\alpha \dot{k}^\alpha\right)_{,\gamma} + m^\gamma \ell_{\alpha;\gamma} \dot{k}^\alpha - \ell_\rho \dot{k}^\rho \left(2\Gamma^1_{12} + \Gamma^3_{10}\right)$$
$$+ m_\rho \dot{k}^\rho \left(2\Gamma^3_{13} - \Gamma^1_{11}\right). \qquad (16.99)$$

Substituting (16.98) in (16.96) we find

$$k^\gamma \Gamma^1_{21,\gamma} = -\Gamma^3_{30}\Gamma^1_{21} + \left[\Gamma^1_{21}\left(\Gamma^1_{32} - 2\Gamma^1_{23}\right) + \Gamma^1_{23}\left(\Gamma^0_{20} - \Gamma^1_{12}\right)\right.$$
$$+ \Gamma^1_{22}\left(2\Gamma^3_{33} + \Gamma^1_{13} - \Gamma^1_{31} + \Gamma^0_{30}\right)$$
$$\left. - m^\gamma \Gamma^1_{23,\gamma} + \overline{m}^\gamma \Gamma^1_{22,\gamma} + A_{(6)} + A_{(7)}\right], \qquad (16.100)$$

where all the terms between the square brackets vanish when $\sigma = 0 = \dot{k}^\alpha$ and $\Gamma^1{}_{23} = 0$. Substituting (16.100) in (16.93) we see that the thesis of the theorem follows when $\Gamma^1{}_{23} = \theta + i\omega = 0$, $\sigma_{22} = \Gamma^1{}_{22} = \Gamma^1{}_{33} = 0$ and $\dot{k}^\alpha = 0$.

Part II: $\theta + i\omega \neq 0$

Then we choose, as the lemma allows, a direction of ℓ^α for which $\Gamma^1{}_{21} = 0$. As follows from (16.93), we then need only show that $\sigma = 0 = \dot{k}^\alpha$ implies $\Gamma^0{}_{20} \equiv \Gamma^3{}_{10} = 0$. Since $\Gamma^1{}_{21} = 0$ was achieved under the assumptions $\sigma = 0 = \dot{k}^\alpha$, we shall not display the shear and acceleration terms in what follows.

Take the equation $0 = R_{22} \equiv 2R_{0212} = 2R_{\rho\alpha\beta\gamma}k^\rho m^\alpha \ell^\beta m^\gamma$; it implies that the contraction of (16.72) with $m^\alpha \ell^\beta m^\gamma$ is zero. After using the assumptions $\Gamma^1{}_{21} = \Gamma^1{}_{22} = \Gamma^2{}_{00} = 0 +$ their complex conjugates we obtain the equation $\Gamma^1{}_{23}\Gamma^0{}_{22} = 0$. Since we are considering the case $\Gamma^1{}_{23} \neq 0$, this means that

$$\Gamma^0{}_{22} \equiv \Gamma^3{}_{12} = 0 \Longrightarrow \Gamma^0{}_{33} \equiv \Gamma^2{}_{13} = 0. \tag{16.101}$$

Then take (16.100) with the current assumptions; the result is

$$m^\gamma \Gamma^1{}_{23,\gamma} = \Gamma^1{}_{23}\left(\Gamma^0{}_{20} - \Gamma^1{}_{12}\right), \tag{16.102}$$

and Eq. (16.89) becomes

$$k^\gamma \Gamma^1{}_{23,\gamma} = -\left(\Gamma^1{}_{23}\right)^2. \tag{16.103}$$

Equations (16.102) and (16.103) imply the integrability condition, which, for non-commuting directional derivatives, has the form

$$e_a{}^\mu \left(e_b{}^\nu \phi_{,\nu}\right)_{;\mu} - e_b{}^\mu \left(e_a{}^\nu \phi_{,\nu}\right)_{;\mu} = -2\Gamma^c{}_{[ab]}e_c{}^\nu \phi_{,\nu} \tag{16.104}$$

for an arbitrary function ϕ. Applying this rule to (16.102) and (16.103) we thus obtain

$$k^\gamma \left(\Gamma^0{}_{20} - \Gamma^1{}_{12}\right)_{,\gamma} = -2\Gamma^1{}_{23}\Gamma^0{}_{20} + \left(\Gamma^2{}_{20} - \Gamma^2{}_{02}\right)\left(\Gamma^0{}_{20} - \Gamma^1{}_{12}\right). \tag{16.105}$$

Using (16.98) and (16.102) in this, we obtain

$$k^\gamma \Gamma^0{}_{20,\gamma} = \Gamma^0{}_{20}\left(-3\Gamma^1{}_{23} + \Gamma^2{}_{20}\right). \tag{16.106}$$

Now, calculating $0 = R_{22} = R^0{}_{202} + R^1{}_{212}$ (in fact, $R^1{}_{212} \equiv 0$ under the current assumptions) and using (9.21) to calculate the scalar components of the Riemann tensor, we obtain

$$m^\gamma \Gamma^0{}_{20,\gamma} = \Gamma^0{}_{20}\Gamma^2{}_{22} - \left(\Gamma^0{}_{20}\right)^2. \tag{16.107}$$

Applying the integrability condition (16.104) to (16.106) and (16.107) we obtain

$$\Gamma^0{}_{20}\left(k^\gamma \Gamma^2{}_{22,\gamma} - m^\gamma \Gamma^2{}_{20,\gamma}\right) = \Gamma^0{}_{20}\left[-\Gamma^0{}_{02}\Gamma^2{}_{20} - 9\Gamma^0{}_{20}\Gamma^1{}_{23}\right.$$
$$\left. - \Gamma^2{}_{02}\left(\Gamma^2{}_{22} - \Gamma^0{}_{20}\right) + \Gamma^2{}_{20}\left(\Gamma^2{}_{22} + \Gamma^0{}_{20}\right)\right]. \tag{16.108}$$

One solution of this is $\Gamma^0{}_{20} = 0$, which is the desired result. In order to verify the implications of the other alternative, we calculate the expression in parentheses on the left-hand side above from the equation $0 = R_{20} \equiv R^1{}_{210} - R^2{}_{220}$, in which we use Eq. (9.21) to calculate the tetrad components of the Riemann tensor. The result is, in consequence of all the assumptions made,

$$k^\gamma \Gamma^2{}_{22,\gamma} - m^\gamma \Gamma^2{}_{20,\gamma} = \Gamma^0{}_{20}\left(\Gamma^1{}_{23} + \Gamma^2{}_{20} + \Gamma^2{}_{02}\right) - \Gamma^0{}_{02}\Gamma^2{}_{20}$$
$$+ \Gamma^2{}_{22}\left(\Gamma^2{}_{20} - \Gamma^2{}_{02}\right). \tag{16.109}$$

Substituting (16.109) into (16.108) we obtain $\Gamma^0{}_{20}\Gamma^1{}_{23} = 0$, which, in view of the currently considered case $\Gamma^1{}_{23} \neq 0$, implies $\Gamma^0{}_{20} = 0$. Thus, $\Gamma^0{}_{20} = 0$ in both branches of (16.108). $\square$

Theorems 16.1–16.3 are summarised in the following theorem:

Theorem 16.4 (The Goldberg–Sachs theorem) *A vacuum spacetime is algebraically special if and only if there exists on it a geodesic and shearfree null vector field k^α. If the Weyl tensor is nonzero, then the field k^α is collinear with the degenerate Debever field.*

·This theorem was first formulated by Goldberg and Sachs (1962).[1] Parts of this result, such as Theorem 16.1 and the other theorems under additional assumptions, had been proven earlier; references to those results are given in the original paper. The theorem is useful in investigating vacuum solutions of Einstein's equations; see Section 20.1.

A few generalisations of Theorems 16.1–16.4 are known (Stephani *et al.*, 2003), but their theses cannot be stated without more elaborate explanations. One generalisation of Theorems 16.2 and 16.3 can be verified rather easily. The field equations we used in proving these two theorems were not really $R_{\alpha\beta} = 0$, but $R_{\alpha\beta}k^\alpha k^\beta = 0 = R_{\alpha\beta}k^\alpha m^\beta = R_{\alpha\beta}m^\alpha m^\beta$ and their complex conjugates; the other equations used in the proof were properties of the curvature tensor. All the equations that we made use of remain valid when $R_{\alpha\beta} = \Phi k_\alpha k_\beta$. However, such a generalisation does not hold for Theorem 16.1 (Stephani *et al.*, 2003).

16.7 * Optical observations in the Universe – part II

The Goldberg–Sachs theorem does not find any direct application in cosmology. Because of the assumptions made in it, it is mainly applied to vacuum solutions, and we will use it in Chapter 20 to derive the Kerr metric. However, some of the results derived along the way in proving it are useful in cosmology, and this is why it was presented in this chapter.

16.7.1 The area distance

In curved spacetime, it is a problem to determine the distance between two objects. Abstract geometric definitions, such as the integral of ds along a uniquely defined spacelike path connecting two worldlines, are not satisfactory. We need a definition that would yield a distance measurable by means of astronomical observations. The reasoning used here is adapted from Ellis (1971), and the definitions of distance go back to Etherington (1933).

---

[1] A few formulae in the later part of the proof in Goldberg and Sachs (1962) have misprints, which have been corrected here.

An observer moving with a 4-velocity u^α would measure the flux, i.e. the energy density, of radiation reaching him to be equal to $\mathcal{F} = T_{\alpha\beta}u^\alpha u^\beta$, where $T_{\alpha\beta}$ is the energy-momentum tensor of radiation, given by (16.15). Thus, neglecting terms of order ε:

$$\mathcal{F} = \frac{\mu}{4\pi}\left(k_\alpha u^\alpha\right)^2. \tag{16.110}$$

On the other hand, combining the conservation equation (16.16) with the surface-area-propagation equation (16.131) (see Exercise 2) we see that $(\mu\delta S)$ is constant along the null geodesics, thus

$$(\mu\delta S)\big|_{s=s_1} = (\mu\delta S)\big|_{s=s_2} \tag{16.111}$$

(recall that δS is the area of an orthogonal cross-section of the bundle of rays). Combining (16.110) and (16.111) we obtain $\mathcal{F} = \text{constant} \times (k_\alpha u^\alpha)^2/\delta S$. The subscript 'e' will denote quantities calculated at the centre; 'O' will denote the quantities at the observer. Remembering that $(k_\alpha u^\alpha)_e$ is constant along each null geodesic and using (16.20), we obtain

$$\mathcal{F}_O = \frac{C_\mathcal{F}}{\delta S_O}\left[\frac{(k_\alpha u^\alpha)_O}{(k_\alpha u^\alpha)_e}\right]^2 = \frac{C_\mathcal{F}}{\delta S_O(1+z_O)^2}, \tag{16.112}$$

where $C_\mathcal{F}$ is constant in each bundle (it depends on the solid angle occupied by the bundle). Thus the flux decreases with increasing redshift. However, δS_O does not necessarily decrease with distance, as can be seen from (16.79). For geodesics, $A_{(1)} = 0$. Assume that $\omega = 0$ and $R_{\mu\nu} = 0$ (vacuum) for simplicity. Then (16.79) shows that the derivative of θ along k^μ is negative, that is, θ decreases and may become negative, too. This observation is expressed in brief by saying that *curvature causes null geodesics to converge*, even in vacuum. The conclusion remains true in a perfect fluid or dust, where $R_{\alpha\beta}k^\alpha k^\beta = \kappa(\epsilon+p)(k_\alpha u^\alpha)^2 > 0$. Consequently, a situation shown in Fig. 16.1 is possible: an observer at Q will see the light source at G to have the same brightness as seen by the observer at P. For Q, the source will appear anomalously bright and anomalously large.[1]

In order to make (16.112) a workable formula, we have to calculate the constant $C_\mathcal{F}$. Imagine then that we surround G with a sphere S with the centre at G and radius r_S.[2] Equation (16.112) shows that, for a given bundle of null geodesics, $(1+z)^2\mathcal{F}\delta S$ is constant along the bundle. On the surface of a sphere, δS is proportional to the solid angle $\delta\Omega_G$ subtended by δS, thus $\delta S_G = r_S^2\delta\Omega_G/(4\pi)$, where r_S is the radius of the sphere S. Assuming that the source G radiates isotropically, and denoting by L its total luminosity

---

[1] In actual observations, the redshift will decrease the flux of photons. The effect of focussing by curvature would be clearly visible only if the observer moved so as to cancel the redshift and if there were no absorption of light between the emitter and the observer.

[2] We assume the radius of the sphere to be sufficiently small that the space inside it can be approximated by the flat tangent space to the manifold. Otherwise, it might be impossible to construct such a sphere – for example, when the bundle has nonzero rotation.

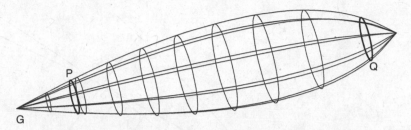

Fig. 16.1. A non-rotating null congruence in vacuum or in a perfect fluid necessarily has its expansion scalar decreasing along the rays. This may (if the distance is sufficiently large) eventually cause the expansion to become negative, which results in refocussing. The observer at Q would see the source G to be anomalously bright and large, if he were able to cancel the redshift. Actually, when the shear is nonzero, the geodesics will be refocussed not to a point, but to a caustic surface.

(i.e. rate of emission of radiation), we have on S $\mathcal{F}_S = L/(4\pi)$. Also, $z = 0$ on S (because S is close to the source), so, taking (16.112) on S, we have

$$C_{\mathcal{F}} = \frac{L}{4\pi} r_S^2 \delta\Omega_G. \tag{16.113}$$

Now, following Etherington (1933) and Ellis (1971), we define the **source area distance** r_G (from the source to the observer) by

$$\delta S_O \stackrel{\text{def}}{=} r_G^2 \delta\Omega_G. \tag{16.114}$$

Finally, combining (16.112)–(16.114), we obtain

$$\mathcal{F}_O = \frac{L}{4\pi} \frac{r_S^2}{(1 + z_O)^2 r_G^2}. \tag{16.115}$$

The quantity $\mathcal{F}_O$ can be measured (this is the flux of energy from a given source through a given surface sheet at the observer's position) and so can z_O; r_S is any unit of distance and can be chosen (for example, it can be the astronomical unit). However, r_G cannot be measured because we cannot get close enough to the source to measure $\delta\Omega_S$. With L being essentially unknown, Eq. (16.115) cannot be observationally tested.

But we can define distance in another way. Imagine a surface sheet of area δS_G placed at the source G, and a bundle of light rays sent from this sheet in such a way that it converges to a single point at the observer's position O (Fig. 16.2). Assume that the sheet is orthogonal to the central ray of the bundle and that the central ray is emitted from G. Let the solid angle filled by this bundle as it reaches O be $\delta\Omega_O$. We can then define the **observer area distance** from G to O by

$$\delta S_G \stackrel{\text{def}}{=} r_O^2 \delta\Omega_O. \tag{16.116}$$

We can measure $\delta\Omega_O$, and we can *in principle* calculate δS_G (for example, assuming that the bundle subtends the whole source, whose geometric size we know). In this way, we can *in principle* measure r_O.

It turns out that r_O is uniquely determined by r_G, as we show in the next subsection.

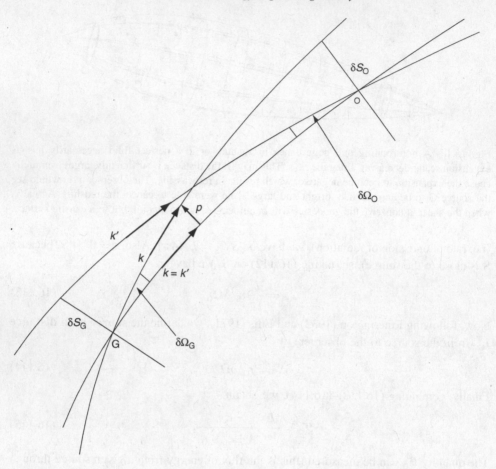

Fig. 16.2.   An illustration of the reciprocity theorem. Explanation is given in the text.

16.7.2 The reciprocity theorem

Figure 16.2 will help in comparing r_O and r_G. The source of light G (suppose, a galaxy) sends a light ray from its middle point that hits the observer O, and a bundle of rays diverging from G that surrounds the central ray GO and, at G, fills the solid angle $\delta\Omega_G$. The bundle has the projected area δS_O at the observer's position. Another bundle of rays is sent from a surface sheet of area δS_G, positioned orthogonally to GO at the central point G. The second bundle converges at the observer, at the endpoint of the ray GO, and fills the solid angle $\delta\Omega_O$. (Figure 16.2 shows a cross-section through one half of each bundle.) Let the affine parameter and the field of null vectors tangent to the first bundle be v and k, respectively, and let the same quantities for the second bundle be v' and k'. Along the ray GO, $k = k'$. Let the field of geodesic deviation defined by the field k be p, and the deviation for the other bundle be p'. Then the following is true:

Theorem 16.5 (the reciprocity theorem)

Assumption:
The bundle k completely surrounds O, i.e. there are rays of the bundle in every direction from O within δS_O.

Thesis:
The area distances r_O and r_G are related by

$$r_G{}^2 = r_O{}^2 (1 + z_O)^2. \tag{16.117}$$

Comment:
The assumption would not be fulfilled if O were placed on a reflecting or absorbing surface, e.g. on a boundary between vacuum and opaque matter.

Proof:
Along the central ray we have $v = v'$ and $k = k'$ (but not $p = p'$). Thus, the geodesic deviation equation (6.56) and the symmetries of the Riemann tensor imply that, along the same ray,

$$p'^\alpha \frac{D^2 p_\alpha}{dv^2} - p^\alpha \frac{D^2 p'_\alpha}{dv^2} = 0. \tag{16.118}$$

Consequently

$$p'^\alpha \frac{D p_\alpha}{dv} - p^\alpha \frac{D p'_\alpha}{dv} = \text{constant} \tag{16.119}$$

along GO. Note that Eq. (6.54), with zero torsion, implies that $k^\beta p^\alpha{}_{;\beta} = p^\beta k^\alpha{}_{;\beta}$. Using this in (16.119) (where $D p^\alpha / dv = p^\alpha{}_{;\beta} k^\beta$), we see that the connection terms cancel out. From now on we will thus replace the covariant derivatives by ordinary derivatives. Taking (16.119) at G (where $p^\alpha = 0$) and at O (where $p'^\alpha = 0$), we obtain

$$\left[p'^\alpha \frac{dp_\alpha}{dv} \right]_G = - \left[p^\alpha \frac{dp'_\alpha}{dv} \right]_O. \tag{16.120}$$

Of the vectors p^α at O we choose those that are orthogonal to the 4-velocity of the observer; of p'^α at G we choose those that are orthogonal to the source 4-velocity. The chosen vectors span 2-dimensional planes tangent to the spacetime at O and G. Consequently, we can choose a pair of the deviation vectors at G that obey

$$\left[\frac{dp_1{}^\alpha}{dv} \frac{dp_{2\alpha}}{dv} \right]_G = 0. \tag{16.121}$$

This is possible, since the vectors $dp^\alpha / dv |_G$, together with $p^\alpha |_G = 0$, form the initial data for the linear set of equations (6.56), and so are in fact a basis of the 2-dimensional tangent plane at G. Condition (16.121) means only that we have chosen an orthogonal basis. We choose any vector p^α at O, and we make use of the assumption: since there are rays of the bundle in every direction from O in the plane orthogonal to GO at O, whatever direction $p^\alpha|_O$ we choose, there will be another direction perpendicular to it. Thus we choose a pair of vectors such that

$$(p_1^\alpha p_{2\alpha})|_O = 0 \tag{16.122}$$

while (16.121) still holds.

These were conditions imposed on the field p^α at both ends of GO.

We now impose the condition on the field p'^α:

$$\left[p_1'^\alpha \frac{dp_{2\alpha}}{dv}\right]_G = 0 = \left[p_2'^\alpha \frac{dp_{1\alpha}}{dv}\right]_G. \tag{16.123}$$

Through (16.120) this implies that

$$\left[p_1^\alpha \frac{dp_{2\alpha}'}{dv}\right]_O = 0 = \left[p_2^\alpha \frac{dp_{1\alpha}'}{dv}\right]_O. \tag{16.124}$$

Since all the vectors lie in 2-planes at G and at O, the first equations of (16.123) and (16.121) imply that $p_1'^\alpha|_G$ is collinear with $dp_1^\alpha/dv|_G$, and then the second of (16.123) says that

$$(p_1'^\alpha p_{2\alpha}')_G = 0. \tag{16.125}$$

By a similar consideration, (16.124) and (16.122) imply that

$$\left[\frac{dp_1'^\alpha}{dv}\frac{dp_{2\alpha}'}{dv}\right]_O = 0. \tag{16.126}$$

In consequence of (16.121), (16.122), (16.125) and (16.126), all scalar products between the vectors p_i^α and dp_i^α/dv at G and O reduce to products of their lengths, denoted $|p_i|$ and $|p_i'|$. Thus

$$\delta S_O = [|p_1||p_2|]_O, \qquad \delta S_G = [|p_1'||p_2'|]_G,$$

$$\delta\Omega_G = \left[\left|\frac{dp_1}{d\ell}\right|\cdot\left|\frac{dp_2}{d\ell}\right|\right]_G, \qquad \delta\Omega_O = \left[\left|\frac{dp_1'}{d\ell}\right|\cdot\left|\frac{dp_2'}{d\ell}\right|\right]_O, \tag{16.127}$$

where ℓ is the radius at which the solid angles $\delta\Omega_G$ and $\delta\Omega_O$ subtend the surfaces δS_G and δS_O, respectively. It is related to v by $d\ell = -k^\rho u_\rho dv$ (see the text after (16.28)). Thus we have from (16.127), (16.123) and (16.124):

$$\delta S_O \delta\Omega_O = \left[|p_1|\left|\frac{dp_1'}{dv}\right||p_2|\left|\frac{dp_2'}{dv}\right|\right]_O \frac{1}{(k^\rho u_\rho)_O{}^2}$$

$$= \left[|p_1'|\left|\frac{dp_2}{dv}\right||p_2'|\left|\frac{dp_1}{dv}\right|\right]_G \frac{1}{(k^\rho u_\rho)_O{}^2}$$

$$= \frac{(k^\rho u_\rho)_G{}^2}{(k^\rho u_\rho)_O{}^2}\delta S_G \delta\Omega_G. \tag{16.128}$$

This is equivalent to (16.117) by (16.20), (16.114) and (16.116). $\square$

The reciprocity theorem implies that

If the observer moves so as to cancel the redshift, then equal projected surface areas at the source G and at the observer O fill equal solid angles at O and G, respectively.

In Fig. 16.1, both r_G and r_O between the endpoints are zero, so (16.117) still holds, but the above conclusion does not.

Recall that an observer at Q would see the source at G to be not only anomalously bright, but also to have an anomalously large angular size. Imagine the observer being placed on the symmetry axis and gradually removed farther from G. The angular size of G would then at first decrease. As the observer moves past the first refocussing point, the angular size would start increasing.

Substituting (16.117) in (16.115) we obtain an expression for r_O:

$$r_O = \sqrt{\frac{L}{4\pi \mathcal{F}_O} \frac{r_S}{(1+z)^2}}. \tag{16.129}$$

This allows us to calculate r_O if we know L from theory and $\mathcal{F}_O$ from measurement. The r_O calculated in this way is sometimes called the **corrected luminosity distance**.

16.7.3 Other observable quantities

Having introduced the two definitions of distance, the redshift, the flux of radiation and the reciprocity theorem, one can define further observationally meaningful quantities. One of them is the number of radiation sources in a given volume. Consider again the bundle k' in Fig. 16.2. Imagine that we move from the surface sheet δS_G (at the observer area distance r_O from O) to a farther sheet $\delta S'$ at the distance $(r_O + d\ell)$. The $d\ell$ is connected to the affine parameter v on the bundle by $d\ell = -k^\rho u_\rho \, dv$; see the text after (16.28). Let $n(x)$ be the number of radiation sources per unit proper volume. The number of sources within the bundle between δS_G and $\delta S'$ will then be

$$dN = -r_O{}^2 \delta\Omega_O \left(n k^\rho u_\rho \right)_G \, dv. \tag{16.130}$$

The integral of this over the volume of the bundle from O to δS_G will be the total number of sources of radiation within this volume. This quantity is considered in observational cosmology, and the procedure of determining it is called **number counts**.

Other observationally meaningful quantities are:

1. The specific flux – the flux per unit frequency interval;
2. The intensity of radiation – the flux per unit solid angle;
3. The specific intensity – the intensity per unit frequency interval;
4. The emission and absorption of light in the intergalactic space, and the optical depth.

While being useful for observations, these quantities do not really involve relativity in their definitions and properties, and are handled by routine mathematics. A good overview source for them is Ellis (1971); we will not discuss them here.

16.8 Exercises

1. Prove that any vector orthogonal to two linearly independent null vectors must be spacelike.
   Hint. Consider the vertex of.the null cone, p_0, and choose the locally Minkowskian coordinates at p_0.
2. Using the methods of Section 15.1, prove that the expansion and rotation of a family of null geodesics have the same interpretation as the corresponding quantities in hydrodynamics. Show that shear changes the shape of a parallelepiped in a plane locally orthogonal to a ray, but preserves its surface area. Verify that the expansion scalar obeys

$$2\theta = \frac{d}{dv} \ln(\delta S), \qquad (16.131)$$

where v is the affine parameter on the null geodesics and δS is the surface area of the propagating light front of a bundle of rays at the parameter value v.

17

Relativistic cosmology II: the Robertson–Walker geometry

17.1 The Robertson–Walker metrics as models of the Universe

The Robertson–Walker (R–W) metrics were derived in Section 10.7 as being spatially homogeneous in the Bianchi sense and at the same time isotropic. In the coordinates of Section 10.7, they are

$$ds^2 = dt^2 - \frac{R^2(t)}{\left(1 + \frac{1}{4}kr^2\right)^2} \left[dr^2 + r^2 \left(d\vartheta^2 + \sin^2 \vartheta \, d\varphi^2\right)\right], \tag{17.1}$$

where $R(t)$ (called the **scale factor**) is a function to be determined from the Einstein equations. The constant k, called the **curvature index**, when nonzero, can be scaled to $+1$ or -1 by the transformations $r = r'/\sqrt{|k|}$, $R(t) = \sqrt{|k|}\widetilde{R}(t)$. After this, the R–W metrics are no longer one continuous family; they become three different metrics and the limiting transition $k \to 0$ becomes impossible. This can cause difficulties when comparing the three classes with each other; see Section 17.4.

Other representations of the R–W metric are also used. Let

$$r = \frac{2r'}{1 + \sqrt{1 - kr'^2}} \qquad \Longleftrightarrow \qquad r' = \frac{r}{1 + \frac{1}{4}kr^2}. \tag{17.2}$$

This changes (17.1) to (primes dropped for better readability):

$$ds^2 = dt^2 - R^2(t) \left(\frac{dr^2}{1 - kr^2} + r^2 \left(d\vartheta^2 + \sin^2 \vartheta \, d\varphi^2\right)\right). \tag{17.3}$$

When $k = +1$, the transformation

$$r = \sin \psi \tag{17.4}$$

changes (17.3) to

$$ds^2 = dt^2 - R^2(t) \left[d\psi^2 + \sin^2 \psi \left(d\vartheta^2 + \sin^2 \vartheta \, d\varphi^2\right)\right], \tag{17.5}$$

where the hypersurfaces $t = \text{constant}$ are 3-dimensional spheres.[1]

---

[1] In (17.4) we assumed that $r \leq 1$. With $k = +1$, the range $0 \leq r \leq 1$ covers half of the 3-sphere. In order to cover the other half, one has to employ a second copy of the coordinate chart introduced in (17.2) – as r goes from 0 to ∞, r' first increases from 0 to $1/\sqrt{k}$, achieved at $r = 2/\sqrt{k}$, and then decreases to 0 as $r \to \infty$.

With $k = -1$, the transformation

$$r = \sinh \psi \tag{17.6}$$

changes (17.3) to

$$ds^2 = dt^2 - R^2(t) \left[d\psi^2 + \sinh^2 \psi \left(d\vartheta^2 + \sin^2 \vartheta \, d\varphi^2 \right) \right]. \tag{17.7}$$

A hypersurface $t = \text{constant}$ in (17.5) and (17.7) is a space of constant curvature: its Riemann tensor is

$$R^{IJ}{}_{KL} = \pm \frac{1}{R(t)} \delta^{IJ}_{KL}, \qquad I, J, K, L = 1, 2, 3;$$

with $+$ for (17.5) and $-$ for (17.7). When $k = 0$, it is seen from (17.1) and (17.3) that the hypersurfaces $t = \text{constant}$ are flat.

Because of these properties, the R–W spacetimes with $k = +1$ are colloquially called the 'closed Universe', those with $k = -1$ are called the 'open Universe' and the one with $k = 0$ is called the 'flat Universe'.

The first author to investigate these spacetimes was A. A. Friedmann[1] in 1922 (the case $k = +1$, in the coordinates of (17.5)) and in 1924 (Friedmann, 1922) (the case $k = -1$; in coordinates different from all those used here, and partly with incorrect conclusions about the evolution (Krasiński and Ellis, 1999)). He solved the Einstein equations for these metrics, with dust source and with the cosmological constant allowed. He himself treated his result as merely a mathematical curiosity, since the expansion of the Universe had not been discovered yet at that time, and no application for those solutions in astronomy could be seen. Both papers were quickly forgotten – most astronomers and physicists had not been ready to accept their implications, while Friedmann was not a sufficiently prominent figure to win attention.[2] Soon after finding these solutions, Friedmann died prematurely and had no chance to claim credit for his discovery when the expansion of the Universe was detected by Hubble (1929).[3] The simplest case $k = 0$ was first introduced by Robertson (1929) in a systematic survey of spacetimes with the, so-called today, Robertson–Walker geometries.

The case $k > 0$ was discussed by Lemaître (1927, 1931),[4] who generalised Friedmann's solution to nonzero pressure and was aware that the result reflects some properties of the observed Universe. Unfortunately, he used to publish his papers in a local Belgian journal, in French, and, being an unknown person at that time, did not win' attention for his results, either. Four years later, when the expansion of the Universe was already well-known, Eddington asked whether exact models of the expanding Universe might be derived from the relativity theory (Royal Astronomical Society Discussion, 1930). It

---

[1] The transliteration of his name from Russian is 'Fridman', but, following the original publications in German, the form 'Friedmann' became standard.

[2] Psychological/social factors of this kind do play a role in the acceptance of new results, also today, very unfortunately.

[3] As already mentioned, Hubble did not believe that the Universe is actually expanding. He insisted that recalculating the redshifts into velocities of recession is merely a convenient mathematical trick (Hubble, 1929, 1936, 1953).

[4] The 1931 paper is a partly updated English version of the earlier one.

was only then that Lemaître brought his paper from 1927 to Eddington's attention; and the English translation (Lemaître, 1931) followed. The recognition of Friedmann's role came gradually later (Krasiński and Ellis, 1999).

The general metric (17.1) was first derived from geometrical considerations by Robertson (1933) and by Walker (1935). Friedmann and Lemaître derived the solutions of Einstein's equations (with slightly different sources) for that metric. In view of this history, the spacetimes with this geometry are often called the Friedmann–Lemaître–Robertson–Walker (FLRW) models, but various subsets of this collection of four names are also in use. It seems appropriate to use the name Robertson–Walker when referring to the general metric (17.1) and the names Friedmann or Lemaître when referring to the explicit solutions.

In each of the coordinate representations, (17.1), (17.3), (17.5) and (17.7), the Einstein tensor for the R–W metric is diagonal. This shows that, if the source in the Einstein equations should be a perfect fluid, then its velocity field can have only the zeroth component, and, because of $u_\alpha u^\alpha = 1$ and $g_{00} = 1$, the velocity must be $u^\alpha = \delta^\alpha{}_0$. This, in turn, means that all those coordinate systems are comoving: each matter particle has fixed spatial coordinates, and its proper time is the time coordinate in spacetime.

The velocity field $u^\alpha = \delta^\alpha{}_0$ has the following properties: $\dot{u}^\alpha = 0$, $\omega_{\alpha\beta} = \sigma_{\alpha\beta} = 0$ and $\theta = 3\dot{R}/R$. The velocity of expansion depends on time, but at each instant it is the same at all points of the space $t = $ constant.

The R–W spacetimes are the only perfect fluid solutions of the Einstein equations for which $\omega_{\alpha\beta} = \sigma_{\alpha\beta} = 0 = \dot{u}^\alpha$ and $\theta \neq 0$. The proof follows from the paper by Stephani (1967a); see also Stephani *et al.* (2003). As we stated in Section 15.3, with $\omega_{\alpha\beta} = \sigma_{\alpha\beta} = 0 = \dot{u}^\alpha$ and a perfect fluid source, the metric must be conformally flat. Theorem 37.17, p. 601 in Stephani *et al.* (2003), says that all conformally flat perfect fluid solutions are those found by Stephani (1967a). Those of them for which $\theta \neq 0$ have $\omega_{\alpha\beta} = \sigma_{\alpha\beta} = 0$, but in general $\dot{u}^\alpha \neq 0$. In the limit $\dot{u}^\alpha = 0$ they reduce to the R–W metrics.

17.2 Optical observations in an R–W Universe

17.2.1 The redshift

In order to apply (16.20) to the R–W spacetimes, we have to know the field of vectors tangent to light rays, k^α. In consequence of the spatial homogeneity of any R–W spacetime, all points within the same space $t = $ constant are equivalent, so the result of any calculation will be independent of the spatial position of the observer. Let us then assume that the observer is at the origin, $r = 0$. A null geodesic sent off radially, with $\dot{\vartheta}_0 = \dot{\varphi}_0 = 0$, will preserve the radial direction $\dot{\vartheta} = \dot{\varphi} = 0$ at all other points. Hence, such a geodesic lies in the surface $\{\vartheta = \text{constant}, \varphi = \text{constant}\}$ and obeys the equation

$$0 = \mathrm{d}t^2 - \frac{R^2(t)}{\left(1 + \frac{1}{4}kr^2\right)^2}\,\mathrm{d}r^2. \tag{17.8}$$

For a ray proceeding *towards* the observer sitting at $r = 0$ we thus have

$$\int_{t_e}^{t_o} \frac{dt}{R(t)} = -\int_{r_e}^{r_o} \frac{dr}{1 + \frac{1}{4}kr^2}. \tag{17.9}$$

It can be verified that v defined by

$$\frac{dt}{dv} = \frac{1}{R(t)} \tag{17.10}$$

is an affine parameter on the geodesic. In this parametrisation, the tangent vector to the geodesic in Eq. (16.20) is

$$k^\alpha = \left[\frac{1}{R}, -\frac{1}{R^2}\left(1 + \frac{1}{4}kr^2\right), 0, 0 \right]. \tag{17.11}$$

The velocity field is $u^\alpha = \delta^\alpha{}_0$ everywhere, so

$$k^\alpha u_\alpha = 1/R, \tag{17.12}$$

and so in (16.20) we have

$$1 + z = R(t_o)/R(t_e). \tag{17.13}$$

When $|R(t_o) - R(t_e)|/R(t_e)$ is small, developing $R(t_o)$ by the Taylor formula around t_e and neglecting terms nonlinear in $(t_o - t_e)$ gives

$$z \approx [\dot{R}(t_e)/R(t_e)](t_o - t_e) \approx \frac{\dot{R}}{R}\delta\ell, \tag{17.14}$$

since $c = 1$ in the coordinates of (17.8), so $t_o - t_e = \delta\ell$ is the distance from the light-source to the observer; $\dot{R} \overset{\text{def}}{=} dR/dt$. The same result follows from (16.29) when we recall that $\sigma_{\alpha\beta} = 0 = \dot{u}^\alpha$ and $\theta = 3\dot{R}/R$ in the R–W models. This is the Hubble law, with the Hubble parameter $H = (c/3)\theta = c\dot{R}/R$ (the factor c appeared because the physical time τ is related to the time coordinate t by $\tau = t/c$).

The Hubble law in fact holds exactly in the R–W models. Note that

$$\ell(t) = R(t)\int_0^{r_e} \frac{dr}{1 + \frac{1}{4}kr^2} \tag{17.15}$$

is the distance of the light-source at the point $r = r_e$ from the observer at $r = 0$. Since the coordinates are comoving, for each definite source of light we have $r_e = \text{constant}$, and then from (17.15) we find

$$\frac{d\ell}{dt} = \frac{\dot{R}}{R}\ell = \frac{H\ell}{c}. \tag{17.16}$$

17.2.2 The redshift–distance relation

Assume that, this time, the origin is at the source G in Fig. 16.2. The element of surface area in the surface $\{t = t_v = \text{constant}, r = r_v = \text{constant}\}$ is, in the coordinates of (17.1),

$$dS = \left(\frac{R(t_v)\, r_v}{1 + \frac{1}{4}kr_v^2}\right)^2 \sin \vartheta\, d\vartheta\, d\varphi.$$

On the other hand, close to the origin $r = 0$, the metric of the space $t = t_v$ is approximately flat, so in that space $dS = r_v^2 \sin \vartheta\, d\vartheta\, d\varphi$, where $d\Omega = \sin \vartheta\, d\vartheta\, d\varphi$ is the element of solid angle. Comparing these quantities with (16.114) we see that in the R–W models

$$r_G = \frac{R_0 \bar{r}}{1 + \frac{1}{4}k\bar{r}^2}, \tag{17.17}$$

where $\bar{r}$ is the r-coordinate of the observer and R_0 is the value of R at the time $t = t_v$, when the ray passes the observer. By the reciprocity theorem (16.117), we then have

$$r_0 = \frac{R_0 \bar{r}}{(1 + z_0)\left(1 + \frac{1}{4}k\bar{r}^2\right)}. \tag{17.18}$$

This provides a relation between $\bar{r}$ and the observable quantity r_0, and (17.13) then provides a relation between R and the observable quantity z. The integral of (17.8) along a radial *outgoing* null geodesic is

$$\int_{t_e}^{t} \frac{dt'}{R(t')} = \int_0^r \frac{d\tilde{r}}{1 + \frac{1}{4}k\tilde{r}^2}. \tag{17.19}$$

We can calculate this integral when $R(t)$ is given explicitly. We will do it in Section 17.4.

The quantities that are used in astronomy to characterise $R(t)$ are the **Hubble parameter** of (17.16)[1] and the **deceleration parameter** q_0:

$$q_0 \overset{\text{def}}{=} -RR_{,tt}/R_{,t}^2\big|_{\text{now}} \equiv -1 - \frac{1}{H_0^2}\frac{dH_0}{dt}. \tag{17.20}$$

The sign of q_0 tells us whether the expansion of the Universe accelerates ($R_{,tt} > 0$) or decelerates, and so is relevant for estimating the value of the cosmological constant.

17.2.3 Number counts

This time, the observer will be at $r = 0$, and r will refer to the radial coordinate of the source G. If we assume that the number of radiation sources is conserved $((nu^\alpha)_{;\alpha} = 0)$, then nR^3 must be constant, so

$$n = n_0 R_0^3 / R^3, \tag{17.21}$$

---

[1] It is most often denoted H_0, to stress that its current value is meant.

where the index 'O' refers to any fixed instant, e.g. the instant of observation. Taking the full solid angle $\delta\Omega_O = 4\pi$ and substituting for the other quantities in (16.130), namely for r_O from (17.18), for $R_O = R(1+z_O)$ from (17.13), for $(k^\rho u_\rho)_G = 1/R$ from (17.12) and for $\mathrm{d}v = -R^2\,\mathrm{d}r/(1+\tfrac14 kr^2)$ from (17.9)–(17.10), we obtain for the total number of sources out to the distance corresponding to $\bar{r}$

$$N(\bar{r}) = 4\pi n_O R_O{}^3 \int_0^{\bar{r}} \frac{r^2\,\mathrm{d}r}{\left(1+\tfrac14 kr^2\right)^3} \overset{\text{def}}{=} 4\pi n_O R_O{}^3 F(\bar{r}). \tag{17.22}$$

The integral is equal to

$$F(\bar{r}) = \begin{cases} \dfrac{1}{2k^{3/2}}(\psi - \sin\psi\cos\psi), & \sin\psi \overset{\text{def}}{=} \dfrac{\tfrac12\sqrt{k}\,\bar{r}}{1+\tfrac14 k\bar{r}^2} & \text{for } k>0, \\[4mm] \bar{r}^3/3, & & \text{for } k=0, \\[4mm] \dfrac{1}{2(-k)^{3/2}}(\sinh\psi\cosh\psi - \psi), & \sinh\psi \overset{\text{def}}{=} -\dfrac{\tfrac12\sqrt{k}\,\bar{r}}{1+\tfrac14 k\bar{r}^2} & \text{for } k<0 \end{cases} \tag{17.23}$$

(this is the same ψ that appears in (17.4)–(17.7)). Equations (17.22)–(17.23) provide (in principle!) a method to determine the sign of k observationally. However, in practice, the determinations of distance in cosmology are too imprecise to distinguish among the three forms of F.

17.3 The Friedmann equations and the critical density

With a perfect fluid source, in a general geometry, the following unknown quantities must be determined from the field equations: six out of ten components of the metric tensor $g_{\alpha\beta}$ (the remaining four can be fixed by choice of the coordinate system), the energy-density, the pressure and three components of the velocity field u^α (the fourth component is determined via the normalisation condition $u_\alpha u^\alpha = 1$). In total, this is 11 quantities to be found from 10 equations $G_{\alpha\beta} = \kappa T_{\alpha\beta}$, and the set is underdetermined. To make it determinate, we have to add one more equation, and usually this is the equation of state.[1]

This indeterminacy survives in the R–W metrics, in spite of their high symmetry. They automatically define a perfect fluid energy-momentum tensor. The energy-density, pressure and velocity field of the fluid are all determined as functions of the scale factor $R(t)$ and of its derivatives. In order to obtain a definite solution, we have to add an equation of state relating the pressure to the energy-density. With all functions depending only on t, this must be a barotropic equation of state $\epsilon = \epsilon(p)$ (see Section 15.5 for an

---

[1] In vacuum, because of $G^{\alpha\beta}{}_{;\beta} = 0$, only six of the Einstein equations are independent (Stephani, 1990, pp. 160–168). In matter, the equations of motion $T^{\alpha\beta}{}_{;\beta} = 0$ are in most cases not obeyed before all the field equations are solved. They are usually simpler than the Einstein equations and are solved first.

explanation of how very special this equation is). The equation of state is a differential equation that determines $R(t)$.

What equation of state is most appropriate for describing the Universe as a whole? At present, the mass-density in the Universe is so small ($< 10^{-28}$ g cm^{-3}) that pressure does not influence the large-scale dynamics of matter and $p = 0$ is an acceptable hypothesis. This is the assumption that Friedmann made in obtaining his solutions (given in Sections 17.4 and 17.6). However, if we want to describe the evolution of the Universe with the early period of high density included, then we have to take pressure into account – and the equation of state cannot be the same at all times. A very brief description of the 'history of the Universe' is given in Section 17.10. We shall consider here the later periods and, following Friedmann (1922), we assume $p = 0$.

Using any algebraic computer program, the tetrad components of the Einstein tensor in the orthonormal tetrad for the metric (17.1) or (17.3)[1] are found, and the Einstein equations are

$$G_{00} = \frac{3k}{R^2} + 3\frac{\dot{R}^2}{R^2} = \kappa\epsilon - \Lambda, \tag{17.24}$$

$$G_{11} = G_{22} = G_{33} = -\left(\frac{k}{R^2} + \frac{\dot{R}^2}{R^2} + 2\frac{\ddot{R}}{R}\right) = \kappa p + \Lambda. \tag{17.25}$$

The equations of motion $T^{\alpha\beta}{}_{;\beta} = 0$, which are integrability conditions for Eqs. (17.24)–(17.25), reduce here to the single equation

$$(\epsilon R^3)^{\cdot} + 3R^2\dot{R}p = 0. \tag{17.26}$$

With $p = 0$, this becomes the mass conservation equation

$$\rho R^3 = \text{constant} \stackrel{\text{def}}{=} 3\mathfrak{M}/(4\pi), \tag{17.27}$$

where $\rho = \epsilon/c^2$ is the mass-density. If $\dot{R} \neq 0$, then (17.25) follows from (17.24) and (17.27). If $\dot{R} = 0$, then the solution of (17.24)–(17.25) is the static 'Einstein Universe' (12.82), and we shall not consider this case. With $p = 0$ and $\dot{R} \neq 0$ we thus have only Eq. (17.24) to solve, which becomes, in consequence of (17.27),

$$R_{,t}^2 = \frac{2G\mathfrak{M}}{c^2 R} - k - \frac{1}{3}\Lambda R^2. \tag{17.28}$$

We can use (17.27) and (17.28) to obtain the formula for the **critical density**. With $\Lambda = 0$ and $\dot{R}/R = H/c$ – the Hubble 'constant', we obtain

$$\rho = \frac{3H^2}{8\pi G} + \frac{3k}{c^2\kappa R^2}. \tag{17.29}$$

---

[1] Since the tetrad components of the Einstein tensor are scalars, and since in the R–W geometry they have to be spatially homogeneous, i.e. independent of the radial coordinate, they come out the same no matter which of the two coordinate representations is used. However, they would be different for (17.5) and (17.7) because there the constant k is scaled to $+1$ or -1, respectively.

The quantities ρ and H are *in principle* directly measurable.[1] Hence, Eq. (17.29) gives us a possibility to determine the sign of k for the real Universe. If the observed density ρ_o is greater than the **critical density**

$$\rho_{cr} \overset{\text{def}}{=} \frac{3H^2}{8\pi G},\tag{17.30}$$

then our Universe is closed ($k > 0$); if $\rho_o < \rho_{cr}$, then it is open; and if $\rho_o = \rho_{cr}$, then it is spatially flat ($k = 0$).

If we take into account only that matter which is visible in telescopes, then $\rho_o \ll \rho_{cr}$. Taking the minimal value of H_0 consistent with observations, $H_0 = 50\,\text{km}\,\text{s}^{-1}\,\text{Mpc}^{-1}$, we obtain $\rho_{cr} = 0.475 \times 10^{-29}\,\text{g}\,\text{cm}^{-3}$, while $\rho_o \leq 10^{-31}\,\text{g}\,\text{cm}^{-3}$ (Lang, 1974, p. 558). Greater values of H_0 increase the discrepancy. However, orbital velocities of stars in galaxies indicate that galaxies contain large amounts of non-luminous matter. Estimates of density of that matter imply $\rho_o \leq 0.2\rho_{cr}$ (Coles and Ellis, 1997). Observations of galaxy clusters show that also the intergalactic space contains non-luminous matter, but including it does not allow one to make ρ_o greater than $0.3\rho_{cr}$. Unfortunately, the debate about the average density of matter in the Universe, a valid scientific question, is compounded by enthusiasts of the *inflationary models* (see Section 17.8), who, more often than not, resort to lobbying (as opposed to scientific dispute) to promote their favourite subject. For these models it is essential that the spatial curvature of the Universe is very close to zero. The current majority view is that the actual density is equal to the critical one, and the invisible part is hidden in the form of 'dark matter' (Padmanabhan, 1993; Peebles, 1993) – that can be anything except matter known from laboratory – or in the form of 'dark energy', i.e. the cosmological constant (see Section 17.9). A serious and critical evaluation of various observational results concerning the amount of matter in the Universe is given by Coles and Ellis (1997).

The history of astronomy teaches us that values of various parameters have sometimes been revised radically – because the final results follow via long chains of assumptions used in interpreting the observations. These assumptions change as knowledge advances. Hence, all such numbers must be treated as temporary. It is possible that future observations will render ρ_o closer to ρ_{cr}. However, the current real knowledge (as opposed to postulates and expectations) gives no grounds to assume that ρ_o is greater than $0.3\rho_{cr}$, and this low density is no problem for any theory but inflation. Moreover, the notion of a critical density applies only to the R–W models. In more general cosmological models (see Chapters 18 and 19) the curvature index k is a function of position. In those models, the local value of matter-density is not connected with the type of evolution. Hence, the whole discussion of the relation between ρ_o and ρ_{cr} concerns only a limited class of models and is not necessarily important for the long-term evolution of our real Universe.

---

[1] The difficult element is measuring the distances. All methods used to measure intergalactic distances rely on various theoretical assumptions that are difficult to verify, and they can be carried out for relatively near galaxies. For very distant galaxies, in astronomical practice the Hubble law is *assumed* to hold exactly, and then unceremoniously used as a measure of distance.

Equation (17.28) can be written as follows:

$$\frac{dR}{dt} = \pm\left(\frac{2G\mathfrak{M}}{c^2 R} - k - \frac{1}{3}\Lambda R^2\right)^{1/2}. \tag{17.31}$$

The general solution of this equation can be expressed in terms of elliptic functions. We will discuss these solutions qualitatively in Section 17.6.

17.4 The Friedmann solutions with $\Lambda = 0$

With $\Lambda = 0$, Eq. (17.31) becomes

$$\frac{dR}{dt} = \pm\left(\frac{2G\mathfrak{M}}{c^2 R} - k\right)^{1/2} \overset{\text{def}}{=} \left(\frac{2\alpha}{R} - k\right)^{1/2}. \tag{17.32}$$

The solutions of this equation are best represented in a parametric way. For $k < 0$ we obtain:

$$R = -\frac{\alpha}{k}(\cosh\omega - 1), \qquad t - t_B = \frac{\alpha}{(-k)^{3/2}}(\sinh\omega - \omega), \tag{17.33}$$

where ω is the parameter and t_B is an arbitrary constant. Usually, the time coordinate is chosen so that $t_B = 0$. For $k = 0$ the solution is

$$R = \left(\frac{9}{2}\alpha\,(t - t_B)^2\right)^{1/3}. \tag{17.34}$$

In this case the constant α can be scaled to any arbitrary value by transformations of r, so its actual value merely defines the unit of distance and has no physical meaning. (Unlike in the previous case and in the next case, where the value of α is related to the curvature of space at a given time.) Finally, for $k > 0$, the solution of (17.32) is

$$R = \frac{\alpha}{k}(1 - \cos\omega), \qquad t - t_B = \frac{\alpha}{k^{3/2}}(\omega - \sin\omega). \tag{17.35}$$

In all three cases we have taken into account the observed fact that at present $\dot{R} > 0$. As we predicted in Section 15.4, each of these solutions has a singularity at $t \to t_B$, at which $R \to 0$ and $\rho \to \infty$. At that instant (in the past), all matter of the model was condensed in one point. The last model has a second such singularity, in the future, at

$$t = t_{FS} = t_B + \frac{2\alpha\pi}{k^{3/2}} = t_B + \frac{2\pi G\mathfrak{M}}{k^{3/2}c^2} = t_B + \frac{8\pi^2 G\rho_o R_o{}^3}{3k^{3/2}c^2}. \tag{17.36}$$

The models (17.33) and (17.34), if they are expanding at the initial instant, will continue to expand for an infinite time. The model (17.35) has a finite lifetime; at $t = t_{FS}$ its existence is terminated. The graph of $R(t)$ corresponding to different values of k is shown in Fig. 17.1, which again explains the dangers connected with rescaling of $|k| \neq 0$ to 1.

Formally, Eqs. (17.35) describe a flattened cycloid, on which t runs through an infinite range, while R oscillates between 0 and $R_{max} = 2\alpha/k = 2G\mathfrak{M}/(kc^2)$. However, $dR/dt \to \infty$ as $t = t_{FS}$, and the integration of (17.32) through this point is not possible. This is why the $k > 0$ Friedmann Universes have only a finite time of existence equal to $(t_{FS} - t_B)$ (note that the value of this quantity depends on k!).

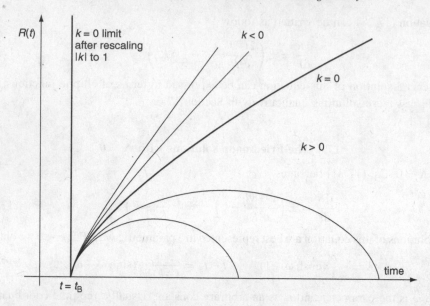

Fig. 17.1. The functions $R(t)$ corresponding to Friedmann models with different values of k. When $k \to 0$, the rescaling of R required to achieve $|k| = 1$ maps the $k = 0$ graph into the vertical straight line at $t = t_B$, and no place is left for the $k < 0$ models. Note that, for models with $k > 0$, the actual value of k determines the lifetime of the model, so *it has physical meaning*. Similarly, for models with $k < 0$, the value of $-k$ determines the asymptotic value of the expansion velocity, $\dot{R}$, as $t \to \infty$. Thus, rescaling k results in tampering with physical parameters.

17.4.1 The redshift–distance relation in the $\Lambda = 0$ Friedmann models

Having found the explicit forms of $R(t)$, we can now integrate Eq. (17.19), to find the observer area distance as a function of the redshift, the Hubble parameter and the deceleration parameter. From (17.32) $R_{,tt} = -\alpha/R^2$, and from (17.20) $R_{,tt} = -q_0 R_{,t}^2/R \equiv -q_0 R H_0^2$. Thus $R = \left[\alpha / \left(q_0 H_0^2 \right) \right]^{1/3}$. Using this, Eq. (17.32) can be rewritten as

$$ H_0^2 = \frac{2\alpha}{R^3} - \frac{k}{R^2} = 2q_0 H_0^2 - \frac{k}{\alpha^{2/3}} q_0^{2/3} H_0^{4/3}. \tag{17.37} $$

Note that the sign of k is the same as the sign of $(2q_0 - 1)$, and $q_0 = 1/2$ when $k = 0$. From here, when $k > 0$,

$$ \alpha = \left(\frac{k}{2q_0 - 1} \right)^{3/2} \frac{q_0}{H_0}, \qquad R = \frac{1}{H_0} \sqrt{\frac{k}{2q_0 - 1}}. \tag{17.38} $$

The integrals in (17.19) are different for each k. For $k > 0$ the result is

$$ \arctan \left(\frac{\sqrt{k}r}{2} \right) = \frac{1}{2} \left[\omega(t) - \omega(t_e) \right]. \tag{17.39} $$

But, from (17.35), $\omega(t) = \arccos(1 - kR/\alpha)$ and $\omega(t_e) = \arccos\{1 - kR/[(1+z)\alpha]\}$. Substituting for α and R from (17.38) we obtain

$$r_O = \frac{q_0 z + (q_0 - 1)\left(\sqrt{2q_0 z + 1} - 1\right)}{H_0 q_0^2 (1+z)^2}. \tag{17.40}$$

Going back to (17.38) and repeating the whole calculation for $k < 0$ we obtain again (17.40), even though the intermediate formulae are different. With $k = 0$, we substitute $q_0 = 1/2$ in (17.40) and obtain

$$r_O = 2\frac{\sqrt{1+z} - 1}{H_0(1+z)^{3/2}}. \tag{17.41}$$

The limit of (17.40) at $q_0 \to 0$ is $r_O = [1/(2H_0)][1 - 1/(1+z)^2]$. This corresponds to $\alpha = 0$, i.e. $\mathfrak{M} = 0$, which is the Minkowski spacetime in coordinates connected with an expanding family of timelike straight lines. Equation (17.40) is called the **Mattig formula** (Mattig, 1958).

Note that $dr_O/dz = 1/H_0 > 0$ at $z = 0$ and becomes negative as $z \to \infty$. Thus the function $r_O(z)$ is increasing at small z and decreasing at large z, achieving a maximum in between. This means that there is refocussing in the Friedmann models with $\Lambda = 0$. This is true for every k, including $k = 0$, for which the maximal r_O occurs at $z = 8$.

17.5 The Newtonian cosmology

We will now use Newton's theory to describe the motion of a cloud of particles that interact only by gravitation. We will make the same assumptions that had led us to the R–W models in relativity: a homogeneous and isotropic matter distribution, $\rho = \rho(t)$, and a spherically symmetric initial distribution of velocities:

$$v_i(t_0) = v(t_0, r)\frac{x_i}{r}. \tag{17.42}$$

Let us consider the motion of the particle on the sphere of radius $r(t)$. Since the distribution of matter is spherically symmetric, the force exerted on this particle by matter that lies outside this sphere is zero (see Exercise 3). The particle thus moves in the gravitational field created by matter inside that sphere, i.e. in the potential

$$V(r) = -\frac{GM}{r} = -\frac{4}{3}\pi G\rho(t)r^2(t). \tag{17.43}$$

Consequently, the equation of motion of the particle is

$$\frac{dv_i}{dt} \equiv \frac{\partial v_i}{\partial t} + v_j\frac{\partial v_i}{\partial x_j} = -\frac{GM}{r^3}x_i = -\frac{4}{3}\pi G\rho(t)x_i(t). \tag{17.44}$$

Using Eq. (17.44) one can easily verify that

$$\frac{d}{dt}\left(\frac{v_i}{v}\right) = 0, \tag{17.45}$$

where $v = \sqrt{v_i v_i}$, i.e. the direction of the velocity vector at every point remains the same for all time. Therefore, using (17.42) we can rewrite (17.44) as follows:

$$\frac{dv}{dt} \equiv \frac{\partial v}{\partial t} + v \frac{\partial v}{\partial r} = -\frac{GM}{r^2} = -\frac{4}{3}\pi G\rho(t)r. \tag{17.46}$$

This is the equation of motion. The equation of continuity $\partial\rho/\partial t + (\rho v_i),_i = 0$, in consequence of $\rho = \rho(t)$, (17.42) and (17.45), takes the form

$$\frac{1}{\rho}\frac{\partial\rho}{\partial t} + v,_r + \frac{2}{r}v = 0. \tag{17.47}$$

We denote

$$\frac{1}{\rho}\frac{\partial\rho}{\partial t} \stackrel{\text{def}}{=} -3F(t). \tag{17.48}$$

Then we have in (17.47) $(r^2 v),_r = 3F(t)r^2$, which is integrated to give

$$v(t, r) = F(t, r)r + \frac{\mathcal{K}(t)}{r^2}. \tag{17.49}$$

We substitute this in (17.46) and obtain

$$\dot{F}r + \frac{\dot{\mathcal{K}}}{r^2} + \left(Fr + \frac{\mathcal{K}}{r^2}\right)\left(F - \frac{2\mathcal{K}}{r^3}\right) = -\frac{4}{3}\pi G\rho(t)r. \tag{17.50}$$

This is an algebraic equation in r whose coefficients are functions of t. The coefficients of different powers of r have to vanish separately, hence

$$\mathcal{K}(t) = 0, \qquad \dot{F} + F^2 = -\frac{4}{3}\pi G\rho(t). \tag{17.51}$$

Consequently, (17.49) becomes

$$v(t, r) = F(t)r, \tag{17.52}$$

and from (17.42) we have $v_i = F(t)x_i$, i.e. the Hubble law.

Equation (17.52) can be integrated again since $v = dr/dt$. We substitute this and (17.48) in (17.52) and obtain $(1/r)dr/dt = -[1/(3\rho)]\partial\rho/\partial t$, which is integrated with the result $r(t) = (A/\rho)^{1/3}$, where A is an arbitrary constant. Hence, for each particle separately

$$\rho r^3 = A = \text{constant} \stackrel{\text{def}}{=} \frac{3M}{4\pi}. \tag{17.53}$$

This is the analogue of (17.27) in the Friedmann solution. Substituting (17.53) in (17.48) and then substituting the result in (17.51) gives $\ddot{r}/r = -GM/r^3$. This is easily integrated with the result

$$\dot{r}^2 = \frac{2GM}{r} - 2K, \tag{17.54}$$

where $K = $ constant. This equation has the same form as (17.32). Thus, it has the same three types of solutions. In relativity, Eq. (17.32) defined the scale factor $R(t)$ that determined the changes of distance (see (17.15)). Here, Eq. (17.54) determines the distance of an arbitrary particle from the origin of the coordinate system. In the R–W models, the constant k determined the sign of the spatial curvature. Here, multiplying (17.54) by m, the mass of a fluid particle, we see that the quantity $(-mK)$ is the total energy of the particle. If $K > 0$, then the energy is negative; then at $r \to GM/K$ the velocity of the particle decreases to zero; the particle turns back towards the origin and falls onto it in a finite time. When $K = 0$, the particle can escape infinitely far from the origin, but its velocity tends to zero as $r \to \infty$. If $K < 0$, then $\dot{r} \underset{r \to \infty}{\to} \dot{r}_0 > 0$.

Equation (17.54) was first found and interpreted by Milne and McCrea (1934), and the authors were surprised that this result had not been known earlier. It is so elementary that it could have been found still in the eighteenth century, if only anybody had allowed for the *possibility* that the Universe is not static (this is one more piece of evidence of the role played by psychology and prejudice in science). However, until Hubble's discovery of the expansion of the Universe, everybody believed that the Universe was unchanging in time.

Although the Milne–McCrea solutions are *formally* identical with the Friedmann solutions, their physical interpretation is not the same. The solutions of (17.54) contain typical Newtonian notions such as absolute space and absolute velocity of matter (relative to space points). In contrast to relativity, they do not imply any law of propagation of light rays and lead to a contradiction with special relativity. Making the Milne–McCrea model consistent with special relativity requires the extension of Newton's theory by new postulates. Such a theory (called 'kinematical relativity') was devised by Milne in later years (Milne 1948), but it did not gain acceptance in the physics community because, unlike relativity, it was well suited only to this particular cosmological model.

17.6 The Friedmann solutions with the cosmological constant

Now we will discuss the full Friedmann equation (17.28). For later convenience we denote $\Lambda = -\lambda$. Equation (17.28) becomes then

$$\dot{R}^2 = \frac{2G\mathfrak{M}}{c^2 R} - k + \frac{1}{3}\lambda R^2. \tag{17.55}$$

The graph of the equation $\dot{R}^2 = 0$ in the plane of the parameters (λ, R) is shown in Fig. 17.2. For $k \leq 0$ the function $\lambda(R)$ determined by this equation is monotonic and negative for all R. With $k > 0$, the function $\lambda(R)$ increases from $-\infty$ at $R = 0$ through 0 (attained at $R = 2G\mathfrak{M}/(c^2 k)$) to the maximum equal to $\lambda \overset{\text{def}}{=} \lambda_{\text{E}} = c^4 k^3/(9G^2\mathfrak{M}^2)$, attained at $R = 3G\mathfrak{M}/(c^2 k) \overset{\text{def}}{=} R_{\text{E}}$, and then monotonically decreases and tends asymptotically to zero as $R \to \infty$.

Since λ is a universal constant, R can vary only along straight lines parallel to the R-axis. Since the λ calculated from (17.55) is never smaller than the λ determined by the $\dot{R} = 0$ curves, the allowed area for R-values is above the corresponding $\dot{R} = 0$ curve, and the extrema of R lie on $\dot{R} = 0$. By definition, $R(t) > 0$. This implies the following:

- (1) For $\lambda < 0$, only models oscillating between $R = 0$ and $R = R_{\text{max}}$ exist. The R_{max} is greater for $k < 0$ and smaller for $k > 0$. The 'cosmological attraction' implied by $\lambda < 0$ will always halt and reverse the expansion of the Universe.

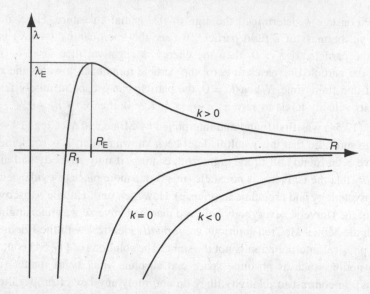

Fig. 17.2.   The curves $\dot{R} = 0$ in the (R, λ) plane. The area accessible for evolution is above each curve. $R_1 = 2G\mathfrak{M}/(c^2 k)$ is the size of the Universe at maximal expansion when $\lambda = 0$. $\lambda_E = c^4 k^3/(3G\mathfrak{M})^2$ is the minimal value of λ at which R changes between 0 and ∞; it corresponds to the maximum of the $k > 0$ curve that occurs at $R = R_E = 3G\mathfrak{M}/(c^2 k)$.

- (2) For $\lambda = 0$ and $k \leq 0$ there exist solutions in which matter expands monotonically from infinite density at $R = 0$ to zero density at $R \to \infty$; they were discussed in Section 17.4.
- (3) For $\lambda = 0$ and $k > 0$ there exists a solution oscillating between $R = 0$ and $R_{max} = 2G\mathfrak{M}/(c^2 k)$.
- (4) For $0 < \lambda < \lambda_E$ the following cases exist:
  - (4a) If $k \leq 0$, the model constantly expands from a singular state at $R = 0$ to zero density at $R \to \infty$, or constantly contracts from $R = \infty$ to $R = 0$. Note that the derivative $\dot{R}$ in (17.55) cannot change sign in this case. Moreover, if $\dot{R} > 0$, then $\dot{R}$ is an increasing function of R. The expansion thus proceeds with acceleration – this is the influence of the 'cosmological repulsion' implied by $\lambda > 0$.
  - (4b) If $k > 0$, then we have two cases:
    - ($4b_A$) If $R < R_E = 3G\mathfrak{M}/(c^2 k)$, then the model is oscillating.
    - ($4b_B$) If $R > R_E$, then the model contracts from zero density at $R = \infty$ to a finite maximal density at $R = R_{min}$ (with $\dot{R}(R_{min}) = 0$), and then expands again to $R \to \infty$.
- (5) For $\lambda = \lambda_E$ there are several possibilities:
  - (5a) If $k \leq 0$, then the model constantly expands or contracts, like in case (4a).
  - (5b) If $k > 0$ and $R < R_E$, then the model either expands from a singular state at $R = 0$ asymptotically to $R = R_E$ at $t \to \infty$, or contracts from the asymptotic

state $R \to R_E$ at $t \to -\infty$ to a singularity at $R = 0$. It can be easily verified that for $\lambda = \lambda_E$ the right-hand side of (17.55) has a double root $R = 3G\mathfrak{M}/(c^2 k)$, and then the time required to reach $R_E = 3G\mathfrak{M}/(c^2 k)$,

$$t = \int_{t_0}^{3G\mathfrak{M}/(c^2 k)} \left(\frac{2G\mathfrak{M}}{c^2 R} - k + \frac{1}{3}\lambda_E R^2 \right)^{-1/2} dR \qquad (17.56)$$

is infinite.

(5c) If $k > 0$ and $R > R_E$, then the model either expands from an asymptotic state $R \to R_E$ at $t \to -\infty$ to $R \to \infty$ at $t \to \infty$, or contracts from $R \to \infty$ at $t \to -\infty$ to $R \to R_E$ at $t \to \infty$.

(5d) For $k > 0$ there also exists the static solution $R \equiv R_E$, which is unstable. This is the 'Einstein Universe' (12.82). It is unstable because a small perturbation of R will cause it to become expanding as in case (5c) or contracting as in case (5b).

- (6) For $\lambda > \lambda_E$, independently of the sign of k, there exist only models that monotonically expand from $R = 0$ to $R \to \infty$ or monotonically contract. Just as in cases (4a) and (5c), the expansion proceeds with acceleration, since $\dot{R}$ increases with increasing R.

These facts are summarised in Figs. 17.3–17.5. Figure 17.3 shows all the possible oscillating models. With λ negative, the maximal size of the oscillating model with $k \leq 0$

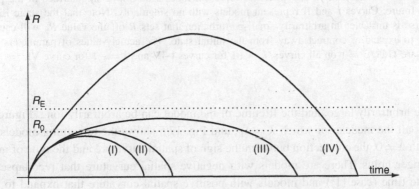

Fig. 17.3. The scale factor as a function of time for recollapsing Friedmann models, for different values of the curvature index and of the cosmological constant. Curve (I): $k > 0, \lambda < 0$; curve (II): $k > 0, \lambda = 0$ (this is one of the models considered in Section 17.4); curve (III): $k > 0, 0 < \lambda < \lambda_E$; curve (IV): $k \leq 0, \lambda < 0$. The horizontal line R_E is the constant value of $R = R_E = 3G\mathfrak{M}/c^2$ in the static Einstein Universe. The horizontal line R_0 is the maximal value of R in the recollapsing $\lambda = 0$ model. As seen from Fig. 17.2, negative λ will always force recollapse, irrespective of the sign of k. However, with $k \leq 0$, an arbitrarily large maximal R can occur with a sufficiently small absolute value of λ. For better visualisation, the parameters of the four curves were chosen as close to each other as possible. On all curves, $G\mathfrak{M}/c^2 = 1$. On curve (I) $k = +1, \lambda = -0.1$; on curve (II) $k = +1, \lambda = 0$; on curve (III) $k = +1, \lambda = +0.1$; on curve (IV) $k = -1, \lambda = -0.1$. With negative λ of sufficiently great absolute value, the lifetime for $k \leq 0$ models may be shorter than in any $k > 0$ model.

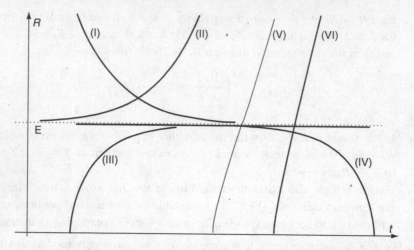

Fig. 17.4.   The graphs of $R(t)$ for those Friedmann models in which $\lambda = \lambda_{\rm E}$. The dotted horizontal line marked E is the static Einstein Universe in which $R = R_{\rm E} = 3G\mathfrak{M}/(c^2 k)$. Curves I–IV represent models that approach the Einstein Universe asymptotically towards the future or towards the past; for all of them $k > 0$. On curve I, $R > R_{\rm E}$ and $R_{,t} < 0$. On curve II, $R > R_{\rm E}$ and $R_{,t} > 0$. On curve III, $R < R_{\rm E}$ and $R_{,t} > 0$. On curve IV, $R < R_{\rm E}$ and $R_{,t} < 0$. Curve V represents the model with $k = 0$, curve VI the model with $k < 0$. Curves V and VI have inflection points at intersections with $R = R_{\rm E}$. Time-reverses of curves V and VI (i.e. collapsing models) are also solutions of Eq. (17.55), but are omitted for clarity. Curves III-VI represent models that have singularities in the past or in the future. Curves I and II represent models with no singularity. Note that the static Einstein Universe is unstable: an arbitrarily small perturbation that sets R off the value $R_{\rm E}$ will cause the model to expand or contract away from the initial state. The actual values of parameters in the figure are $G\mathfrak{M}/c^2 = 1$ on all curves, $k = +1$ for curves I–IV and $k = -1$ for curve VI.

can be arbitrarily large, and the lifetime of the model can be arbitrarily long. Figure 17.4 shows all the models for which $\lambda = \lambda_{\rm E}$, and Fig. 17.5 shows the remaining models.

With $\lambda \neq 0$, the connection between the sign of spatial curvature and the type of motion no longer holds. There are models with negative spatial curvature that recollapse after a finite time (case (1)) and models with positive spatial curvature that expand to reach infinite extent (cases $(4b_{\rm B})$, (5c) and (6)). The latter do not always attain a singular state.

The fact that, with $\Lambda < 0$, models expanding with acceleration may exist will be important in the discussion of horizons in the next section.[1]

The discussion introduced above was first presented by Friedmann in his original papers (Friedmann, 1922); it can also be found in the book by Robertson and Noonan (1968) and, in a more extended form, in Rindler (1980). Friedmann's original discussion was incomplete because he considered only the case $k = +1$, and did not know the case $k = 0$.

---

[1] Currently, it is believed that our observed Universe is accelerating; see Section 17.9.

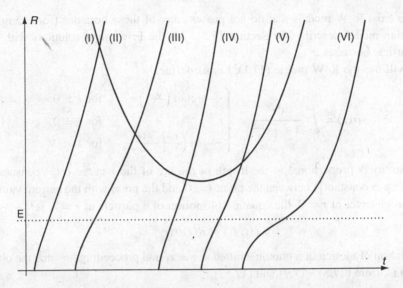

Fig. 17.5. The graphs of $R(t)$ for the remaining Friedmann models, for different values of the curvature index and of the cosmological constant. The horizontal line E is the constant value of $R = R_E = 3G\mathfrak{M}/(c^2k)$ in the static Einstein Universe. On curves I–III $0 < \lambda < \lambda_E$; on curves IV–VI $\lambda > \lambda_E$. On curve I $k > 0$ and $R > R_E$, on curve II $k < 0$, on curve III $k = 0$. On curves IV, V and VI k is, respectively, negative, zero and positive. Curves II and III have their inflection points for $R > R_E$; curves IV–VI have their inflection points for $R < R_E$. Positive λ implies cosmic repulsion that opposes gravitational attraction. Beyond the inflection point the repulsion definitely prevails and sets the Universe into accelerated expansion. With large initial R, the repulsion prevents collapse even for $k > 0$.

17.7 Horizons in the Robertson–Walker models

This section is a brief presentation of the results of Rindler (1956).

Every observer in the Universe receives information about distant objects via null geodesics. It turns out that in most of the Robertson–Walker models there exist, for every observer, such objects, from which he/she has not yet received any signal. Then, in some of the R–W models (those which expand with acceleration) objects exist from which a given observer has not received and will never receive any signal. The boundaries separating objects already observed from those not yet observed, and objects observable from unobservable, are called *horizons*. This definition will be made precise below.

The **event horizon** for an observer A is the hypersurface in the spacetime that divides the collection of all events into two nonempty classes: those events that have been, are being or will be observed by A, and those that A has never observed and will never be able to observe. Not every R–W spacetime has event horizons.

The **particle horizon** for an observer A at the instant t_0 is a 2-dimensional surface in the space $t = t_0$ that divides all particles (i.e. worldlines of matter) into two nonempty classes: those that had been observed by A up to $t = t_0$, and those that A has not yet observed.

There exist R–W models that do not possess any of these horizons (for example, the Newtonian model described in Section 17.5). All the Friedmann solutions with $\Lambda = 0$ have particle horizons.

We will use the R–W metric (17.1). Let us define

$$\sigma(r) \overset{\text{def}}{=} \int_0^r \frac{\mathrm{d}r'}{1 + \frac{1}{4}kr'^2} = \begin{cases} \frac{2}{\sqrt{k}} \arctan\left(\frac{\sqrt{k}r}{2}\right) & \text{for } k > 0, \\ r & \text{for } k = 0, \\ \frac{1}{\sqrt{-k}} \ln\left(\frac{1 + \frac{1}{2}\sqrt{-k}r}{1 - \frac{1}{2}\sqrt{-k}r}\right) & \text{for } k < 0. \end{cases} \tag{17.57}$$

This quantity is proportional to the length of the arc of the r-curve ($\{t = \text{constant}, \vartheta = \text{constant}, \varphi = \text{constant}\}$) between the point $r = 0$ and the point with the current value of r.

For an observer at $r = 0$, the equation of motion of a particle at $r = r_1$ is

$$\ell(t, r_1) = R(t)\sigma(r_1). \tag{17.58}$$

The equation of motion of a photon emitted at $r = r_1$ and proceeding *towards* the observer at $r = 0$ is, from (17.8), (17.9) and (17.57),

$$\sigma(r) = \sigma(r_1) - \int_{t_1}^{t} \frac{\mathrm{d}\tau}{R(\tau)}. \tag{17.59}$$

With $k < 0$ we have $\sigma(r) \to \infty$ for $r \to 2/\sqrt{-k}$, hence the range $r \in [0, 2/\sqrt{-k})$ covers the whole space $t = \text{constant}$. For $k > 0$ we have $\sigma \to \pi/\sqrt{k}$ for $r \to \infty$. Hence, $r = \infty$ corresponds to the point that is antipodal to $r = 0$ on each sphere $t = \text{constant}$. The r-curve can, of course, be continued beyond this point, although the coordinate r does not cover the extra stretch. The extended line may wind multiply around the sphere $t = \text{constant}$. On such multiply wound curves we will define σ to be $\sigma(r) = n\pi/\sqrt{k} + \widetilde{\sigma}(r)$, where n is the number of passages of the curve through the poles $r = 0$ and $r = \infty$, and $\widetilde{\sigma}(r)$ is the quantity calculated by (17.57) between the final point r and the last pole passed. Hence, also in the model with positive curvature, a point in the space $t = \text{constant}$ can be assigned to every value of σ.

The reasoning below applies to models of infinite time of existence. For recollapsing models, modifications are necessary and they will not be discussed here. For example, the limit $t \to \infty$ has to be replaced by $t \to t_{FS}$, where t_{FS} is the instant of the final singularity, and 'finite limit' has to be replaced by 'value smaller than $\pi/\sqrt{k}$'.

With $k \leq 0$, the necessary and sufficient condition for the existence of an event horizon is the convergence of the integral $\int_{t_1}^{t} \mathrm{d}\tau/R(\tau)$ to a finite limit at $t \to \infty$. Then, there exist particles at $r = r_H$ (with $\sigma(r_H) < \infty$) such that the photon emitted from there at $t = t_H$ will not reach the observer at $r = 0$ (where $\sigma(0) = 0$) even after an infinite time. From (17.59) we obtain for r_h (the minimal value of r_H)

$$\sigma(r_h) = \int_{t_h}^{\infty} \frac{\mathrm{d}\tau}{R(\tau)} < \infty. \tag{17.60}$$

With $k > 0$, the event horizon exists if $\sigma(r_h)$ defined by (17.60) obeys $\sigma(r_h) \leq \pi/\sqrt{k}$. With $\sigma(r_h) > \pi/\sqrt{k}$, each photon sent from $r = r_h$ will manage, in a finite time, to travel farther than half of the circumference of the Universe, and, in consequence of this, the light signal from the emission event will reach the observer at $r = 0$ on one route or another.

For the recollapsing models with $k > 0$, the event horizon may exist even if $\sigma(r_{\mathrm{H}}) \overset{\text{def}}{=} \int_{t_{\mathrm{H}}}^{t_{\mathrm{FS}}} \mathrm{d}\tau/R(\tau) > \pi/\sqrt{k}$ (see the example below).

In consequence of the isotropy of the R–W models, if there exists r_{h} obeying (17.60) (or the corresponding condition for $k > 0$), then the event horizon exists for every direction of observation. In consequence of their homogeneity, the origin $r = 0$ may be placed at any point of the space $t = $ constant. Hence, if there exists an event horizon for one given observer, then event horizons exist for all observers.

When the event horizon exists, the events whose coordinate r obeys $r > r_{\mathrm{h}}$ (when $k \leq 0$) or $r_{\mathrm{h}} < r \leq \pi/\sqrt{k}$ (when $k > 0$) will never be observed by the observer at $r = 0$. How can this be explained intuitively? If $R(t)$ increases sufficiently rapidly, then the spatial distance between the observer at $r = 0$ and the light source at $r = r_1 > r_{\mathrm{H}}$, given by (17.58), increases so fast that the light cannot overcome this 'swelling of space', even though it keeps going towards the observer. Eddington once explained this as follows: imagine a runner who runs on an expanding race track whose finish line moves away faster than the maximum speed at which the runner can run. In the Friedmann models with $\Lambda = 0$, $\dot{R}$ is a decreasing function. Consequently, with $k \leq 0$, every event will eventually become visible for every observer. From (17.33)–(17.34) it follows that $\int_{t_1}^{t_2} \mathrm{d}\tau/R(\tau) \underset{t_2 \to \infty}{\to} \infty$, i.e. the event horizons do not exist. However, with $k > 0$, $\omega = 2\pi$ at the final singularity, and then

$$\sigma(r_1) = \int_{t_1}^{t(2\pi)} \frac{\mathrm{d}\tau}{R(\tau)} = \int_{\omega(t_1)}^{2\pi} \frac{1}{k}\mathrm{d}\omega = \frac{2\pi - \omega(t_1)}{k}. \tag{17.61}$$

Hence, if $\omega(t_1) > \pi$ (i.e. $t_1 > \pi\alpha/k$), then events at $t > t_1$ and $r > r_1$ will not become visible for the observer at $r = 0$ before the final singularity occurs. The event horizon will thus exist, although the integral in (17.61) is smaller than $\pi/\sqrt{k}$ for $t_1 < \pi\alpha/k$.

From (17.59) it is also seen that, if a particle was initially in the field of view (i.e. inside the event horizon) for the observer A, then it will remain visible to him for ever. This is because, if the equation

$$0 = \sigma(r_1) - \int_{t_1}^{t_0} \frac{\mathrm{d}\tau}{R(\tau)} \tag{17.62}$$

has a solution for t_1 at a given r_1 and a given reception time t_0, then it will have a solution for all times $t > t_0$.

Proof: Since $\sigma(r_1)$ does not depend on t, the solution of the problem consists in finding an integration interval in which the integral in (17.62) has a given value. From Fig. 17.6 one sees that if one such interval $[t_1, t_0]$ exists, then for every $t_0' > t_0$ there will exist a $t_1' > t_1$ such that

$$\int_{t_1'}^{t_0'} \frac{\mathrm{d}\tau}{R(\tau)} = \int_{t_1}^{t_0} \frac{\mathrm{d}\tau}{R(\tau)}.$$

(if $1/R(t)$ is continuous, but this we are assuming all the time). $\square$

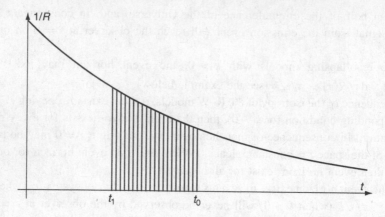

Fig. 17.6. An illustration to (17.62). If the integral of a continuous function $1/R$ over the interval $[t_1, t_0]$ equals $\sigma(r_1)$, then for every $t'_0 > t_0$ there exists a $t'_1 > t_1$ such that the integral of the same function over $[t'_1, t'_0]$ has the same value $\sigma(r_1)$.

If the event horizon exists (i.e. $\sigma(r_1) < \infty$ for $t_0 \to \infty$ in (17.62)), then t_1 goes to a finite limit t_{out} as $t_0 \to \infty$. This means that, although the particle at $r = r_1$ will always be visible to the observer A, he/she will see it only up to the instant $t = t_{\text{out}}$ on the particle's clock. The signal sent out from $r = r_1$ at $t = t_{\text{out}}$ will reach A after an infinite time ($t_0 \to \infty$), i.e. never. This, in turn, means that the worldline of the particle at $r = r_1$ will intersect the event horizon of A at the instant $t = t_{\text{out}}$. Moreover, a $t_{\text{out}} < \infty$ exists for every $r_1 < r_{\text{h}}$. Consequently, when the event horizon exists, each particle different from A will be visible to A only for a finite period of its history and will eventually escape from A's field of view through the horizon. The crossing of the horizon occurs in the infinite future of A. The farther away the particle, the shorter the period of its history that will be visible to A.

The necessary and sufficient condition for the existence of a particle horizon is the convergence of the integral in (17.62) to a finite limit at $t_1 \to 0$. (In those models that have no initial singularity, it is the convergence to a finite limit at $t \to -\infty$. The analysis of this second case is similar and will be omitted here.) When the integral is finite at $t_1 \to 0$,

$$\sigma(r_1) = \int_0^{t_0} \frac{d\tau}{R(\tau)} \stackrel{\text{def}}{=} \phi(t_0) \tag{17.63}$$

determines the farthest particles from which the observer at $r = 0$ could have received a light signal up to t_0. If $\sigma(r_1) \xrightarrow[t_1 \to 0]{} \infty$ in (17.62), then the observer at $r = 0$ could receive signals from all other particles for any t_0. At fixed t_0, (17.63) is an equation of a 2-dimensional sphere. Since $1/R > 0$, the function $\phi(t)$ is increasing. This means that, if the particle horizon exists, then with time still more particles come within it. Whether every particle will eventually enter the particle horizon depends on whether the integral in (17.63) has a finite limit at $t_0 \to \infty$. If it has, then, according to the previous definition, the event horizon exists and some particles will never enter A's field of view.

The first signal received by an observer A from each particle is the signal sent out at the initial singularity at $t = 0$. However, by Eq. (17.13), the signal sent out at $t_e = 0$ (for which $R(t_e) = 0$) is received with an infinite redshift. This means that *in the Robertson–Walker models the initial singularity is not observable.*[1]

In the models that contract from $R = \infty$ at $t = -\infty$ and possess a particle horizon, Eq. (17.13) gives $1 + z = 0$ at $t_e = -\infty$. Since $1 + z = \lambda_o/\lambda_e$, in these models the first signal received from each particle has zero wavelength at the observation point, i.e. is infinitely blueshifted.

In the Friedmann models with $\Lambda = 0$ the particle horizons do exist.

The two kinds of horizon were defined above for observers comoving with matter. However, one can also consider observers moving independently of the matter background. In some cosmological models there exist horizons also for such observers: irrespective of how fast the observer moves towards a given worldline of matter, he/she will never see any event on that worldline if the line is sufficiently far from the observer's starting point.

Rindler's paper contains several examples of horizons in various R–W models; readers are advised to consult that paper for further information.

There exists one more type of horizon: the *apparent horizon* (see Sections 14.11 and 16.4). In a Universe expanding from a singularity, it is a closed hypersurface at which both the inward- and the outward-going light rays are diverging, i.e. at which the expansion scalar of any bundle of emitted light rays is positive. This notion plays no useful role in the R–W cosmology; it applies mainly to black hole models. Because of the peculiar properties of the R–W models, it is actually easier to interpret an apparent horizon in the more general Lemaître–Tolman model. However, we will calculate its position in the Friedmann models for completeness, as an exercise. The calculation is simpler in the coordinates of (17.3). The tangent vector field to a bundle of null geodesics emanating radially from a sphere is found by transforming (17.11) with use of (17.2). Dropping primes, the result is

$$k^\alpha = \left[\frac{1}{R}, \frac{\varepsilon}{R^2}\sqrt{1 - kr^2}, 0, 0\right],\tag{17.64}$$

where $\varepsilon = +1$ for outward-going and $\varepsilon = -1$ for inward-going rays. The expansion scalar of this bundle is

$$2\theta = k^\alpha{}_{;\alpha} = \frac{1}{\sqrt{-g}}(\sqrt{-g}k^\alpha)_{,\alpha}.\tag{17.65}$$

Calculating this and substituting for $R_{,t}$ from (17.32) with the $+$ sign (we consider an expanding Universe) we find that $\theta > 0$ is equivalent to

$$\frac{r\sqrt{2\alpha/R - k}}{\sqrt{1 - kr^2}} + \varepsilon > 0.\tag{17.66}$$

---

[1] This is not a universal property of all cosmological models; in Chapter 18 we will see that a typical situation in the Lemaître–Tolman cosmological model is an infinite blueshift for rays emitted from the initial singularity.

This is fulfilled for an outward-going bundle ($\varepsilon = +1$), and it will be fulfilled also for an inward-going bundle ($\varepsilon = -1$) when

$$R < 2\alpha r^2 = \frac{2G\mathfrak{M}}{c^2} r^2. \tag{17.67}$$

This always holds in a vicinity of the Big Bang singularity; this is a region of past-trapped surfaces. Its envelope, where $R = 2\alpha r^2$, is the past apparent horizon. Note from (17.17) and (17.2) that, in the coordinates we are now using, rR is the source area distance from the singularity at $R = 0$. Then, by analogy between (17.28) and the Newtonian equation (17.54), note from (17.27) that $\mathfrak{M}r^3$ is equal to the mass contained within the source area distance rR from the singularity. Thus (17.67) can be written in the equivalent form $r_G < 2m$, where $m \stackrel{\text{def}}{=} G\mathfrak{M}r^3/c^2$. This is, not accidentally, similar to the condition defining the interior of the horizon in the Schwarzschild solution. The meaning of this will become clearer in the Lemaître–Tolman model; see Section 18.8.

17.8 The inflationary models and the 'problems' they solved

(The views expressed in this section are A. K.'s. J. P. should not be held responsible for them.)

The R–W metrics are oversimplified models of the Universe. When taken literally, they can lead to puzzles or problems, most of which do not exist in the L–T and Szekeres models of Chapters 18 and 19. Two of such 'problems' gave rise to a booming field of activity in cosmology – the inflationary models. They are of marginal interest for basic relativity, but gained such prominence that we have to mention them.

The two problems are the 'flatness problem' and the 'horizon problem' (Guth, 1981). In brief, the 'flatness problem' is this. With Λ allowed,[1] the density at $k = 0$ is found from (17.27) and (17.28) to be $c^2 \kappa \rho_{cr}^\Lambda = 3\dot{R}^2/R^2 + \Lambda$. Consider the quantity

$$\widetilde{\Omega} \stackrel{\text{def}}{=} \frac{\rho_{cr}^\Lambda}{\rho} - 1 = -\frac{kR}{2\alpha}, \tag{17.68}$$

where α is defined in (17.32). Since in the ever-expanding models $R \to \infty$ as $t \to \infty$, the parameter $\widetilde{\Omega}$ will tend to infinity, unless $k = 0$. To estimate the value of $\widetilde{\Omega}$ by today, we take the $\Lambda = 0$ Friedmann model,[2] so that (17.27) holds. We will calculate how $\widetilde{\Omega}$ has changed since the instant when the average density in the Universe was equal to that in the atomic nucleus, $\rho_N = 10^{14} \, \text{g cm}^{-3}$; the present density is assumed to be $\rho_0 \approx 10^{-31} \, \text{g cm}^{-3}$. Using (17.27) we obtain $\widetilde{\Omega}_0/\widetilde{\Omega}_N = (\rho_N/\rho_0)^{1/3} = 10^{15}$.

At present $\rho_{cr}/\rho_0 \approx 0.475 \times 10^2$ (see after (17.30)), which means $\widetilde{\Omega}_0 \approx 46.5$. This discrepancy seems large. However, in order that the present value $\widetilde{\Omega}_0$ can be so small, at the epoch of nuclear density it had to equal $\widetilde{\Omega}_N = 10^{-15}\widetilde{\Omega}_0 \approx 4.65 \times 10^{-14}$. Hence, during the first moments of the existence of the Universe the density had to be very close to the

---

[1] We include the cosmological constant for later reference, although the original paper (Guth, 1981) assumed $\Lambda = 0$.

[2] In the original paper (Guth, 1981), the radiation source with $\Lambda = 0$ and $\epsilon = 3p$ was considered, but the conclusion was similar.

critical density, or else the present discrepancy between these two quantities would be huge.[1] Why had the initial conditions of the evolution of the Universe been set up in such a way that it was described by the flat Friedmann Universe with such a high precision?

The 'horizon problem' needs to be explained first qualitatively; see Fig. 17.7. It is conventionally assumed that the emission process of the cosmic microwave background (CMB) took place instantly, at the **last scattering** instant (also called the **recombination** instant), $t_{ls} \approx 3 \times 10^5$ years after the Big Bang (BB) (Lang, 1999, Vol. II). A light cone with its vertex at the BB encompasses a certain volume at $t = t_{ls}$; Fig. 17.7 shows two such cones. Particles that are still outside P's light cone at t_{ls} could not have exchanged any interaction with P because the interaction would have had to propagate faster than light. The past light cone of a contemporary observer, with the vertex at O, encompasses at t_{ls} a much larger volume than does any BB cone. Yet the radiation received by O from points in the $t = t_{ls}$ space has the same temperature, as if the various sources had interacted before, to achieve equilibrium. How is this possible? This is the 'horizon problem'.

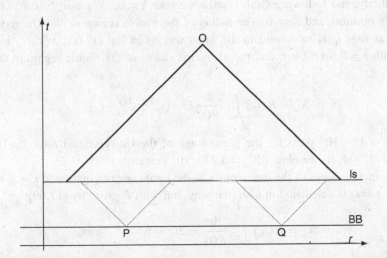

Fig. 17.7. The 'horizon problem' in a Robertson–Walker Universe – an illustrative sketch in comoving coordinates. The line BB is the Big Bang set, the line ls is the *last scattering hypersurface* – the time by which the CMB radiation was emitted. O is the observer today. The past light cone of the observer O encompasses a much larger volume at $t = t_{ls}$ than the volume inside any single light cone emitted at the Big Bang (two of which are shown). Hence, the different regions we see today had had no chance to come into causal contact prior to the emission of the CMB radiation, yet the temperature of the CMB radiation coming from them is nearly the same. Inflationary models avoid this problem by replacing the part of spacetime prior to t_{ls} with a model that expands much faster than any matter-filled R–W model. In effect, inflation makes the light cones at P and Q much wider – encompassing the whole interior of O's past light cone at $t = t_{ls}$.

---

[1] The inflationary models consider still earlier periods, where the fine-tuning is by several orders of magnitude finer. We want to be conservative and do not go too far from laboratory physics.

To formulate the problem in numbers, we have to assume a model of matter for the period prior to t_{ls}. After t_{ls} we assume dust. It is usually assumed that the Universe prior to t_{ls} was **radiation dominated**, i.e. that the energy-density of radiation was larger than the energy-density contained in the massive particles. Since radiation is supposed to obey the Maxwell equations, and the energy-momentum tensor of a Maxwell field is traceless (see (13.10)), it is assumed that a radiation-dominated matter is a perfect fluid with the property $T^\rho{}_\rho = 0$. This implies the equation of state $\epsilon = 3p$, and then the equation of motion (17.26) becomes $\kappa \epsilon R^4 = \epsilon_0 = $ constant. Substituting this in (17.24) with $\Lambda = 0$ we obtain an equation for R that is easily integrated with the result

$$
R = \begin{cases} \sqrt{\epsilon_0/(3k) - k\left(t - t_B - \dfrac{1}{k}\sqrt{\epsilon_0/3}\right)^2} & \text{when} \quad k \neq 0, \\[2mm] \sqrt{2\sqrt{\epsilon_0/3}\,(t - t_B)} & \text{when} \quad k = 0. \end{cases}
\tag{17.69}
$$

For simplicity, and following Guth (1981), we take $k = 0$.[1] We also take $r = 0$ for the observer's position, and $t_B = 0$. The radius of the visible region of the Universe for an observer at time t_0 is calculated in the same way as in Eq. (17.63), with t_{ls} instead of $t = 0$. With $k = 0$, $\sigma(r_1) = r_1$ and the geometric radius of the visible region in the $t = t_{ls}$ space is

$$
\mathcal{R}^0_{ls} = R(t_{ls}) \int_{t_{ls}}^{t_0} \frac{\mathrm{d}t}{R(t)} = 3\left(t_{ls}^{2/3} t_0^{1/3} - t_{ls}\right).
\tag{17.70}
$$

Taking $t_0 = 15 \times 10^9$ years for the present age of the Universe and $t_{ls} = 3 \times 10^5$ years (Lang, 1999, Vol. II) we obtain $\mathcal{R}^0_{ls} = 3.23 \times 10^7$ years.

The geometric radius of the intersection of the light cone originating at $t = t_B = 0$ with the space $t = t_{ls}$ is calculated in a similar way, but with R given by (17.69):

$$
\mathcal{R}^r_{ls} = R(t_{ls}) \int_0^{t_{ls}} \frac{\mathrm{d}t}{R(t)} = 2t_{ls} = 6 \times 10^5 \text{ years.}
\tag{17.71}
$$

The observed region thus has the radius 53.8 times larger, and contains 1.56×10^5 times as many particles.[2]

The solution of these two problems, proposed by Guth (1981), was that the equation of state of matter during the early phase of evolution (between 10^{-34} and 10^{-32} s after the BB) was such that the Universe expanded exponentially ('inflated'), so that the radius of a light cone originating at the BB was, at t_{ls}, much larger than the radius $\mathcal{R}^0_{ls}$ of the visible region of the Universe. In Guth's original paper, the de Sitter metric (8.85) was proposed as the model of the inflation process. This metric can formally be treated as

---

[1] The difference with the cases $k \neq 0$ is smaller than the uncertainties resulting from statistical errors in determining the astrophysical quantities, and from controversies between different schools in interpreting the observations.
[2] Guth (1981) compared the two radii at an earlier time than t_{ls} and obtained a much more impressive ratio of particle numbers, 10^{83}.

a perfect fluid solution of Einstein's equations with the equation of state $\epsilon = -p$.[1] In later papers, the source was assumed to be a scalar field, i.e. an entity $\phi(x)$ whose field equation is $g^{\alpha\beta}\phi_{;\alpha\beta} - \partial V/\partial\phi = 0$, where the potential $V(\phi)$ is chosen so as to give ϕ the desired behaviour. To cause inflation, it is enough that the energy-momentum tensor

$$T_{\alpha\beta} = (\epsilon + p)u_\alpha u_\beta - pg_{\alpha\beta}, \qquad u_\alpha = \phi_{,\alpha}/\sqrt{\phi_{,\rho}\phi^{;\rho}},$$

$$\epsilon = \phi_{,\rho}\phi^{;\rho}/2 - V, \qquad p = \phi_{,\rho}\phi^{;\rho}/2 + V \qquad (17.72)$$

has the property $p < -\epsilon/3$ (Peebles, 1993).

Thus, the two 'problems' were solved within the class of R–W models by postulating a new equation of state. They turned out to be problems of choosing a model rather than fundamental problems of cosmology. In effect, they just implied that the matter of the very early Universe must have obeyed an equation of state different from $p = 0$ and $p = \epsilon/3$.

It is a mystery why the inflationary idea has become such a huge success with cosmologists. It gave rise to hundreds if not thousands of papers, and the subject cannot be avoided by anyone interested in cosmology. This is so in spite of two circumstances:

1. The two 'problems' that the inflationary models were designed to solve had not been perceived as much of a problem before. So much so that Guth (1981) felt compelled to argue at length to convince readers that they were indeed problems.
2. Inflationary models created more problems than they solved.

The 'flatness problem' is completely transformed if we consider the Lemaître–Tolman (L–T) and Szekeres models (see Chapters 18 and 19). In those models, the curvature index is not a constant, but a function of a spatial variable. If we find that it is unusually close to zero, then this means that we live in a privileged position in space. The 'horizon problem' finds a simple solution in the L–T model without recourse to inflation; see Section 18.17. Besides, inflation would have solved this problem if it were able to provide a model in which the Universe is initially inhomogeneous and acquires the uniform temperature of the CMB radiation in the course of its evolution. No such model has been presented. All the inflationary models are homogeneous all the time, in the best case they are anisotropic, of various Bianchi types. A great majority of them has a R–W geometry all the way. So all inflation has done so far in solving the 'horizon problem' is to say: these models expand so fast that they *might* homogenise the CMB radiation!

We will not go into details; overviews of the inflationary scenario may be found in textbooks on cosmology (Peebles, 1993; Padmanabhan, 1993). The main virtue of the inflationary idea is the collection of speculations and hypotheses that it inspired (e.g. concerning the formation of structures in the Universe) rather than solid calculated results.

---

[1] Note, however, that the corresponding energy-momentum tensor $T_{\alpha\beta} = -pg_{\alpha\beta}$ does not define any velocity field. This gave rise to the following question: how are the local directions of flow defined after inflation stops? This was listed by Guth (1981) as one of the problems – *created by the idea of inflation* – that needed to be solved.

Among the problems that the idea of inflation created are these:

1. The so-called 'graceful exit' problem – once inflation sets in (its onset being a problem in itself), it cannot end.
2. The cosmological constant problem – Λ is determined by the value of the scalar field at the end of the inflationary period, and it could be arbitrary. Why is it so close to zero?
3. Inflation overdid the solution of the 'flatness problem' – it implies that k must be very close to zero, i.e. that the density of matter in the Universe must be very near the critical one. Since observations of luminous matter cannot honestly account for more than 20% of the critical density (Coles and Ellis, 1997), inflation had to postulate the existence of 'dark matter' that makes up for the discrepancy. For reasons that we will not discuss here, this should not be any kind of matter known from laboratory, but matter consisting of hypothetical particles considered in high-energy physics (Peebles, 1993; Padmanabhan, 1993). However, such matter has not been observed directly.
4. Currently, it is assumed that the missing matter is made up for by the cosmological constant (see the next section), and the numbers now favoured by astronomers are: 30% of the total density for ordinary (but not necessarily luminous) matter and 70% for the cosmological constant (Perlmutter *et al.*, 1999). This creates another problem: the cosmological constant does not change with time, while the mass density falls with time by the law $\rho \propto R^{-3}$. They cannot be so close to each other at all times, which means that we live in a special epoch. This is a problem very much like the 'flatness problem'.

Finally, let us mention a problem that does not seem to bother cosmologists. Inflation is supposed to have taken place between 10^{-34} and 10^{-32} s after the BB. Note that, both in the dust era ($p = 0$ for $t \geq 3 \times 10^5$ years) and in the radiation era ($p = \epsilon/3$ for $t \leq 3 \times 10^5$ years), the matter density obeys $\rho (t - t_{\rm B})^2 = $ constant. Thus, taking $\rho_0 = 10^{-31}\,{\rm g\,cm}^{-3}$ for the present density, $t_0 = 15 \times 10^9$ years for the present age of the Universe and 1 year $= 3.156 \times 10^7$ s, we conclude that during the inflation period the density must have been greater than $2.24 \times 10^{68}\,{\rm g\,cm}^{-3}$. This is far outside the range of laboratory physics and of any astronomical observations. Thus, if we wish to stick with the tradition that physics is an empirical science, inflation cannot be called a physical theory.

Some more problems created by inflation were discussed by Rothman and Ellis (1987).

17.9 The value of the cosmological constant

Perlmutter and collaborators (1999) used a collection of high-redshift supernovae to find the best-fit cosmological model in the collection that included different values of the present density and of the cosmological constant. The underlying idea was that supernovae of a certain kind, called type Ia, are 'standard candles', i.e. that their absolute luminosity at maximum is the same for all of them. Thus, knowing L and measuring $\mathcal{F}_0$ and z_0

in (16.115), one can calculate the source area distances r_G. (In astronomy, a measure of $\mathcal{F}_O$, is used, called *apparent magnitude*, m, and defined by $m = -2.5\log_{10}\mathcal{F}_O + \text{constant.}$) Then, (16.117) and (17.40) provide a relation between r_G and the (H_0, q_0) parameters. By plotting magnitude against redshift, while knowing the value of H_0, one can determine q_0 and thus discern between different evolution laws. This gives, in principle, the possibility to determine q_0 *if we know all the other parameters with sufficient precision.*

The procedure used by the Perlmutter team requires advanced knowledge of astronomy, and we cannot go into any of its details in this book. The authors took several disturbances into account, such as extragalactic extinction of light from the supernovae and the possibility of gravitational lenses in the line of sight or evolutionary changes in the supernovae-to-be (called *progenitors*). The result they obtained is that $\Lambda = 0$ is strongly inconsistent with the measurements. With the Hubble parameter $H_0 = 63\,\text{km s}^{-1}\text{Mpc}^{-1}$, the best fit implies Λ accounting for more than 50% of the observed density. With $k = 0$, this gives $\Omega_M = 0.28^{+0.05}_{-0.04}$ as the percentage of mass in the form of matter, the remaining 0.72 being the cosmological constant. An additional bonus is the determination of the age of the Universe, $t_0 = (14.5 \pm 1.0) \times 10^9$ years.

This result will undoubtedly be checked by other teams.[1] Meanwhile, we can take it with caution – supernova explosions are rare and distant events. It is rather bold to assume that we know enough about the mechanism of the explosion, and about the propagation of light from it to us, to use the brightness measurements for such intricate calculations. However, the quoted result has been welcomed by the astronomical community, and the now-standard model of the Universe is the Friedmann model with $k = 0$ and $-c^2\Lambda = 0.72\rho_{cr}$.

Meanwhile, Celerier (2000) interpreted the same observations in terms of the Lemaître–Tolman model (see Chapter 18). She found that they can be reproduced by an inhomogeneous mass distribution, with no implications for the value of Λ, and $\Lambda = 0$ not being excluded.

17.10 The 'history of the Universe'

The realisation that the Universe might have a history was coming only gradually to the minds of astronomers and physicists. The Hubble (1929, 1936) discovery of the expansion of the Universe and the realisation that the Friedmann (1922)–Lemaître (1927) solutions of Einstein's equations account for it set this line of thinking in motion. Since the Universe had been denser in the past, it must have been hotter as well. Some time ago then, it should have been sufficiently hot that all atoms were ionised. When the temperature dropped below the ionisation temperature, radiation had to be emitted. Expansion of the Universe should have cooled the radiation. Assuming that it had a black-body spectrum all the

---

[1] The subject of 'dark energy' is relatively new and can be followed only through papers and conference proceedings. An overview of the situation can be found in Klapdor-Kleingrothaus (2001).

time, the evolution of temperature can be calculated. The intensity I of the black-body radiation as a function of frequency ν is given by

$$I(\nu) = \frac{2h\nu^3}{c^2 \{\exp[h\nu/(kT) - 1]\}}, \tag{17.73}$$

where h and k are the Planck and Boltzmann constants, respectively, and T is the temperature of the radiation. The received frequencies of radiation obey (16.20), so $\nu(1 + z) = \nu_e = $ constant along each ray. Consequently, to keep the form of the function $I(\nu)$ in (17.73) unchanged, we must have $T(1 + z) = $ constant. That radiation should still be present in the Universe today. This idea first occurred to Gamow (1948) and collaborators (Alpher and Herman, 1948). The discovery came in 1965: the radiation exists and has the temperature 2.73 K (Dicke *et al.*, 1965; Penzias and Wilson, 1965).

The speculation went on. Atomic nuclei are built of protons and neutrons. Of these, only protons are stable. All the known atomic nuclei could, in principle, be built by adding protons and electrons consecutively one by one to the nucleus of hydrogen (the proton). Still farther in the past, the Universe must have been hot enough to crash all heavier nuclei; only protons, neutrons and loose electrons could survive. Could it have been that matter originally consisted only of protons and electrons, and heavier nuclei came into existence through collisions between these? This idea again occurred to Gamow (1948) and Alpher and Herman (1948). Simulations, with use of computers, indicated that something like this was going on, but only the first few nuclei of the Mendeleev table could be created in this way: about 25% of the mass of the Universe would be converted into helium ^{4}He. Tiny traces of deuterium, helium ^{3}He, lithium, beryllium and boron could be created, but the falling temperature of the Universe would stop any further synthesis (Wagoner, Fowler and Hoyle, 1967). (Heavier elements are synthesised later, in the stars (Fowler, 1967), but this process is not within the domain of cosmology.) These calculated proportions of nuclides were confirmed by observations (Boesgaard and Steigman, 1985).

Thinking along these lines, cosmologists reconstructed the possible sequence of events in the evolution of the Universe. Some of the conclusions have been verified observationally (like the ones mentioned above); some others remain speculations. We will not go into details since they are not within the domain of relativity; they can be found in books on astrophysical cosmology (e.g. Padmanabhan, 1993). Here we give only a short list of the most important events.

The leading motive in theoretical cosmology was this: as we proceed backwards in time, approaching the singularity, the density and temperature of matter become arbitrarily high. Thus, whatever processes we know or can think of that should take place at high temperatures, must have taken place in the early Universe. Why the BB explosion occurred and what preceded it are questions that cannot be answered by means of the currently existing physics or mathematics. Thus, they are usually not asked and we take the BB as a given thing.

The natural and interesting question is: at precisely what times the consecutive stages of evolution took place. The numbers given in different sources are different. The selection of numbers and references given below is random and we make no claim to being precise or up to date.

Assuming that $k = 0$ and $\rho/\rho_{cr} = 0.27$, $-c^2\Lambda/\rho_{cr} = 0.73$, the values currently favoured by astronomers, the BB explosion took place $(14.5 \pm 1.0) \times 10^9$ years ago (Perlmutter *et al.*, 1999).

During the first 10^{-34} s, the Universe had a temperature of about 10^{27} K and was supposedly described by a 'Grand Unified Theory' (GUT) that unites the strong nuclear, weak nuclear and electromagnetic interactions (Lang, 1999, Vol. II). The theory is still in the making. The elementary particles we know today did not necessarily exist then, and matter might have been composed of loose quarks and gluons.

Between 10^{-34} and 10^{-32} s after the BB, inflation took place (see Section 17.8). No set of observations necessarily requires inflation for its explanation, but inflation is widely believed to be a proven theory.[1] At the end of the inflation period, the elementary particles we know today should have come into existence, together with seeds for structure formation. The reasons, mechanism and exact instant of creation of the seeds are still a matter of debate (Lang, 1999, Vol. II). The BB did not have to be spatially homogeneous (see Chapter 18), and the inhomogeneities may have existed in the Universe from the very beginning, but this idea is not popular in the astronomical community.

About 1 s after the BB, neutrinos decoupled and should thereafter propagate freely through space, similarly to the CMB photons. During the next few seconds, protons, neutrons, electrons, positrons and photons existed in thermal equilibrium, at temperatures $T \geq 10^{10}$ K (Misner, Thorne and Wheeler, 1973).

The formation of light atomic nuclei occurred between 2 and 1000 s after the BB (Misner, Thorne and Wheeler, 1973). At the end of this period, the temperature dropped to about 10^9 K.

Later, the Universe continued to be a radiation-dominated plasma, but the radiation mass-density ρ_r (which obeyed $\rho_r R^4 = $ constant, with $R \propto 1/T$, as follows from (17.73) and (17.13)) was decreasing faster than the mass-density of massive particles, ρ_m (obeying $\rho_m R^3 = $ constant). About 3×10^5 years after the BB (Lang, 1999, Vol. II), and at temperatures of the order 10^3 K, ρ_r became smaller than ρ_m and radiation decoupled from matter, having too little energy to ionise the atoms that had already captured their electrons. It evolved into the CMB radiation of temperature 2.73 K that is observed today.

Later, structures like galaxies, galaxy clusters, superclusters and still larger condensations, and voids, must have formed. Even though the process of structure formation should have proceeded by gravitation and should be well within the domain of classical gravitation theory, it is still poorly understood. There exists no quantitative account of the emergence of structures, only a general firm belief that structures came about by gravitational magnification of fluctuations created very early, nobody knows how (see above). The Lemaître–Tolman model offers some interesting possibilities in this respect;

---

[1] Observations confirming the R–W models are sometimes appropriated as confirmations of inflation.

see Sections 18.5 and 18.6. There is even no general agreement on what came earlier: whether stars condensed out of galaxies that had been formed before, or galaxies came into being by pulling pre-existing stars together.

As this very brief account shows, the only input from relativity to this subject was the idea that the Universe began hot and dense, and then cooled as it expanded. Everything else is thermodynamics and particle physics. The fact that observations confirm some of the results does not uniquely point to the R–W models. Even in the R–W class, observations are not precise enough to define the evolution uniquely, so there is certainly enough room to consider more general models. It is often claimed that the high isotropy of the CMB radiation (with maximal temperature anisotropies of $\Delta T/T \approx 2.93 \times 10^{-6}$ at the angular scale of about $0.9°$; see Hu (2004)) *proves* that our Universe has an R–W geometry. Such statements are, however, meaningless without a quantitative account of interaction between the CMB radiation and inhomogeneities in matter distribution. Existing estimates (see Section 18.12) show that the interaction is weak, and no temperature anisotropies larger than 10^{-5} should ever have been expected.

The above remarks were meant to convince the reader that, by considering more general (L–T and Szekeres) cosmological models, one does not in any way deny the confirmed successes of the R–W cosmology. The more general models should fill in the finer details that cannot be captured in the R–W geometries, like the structure formation. The R–W models still remain valid as a rough first approximation to a more detailed description. Very unfortunately though, there is a tendency in the astronomical community to treat any departure from the R–W class of geometries as a blasphemy that requires instant punishment.

17.11 Invariant definitions of the Robertson–Walker models

It is often difficult to recognise a known metric when it is represented in an unfamiliar coordinate system. Metrics of high symmetry can be disguised in particularly elaborate ways; an example is the Schwarzschild solution represented in the Lemaître–Novikov coordinates. The same applies to the R–W metrics, so we shall list here some invariant criteria by which to recognise them. One invariant criterion, that has already been presented in Chapter 10, is by the symmetry group, but it requires solving the Killing equations and is therefore rather laborious.

The following set of properties is a necessary and sufficient condition for a spacetime to be in the R–W class:

1. The metric obeys Einstein's equations with a perfect fluid source.[1]
2. The velocity field of the perfect fluid has zero rotation, shear and acceleration.

---

[1] Some authors considered the R–W metrics with nonperfect fluid and/or shearing/rotating/accelerating contributions to the energy-momentum tensor. Those contributions cancel each other and leave no trace in the metric tensor – which means that, from the point of view of the 4-dimensional geometry, they are undetectable. Hence, considering them is in contradiction to the Ockham razor principle.

The necessity of these properties follows by calculation. The proof of their sufficiency follows from (15.43)–(15.51). Those equations imply that a perfect fluid solution with $\sigma = \omega = 0 = \dot{u}^{\alpha}$ must be conformally flat. All conformally flat perfect fluid solutions were found by Stephani (1967a) (Stephani *et al.*, 2003, Theorem 37.17, p. 601). They have $\sigma = \omega = 0$, but in general $\dot{u}^{\alpha} \neq 0$. Specialising them to the case $\dot{u}^{\alpha} = 0$ we obtain the R–W models.

In this text, we are concerned only with expanding and nonempty R–W models for which $\theta \neq 0 \neq \epsilon + p$, and these inequalities are always tacitly assumed. The R–W model with $\theta = 0$ is the Einstein static Universe; the ones with $\epsilon + p = 0$ are the de Sitter vacuum Universes (the Minkowski spacetime is included among them as the limit $\Lambda = 0$); neither is appropriate for describing the observed Universe.

Condition 1, that is, the perfect fluid source, is essential. Examples are known (Krasiński, 1997) of solutions for which $\omega = \sigma = 0 = \dot{u}^{\alpha}$, but which are not R–W because the source is not a perfect fluid.

There is another invariant definition of the R–W spacetimes, which makes no use of the field equations. The following set of properties is a necessary and sufficient condition for a spacetime to be R–W:

3. The spacetime admits a foliation into spacelike hypersurfaces of constant curvature.
4. The congruence of lines orthogonal to the leaves of the foliation consists of shearfree geodesics.
5. The expansion scalar of the geodesic congruence has its gradient tangent to the geodesics.

17.12 Different representations of the R–W metrics

Several representations of the R–W metrics are met in the literature. The forms that appear most often are (17.1), (17.3), (17.5) and (17.7). Another frequently used representation is:

$$\mathrm{d}s^2 = \mathrm{d}t^2 - R^2(t)\left[\mathrm{d}r^2 + f^2(r)\left(\mathrm{d}\vartheta^2 + \sin^2\vartheta\,\mathrm{d}\varphi^2\right)\right], \tag{17.74}$$

where

$$f(r) = \begin{cases} \sin r & \text{for } k > 0, \\ r & \text{for } k = 0, \\ \sinh r & \text{for } k < 0. \end{cases} \tag{17.75}$$

The three cases of (17.75) can be written in one formula as

$$\mathrm{d}s^2 = \mathrm{d}t^2 - R^2(t)\left(\mathrm{d}r^2 + \frac{1}{k}\sin^2(\sqrt{k}r)\left(\mathrm{d}\vartheta^2 + \sin^2\vartheta\,\mathrm{d}\varphi^2\right)\right). \tag{17.76}$$

The range of r is finite or infinite, depending on the sign of k and on the coordinates used, but this is easy to recognise in each case. Note that in the case $k > 0$ the coordinates of (17.3) cover only half of the 3-sphere $t = \text{constant}$ ($0 \leq r < 1/k^{1/2}$) and are unsuitable for considering the geometry in the vicinity of the equator $r = k^{-1/2}$.

Rather unfamiliar is the form of the R–W metric that results from solutions with two commuting Killing vector fields. The easiest way to find it is to demand that the metric be independent of y and z while the spaces $t = $ constant have constant curvature. Then

$$ds^2 = dt^2 - R^2(t)\left[dx^2 + f_{,x}{}^2 dy^2 + f^2(x)dz^2\right], \tag{17.77}$$

where

$$f(x) = \begin{cases} \sin(\sqrt{k}x) & \text{for } k > 0, \\ x & \text{for } k = 0, \\ \sinh(\sqrt{-k}x) & \text{for } k < 0. \end{cases} \tag{17.78}$$

Yet another form of the R–W metrics follows as a limit of plane symmetric solutions. Then

$$ds^2 = dt^2 - R^2(t)\left[dx^2 + e^{2Cx}\left(dy^2 + dz^2\right)\right], \tag{17.79}$$

where C is a constant. When $C = 0$, this is evidently the flat R–W metric; when $C \neq 0$, it is the $k < 0$ R–W metric. The $k > 0$ R–W metric is incompatible with plane symmetry.

The R–W models result in quite unfamiliar form from the Goode–Wainwright (G–W) representation of the Szekeres models, which are discussed in Section 19.8. One of the G–W forms is:

$$ds^2 = dt^2 - S^2\left[W^2 f^2 \nu_{,z}{}^2 dz^2 + e^{2\nu}\left(dx^2 + dy^2\right)\right], \tag{17.80}$$

where

$$W^2 = \frac{1}{\varepsilon - kf^2}, \tag{17.81}$$

$$e^{\nu} = \frac{f(z)}{a(z)(x^2 + y^2) + 2b(z)x + 2c(z)y + d(z)},$$

ε and k are arbitrary constants, and $S(t), f(z), a(z), b(z), c(z)$ and $d(z)$ are arbitrary functions subject to

$$ad - b^2 - c^2 = \varepsilon/4. \tag{17.82}$$

The slices $t = $ constant of the metric (17.80)–(17.82) are spaces of constant curvature equal to k/S^2. The t-coordinate lines are shearfree geodesics with the expansion scalar depending only on t. This is a characteristic property of the R–W spacetimes (see points 3, 4 and 5 in the previous section).

The other G–W form is:

$$ds^2 = dt^2 - S^2\left[A^2 dz^2 + e^{2\nu}\left(dx^2 + dy^2\right)\right], \tag{17.83}$$

where

$$e^{\nu} = \frac{1}{1 + \frac{1}{4}k(x^2 + y^2)},$$

$$A = e^{\nu}\left\{a(z)\left[1 - \frac{1}{4}k(x^2 + y^2)\right] + b(z)x + c(z)y\right\}, \tag{17.84}$$

$S(t)$, $a(z)$, $b(z)$ and $c(z)$ are arbitrary functions and k is an arbitrary constant. This metric has exactly the same properties as (17.80)–(17.82).

17.13 Exercises

1. Verify that the Einstein tensor for the R–W metric is diagonal in all the coordinate representations (17.1), (17.3), (17.5) and (17.7).
2. Prove that a null geodesic sent off radially from any point in an R–W spacetime, i.e. with $\dot{\vartheta}_0 = \dot{\varphi}_0 = 0$ at the initial point, will remain radial along its whole length.
3. Let $\rho(t, r)$ be any spherically symmetric (possibly inhomogeneous) finite distribution of matter (i.e. $\rho(t, r) = 0$ for $r > r_0$, $r_0 < \infty$). Let A be a point inside this distribution located at $r = r_1$. Show that the total (Newtonian) gravitational force exerted on the point A by matter outside the sphere $r = r_1$ is zero.
4. Verify that the various forms of the R–W metrics given in Section 17.11 do fulfil the invariant criteria.

   Note. It is not advisable to try to prove the equivalence of the G–W forms to R–W by explicit coordinate transformations.

18

Relativistic cosmology III: the Lemaître–Tolman geometry

18.1 The comoving–synchronous coordinates

The curvature coordinates introduced in Section 14.1 were convenient for investigating the Schwarzschild solution. However, for other purposes other coordinates might be more useful, the **comoving coordinates** are among them. They can be introduced whenever there exists a timelike vector field u^α in the spacetime, for example, the 4-velocity of matter. By definition, they are coordinates in which the (contravariant) vector field u^α has only the time coordinate, thus $u^{\alpha'} \propto \delta^{\alpha'}{}_0$. They exist always – one has to solve the set of three equations $u^\alpha x^{\alpha'}{}_{,\alpha} = 0$ for $\alpha' = 1, 2, 3$. Then, the transformations preserving the comoving coordinates are $x^I = f^I(x^{J'})$, $x^0 = f^{0'}(x^{J'})x^{0'} + g(x^{J'})$.

If the vector field u^α has zero rotation, then the comoving coordinates can be chosen so that, in addition, they are **synchronous**, that is, in these coordinates the metric tensor has no time–space components. Here is the proof that vanishing rotation is a necessary condition.

Suppose that comoving–synchronous coordinates $\{x^{\alpha'}\}$ exist. Then $u_{\alpha'} = \lambda \delta^0{}_{\alpha'}$ and, on changing from $\{x^{\alpha'}\}$ to any other coordinates $\{x^\alpha\}$, we obtain $u_\alpha = f_{,\alpha}\lambda$, where $f = x^{0'}$. Because of $u_\alpha u^\alpha = 1$ we have $f_{,\rho} u^\rho = 1/\lambda$, and then $\dot{u}_\alpha = f_{,\alpha}\lambda_{,\rho}u^\rho + \lambda f_{,\alpha\rho}u^\rho - \frac{1}{2}u^\mu u^\rho g_{\rho\mu,\alpha}$. But $u^\mu u^\rho g_{\rho\mu,\alpha} \equiv 2u^\rho u_{\rho,\alpha} = 2u^\rho (\lambda f_{,\rho\alpha} + \lambda_{,\alpha}f_{,\rho})$. From all this, we find $\omega_{\alpha\beta} = 0$. $\square$

The proof that $\omega_{\alpha\beta} = 0$ is also a sufficient condition for the existence of synchronous coordinates is somewhat complicated. If $\omega_{\alpha\beta} = 0$, then also $\sqrt{-g}w^\alpha = \epsilon_{\alpha\beta\gamma\delta}u_\beta u_{\gamma,\delta} = 0$. This can be written as $u \wedge \mathrm{d}u = 0$, where $u \overset{\text{def}}{=} u_\alpha \mathrm{d}x^\alpha$. This means that there exists a 1-form τ such that $\mathrm{d}u = \tau \wedge u$. Then there exist such functions f and λ that $u = \lambda \mathrm{d}f$, i.e. $u_\alpha = \lambda f_{,\alpha}$ (Flanders, 1963, Section 7.2). Choosing $x^{0'} = f$ as the time coordinate, we see that it is synchronous.

The comoving–synchronous coordinates will be used throughout the remaining part of this chapter.

18.2 The spherically symmetric inhomogeneous models

In a spherically symmetric spacetime in which the source in Einstein's equations is a perfect fluid, rotation is necessarily zero (see Exercise 1). Hence, the comoving–synchronous coordinates can be introduced, in which, in consequence of (8.51), the metric has the form[1]

$$ds^2 = e^{C(t,r)} dt^2 - e^{A(t,r)} dr^2 - R^2(t,r) \left(d\vartheta^2 + \sin^2\vartheta \, d\varphi^2 \right), \qquad (18.1)$$

and the velocity field is

$$u^\alpha = e^{-C/2} \delta^\alpha{}_0. \qquad (18.2)$$

Note that R is connected with S, the area of the surface $\{t = \text{constant}, r = \text{constant}\}$, by the Euclidean relation $S = 4\pi R^2$. Hence, R is called the **areal radius**. (N.B. this is the source area distance, defined in (16.114), between an observer at an arbitrary location and the centre at $R = 0$.)

The Einstein equations for this metric (in coordinate components), with the cosmological constant taken into account, are

$$G^0{}_0 = e^{-C} \left(\frac{R_{,t}^2}{R^2} + \frac{A_{,t} R_{,t}}{R} \right) - e^{-A} \left(2\frac{R_{,rr}}{R} + \frac{R_{,r}^2}{R^2} - \frac{A_{,r} R_{,r}}{R} \right) + \frac{1}{R^2}$$

$$= \kappa\epsilon - \Lambda, \qquad (18.3)$$

$$G^1{}_0 = e^{-A} \left(2\frac{R_{,tr}}{R} - \frac{A_{,t} R_{,r}}{R} - \frac{R_{,t} C_{,r}}{R} \right) = 0, \qquad (18.4)$$

$$G^1{}_1 = e^{-C} \left(2\frac{R_{,tt}}{R} + \frac{R_{,t}^2}{R^2} - \frac{C_{,t} R_{,t}}{R} \right) - e^{-A} \left(\frac{R_{,r}^2}{R^2} + \frac{C_{,r} R_{,r}}{R} \right) + \frac{1}{R^2}$$

$$= -\kappa p - \Lambda, \qquad (18.5)$$

$$G^2{}_2 = G^3{}_3$$

$$= \frac{1}{4} e^{-C} \left(4\frac{R_{,tt}}{R} - 2\frac{C_{,t} R_{,t}}{R} + 2\frac{A_{,t} R_{,t}}{R} + 2A_{,tt} + A_{,t}^2 - C_{,t} A_{,t} \right)$$

$$- \frac{1}{4} e^{-A} \left(4\frac{R_{,rr}}{R} + 2\frac{C_{,r} R_{,r}}{R} - 2\frac{A_{,r} R_{,r}}{R} + 2C_{,rr} + C_{,r}^2 - C_{,r} A_{,r} \right)$$

$$= -\kappa p - \Lambda. \qquad (18.6)$$

Equation (18.3), multiplied by $R^2 R_{,r}$, may be rewritten as

$$\left(R + e^{-C} R R_{,t}^2 - e^{-A} R R_{,r}^2 + \frac{1}{3}\Lambda R^3 \right)_{,r} - R \left(e^{-C} R_{,t}^2 \right)_{,r} + e^{-C} A_{,t} R R_{,t} R_{,r}$$

$$= \kappa\epsilon R^2 R_{,r}. \qquad (18.7)$$

But, in consequence of (18.4), the second and third terms on the left-hand side sum up to zero, so in fact (18.7) is

$$\left(R + e^{-C} R R_{,t}^2 - e^{-A} R R_{,r}^2 + \frac{1}{3}\Lambda R^3 \right)_{,r} = \kappa\epsilon R^2 R_{,r}. \qquad (18.8)$$

---

[1] Several different notations are used in the literature for the components of this metric. We have adopted here a notation that will, we hope, avoid clashes.

Multiplying it by $R^2 R,_t$, we transform Eq. (18.5) to the equivalent form

$$\left(R + e^{-C}RR,_t^2 - e^{-A}RR,_r^2 + \frac{1}{3}\Lambda R^3\right),_t = -\kappa pR^2 R,_t. \tag{18.9}$$

From (18.8) we can now recognise that the quantity

$$m \overset{\text{def}}{=} \frac{c^2}{2G}\left(R + e^{-C}RR,_t^2 - e^{-A}RR,_r^2 + \frac{1}{3}\Lambda R^3\right) \tag{18.10}$$

has all the properties we expect from mass: assuming $R = 0$ at $r = r_0$ and integrating (18.8) from r_0 to a current r in the hypersurface of constant t, we obtain $m(r) = \int_{r_0}^{r} 4\pi\epsilon c^2 R^2 R,_r \, dr' \equiv \int_0^{R(r)} 4\pi\epsilon c^2 u^2 \, du$. Then, Eq. (18.9) can be read as the energy-conservation equation: the time-derivative of mass equals volume-work.

In the derivation of (18.8) and (18.9) we assumed $R,_r \neq 0 \neq R,_t$. The case $R,_t = 0$ is not interesting for cosmology: it requires, from (18.4), that either $R,_r = 0$, which leads to the Nariai (1950) solution, or $A,_t = 0$, which, by virtue of (18.3), has constant density. But with $R,_r = 0$, Eqs. (18.3)–(18.6) do admit an interesting solution; see Section 19.4.

At the surface of a spherical star, where $p = 0$, Eq. (18.9) reduces to $m,_t = 0$, i.e. the total mass of a star immersed in vacuum is constant.

The definition of mass given above (in the case $\Lambda = 0$) was first introduced by Lemaître (1933a), and then re-derived by Podurets (1964), but in the literature it is most often credited to Misner and Sharp (1964).

18.3 The Lemaître–Tolman model

In order to solve (18.3)–(18.6), we have to assume an equation of state. A relation of the type $\epsilon = f(p)$ does not seem to be the right one in inhomogeneous models since it implies the entropy per particle to be a universal constant (see the end of Section 15.5). Lacking any other workable idea, the most natural equation of state is $p = 0$, i.e. evolution by gravitation only. We shall assume this now.

From (12.78), with $p = 0$ the fluid will move along timelike geodesics. Acceleration being equal to zero then implies $C,_r = 0$ in (18.1) (see Exercise 2), and the transformation $t' = \int e^{C/2} \, dt$ leads to $C = 0$ in the new coordinates. Then (18.4) becomes

$$\left(e^{-A/2}R,_r\right),_t = 0. \tag{18.11}$$

The case $R,_r = 0$ has already been set aside for separate consideration; see Section 19.4. With $R,_r \neq 0$, Eq. (18.11) is integrated with the result

$$e^A = \frac{R,_r^2}{1 + 2E(r)}, \tag{18.12}$$

where $E(r)$ is an arbitrary function; for the metric (18.1) to have the right signature it is necessary that E obeys $E \geq -1/2$ for all r.[1]

---

[1] The value $E = -1/2$ is admissible provided that $R,_r = 0$ at the same location. This is a *neck* or *wormhole* – see Section 18.10 and Hellaby and Lake (1985).

Using $C = 0$ and (18.12) to eliminate $R_{,r}{}^2$ from (18.5), we obtain

$$2\frac{R_{,tt}}{R} + \frac{R_{,t}{}^2}{R^2} - \frac{2E}{R^2} + \Lambda = 0. \tag{18.13}$$

Since we assumed that $R_{,t} \neq 0$, we multiply (18.13) by $R^2 R_{,t}$. The resulting equation has $(\partial/\partial t)\left(RR_{,t}{}^2 - 2ER + \frac{1}{3}\Lambda R^3\right) = 0$, and the integral

$$R_{,t}{}^2 = 2E(r) + \frac{2M(r)}{R} - \frac{1}{3}\Lambda R^2, \tag{18.14}$$

$M(r)$ being one more arbitrary function. With $\Lambda = 0$, this equation is formally identical to the Newtonian equation of radial motion in a Coulomb potential. In this Newtonian analogy, $c^2 M(r)/G$ plays the role of the active gravitational mass within an $r = $ constant shell, and $c^2 E(r)/G$ plays the role of the total energy within the same shell.

The solution of (18.14) will contain one more arbitrary function, $(t_B(r))$, that will appear in the combination $(t - t_B(r))$. It is called the **bang-time function**, and in the case $\Lambda = 0$ we will see below that it does indeed define the time coordinate of the Big Bang singularity, which is in general position-dependent. With an arbitrary value of Λ, the Big Bang will not always occur, just as in the Friedmann models.

With $C = 0$ and Eqs. (18.11) and (18.13) fulfilled, Eq. (18.6) becomes an identity, while (18.3) provides the definition of the mass density:

$$\frac{8\pi G}{c^4}\epsilon = \frac{2M_{,r}}{R^2 R_{,r}}. \tag{18.15}$$

The mass density ϵ/c^2 becomes infinite where $R = 0 \neq M_{,r}$ and where $R_{,r} = 0 \neq M_{,r}$. The first of these is the Big Bang, which occurs necessarily whenever $\Lambda = 0$. The second is a shell crossing singularity, where the mass density goes to infinity and changes sign to become negative. At those points, the radial geodesic distance between the point $\{t_0, r_0, \vartheta_0, \varphi_0\}$ and the point $\{t_0, r_0 + dr, \vartheta_0, \varphi_0\}$, equal to $\sqrt{|g_{rr}|}\,dr$, becomes zero, which means that shells of different values of the r-coordinate coincide. This singularity can be avoided with an appropriate choice of $M(r)$, $E(r)$ and $t_B(r)$; see Section 18.10. In the Friedmann limit, defined below, the shell crossing coincides with the Big Bang.

The final solution of Einstein's equations is

$$ds^2 = dt^2 - \frac{R_{,r}{}^2}{1 + 2E(r)}dr^2 - R^2(t, r)\left(d\vartheta^2 + \sin^2\vartheta\,d\varphi^2\right), \tag{18.16}$$

where $R(t, r)$ is determined by (18.14). It was first found and interpreted by Lemaître (1933a). Then, more of its properties were discussed by Tolman (1934) and by Bondi (1947). It will be called here the **Lemaître–Tolman (L–T) model**.[1] Many papers have

---

[1] Even though Tolman made it clear that he was discussing Lemaître's solution, and Bondi cited Tolman, for some reason, over many years, this solution had been known in the literature as the 'Tolman' or 'Tolman–Bondi' model. It is called here 'Lemaître–Tolman' only to avoid confusion with the Friedmann–Lemaître models, otherwise there would be no reason to add anybody's name to Lemaître's.

been published in which various properties of the L–T model were discussed. The book by Krasiński (1997) contains an overview of these, complete until 1994.

The mass $c^2 M/G$ in (18.14) and (18.15) is the **active gravitational mass** that generates the gravitational field. It is not equal to $c^2 N/G$ – the sum of masses of particles that formed the gravitating body (compare Eqs. (14.124)–(14.126)). Suppose that matter fills a sphere V with the centre at the centre of symmetry of the space, $r = r_0$, and the surface at $r = r_S$. Then $c^2 N/G$ is, from (18.16),

$$c^2 N/G = \int_V \left(\epsilon/c^2\right) \sqrt{-g}\, d_3 V \equiv 4\pi \int_{r_0}^{r_S} \frac{\left(\epsilon/c^2\right) R^2 R_{,r}}{\sqrt{1+2E(r)}}\, dr, \tag{18.17}$$

while the active gravitational mass M is, from (18.15):

$$c^2 M/G = 4\pi \int_{r_0}^{r_S} \left(\epsilon/c^2\right) R^2 R_{,r}\, dr. \tag{18.18}$$

Depending on the sign of E, M may be larger than N (when $E > 0$), smaller than N (when $E < 0$) or equal to it (when $E = 0$). If $M < N$, then $(N - M)$ is called the **relativistic mass defect** – the general-relativistic analogue of the mass defect known from nuclear and elementary particle physics. In a bound system $(E < 0)$, part of the energy contained in the component particles had to be shed, and this lost energy is responsible for the mass defect. In the opposite case $(E > 0)$, the system is unbound and its excess energy sums up with the energy equivalent to the sum of masses of components. When $E = 0$, the system is 'marginally bound' – no energy has been shed to form it, and there is no excess energy. This interpretation of E was first given by Bondi (1947).

With $\Lambda \neq 0$, the explicit solutions of (18.14) involve elliptic functions. They were discussed by Lemaître (1933a) and Omer (1965). When $\Lambda = 0$, they are as follows:

When $E(r) < 0$:

$$R(t, r) = -\frac{M}{2E}(1 - \cos \eta),$$
$$\eta - \sin \eta = \frac{(-2E)^{3/2}}{M}\, (t - t_B(r)). \tag{18.19}$$

When $E(r) = 0$:

$$R(t, r) = \left[\frac{9}{2} M(r)\, (t - t_B(r))^2\right]^{1/3}. \tag{18.20}$$

When $E(r) > 0$:

$$R(t, r) = \frac{M}{2E}(\cosh \eta - 1),$$
$$\sinh \eta - \eta = \frac{(2E)^{3/2}}{M}\, (t - t_B(r)). \tag{18.21}$$

Actually, the conditions for the occurrence of each case are $E/M^{2/3} <, = $ or > 0, since the conditions of regularity of the spacetime at the centre of symmetry require that $E = 0$ at $M = 0$; see Section 18.4.

The solutions (18.19)–(18.21) have the same algebraic form as the Friedmann solutions (17.33)–(17.35). Indeed, the Friedmann solutions (17.33)–(17.35) follow from (18.19)–(18.21) when

$$M = M_0 r^3, \qquad E = -\frac{1}{2}kr^2, \qquad t_{\mathrm{B}} = \text{constant}. \tag{18.22}$$

This limit is coordinate-dependent. We will show that the invariant condition is $\epsilon,_r = 0$. In each interval of r in which $M,_r$ does not change sign, $M(r)$ can be used as the independent variable instead of r, since Eqs. (18.15) and (18.16) are covariant under the coordinate transformations $r = f(r')$. Using M as the radial coordinate, we rewrite (18.15):

$$\kappa\epsilon = \frac{6}{(R^3),_M} \iff R^3 - R^3(M_0) = \int_{M_0}^{M} \frac{6}{\kappa\epsilon(\widetilde{M})} \, \mathrm{d}\widetilde{M}. \tag{18.23}$$

Then, the condition of spatial homogeneity is $\epsilon,_M = 0$, i.e. $(R^3),_{MM} = 0$. Working it out in detail, one finds that this is equivalent to

$$E/M^{2/3} = \text{constant}, \qquad t_{\mathrm{B}} = \text{constant}, \tag{18.24}$$

and these two equations define the Friedmann limit invariantly (see Exercise 3). The functions $(E/M^{2/3}),_M$ and $t_{\mathrm{B},M}$ generate, respectively, the increasing and the decreasing perturbation of the background Friedmann model (see Silk (1977) and Section 18.19).

The function $E(r)$ has one more interpretation. In the subspace $t = $ constant of (18.16), R depends only on r, so it can be used as the radial coordinate. Taking the orthonormal tetrad defined by the 3-metric of that subspace $(e^1 = \mathrm{d}R/\sqrt{1+2E}, e^2 = R \, \mathrm{d}\vartheta, e^3 = R \sin \vartheta \, \mathrm{d}\varphi)$, we find the tetrad components of the 3-dimensional Riemann tensor to be

$$R_{1212} = -\frac{E,_R}{R} = R_{1313}, \qquad R_{2323} = -\frac{2E}{R^2}. \tag{18.25}$$

Thus, with $E = 0$, every space $t = $ constant is flat. If $E/R^2 = $ constant, then the curvature is constant and its sign is the sign of $-E$. Consequently, $(-E)$ is a measure of the curvature of the subspaces $t = $ constant. Unlike in the Friedmann models, that curvature is *local* – it depends on r, and in particular it may be positive in one neighbourhood of the space, but negative elsewhere. This shows that the distinction between the R–W models of different values of k is a peculiarity of the R–W class, and not a property of the physical Universe – the same spacetime can be approximately like the $k > 0$ Friedmann model in one neighbourhood and like the $k < 0$ Friedmann model in another. Thus, Eqs. (18.19)–(18.21) do not necessarily describe different cosmological models – they can hold in different regions of the same spacetime.

Equation (18.15) implies that, with $M,_r = 0$, the L–T model becomes vacuum. Being spherically symmetric, when $\Lambda = 0 = M,_r$, it must coincide with the Schwarzschild solution or its extension through the event horizon – and indeed it does, as shown in Section 14.12 (Eqs. (14.116)–(14.117) and Exercise 10 in Chapter 14). See there for a discussion.

Other coordinate representations of the L–T model are rarely used. Gautreau (1984) used a hybrid of the comoving and curvature coordinates; see Section 18.7. The proper curvature coordinates are rather useless for this model (see Exercise 4) – in them, the simple statements $M_{,t} = E_{,t} = 0$ become complicated differential equations.

Another example: Stoeger, Ellis and Nel (1992) tested their *observational cosmology programme* on the L–T model. The basis of the programme is coordinates connected with the observer's past null cone. Transforming the L–T solution to such coordinates requires the integration of the equations of null geodesics, which proved to be an impossible task. Therefore, the authors re-derived the model from the Einstein equations using the observational coordinates from the beginning. The L–T metric in these coordinates is (in the original notation):

$$ds^2 = A^2(\eta)\left(-dw^2 + 2\,dw\,dy\right) + C^2(w, y)\left(d\vartheta^2 + \sin^2\vartheta\,d\varphi^2\right), \tag{18.26}$$

where $\eta \overset{\text{def}}{=} w - y$, $A(\eta)$ is a function related to the observed redshift $z(\eta)$ by $A(\eta) = A_0/[1 + z(\eta)]$, $A_0 = $ constant and C is determined by

$$\frac{\partial C}{\partial y} = A(\eta)W(y), \qquad \frac{\partial C}{\partial \eta} = A\sqrt{\frac{2\omega_0(y)}{C} + W^2(y) - 1}, \tag{18.27}$$

$W(y)$ and $\omega_0(y)$ being arbitrary functions; W^2 is the analogue of $(1 + 2E)$. The matter-density is $\kappa\epsilon = \epsilon_0(y)/\left(AC^2\right)$, and $\epsilon_0(y)$ is one more arbitrary function. The authors explained how the functions are related to observable quantities, but the connection is not simple, and readers are advised to refer to the original paper. The function $z(y)|_{w=\text{const}}$ can in principle be read out from observations, but in practice the procedure would require cosmological parallax distances to be measured and this is unrealistic with the present-day technology. For comparison, the R–W metric in the observational coordinates is

$$ds^2 = R^2(\eta)\left[dw^2 - 2\,dw\,dy - k^{-2}\sin^2(ky)\left(d\vartheta^2 + \sin^2\vartheta\,d\varphi^2\right)\right]. \tag{18.28}$$

18.4 Conditions of regularity at the centre

We found that the mass density ϵ/c^2 becomes infinite at those locations where $R = 0 \neq M_{,r}$. Thus, not the whole set $R = 0$ is a singularity. Part of this set is the centre of symmetry, and we will now formulate the conditions that the arbitrary functions in the L–T model have to obey in order that the centre of symmetry remains nonsingular.[1]

Let $r = r_c$ be the radial coordinate of the centre of symmetry, where $R(t, r_c) = 0$ for all $t > t_B(r_c)$, and let $\epsilon(t, r_c)/c^2 = \rho(t, r_c) \overset{\text{def}}{=} \alpha(t) < \infty$. Assuming that $\rho(t, r)$ is not only finite at $r = r_c$, but also continuous in a neighbourhood of $r = r_c$, $\alpha(t) < \infty$, implies that

$$M(r_c) = 0. \tag{18.29}$$

---

[1] One could consider models with a permanently existing singularity at the centre of symmetry, but they do not seem to be relevant for astrophysics.

Applying the de l'Hôpital rule and using (18.23), we then find

$$\lim_{r \to r_c} \frac{R}{M^{1/3}} = \left(\lim_{r \to r_c} \frac{R^3}{M} \right)^{1/3} = \left(\lim_{r \to r_c} (R^3)_{,M} \right)^{1/3} = \left(\frac{6}{\kappa \epsilon(t, r_c)} \right)^{1/3}. \tag{18.30}$$

Thus R must behave in the neighbourhood of $r = r_c$ as

$$R = \beta(t) M^{1/3} + O_{1/3}(M), \qquad \beta = \left(\frac{6}{\kappa \alpha} \right)^{1/3}, \tag{18.31}$$

where the symbols $O_a(M)$ will denote quantities with the property

$$\lim_{M \to 0} \frac{O_a(M)}{M^a} = 0. \tag{18.32}$$

Thus, if $\rho(t, r_c) = 0$, then $\lim_{r \to r_c}(R/M^{1/3}) = \infty$.

Note (from (18.20)) that (18.31) is always fulfilled when $E = 0$. For the other two models, from (18.19) and (18.21), Eq. (18.31) implies that

$$\beta(t) + O_0(M) = \begin{cases} -\frac{M^{2/3}}{2E}(1 - \cos \eta) & \text{for } E < 0, \\ \frac{M^{2/3}}{2E}(\cosh \eta - 1) & \text{for } E > 0. \end{cases} \tag{18.33}$$

Thus, in order to allow for a well-defined value of η at $r = r_c$ (which is necessary for the density to have a well-defined value at the centre at all times), E must behave in the neighbourhood of the centre as

$$E = -CM^{2/3} + O_{2/3}(M), \qquad C\beta = 1 - \cos \eta(t, r_c). \tag{18.34}$$

Finally, from (18.20) and (18.21), we see that $t_B(r_c)$ must have a finite value, which can be formally written as

$$t_B = \tau + O_0(M). \tag{18.35}$$

Equations (18.29)–(18.35) will be assumed from now on.

18.5 Formation of voids in the Universe

Voids are large (approximately 60 Mpc in radius) volumes in the intergalactic space with a very low matter-density. Their observational discovery at the end of the 1970s (Gregory and Thompson, 1978) was a surprise because it contradicted the universal belief in the cosmological principle. In fact, the first papers indicating that voids should be ubiquitous were published in the early 1930s, but had not been understood properly.

The first such indication was given by Tolman (1934) and Sen (1934). Tolman's main result was the proof that the Einstein and Friedmann models are unstable against the growth of inhomogeneities. This is derived as follows. Let the initial conditions at $t = t_1$ be chosen so that $R = R_{LT}(t_1, r)$ in the L–T model is the same function as $rR = rR_F(t_1)$ in the Friedmann model and $R_{LT,r}(t_1, r) = rR_{F,r}(t_1)$. This does not uniquely determine the evolution of the L–T model, since the radial coordinate can be chosen arbitrarily.

(In fact, the condition $R_{LT}(t_1, r) = rR_F(t_1)$ defines the relation between the coordinate systems.) Hence, $R_{,tt}$ can still be chosen at will, so we can assume that the densities at t_1 are different. From the equations assumed it follows then that $(R_{,tr}/R_{,r})(t_1)|_{LT} = (R_{,t}/R)(t_1)|_F$. Since $R_{,t}$ is a measure of the velocity of expansion, the above implies a perturbation of the initial density in the Friedmann model, with unperturbed initial velocity.

From (18.15) we find

$$\left[\frac{\partial^2}{\partial t^2} \ln \epsilon\right]_{LT}(t_1) = \left[-2\frac{R_{,tt}}{R} + 2\frac{R_{,t}^2}{R^2} - \frac{R_{,ttr}}{R_{,r}} + \frac{R_{,tr}^2}{R_{,r}^2}\right]_{LT}(t_1). \tag{18.36}$$

Using (18.14) to find $R_{,tt}$ and $R_{,ttr}$ and then using (18.15), we simplify (18.36) to

$$\left[\frac{\partial^2}{\partial t^2} \ln \epsilon\right]_{LT}(t_1) = \left[\frac{1}{2}\kappa\epsilon + \Lambda + 2\frac{R_{,t}^2}{R^2} + \frac{R_{,tr}^2}{R_{,r}^2}\right]_{LT}(t_1). \tag{18.37}$$

A similar equation holds for the Friedmann model:

$$\left[\frac{\partial^2}{\partial t^2} \ln \epsilon\right]_F(t_1) = \left[\frac{1}{2}\kappa\epsilon + \Lambda + 3\frac{R_{,t}^2}{R^2}\right]_F(t_1). \tag{18.38}$$

Subtracting (18.38) from (18.37) and using the assumptions, we find:

$$\frac{\partial^2}{\partial t^2}(\ln \epsilon_{LT} - \ln \epsilon_F) = \frac{1}{2}\kappa(\epsilon_{LT} - \epsilon_F). \tag{18.39}$$

Thus, wherever the density of the L–T model is greater or smaller than the density of the corresponding Friedmann model, the difference will be increasing in time. This means that an L–T model with initial condensations *or voids* will be evolving away from the background Friedmann model. This was as close as possible at that time to predicting that the Friedmann models are unstable against the formation of condensations and voids. This is how Tolman himself formulated the prediction:

'... at those values of r where the density in the distorted model is different from that in the Friedmann model, there is at least an initial tendency for the differences to be emphasised ... in cases where condensation is taking place ... the discrepancies will continue until we reach a singular state involving infinite density or reach a breakdown in the simplified equations.'

Sen carried out a complementary study of stability of the Einstein and Friedmann models with respect to the L–T perturbation – he assumed the initial density to be unperturbed, with the velocity distribution being non-Friedmannian at the initial time. By a similar method to Tolman's, he concluded explicitly that 'the models are unstable for initial rarefaction'.

The book by Krasiński (1997) contains a complete overview of studies of void formation done on the basis of the L–T model until 1994. They were mostly motivated by astronomical observations and used various shapes of the arbitrary functions to investigate observable effects. The most complete study (Maeda, Sasaki and Sato, 1983; Maeda and Sato, 1983; Sato and Maeda, 1983), extensively summarised by Sato (1984), considered the long-term evolution of voids.

A new approach was proposed by Krasiński and Hellaby (2002, 2004a, 2005), which will be presented in the next section.

18.6 Formation of other structures in the Universe

Formation of galaxies (then called 'nebulae') was first considered by Lemaître (1933b). He showed that the initial mass distribution may be set up so that a region of comoving size $r = r_0$ around the centre will recollapse while the region $r > r_0$ will keep expanding forever. In this situation, the curvature of the space is positive everywhere and the expansion of the outer region is caused by the cosmological repulsion.

Bonnor's (1956) model was a Friedmann dust tube around the centre of symmetry, surrounded by an L–T transition zone, and that in turn surrounded by another Friedmann dust region so that, at any $t = t_0 = $ constant, the density in the outer region is different from that in the inner region. The boundaries of the Friedmann regions were assumed to be comoving. If both Friedmann regions have positive spatial curvature and the density in the inner region is higher than that in the outer region, then the inner region will start to recollapse earlier than the background and will form a condensation. Bonnor assumed that the condensation has the mass of a typical galaxy, that is, it contains $N \cong 3 \times 10^{67}$ nucleons, and that it formed at $t_i \cong 1000$ years after the Big Bang. Then the following problem arose: if such a condensation formed as a statistical fluctuation in a homogeneous background, then the initial density contrast is $\delta\epsilon/\epsilon = |\epsilon_c - \epsilon_b|/\epsilon_b \cong N^{-1/2} \cong 10^{-34}$, where ϵ_c is the density in the condensation and ϵ_b is the background density. However, in order to develop into a galaxy of typical density, the initial perturbation at t_i would have to be of the order $\delta\epsilon/\epsilon \cong 10^{-5}$. On the other hand, if a perturbation of the order of 10^{-5} is to arise as a statistical fluctuation, then it can involve only 10^{10} particles.

If the outer region has negative curvature while the inner region has positive curvature, then the initial perturbation has to be about 10 times larger than in the preceding case. If both Friedmann regions have negative curvature, a galaxy cannot form at all. The cosmological constant does not help in a model that begins from a Big Bang. Hence, two possibilities are left: either $\Lambda \neq 0$ and the Universe begins as an instability in the Einstein Universe in the asymptotic past (then there is an arbitrary amount of time available for the statistical fluctuations to grow) or there exists a mechanism for producing large perturbations.

The current thinking is that initial fluctuations of density were generated by quantum fluctuations of the scalar field that drives the inflation and indeed are of the order 10^{-5} (Padmanabhan, 1996). However, one of the results of Krasiński and Hellaby (2004a) is that density fluctuations alone cannot be responsible for the generation of structures: the velocity distribution at the initial time must be taken into account. In particular, an initial condensation can evolve into a void (Mustapha and Hellaby, 2001).

The Lemaître–Tolman geometry

The book by Krasiński (1997) contains an overview of other discussions of structure formation in the L–T model. We shall present here a more recent approach (Krasiński and Hellaby, 2002, 2004, 2005).

18.6.1 Density to density evolution

The mass M will be used as the radial coordinate. We specify the density distributions at the instants $t = t_i, i = 1, 2$:

$$\rho_i(M) = \rho(t_i, M) \equiv \epsilon(t_i, M)/c^2, \tag{18.40}$$

and calculate the corresponding functions $R(t_i, M)$ from Eq. (18.23). Throughout the whole consideration we assume that $R_{,t}(t_1, M) > 0$, i.e. that matter is expanding at the initial instant. This assumption is dictated by the intended application of the results (structure formation in the Universe), but a similar investigation could be done for collapsing matter. For definiteness, we also assume $t_2 > t_1$ throughout. The cases $E > 0$ and $E < 0$ have to be considered separately.[1]

For $E > 0$ we write out the evolution equations (18.21) at t_1 and t_2:

$$R_i(M) = R(t_i, M) = \frac{M}{2E}(\cosh \eta_i - 1),$$

$$\sinh \eta_i - \eta_i = \frac{(2E)^{(3/2)}}{M}(t_i - t_B(M)), \tag{18.41}$$

and $R_2(M) > R_1(M)$ in consequence of $\rho(t_2, M) < \rho(t_1, M)$. Solving for t_B at t_1 we obtain

$$t_B = t_1 - \frac{M}{(2E)^{3/2}}\left[\sqrt{(1 + 2ER_1/M)^2 - 1} - \operatorname{arccosh}(1 + 2ER_1/M)\right]. \tag{18.42}$$

We substitute this in (18.41) at t_2 and obtain

$$\sqrt{(1 + 2ER_2/M)^2 - 1} \; -\operatorname{arcosh}(1 + 2ER_2/M)$$
$$-\sqrt{(1 + 2ER_1/M)^2 - 1} + \operatorname{arcosh}(1 + 2ER_1/M)$$
$$= \frac{(2E)^{3/2}}{M}(t_2 - t_1). \tag{18.43}$$

This equation defines $E(M)$, and, given $E(M)$, the previous one defines $t_B(M)$. These two functions then define the L–T evolution from $\rho(t_1, M)$ to $\rho(t_2, M)$. In fact, this is already a solution of our problem, but we have to answer the following question: does

---

[1] The case $E = 0$ is highly exceptional because then just one of the distributions $\rho_i(t, M)$ uniquely determines an L–T model. Choosing the other $\rho_i(t, M)$, it is practically impossible to hit upon such a pair that is connected by the $E = 0$ evolution. However, the $E = 0$ model will appear as an intermediate limiting case.

Eq. (18.43) have any solutions, and, if so, is the solution unique? For ease of calculation, let us denote

$$x \stackrel{\text{def}}{=} 2E/M^{2/3}, \qquad a_i \stackrel{\text{def}}{=} R_i/M^{1/3}, \quad i = 1, 2;$$

$$\begin{aligned} \psi_H(x) \stackrel{\text{def}}{=} & \sqrt{(1+a_2 x)^2 - 1} - \text{arcosh}(1 + a_2 x) \\ & - \sqrt{(1+a_1 x)^2 - 1} + \text{arcosh}(1 + a_1 x) \\ & - (t_2 - t_1) x^{3/2}, \end{aligned} \tag{18.44}$$

Our problem is this: for what values of the parameters $a_2 > a_1$ and $t_2 > t_1$ does the equation $\psi_H(x) = 0$ have a solution $x \neq 0$?[1] From the properties of the function $\psi_H(x)$ (Krasiński and Hellaby, 2002, and Exercise 5) it follows that (18.44) has a solution if and only if

$$t_2 - t_1 < \frac{\sqrt{2}}{3} \left(a_2^{3/2} - a_1^{3/2} \right), \tag{18.45}$$

and then the solution is unique. This inequality says that the expansion between t_1 and t_2 must have been faster than in the $E = 0$ model.

The result above shows only the existence of a solution for a given value of M. Some initial conditions may lead to shell crossings, and these are not excluded by (18.45) – one must check for their presence separately. But the criteria for the occurrence of shell crossings are known – see Section 18.10. If they occur before t_1, then they are no problem – the L–T model cannot describe those epochs anyway.

For $E < 0$, the inverse of the function cos in the range $[0, \pi]$ is different from that in the range $[\pi, 2\pi]$. Consequently, the L–T model evolving between two given states must be considered separately for the case when the final state is still expanding ($\eta \in [0, \pi]$ in (18.19)) and for the case when the final state is already recollapsing ($\eta \in [\pi, 2\pi]$).

For the still-expanding final state, the variables and the function whose zero has to be found are the a_i from (18.44) and

$$x \stackrel{\text{def}}{=} -2E/M^{2/3},$$

$$\begin{aligned} \psi_X(x) \stackrel{\text{def}}{=} & \arccos(1 - a_2 x) - \sqrt{1 - (1 - a_2 x)^2} \\ & - \arccos(1 - a_1 x) + \sqrt{1 - (1 - a_1 x)^2} - (t_2 - t_1) x^{3/2}. \end{aligned} \tag{18.46}$$

The reasoning is analogous to that for (18.44), but this time the arguments of arccos must have absolute values not greater than 1. This implies $x \leq 2/a_i$ for both i, so, since $a_2 > a_1$,

$$0 \leq x \leq 2/a_2. \tag{18.47}$$

---

[1] $x = 0$ is a solution of $\psi_H(x) = 0$, but it corresponds to $E = 0$. In fact, we are looking for a zero of the function $\psi_H(x)/x^{3/2}$.

The two square roots in (18.46) will then also exist. Equation (18.47) is equivalent to the requirement that $(R_{,t})^2$ (in (18.14) with $\Lambda = 0$) is non-negative at both t_1 and t_2.

Again, elementary reasoning (Krasiński and Hellaby, 2002) shows that $\psi_X(x) = 0$ has a nonzero solution if and only if

$$\frac{\sqrt{2}}{3}\left(a_2^{3/2} - a_1^{3/2}\right) < t_2 - t_1 \leq$$

$$(a_2/2)^{3/2}\left[\pi - \arccos(1 - 2a_1/a_2) + 2\sqrt{a_1/a_2 - (a_1/a_2)^2}\right]. \tag{18.48}$$

With (18.48) fulfilled, the solution is unique. The first inequality means that the model must have expanded between t_1 and t_2 slower than the $E = 0$ model; the second one means that the final state is still earlier than the instant of maximal expansion. The condition $E \geq -1/2$ does not follow from (18.48) and has to be verified separately.

For the recollapsing final state, the variables are defined as in (18.48), but the function whose zeros are sought is different:

$$\psi_C(x) \stackrel{\text{def}}{=} \pi + \arccos(-1 + a_2 x) + \sqrt{1 - (1 - a_2 x)^2}$$

$$- \arccos(1 - a_1 x) + \sqrt{1 - (1 - a_1 x)^2} - (t_2 - t_1)x^{3/2}. \tag{18.49}$$

The solution of $\psi_C(x) = 0$ exists if and only if

$$t_2 - t_1 \geq (a_2/2)^{3/2}\left[\pi - \arccos(1 - 2a_1/a_2) + 2\sqrt{a_1/a_2 - (a_1/a_2)^2}\right], \tag{18.50}$$

and then it is unique (Krasiński and Hellaby, 2002).

The results obtained above can be summarised in the following:

Theorem 18.1 *Given any two instants t_1 and $t_2 > t_1$, and any two spherically symmetric density profiles $0 < \rho_2(M) < \rho_1(M)$ defined over the same range of M, an L–T model can be found that evolves from ρ_1 to ρ_2 in time $(t_2 - t_1)$. The inequalities (18.45), (18.48) and (18.50) will tell which class of L–T evolution applies at each value of M. The possibilities of shell crossings or excessively negative energies are not excluded, and must be separately checked for.*

18.6.2 Velocity to density evolution

A measure of velocity of expansion is $R_{,t}(t, M)$. A more convenient measure is $R_{,t}(t, M)/M^{1/3}$ because it becomes constant in the Friedmann limit, and so is also a measure of inhomogeneity of the spacetime. Suppose now that the initial state of the Universe is specified by

$$b_1(M) = R_{,t}(t_1, M)/M^{1/3}, \tag{18.51}$$

while the final state is specified, as before, by a density distribution $\rho(t_2, M)$. Again, the two signs of E must be considered separately.

For $E > 0$, with the variables defined as in (18.44) and (18.51), the equation whose solutions are sought is $\Phi_H(x) = 0$, where

$$\Phi_H(x) \overset{\text{def}}{=} \sqrt{(1 + a_2 x)^2 - 1} - \sqrt{\left(\frac{b_1^2 + x}{b_1^2 - x}\right)^2 - 1}$$

$$- \operatorname{arcosh}(1 + a_2 x) + \operatorname{arcosh}\left(\frac{b_1^2 + x}{b_1^2 - x}\right) - x^{3/2}(t_2 - t_1). \tag{18.52}$$

Now the necessary and sufficient condition for the existence of solutions of (18.52) consists of two inequalities (Krasiński and Hellaby, 2004a):

$$t_2 - t_1 < \frac{\sqrt{2}}{3} a_2^{3/2} - \frac{4}{3 b_1^3}, \qquad 2/a_2 < b_1^2. \tag{18.53}$$

The second one is a necessary condition for the existence of a $t_2 > t_1$ obeying the first, and is equivalent to $R_2 > R_1$. It came in place of $\rho(t_2, M) < \rho(t_1, M)$ that we assumed when considering a density to density evolution. With both inequalities fulfilled, the solution of $\Phi_H(x) = 0$ is unique. Equation (18.53) is equivalent to (18.45).

For $E < 0$, as before, the cases when the final state is still expanding and recollapsing have to be considered separately. When the final state is still expanding, the function whose zeros are to be found is

$$\Phi_X(x) \overset{\text{def}}{=} \sqrt{1 - \left(\frac{b_1^2 - x}{b_1^2 + x}\right)^2} - \sqrt{1 - (1 - a_2 x)^2}$$

$$+ \arccos(1 - a_2 x) - \arccos\left(\frac{b_1^2 - x}{b_1^2 + x}\right) - x^{3/2}(t_2 - t_1), \tag{18.54}$$

and the equation $\Phi_X(x) = 0$ has a solution if and only if

$$\frac{\sqrt{2}}{3} a_2^{3/2} - \frac{4}{3 b_1^3} < t_2 - t_1 \le$$

$$(a_2/2)^{3/2} \left[\pi + \frac{b_1 \sqrt{2 a_2}}{a_2 b_1^2 / 2 + 1} - \arccos\left(\frac{a_2 b_1^2 / 2 - 1}{a_2 b_1^2 / 2 + 1}\right) \right]. \tag{18.55}$$

This is equivalent to (18.48). With the inequalities fulfilled, the solution of the corresponding equation is unique (Krasiński and Hellaby, 2004a).

When the final state is recollapsing, the appropriate function is

$$\Phi_C(x) \overset{\text{def}}{=} \sqrt{1 - \left(\frac{b_1^2 - x}{b_1^2 + x}\right)^2} + \sqrt{1 - (1 - a_2 x)^2} + \pi$$

$$- \arccos\left(\frac{b_1^2 - x}{b_1^2 + x}\right) + \arccos(-1 + a_2 x) - x^{3/2}(t_2 - t_1), \tag{18.56}$$

and the condition for the existence of the appropriate evolution is

$$t_2 - t_1 \geq (a_2/2)^{3/2} \left[\pi + \frac{b_1 \sqrt{2a_2}}{a_2 b_1^2/2 + 1} - \arccos \left(\frac{a_2 b_1^2/2 - 1}{a_2 b_1^2/2 + 1} \right) \right]. \tag{18.57}$$

All derivations can be found in Krasiński and Hellaby (2004a).

Using the methods of this section, examples of evolution of various initial configurations to galaxy clusters, voids and galaxies with central black holes were found (Krasiński and Hellaby, 2002, 2004, 2005).

The approach presented in this section has not been entirely successful so far. If the distributions of velocity at t_1 and of density at t_2 are assumed, then the density distribution at t_1 is determined, and has to obey the observational constraints. It is sensitive to the shape of the velocity profile at t_1, and the optimal shape has not been identified yet.

18.6.3 Velocity to velocity evolution

The given quantities are now the velocity distributions at t_1 and t_2:

$$b_i(M) = R_{,t}(t_i, M)/M^{1/3}, \qquad i = 1, 2; \tag{18.58}$$

the definition of $x(x = \pm 2E/M^{2/3})$ being in each case the same as in the previous cases. We assume that $t_2 > t_1$ and that the model is expanding at t_1, thus $b_1 > 0$. This way of specifying the data is probably not useful for astrophysics, but is included here for academic completeness.

For $E > 0$, the function whose zeros are to be found is

$$\chi_H(x) \overset{\text{def}}{=} \sqrt{\left(\frac{b_2^2 + x}{b_2^2 - x} \right)^2 - 1} - \sqrt{\left(\frac{b_1^2 + x}{b_1^2 - x} \right)^2 - 1}$$

$$- \operatorname{arcosh} \left(\frac{b_2^2 + x}{b_2^2 - x} \right) + \operatorname{arcosh} \left(\frac{b_1^2 + x}{b_1^2 - x} \right) - x^{3/2} (t_2 - t_1). \tag{18.59}$$

The necessary and sufficient condition for the existence of an $E > 0$ evolution between the two states is the set of two inequalities

$$0 < b_2 < b_1, \qquad t_2 - t_1 > \frac{4}{3} \left(\frac{1}{b_2^3} - \frac{1}{b_1^3} \right). \tag{18.60}$$

The second inequality becomes more intelligible when it is rewritten as

$$b_2^3 > \frac{b_1^3}{1 + \frac{3}{4} b_1^3 (t_2 - t_1)}, \tag{18.61}$$

which means that the expansion at t_2 is faster than it would be in the $E = 0$ model (for which $b^3 = 4/[3(t - t_B)]$, giving an equality in (18.61)).

For an $E < 0$ evolution between t_1 and t_2 with the final state still expanding, the function χ is

$$\chi_X(x) \stackrel{\text{def}}{=} -\sqrt{1 - \left(\frac{b_2^2 - x}{b_2^2 + x}\right)^2} + \sqrt{1 - \left(\frac{b_1^2 - x}{b_1^2 + x}\right)^2}$$

$$+ \arccos\left(\frac{b_2^2 - x}{b_2^2 + x}\right) - \arccos\left(\frac{b_1^2 - x}{b_1^2 + x}\right) - x^{3/2}(t_2 - t_1). \quad (18.62)$$

A solution of $\chi_X(x) = 0$ exists if and only if

$$0 < b_2 < b_1, \qquad t_2 - t_1 < \frac{4}{3}\left(\frac{1}{b_2^3} - \frac{1}{b_1^3}\right), \quad (18.63)$$

which now implies that the expansion between t_1 and t_2 must have been slower than it would be in an $E = 0$ model.

These two cases together exhaust all possibilities for $b_2 > 0$. When the final state is already recollapsing, we have $b_2 < 0$. Then the evolution exists for any values of $b_1 > 0$ and $b_2 < 0$. The function χ is here

$$\chi_C(x) \stackrel{\text{def}}{=} \sqrt{1 - \left(\frac{b_1^2 - x}{b_1^2 + x}\right)^2} - \sqrt{1 - \left(\frac{x - b_2^2}{x + b_2^2}\right)^2} + \pi$$

$$+ \arccos\left(\frac{x - b_2^2}{x + b_2^2}\right) - \arccos\left(\frac{b_1^2 - x}{b_1^2 + x}\right) - x^{3/2}(t_2 - t_1), \quad (18.64)$$

and with $t_2 > t_1$ the equation $\chi_C(x) = 0$ always has an $x > 0$ solution.

18.7 The influence of cosmic expansion on planetary orbits

The first formally correct study of the problem of expansion of orbits was carried out by Einstein and Straus (1945).[1] They showed that the Schwarzschild solution can be matched to any Friedmann model. This implies that the planetary orbits are *in this configuration* not influenced by the expansion of the Universe. The Schwarzschild mass m is related to the Friedmann mass integral $\mathfrak{M}$ from (17.27) by

$$m = \frac{G\mathfrak{M}r_0^3}{c^2\left(1 + \frac{1}{4}kr_0^2\right)^3} \stackrel{\text{def}}{=} \mu(r_0), \quad (18.65)$$

where r_0 is the radius of the Schwarzschild vacuole in the coordinates of (17.1). The geodesic radius of the vacuole, $R(t)\int_0^{r_0} r(1 + \frac{1}{4}kr^2)^{-1}\,dr$, expands together with the Universe.

This relation was derived by Einstein and Straus from the Einstein equations in a rather outlandish notation. It can be derived in a simple way from the Lemaître–Novikov

---

[1] Some papers were published by other authors earlier, but they were based on questionable assumptions (Krasiński, 1997).

representation of the Schwarzschild solution, Eqs. (14.116)–(14.117), and from the Friedmann limit of Eqs. (18.16) and (18.14) (with $\Lambda = 0$). Assuming that the coordinates on both sides of the hypersurface $r = r_b$ are the same, we conclude, from the matching conditions of Section 12.17, that the $R(t, r_b)$ in (14.116) must be the same (*as a function of t*) as the $R(t, r_b)$ in (18.16). Consequently, $R_{,t}(t, r_b)$ must be the same in both metrics, and then (18.14) implies that $E(r_b)$ is the same on both sides, and $m = M(r_b)$, where $M(r_b)$ is the L–T mass contained within the $r = r_b$ hypersurface of (18.16).

So far we have considered the matching conditions between the Schwarzschild metric and a general L–T model. Now we take the Friedmann limit of (18.14) and (18.16). Hence, by (18.22), $M(r_b) = \left(G\mathfrak{M}/c^2\right) r_b^{\,3}$, while the Friedmann limit of the metric (18.16) is

$$\mathrm{d}s^2 = \mathrm{d}t^2 - R_F^{\,2}(t) \left(\frac{\mathrm{d}r_F^{\,2}}{1 - k r_F^{\,2}} + r_F^{\,2} \left(\mathrm{d}\vartheta^2 + \sin^2 \vartheta\, \mathrm{d}\varphi^2 \right) \right), \tag{18.66}$$

and the transformation leading from this to the coordinates of (17.1) is (17.2). Hence, denoting by r_0 the value of r corresponding to $r_F = r_b$, we obtain (18.65). This says that the Schwarzschild mass at the centre of symmetry must be equal to the Friedmann mass removed from within the sphere $r = r_b$. However, contrary to superficial appearance, Eq. (18.65) is not invariant under coordinate transformations in the Friedmann metric – because the constant $\mathfrak{M}$ is not a scalar, but can be scaled by coordinate transformations.

Einstein and Straus' result was for many years taken as the general implication of relativity. However, (18.65) need not be fulfilled if the Einstein–Straus configuration is taken only at a single moment $t = t_0$ as an initial condition for an L–T model. Then, the results of other papers (Sato, 1984 and papers cited therein; Lake and Pim, 1985) imply that if $m < \mu(r_0)$, then the boundary of the vacuole will expand faster than the Friedmann background, whereas if $m > \mu(r_0)$, then initial conditions may be set up so that the vacuole will start to collapse. This indicates that the Einstein–Straus configuration is unstable against perturbations of the condition (18.65); that is, it is an exceptional situation.

The same problem was studied by quite a different method by Gautreau (1984). He based his study on the subcase $E = 0$ of the L–T model that he derived in a hybrid of comoving and curvature coordinates, in which the metric is non-diagonal. In these coordinates, R is the radial coordinate defined by the curvature radius of the orbits of the symmetry group. These orbits do not participate in the cosmic expansion and therefore R of any single orbit can be used as a standard of length.

In Gautreau's coordinates, the $E = 0$ L–T solution takes the form:

$$\mathrm{d}s^2 = A\,\mathrm{d}t^2 - \frac{1}{A} \left(\mathrm{d}R - \sqrt{Z + 2\mu - \frac{1}{3}\Lambda R^2}\ \mathrm{d}t \right)^2 - R^2 \left(\mathrm{d}\vartheta^2 + \sin^2 \vartheta\, \mathrm{d}\varphi^2 \right), \tag{18.67}$$

where

$$A \stackrel{\text{def}}{=} 1 - \frac{2\mu + Z}{R} + \frac{1}{3}\Lambda R^2. \tag{18.68}$$

The function Z is given by an equation equivalent to (18.213), derived in Exercise 4:

$$R^{1/2}\frac{\partial Z}{\partial t} + \varepsilon \left(Z + 2\mu + \frac{1}{3}\Lambda R^3 \right)^{1/2} \frac{\partial Z}{\partial R} = 0, \qquad (18.69)$$

and $\mu = $ constant is the mass of the star. This is the L–T model with $E = 0$ and $M = \mu + Z/2$. The metric (18.67) applies only outside the star, for $R \geq R_b = $ constant > 0; the metric for the inside of the star is not considered. By investigating the equations of timelike geodesics in (18.67), Gautreau showed that circular orbits do not exist. This is in fact a purely Newtonian phenomenon: in the model (18.67)–(18.69), the smoothed-out cosmic matter-density extends throughout the planetary system, and, as a result of cosmic expansion, matter streams out of every sphere $R = $ constant. Hence, each planet moves under the influence of a gravitational force that is decreasing with time, so the orbit must spiral out. Gautreau derived the Newtonian formula for the rate of change of orbital radius, $dR/dt = 8\pi R^4 H\overline{\rho}/(2\mu)$, where R is the orbital radius, H is the Hubble parameter and $\overline{\rho}$ is the mean cosmic density of matter. The effect is thus greater for orbits of greater radius; for Saturn it is $(dR/dt)_S = 6 \times 10^{-18}$ m per year. This is obviously unmeasurable (one proton diameter per 1000 years!). For a star at the edge of the Andromeda galaxy the effect would be $(dR/dt)_{gal} = 1100$ km per year. From the 'practical' point of view Gautreau's result thus implies that planetary orbits do not react to the expansion of the Universe. However, it is important to know that in principle the effect is nonzero. In the Einstein–Straus approach, it was exactly zero. As explained above, the model of Einstein and Straus is unstable against the perturbations of (18.65), and hence is less realistic than Gautreau's.

18.8 * Apparent horizons in the L–T model

As defined in Section 16.4, an apparent horizon is the outer envelope of a region of closed trapped surfaces, while a closed trapped surface S_t is one from which it is impossible to send a diverging bundle of light rays – because both the outward-directed and the inward-directed bundles immediately converge: $k^\mu{}_{;\mu} \leq 0$ at S_t.

Since the L–T model is spherically symmetric, the apparent horizon must be a sphere around the centre of symmetry. Hence, it suffices to consider families of null geodesics sent orthogonally from a surface $r = $ constant. We must identify the surface at which $k^\mu{}_{;\mu}$ becomes zero for all future-directed null geodesics. For this, we shall use the method of Szekeres (1975b). Since the surface we are looking for is a sphere, its normal rays will be radial. From (18.16), the tangent vectors to radial null curves obey $k^0 - \varepsilon R_{,r}/\sqrt{1+2E}\, k^1 = 0$, $k^2 = k^3 = 0$, where $\varepsilon = +1$ for outward-directed and $\varepsilon = -1$ for inward-directed curves. Because of spherical symmetry, these curves must be geodesics. Now consider a bundle of null geodesics originating at a surface $S_{t,r}$ given by $\{t = t_s, r = r_s\}$. At these constant values of t and r, the affine parameter on the null geodesics may be chosen so that

$$k^0 = \frac{R_{,r}}{\sqrt{1+2E}}, \qquad k^1 = \varepsilon \qquad \text{on } S_{t,r}. \qquad (18.70)$$

The divergence of this field on $S_{t,r}$ is then

$$k^\mu{}_{;\mu} = k^\mu{}_{,\mu} + \begin{Bmatrix} \mu \\ \rho\nu \end{Bmatrix} k^\rho k^\nu$$

$$= k^0{}_{,t} + k^1{}_{,r} + \frac{R_{,r}}{\sqrt{1+2E}}\left(\frac{R_{,tr}}{R_{,r}} + 2\frac{R_{,t}}{R}\right) + \varepsilon\left(-\frac{E_{,r}}{1+2E} + \frac{R_{,rr}}{R_{,r}} + 2\frac{R_{,r}}{R}\right). \quad (18.71)$$

Since k^μ is null, geodesic and affinely parametrised, we have $k_\mu k^\mu = 0$ and $k^\mu{}_{;\nu} k^\nu = 0$. Differentiating the first equation by t, and taking the second one with the index $\mu = 1$, we obtain

$$k^0{}_{,t} - \frac{R_{,tr}}{\sqrt{1+2E}} - \varepsilon\frac{R_{,r}}{\sqrt{1+2E}} k^1{}_{,t} = 0, \quad (18.72)$$

$$\frac{R_{,r}}{\sqrt{1+2E}} k^1{}_{,t} + \varepsilon k^1{}_{,r} + 2\varepsilon\frac{R_{,tr}}{\sqrt{1+2E}} + \frac{R_{,rr}}{R_{,r}} - \frac{E_{,r}}{1+2E} = 0. \quad (18.73)$$

We add (18.73) multiplied by ε to (18.72), and thereby eliminate $k^1{}_{,t}$. The resulting equation is used to eliminate $k^0{}_{,t} + k^1{}_{,r}$ from (18.71), and the result is

$$2\frac{R_{,r}}{R}\left(\frac{R_{,t}}{\sqrt{1+2E}} + \varepsilon\right) = 0. \quad (18.74)$$

One solution of this equation is $R_{,r} = 0$, but this is either a shell crossing singularity or a neck. What happens there will be investigated in Section 18.10. The generic solution of (18.74) is

$$\frac{R_{,t}}{\sqrt{1+2E}} = -\varepsilon. \quad (18.75)$$

For outward-directed geodesics $\varepsilon = +1$, so the solution of (18.75) will exist only in collapsing models, in which $R_{,t} < 0$. For inward-directed geodesics $\varepsilon = -1$, and the solution of (18.75) will exist only in expanding models. In each case, $R_{,t}$ must have the sign of $-\varepsilon$. Using the evolution equation (18.14), we have in (18.75)

$$\sqrt{2E + \frac{2M}{R} - \frac{1}{3}\Lambda R^2} = \sqrt{1+2E}. \quad (18.76)$$

With $\Lambda = 0$, the solution of this is

$$R = 2M. \quad (18.77)$$

In the Schwarzschild limit, $M = $ constant, the apparent horizon becomes identical to the event horizon. In a general L–T spacetime, the $r = $ constant shell obeying (18.77) is just falling into its own Schwarzschild horizon.

Using the result of Exercise 2 in Chapter 16, the formula for the apparent horizon can be derived in an equivalent way. A bundle of light rays sent outwards from a spherical surface forms, at any time, a spherical light front. The apparent horizon is where the surface area of this front starts decreasing. Since this surface area is proportional to R^2, it suffices to identify the location where $R(t, r)$ starts decreasing as we go along the null geodesics (Krasiński and Hellaby, 2004b).

The evolution equation (18.14) with $\Lambda = 0$ can be written as

$$\dot{R} = \ell \sqrt{\frac{2M}{R} + 2E}, \tag{18.78}$$

where $\ell = +1$ for expanding models and $\ell = -1$ for collapsing models. The radial null geodesics, from (18.16), are given by

$$\left. \frac{dt}{dr} \right|_n = \frac{jR_{,r}}{\sqrt{1+2E}}, \tag{18.79}$$

where $j = +1$ for outgoing rays and $j = -1$ for incoming rays, whose solution we write as $t = t_n(r)$. Along this ray we have

$$R_n = R(t_n, r),$$

$$(R_n)_{,r} = \dot{R} \left. \frac{dt}{dr} \right|_n + R_{,r} = \left(\ell j \frac{\sqrt{2M/R + 2E}}{\sqrt{1+2E}} + 1 \right) R_{,r}. \tag{18.80}$$

The apparent horizon (AH) is the hypersurface in spacetime where R stops increasing along the rays:

$$(R_n)_{,r} = 0 \Rightarrow \sqrt{\frac{2M}{R} + 2E} = -\ell j \sqrt{1+2E} \Rightarrow \ell j = -1, \qquad R = 2M. \tag{18.81}$$

Inside the future apparent horizon, all light rays inevitably proceed towards the final singularity ($(R_n)_{,r} < 0$ necessarily). The existence of such a region was predicted by Bondi (1947) and later by Barnes (1970). Note that, in every L–T model that collapses, the dust matter must enter the future apparent horizon before it hits the final singularity at $R = 0$, and in every L–T model that expands the dust matter remains inside the past apparent horizon for a while after leaving the Big Bang.

We now wish to establish whether the apparent horizon is timelike, spacelike or null. For this purpose, we find dt/dr along the AH by differentiating (18.81):

$$\dot{R} dt + R_{,r} dr = 2M_{,r} dr. \tag{18.82}$$

The result is

$$t'_{\text{AH}} = \left. \frac{dt}{dr} \right|_{\text{AH}} = \frac{2M_{,r} - R_{,r}}{\dot{R}} = \frac{2M_{,r} - R_{,r}}{\ell\sqrt{2M/R + 2E}}, \tag{18.83}$$

and, since $R = 2M$ on the AH,

$$\left. \frac{dt}{dr} \right|_{\text{AH}} = \frac{\ell(2M_{,r} - R_{,r})}{\sqrt{1+2E}}. \tag{18.84}$$

In the vacuum case $M_{,r} = 0$, we have $dt/dr|_{\text{AH}} = dt/dr|_n$ since $\ell j = -1$, i.e. the AH is null. Note that $M_{,r} = 0$ could be only local, so the AH would be null only in that region.

For hyperbolic regions, with $E \geq 0$ along each dust worldline, there is either only expansion or only collapse, i.e. only one AH (either the future AH^+ or the past AH^-)

can occur. The two AHs can thus cross only in an elliptic $E < 0$ region. At the moment of maximum expansion, where $R_{,t} = 0$, the maximal value of R is $R = -M/E$. In elliptic regions, with $-1 < 2E < 0$, this $R_{\max}$ is thus always greater than $2M$, so the dust source always escapes from the past AH before it falls into the future AH. However, at those locations where $2E = -1$, the maximal R equals $2M$, which means that the past AH touches the future AH. Such a location is a *neck*, the nonvacuum analogue of the Kruskal wormhole. We shall come back to it in Section 18.10.

To establish whether the AH is timelike, null or spacelike, we compare the slope of the AH^+ with the outgoing light ray (or the slope of the AH^- with the incoming light ray), i.e. $\ell j = -1$:

$$B \stackrel{\text{def}}{=} \left.\frac{dt}{dr}\right|_{AH} \bigg/ \left.\frac{dt}{dr}\right|_{n} = -\ell j \left(1 - \frac{2M_{,r}}{R_{,r}}\right) = 1 - \frac{2M_{,r}}{R_{,r}}. \tag{18.85}$$

Now, since the conditions for no shell crossings (Section 18.10) require $M_{,r} \geq 0$ where $R_{,r} > 0$ and vice versa, we have

$$
\begin{aligned}
B_{\max} = 1, &\rightarrow AH^+ \text{ outgoing null} & (M_{,r} = 0, \rho = 0), \\
1 > B > -1, &\rightarrow AH^+ \text{ spacelike} & (\text{for most } M_{,r}), \\
B = -1, &\rightarrow AH^+ \text{ ingoing null} & (\text{large } M_{,r}/R_{,r}), \\
-1 > B > -\infty, &\rightarrow AH^+ \text{ ingoing timelike} & (\text{very large } M_{,r}/R_{,r}),
\end{aligned}
\tag{18.86}
$$

so an outgoing timelike AH^+ is not possible. Thus outgoing light rays that reach the AH^+ fall inside the AH^+, except where $M_{,r} = 0$, in which case they move along it. The AH^+ is timelike if[1]

$$\ell \frac{dt_{AH}}{dM} > \frac{1}{\sqrt{1 + 2E}}. \tag{18.88}$$

The argument is similar for light rays at the AH^-, except that 'ingoing' should be swopped with 'outgoing'. If ingoing light rays reach the AH^-, they pass out of it or run along it.

The apparent horizons in the Friedmann models with $\Lambda = 0$ are always timelike, with the parameter $B = -2$ (see Exercise 18.8).

The apparent horizons in elliptic regions require special attention, since two AHs are present in the same spacetime.

We shall first consider the expansion phase of an elliptic model, where $0 \leq \eta \leq \pi$ and $E < 0$, so we have $\ell = +1$ and only the AH^- is present. Since $R = 2M$ on an AH, we have from (18.19):

$$\cos \eta_{AH} = 1 + 4E, \tag{18.89}$$

---

[1] Note that Eq. (18.88) can be equivalently written as follows:

$$\ell \left.\frac{dt}{dr}\right|_{AH} > \frac{M_{,r}}{\sqrt{1 + 2E}} = N_{,r}, \tag{18.87}$$

where N is the sum of rest masses within the $r = $ constant sphere given by (18.17).

and thus, along a given worldline, the proper time of passing through the AH⁻, counted from the Big Bang time t_B, can be calculated from (18.19) with $R = 2M$ to be

$$t_{AH^-} - t_B = M \frac{\arccos(1+4E) - 2\sqrt{-2E(1+2E)}}{(-2E)^{3/2}}. \tag{18.90}$$

Note that $t_{AH^-} - t_B = 0$ only at the centre of symmetry, where $M = 0$. This is the only place where the AH⁻ can touch the Big-Bang set.[1] The function $F = t_{AH^-} - t_B$ of the argument $f = 2E$, defined in (18.90), has the following properties,

$$F(-1) = M\pi, \qquad F(0) = 4M/3, \qquad \frac{dF}{df} < 0 \qquad \text{for} -1 < f < 0, \tag{18.91}$$

i.e. it is decreasing. These properties mean that the AH does not touch the Big Bang anywhere except at a centre, $M = 0$, even if $E = 0$. Along all worldlines with $E > -1/2$, even parabolic worldlines for which $E = 0$ and $M \neq 0$, the dust particles emerge from the AH⁻ a finite time after the Big Bang ($\eta = 0$) and a finite time before maximum expansion ($\eta = \pi$).

So, although (18.89) shows that wordlines with larger E exit the AH⁻ at a later stage of evolution, this need not correspond to a later time t, or even to a longer time $(t - t_B)$ since the bang. It is not at all necessary that E is a monotonically decreasing function of r in an elliptic region; in general it can increase and decrease again any number of times.

A shell of parabolic ($E = 0$) worldlines occurs at the boundary between elliptic and hyperbolic regions, where $E \to 0$, but $E_{,r} \neq 0$ and $M > 0$. From (18.89) and (18.90) with $\ell = +1$, we see that

$$\eta \xrightarrow[E \to 0]{} 0, \qquad t_{AH^-} - t_B \xrightarrow[E \to 0]{} \frac{4M}{3}$$

$$\frac{dt_{AH^-}}{dM} \xrightarrow[E \to 0]{} \frac{4}{3} - \frac{4M}{5} \frac{dE}{dM} + \frac{dt_B}{dM}, \tag{18.92}$$

so the AH⁻ never touches the bang here, despite η being zero.

The crunch time, i.e. the time of the final singularity, where $\eta = 2\pi$, is found from (18.19) to be

$$t_C(r) = t_B(r) + \frac{2\pi M}{(-2E)^{3/2}}, \tag{18.93}$$

so it diverges wherever $E \to 0$ at $M \neq 0$. This divergence shows that either the bang time or the crunch time recedes to infinity, or both of them do. The third possibility will be illustrated in the next section. We also see the slope and the causal nature of the AH⁻ are uncertain here.

---

[1] But whether the coincidence of the AH⁻ and the Big Bang at $M = 0$ is real or only a spurious coordinate effect depends on the precise shapes of the functions $E(r)$ and $t_B(r)$. A null or timelike segment of the Big Bang/Big Crunch singularity can stick out at the centre, and it is invisible in the comoving coordinates because they have a singularity there that squeezes this segment into a point. A signature of such a singularity is a family of different light cones with vertices at the same point of the coordinate grid. See Section 18.14.

The behaviour of the AH^+ during collapse can be obtained form the above by replacing $(t - t_B)$ with $(t_C - t)$, η with $(2\pi - \eta)$, flipping the signs of ℓ and j, and swopping 'ingoing' with 'outgoing'. However, keeping t_B as our arbitrary function, we have $\pi \leq \eta \leq 2\pi$. Equation (18.89) still applies, but instead of (18.90) we now obtain

$$(t - t_B)_{AH^+} = M\frac{\pi + \arccos(-1 - 4E) + 2\sqrt{-2E(1 + 2E)}}{(-2E)^{3/2}}, \tag{18.94}$$

and (18.87) applies with $\ell = -1$.

The corresponding results in an expanding $E = 0, E_{,r} = 0$ L–T model follow from (18.20) with $R = 2M$. Then, Eq. (18.92) becomes

$$\frac{dt_{AH^-}}{dM} = \frac{4}{3} + \frac{dt_B}{dM}. \tag{18.95}$$

The AH^- may still exhibit all possible behaviours of (18.86) with 'outgoing' and 'ingoing' interchanged. In a collapsing parabolic model, time-reversed results apply.

In expanding hyperbolic regions, using the same methods as for expanding elliptic regions, we find that the behaviour of the AH is qualitatively the same. There is of course only one AH, no maximum expansion, and loci where $E = -1/2$ are not possible, but the results for origins and for the parabolic limit both carry over. Collapsing hyperbolic regions are essentially like collapsing elliptic regions.

18.9 * Black holes in the evolving Universe

This process was first described by Oppenheimer and Snyder (1939). They discussed the collapse of a Friedmann dust cloud matched to the Schwarzschild solution. They found that the collapse to the Schwarzschild horizon $r = 2m$ takes a finite proper time for each infalling particle, but an infinite time for a static distant observer in the Schwarzschild region who would see the star gradually redden. For a comoving observer on the surface of an object of the mass of the Sun, but rarefied so that the initial density equals that of water, the time needed to reach the horizon would be of the order of one day. As the horizon is approached, light can escape outwards within a cone around the radial direction that becomes progressively narrower and closes completely at the horizon.

The novel idea in the Oppenheimer–Snyder paper was to study the black hole in the process of its formation. The theory of black holes developed from the 1960s onwards is based on stationary vacuum solutions; that is, it describes black holes that have always existed and are observed from afar. The L–T model allows for an even more sophisticated approach (see below).

Bondi (1947) observed that if matter in the L–T model is collapsing with a great velocity so that $[1 + 2E(r)]^{1/2} + \partial R/\partial t < 0$, then along a light ray emitted away from the centre with the tangent vector k^μ the quantity $k^\mu R_{,\mu}$ is negative, that is, the ray is forced to move inwards. A necessary condition for this is $R < 2M$; that is, a sufficiently high matter-density over a sufficiently large region. With hindsight, this was a prediction that black holes would form under certain conditions.

Several properties of the apparent horizons and black holes in an L–T model can be explained using the simple example discussed below. The parameter values used in it are unrealistic, having been chosen in such a way that all the figures are easily readable.

We take an $E < 0$ L–T model whose Big Bang function $t_B(M)$ is

$$t_B(M) = -bM^2 + t_{B0}, \tag{18.96}$$

and whose Big Crunch function is

$$t_C(M) = aM^3 + T_0 + t_{B0}, \tag{18.97}$$

where T_0 is the lifetime of the central worldline ($M = 0$). The values of the parameters used in the figures will be $a = 2 \times 10^4$, $b = 200$, $t_{B0} = 5$ and $T_0 = 0.05$. They were chosen so as to make the figures readable and illustrative; they are unrelated to any astrophysical quantities. Since $t = t_C$ at $\eta = 2\pi$, we find from Eq. (18.19):

$$E(M) = -\frac{1}{2}\left(\frac{2\pi M}{t_C - t_B}\right)^{2/3} = -\left(\frac{\pi^2}{2}\right)^{1/3} \frac{M^{2/3}}{(aM^3 + bM^2 + T_0)^{2/3}}. \tag{18.98}$$

As $M \to \infty$, we have $t_B \to -\infty$, $t_C \to +\infty$ and $E \to 0$. Hence, the space contains infinite mass and has infinite volume. Unlike in the Friedmann models, positive space curvature does not imply finite volume; this has been known since long ago (Bonnor, 1985; Hellaby and Lake, 1985).

The main features of this model are shown in Fig. 18.1. Note that the AH$^+$ first appears at a finite distance from the centre, where the function $t_{AH^+}(M)$ has its minimum, and at a time $t_{hs} < t_C(0)$.

At all times after the crunch first forms, $t > t_C(0)$, the mass M_S already swallowed up by the singularity is necessarily smaller than the mass M_{BH} that had disappeared into the AH$^+$. The mass M_S cannot even be estimated by astronomical methods. The situation is reversed in time for the Big Bang singularity and the AH$^-$.

The 3-dimensional diagram of Fig. 18.2 shows the value of R at each t and each M. Figure 18.3 shows the 'topographic map' of the surface from Fig. 18.2. It contains contours of constant R (the thinner curves) inscribed into Fig. 18.1. It also shows several outgoing radial null geodesics. Each geodesic has a vertical tangent at the centre. This is a consequence of using M as the radial coordinate. Since $dt/dM = (\partial R/\partial M)/\sqrt{1 + 2E}$ on each geodesic and $R \propto M^{2/3}$ close to the centre, $dt/dM \propto M^{-1/3}$ and $dt/dM \to \infty$ as $M \to 0$. Each geodesic proceeds to higher values of R before it meets the apparent horizon AH$^+$, where it is tangent to an $R = $ constant contour, and then proceeds towards smaller R values. The future event horizon consists of those radial null geodesics that approach the AH$^+$ asymptotically. In Fig. 18.3, it lies between geodesics 5 and 6 (counted from the lower right); we shall discuss its location in more detail below.

Geodesic 5 emanates from the centre $M = 0$, where the Big Bang function has a maximum. The tangent to the geodesic is horizontal there. Geodesics to the lower right of this one all begin with a vertical tangent. Likewise, the geodesics meet the Big Crunch with their tangents being vertical.

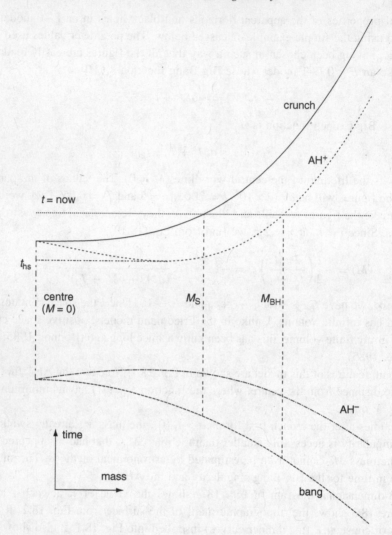

Fig. 18.1. Evolution leading to a black hole in the $E < 0$ L–T model of Eqs. (18.96)–(18.97). The current state is defined at the instant $t = t_2 =$ now. Worldlines of dust particles are vertical straight lines, each has a constant mass-coordinate. Intersections of the line $t = t_2$ with the lines representing the Big Crunch and the future apparent horizon determine the masses M_S and M_{BH}, respectively.

By the time the crunch has formed at $t = t_C(0)$, the future apparent horizon already exists (see Figs. 18.1 and 18.3). The shells of constant values of M first go through the AH, and then hit the singularity at $t = t_C(M)$. We assume that at the time $t = t_2$ the singularity has accumulated the mass M_S, while the mass hidden inside the apparent horizon at the same time is $M_{BH} > M_S$. Both of them grow with time. From the definitions of M_S and M_{BH} it follows that

$$t_2 = t_C(M_S) = t_{AH^+}(M_{BH}). \tag{18.99}$$

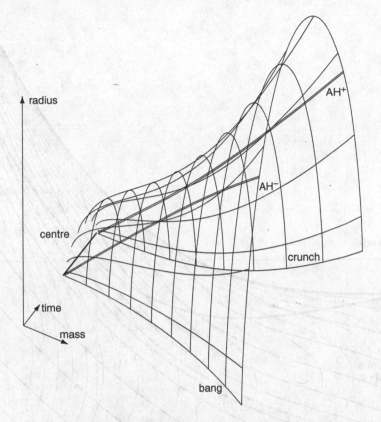

Fig. 18.2. A 3-dimensional graph of the black hole formation process from Fig. 18.1: the areal radius as a function of M and t. Each shell of constant mass evolves in a plane given by $M =$ constant. It starts at $R = 0$, then escapes from the past apparent horizon AH$^-$, then reaches maximum R, then falls into the future apparent horizon AH$^+$, and finally hits the Big Crunch. Note that the surface intersects the $R = 0$ plane perpendicularly all along the $R = 0$ contour. The apparent horizons are intersections of the $R(M, t)$ surface with the plane $R = 2M$. (In the $k > 0$ Friedmann model, the bang and the crunch would be parallel straight lines, and all the $R(t)|_{M=\text{const}}$ curves would be identical.)

Even though the model has a rather simple geometry, locating the event horizon is quite a complicated task that requires complete knowledge of the whole spacetime, including the null infinity. Hence, in a real Universe, where our knowledge (mostly incomplete and imprecise) is limited to a relatively small neighbourhood of our past light cone and our past worldline, the event horizon simply cannot be located by astronomical observations.

The future event horizon (EH) is formed by those null geodesics that fall into the future apparent horizon 'as late as possible', i.e. approach it asymptotically. Hence, in order to locate the event horizon, we must issue null geodesics backwards in time from the 'future endpoint' of the AH$^+$. This cannot be done in the (M, t) coordinates used so far because the spacetime and the AH$^+$ are infinite. Hence, we must first compactify the spacetime. The most convenient compactification for considering null geodesics at a null infinity is,

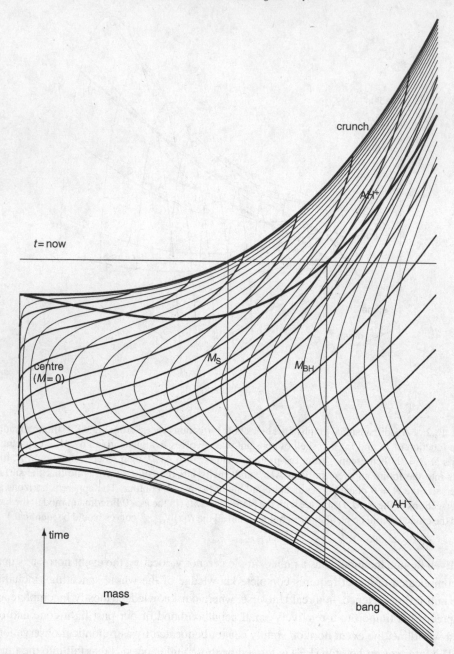

Fig. 18.3. Contours of constant R-value (the thinner lines) and outgoing radial null geodesics inscribed into the spacetime diagram of Fig. 18.1. The R-values on consecutive contours differ by always the same amount.

theoretically, the Penrose (1964) transform that spreads the null infinities into finite sets. However, in order to find a Penrose transformation, one must first choose null coordinates, and in the L–T model this has so far proven to be an impossible task (Stoeger, Ellis and Nel, 1992; Hellaby, 1996a). Hence, we will use a less convenient compactification that will squeeze the null infinities into single points in the 2-dimensional (time–radius) spacetime diagram. It is provided by the transformation

$$M = \tan \mu, \qquad t = \tan \tau. \qquad (18.100)$$

In these coordinates, the $\mathbb{R}^1 \times \mathbb{R}^1_+$ space of Figs. 18.1–18.3 becomes the finite $[-\pi/2, \pi/2] \times [0, \pi/2]$ rectangle. See Fig. 18.4. The upper curve in Fig. 18.4 is the Big Crunch singularity; the future apparent horizon runs so close to it that it seems to coincide with it.[1] The horizontal line is the $\tau =$ now line. The lower curve is the Big Bang and the past apparent horizon, again running one on top of the other. The point on the τ-axis where the three lines meet is the image of the $M = 0$ line of Fig. 18.1, squeezed here into a point because of the scale of this figure.

The theoretical method to locate the future event horizon would now be to run a radial null geodesic backwards in time from the point $(\mu, \tau) = (\pi/2, \pi/2)$, i.e. from the image of the future end of the AH^+. However, for the most part, the AH^+ runs so close to the crunch singularity, and the geodesics intersecting the AH^+ are so nearly tangent to AH^+, that numerical instabilities crash any such geodesic into the singularity instantly. This happens all the way down to $\mu = 1.1$ at double precision. The calculation succeeded, with double precision, only at $\mu = 1.0$, and a null geodesic could be traced from there to the centre at $\mu = 0$. On the scale of Fig. 18.4, this whole geodesic seems to coincide with the crunch and the AH^+. However, it is well visible if one closes in on the image of the area shown in Fig. 18.1; the closeup is shown in the inset.

The event horizon is transformed back to the (M, t) coordinates and written into the frame of Figs. 18.1–18.3 in Fig. 18.5. As stated earlier, the event horizon is located between geodesics 5 and 6 from the lower right in Fig. 18.3. By accident (caused by our choice of numerical values in this example), the EH hits the centre very close to the central point of the Big Bang, but does not coincide with it.

This whole construction should make it evident that there is no chance to locate the event horizon by astronomical observations, even approximately. It only makes sense, in the observational context, to speak about an upper limit on the mass inside the *apparent horizon*.

18.10 * Shell crossings and necks/wormholes

The mass density in the L–T model becomes infinite where $R_{,r} = 0 \neq M_{,r}$. This singularity is called **shell crossing** because at those locations the radial distance between two adjacent

---

[1] From Eqs. (18.93)–(18.94), the time difference between the crunch and the AH^+ goes to infinity when $M \to \infty$. However, the *ratio* of this time difference to the crunch time goes to zero, which explains why the two curves in Fig. 18.4 meet at the image of the infinity. The same is true for the Big Bang and the AH^-.

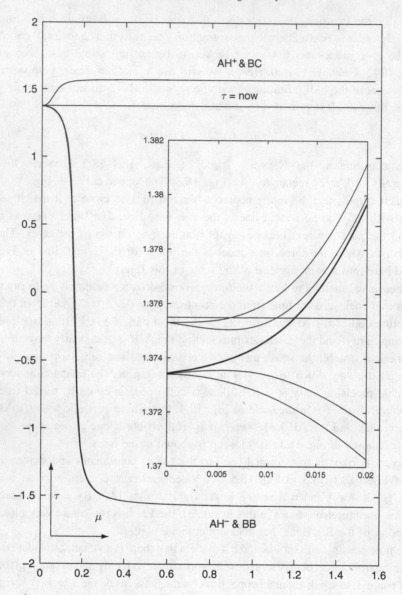

Fig. 18.4. The spacetime diagram of Fig. 18.1 compactified according to Eq. (18.100). The region shown in Fig. 18.1 is squeezed into the point where the three lines meet at the τ-axis. The worldlines of dust are still vertical straight lines here. The upper curve is the future apparent horizon (AH$^+$) and the Big Crunch (BC) singularity, they seem to coincide at the scale of this picture. The lower curve is the past apparent horizon (AH$^-$) and the Big-Bang (BB) singularity, again coinciding only spuriously. The horizontal straight line is the $\tau =$ now time. Inset: a closeup view of the image (in the coordinates (μ, τ)) of the region shown in Fig. 18.1. The thicker line is the event horizon. It does not really hit the central point of the Big Bang; the apparent coincidence is just an artefact of the scale.

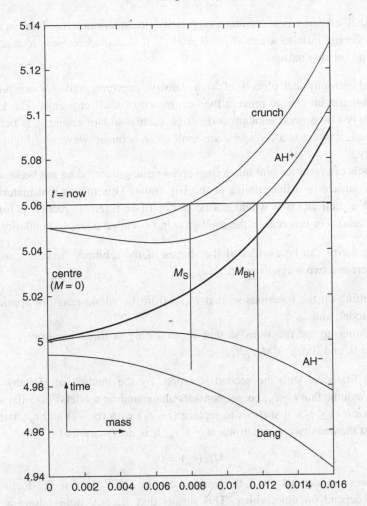

Fig. 18.5. The event horizon (thicker line) written into the frame of Fig. 18.1. Its intersection with the $M = 0$ axis does not coincide with the central point of the Big Bang; this is only an illusion created by the scale.

shells that have different values of r becomes zero. If $R_{,r}$ changes sign there, then the mass density on the other side of the shell crossing becomes negative.

The tetrad components of the Riemann tensor (in the tetrad $e^0 = dt, e^1 = (R_{,r}/\sqrt{1+2E})\,dr, e^2 = R\,d\vartheta, e^3 = R\sin\vartheta\,d\varphi$) are:

$$R_{0101} = \frac{2M}{R^3} - \frac{M_{,r}}{R^2 R_{,r}}, \qquad R_{0202} = R_{0303} = -\frac{M}{R^3} = \frac{1}{2}R_{2323},$$

$$R_{1212} = R_{1313} = \frac{M}{R^3} - \frac{M_{,r}}{R^2 R_{,r}}.$$

(18.101)

Thus, the shell crossing is a curvature singularity (the quantities listed above are scalars and some become infinite where $R,_r = 0 \neq M,_r$). It is considered less 'dangerous' than the Big Bang for two reasons:

1. In real astrophysical objects of high density, pressure gradients are present, and these should be able to prevent the occurrence of shell crossings. The L–T model is simply not general enough to describe such a situation, and it is believed that the shell crossing is a zero-pressure limit of an acoustic wave – of high, but finite, density.
2. A bundle of geodesics sent into a shell crossing singularity does not become focussed into a surface or a line, unlike at the Big Bang. This means that material objects hitting a shell crossing would not be crushed (we refer the reader to Joshi (1993) for details). For this reason the shell crossing is called a *weak* singularity.

This singularity can be avoided if the shapes of the arbitrary functions are properly chosen. There are two ways to avoid it:

(1) by setting up the functions so that $R,_r \neq 0$ in the whole range of applicability of the model; and
(2) by setting up the functions so that $R,_r = 0$ only at those locations $r = r_w$ where $M,_r = 0$, and $\lim_{r \to r_w} |M,_r / R,_r| < \infty$.

We shall first deal with the second situation, by the method of Hellaby and Lake (1985). We assume that $t \geq t_B$, i.e. we consider an expanding model. For collapse towards the Big Crunch ($t \leq t_C$), it suffices to replace $(t - t_B)$ with $(t_C - t)$ and $t_{B,r}$ with $(-t_{C,r})$. In order that the density remains finite at $r = r_w$, it is necessary that

$$M,_r (r_w) = 0. \qquad (18.102)$$

Since M does not depend on t, the density can be finite only if the locus of $R,_r = 0$ does not depend on time, either. This means that $R,_r = 0$ holds along a worldline of dust. Equation (18.102) is only a necessary condition for finite mass density. The sufficient condition is that the limit of $M,_r / R,_r$ at $r = r_w$ is finite. This condition becomes particularly simple if we choose M as the radial coordinate, it is then

$$\lim_{M \to M(r_w)} (R^3),_M > 0. \qquad (18.103)$$

From (18.20) we then easily see that for the $E = 0$ model, if $R,_r = 0 = M,_r$ at $r = r_w$, then automatically $t_{B,r}(r_w) = 0$. For the other two models, we find from (18.19) and (18.21)

$$R,_r = \left(\frac{M,_r}{M} - \frac{E,_r}{E} \right) R + \left[\left(\frac{3}{2} \frac{E,_r}{E} - \frac{M,_r}{M} \right)(t - t_B) - t_{B,r} \right] R,_t. \qquad (18.104)$$

Since $R,_t = \pm\sqrt{2M/R + 2E}$ is a different function of time than R, the equation $R,_r = 0 = M,_r$ can hold over all times only if the coefficients of $R,_t$ and R vanish identically, i.e.

$$E,_r = 0, \qquad t_{B,r} = 0. \qquad (18.105)$$

Thus the necessary conditions for the absence of a shell crossing at such $r = r_w$ at which $R_{,r} = 0$ are (18.102), (18.104) and (18.105).[1] If these conditions hold and $1 + 2E(r_w) \neq 0$, then Eq. (18.16) shows that the metric component g_{11} goes to zero. However, since we have already assumed that $M_{,r}/R_{,r}$ is finite at $r = r_w$, Eqs. (18.101) show that the curvature is nonsingular there. Hence, with (18.102), (18.103) and (18.105) all fulfilled, while $1 + 2E(r_w) \neq 0$, the locus $r = r_w$ is only a coordinate singularity. It can be removed by a coordinate transformation that also removes the property $R_{,r} = 0$, i.e. makes at least one of the functions $M_{,r}$, $E_{,r}$ and $t_{B,r}$ non-zero at r_w.

The situation is different when $E(r_w) = -1/2$. Equations (18.90) and (18.94) show then that the past and future apparent horizons meet at $r = r_w$, $t = t_w = t_{AH}(r_w)$, which is the instant of maximal expansion. An outgoing radial null geodesic that would leave the past AH through the hypersurface $r = r_w$ would immediately fall inside the future AH. An outgoing radial null geodesic that is already out of the past AH with $r < r_w$ must enter the future AH with $t > t_w$, $r < r_w$. Thus, no light ray can be sent from the region $\{t_{AH^-} \leq t \leq t_{AH^+}, r < r_w\}$ to the region $\{t_{AH^-} \leq t \leq t_{AH^+}, r > r_w\}$. In this case, the locus $r = r_w$ is called a **neck** or a **wormhole**, and it is a generalisation of the Kruskal–Szekeres 'throat' at $r = 2m$ of the Schwarzschild solution. In the vacuum limit ($M_{,r} = 0$ over an extended region), the neck goes over into the Kruskal–Szekeres throat. In the nonvacuum case, the neck need not be mirror-symmetric with respect to $(r - r_w)$. The possibility of existence of such a neck was first noted by Barnes (1970). Then, a generalisation of it appeared in numerical studies of a model with nonzero pressure gradients by Suto et al. (1984). For illustrative examples of necks see Hellaby (1987), Hellaby and Krasiński (2002), Krasiński and Hellaby (2004b) and Fig. 18.6.[2]

Now we come to the conditions for avoiding shell crossings at those points where $M_{,r} \neq 0$. We wish to translate the condition $R_{,r} \neq 0$ into properties of the functions $M(r)$, $E(r)$ and $t_B(r)$. The cases $R_{,r} > 0$ and $R_{,r} < 0$ have to be considered separately. We will write out the conditions only for $R_{,r} > 0$. For $R_{,r} < 0$, the inequalities (18.106), (18.110), (18.111), (18.116) and (18.117) must have their senses inverted.

In all cases, in those regions where $R_{,r} > 0$, in order that the mass density is positive, we must have

$$M_{,r} > 0. \tag{18.106}$$

From here on, the three types of models have to be considered separately.

18.10.1 E < 0

Making use of (18.14) for expansion ($R_{,t} > 0$) and of (18.19), we rewrite (18.104) as

$$\frac{R_{,r}}{R} = \left(\frac{M_{,r}}{M} - \frac{E_{,r}}{E}\right) + \left(\frac{3}{2}\frac{E_{,r}}{E} - \frac{M_{,r}}{M}\right)\Phi_1(\eta) - \frac{(-2E)^{3/2}}{M}t_{B,r}\Phi_2(\eta) \overset{\text{def}}{=} f(\eta), \tag{18.107}$$

---

[1] Sufficient conditions more specific than (18.103) are difficult to formulate.

[2] The functions used in drawing Fig. 18.6 are as follows: $M(r) = M_w + d(r - r_w)^2$, $t_B(r) = t_{B0} - bd^2(r - r_w)^2$, $E(r) = -(1/2)M_w/[M_w + d(r - r_w)^2] + ad(r - r_w)^2$ and $t_C(r) = 2\pi M(r)/(-2E(r))^{3/2} + t_B(r)$ – the crunch time; the parameters are $M_w = 20.0$, $d = 0.1$, $r_w = 2.0$, $t_{B0} = 10.0$, $b = 2000.0$ and $a = 1.0$.

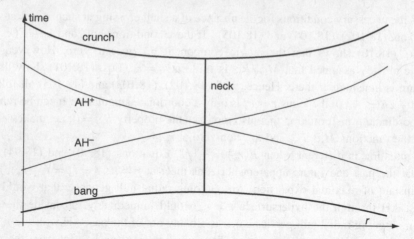

Fig. 18.6. An example of a spacetime region around a neck. The neck is the vertical straight line, which represents a world-tube with spherical cross-sections $r = r_w$. The two apparent horizons touch each other at the neck, at the instant of maximal expansion of the model. Every future-directed radial null geodesic that touches the AH⁻ is forced to leave the AH⁻; every such geodesic that touches the AH⁺ is forced to fall into the AH⁺. This is why there is no communication between the region outside both AHs to the left of the neck and the corresponding region to the right of the neck. The functions in the figure are mirror-symmetric with respect to the neck, but in general they would not be.

where

$$\Phi_1(\eta) \overset{\text{def}}{=} \frac{\sin \eta(\eta - \sin \eta)}{(1 - \cos \eta)^2}, \qquad \Phi_2(\eta) \overset{\text{def}}{=} \frac{\sin \eta}{(1 - \cos \eta)^2}. \tag{18.108}$$

The function $f(\eta)$ should be strictly positive in the whole range $\eta \in [0, 2\pi]$. Note that

$$\lim_{\eta \to 0} \Phi_1(\eta) = 2/3, \qquad \lim_{\eta \to 2\pi} \Phi_1(\eta) = -\infty,$$

$$\lim_{\eta \to 0} \Phi_2(\eta) = \infty, \qquad \lim_{\eta \to 2\pi} \Phi_2(\eta) = -\infty, \tag{18.109}$$

$$\lim_{\eta \to 2\pi} \frac{\Phi_1(\eta)}{\Phi_2(\eta)} = 2\pi.$$

Hence, taking (18.107) in the limit $\eta \to 0$, we see that the last term becomes unbounded, and it will be positive only if

$$t_{B,r} < 0. \tag{18.110}$$

Now we take (18.107) in the limit $\eta \to 2\pi$. Then the last two terms become unbounded. Factoring out Φ_2, which goes to $-\infty$, and demanding that the (bounded) coefficient is negative in the limit, we obtain:

$$2\pi \left(\frac{3}{2} \frac{E_{,r}}{E} - \frac{M_{,r}}{M} \right) - \frac{(-2E)^{3/2}}{M} t_{B,r} < 0. \tag{18.111}$$

Equations (18.106), (18.110) and (18.111) are the necessary conditions for the absence of shell crossings in the case $R_{,r} > 0$. They are also sufficient (see Exercise 10). The meaning of (18.111) is that the crunch time must be an increasing function of r.

Figure 18.3 can now be used to illustrate how Eqs. (18.110) and (18.111) work. Recall that mass is used as the radial coordinate there, so $M_{,r} = 1 > 0$ is assured. Since the worldlines of the dust source are constant-M lines (i.e. vertical straight lines), they can never intersect in a t–M diagram, even at shell crossings. A shell crossing would show in this picture as a point where a constant-R contour has a horizontal tangent. (Apart from the central points of the two singularities, $M = 0$ at $R = 0$, no such points exist in this example, because the functions $E(M)$ and $t_{\rm B}(M)$ were chosen appropriately.) The centre of symmetry $M = 0$, the Big Crunch and the Big Bang together form the $R = 0$ contour. Adjacent contours of small constant R must have a similar shape. Hence, if either of the two equations were not fulfilled, either the upper branch or the lower branch of some contours would be a non-monotonic function, whose derivative by M would change sign somewhere. At the changeover points, the tangents to the contours would be horizontal, and these would be the shell crossings.

18.10.2 $E = 0$

We calculate from (18.20):

$$\frac{R_{,r}}{R} = \frac{M_{,r}}{3M} - \frac{\sqrt{2M}\, t_{{\rm B},r}}{R^{3/2}}. \tag{18.112}$$

As $R \to 0$, the second term dominates, so, in order that $R_{,r} > 0$ everywhere, we must have $t_{{\rm B},r} < 0$. Together with (18.106) this is then seen to be the necessary and sufficient condition.

18.10.3 $E > 0$

The analogue of (18.107)–(18.108) is here

$$\frac{R_{,r}}{R} = \left(\frac{M_{,r}}{M} - \frac{E_{,r}}{E}\right) + \left(\frac{3}{2}\frac{E_{,r}}{E} - \frac{M_{,r}}{M}\right)\Phi_3(\eta) - \frac{(2E)^{3/2}}{M} t_{{\rm B},r}\Phi_4(\eta), \tag{18.113}$$

where

$$\Phi_3(\eta) \stackrel{\text{def}}{=} \frac{\sinh\eta(\sinh\eta - \eta)}{(\cosh\eta - 1)^2}, \qquad \Phi_4(\eta) \stackrel{\text{def}}{=} \frac{\sinh\eta}{(\cosh\eta - 1)^2}. \tag{18.114}$$

The following properties of $\Phi_3(\eta)$ and $\Phi_4(\eta)$ are useful in calculations:

$$\lim_{\eta \to 0} \Phi_3(\eta) = 2/3, \qquad \lim_{\eta \to \infty} \Phi_3(\eta) = 1,$$

$$\lim_{\eta \to 0} \Phi_4(\eta) = \infty, \qquad \lim_{\eta \to \infty} \Phi_4(\eta) = 0,$$

$$\frac{{\rm d}\Phi_3}{{\rm d}\eta} > 0 \quad \text{for } \eta > 0, \qquad \frac{{\rm d}\Phi_4}{{\rm d}\eta} < 0 \quad \text{for } \eta > 0. \tag{18.115}$$

Now take (18.113) at $\eta \to 0$. The last term dominates, and it will be positive if

$$t_{B,r} < 0. \tag{18.116}$$

Taking (18.113) at $\eta \to \infty$ we easily obtain

$$E_{,r} > 0. \tag{18.117}$$

Equations (18.106), (18.116) and (18.117) are necessary conditions for $R_{,r} > 0$. To see that they are also sufficient, it suffices to rewrite (18.113) in the form

$$\frac{R_{,r}}{R} = \frac{M_{,r}}{M} (1 - \Phi_3) + \frac{E_{,r}}{E} \left(\frac{3}{2}\Phi_3 - 1 \right) - \frac{(2E)^{3/2}}{M} t_{B,r} \Phi_4(\eta) \tag{18.118}$$

and take note of (18.115).

18.11 The redshift

In several applications of a cosmological model, we need to calculate the redshift of light emitted by a source at a given time and location and received by an observer located down a null geodesic from the source. In fact, we already have a ready-to-use formula, Eq. (16.20), but the field of tangent vectors to the geodesic, k^α, must be affinely parametrised. Transforming it to such a parametrisation is usually not easy, so, for numerical calculations, it is often more convenient to use other methods.

One of them is the following method (copied from Bondi (1947)). From (18.16), for a radial null geodesic proceeding towards the observer, we find

$$\frac{dt}{dr} = -\frac{R_{,r}(t, r)}{\sqrt{1 + 2E(r)}}. \tag{18.119}$$

Let two light rays be emitted in the same direction, the second one later by a small time-interval τ. Let the equation of the first ray be

$$t = T(r) \tag{18.120}$$

and that of the second ray $t = T(r) + \tau(r)$. Both rays must obey (18.119), so

$$\frac{dT}{dr} = -\frac{R_{,r}(T(r), r)}{\sqrt{1 + 2E(r)}}, \qquad \frac{d(T + \tau)}{dr} = -\frac{R_{,r}(T(r) + \tau(r), r)}{\sqrt{1 + 2E(r)}}. \tag{18.121}$$

Since $\tau(r)$ was assumed small, we have, to first order in τ,

$$R_{,r}(T(r) + \tau(r), r) = R_{,r}(T(r), r) + \tau(r) R_{,tr}(T(r), r). \tag{18.122}$$

Using (18.122) and the first of (18.121) in the second of (18.121) we obtain

$$\frac{d\tau}{dr} = -\tau(r) \frac{R_{,tr}(T(r), r)}{\sqrt{1 + 2E(r)}}. \tag{18.123}$$

If τ at emission is the period of a wave, then, comparing it with the corresponding period at the point of observation, we obtain

$$\frac{\tau(r_{\text{obs}})}{\tau(r_{\text{em}})} = 1 + z(r_{\text{em}}). \tag{18.124}$$

Keeping the observer at a fixed position and considering the sources at two distances, r_{em} and $(r_{\text{em}} + dr)$, we find by differentiating the above that $(d\tau/dr)/\tau = -(dz/dr)/(1+z)$. Using this in (18.123):

$$\frac{1}{1+z}\frac{dz}{dr} = \frac{R,_{tr}(T(r),r)}{\sqrt{1+2E(r)}}. \tag{18.125}$$

Hence, the redshift may be calculated numerically from:

$$\ln(1 + z(r)) = \int_{r_{\text{em}}}^{r_{\text{obs}}} \frac{R,_{tr}(T(r),r)}{\sqrt{1+2E(r)}} dr. \tag{18.126}$$

This formula is equivalent to (16.20) (see Exercise 11), but here the parameter on the null geodesic is just the coordinate r.

Exercise 11 also shows how to introduce the affine parametrisation for the field k^α. That parametrisation can be introduced by the general method presented in Section 5.2, and we shall do it here.

Let us choose the parametrisation in (18.119) so that

$$k^1 = -1, \qquad k^0 = \frac{R,_r}{\sqrt{1+2E}}. \tag{18.127}$$

The first equation above says that the parameter on the geodesic is

$$\tau = -r. \tag{18.128}$$

Since this parametrisation is not affine, k^α obeys, according to (5.11), $k^\alpha{}_{;\beta} k^\beta = -(d\ln\lambda/d\tau)k^\alpha$. Substituting here the k^α from (18.127) we find the coefficient to be

$$-\frac{1}{\lambda}\frac{d\lambda}{d\tau} = 2\frac{R,_{tr}}{\sqrt{1+2E}} - \frac{R,_{rr}}{R,_r} + \frac{E,_r}{1+2E}. \tag{18.129}$$

Now the meaning of all the symbols has to be precisely understood. The differential equation (18.129) holds along a single null geodesic, so all quantities in it have to be taken along that geodesic. In particular, wherever any function depends on t, the coordinate t in it must be replaced by the $T(r)$ of (18.120). The derivatives of R in (18.129) had been taken before the quantities were specialised to the null geodesic, so they are partial derivatives. Hence, $R,_{rr}$ is not the total derivative of $R,_r$ by r, but just the partial derivative by the second argument of $R,_r$. It is related to the total derivative, denoted D/dr, by

$$\frac{DR,_r(T(r),r)}{dr} = R,_{tr}(T(r),r)\frac{dT}{dr} + R,_{rr}(T(r),r). \tag{18.130}$$

Substituting from (18.121) for dT/dr we obtain from the above

$$R_{,rr}(T(r),r) = \frac{DR_{,r}(T(r),r)}{dr} + R_{,tr}(T(r),r)\frac{R_{,r}(T(r),r)}{\sqrt{1+2E(r)}}. \tag{18.131}$$

Substituting this in (18.129) and recalling that $\tau = -r$ we obtain

$$\frac{1}{\lambda}\frac{d\lambda}{dr} = \frac{R_{,tr}(T(r),r)}{\sqrt{1+2E(r)}} - \frac{1}{R_{,r}(T(r),r)}\frac{DR_{,r}(T(r),r)}{dr} + \frac{E_{,r}}{1+2E(r)}. \tag{18.132}$$

Now this can be integrated with the result

$$\lambda = C\frac{\sqrt{1+2E}}{R_{,r}(T(r),r)}\exp\left(\int \frac{R_{,tr}(T(r),r)}{\sqrt{1+2E(r)}}\,dr\right), \tag{18.133}$$

where C is an arbitrary constant. From (5.12), the parameter τ of (18.127)–(18.128) is related to the affine parameter s by $d\tau/ds = \lambda/C$, so the field k^{α} of (18.127)–(18.128) is related to the same field in the affine parametrisation, denoted $\tilde{k}^{\alpha}$, by $\tilde{k}^{\alpha} = (\lambda/C)k^{\alpha}$. Consequently, the components of the affinely parametrised field are

$$\tilde{k}^0 = \exp\left(\int \frac{R_{,tr}(T(r),r)}{\sqrt{1+2E(r)}}\,dr\right),$$

$$\tilde{k}^1 = -\frac{\sqrt{1+2E}}{R_{,r}(T(r),r)}\exp\left(\int \frac{R_{,tr}(T(r),r)}{\sqrt{1+2E(r)}}\,dr\right). \tag{18.134}$$

18.12 The influence of inhomogeneities in matter distribution on the cosmic microwave background radiation

Saez and collaborators (Arnau *et al.*, 1993, 1994; Saez *et al.*, 1993) studied the interaction of the cosmic microwave background (CMB) with inhomogeneities in matter distribution placed in the path of the CMB rays. Since it was a major numerical project, its results will be only briefly reported here.

The CMB radiation has, to a high accuracy, the black-body spectrum (17.73). During the period to which the L–T model can be applied (after radiation had decoupled from matter), photons do not interact with matter and it is assumed that their number is conserved. In a Friedmann model, this conservation implies $d[I(\nu(t))R^3(t)]/dt = 0$. Then, again in a Friedmann model, Eqs. (16.20) and (17.13) imply $\nu R = $ constant. Thus, *assuming* that the black body spectrum is preserved in time, we conclude from (17.73) and (17.13) that the temperature of the radiation changes from the emission instant t_e to the observation instant t_o by the law

$$T(t_o) = \frac{T(t_e)}{1+z}. \tag{18.135}$$

This can be generalised to the L–T model by observing that in small neighbourhoods it behaves like a Friedmann model, so Eq. (18.135) can be applied to nearby points on the same null geodesic, with parameter values r and $(r + dr)$, thus $d(\ln T)/dr = -d[\ln(n + z)]/dr$. In this way, differences in temperature at reception between light rays going along different paths can be calculated. The quantity used in calculations is the **temperature contrast** $\Delta T/T$, where T is the temperature along a ray that has been propagating through a Friedmann region all the way and ΔT is the difference between T and the observed temperature in a given direction.

This approach was first applied by Raine and Thomas (1981). They considered a large-scale but small-amplitude condensation in the path of light rays of the microwave background reaching an observer. The rays were assumed to be emitted and received in a Friedmann region. The equations of null geodesics were integrated numerically in order to calculate the temperature at the reception point and compare it with the temperature at endpoints of rays propagating all the way through a Friedmann medium. The temperature variation was calculated as a function of the direction of observation.

The most comprehensive study by this method was done by Arnau, Fullana, Monreal and Saez (Arnau *et al.* 1993, 1994; Saez *et al.*, 1993). They computed numerically the dependence of the temperature contrast on the direction of observation for the microwave background radiation for a model consisting of a Friedmann background, with an arbitrary density parameter $\Omega = \rho/\rho_{cr}$ (where ρ is the actual mean mass density in the Universe and ρ_{cr} is the critical density defined by (17.30)), and with a localised L–T perturbation superimposed on it. The numerical code allowed them to fit a model with a few free parameters (describing the density profile of the condensation, the background density parameter and the Hubble parameter) to the observed characteristics of the Great Attractor or of the Virgo cluster, and then to calculate the anisotropy in temperature of the background radiation caused by these condensations. The authors calculated the temperature anisotropies in the background radiation produced by models of the Great Attractor with different density and velocity profiles, and with different density parameters of the Friedmann background. Specifically, they calculated the effects of the following:

1. Various velocity profiles in the attractor model with a fixed background density parameter and fixed distance from the observer.
2. Various background density parameters with a fixed velocity profile and the distance adjusted so as to produce the largest effect.
3. Various distances to the condensation with a fixed velocity profile and a fixed background density parameter.

The result was that the maximal anisotropy to be expected is up to 3×10^{-5} (when $\Omega = 0.15$), at the angular scale of $10°$. This agrees with the current measurements, which give 2×10^{-5} (WMAP, 2004).[1]

---

[1] The value of this anisotropy invites a comment. The high degree of isotropy of the CMB radiation has frequently been used as an argument in favour of the Friedmann models, and the 'reasoning' was that inhomogeneities in matter distribution would leave an imprint on the radiation. However, for a long time these claims had not been backed up by any calculations. The results of Saez *et al.* show that no trace of such an imprint could show up until the precision of measurements of temperature anisotropies reached the level of 10^{-6}, which happened only in 1993 (Mather *et al.*, 1993).

18.13 Matching the L–T model to the Schwarzschild and Friedmann solutions

As already observed in Section 18.3, the Schwarzschild solution is the vacuum limit of the L–T model that results when $M,_r = 0$. For various purposes, a spacetime model is sometimes considered consisting of the L–T solution in the region $r \leq r_b$ matched to the Schwarzschild solution in the region $r \geq r_b$. The Schwarzschild solution should then be represented in the Lemaître–Novikov coordinates, as in (14.116)–(14.117). As shown in Section 12.17, the matching conditions to (18.16) require that

$$R_{LT}(t, r_b) = R_S(t, r_b), \qquad R,_r /N|_{LT} = R,_r /N|_S, \qquad (18.136)$$

where S stands for 'Schwarzschild' and LT for 'Lemaître–Tolman', and

$$N = \frac{R,_r}{\sqrt{1+2E}} \qquad (18.137)$$

in both metrics. As a result, $R,_r$ cancels out in the equations, and the matching conditions reduce just to the continuity of E, R and $R,_t$, i.e.

$$E(r_b)|_{LT} = E(r_b)|_S, \qquad t_B(r_b)|_{LT} = t_B(r_b)|_S,$$
$$M(r_b) = m, \qquad (18.138)$$

where m is the mass parameter of the Schwarzschild solution. What is obtained after the matching is in fact a single L–T model with $M(r)$ going over into the constant function $M = m$ at $r = r_b$.

Sometimes, the matching between the L–T and Friedmann spacetimes is considered; for example, when discussing the formation of galaxy clusters or voids (Krasiński and Hellaby, 2002, 2004, 2005, and Section 18.6). It is then assumed that at a certain distance from the centre of the cluster or void the L–T metric goes over into an unperturbed Friedmann model. Equations (18.136)–(18.138) must apply also then, with the subscript 'S' now referring to Friedmann. In particular, the last equation of (18.138) now reads $M_{LT}(r_b) = M_S(r_b)$, where $M_S(r_b)$ is the Friedmann mass function calculated at $r = r_b$. This means that the Friedmann mass removed from within $r = r_b$ must be the same as the L–T mass filling that sphere. This is a necessary and sufficient condition for matching because the other two equations in (18.138) can always be fulfilled: the first determines the value of the Friedmann curvature index k, the second can be fulfilled by time-translation in the Friedmann spacetime.

As observed by Ribeiro (1992a), this equality of masses implies that in such a configuration any region of density higher or lower than that in the Friedmann background must be compensated by a region of a lower or higher density, respectively.

18.14 * General properties of the Big Bang/Big Crunch singularities in the L–T model

At all points apart from the centre of symmetry, both these singularities are spacelike. This is verified as follows: since the singularities are parts of the set $R = 0$, calculate the normal vector field to any hypersurface $R =$ constant. It has components $(R_{,t}, R_{,r}, 0, 0)$, so $g^{\mu\nu} n_{,\mu} n_{,\nu} = g^{00} R_{,t}^2 + g^{11} R_{,r}^2$, which, using (18.14) and (18.16), becomes

$$g^{\mu\nu} n_{,\mu} n_{,\nu} = \frac{2M}{R} - 1. \tag{18.139}$$

This is negative for $R > 2M$, zero at the apparent horizon and positive for $R < 2M$. Consequently, the vector field n_μ is, respectively, spacelike, null or timelike, i.e. the hypersurface $R =$ constant is timelike, null or spacelike. Thus, as we approach the nonsingular centre of symmetry $R \to 0$, the regularity condition (18.31) implies that $2M/R \to 0$, so the nonsingular part of the set $R = 0$ is timelike. As we approach the singularity $R \to 0$ along a line $M > 0$, the expression (18.139) tends to $+\infty$, so the part of $R = 0$ in which $M > 0$ is spacelike.

The orientation of the central part of the singularity, where the line of centre of symmetry hits the Big Bang or Big Crunch set, is not simple to determine, since the comoving coordinates have a singularity there, too, and they do not allow one to see the geometric relations clearly. It had taken quite an effort to notice that the locus in which the two parts of the set $R = 0$ meet need not be a single point, but, under certain circumstances, may be a finite segment of a timelike or null curve.

The first to note the possibility of this peculiar behaviour (during a numerical investigation of an $E = 0$ model) were Eardley and Smarr (1979), and they called the phenomenon **shell focussing**. They stated that the segment of the centre of symmetry in which shell focussing occurs can only be null. However, the criteria for the occurrence of a shell focussing, which they gave without derivation, do not look entirely credible because they refer to the limit of t_B/M as $r \to r_c$, while the value of t_B is coordinate-dependent. Hellaby, in his unpublished Ph.D. Thesis,[1] found that the segment in question can be timelike as well.

If the singularity is all spacelike, then its intersection with the worldline of the centre of symmetry can be imagined as a single point – no future-directed light ray can leave the singular set. However, if the singularity contains a timelike or null segment, then, as will be shown below, that segment is a common vertex of an infinite family of distinct light cones. The conclusion is that the non-spacelike segment of the singularity is an extended arc of a curve that is mapped into a single point in the comoving coordinates, which have their own singularity there.

The discussion below was inspired by the paper of Christodoulou (1984) and the unpublished work of Hellaby and Lake (1988), but is a great simplification of both. Take the $E = 0$ L–T model with the function $M(r)$ chosen as follows:

$$M(r) = M_0 r^3, \tag{18.140}$$

---

[1] We are grateful to C. Hellaby for giving us access to his thesis.

where M_0 is a constant. (This is a definition of the coordinate r.) Since we will be discussing collapse, we set in (18.20):

$$R(t, r) = \left\{ \frac{9}{2} M(r) \left[t_B(r) - t \right]^2 \right\}^{1/3} \tag{18.141}$$

so that t increases from an initial value to $t = t_B(r)$, which is now the final (Big Crunch) singularity. Now choose $t_B(r)$:

$$t_B(r) = ar^2, \tag{18.142}$$

where $a > 0$ is a constant, so that the Big Crunch starts at $r = 0$ with $t = 0$ and then proceeds to greater r-values as t increases (see Fig. 18.7). We find then that the equation of an outgoing radial null geodesic is

$$\left(\frac{dt}{dr} \right)_n = \left(\frac{9M_0}{2} \right)^{1/3} \frac{\frac{7}{3} ar^2 - t}{(ar^2 - t)^{1/3}}. \tag{18.143}$$

The set where $t = 7ar^2/3$ would be a shell crossing singularity, but it lies to the future of the Big Crunch, and hence does not belong to the physical spacetime. It is seen that, as we approach the Big Crunch singularity along any line that terminates at $t = ar^2$ with a nonzero r-value, the derivative dt/dr becomes infinite, i.e. the geodesics hit the Big Crunch set vertically in a (t, r) diagram. As we approach any nonsingular point of the centre of symmetry line, $\{r = 0, t < t_B\}$, the derivative dt/dr has a finite nonzero limit. In order to note what happens as we approach the intersection of the centre of symmetry with the Big Crunch along an arbitrary curve $t = f(r)$, write (18.143) in the form

$$\left(\frac{dt}{dr} \right)_n = \left(\frac{9M_0}{2} \right)^{1/3} r^{4/3} \frac{\frac{7}{3} a - t/r^2}{(a - t/r^2)^{1/3}}, \tag{18.144}$$

and consider the various possible limits of t/r^2 at $r \to 0$ along different curves. The limit of t/r^2 may be zero, finite non-zero, or infinite. In the first two cases, (18.144) clearly shows that

$$\lim_{r \to 0, t \to 0} \left(\frac{dt}{dr} \right)_n = 0. \tag{18.145}$$

In the third case, when the limit of t/r^2 is infinite, it is useful to rewrite (18.144) in yet another form

$$\left(\frac{dt}{dr} \right)_n = \left(\frac{9M_0}{2} \right)^{1/3} t^{2/3} \frac{7ar^2/(3t) - 1}{(ar^2/t - 1)^{1/3}}. \tag{18.146}$$

The infinite limit of t/r^2 means zero limit of r^2/t, so (18.145) again holds. Thus, the limiting slope of a light ray emitted from the point p_0 of coordinates $(t, r) = (0, 0)$ consistently comes out to be zero for all nonsingular paths of approach to p_0 (see Fig. 18.7), but when this point is approached along the Big Crunch singularity, the tangent to the light ray is vertical. This suggests that there is some structure hidden in p_0.

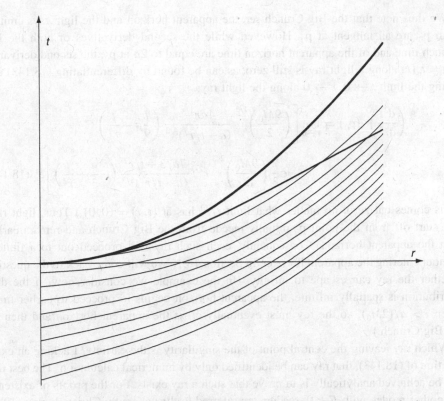

Fig. 18.7. Outgoing radial light rays in the neighbourhood of the central singularity in the space-time (18.141). The vertical arrow is the axis of symmetry, the thicker parabola is the Big Crunch singularity and the lighter curve is the apparent horizon. At all points except at the centre, the rays hit the singularity vertically (the figure does not show this). At all points except at the singularity, the rays hit the axis of symmetry with non-horizontal tangents. When the common point of the axis and of the singularity is approached along any curve except the singularity itself, the direction of the outgoing light ray tends to the horizontal. Light rays emanating from the central point of the singularity are tangent to the singular set and to the apparent horizon, but d^2t/dr^2 calculated along those rays at $(t, r) = (0, 0)$ is still zero, while $d^2t_B/dr^2(0) = d^2t_{AH}/dr^2 \neq 0$, so the rays can recede from the singularity out to a finite distance, still remaining before the apparent horizon. One such ray, the uppermost in the figure, is shown.

The equation of the apparent horizon in the spacetime considered is

$$t_{AH}(r) = ar^2 - \frac{4}{3}M_0 r^3. \tag{18.147}$$

Thus, the apparent horizon surface is tangent to the Big Crunch at $r = 0$, and initially proceeds towards larger t, $t_{AH}(r)$ becomes a decreasing function of r only at $r = a/(2M_0) > 0$ (that region is not visible in Fig. 18.7, which covers only the region of $r < 0.2a/M_0$).

We thus note that the Big Crunch set, the apparent horizon and the light rays emitted from p_0 are all tangent at p_0. However, while the second derivatives of both the Big Crunch time and of the apparent horizon time are equal to $2a$ at p_0, the second derivative of $t = t_n(r)$ along a light ray is still zero, as can be found by differentiating (18.143) and taking the limit $t \to 0$, $r \to 0$ along the light ray:

$$\left(\frac{d^2 t}{dr^2}\right)_n (p_0) = \lim_{r \to 0} \left[\left(\frac{9M_0}{2}\right)^{1/3} \frac{4ar^{4/3}}{(a - t/r^2)^{4/3}} \left(\frac{7}{9}a - \frac{t}{r^2}\right) \right.$$

$$\left. + \frac{2}{3}\left(\frac{9M_0}{2}\right)^{2/3} r^{2/3} \frac{7a/3 - t/r^2}{(a - t/r^2)^{5/3}} \left(-\frac{a}{3} + \frac{t}{r^2}\right) \right]. \quad (18.148)$$

(This comes out zero, no matter what limit t/r^2 has at $(t, r) = (0, 0)$.) Thus, light rays that start off from the point p_0 initially recede from the Big Crunch and remain earlier than the apparent horizon. Consequently, each such ray will proceed out to a finite r without entering the apparent horizon. In Section 18.16 we will come back to the question whether the ray can escape to infinity. (In the example we consider now, if the dust distribution is spatially infinite, the apparent horizon begins to proceed to earlier times when $r > a/(2M_0)$, so the ray must eventually cross the apparent horizon and then hit the Big Crunch.)

Which ray leaving the central point of the singularity is the earliest? Lacking an exact solution of (18.143), that ray can be identified only by numerical calculation. The best that can be achieved analytically is to prove that such a ray exists. For the proofs of existence in another model, with $E < 0$, readers are referred to the papers by Christodoulou (1984) and Newman (1986).

Suppose that L is any such ray emitted from $p_0 = (0, 0)$. Outgoing rays cannot intersect at $(t, r) \neq (0, 0)$, and cannot be emitted into the spacetime from any other point of the Big Crunch. It follows that a ray passing through any point that lies to the future of L was also emitted from p_0. Thus there is an infinite family of light rays emanating from the point p_0 and initially receding from the Big Crunch and the apparent horizon. It follows that p_0 only looks to be a single point in the comoving coordinates; in reality it must be a timelike or null set.

We have considered just one example of an L–T model, constructed deliberately so that it shows this unusual behaviour. There are L–T models in which the central point of the Big Crunch singularity cannot emit any light ray because that singularity is all spacelike.

A time-reversed behaviour can occur at the central point of the Big Bang.

It is seen from (18.119) that the tangent to the light ray becomes horizontal wherever $R_{,r} = 0$. Using (18.104) for the $E > 0$ and $E < 0$ models and (18.20) for the $E = 0$ model we find that the terms not involving $t_{B,r}$ go to zero at any point of the singular set $t = t_B$, independently of the path of approach.[1] If $t_{B,r} \neq 0$ at $r = r_1$, then the term $t_{B,r}R_{,t}$ in (18.104) and the corresponding term in (18.20) will become infinite as $r \to r_1$ because

---

[1] To see this, note that $R_{,t} = \ell\sqrt{2M + 2ER}/\sqrt{R}$, $\ell = \pm 1$ and that $\lim_{t \to t_B} (t - t_B)/\sqrt{R} = 0$ (use the de l'Hôpital rule).

$R_{,t} \xrightarrow[r \to r_1]{} \infty$. Hence, a necessary condition for the light ray to have a horizontal tangent at a point p_1 located in the Big Bang/Big Crunch singularity is $t_{B,r}(p_1) = 0$ (the same is true for the $E = 0$ model). Whether it is also sufficient cannot be said in general because the limit of $t_{B,r} R_{,t}$ as $p \to p_1$ may depend on the path of approach to the point p_1. If we approach it along the $r = r_1$ line, then the limit is always zero.

The tangent to the light ray being vertical or horizontal in the (t, r) graph has physical consequences. If the tangent is horizontal at the emission point, then the redshift observed at any other point is infinite. If the tangent is vertical, then an observer receiving the ray elsewhere will see it infinitely *blueshifted*. This was first noted without a proof by Szekeres (1980), and then calculated by Hellaby and Lake (1984) by the method of developing the quantities present in Eq. (16.20) into truncated Taylor series in R around $R = 0$. However, the calculation turned out to be rather complicated, so we shall explain this result by intuitive arguments only.

Suppose that the tangent to the ray at the emission point is horizontal. This means that the ray proceeds along the surface of constant time, so there will be an infinite number of cycles of an electromagnetic wave in a unit of time. This just means that the frequency at the emission point is then infinite, which, by (16.20), implies an infinite redshift.

Now let the tangent at the emission point be vertical. In the comoving coordinates, the velocity of the dust is vertical everywhere, i.e. it is tangent to the light cone at emission. Thus, the dust is leaving the Big Bang with the velocity of light. Consequently, an observer sitting on a dust particle would be riding on the crest of a light wave and would see zero frequency. By (16.20), this means that the ratio of the observed frequency to the frequency at emission is infinite, so at the observation point there would be an infinite blueshift. However, this infinity is just a consequence of the frequency of the emitter being zero – the frequency at the observation point is finite. Therefore, one should not expect any unusually strong influence of the radiation on matter in this case.

In the Friedmann models, the light rays are horizontal at the Big Bang, i.e. the Big Bang is seen with an infinite redshift by any observer. Since the Big Bang in the L–T model is in general non-simultaneous, most light rays leave it vertically, which implies an instability of the Friedmann models – one more, in addition to the instability against the formation of condensations and voids.

18.15 * Extending the L–T spacetime through a shell crossing singularity

The shell crossing singularity is considered a less serious breakdown of Einstein's theory than the Big Bang – because it is believed to be just an artefact of the zero pressure gradient in the L–T model. In models with nonzero pressure gradients (which are so far unknown), it is expected that the shell crossing will be replaced by a region of high but finite density.

In this section, we shall present one more reason why a shell crossing is relatively harmless: as observed by Newman (1986), the L–T spacetime can be extended through it. The extension is provided by the Gautreau coordinates of Section 18.7. For the general

L–T model (with arbitrary E) the Gautreau coordinates are $(\tau, R, \vartheta, \varphi)$, where $\tau = t$ and R is the areal radius function obeying (18.14). Thus, in (18.16) we have

$$R,_r \, \mathrm{d}r = \mathrm{d}R - R,_t \, \mathrm{d}t = \mathrm{d}R - \ell\sqrt{2E + \frac{2M}{R} + \frac{1}{3}\Lambda R^2} \, \mathrm{d}\tau, \tag{18.149}$$

$$\mathrm{d}s^2 = \frac{1}{1+2E}\left[\left(1 - \frac{2M}{R} - \frac{1}{3}\Lambda R^2\right)\mathrm{d}\tau^2 \right.$$

$$\left. + 2\ell\sqrt{2E + \frac{2M}{R} + \frac{1}{3}\Lambda R^2} \, \mathrm{d}\tau\,\mathrm{d}R - \mathrm{d}R^2\right]$$

$$- R^2\left(\mathrm{d}\vartheta^2 + \sin^2\vartheta\,\mathrm{d}\varphi^2\right), \tag{18.150}$$

where now the statement that M and E are functions of r translates into the partial differential equations:

$$\frac{\partial M}{\partial \tau} + \ell\sqrt{2E + \frac{2M}{R} + \frac{1}{3}\Lambda R^2} \, \frac{\partial M}{\partial R} = 0, \tag{18.151}$$

and a similar equation for E. In these coordinates there is no trace of the shell crossing singularity, which means that the metric can be extended through it. However, the derivatives of the metric components g_{00}, g_{01} and g_{11} by R will now be singular at the shell crossing because

$$\frac{\partial M}{\partial R} = \frac{\partial M}{\partial r}\frac{\partial r}{\partial R} = \frac{\partial M}{\partial r}\bigg/R,_r \tag{18.152}$$

(all quantities to be taken at $r = r(\tau, R)$). Thus, the extension is of class C^0 (continuous but non-differentiable), and the curvature tensor still has a singularity at the set $R,_r(\tau, R) = 0$.

The components of the velocity field in the (τ, R) coordinates are

$$u^0 = 1, \qquad u^1 = R,_t = \ell\sqrt{2E + \frac{2M}{R} + \frac{1}{3}\Lambda R^2}, \qquad u^2 = u^3 = 0. \tag{18.153}$$

The u^1 is non-differentiable at the shell crossing, but continuous. Thus the derivative of u^1 (and consequently the Christoffel symbols) will have a finite discontinuity there. The singularity of the flow makes itself visible in the behaviour of the geodesic deviation of the velocity field. As seen from (6.56), to offset the infinite values of the curvature at the shell crossing, the geodesic deviation δx^α must be zero there, which means that different flow lines intersect. However, they continue behind the intersection points. Figures 18.8 and 18.9 show an example of a shell crossing in an $E = 0$ L–T model in the comoving coordinates (Fig. 18.8) and in the Gautreau coordinates (Fig. 18.9). As seen from Fig. 18.9, the region behind the shell crossing contains three superposed flows of dust: the streams that entered through the left wall, through the right wall and through the lower vertex.[1]

---

[1] This kind of extension was considered by Clarke and O'Donnell (1992), but it is not clear from their figure what quantities are measured on the coordinate axes.

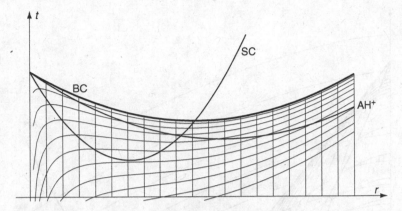

Fig. 18.8. A shell crossing (SC) singularity in an $E = 0$ L–T model in the comoving coordinates. (The model used to draw the figure is the same as will appear in Section 18.16, Eqs. (18.154)–(18.159), but with different parameter values.) The vertical straight lines are flow lines of the dust, terminating in the Big Crunch (BC) singularity. AH$^+$ is the apparent horizon. The other curves in the figure are contours of constant values of R. In the comoving coordinates, the flow lines cannot intersect in a spacetime diagram, even at shell crossings. A shell crossing is a location where $R_{,r} = 0$, i.e. where the contour of constant R has a horizontal tangent – note that this is indeed so. The figure should help understand why the necessary and sufficient condition for avoiding a shell crossing is that the BC is an increasing function of r: note that the centre of symmetry and the BC singularity together form the $R = 0$ contour. Thus, since the contours of small constant R-values must approximately follow the shape of the $R = 0$ contour, if $t_{\mathrm{BC}}(r)$ is not an increasing function, then the other $R = $ constant contours must have a horizontal tangent somewhere. At the point where t_{BC} becomes an increasing function, the shell crossing set meets the BC and at larger r-values is later than the BC, i.e. not in the physical region of the spacetime. Compare this with Fig. 18.3, where there are no shell crossings.

18.16 * Singularities and cosmic censorship

The cosmic censorship hypothesis (CCH) does not in fact belong to the field of cosmology. However, the L–T model has been used many times as a testing ground for various formulations of the CCH. After reading this section, readers may have a false impression that the subject of CCH is buried under counterexamples. In reality, the debate is far from settled. The emphasis on counterexamples in this text does not reflect the true state of affairs and results from the fact that we are more interested here in applications of the L–T model than in settling the CCH issue.

In its original formulation by Penrose (1969), the CCH said that relativity does not allow **naked singularities**. All singularities were supposed to be hidden inside horizons. A singularity can be **locally naked**, when light rays proceed from it out to a finite distance and turn back, or **globally naked**, when the light rays it emitted can escape to infinity. The CCH would preferably prohibit both kinds of naked singularity.

It would certainly be desirable if the CCH were correct because a naked singularity could emit matter or radiation in unpredictable quantities at unpredictable times. However,

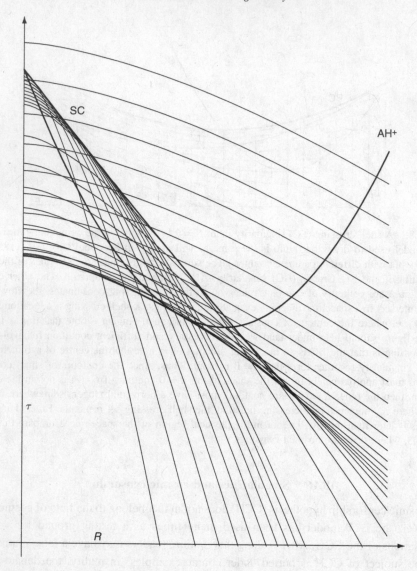

Fig. 18.9. The situation from Fig. 18.8 in the (τ, R) coordinates of Gautreau. The Big Crunch
singularity is at the vertical axis, where $R = 0$. The shell crossing (SC) is the wedge-shaped pair
of curves; the tip of the wedge is the minimum of t_S from Fig. 18.8 and the lower branch of the
wedge is the part to the right of the minimum of t_S in Fig. 18.8. More flow lines are drawn here
than in Fig. 18.8 – the extra lines are the leftmost nine (which would lie, evenly spaced, to the
left of the leftmost line in Fig. 18.8) and the uppermost one (which would lie to the right of the
rightmost line in Fig. 18.8). Inside the wedge, three streams of matter are seen: the particles that
entered through the lower face, the particles that entered through the upper face and the particles
that entered through the tip. Note that this diagram does not faithfully represent all the features of
Fig. 18.8. For example, the flow lines that intersect the right face of the SC here do not intersect
the SC in Fig. 18.8. The function $t(r)$ given by $R(t, r) = $ constant is not one-to-one, i.e. for every
point (t_1, r_1) in Fig. 18.8 in the SC to the left of the minimum, with the value R_1 of R, there is a
point (t_1, r_2) to the right of the SC that has the same value of R.

the history of the CCH turned out to be turbulent. A counterexample to this earliest formulation was provided by Yodzis, Seifert and Müller zum Hagen (1973), then the hypothesis was modified to exclude 'non-generic' cases, then another counterexample to the modified version appeared, followed by a new modification, and so the story went. We shall not follow it here; more information can be found in the book by Joshi (1993). Instead, we will describe some of the applications of the L–T model to the discussion of properties of singularities, the CCH among them.

Our example of a naked shell crossing singularity closely follows the method of Yodzis, Seifert and Müller zum Hagen (1973). However, the functions used here are different because the original example was, apparently needlessly, quite complicated. The idea is to construct a ball of L–T dust of finite radius, matched to the Schwarzschild solution, and then to design a shell crossing singularity in it so that it reaches the surface of the ball before the surface crosses the future apparent horizon at $R = 2M$. At the surface, the apparent horizon coincides with the Schwarzschild event horizon, so the singularity thus constructed will be naked.

We take the collapsing $E = 0 = \Lambda$ L–T model (in which $t < t_B$ and t increases towards $t = t_B$), and we find the locus of $R_{,r} = 0$ in it:

$$t = t_S(r) \overset{\text{def}}{=} t_B(r) + 2\frac{M}{M_{,r}} t_{B,r}. \tag{18.154}$$

Following Yodzis *et al.*, we choose the simple function $M(r) = \mu r$, where μ is a constant (this means only that we choose M/μ as the radial coordinate; it does not limit the generality of the example). The centre of symmetry is at $r = 0$, and the regularity condition (18.31) is fulfilled. The surface of the L–T ball will be at $r = b$, and for $r > b$ we have the Schwarzschild solution with the mass parameter $m = \mu b$ (it must be represented in the Lemaître–Novikov coordinates with $E = 0$ and with the appropriate value of $t_B(b)$).

Next, we choose such a form of the function $t_S(r)$ in (18.154) that it has a minimum at some r:

$$t_S(r) = a(r - c)^2 + d, \tag{18.155}$$

where a, c and d are arbitrary positive parameters, whose values will be discussed later. The minimum of $t_S(r)$ is at $r = c$ and it equals d, which means that for $t < d$ the shell crossing does not exist yet. We equate (18.154) and (18.155) and solve the resulting differential equation for $t_B(r)$. In order to fulfil the regularity condition (18.35), the integration constant must be zero. The result is

$$t_B(r) = \frac{a}{5} r^2 - \frac{2}{3} acr + ac^2 + d. \tag{18.156}$$

The time by which the shell crossing hits the surface of the ball is

$$t = t_S(b) = a(b - c)^2 + d. \tag{18.157}$$

The time by which the surface of the ball crosses the apparent horizon is found by solving the equation $R = 2M$ for t; it is

$$t_H(b) = t_B(b) - \frac{4}{3}\mu b = \frac{a}{5}b^2 - \frac{2}{3}b(2\mu + ac) + ac^2 + d, \qquad (18.158)$$

and this must be greater than $t_S(b)$, which means that

$$c > \frac{3}{5}b + \frac{\mu}{a}. \qquad (18.159)$$

This allows us to choose the mass parameter μ and the location of the surface of the ball, $r = b$, quite arbitrarily. Also the instant when the shell crossing first appears, $t = d$, is arbitrary – Eq. (18.159) is the only condition to be fulfilled. If it is fulfilled, then the singularity reaches the surface of the ball at $t = t_S(b)$ given by (18.157), while the surface is still outside the Schwarzschild event horizon. Consequently, the singularity can send a null geodesic out to the future null infinity; see Fig. 18.10. In fact, for better physical interpretation, it is good to assume that the minimum of $t_S(r)$ is still inside the ball, so that the shell crossing begins inside the ball and not at the surface. For this, it is necessary that $c < b$. This can be fulfilled together with (18.159) only if $a > 5\mu/(2b)$. This still leaves the parameters μ, b and d arbitrary.

The Schwarzschild solution represented in the Lemaître–Novikov coordinates must have its $R(t, r)$ equal at $r = b$ to the $R(t, r)$ of (18.154)–(18.156). Hence, $E_S(b) = 0$, $t_{BS}(b) = t_B(b)$ and $m = \mu b$, where the subscript 'S' stands for 'Schwarzschild'. Thus, it is simplest to choose $E_S(r) = 0$ (which, in the Schwarzschild limit, is just a choice of coordinates; see Exercise 10 in Chapter 14), and then $t_{BS}(r)$ can be given any form by a transformation of r alone. A simple form is $t_{BS}(r) = \alpha r$, with $\alpha = t_B(b)/b$. Then, the event horizon, $R = 2m$, is given by $t = \alpha r - 4m/3$, i.e. the singularity $t = t_{BS}(r)$ and the event horizon are parallel straight lines in the (t, r) diagram, as in Fig. 18.10. (With arbitrary r, the event horizon would still be a parallely displaced copy of the singularity, but the graphs of both sets are in general curves.) The null geodesic equation in the Schwarzschild spacetime must then be integrated numerically.

Now we will show that a shell focussing singularity can be globally naked. The method of proof is similar to the one used for shell crossings: the surface of the L–T dust ball has to be chosen at such a location that the ray emitted from the central point of the Big Crunch singularity hits it before the surface crosses the apparent horizon. Then the ray can continue out to infinity in the Schwarzschild region. Figure 18.7 shows that this is possible, since that ray remains before the apparent horizon for some time, but we will prove it formally.

We will continue to consider the same example as in Section 18.14, i.e. the model defined by (18.140)–(18.142). We showed that the apparent horizon in it, (18.147), and the Big Crunch, (18.142), have a zero of second order at $r = 0$, while the function $t_n(r)$ that obeys the equation of a radial null geodesic has a zero of higher order there.

Consider the function $f(r) \overset{\text{def}}{=} t_{AH}(r) - t_n(r)$. It is zero at $r = 0$ and its first derivative by r is also zero at $r = 0$, while $d^2t/dr^2(0) = 2a > 0$. From (18.148) we see that d^2t_n/dr^2 is

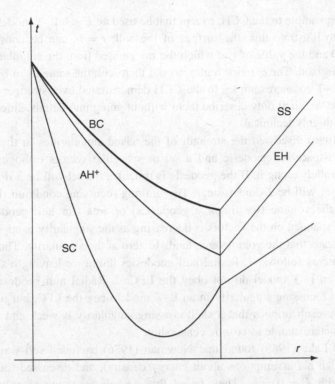

Fig. 18.10. A schematic illustration of a naked shell-crossing singularity of the type of Yodzis *et al.* (1973). The vertical line is the worldline of the surface of the dust ball; the area to the right of it is the Schwarzschild spacetime. BC is the Big Crunch singularity, where $t = t_B(r)$; outside the ball it goes over into the Schwarzschild singularity denoted SS. AH^+ is the future apparent horizon; outside the ball it goes over into the Schwarzschild event horizon EH. SC is the shell crossing singularity that appears at $t = d$ and grows inwards and outwards. The seemingly straight line below EH is the light ray emitted when the shell crossing reaches the surface of the body – it proceeds unimpeded to the future null infinity. For the shell crossing singularity to exist it is necessary that the Big Crunch function inside the body is decreasing with r (for an increasing $t_B(r)$, the shell crossing is later than the Big Crunch).

continuous at $r = 0$, and so is $\left(d^2 t_{AH}/dr^2\right) = 2a - 8M_0 r$. Consequently, $d^2 f/dr^2$, which equals $2a > 0$ at $r = 0$, must remain strictly positive for $0 \leq r \leq r_1$, where $r_1 > 0$. This means that df/dr (which equals 0 at $r = 0$) is a strictly increasing function for $0 < r \leq r_1$, and so, being continuous, must be positive for $0 < r \leq r_2$, where $r_2 > r_1$. This, in turn, means that $f(r)$, which is continuous, is increasing for $0 < r \leq r_2$, and so must be positive for $0 < r \leq r_3$, where $r_3 > r_2$. Thus, at $r = r_3$, $f(r_3) > 0$, i.e. $t_n(r_3) < t_{AH}(r_3)$. Choose the surface of the dust ball at $r = b = r_3$, and match the L–T model to the Schwarzschild space-time there. The ray emitted from the central point of the Big Crunch singularity will reach the surface of the ball at $r = b$ while $t_n(b) < t_{AH}(b)$, and then will continue unimpeded to infinity in the Schwarzschild region. This was the essential point of Christodoulou's

(1984) counterexample to the CCH, except that he used an $E < 0$ L–T model. Figure 18.7 can be used to illustrate this: the surface of the ball, $r = b$, can be chosen anywhere between $r = 0$ and the value of r at which the ray emitted from the singularity intersects the apparent horizon. The exterior region would then look the same as in Fig. 18.10.

The other L–T counterexamples to the CCH demonstrated ever stronger violations of the hypothesis. We shall only describe them without going into details, since most of the arguments are highly technical.

Newman (1986) discussed the strength of the naked singularities in the L–T model. Imagine a non-spacelike geodesic and a set of spacelike vectors orthogonal to it and propagated parallely along it. If the geodesic is timelike, the set will be 3-dimensional; if it is null, the set will be 2-dimensional. The limiting focussing condition (LFC) means, roughly, that the volume (for timelike geodesics) or area (for null geodesics) of the parallelepiped spanned on the vectors is decreasing as the singularity is approached. The strong LFC means that the volume/area tends to zero at the singularity. The conclusions of the paper are as follows: 1. Radial null geodesics hitting or leaving a shell crossing singularity in an L–T model do not obey the LFC. 2. Radial null geodesics hitting or leaving a shell focussing singularity in an L–T model obey the LFC, but not the strong LFC. The first result implies that a shell crossing singularity is weak and thus is not a conclusive counterexample to cosmic censorship.

Waugh and Lake (1988) found that Newman (1986) excluded self-similar configurations[1] (through his assumptions about energy density), and discussed the self-similar L–T model with $E = 0 = \Lambda$. They found that a shell focussing singularity may be formed in it which will be globally naked and strong. It is strong in the sense that $\lim_{\lambda \to 0}(\lambda^2 R_{\alpha\beta} k^\alpha k^\beta) \neq 0$, where k^α is the tangent vector to a null geodesic hitting the singularity, λ is the affine parameter on the geodesic with $\lambda = 0$ at the singularity and $R_{\alpha\beta}$ is the Ricci tensor of the L–T spacetime.

Gorini, Grillo and Pelizza (1989) identified another subcase of the L–T models that was not captured by Newman and showed that it develops a globally naked shell focussing singularity that obeys the strong LFC, so it is to be taken as a serious counterexample to (one particular formulation of) the cosmic censorship postulate.

Grillo (1991) showed by an example that a $\Lambda = 0$ L–T model with $E < 0$ and with a certain definite choice of $E(r)$, $M(r)$ and $t_B(r)$ will form a locally naked shell focussing singularity that satisfies the strong LFC.

A few authors (Waugh and Lake, 1988, 1989; Lemos, 1991; Eardley, 1974b; Dyer, 1979; Eardley and Smarr, 1979) found the conditions for the occurrence of a strong globally naked singularity in self-similar L–T models with $\Lambda = 0$. For $E \neq 0$, Lemos (1991) found the conditions numerically; for $E = 0$ an exact result can be derived with reference to the papers by Waugh and Lake (1989) and Dyer (1979). (In a self-similar spacetime, the Big Crunch function must have the form $t_C(r) = Br$, where $B > 0$ is a constant. Then, a light ray can escape from the central point p_0 of the Big Crunch only

---

[1] A self-similar spacetime is one that admits a conformal Killing vector field such that the function λ in (8.55) is constant.

if the slope of the Big Crunch function, B, is larger than the slope of the ray at p_0.) A self-similar L–T spacetime has the metric (Dyer, 1979, and Exercise 12):

$$ds^2 = dt^2 - \frac{(S - sS')^2}{1 + 2E} dr^2 - r^2 S^2 \left(d\vartheta^2 + \sin^2\vartheta \, d\varphi^2\right), \tag{18.160}$$

where E is a constant and S is a function of the variable $s = t/r$, the prime denotes the derivative with respect to s, and S obeys the equation

$$S'^2 = 2E + 2/S \tag{18.161}$$

(because of self-similarity, the function $M(r)$ must obey $M(r)/r = C = $ constant, and then the coordinate freedom was used to set $C = 1$).

Now consider the orbits of the self-similarity transformations, $t' = e^\tau t, r' = e^\tau r, \vartheta = $ constant, $\varphi = $ constant, i.e. the curves given by $r = Dt, \vartheta = $ constant, $\varphi = $ constant. They all meet at the point p_0 with coordinates $(t, r) = (0, 0)$. They are timelike, null or spacelike according as the expression $\chi = 1 - D^2(S - sS')^2/(1 + 2E)$ is positive, zero or negative. With S and E given, the equation $\chi = 0$ may or may not have a solution for D. If it has no solution, then all the orbits are spacelike and no light ray can escape the point p_0.[1] Further (see Exercise 13), if $\chi = 0$ along the orbit, then the orbit is also geodesic (but neither t nor r is an affine parameter). Consequently, an orbit with $\chi = 0$ is also a path of a light ray escaping the central point of the singularity. We shall now find a condition for its existence in the $E = 0$ model.

The solution of (18.161) with $E = 0$ is $S = [(3/\sqrt{2})(s - B)]^{2/3}$, where $B > 0$ is a constant (so that the singularity is at $t = Br$), and for further considerations we choose the new variable u so that

$$s - B = \frac{1}{6}u^3, \qquad S = \frac{1}{2}u^2, \qquad S' = 2/u. \tag{18.162}$$

As shown above, the first ray to escape the singularity obeys $t/r \equiv s = S - sS'$. Substituting from the above and solving for B we obtain

$$B = \frac{u^3(1 - u)}{6(u + 2)}, \tag{18.163}$$

and then the first of (18.162) yields

$$s = \frac{u^3}{2(u + 2)}. \tag{18.164}$$

Now we interpret (18.163) and (18.164) as the parametric equations of a curve in the (B, s) plane; see Fig. 18.11. We are interested in the smallest value of $B > 0$. We see that it corresponds to the local minimum of the function $B(u)$. (In the other branch of the graph, the smallest value of B does not exist. That branch corresponds to rays proceeding towards the central point of the Big Bang in an expanding model.) By elementary methods

---

[1] The orbits cannot be all timelike because the Big Crunch singularity also begins at p_0 and is spacelike in every neighbourhood of p_0, however small. Hence, any future-directed timelike curve leaving p_0 would then immediately hit the Big Crunch.

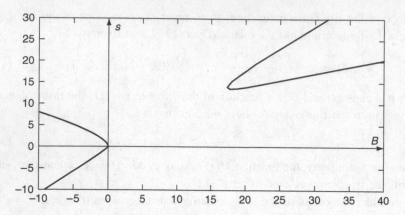

Fig. 18.11.   The graph of solutions of the equation $s = S - sS'$ that determines the first ray to escape from the central point of the Big Crunch. The left branch is not really tangent to the vertical axis, it goes to small positive values of B. Values of B on the left branch correspond to a model expanding away from the Big Bang, in which the incoming rays hit the central point of the BB.

we find that $B,_u = 0$ at $u_\epsilon = -1 + \epsilon\sqrt{3}$, where $\epsilon = \pm 1$. Since $B,_{uu} < 0$ at $u = u_-$ and $B,_{uu} > 0$ at $u = u_+$, we conclude that the minimum is achieved at $u = -1 - \sqrt{3}$, which corresponds to the value of $B = B_m = 26/3 + 5\sqrt{3}$. This is the minimum slope of the Big Crunch function, below which no light ray can escape from the central point of the Big Crunch. For $B > B_m$, the singularity is globally naked.

Dwivedi and Joshi (1992) identified a class of the functions $M(r)$ which generate a strong naked singularity at the centre. In this class, the first and second derivatives of the initial density distribution by r are zero at $r = 0$. Most of the earlier papers considered special cases of this.

In another paper, the same authors (Joshi and Dwivedi, 1993) showed that among the L–T variety of models one can find a subset in which there will be an at least locally naked strong curvature singularity at the centre. The class includes the configurations of Eardley and Smarr (1979) and of Christodoulou (1984). A continuous family of non-spacelike curves can be emitted from the singularity.

The status of the CCH in 1993 was reviewed by Clarke (1993) and, more extensively, by Joshi (1993).

Deshingkar, Joshi and Dwivedi (1999) defined a shell focussing singularity (SFS) to be strong if there exists a timelike or null geodesic γ terminating in the SFS such that all the geodesic deviation fields V^α along γ obey $\mathcal{V} \overset{\text{def}}{=} (s_0 - s)^2 R_{\alpha\beta} V^\alpha V^\alpha = \text{constant} > 0$, where $R_{\alpha\beta}$ is the Ricci tensor, s is the affine parameter along γ and $s = s_0$ is the location of the singularity. In terms of this definition, SFSs are *always strong*. The reason why earlier papers identified some of these singularities as weak was the selection of a special family of curves to calculate $\mathcal{V}$. The property of being strong is stable against perturbations of symmetry.

18.17 Solving the 'horizon problem' without inflation

The 'horizon problem' was explained in Section 17.8 and Fig. 17.7. The inflationary models solve it by postulating that at some time before last scattering the Universe had been modelled by geometries in which the matter source was a scalar field, acting like a repulsive cosmological constant of large value. In that case, the Universe would have expanded exponentially, by the law $R \propto e^{\Lambda t}$.[1]

Another solution was invented by Celerier and Schneider (1998). They assumed that the L–T Big Bang function $t_B(r)$ has a local minimum at $r = 0$ and is increasing with r so that there is a shell crossing at some $t > t_B(r)$ for every $r > 0$. As seen from (18.16), at a shell crossing, where $R,_r = 0$, a radial null geodesic has a horizontal tangent in the comoving coordinates. Consequently, all the curves passing through a shell crossing are timelike or null at the intersection point. The time of shell crossing in the $E = 0$ model is $t \overset{\text{def}}{=} t_S(r) = t_B + 2Mt_{B,r}/M,_r$, where $t_{B,r} > 0$. Choosing the r-coordinate so that $M = M_0 r^3$, and assuming that $t_{S,r} > 0$, i.e. $5t_{B,r} + 2rt_{B,rr} > 0$, we obtain that the curve $t = t_S(r)$ is timelike everywhere.

Now consider a radial null geodesic k_1 (given by the equation $t = t_n(r)$) sent backwards in time from the observer's position at $t = t_p$, $r = 0$; see Fig. 18.12. It obeys $(dt/dr)_n = -R,_r < 0$ in the whole spacetime region that is later than $t_S(r)$. Since the curve $t = t_S(r)$ started also from $r = 0$ at an earlier time, and has a strictly positive derivative everywhere, the two curves must intersect somewhere, and $t_n(r)$ will have a zero derivative there. Let the point of intersection p_{is} have the coordinates (t_{is}, r_{is}). Consider another radial null geodesic k_2 sent backwards in time from a point p_2 of coordinates (t_2, r_2), where $t_2 > t_{is}$, $r_2 < r_{is}$, towards the centre of symmetry at $r = 0$. The function $t = t_{n2}(r)$ describing that geodesic has a strictly positive derivative at every point at which $t_{n2}(r) > t_S(r)$. Assume that the curves $t_S(r)$ and $t_{n2}(r)$ are both differentiable for $r \in [0, r_2]$ and that the mass density is nowhere zero. It is then a simple exercise to prove that k_2 cannot intersect $t_S(r)$ at any r obeying $0 \leq r \leq r_2$. In brief, the argument is that at an intersection point q of coordinates (t_q, r_q) the derivative $t'_{n2}(r)$ would have to be zero and would have to go down to zero from positive values as r is decreasing. This means, when $t_{n2}(r)$ would be approaching $t_{n2}(r_q)$, the difference $(t_{n2}(r) - t_S(r))$ would have to be increasing, and the intersection could not occur (see Exercise 14 for more details). Consequently, k_2 will reach the centre at $r = 0$ at a later time than the central point of the Big Bang singularity, $t_B(0)$ (which coincides with the central point of the shell crossing, $t_S(0)$).

Consequently, there was a period in the past of the observer O (with time coordinates $t < T$) such that a signal sent out from the point $(t_1, 0)$ with $t_1 < T$ had enough time to proceed to the neighbourhood of the shell crossing, and then bounce back towards the observer, to reach him at $t = t_p$. Such a period (i.e. such a value of T) exists for every $t_p > t_B(0)$. Hence, all the regions that the observer sees at t_p could have been in causal

---

[1] As observed by Celerier and Szekeres (2002), this not so much *solves*, but rather *postpones* the horizon problem. At a later time, the observer would still be able to see regions that had not been in causal contact at $t = t_{is}$.

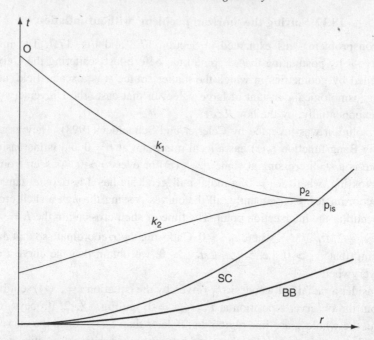

Fig. 18.12. A solution of the 'horizon problem' in an $E = 0$ L–T model with an increasing Big Bang (BB) function $t_B(r)$ (the form used in the picture is $t_B(r) = br^2$, where $b = $ constant). The null radial geodesic k_1 sent backwards in time from the observer's position O hits the shell crossing SC at the point p_{is} of coordinates $(t, r) = (t_{is}, r_{is})$. From a point p_2 on k_1, of coordinates (t_2, r_2), with $t_2 > t_{is}$, $r_2 < r_{is}$, another radial null geodesic k_2 is sent backwards in time towards the centre at $r = 0$. As explained in the text, k_2 must hit the centre at a time later than the BB at $r = 0$ (the shell crossing set is in this example tangent to the BB set at $r = 0$). Consequently, all areas of the sky that the observer O sees at present had a chance to be in causal contact with a common source in the past. This solves the 'horizon problem' for as long as the observer has in the field of view the shell crossing set $t = t_S(r)$ with $t'_S(r) > 0$; the solution is permanent if $t'_S(r) > 0$ for all $r > 0$.

contact with a common source located in the observer's past. This solution of the 'horizon problem' is not just a postponement; with $t'_S(r) > 0$ for all $r > 0$, the problem is avoided permanently.

Note that the two rays in the construction remain in the region $t > t_S(r)$, i.e. they never hit the shell crossing set. Hence, they remain all the time in the physical region of the spacetime.

18.18 * The evolution of $R(t, M)$ versus the evolution of $\rho(t, M)$

In the Friedmann models, the relation between the areal radius R and the mass density is simple: in any given model, the greater R, the smaller ϵ. In an L–T model, no such relation exists because of the possible existence of local condensations, as we will prove here.

From (18.23), at any fixed time t:

$$R^3(M) - R_0{}^3 = \frac{6}{\kappa} \int_{M_0}^{M} \frac{\mathrm{d}u}{\epsilon(u)}, \tag{18.165}$$

where R_0 is a (possibly time-dependent) constant of integration and M_0 is another constant. Equation (18.165) means that the value of R at any M depends on the values of ϵ in the whole range $[M_0, M]$. Also the inverse relation in (18.23) is non-local – to find $\epsilon(t, M)$ we need to know R in an open neighbourhood of the value of M.

Assume now that $\epsilon_2 < \epsilon_1$ over the whole of $[M_0, M]$. Then

$$R_2{}^3(M) - R_1{}^3(M) - \left(R_0{}^3(t_2) - R_0{}^3(t_1)\right) = \frac{6}{\kappa} \int_{M_0}^{M} \left(\frac{1}{\epsilon_2(u)} - \frac{1}{\epsilon_1(u)}\right) \mathrm{d}u > 0. \tag{18.166}$$

Hence, $R_2{}^3(M) > R_1{}^3(M)$ if

$$R_0{}^3(t_2) - R_0{}^3(t_1) + \frac{6}{\kappa} \int_{M_0}^{M} \left(\frac{1}{\epsilon_2(u)} - \frac{1}{\epsilon_1(u)}\right) \mathrm{d}u > 0. \tag{18.167}$$

This will hold if, for example, $R_0(t_i) = 0$, $i = 1, 2$, in addition to $\epsilon_2 < \epsilon_1$.

The converse implication is simply not true. This is easy to understand on physical grounds: $R(t_2, M) > R(t_1, M)$ for all $M \in [M_0, M_1]$ means that every shell of constant $M = \widetilde{M}$ has a larger radius at t_2 than it had at t_1. However, the neighbouring shells may have moved closer to $\widetilde{M}$ at t_2 than they were at t_1. If they did, then a local condensation around $\widetilde{M}$ was created, which may result in $\epsilon(t_2, \widetilde{M})$ being larger than $\epsilon(t_1, \widetilde{M})$. This does not happen in the Friedmann limit, where local condensations are excluded by the symmetry assumptions.

A sufficient condition for $\epsilon_2(M) < \epsilon_1(M)$, $M \in [M_0, M_1]$, is

$$\frac{1}{(R_1{}^3)_{,M}} > \frac{1}{(R_2{}^3)_{,M}} \qquad \text{for all } M \in [M_0, M_1]. \tag{18.168}$$

If $R_{,M} > 0$ for all $M \in [M_0, M_1]$ at both t_1 and t_2 (i.e. there are no shell crossings in $[M_0, M_1]$), then this is equivalent to $(R_2{}^3)_{,M} > (R_1{}^3)_{,M}$ for all $M \in [M_0, M_1]$. Incidentally, this implies $R_2 > R_1$ for all $M \in [M_0, M_1]$ if $R_2(M_0) > R_1(M_0)$.

18.19 * Increasing and decreasing density perturbations

Formation of structures in the Universe used to be described by approximate perturbations of the Robertson–Walker models. In that approach, two classes of perturbations had been identified: those that increase with time and those that decrease with time. The same classification can be done in the L–T and Szekeres models (for the latter see Section 19.8). For the L–T model with $\Lambda = 0$, this approach was first applied by Silk (1977). His results are re-derived here, by a somewhat different method.

The quantities used in studying perturbations of the Friedmann models are the density contrast $\epsilon_{,r}/\epsilon$ and the curvature contrast $R^{(3)}{}_{,r}/R^{(3)}$, where $R^{(3)}$ is the scalar curvature of the hypersurfaces $t = $ constant. Both these functions vanish identically in the Friedmann limit. From (18.15) and (18.25) we find

$$\frac{\epsilon_{,r}}{\epsilon} = \frac{M_{,rr}}{M_{,r}} - 2\frac{R_{,r}}{R} - \frac{R_{,rr}}{R_{,r}}, \tag{18.169}$$

$$R^{(3)} = -\frac{4(RE)_{,r}}{R^2 R_{,r}}, \tag{18.170}$$

$$\Delta K \stackrel{\text{def}}{=} \frac{R^{(3)}_{,r}}{R^{(3)}} = \frac{(RE)_{,rr}}{(RE)_{,r}} - 2\frac{R_{,r}}{R} - \frac{R_{,rr}}{R_{,r}}$$

$$= \frac{E_{,rr}/E - 2R_{,r}{}^2/R^2 - (R_{,rr}/R_{,r})(E_{,r}/E)}{R_{,r}/R + E_{,r}/E}. \tag{18.171}$$

Since a perturbation with $E = 0$ would be rather exceptional, we assume $E \neq 0$. For $E > 0$, using (18.113)–(18.114) and (18.21), we find

$$\alpha(r) \stackrel{\text{def}}{=} \frac{M_{,r}}{M} - \frac{E_{,r}}{E}, \qquad \beta(r) \stackrel{\text{def}}{=} \frac{3}{2}\frac{E_{,r}}{E} - \frac{M_{,r}}{M},$$

$$\gamma(r) \stackrel{\text{def}}{=} \frac{|2E|^{3/2}}{M} t_{B,r}, \tag{18.172}$$

$$\frac{R_{,rr}}{R_{,r}} = \alpha_{,r}\frac{R}{R_{,r}} + \alpha + \left(\frac{R\beta_{,r}}{R_{,r}} + \beta\right)\frac{\sinh\eta\,(\sinh\eta - \eta)}{(\cosh\eta - 1)^2}$$

$$+ \frac{R\beta}{R_{,r}}\frac{2\eta + \eta\cosh\eta - 3\sinh\eta}{(\cosh\eta - 1)^3}[\beta(\sinh\eta - \eta) - \gamma]$$

$$- \left(\frac{R\gamma_{,r}}{R_{,r}} + \gamma\right)\frac{\sinh\eta}{(\cosh\eta - 1)^2}$$

$$+ \frac{R\gamma}{R_{,r}}\frac{\cosh\eta + 2}{(\cosh\eta - 1)^3}[\beta(\sinh\eta - \eta) - \gamma]. \tag{18.173}$$

We need to know the behaviour of $\epsilon_{,r}/\epsilon$ and ΔK as $\eta \to 0$ and $\eta \to \infty$ in two cases: when $t_{B,r} = 0$ and when $[(2E)^{3/2}/M]_{,r} = 0$. For this purpose, we must know the behaviour in these limits of a few functions present in (18.169)–(18.171). Two of them are $\Phi_3(\eta)$ and $\Phi_4(\eta)$, defined in (18.114); their relevant properties are given in (18.115). The other functions are (a coefficient at ∞ is meant to determine the sign):

$$\lim_{\eta\to 0}\frac{R_{,r}}{R} = \frac{M_{,r}}{3M}, \qquad \lim_{\eta\to\infty}\frac{R_{,r}}{R} = \frac{E_{,r}}{2E} \qquad \text{when } t_{B,r} = 0,$$

$$\lim_{\eta\to 0}\frac{R_{,r}}{R} = -t_{B,r}\times\infty, \tag{18.174}$$

$$\lim_{\eta\to\infty}\frac{R_{,r}}{R} = \frac{M_{,r}}{3M} \qquad \text{when } [(2E)^{3/2}/M]_{,r} = 0, \tag{18.175}$$

$$\psi_3(\eta) \stackrel{\text{def}}{=} \frac{(2\eta + \eta\cosh\eta - 3\sinh\eta)(\sinh\eta - \eta)}{(\cosh\eta - 1)^3},$$

$$\lim_{\eta \to 0} \psi_3(\eta) = 0, \qquad \lim_{\eta \to \infty} \psi_3(\eta) = 0, \qquad (18.176)$$

$$\psi_4(\eta) \overset{\text{def}}{=} \frac{\cosh \eta + 2}{(\cosh \eta - 1)^3},$$

$$\lim_{\eta \to 0} \psi_4(\eta) = \infty, \qquad \lim_{\eta \to \infty} \psi_4(\eta) = 0, \qquad (18.177)$$

$$\lim_{\eta \to 0} \frac{R_{,r}}{R} \frac{(\cosh \eta - 1)^2}{\sinh \eta} = -\frac{(2E)^{3/2}}{M} t_{B,r}, \qquad \lim_{\eta \to 0} \frac{R_{,r}}{R}(\cosh \eta - 1)^2 = 0. \qquad (18.178)$$

With $t_{B,r} = 0$, we now find the following limits:

$$\lim_{\eta \to 0} \frac{R_{,rr}}{R_{,r}} = \frac{M_{,rr}}{M_{,r}} - \frac{2}{3}\frac{M_{,r}}{M}, \qquad \lim_{\eta \to \infty} \frac{R_{,rr}}{R_{,r}} = \frac{E_{,rr}}{E_{,r}} - \frac{E_{,r}}{2E}. \qquad (18.179)$$

With these, one finds that $\epsilon_{,r}/\epsilon \xrightarrow[\eta \to 0]{} +\infty$, while

$$\lim_{\eta \to 0} \Delta K = \frac{\dfrac{E_{,rr}}{E} - \dfrac{2M_{,r}^2}{9M^2} - \dfrac{E_{,r}}{E}\left(\dfrac{M_{,rr}}{M_{,r}} - \dfrac{2M_{,r}}{3M}\right)}{M_{,r}/(3M) + E_{,r}/E}. \qquad (18.180)$$

As $\eta \to \infty$, we obtain

$$\lim_{\eta \to \infty} \frac{\epsilon_{,r}}{\epsilon} = \frac{M_{,rr}}{M_{,r}} - \frac{E_{,rr}}{E_{,r}} - \frac{E_{,r}}{2E}, \qquad \lim_{\eta \to \infty} \Delta K = 0. \qquad (18.181)$$

This shows that with $t_{B,r} = 0$ the perturbation of the Friedmann density vanishes at the initial singularity and tends to a finite limit as $t \to \infty$, i.e. it is increasing. The perturbation of curvature displays the exactly opposite behaviour.

With $[(2E)^{3/2}/M]_{,r} = 0$, the limits are

$$\lim_{\eta \to 0} \frac{\epsilon_{,r}}{\epsilon} = +t_{B,r} \times \infty = \lim_{\eta \to 0} \Delta K, \qquad \lim_{\eta \to \infty} \frac{\epsilon_{,r}}{\epsilon} = 0,$$

$$\lim_{\eta \to \infty} \Delta K = \frac{\dfrac{E_{,rr}}{E} - \dfrac{2M_{,r}^2}{9M^2} - \dfrac{E_{,r}}{E}\left(\dfrac{M_{,rr}}{M_{,r}} - \dfrac{2M_{,r}}{3M}\right)}{M_{,r}/(3M) + E_{,r}/E}, \qquad (18.182)$$

so the perturbations generated by $t_{B,r}$ are decreasing.

Now we repeat the calculations done in (18.171)–(18.182) for the $E < 0$ model. Since infinities will now appear both in $R_{,rr}/R_{,r}$ and in $R_{,r}/R$; we have to substitute for these quantities before taking the limits. Using (18.15), (18.19), (18.107)–(18.109) and (18.172) we find

$$\frac{\epsilon_{,r}}{\epsilon} = \frac{M_{,rr}}{M_{,r}} - \frac{\alpha_{,r} R}{R_{,r}} - 3\alpha - \left(\frac{R\beta_{,r}}{R_{,r}} + 3\beta\right)\frac{\sin \eta(\eta - \sin \eta)}{(1 - \cos \eta)^2}$$

$$- \frac{R\beta}{R_{,r}} \times \frac{-2\eta - \eta\cos \eta + 3\sin \eta}{(1 - \cos \eta)^3}[\beta(\eta - \sin \eta) - \gamma]$$

$$+ \left(\frac{R\gamma_{,r}}{R_{,r}} + \gamma\right)\frac{\sin \eta}{(1 - \cos \eta)^2}$$

$$- \frac{\gamma R}{R_{,r}} \frac{\cos \eta + 2}{(1 - \cos \eta)^3} [\beta(\eta - \sin \eta) - \gamma], \tag{18.183}$$

$$\Delta K = -2 \frac{R_{,r}}{R} + \frac{E_{,rr}/E - (E_{,r}/E)(\epsilon_{,r}/\epsilon - M_{,rr}/M_{,r})}{R_{,r}/R + E_{,r}/E}. \tag{18.184}$$

Here we give some useful limits. For $t_{B,r} = 0$:

$$\lim_{\eta \to 0} \frac{R_{,r}}{R} = \frac{M_{,r}}{3M}, \qquad \lim_{\eta \to 2\pi} \frac{R_{,r}}{R} = -\left(\frac{3E_{,r}}{2E} - \frac{M_{,r}}{M} \right) \times \infty; \tag{18.185}$$

and for $[(2E)^{3/2}/M]_{,r} = 0$:

$$\lim_{\eta \to 0} \frac{R_{,r}}{R} = -t_{B,r} \times \infty, \qquad \lim_{\eta \to 2\pi} \frac{R_{,r}}{R} = t_{B,r} \times \infty. \tag{18.186}$$

Other functions and limits are:

$$\psi_1(\eta) \overset{\text{def}}{=} \frac{(-2\eta - \eta \cos \eta + 3 \sin \eta)(\eta - \sin \eta)}{(1 - \cos \eta)^3},$$

$$\lim_{\eta \to 0} \psi_1(\eta) = 0, \qquad \lim_{\eta \to 2\pi} \psi_1(\eta) = -\infty, \tag{18.187}$$

$$\psi_2(\eta) \overset{\text{def}}{=} \frac{\cos \eta + 2}{(1 - \cos \eta)^3}, \qquad \lim_{\eta \to 0} \psi_2(\eta) = \lim_{\eta \to 2\pi} \psi_2(\eta) = \infty. \tag{18.188}$$

In calculating the limits of the contrasts with $\gamma = 0$, we observe that $\Phi_1(\eta)/\psi_1(\eta) \underset{\eta \to 2\pi}{\longrightarrow} 0$, so the behaviour of the whole function at $\eta \to 2\pi$ is determined by the term containing ψ_1. Other useful limits are:

$$\lim_{\eta \to 0} \frac{R_{,r}}{R} \frac{(1 - \cos \eta)^2}{\sin \eta} = -\frac{(-2E)^{3/2}}{M} t_{B,r} = -\gamma, \tag{18.189}$$

$$\lim_{\eta \to 0} \frac{R_{,r}}{R} (1 - \cos \eta)^2 = 0 = \lim_{\eta \to 2\pi} \frac{R_{,r}}{R} (1 - \cos \eta)^2, \tag{18.190}$$

$$\lim_{\eta \to 2\pi} \frac{R_{,r}}{R} \frac{(1 - \cos \eta)^2}{\sin \eta} = 2\pi\beta - \gamma. \tag{18.191}$$

Now, when $t_{B,r} = 0$ ($\Longrightarrow \gamma = 0$), we find $\lim_{\eta \to 0} \epsilon_{,r}/\epsilon = 0$ and $\lim_{\eta \to 2\pi} \epsilon_{,r}/\epsilon = +\infty$. Thus, the gradient of $[(-2E)^{3/2}/M]$ generates a perturbation of the Friedmann model that vanishes at the initial singularity and increases without limits at the final singularity.

With $[(-2E)^{3/2}/M]_{,r} = 0$, the density contrast tends to infinity at both singularities. In this case, the time-difference between the bang and the crunch singularities is constant: $t_C - t_B = 2\pi/B$, where $B \overset{\text{def}}{=} (-2E)^{3/2}/M = $ constant. Thus, it is impossible to fulfil both of (18.110)–(18.111) and shell crossings are unavoidable.

For ΔK we obtain the following results. With $t_{B,r} = 0$ ($\gamma = 0$):

$$\lim_{\eta \to 0} \Delta K = -\frac{2M_{,r}}{3M} + \frac{E_{,rr}/E + E_{,r} M_{,rr}/(EM_{,r})}{M_{,r}/(3M) + E_{,r}/E}, \tag{18.192}$$

$$\lim_{\eta \to 2\pi} \Delta K = \beta \times \infty$$

(only the first term in (18.184) is infinite).

With $[(-2E)^{3/2}/M]_{,r} = 0$ ($\beta = 0$), the limits are

$$-\lim_{\eta \to 0} \Delta K = \lim_{\eta \to 2\pi} \Delta K = t_{B,r} \times \infty. \tag{18.193}$$

Thus, the gradient of $[(-2E)^{3/2}/M]$ generates increasing perturbations of the Friedmann curvature contrast, while the gradient of t_B generates perturbations that increase without limits at both singularities.

18.20 * L&T curio shop

In this section, a selection of material will be presented that did not fit under any other heading in this chapter. A complete overview of such papers until 1994 can be found in Krasiński (1997).

18.20.1 Lagging cores of the Big Bang

Novikov (1964a) used a similar setup to that of Bonnor (1956) (see Section 18.6) to describe the *lagging cores of Big Bang* (BB). He considered the flat Friedmann model with several Schwarzschild holes in it, matched into the background. Inside each hole there was a small expanding Friedmann region that had its local BB later than the general BB of Friedmann Universe. Particles and radiation emitted from such lagging cores of BB should in principle be observable. The hope was that this could explain the source of energy in quasars, but this attempt has had almost no following, except the paper mentioned below.

Neeman and Tauber (1967) considered more possibilities for matter inside the core: it could be either the Friedmann dust or the so-called stiff fluid (equation of state $\epsilon = p$), both with or without $\Lambda = 0$. However, in every case there were difficulties to reconcile the parameters of the model with the results of observations in a self-consistent way.

Miller (1976) described a Universe with the L–T geometry and with $E = 0 = \Lambda$, $t_{B,r} < 0$ gradually emerging from the BB singularity. With the BB first occurring far from the centre and proceeding inwards, one can assign mass to the singularity and interpret it as the mass to be emitted from it. The mass of the singularity is the function $M(r)$ from (18.14) calculated at $r = r_B(t)$, where $r_B(t)$ is the inverse function to $t_B(r)$; it is well behaved on the singularity. When the BB reaches $r = 0$, the singularity may either disappear or continue to exist with a negative mass and still emitting matter. In the second case it becomes a permanent timelike singularity.

18.20.2 Strange or non-intuitive properties of the L–T model

Novikov (1962a) showed that, with some choices of the arbitrary functions in the L–T model, adding some amount of rest mass to a source may decrease the active gravitational mass. The model used (a Friedmann region going over into the Schwarzschild spacetime through an interpolating L–T region) was specified numerically, and the effect was demonstrated only for a definite amount of mass added. Novikov's interpretation: 'The (negative) potential energy of gravitational interaction between the medium added and the one present previously is greater than the energy corresponding to the mass of the medium added. This leads to a decrease of the total gravitational mass.' The effect is an example of what happens beyond the neck (see Section 18.10): adding more rest mass results in reducing the total active mass. By another numerically specified example Novikov showed that an infinite amount of added rest mass may leave the active mass still finite.

Hellaby and Lake (1985) presented a few more examples that run counter to the intuitions based on the properties of the R–W models:

1. A model with $E > 0$ everywhere that has a globally closed space.
2. A model with $E < 0$ everywhere that has an open space.
3. A region of negative curvature placed between two regions of positive curvature, yet it does not develop shell crossings.

The first situation occurs when, for example:

$$M = M_0 \gamma^m, \qquad E = E_0 \gamma^{2m/3}, \qquad t_B = -a_0 \gamma^p, \tag{18.194}$$

where M_0, E_0, a_0, m and p are positive constants, and

$$\gamma(r) \overset{\text{def}}{=} 3\sin(\pi r/l) + 2\sin(3\pi r/l), \tag{18.195}$$

$l > 0$ being one more constant. Note that $\gamma(0) = 0 = \gamma(l)$, and $\gamma(r) > 0$ for all $0 < r < l$. The function $\gamma(r)$ has a (positive) local minimum at $r = l/2$ and two local maxima, at $r = r_1 = l \arcsin[\sqrt{3}/(2\sqrt{2})]$ and $r = r_2 = l\pi - r_1$. For this model, $(2E)^{3/2}/M = (2E_0)^{3/2}/M_0 \overset{\text{def}}{=} C = \text{constant}$, and, from (18.19):

$$R(t, r) = \frac{M_0}{2E_0} \gamma^{m/3}(\cosh \eta - 1), \qquad \sinh \eta - \eta = C(t + a_0 \gamma^p). \tag{18.196}$$

Thus, for all $t > t_B$, $R = 0$ at $r = 0$ and at $r = l$, and $R > 0$ for $0 < r < l$. Consequently, $R(t, r)$ must have, at every fixed t, at least one local maximum. We have

$$R_{,r} = \left[\frac{m}{3} \gamma^{m/3-1}(\cosh \eta - 1) + Ca_0 p \gamma^{p-1} \frac{\sinh \eta}{\cosh \eta - 1}\right] \gamma_{,r}. \tag{18.197}$$

Since both terms in square brackets are positive for $r \in (0, l)$, $R_{,r}$ can vanish only where $\gamma_{,r}$ does, i.e. the maximum of $R(t, r)$ at fixed t must coincide with one of the extrema of $\gamma(r)$. Since $R = 0$ at two different values of r for each t, each space $t = \text{constant}$ has two origins and a finite radius. Still, at every r the space expands to infinite size as $t \to \infty$. Hence, the model has closed spaces, but is ever-expanding – a situation that cannot happen in the Friedmann models with $\Lambda = 0$.

The second situation occurs in the spacetime discussed in Section 18.9. Hellaby and Lake's example had similar qualitative properties.

The third situation, an ever-expanding region sandwiched between two recollapsing regions, occurs with the following functions:

$$E = E_0 r^2 (l - r)^2 (br - 1)[b(l - r) - 1],$$
$$M = M_0 r^3 (l - r)^3, \qquad t_B = -a_0 r(l - r),$$

(18.198)

where $l > 0$ and $b > 2/l$ are constants. The three functions have vanishing derivatives by r at the same three values: $r = 0, r = l/2$ and $r = l$. All three vanish at $r = 0$ and $r = l$, while $E(r)$ has two additional zeros, at $r = r_1 \stackrel{\text{def}}{=} 1/b$ and at $r = r_2 \stackrel{\text{def}}{=} l - 1/b$. We have $M(r) > 0$ for $0 < r < l$, $E(r) > 0$ for $r_1 < r < r_2$ and $E(r) < 0$ for $0 < r < r_1$ and for $0 < r < r_2$. From the behaviour of $E(r)$ it follows that it has minima in $(0, r_1)$ and (r_2, l), and a maximum in (r_1, r_2) at $r = l/2$. In order to render the model physical, we have to ensure that $E(r) \geq -1/2$ for all $r \in [0, l]$ and that there are no shell crossings.

The minima of $E(r)$ occur at

$$r = r_{3,4} = \frac{l}{2} \pm \frac{1}{6b}\sqrt{3\left(3b^2 l^2 - 8bl + 8\right)}$$

(18.199)

and at both of them E has the same value $E_{\min} = -[4E_0/(27b^4)](bl - 1)^3$. Thus, the condition $E \geq -1/2$ implies that

$$E_0 < \frac{27}{8}\frac{b^4}{(bl - 1)^3}.$$

(18.200)

With the functions (18.198), the no-shell-crossing conditions in the $E > 0$ region are fulfilled, we have $t_{B,r} < 0, M_{,r} > 0$ and $E_{,r} > 0$ for $r < l/2$; $t_{B,r} = M_{,r} = E_{,r} = 0$ for $r = l/2$; and $t_{B,r} > 0, M_{,r} < 0, E_{,r} < 0$ for $r > l/2$. In the $E < 0$ region, (18.111) has to be fulfilled for $r < l/2$, and its opposite for $r > l/2$. No additional restrictions on the constants come from this requirement, since these conditions are fulfilled automatically; the left-hand side of Eq. (18.111) is equivalent to

$$\left[3\pi b^2 + \frac{(2E_0)^{3/2}}{M_0}a_0\left(b^2 r^2 + b^2 lr + bl - 1\right)^{3/2}\right](2r - l).$$

(18.201)

The trinomial in r is positive for all $r \geq 0$, the expression in square brackets is thus positive, while the factor $(2r - l)$ ensures the correct overall sign of the whole expression in the ranges $r < l/2$ and $r > l/2$.

Just like the spacetime (18.194)–(18.196), the one specified in (18.198) has two centres, one at $r = 0$ and the other at $r = l$. $E(r)$ is negative in the vicinity of each centre and positive around $r = l/2$. Thus, each of the two $E < 0$ regions recollapses onto a different centre, and this is why the ever-expanding region between them can exist without shell crossings ever being created.

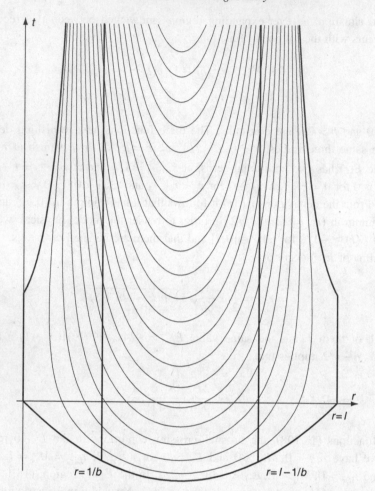

Fig. 18.13. A map of the graph showing the evolution of the (t, r) subspace in the model given by (18.198). The thin lines are contours of constant values of R. The differences in R between consecutive contours are the same everywhere. The curve at the bottom is the Big Bang, the two curves bordering the surface on the right and on the left together form the Big Crunch. The two thick lines, at $r = r_1 = 1/b$ and at $r = r_2 = l - 1/b$, separate the region with $E(r) > 0$ (between them) from the regions with $E(r) < 0$. For $r < r_1$ and $r > r_2$, the lifetime of the model is finite, although it tends to infinity as r approaches r_1 or r_2. The short straight segments at $r = 0$ and at $r = l$ are the two centres of spherical symmetry.

Figure 18.13 shows the evolution of the (t, r) surfaces in this model. At $r = 0, r = r_1 = 1/b, r = r_2 = l - 1/b$ and at $r = l$ the $r = $ constant shells evolve by the $E = 0$ law. For every $0 < r < r_1$ and for every $r_2 < r < l$, the evolution proceeds by the $E < 0$ law, and that part of the space recollapses after a finite time T, with $T \to \infty$ as $r \to r_1$ or $r \to r_2$. For $r_1 < r < r_2$, the evolution proceeds by the $E > 0$ law.

18.20.3 Chances to fit the L–T model to observations

Mészáros (1986) considered the implications of the assumption that the observed part of the actual Universe is described by a finite-volume portion of the L–T model with a simultaneous Big Bang. There are two chances of confirming this assumption observationally: (1) by observing the outer surface of the L–T sphere; (2) by detecting the *cosine anisotropy* in the matter-density and in the Hubble parameter. There is no observational evidence for point (1). As for point (2), Mészáros noted the following: (I) Observations of matter-density are inconclusive. (II) The Hubble parameter and the temperature of the microwave background radiation do reveal cosine anisotropies. They are usually explained away as resulting from the motion of the Solar System around the Galaxy. However, the directions and magnitudes of the velocities calculated from the two effects are inconsistent. This inconsistency is a problem in a Robertson–Walker model, but it can be accounted for on the basis of an L–T model. Unfortunately, the precision and interpretation of the observations are questionable. The quadrupole anisotropy of the background radiation is much simpler to explain in an L–T model than in a Friedmann model. Mészáros concluded that there are more arguments for the L–T models than against them, but present observations are inconclusive.

18.20.4 An 'in one ear and out the other' Universe

Hellaby (1987) presented a few illuminating examples of nontrivial L–T geometries and topologies. One of them, called 'In one ear and out the other', has the geometry shown in Fig. 18.14. The coordinate r is allowed to assume negative values. For $r \leq -E_1$, where $E_1 > 0$ is a constant, the model has $E \geq 0$ ($E = 0$ at $r = -E_1$) and collapses from past infinity to the Big Crunch that has a finite value at all $0 > r > -\infty$. For $-E_1 < r < E_2$ (where $E_2 > 0$ is another constant), E is negative, so the model has both a Big Bang and a Big Crunch at finite times. For $r \geq E_2$, the model has $E \geq 0$ again (with $E = 0$ at $r = E_2$), and expands from the Big Bang in the finite past to infinity.[1] In the range $r \leq 0$, the L–T arbitrary functions are

$$M(r) = \frac{1}{2}\left(M_0 - M_1 r^3\right), \qquad E(r) = -\frac{1}{2}\left(1 - r^2/E_1^2\right),$$

$$t_C(r) = \frac{1}{2}\pi M_0 \left(1 + \frac{3}{2}r^2/E_1^2\right). \tag{18.202}$$

The function $t_C(r)$ is the crunch time, and the bang time is

$$t_B(r) = t_C(r) - \frac{\pi\left(M_0 - M_1 r^3\right)}{\left(1 - r^2/E_1^2\right)^{3/2}}. \tag{18.203}$$

Thus, as seen in Fig. 18.14, the bang time goes down to $-\infty$ as $r \to E_1$.

---

[1] The configuration presented here is a slight generalisation of that used by Hellaby, who considered the case $(E_1, M_1) = (E_2, M_2)$.

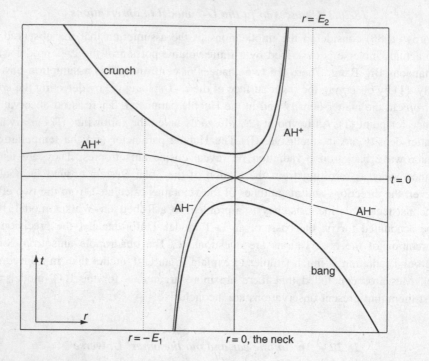

Fig. 18.14. The spacetime diagram for the L–T model of Eqs. (18.202)–(18.205). It has past timelike infinity for $r \leq -E_1 < 0$, a finite lifetime for $-E_1 < r < E_2$ and future timelike infinity for $r \geq E_2 > 0$. There is a neck at $r = 0$, where the past and future apparent horizons touch each other. In the figure, the time along AH^+ tends to $-\infty$ as $r \to -\infty$, and so does the time along AH^- as $r \to +\infty$. Whether these times go to $+\infty$ or $-\infty$ depends on the numerical balance between the various constants; see Exercise 15.

In the range $r \geq 0$, the functions are

$$M(r) = \frac{1}{2}\left(M_0 + M_2 r^3\right), \qquad E(r) = -\frac{1}{2}\left(1 - r^2/E_2^2\right),$$

$$t_B(r) = -\frac{1}{2}\pi M_0 \left(1 + \frac{3}{2}r^2/E_2^2\right). \tag{18.204}$$

The crunch time corresponding to the Big Bang at $t = t_B(r)$ is

$$t_C(r) = t_B(r) + \frac{\pi\left(M_0 + M_2 r^3\right)}{\left(1 - r^2/E_2^2\right)^{3/2}}, \tag{18.205}$$

and it goes up to $+\infty$ as $r \to E_2$. There is a neck at $r = 0$, where the past and future apparent horizons touch each other.

As we observed in Section 18.8, Eq. (18.86), an outgoing future apparent horizon can never be timelike, and can be null only in vacuum. Since we are considering here a model that is not vacuum anywhere, the future apparent horizon in Fig. 18.14 is everywhere spacelike. This observation is sufficient to see that no light ray can be sent from the left part of the spacetime between AH^- and AH^+ to the right part between AH^- and AH^+: a light ray that would touch any point of the future apparent horizon would immediately pass inside it. This is in contrast to the Schwarzschild spacetime, in which, as seen from the Kruskal diagram (Fig. 14.6), it is possible to send light rays along the apparent/event horizon. Thus, as stated by Hellaby (1987), in the L–T spacetime the communication between the opposite sides of the neck is even worse than in the Schwarzschild spacetime. However, it is possible to receive a light ray that originated behind the neck before the AH^-.

18.20.5 A 'string of beads' Universe

Another of Hellaby's (1987) illuminating examples is a 'string of beads' model. This is a chain of recollapsing L–T regions, each of which begins in a separate Big Bang (BB), then they become connected by necks, and then split to end their lives in separate Big Crunches (BCs), see Fig. 18.15.[1] In our example, for simplicity, the BC is a mirror image of the BB. In general, in order to have necks without shell crossings, the minima of the BC must have the same radial coordinates as the maxima of the BB, but otherwise the BB and BC can be arbitrary. The L–T functions in Fig. 18.15 are:

$$t_B(r) = -\frac{\pi M}{(-2E)^{3/2}}, \qquad M(r) = M_0 + M_1 e^{-ar^2} \cos(br),$$

$$E(r) = -\frac{1}{2}\left(1 - E_1 e^{-ar^2} \sin^2(br)\right),$$

(18.206)

where M_0, a, b and E_1 are arbitrary positive constants, subject to the conditions that $M > 0$ and $-1/2 \leq E < 0$ everywhere. The corresponding crunch function is $t_C(r) = \pi M/(-2E)^{3/2}$. The space extends infinitely far in all directions. This Universe has no centre of symmetry. It begins its existence in a series of (non-simultaneous!) explosions, each of which creates a bubble of spacetime isolated from the others. Inside each bubble, the BB is going on for some time, and it ends when the bubble becomes connected to a neighbouring bubble by a neck. At a certain time t_e (different for each bubble), the BB expires. After the last BB is over, the Universe becomes a chain of spacetime regions connected by necks. Later, the BC sets in, also at a different time for each bubble. The Universe splits again into unconnected bubbles, each of which gradually collapses into its own final singularity.

18.20.6 Uncertainties in inferring the spatial distribution of matter

Partovi and Mashhoon (1984) calculated various cosmologically relevant parameters for two types of a spherically symmetric inhomogeneous spacetime: one with a perfect fluid

---

[1] Again, our example is a slight generalisation of that presented by Hellaby, who considered 'beads' isometric.

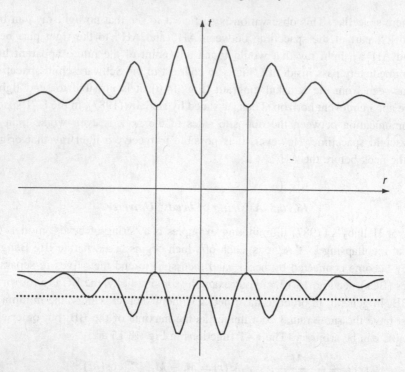

Fig. 18.15. A 'string of beads' Universe. The lower curve is the Big Bang (BB); the upper curve is the Big Crunch (BC). The Universe begins its existence as a chain of unconnected bubbles. The dotted line shows the time by which four bubbles already exist, and for the fifth and sixth the BB just begins. The horizontal line shows the time by which the two central bubbles have completed their BBs. The vertical lines show the positions of three of the necks. (At the local minima of the BB, which are the local maxima for the BC, there are no necks because $E \neq -1/2$ there.)

source and a barotropic equation of state, the other with a dust source. The quantities are calculated in terms of successive powers of the redshift around the values in a R–W background. It was found that the leading terms in z are identical to the R–W terms, while the next following terms are formally modified, but in such a way that the inhomogeneous corrections cannot be properly read out from the formulae, and may lead to assignment of incorrect values to the standard cosmological parameters. The conclusion is that up to those following terms these inhomogeneous models are observationally indistinguishable from the R–W models.

Kurki-Suonio and Liang (1992) used the L–T model to demonstrate the ambiguity in reading out the spatial distribution of matter from redshift measurements. What the astronomers can directly measure is the number of light sources in a given solid angle and a given redshift interval, i.e. the density of matter in the **redshift space**. Lacking any information about its geometry, the redshift space, in which z is the radial coordinate, is assumed flat. Let us assume that the observer is at the centre of symmetry at $z = 0$, and let $N(z)$ be the mass contained in the sphere of radius z around the observer in the redshift space. This mass is the sum of rest masses defined in (18.17), and, from (18.18), is related to the mass function $M(r)$ of the L–T model by $N_{,r} = M_{,r}/\sqrt{1+2E}$. We define the mass density in the redshift space, $\widehat{\rho}(z)$, by

$$\frac{c^2 H_0^3}{G} N(z) = \int_0^z 4\pi z'^2 \widehat{\rho}(z')\mathrm{d}z'. \tag{18.207}$$

The coefficients are there in order to provide convenient units; H_0 is the current value of the Hubble parameter at the observer's position. Thus

$$\frac{c^2 H_0^3}{G} N_{,z} = 4\pi z^2 \widehat{\rho}(z). \tag{18.208}$$

But $N_{,z} = N_{,r}/z_{,r}$. Substituting for $N_{,r}$ from (18.17)–(18.18) and for $z_{,r}$ from (18.125) we obtain

$$\widehat{\rho}(z) = \frac{2H_0^3 M_{,r}}{c^2 \kappa (1+z)z^2 R_{,tr}} = \frac{H_0^3 R^2 R_{,r}}{(1+z)z^2 R_{,tr}} \rho(r), \tag{18.209}$$

where $\rho(r)$ is the physical mass density. The values of t and r refer to the time and position of emission. Equation (18.209) connects one observed quantity, $\widehat{\rho}(z)(1+z)z^2$, to two unknown functions, $M(r)$ and $R(t,r)$, and obviously cannot determine both. The authors investigated numerically the dependence of these functions on $\widehat{\rho}(z)$ with various shapes of $t_B(r)$, $E(r)$ and initial density distributions $\rho(t,r)$. They assumed no shell crossings ($R_{,r} > 0$) and redshifts monotonically growing with distance ($R_{,tr} > 0$). The results, greatly abbreviated, are as follows:

1. If $t_B(r) = $ constant, then the amplitudes of overdensity in the redshift space are larger than the amplitudes of proper density by 40% or more. The overdensity regions in proper density have larger sizes than their images in the redshift space.
2. When $M(r)/r^3 = $ constant and $E(r)/r^2 = $ constant, that is, when the bang-time function is the only factor generating inhomogeneity, overdensities in the redshift space translate into *underdensities* in proper density, of larger size and much larger amplitude.
3. Since the two effects influence observations in opposite directions, they can momentarily cancel each other, producing a homogeneous density in the redshift space while the Universe is highly inhomogeneous. Conversely, the Universe may be momentarily very smooth in density, but can appear highly inhomogeneous in the redshift space because of the effects of inhomogeneous velocities. To illustrate this last possibility, Kurki-Suonio and Liang fitted the velocity function in the L–T model so that the proper density is constant on the past light cone, while the density in redshift space is the same as in one of the deep-redshift observational surveys. The proper density at the present time then has smaller amplitudes in the peaks and a less regular distribution of the peaks. The authors concluded that using the R–W relation between the redshift and the comoving distance to describe inhomogeneities is 'fundamentally self-inconsistent'. They said, further: 'What we see on the past light cone are only momentary "images".'

Kurki-Suonio and Liang's paper contains several graphs and spacetime diagrams that illustrate the conclusions more clearly. In particular, they show how the bang-time non-simultaneity generates decaying inhomogeneities while variations in initial energy generate growing inhomogeneities, even though these processes are not discussed in the paper.

18.20.7 Is the matter distribution in our Universe fractal?

Ribeiro (1992a) defined a fractal distribution of matter by requiring that the number of objects N_c and luminosity distance d_1 to the surface of the sphere containing the N_c objects are connected by $N_c = \sigma d_1^D$, where σ is a constant and D is the fractal dimension of the distribution.

In a sequel paper (Ribeiro, 1992b), the author calculated the density versus distance function along the past light cone in the $k = 0$ Friedmann model. This function is not constant and does not seem to agree with the observations. Ribeiro stressed that cosmological models should be tested in precisely this way. For the $k = 0$ Friedmann model this function does not possess any features of a fractal distribution of mass.

In the third paper, Ribeiro (1993) showed that the density versus distance relation measured along the past light cone in the $k \neq 0$ Friedmann models is not homogeneous either, nor is it fractal. However, the functions $E(r)$, $M(r)$ and $t_B(r)$ in the L–T model *can* be chosen so that $N_c = \sigma d_1^D$. Ribeiro found several examples of such functions by numerical experiments. In most of the examples, either the Hubble law is contradicted or the Hubble constant is smaller than the astronomers currently believe it should be. The only example that yields acceptable parameters is $2E(r) = \sinh^2 r$, $2M(r) = \alpha r^p$, $t_0(r) = \ln(e^{\beta_0} + \eta_1 r)$. If $\alpha = 10^{-4}$, $p = 1.4$, $\beta_0 = 3.6$ and $\eta_1 = 1000$; then the implied fractal dimension is $D = 1.3$, the Hubble constant is $H = 61\,\mathrm{km\ s^{-1}\,Mpc^{-1}}$, and $\sigma = 5.4 \times 10^5$. Even better agreement with observations follows when $\eta_1 = 0$; then $D = 1.4$ and $H = 80\,\mathrm{km\ s^{-1}\,Mpc^{-1}}$. Ribeiro emphasised that no positive observational evidence of spatial homogeneity of the Universe is available. Should it be available, it would actually rule out the Friedmann models because the light cones in them intersect hypersurfaces of different densities.

18.20.8 General results related to the L–T models

Novikov (1962b) proved the following theorems for a general spherically symmetric spacetime: 1. If $\lim_{r \to r_0} R(t_0, r) = \infty$ for a certain t_0 (the r_0 may be infinite), and for $r > r_1$ the matter-density $\epsilon \geq A/(8\pi R^2)$, where $A = \text{constant} > 1$, then there exists an r_2 such that the points with $r > r_2$ and $t = t_0$ are in the T-region.[1] 2. For every $r = r_2$ there exist such t_1 and t_2 (dependent on r_2) that the points with (t, r_2) where $t_1 < t < t_2$ are in the T-region.[2] 3. There exists a solution for which the volume of a finite mass

---

[1] For the definition of the T- and R-regions, see Section 14.1. This means that with sufficiently high density the spheres $t = \text{constant}$, $r = \text{constant}$ will not be static to any observer.

[2] This means that each matter particle remains in a T-region for a period of time – in the vicinity of the Big Bang singularity.

goes through a nonsingular minimum and the minimum is simultaneous for the whole space. 4. The same is possible for an infinite mass. 5. There exists a solution in which a volume minimum occurs at small distances from the centre, and, at the same instant, a volume maximum occurs at large distances from the centre. 6. If $M(r) = \int \epsilon \, dV$ grows with r faster than the affine radius $l = \int_0^r (-g_{11})^{1/2} \, dr'$, and does so for a sufficiently large interval of l, then the configuration cannot be static. 7. If $M(r)$ grows faster than $l(r)$ over a sufficiently large interval of monotonic growth of R, then the volumes of different shells cannot go through an extremum simultaneously. 8. If the surface of a spherical ball is in a T-region and is shrinking, then, in a finite proper time, it will shrink to a point; if the surface is in a T-region and is expanding, then it started its expansion from a point. Finally, Novikov calculated the position of the boundary between the R- and T-regions in a Friedmann model and concluded that the radio sources observed must lie in the T-region. Consequently, they seem to have emerged from a point. The properties listed above reveal what should be expected from exact solutions generalising the L–T model for pressure with nonzero spatial gradient.

A later paper by Novikov (1964b) is a continuation of the one described above. One of its new results is that in the Schwarzschild solution the boundary between the R- and T-regions is the $r = 2m$ event horizon, and such boundaries also exist for the Kottler (i.e. Schwarzschild with $\Lambda \neq 0$) and Reissner–Nordström solutions. Another result was discussed in Section 14.12 (Eqs. (14.116)–(14.117) and Exercise 10 in Chapter 14).

In a series of papers, Dautcourt (1980, 1983a,b, 1985) developed a new approach to integrating the Einstein equations that is better suited to incorporating observational results into cosmological models. In this approach, the initial data are given on the past light cone of a single event in the spacetime, and then the Einstein equations are integrated off that cone to the future. The first paper (Dautcourt, 1980) is a short exposition of the ideas of the other papers. In the second paper (Dautcourt, 1983b) the author gave a prescription for integrating the Einstein equations in vacuum and for dust with the initial data given on the past light cone of an observer. The Einstein Universe and the Friedmann models were used as illustrations. In the third paper (Dautcourt, 1983a) the author described the ideal observations needed to determine the initial data. In practice, several of those observations are beyond the limits of current technology. In the last paper (Dautcourt, 1985), the author noted that the observable quantities introduced in the preceding papers depend on the 4-velocity of the observer, and proposed a preferred set of coordinates with a simple relation to the observable quantities and a simple transformation law under a Lorentz rotation of the observer. Dautcourt's approach was later developed by Ellis and coworkers; the paper by Stoeger, Ellis and Nel (1992) mentioned in Section 18.3 is one entry in that series.

18.21 Exercises

1. Prove that if the spacetime is spherically symmetric and the metric obeys the Einstein equations with a perfect fluid source, then the rotation is necessarily zero.
2. Prove that $\dot{u}^\alpha = 0$ in (18.1)–(18.2) implies $C_{,r} = 0$.

3. Prove that Eqs. (18.24) are the necessary and sufficient conditions for the L–T model to reduce to the Friedmann limit.

   Hint. Use M as the radial coordinate. Calculate $(R^3)_{,MM}$ in (18.23) separately for each sign of E. With $E \neq 0$, observe that the Jacobian $\partial(M, \eta)/\partial(M, t) = \eta_{,t} \neq 0$. Hence, M and η can be chosen as the independent variables. Eliminate $(t - t_B)$ from $(R^3)_{,MM}$ using Eqs. (18.19) and (18.21). What will result will be a polynomial in functions of η with coefficients depending on M only. Equate to zero the coefficients of independent functions of η. In order to show that (18.24) do indeed define the Friedmann model, change the radial variable to $r = (M/M_0)^{1/3}$ and substitute whatever you have found in the metric (18.16), then compare the result with (17.1) and (17.33)–(17.35).

4. Transform the L–T metric to the curvature coordinates.

   Hint. Note that, from (18.14),

$$R_{,r}\, dr = dR - R_{,t}\, dt = dR - l\sqrt{2E + \frac{2M}{R} - \frac{1}{3}\Lambda R^2}\; dt, \tag{18.210}$$

   where $l = +1$ for expanding and $l = -1$ for collapsing models. Then transform t by $t = f(T, R)$ and demand that $g_{TR} = 0$. This implies:

$$\frac{\partial f}{\partial R} = -\frac{1}{1 - 2M/R + \frac{1}{3}\Lambda R^2}\sqrt{2E + \frac{2M}{R} - \frac{1}{3}\Lambda R^2}. \tag{18.211}$$

   Now M and E are functions of $r(T, R)$ and the metric becomes

$$ds^2 = \frac{f_{,T}^2}{1 + 2E}\left(1 - \frac{2M}{R} + \frac{1}{3}\Lambda R^2\right) dT^2 - \frac{1}{1 - 2M/R + \frac{1}{3}\Lambda R^2}\, dR^2$$
$$- R^2\left(d\vartheta^2 + \sin^2\vartheta\, d\varphi^2\right). \tag{18.212}$$

   The functions E and M obey the equations

$$0 = M_{,T}\, T_{,t} + M_{,R}\, R_{,t}, \quad 0 = E_{,T}\, T_{,t} + E_{,R}\, R_{,t}, \tag{18.213}$$

   where $T_{,t} = \partial T/\partial t$ is found as an element of the matrix inverse to $\begin{bmatrix} f_{,t} & f_{,R} \\ r_{,T} & r_{,R} \end{bmatrix}$.

5. Prove that Eq. (18.45) is the necessary and sufficient condition for the existence of a solution of $\psi_H(x) = 0$ at $x \neq 0$, where ψ_H is given by (18.44). Prove also that only one such value of x exists.

   Hint. Calculate $d[\psi_H(x)]/dx$ and consider how it changes.

6. Prove that (18.48) are the necessary and sufficient condition for the existence of a solution of $\psi_X(x) = 0$ at $x \neq 0$, where ψ_X is given by (18.46). Prove also that only one such value of x exists.

7. Verify that Eqs. (18.60) and (18.63) are indeed the sets of necessary and sufficient conditions for the existence of the appropriate evolutions. Interpret the results (i.e. why the time-difference has this time to be sufficiently *large* in (18.60) and sufficiently *small* in (18.63)).

8. Prove that for a Friedmann model with $\Lambda = 0$ (in which also $p = 0$) the parameter B given by (18.85) is always equal to -2, and so the apparent horizons in it are timelike.

9. Derive Eq. (18.104) for the $E > 0$ and $E < 0$ L–T models.

10. Prove that Eqs. (18.106), (18.110) and (18.111) are sufficient conditions for avoiding a shell crossing when $R_{,r} > 0$.

Hint. Note that (18.110) and (18.111) imply

$$\frac{M_{,r}}{M} > \frac{3}{2}\frac{E_{,r}}{E}, \tag{18.214}$$

i.e. that the coefficient of $\Phi_1(\eta)$ in (18.107) is negative. Then write Eq. (18.111) as follows:

$$f(\eta) = \frac{M_{,r}}{3M} + \left(\frac{3}{2}\frac{E_{,r}}{E} - \frac{M_{,r}}{M}\right)\left(-\frac{2}{3} + \Phi_1(\eta)\right) - \frac{(-2E)^{3/2}}{M}t_{B,r}\Phi_2(\eta). \tag{18.215}$$

Next, note that $f(\eta)$ is evidently positive for all $\eta \geq \pi$ in consequence of (18.106), (18.110) and (18.214). Thus, it remains to investigate $f(\eta)$ in the range $[0, \pi)$. The first and third terms in (18.215) are positive. The first factor in the second term is negative. Show that

$$\frac{d\Phi_1}{d\eta} = \frac{3\sin\eta - \eta\cos\eta - 2\eta}{(1 - \cos\eta)^2} \tag{18.216}$$

is negative for $\eta \in (0, \pi]$. Since $\lim_{\eta\to 0}\Phi_1 = 2/3$ and $\Phi_1(\pi) = 0$, this means that $\Phi_1(\eta) < 2/3$ for $\eta \in (0, \pi]$. This proves that the middle term in (18.215) is positive, too, for $\eta \in [0, \pi)$.

11. Prove that Eq. (18.126), is equivalent to (16.20).

Hint. With the coordinates of (18.16), Eq. (16.20) can be written as $1 + z = (k^0)_e/(k^0)_o$. Keeping the observer at a fixed position, take the logarithmic derivative of this by r and compare the result with (18.125). The conclusion is

$$\frac{(k^0)_{,r}}{k^0} = \frac{R_{,tr}(T(r), r)}{\sqrt{1 + 2E(r)}}. \tag{18.217}$$

Integrating this and using (18.119) we obtain (18.134). Verify that this field is affinely parametrised (note that the integral is calculated along a null geodesic, so the integrand is a function of r only).

12. The metric (18.160) is invariant under the transformations $t' = e^{\tau}t$, $r' = e^{\tau}r$, where τ is the group parameter. Verify that their generators, $k^{\mu} = t\delta^{\mu}{}_0 + r\delta^{\mu}{}_1$, obey (8.55) with $\lambda = 2$.

13. Verify that if the curve $t = Dr$, $\vartheta = $ constant, $\varphi = $ constant in the spacetime with the metric (18.160) is null, then it is necessarily a geodesic (but the parametrisation by r or t is non-affine!).

14. Prove that the light ray k_2 sent backwards in time towards the centre of symmetry from the point p_2 in Fig. 18.12 can never intersect the shell crossing set if the following assumptions are made:

(a) The functions $t = t_{n2}(r)$ (describing the light ray k_2) and the shell-crossing function $t = t_S(r)$ are differentiable for $r \in [0, r_2]$.

(b) $t'_S(r) > 0$ for $r \in [0, r_2]$.

(c) $R_{,r} > 0$ for all $t > t_S(r)$.

(d) The mass density is nowhere zero.

Hint. Here is one suggestion of a proof. Since $t'_S(r) > 0$ for $r \in [0, r_2]$, there exists a minimal value $M > 0$ of $t'_S(r)$ in this range, so $t'_S(r) \geq M$ for $r \in [0, r_2]$.

If k_2 should intersect $t_S(r)$ at some $r_1 \in [0, r_2]$, then $t'_{n2}(r_2) = 0$ and $t'_{n2} \xrightarrow[r\to r_1]{} 0$, while $t'_{n2}(r) > 0$ for $r > r_1$ (by virtue of assumption (c)). Hence, for every $0 < \epsilon < M$ there exists such a $\delta > 0$ that $0 \leq t'_{n2}(r) < \epsilon < M$ if $r_1 \leq r < r_1 + \delta$.

Since $t_{n2}(r)$ starts from (t_2, r_2) with a value $t_{n2}(r_2) > t_S(r_2)$, assume that $t_{n2}(r_1 + \delta) > t_S(r_1 + \delta)$ and compare $t_{n2}(r_1)$ with $t_S(r_1)$. By virtue of the statements made we have

$$t_S(r_1) < t_S(r_1 + \delta) - M\delta, \qquad t_{n2}(r_1) > t_{n2}(r_1 + \delta) - \epsilon\delta,$$

$$t_{n2}(r_1) - t_S(r_1) > t_{n2}(r_1 + \delta) - t_S(r_1 + \delta) + (M - \epsilon)\delta$$

$$> t_{n2}(r_1 + \delta) - t_S(r_1 + \delta).$$

Thus, whenever k_2 would be approaching the horizontal direction, it would be receding from the shell crossing rather than intersecting it.

In order that k_2 can intersect the shell crossing set, it would have to either (1) change abruptly from past-directed ingoing to past-directed outgoing, which is prohibited by assumption (a), or (2) become vertical and then past-directed outgoing. However, at the point where the tangent to k_2 would be vertical, $R_{,r}$ would go to infinity, making the mass density zero, which is prohibited by assumption (d).

Hence, k_2 and $t_S(r)$ cannot intersect between $r = 0$ and $r = r_2$.

15. Calculate the equations $t = t_{AH}(r)$ of the apparent horizons for the model (18.202)–(18.205). Show that for the past AH $t_{AH} \xrightarrow[r \to +\infty]{} \varepsilon_r \infty$, where $\varepsilon_r = \pm 1$ and the sign of ε_r is the sign of $(2M_2 E_2{}^3 - \pi M_0/2)$. Show that for the future AH $t_{AH} \xrightarrow[r \to -\infty]{} \varepsilon_l \infty$, where $\varepsilon_l = \pm 1$ and its sign is the sign of $(\pi M_0/2 - 2M_1 E_1{}^3)$.

16. Let the mass density in the L–T model (18.15) be a differentiable funciton of r. Prove that the spatial extrema of density in the $E \neq 0$ models are comoving only when the following equations are fulfilled at the extremal values:

$$\left(\frac{|2E|^{3/2}}{M} \right)_{,r} = \left(\frac{|2E|^{3/2}}{M} \right)_{,rr} = t_{B,r} = t_{B,rr} = 0,$$

$$\left[\left(\frac{M^3}{E^3} \right)_{,r} \frac{1}{M_{,r}} \right]_{,r} = 0. \tag{18.218}$$

Find the corresponding condition for the $E = 0$ model.

Note. The fact that the extrema are comoving only under the conditions (18.218) means that in general they are not comoving. Thus, they move across the flow lines of dust. This phenomenon is known in cosmology as **density waves**. Their existence in relativistic models was predicted by Ellis, Hellaby and Matravers (1990) by methods of the linearised Einstein theory. Note also that the first four conditions mean that at the extremum the L–T model has a second-order contact with the Friedmann model that approximates it.

Hint. Choose M as the spatial coordinate. The first four conditions do not change then, but the last one takes the more readable form $(M^3/E^3)_{,MM} = 0$. The extrema are at those values of M where $\epsilon_{,M} = 0$, i.e. $(R^3)_{,MM} = 0$. Then proceed as in Exercise 3. If the extrema are comoving, then the values of M obeying (18.218) do not depend on t.

19

Relativistic cosmology IV: generalisations of L–T and related geometries

19.1 The plane- and hyperbolically symmetric spacetimes

The L–T models considered in the previous chapter have 2-spheres as orbits of their symmetry groups. Do there exist spacetimes whose symmetry orbits are surfaces of zero or negative constant curvature?

A surface of zero curvature has the local geometry of a plane, whose symmetries in the Cartesian coordinates (x, y, z) are

$$(x', y', z') = (x + a_1, y, z), \qquad (x', y', z') = (x, y + a_2, z),$$
$$(x', y', z') = (x \cos a_3 + y \sin a_3, -x \sin a_3 + y \cos a_3, z), \tag{19.1}$$

where a_1, a_2 and a_3 are the group parameters; a_1 and a_2 are arbitrary, while $0 \le a_3 \le 2\pi$. Proceeding just as in Section 8.9, we find that a 4-dimensional metric for which (19.1) is the symmetry group is

$$ds^2 = \alpha(t, r)dt^2 + 2\beta(t, r)dt\,dr + \gamma(t, r)dr^2 + \delta(t, r)\left(dx^2 + dy^2\right). \tag{19.2}$$

Spacetimes with this symmetry group are called **plane symmetric**.

Then, proceeding as in Section 18.1, we prove that in such a spacetime rotation must be zero, so comoving–synchronous coordinates can be introduced, in which the metric becomes similar to (18.1):

$$ds^2 = e^{C(t,r)}\,dt^2 - e^{A(t,r)}\,dr^2 - R^2(t, r)\left(dx^2 + dy^2\right). \tag{19.3}$$

The curvature tensor of a sphere, $R^2\left(d\vartheta^2 + \sin^2 \vartheta\, d\varphi^2\right)$, $R = \text{constant}$, is

$$R^{\alpha\beta}{}_{\gamma\delta} = R^{-2}\delta^{\alpha\beta}_{\gamma\delta}, \tag{19.4}$$

where $1/R^2 > 0$ is the constant scalar curvature. We can make the curvature negative by the complex transformation $\{\vartheta = i\vartheta', R = iR'\}$. Thereafter, the 2-metric becomes (with primes dropped)

$$ds^2 = R^2\left(d\vartheta^2 + \sinh^2 \vartheta\, d\varphi^2\right), \tag{19.5}$$

and its curvature tensor is $R^{\alpha\beta}{}_{\gamma\delta} = -R^{-2}\delta^{\alpha\beta}_{\gamma\delta}$. We choose $(x^2, x^3) = (\vartheta, \varphi)$ as coordinates in the spacetime. The Killing fields are then

$$\underset{(1)}{k^{\alpha}} = \sin\varphi\,\delta^{\alpha}{}_2 + \cos\varphi\coth\vartheta\,\delta^{\alpha}{}_3,$$

$$\underset{(2)}{k^{\alpha}} = \cos\varphi\,\delta^{\alpha}{}_2 - \sin\varphi\coth\vartheta\,\delta^{\alpha}{}_3, \qquad \underset{(3)}{k^{\alpha}} = \delta^{\alpha}{}_3, \tag{19.6}$$

and we require these to be the Killing fields for the whole spacetime. Solving the Killing equations with (19.6) we find the metric

$$ds^2 = \alpha(t, r)dt^2 + 2\beta(t, r)dt\,dr + \gamma(t, r)dr^2 + \delta(t, r)\left(d\vartheta^2 + \sinh^2\vartheta\,d\varphi^2\right), \tag{19.7}$$

where $\overset{*}{\alpha}, \beta, \gamma$ and δ are arbitrary functions. These spacetimes are called **hyperbolically symmetric** because the orbits of their symmetry group have several properties in common with two-sheeted hyperboloids.

Once again, we prove that in such a spacetime rotation must be zero, and we introduce the comoving–synchronous coordinates, in which

$$ds^2 = e^{C(t,r)}\,dt^2 - e^{A(t,r)}\,dr^2 - R^2(t, r)\left(d\vartheta^2 + \sinh^2\vartheta\,d\varphi^2\right). \tag{19.8}$$

The surface with the metric (19.5) cannot be embedded in a Euclidean 3-space – the 3-metric must have the signature $(+ + -)$. To verify this, consider (19.5) embedded in a flat 3-space, draw an orthogonal geodesic through each point of the surface, take the affine parameter r on the geodesics as the x^1-coordinate in space, let $r = R$ on the initial surface and let the surfaces $R \neq r = $ constant have the geometry of (19.5) with different values of R. In such coordinates, the 3-space metric will be

$$ds_3{}^2 = \varepsilon\,dr^2 + f^2(r)\left(d\vartheta^2 + \sinh^2\vartheta\,d\varphi^2\right), \tag{19.9}$$

where $\varepsilon = \pm 1$, to allow for different signatures. The Riemann tensor for the metric (19.9) can vanish only when $\varepsilon = -1$, and $f_{,r}(r) = 1$.

The symmetry groups of the spherically, plane- and hyperbolically symmetric spacetimes have 2-dimensional orbits. However, the groups themselves are 3-dimensional, so their algebras can be classified by the Bianchi method of Chapter 10. The algebras of spherical, plane and hyperbolic symmetry are of Bianchi types IX, VII_0 and VIII, respectively.

It is convenient to represent these spacetimes by a single formula:

$$ds^2 = \alpha(t, r)dt^2 + 2\beta(t, r)dt\,dr + \gamma(t, r)dr^2$$
$$+ \delta(t, r)\left[d\vartheta^2 + (1/\varepsilon)\sin^2(\sqrt{\varepsilon}\vartheta)d\varphi^2\right], \tag{19.10}$$

where $\varepsilon = \pm 1, 0$. Then, with $\varepsilon = +1$, the metric is spherically symmetric; with $\varepsilon = -1$ we recover the hyperbolically symmetric metric via the identity $\sin(i\vartheta) = i\sinh\vartheta$, and the limit $\varepsilon \to 0$ of (19.10) is plane symmetric. The three cases will be collectively called G_3/S_2-**symmetric** spacetimes, G_3 standing for 3-dimensional groups and S_2 for 2-dimensional orbits.

In the comoving–synchronous coordinates that exist for each of the three metrics we will use a variant of the notation of (19.10):

$$ds^2 = e^{C(t,r)} dt^2 - e^{A(t,r)} dr^2 - R^2(t,r)\left[d\vartheta^2 + f^2(\vartheta)d\varphi^2\right], \qquad (19.11)$$

where

$$f(\vartheta) = \sin\vartheta \qquad \text{for spherical symmetry,}$$

$$f(\vartheta) = \vartheta \qquad \text{for plane symmetry,} \qquad (19.12)$$

$$f(\vartheta) = \sinh\vartheta \qquad \text{for hyperbolic symmetry.}$$

19.2 G_3/S_2-symmetric dust solutions with $R,_r \neq 0$

Just as for the spherically symmetric models, the case $R,_r = 0$ has to be considered separately; we will present it in Section 19.4.

The most general set of solutions for the metric (19.11)–(19.12) with $R,_r \neq 0$, a dust source and nonzero cosmological constant was found by Ellis (1967, his case IIaiii). They are given by

$$ds^2 = dt^2 - \frac{R,_r^{\,2}(t,r)}{\varepsilon + 2E(r)} dr^2 - R^2(t,r)\left[d\vartheta^2 + f^2(\vartheta)d\varphi^2\right], \qquad (19.13)$$

where $E(r)$ is an arbitrary function, while $R(t,r)$ and the matter-density are determined by (18.14) and (18.15). The required signature $(+ - - -)$ puts different limits on $E(r)$ for each ε. With $\varepsilon = 0$, $E(r)$ must be non-negative everywhere, and can vanish only at isolated values of r, provided that $R,_r = 0$ and $R,_r^{\,2}/(2E)$ has a finite limit at the same values. With $\varepsilon = -1$, $E(r)$ must be strictly positive everywhere.

In the Friedmann limit $R(t,r) = g(r)S(t)$; then the r-coordinate can be chosen so that $R(t,r) = rS(t)$ and it follows further that $-2E/r^2 = k =$ constant, k being the Friedmann curvature index, $M(r)/r^3 = M_0 =$ constant, M_0 being the Friedmann mass integral. Thus, only $\varepsilon = +1$ allows all three Friedmann limits. The requirement of the Lorentzian signature forces the Friedmann limit to have $k < 0$ when $\varepsilon = 0$ or $\varepsilon = -1$.

The physical and geometrical interpretation of the plane- and hyperbolically symmetric solutions is less clear than that of the L–T model. In particular, it cannot be uniquely determined where exactly the mass $M(r)$ is contained. With plane symmetry, one can imagine it to be the mass contained within a column of height r along the r-direction, its base being a rectangle $\{\vartheta_1 \leq \vartheta \leq \vartheta_2, \varphi_1 \leq \varphi \leq \varphi_2\}$. For brevity, we will be saying that $M(r)$ is contained *within* the surface $r =$ constant.

19.3 G_3/S_2-symmetric dust in electromagnetic field, the case $R,_r \neq 0$

19.3.1 Integrals of the field equations

The most general solution of the Einstein–Maxwell equations with $R,_r \neq 0$ was found by Bronnikov and Pavlov (1979). They considered charged dust under the assumption that the metric tensor and the electromagnetic tensor both have a G_3/S_2 group of isometries, with the cosmological constant allowed. We shall allow also for magnetic charges.

With the assumed symmetries, in the coordinates of (19.11)–(19.12), the electromagnetic tensor can have at most two nonzero components, F_{tr} (the electric field) and $F_{\vartheta\varphi}$ (the magnetic field). The $F_{\vartheta\varphi} \neq 0$ is due to a distribution of magnetic monopoles – admittedly a nonrealistic source, but it is useful to know what the theory implies for it.

The Einstein–Maxwell equations are in this case (compare them with Eqs. (13.6), (13.10), (13.12), (13.15) and (13.16)):

$$R_{\alpha\beta} - \frac{1}{2}g_{\alpha\beta}R + \Lambda g_{\alpha\beta} = \frac{8\pi G}{c^4}\epsilon u_\alpha u_\beta + \frac{2G}{c^4}\left(F_\alpha{}^\mu F_{\mu\beta} + \frac{1}{4}g_{\alpha\beta}F_{\mu\nu}F^{\mu\nu}\right), \quad (19.14)$$

$$F^{\mu\nu}{}_{;\nu} = (4\pi/c)\rho_e u^\mu, \quad (19.15)$$

$$*F^{\mu\nu}{}_{;\nu} = (4\pi/c)\rho_m u^\mu. \quad (19.16)$$

Equation (19.14) is (13.16) rewritten for dust. Equation (19.15) is (13.6) with the current generating the field being composed of electric charges attached to the dust particles. Equation (19.16) is the generalisation of (13.15) for the current of magnetic charges attached to dust particles. The ρ_e and ρ_m are the densities of the electric and magnetic charge, respectively.

Equation (19.15) is equivalent to $\left(\sqrt{-g}F^{\mu\nu}\right)_{,\nu} = (4\pi/c)\sqrt{-g}\rho_e u^\mu$. Taking the various values for the index μ we find

$$F^{01} = Q_e(r)e^{-(A+C)/2}/R^2, \quad (19.17)$$

$$Q_{e,r} = (4\pi/c)\rho_e e^{A/2}R^2, \quad (19.18)$$

where $Q_e(r)$ is an arbitrary function; it is the electric charge within the r-surface as (19.18) shows. Doing the same with (19.16) we find

$$*F^{01} = Q_m(r)e^{-(A+C)/2}/R^2, \quad (19.19)$$

$$Q_{m,r} = (4\pi/c)\rho_m e^{A/2}R^2, \quad (19.20)$$

where now $Q_m(r)$ is the magnetic charge within the r-surface. Equation (19.19) translates into

$$F_{23} = f(\vartheta)Q_m(r). \quad (19.21)$$

Now the coordinate components of the Einstein equations with the metric (19.11) and the electromagnetic tensor (19.17)–(19.21) become

$$G_{00} = \frac{8\pi G}{c^4}\epsilon e^C + \frac{G}{c^4}\frac{Q_e{}^2 + Q_m{}^2}{R^4}e^C - \Lambda e^C, \quad (19.22)$$

$$G_{01} = 0, \quad (19.23)$$

$$G_{11} = -\frac{G}{c^4}\frac{Q_e{}^2 + Q_m{}^2}{R^4}e^A + \Lambda e^A, \quad (19.24)$$

$$G_{22} \equiv G_{33}/f^2(\vartheta) = \frac{G}{c^4}\frac{Q_e{}^2 + Q_m{}^2}{R^2} + \Lambda R^2. \tag{19.25}$$

For $G_{\alpha\beta}$ we find

$$G_{00} = \frac{\varepsilon e^C}{R^2} + \frac{R_{,t}{}^2}{R^2} + \frac{A_{,t}\, R_{,t}}{R} + e^{C-A}\left(-\frac{R_{,r}{}^2}{R^2} - 2\frac{R_{,rr}}{R} + \frac{A_{,r}\, R_{,r}}{R}\right), \tag{19.26}$$

$$G_{01} = -2\frac{R_{,tr}}{R} + \frac{A_{,t}\, R_{,r}}{R} + \frac{R_{,t}\, C_{,r}}{R}, \tag{19.27}$$

$$G_{11} = -\frac{\varepsilon e^A}{R^2} + \frac{R_{,r}{}^2}{R^2} + \frac{C_{,r}\, R_{,r}}{R} + e^{A-C}\left(-\frac{R_{,t}{}^2}{R^2} - 2\frac{R_{,tt}}{R} + \frac{C_{,t}\, R_{,t}}{R}\right), \tag{19.28}$$

$$G_{22} = \frac{1}{4}e^{-C}\left(-4RR_{,tt} + 2RC_{,t}\, R_{,t} - 2RA_{,t}\, R_{,t} - R^2 A_{,t}{}^2 - 2R^2 A_{,tt} + R^2 C_{,t}\, A_{,t}\right)$$

$$+ \frac{1}{4}e^{-A}\left(4RR_{,rr} + 2RC_{,r}\, R_{,r} - 2RA_{,r}\, R_{,r} + R^2 C_{,r}{}^2 + 2R^2 C_{,rr} - R^2 C_{,r}\, A_{,r}\right). \tag{19.29}$$

Equations (19.14)–(19.16) imply the following:

$$\left(\epsilon u^\beta\right)_{;\beta} = 0, \tag{19.30}$$

$$\epsilon u^\alpha{}_{;\beta}\, u^\beta = -(1/c)\left(\rho_e F^\alpha{}_\beta + \rho_m {*}F^\alpha{}_\beta\right) u^\beta. \tag{19.31}$$

In order to derive these equations, apply $\left(R^{\alpha\beta} - \frac{1}{2}g^{\alpha\beta}R\right)_{;\beta} = 0$ to the right-hand side of (19.14) and note that (19.16) is equivalent to

$$F_{\mu\nu;\beta} + F_{\nu\beta;\mu} + F_{\beta\mu;\nu} = -\frac{4\pi}{c}\sqrt{-g}\,\rho_m u^\lambda \epsilon_{\lambda\mu\nu\beta}. \tag{19.32}$$

Equation (19.30) follows when $T^{\alpha\beta}{}_{;\beta} = 0$ is contracted with u_α, Eq. (19.31) follows when (19.30) is used in $T^{\alpha\beta}{}_{;\beta} = 0$. The remaining terms cancel in consequence of (19.32) and the antisymmetry of $F_{\mu\nu}$.

Equations (19.30) (the conservation of mass) and (19.31) (the **Lorentz force** acting on charges in motion and pushing them off geodesic trajectories) are quite general and independent of any symmetry properties of spacetime. Applying (19.31) to our metric (19.11)–(19.12) we obtain

$$\epsilon C_{,r} = (\rho_e Q_e + \rho_m Q_m)\frac{2e^{A/2}}{cR^2} \equiv \frac{QQ_{,r}}{2\pi R^4}, \tag{19.33}$$

where

$$Q^2 \overset{\text{def}}{=} Q_e{}^2 + Q_m{}^2. \tag{19.34}$$

Applying (19.30) to our metric (19.11)–(19.12) we then obtain

$$\frac{\kappa}{2}\epsilon R^2 e^{A/2} = \frac{G}{c^4}N_{,r}, \tag{19.35}$$

where $N_{,r}$ is an arbitrary function of integration. We see that N so defined is, up to the choice of units, the sum of rest masses within the r-surface. The coefficient defining the units was chosen so that the notation agrees with that in the pioneering paper (Bronnikov, 1983). From (19.18), (19.20) and (19.35) we now see that the ratios $Q_{e,r}/N_{,r} = \rho_e/(c\epsilon)$ and $Q_{m,r}/N_{,r} = \rho_m/(c\epsilon)$ are both time-independent.

Using now (19.33) and (19.27) we obtain from the equation $G_{01} = 0$

$$2e^{-A/2}R_{,tr} - e^{-A/2}A_{,t}R_{,r} = \frac{2R_{,t}}{cR^2}\left(\frac{\rho_e}{\epsilon}Q_e + \frac{\rho_m}{\epsilon}Q_m\right). \qquad (19.36)$$

(It is here that the case $R_{,r} = 0$ has to be set aside for separate investigation. With $R_{,r} = 0$, the equation $G_{01} = 0$ reduces to $R_{,t}C_{,r} = 0$ and cannot be used to determine $R_{,r}$ as we do below.) Since the expression in parentheses is independent of t, this can be integrated with the result

$$e^{-A/2}R_{,r} = \Gamma(r) - \frac{\rho_e Q_e + \rho_m Q_m}{c\epsilon R}, \qquad (19.37)$$

where $\Gamma(r)$ is an arbitrary function of integration. Using (19.18), (19.20), (19.34) and (19.35) we now find

$$\frac{\rho_e Q_e + \rho_m Q_m}{c\epsilon} \equiv \frac{QQ_{,r}}{N_{,r}} = QQ_{,N}. \qquad (19.38)$$

With this, (19.37) becomes

$$e^{-A/2}R_{,r} = \Gamma(r) - \frac{QQ_{,N}}{R}, \qquad (19.39)$$

and now the first part of (19.33) becomes

$$C_{,r} = 2\frac{e^{A/2}}{R^2}QQ_{,N}. \qquad (19.40)$$

Using (19.39) and (19.40) to eliminate $R_{,r}$ and $C_{,r}$ from (19.28), we obtain from the G_{11} equation

$$e^{-C}\left(2\frac{R_{,tt}}{R} + \frac{R_{,t}^2}{R^2} - \frac{C_{,t}R_{,t}}{R}\right) - \frac{\Gamma^2 - \varepsilon}{R^2} + \frac{Q^2\left(Q_{,N}^2 - G/c^4\right)}{R^4} + \Lambda = 0. \qquad (19.41)$$

Multiplying by $R^2R_{,t}$, we easily find that the integral of this is

$$e^{-C}R_{,t}^2 = \Gamma^2 - \varepsilon + \frac{2M(r)}{R} + \frac{Q^2\left(Q_{,N}^2 - G/c^4\right)}{R^2} - \frac{1}{3}\Lambda R^2, \qquad (19.42)$$

where $M(r)$ is an arbitrary function. Comparing this with (18.14) we see that $(\Gamma^2 - \varepsilon)/2$ plays here the role of the L–T energy function $E(r)$.

The function $M(r)$ is the **effective mass** that drives the evolution, but, as we will see, it is a combination of mass and charge that *need not be positive*. In order to see this, we will now compare (19.42) with the Newtonian limit, assuming $\Lambda = 0$. Let $[x]$ denote the physical dimension (unit) of x. The dimensions of the quantities appearing in (19.42) are: $[R] = [\text{length}]$, $[e^{C/2}dt] \equiv [ds] = [c] \times [\text{time}]$, $[M] = [G] \times [\text{mass}]/[c^2]$, $[N] = [\text{mass}] \times [c^2]$ and $[Q] = [\text{charge}] \equiv [\sqrt{\text{mass}}] \times [\text{length}^{3/2}]/[\text{time}]$. The function $(\Gamma^2 - \varepsilon)$ is dimensionless, but, for consistency, must be assumed to have the form $2\mathcal{E}/c^2$, where

$[\mathcal{E}] = [\text{velocity}^2]$. In order to find the Newtonian limit of (19.42), we multiply it by c^2 and let $c \to \infty$. Denoting the Newtonian time by τ, we obtain

$$R_{,\tau}^2 = 2\mathcal{E} + 2Gm(r)/R.$$

But the Newtonian equation of motion of spherically symmetric charged dust is

$$r_{,\tau}^2 = 2\mathcal{E}(r) + 2[G\mathcal{M}(r) - (\rho_e(r)/\rho_\mu(r))Q(r)]/r,$$

where $\mathcal{M}(r)$ is the total mass within the sphere of radius r, $Q(r)$ is the total charge within the same sphere, ρ_e is the charge density and ρ_μ is the mass density. Thus, the $M(r)$ in (19.42) corresponds to the Newtonian $[G\mathcal{M}(r) - (\rho_e(r)/\rho_\mu(r))Q(r)]$, and, as announced, does not have to be positive. This relation will appear in other formulae later on.

In order to verify the G_{22} equation, we have to find $C_{,t}$ from the G_{11} equation and $A_{,t}$ from the $G_{01} = 0$ equation. Substituting these, then finding $A_{,r}$ from (19.39), using the r-derivative of (19.42) to eliminate $R_{,t} R_{,tr}$ and again using (19.39) to eliminate e^{-A}, we obtain

$$QQ_{,N}\left(-\frac{G}{c^4}\Gamma N_{,r} + (M + QQ_{,N}\Gamma)_{,r}\right) = 0. \tag{19.43}$$

One solution of this is $Q_{,N} = 0$, i.e. a constant total charge. We will deal with this simpler case later. When $Q_{,N} \neq 0$,

$$\frac{G}{c^4}\Gamma N_{,r} = (M + QQ_{,N}\Gamma)_{,r}. \tag{19.44}$$

The quantity $\mathcal{M} \overset{\text{def}}{=} M + QQ_{,N}\Gamma$ will appear again in Section 19.3.2, where, by matching our metric to the Reissner–Nordström solution, we will find that $\mathcal{M}$ is the active gravitational mass. Thus, via (19.44), Γ determines by how much $\mathcal{M}$ increases when a unit of rest mass is added to the source, i.e. Γ is a measure of the gravitational mass defect/excess. Solutions with $\Gamma = 0$ are known, this is the Datt–Ruban class of Subsection 19.4. Negative Γ can also occur; see the comments on the Novikov (1962a) paper in Section 18.20.2. This happens when the distribution of matter contains necks (see Section 18.10) or fills more than half the volume of a closed 3-space, like in the Friedmann $k > 0$ model matched to the Schwarzschild spacetime. However, the case $\Gamma > 0$ corresponds to the most ordinary configuration.

Note the similarity of the formula $M = \mathcal{M} - QQ_{,N}\Gamma$ to the Newtonian relation $Gm(r) = G\mathcal{M}(r) - (\rho_e/\rho_\mu)Q$.

The final equation to take into account is the G_{00} equation that defines the mass density of the dust. It reproduces (19.35). Using (19.39) to eliminate $e^{A/2}$ from (19.35), we obtain ϵ in a form analogous to (18.15):

$$\kappa\epsilon = \frac{2GN_{,r}}{c^4 R^2 R_{,r}}\left(\Gamma - \frac{QQ_{,N}}{R}\right). \tag{19.45}$$

Finally then, the Einstein–Maxwell equations for charged dust reduced to a set that defines the metric functions $A(t, r)$, $C(t, r)$ and $R(t, r)$ implicitly. The prescription for constructing a solution is this:

1. Choose the arbitrary functions $M(r)$, $Q_e(r)$, $Q_m(r)$ and $\Gamma(r)$, and then solve (19.44) to find $N(r)$.
2. Given these, express $e^{A/2}$ through R via (19.39).
3. The set of equations (19.40) and (19.42) then defines e^C and R.

In the following, we will assume $Q_m = 0$, i.e. no magnetic charges, in agreement with experimental knowledge. This does not influence the solutions in a significant way – it just changes the interpretation of some functions: with $Q_m = 0$ we have $Q = Q_e$. If, further, $Q_{,N} = 0$, then $Q_e =$ constant, i.e. $\rho_e = 0$ from (19.18). Thus in this case the whole charge is contained in the set $r = 0$ – it is the electrically neutral dust moving in the exterior electric field of a charge concentrated at $r = 0$.

Note, from (19.40), that when $Q_{,N} = 0$, the dust moves on geodesics, whereas when $Q_{,N} \neq 0$ it does not. Even so, note from (19.42) and (18.14) that a constant charge Q modifies the geometry, and the neutral dust in electric field moves on geodesics in a different geometry than does a neutral dust in pure gravitational field. This means that electromagnetic field exerts an influence even on electrically neutral matter via the Einstein equations. This is a purely relativistic effect. However, a constant charge at $r = 0$ does not modify the mass, and with $Q_{,N} = 0$ Eq. (19.45) reduces back to (18.15).

Following Bronnikov and Pavlov (1979) let us note from Eq. (19.42) that, with $\Lambda = 0$, the sign of ε puts certain limits on the possible types of evolution of $R(t, r)$. When R increases, the third and fourth terms on the right of (19.42) decrease. Nothing can prevent R from growing to infinity when $\varepsilon \leq 0$. Thus these two cases cannot contain the recollapsing L–T or Friedmann models as subcases. When $\varepsilon = 0$, $R_{,t}$ will decrease to zero at $R \to \infty$ only if $\Gamma = 0$. However, with $\Gamma = 0$, Eq. (19.44) implies $M_{,r} = 0$, i.e. the vacuum limit. Thus, the plane symmetric charged models do not contain the zero-energy L–T (or $k = 0$ Friedmann) models in a limit, although the evolution of the charged model can have qualitative features of the $k = 0$ model. Only in the spherically symmetric case, $\varepsilon = +1$, are all three types of evolution possible, and all three L–T models are contained as subcases.

Special cases of the Bronnikov–Pavlov class presented here were discovered and discussed by other authors in earlier papers – see Krasiński (1997) for a complete list. Of the more important subcases, the electrically neutral case for all three symmetries was solved by Ellis (1967), the spherically symmetric case with no magnetic charge was solved and discussed by Vickers (1973), the $\Lambda = 0$ subcase of the Vickers case was solved and discussed by Markov and Frolov (1970), the case with zero density of magnetic monopoles and zero cosmological constant for all three symmetries was investigated by Shikin (1974), and the further subcase with zero magnetic charge and zero electric charge density was discussed in an earlier paper by Shikin (1972). In this chapter, we will mention only the most important physical contributions.

With vanishing charges, $Q_e = Q_m = 0$, in the spherically symmetric case $\varepsilon = +1$, (19.31) reduces to the equation of a geodesic, while Eqs. (19.39), (19.40) and (19.42) reduce to those defining the L–T model, where $2E = \Gamma^2 - 1$.

19.3.2 Matching the charged dust metric to the Reissner–Nordström metric

From now on, we will limit ourselves to the case of spherical symmetry ($\varepsilon = +1$) and zero magnetic charge ($Q_m = 0$).

Matching a matter solution to an electrovacuum[1] metric can help in interpreting the arbitrary functions and constants. For spherical symmetry and zero magnetic charge, the appropriate electrovacuum metric is the Reissner–Nordström (R–N) solution[2] presented in Section 14.4. We will allow for a nonzero cosmological constant.

We first transform the R–N metric with Λ, Eqs. (14.40)–(14.41), to the appropriate coordinates. For the (t, r) coordinates of (14.40) we will write (τ, R_{RN}), and we introduce the symbol

$$h \stackrel{\text{def}}{=} 1 - \frac{2m}{R_{RN}} + \frac{e^2}{R_{RN}{}^2} + \frac{1}{3}\Lambda R_{RN}{}^2 \tag{19.46}$$

in place of $e^{2\nu}$. We demand that the new coordinates (t, r) are still orthogonal, so that $g_{tr} = 0$, i.e.

$$h\tau_{,t}\,\tau_{,r} - \frac{1}{h}R_{RN,t}R_{RN,r} = 0. \tag{19.47}$$

The function $C_{RN}(t, r)$ is

$$e^{C_{RN}} = h\tau_{,t}{}^2 - \frac{1}{h}R_{RN,t}{}^2. \tag{19.48}$$

We now solve (19.47)–(19.48) for $\tau_{,r}$ and calculate

$$(g_{11})_{RN} \equiv h\tau_{,r}{}^2 - \frac{1}{h}R_{RN,r}{}^2 = -\frac{e^C R_{RN,r}{}^2}{he^C + R_{RN,t}{}^2}. \tag{19.49}$$

The component $(g_{00})_{RN}$ is not fully determined at this point; we will determine it later. Since τ is defined by the partial differential equation (19.47), it still involves an arbitrary function of one variable.

We wish to match the charged dust metric of Section 19.3 to the R–N solution given above across a hypersurface $r = r_b$. This requires that the induced 3-metric and the second fundamental form of this hypersurface are the same for both spacetime metrics (by Section 12.17). The coordinates are adapted to the boundary, so we may use (7.96).

Continuity of the 3-metric requires that

$$e^{C(t,r_b)} = e^{C_{RN}(t,r_b)}, \qquad R(t, r_b) = R_{RN}(t, r_b). \tag{19.50}$$

The transformations that keep the metric diagonal are still allowed. Transforming t by $t' = \int e^{\alpha(t)}dt$, where $\alpha = C_{RN}(t, r_b) - C(t, r_b)$, we fulfil the first of (19.50), while g'_{11} and g'_{01} are not changed.

---

[1] An **electrovacuum metric** is a solution of the Einstein–Maxwell equations such that the mass density ϵ and charge density ρ_e are both zero, but there may be an electromagnetic field in the spacetime generated by charges placed outside the region considered.

[2] A generalisation of the Reissner–Nordström solution for magnetic charge is known (Stephani *et al.*, 2003), but it is geometrically insignificant. The only change is replacing e^2 in the metric by $e^2 + q^2$, where q is the magnetic charge.

On the surface $r = r_b$, $R_{RN}(t, r_b)$ must be the same function of t as $R(t, r_b)$. Consequently, $R_{,t}(t, r_b)$ must be the same in both metrics, and so $R_{RN}(t, r_b)$ must obey

$$e^{-C_{RN}(t,r_b)} R_{RN,t}{}^2(t, r_b) = \Gamma^2(r_b) - 1 + \frac{2M(r_b)}{R_{RN}(t, r_b)} - \frac{GQ^2(r_b)}{c^4 R_{RN}{}^2(t, r_b)}$$

$$-\frac{1}{3}\Lambda R_{RN}{}^2(t, r_b). \tag{19.51}$$

The unit vector normal to the hypersurface $r = r_b$ has components

$$X^\mu = \left(0, e^{-A/2}, 0, 0\right) \equiv \left(0, \left(\Gamma - \frac{QQ_{,N}}{R}\right)/R_{,r}, 0, 0\right) \tag{19.52}$$

for the interior metric, and, from (19.49),

$$X^\mu_{RN} = \left(0, \frac{\sqrt{h + e^{-C_{RN}} R_{RN,t}{}^2}}{R_{RN,r}}, 0, 0\right) \tag{19.53}$$

for the R–N solution. From the continuity of the second fundamental form we have $R_{,r} X^r|_{r=r_b} = R_{RN,r} X^r_{RN}|_{r=r_b}$ and $\left(e^C\right)_{,r} X^r|_{r=r_b} = \left(e^{C_{RN}}\right)_{,r} X^r_{RN}|_{r=r_b}$. The first condition says that

$$\left(h + e^{-C_{RN}} R_{RN,t}{}^2\right)_{r=r_b} = \left(\Gamma - \frac{QQ_{,N}}{R}\right)^2_{r=r_b}, \tag{19.54}$$

which ensures the continuity of $g_{11} = -e^A$ across $r = r_b$, even though we have not required this. Substituting for $R_{RN,t}{}^2$ from (19.51) and for h from (19.46), and then comparing the coefficients, we obtain

$$e = \frac{\sqrt{G}}{c^2} Q(r_b), \qquad m = (M + QQ_{,N}\Gamma)_{r=r_b}. \tag{19.55}$$

The continuity of the second fundamental form imposes one last condition, on $C_{,r}$. Using (19.40), (19.39) and (19.49), the condition is

$$C_{RN,r} \left.\frac{\Gamma - QQ_{,N}/R}{R_{RN,r}}\right|_{r=r_b} = 2\frac{QQ_{,N}}{R^2}. \tag{19.56}$$

We have no expression yet for $C_{RN,r}$, and we will find it from the field equations now. We know that e^A is continuous at $r = r_b$, so we can use (19.39) for $A_{RN}(t, r_b)$. Substitute this in (19.27), and take the equation $G_{01} = 0$ at $r = r_b$. The result is

$$C_{RN,r}(t, r_b) = \left.\frac{2R_{RN,r} QQ_{,N}}{R^2 (\Gamma - QQ_{,N}/R)}\right|_{r=r_b}, \tag{19.57}$$

and it shows that (19.56) is fulfilled.

Thus, (19.55) are the only limitations imposed on the charged dust metric by the matching conditions. This matching was first discussed by Vickers (1973). The second of (19.55) once more reveals the connection between the active gravitational mass m and the effective mass M.

19.3.3 Prevention of the Big Crunch singularity by electric charge

It turns out that the presence of electric charges has a strong influence on the evolution of the dust. In discussing this, we shall follow Vickers (1973), with a few corrections. In the present section we will deal only with the Big Bang/Big Crunch (BB/BC) singularities. The shell crossings will be discussed separately in Section 19.3.6. Also, we assume $\Lambda = 0$, since with nonzero Λ the situation becomes more complicated.

When $Q_{,r} \neq 0$, the avoidance of the Big Bang singularity is not guaranteed. In the following we shall denote $E(r) = (\Gamma^2 - 1)/2$.

The presence or absence of a singularity is detected by investigating the roots of the right-hand side of (19.42), which, for this purpose, is more conveniently written as

$$e^{-C} R^2 R_{,t}{}^2 = 2E(r)R^2 + 2M(r)R + Q^2 \left(Q_{,N}{}^2 - G/c^4 \right) \overset{\text{def}}{=} W(R). \tag{19.58}$$

At each root of $W(R)$, the sign of $R_{,t}$ changes, and evolution is possible only in those regions where $W(R) \geq 0$. The following cases occur:

(a) When $E < 0$, $W(R)$ has roots only if

$$M^2 \geq 2EQ^2 \left(Q_{,N}{}^2 - G/c^4 \right). \tag{19.59}$$

With no roots, $W(R)$ would be negative at all R, so (19.59) is the condition for the existence of a solution of (19.58). The roots are

$$R_{\pm} = -\frac{M}{2E} \pm \frac{1}{2E} \sqrt{M^2 - 2EQ^2 \left(Q_{,N}{}^2 - G/c^4 \right)}, \tag{19.60}$$

and $W(R) > 0$ between them. Nonsingular solutions will exist when both $R_{\pm} > 0$ (with $R_{\pm} < 0$, no solution exists at all, with $R_- R_+ < 0$, $R = 0$ is in the allowed range.) This is equivalent to

$$Q_{,N}{}^2 < G/c^4 \quad \text{and} \quad M > 0. \tag{19.61}$$

We will interpret these conditions later on in this section.

If there is equality in (19.59), then $W(R) < 0$ for all $R \neq R_- = R_+$, and $W(R_\pm) = 0$. Then $R = R_\pm$ and the model is static. If, in addition, $Q_{,N}(r_b) = 0$ (meaning that $\rho_e(r_b) = 0$) and $\Gamma(r_b) = 0$, then $E = -1/2$, and in this case the exterior R–N metric is the extreme one, with $e^2 = m^2$, as seen from (19.59) and (19.55).

With (19.61) fulfilled, R oscillates between a minimum and a maximum, never going down to zero.

(b) When $E = 0$, singularity is avoided if and only if $M > 0$ and $Q_{,N}{}^2 < G/c^4$. Collapse is then halted and reversed once and for all.

(c) When $E > 0$, $W(R) > 0$ either everywhere (if there are no roots) or beyond the roots. There will be no roots when $M^2 < 2EQ^2 \left(Q_{,N}{}^2 - G/c^4 \right)$, in which case $W(R) > 0$ for all R including $R = 0$, and the model can run into the singularity. Thus (19.59) is here one of the necessary conditions for the existence of nonsingular solutions. With (19.59) fulfilled, $W(R)$ has two roots, and at least one of them has to be positive if singularity is to be avoided. With $M > 0$, we have $R_- < 0$ always and $R_+ > 0$ if and only if $Q_{,N}{}^2 < G/c^4$. With $M < 0$, $R_+ > 0$ and $R_- < R_+$ always, so non-singular solutions exist with no further conditions, provided that $R > R_+$ initially. Collapse is then halted and reversed as in case (b).

Now we will interpret the conditions (19.61). The inequality $Q_{,N}{}^2 < G/c^4$, with $Q_m = 0$ and $Q_e = Q$, translates into $|\rho_e| < \sqrt{G}\epsilon/c$. Thus, with $M < 0$, the BB/BC singularity is avoided only if $E > 0$ and (19.59) is fulfilled, with no further conditions on ρ_e. This is a non-relativistic bounce, since it occurs also in Newton's theory, under the same conditions.[1] With $M > 0$, a nonsingular bounce will occur only if $|\rho_e| < \sqrt{G}\epsilon/c$, i.e. if the absolute value of the charge density is sufficiently *small* (but nonzero) compared with the mass density. This is a purely relativistic effect that does not occur in the Newtonian limit: with $M > 0$, $R = 0$ is always in the allowed range of Newtonian solutions. The physical interpretation of the relativistic bounce is this: as seen from (19.42), the charges provide a correction to the effective mass M, so that it becomes $\overline{M} = M + (1/2)Q^2 \left(Q_{,N}{}^2 - G/c^4\right)/R$. This correction is negative at small charge density (when $Q_{,N}{}^2 < G/c^4$), so it weakens gravitation, thus helping the dust to bounce. However, at large charge density ($Q_{,N}{}^2 > G/c^4$), charges enhance the effective mass and thus oppose bounce. Nevertheless, in the latter case, the Newtonian electrostatic repulsion can prevail, provided that $M < 0$ at the same time. Recall a similar phenomenon that we encountered while discussing the motion of particles in the Reissner–Nordström spacetime (Section 14.16), where an electric charge in the source of the gravitational field creates effective antigravitation, provided the charge is small enough compared with the mass.

With $Q_{,N} = 0$ the singularity is avoided in every case when a solution exists. Thus, for neutral dust moving in an exterior electric field, the BC/BB singularity never occurs. This was first found by Shikin (1972). This is a purely relativistic effect.

We have proven that in some cases a solution of (19.42), for which $R \neq 0$ initially, does not go down to 0. However, if the charged dust occupies a volume around the centre of symmetry $R = 0$, then, at any time, there are dust particles with all values of R, including $R = 0$. We will find in the next section the conditions to be obeyed in order that the centre is nonsingular. For the particles at $R \neq 0$, the initial conditions can be set up so that the dust is collapsing at $t = t_i$. Thus, the inner turning points given by (19.60) will exist arbitrarily close to the centre. This has consequences for the shell crossings discussed in Section 19.3.6.

If $EM < 0$ and $Q_{,N}{}^2 = G/c^4$, then (19.58) has a time-independent solution $R = -M/E$. In this case, the electrostatic repulsion just balances the gravitational attraction and the whole configuration is static – but unstable. Any small (necessarily negative) perturbation of this value of R will send the dust into collapse that will terminate at $R = 0$.

At the surface of the charged sphere Eq. (19.58) will coincide with (14.176) (up to a transformation of time) if $J_0 = 0$ (radial motion) and $e = \sqrt{G}Q/c^2$, $q/\mu = c^2 Q_{,N}/\sqrt{G}$,

---

[1] It should not be surprising that there are no further conditions on ρ_e when $M < 0$ and $E > 0$, since these two inequalities already imply that the absolute value of the charge density must be large.

$m = M - qe\Gamma/\mu$. Consequently, the surface of a charged dust sphere obeys the equation of radial motion of a charged particle in the Reissner–Nordström (R–N) spacetime. As shown in Section 14.16, for such a particle, if $(q/\mu)^2 < 1$, then the reversal of fall to escape can occur only inside the inner R–N horizon, at $R < r_- = m - \sqrt{m^2 - e^2}$. Thus, the surface of a collapsing charged dust sphere must continue to collapse until it crosses the inner horizon at $R = r_-$, and can bounce at $R < r_-$. Then, as seen from Fig. 14.10, it cannot re-expand back into the spacetime region I from which it collapsed, since this would require motion backwards in time. The surface would thus continue through the tunnel between the singularities and re-expand into another copy of the asymptotically flat region.

The bounce at small charge density $(Q,_N{}^2 < G/c^4)$ would be more interesting physically, since the real Universe has no detectable net charge, so only small charges could exist in it. We saw that *an arbitrarily small uncompensated charge can prevent the BB/BC singularity*. Unfortunately, Ori (1990, 1991) proved that precisely in this case a shell crossing is unavoidable, and it will block the passage through the tunnel. We will derive this result in Section 19.3.6. Thus, a nonsingular bounce through the R–N tunnel is impossible when $Q,_N{}^2 < G/c^4$ everywhere.

19.3.4 * Charged dust in curvature and mass-curvature coordinates

It is instructive to transform the metric given by (19.11) with $f(\vartheta) = \sin\vartheta$, (19.39), (19.40) and (19.42) to coordinates in which the function $R(t, r)$ is the radial coordinate.

We note that $R,_r dr = dR - R,_t dt$, and then we take t to be a function of the new coordinates: $t = f(\tau, R)$. Thus

$$R,_r dr = dR - R,_t (f,_\tau d\tau + f,_R dR),$$
$$dt = f,_\tau d\tau + f,_R dR, \tag{19.62}$$

and the new metric components, using (19.42), are found to be

$$g_{\tau\tau} = \frac{e^C f,_\tau{}^2 \Delta}{(\Gamma - QQ,_N/R)^2}, \tag{19.63}$$

$$g_{\tau R} = \frac{e^C f,_\tau f,_R \Delta + f,_\tau R,_t}{(\Gamma - QQ,_N/R)^2}, \tag{19.64}$$

$$g_{RR} = \frac{e^C f,_R{}^2 \Delta - 1 + 2f,_R R,_t}{(\Gamma - QQ,_N/R)^2}, \tag{19.65}$$

$$\Delta \stackrel{\text{def}}{=} 1 - \frac{2\mathcal{M}}{R} + \frac{GQ^2}{c^4 R^2} + \frac{1}{3}\Lambda R^2, \tag{19.66}$$

where we have defined

$$\mathcal{M} \stackrel{\text{def}}{=} M + QQ,_N \Gamma. \tag{19.67}$$

Note, from (19.62), that the transformation for collapsing dust $(R,_t < 0)$ is *different* from that for expanding dust $(R,_t > 0)$. The transformation from (τ, R) to (t, r), inverse to (19.62), is analogous to introducing, in the Reissner–Nordström region $r > r_+ = m + \sqrt{m^2 - e^2}$, coordinates comoving with the congruence of charged particles that are radially collapsing or expanding, respectively. Figure 14.10 shows that in this way we obtain a coordinate map that covers regions I and II for the collapsing congruence or regions I and IV for the expanding congruence. The extension is to the future or to the past, respectively.

This can now be specialised in two ways. One possibility is to choose the proper curvature coordinates, in which $g_{\tau R} = 0$. This representation of the Vickers metric has not found any application in the literature so far, but it is instructive. From $g_{\tau R} = 0$ we find

$$f_{,R} = -e^{-C} R_{,t}/\Delta, \tag{19.68}$$

and substituting it in (19.65) we obtain

$$g_{RR} = -1/\Delta, \tag{19.69}$$

The resemblance to the Reissner–Nordström metric is now evident, and the quantity $\mathcal{M}$ that plays the role of mass is indeed the one identified by Vickers (1973); see Eq. (19.55). The component $g_{\tau\tau}$ is given by (19.63), with $f_{,\tau}$ to be found from the Einstein equations.

The other possible specialisation of (19.63)–(19.65) is to choose the $\mathcal{M}$ defined in (19.67) as the new τ-coordinate. These **mass-curvature coordinates** were first introduced by Ori (1990). The surfaces $\mathcal{M} = $ constant are timelike, so none of the coordinates is time and the metric cannot be diagonal. Since $\mathcal{M}$, Q, Γ and N depend only on r in the original coordinates, we have $Q = Q(\mathcal{M})$, $\Gamma = \Gamma(\mathcal{M})$ and $N = N(\mathcal{M})$. The Jacobi matrices of the transformations $(t, r) \leftrightarrow (\mathcal{M}, R)$ are

$$\frac{\partial(t, r)}{\partial(\mathcal{M}, R)} = \begin{bmatrix} f_{,\mathcal{M}} & f_{,R} \\ r_{,\mathcal{M}} & r_{,R} \end{bmatrix}, \tag{19.70}$$

$$\frac{\partial(\mathcal{M}, R)}{\partial(t, r)} = \begin{bmatrix} 0 & \mathcal{M}_{,r} \\ R_{,t} & R_{,r} \end{bmatrix}. \tag{19.71}$$

These matrices is must be inverse to each other. Hence:

$$f_{,\mathcal{M}} = -\frac{R_{,r}}{R_{,t}\mathcal{M}_{,r}}, \qquad f_{,R} = 1/R_{,t}, \qquad r_{,\mathcal{M}} = 1/\mathcal{M}_{,r}, \qquad r_{,R} = 0. \tag{19.72}$$

In the coordinates $(x^0, x^1) \stackrel{\text{def}}{=} (\mathcal{M}, R)$ the velocity field still has only one contravariant component:

$$u'^1 = \pm\sqrt{\Gamma^2 - 1 + \frac{2M}{R} + \frac{Q^2(Q_{,N}^2 - G/c^4)}{R^2} - \frac{1}{3}\Lambda R^2}. \tag{19.73}$$

We define the auxiliary quantities

$$u \stackrel{\text{def}}{=} \Gamma - QQ_{,N}/R, \tag{19.74}$$

$$\alpha \stackrel{\text{def}}{=} e^{C/2}/u, \qquad \alpha f_{,\mathcal{M}} \stackrel{\text{def}}{=} F(\mathcal{M}, R), \tag{19.75}$$

and, using (19.66), (19.73) and (19.42), we obtain in (19.63)–(19.65):

$$g_{\mathcal{MM}} = F^2 \Delta \overset{\text{def}}{=} \mathcal{A}, \tag{19.76}$$

$$g_{\mathcal{MR}} = F\alpha u^2/R,_t = Fu/u'^1 \overset{\text{def}}{=} \mathcal{B}, \tag{19.77}$$

$$g_{\mathcal{RR}} = 1/(u'^1)^2 \overset{\text{def}}{=} \mathcal{C}. \tag{19.78}$$

The function $F(\mathcal{M})$ is to be found from the field equations. Equations (19.76)–(19.78) differ from those of Ori (1990) only by notation.

For some further calculations it is useful to note that

$$\mathcal{AC} - \mathcal{B}^2 = -F^2. \tag{19.79}$$

Using (19.72), (19.17) and (19.75) we find that the only nonvanishing components of the electromagnetic tensor in the $(\mathcal{M}, R)$ coordinates are

$$F^{\mathcal{MR}} = -F^{R\mathcal{M}} = \frac{Q}{FR^2}, \qquad F_{\mathcal{MR}} = -F_{R\mathcal{M}} = -\frac{FQ}{R^2}. \tag{19.80}$$

Further, using (19.17), (19.18), (19.35), (19.39), (19.44), (19.42) and (19.73) we find for the charge density and energy-density

$$\frac{4\pi\rho_e}{c} = \frac{uQ,_r}{R^2 R,_r} \equiv -\frac{uQ,_{\mathcal{M}}}{R^2 f,_{\mathcal{M}} R,_t} \equiv -\frac{Q,_{\mathcal{M}}}{R^2 F u'^1}, \tag{19.81}$$

$$\kappa\epsilon = \frac{2u\mathcal{M},_r}{\Gamma R^2 R,_r} \equiv -\frac{2\alpha u}{\Gamma R^2 F R,_t} \equiv -\frac{2}{\Gamma R^2 F u'^1}. \tag{19.82}$$

Equations (19.76)–(19.78) and (19.80)–(19.82) show that $F(\mathcal{M}, R)$ is the only unknown function. Denoting

$$u^R \overset{\text{def}}{=} u'^1 \equiv \pm\sqrt{u^2 - \Delta} \tag{19.83}$$

(the $+$ sign for expansion, $-$ for collapse), we obtain

$$u_{\mathcal{M}} = uF, \qquad u_R = 1/u^R, \tag{19.84}$$

$$g^{\mathcal{MM}} = -\frac{1}{(Fu^R)^2}, \qquad g^{\mathcal{MR}} = \frac{u}{Fu^R}, \qquad g^{RR} = -\Delta. \tag{19.85}$$

Dividing (19.44) by $\mathcal{M},_r$ and using the definition of $\mathcal{M}$, Eq. (19.67), we find that in the $(\mathcal{M}, R)$ coordinates Eq. (19.44) reads

$$\frac{G}{c^4} \Gamma N,_{\mathcal{M}} = 1. \tag{19.86}$$

The function F can be found from the equations of motion (19.31). With $\rho_m = Q_m = 0$, using (19.38), they read

$$u^\alpha;_\beta u^\beta = -\frac{\rho_e}{c\epsilon} F^{\alpha\beta} u_\beta \equiv -Q,_N F^{\alpha\beta} u_\beta, \tag{19.87}$$

and imply just one equation:

$$F_{,R} = -\frac{u^R{,}_{\mathcal{M}}}{u\,(u^R)^2}.$$ (19.88)

Using (19.74), (19.83) and (19.86), we transform (19.88) to

$$F_{,R} = -\frac{1}{(u^R(\mathcal{M},R))^3}\left[\Gamma_{,\mathcal{M}} + \frac{1}{R\Gamma}\left(1 - \frac{c^4}{G}(Q_{,N}{}^2 + QQ_{,NN})\right)\right].$$ (19.89)

This coincides, except for notation, with Ori's (1990) result. Note that we have *not* assumed $\Lambda = 0$. As Ori (1990) stressed, Eqs. (19.76)–(19.78), (19.84) and (19.89) determine the metric explicitly, in contrast to the representation by Vickers used in Section 19.3, where the Einstein–Maxwell equations were reduced to a set of two differential equations.

With $\Lambda = 0$, the integral of (19.89) is of the form $\int \frac{(ax+b)x^2}{(cx^2+dx+e)^{3/2}}\,\mathrm{d}x$ and is elementary, though complicated; it requires separate treatment of various subcases. The full list of results is given in Ori's paper.

It can now be verified that the Einstein–Maxwell equations are all fulfilled. The identity $Q_{,N} \equiv Q_{,\mathcal{M}}/N_{,\mathcal{M}} = G\Gamma Q_{,\mathcal{M}}/c^4$, resulting from (19.86), is helpful in the calculations.

Note, from (19.76)–(19.78) and (19.82), that the metric and the mass density are insensitive to the sign of u^R. As explained in the remark after (19.66), $u^R > 0$ and $u^R < 0$ correspond to different maps with different domains. Thus, integrating (19.89) from R_1 to $R > R_1$ with $u^R > 0$, we integrate forward in time, whereas calculating the same integral with $u^R < 0$, we integrate backwards in time.

19.3.5 Regularity conditions at the centre

Just as in the L–T model, the set $R = 0$ in charged dust consists of the BB/BC singularity (which we showed to be avoidable) and the centre of symmetry, which may or may not be singular. We will now derive the conditions for the absence of the central singularity. We assume no magnetic charges.

Let $r = r_c$ correspond to the centre of symmetry. From (19.11), (19.12) and (19.35) we see that $N(r) = 4\pi \int_{\mathcal{V}} \epsilon\sqrt{-g}\,\mathrm{d}_3 x$, where $\mathcal{V}$ is a sphere centred at $r = r_c$ in a $t = $ constant space. Thus, if ϵ has no singularity of the type of the Dirac delta at the centre, $N(r)$ must obey $N(r_c) = 0$. Similarly, Eqs. (19.11)–(19.12) and (19.18) show that if there is no delta-type singularity of ρ_e at the centre then the electric charge must obey $Q(r_c) = 0$. With both ϵ and ρ_e being nonsingular at r_c, the ratio $\rho_e(r_c)/\epsilon(r_c)$ is nonsingular, and (19.18) with (19.35) show that $\lim_{r\to r_c} Q/N = \lim_{r\to r_c} Q_{,N} = \rho_e(r_c)/[c\epsilon(r_c)]$ is finite (possibly zero). Then, (19.86) implies that $\Gamma(r_c) \neq 0$ and $\lim_{r\to r_c} N/\mathcal{M} = $ constant $\neq 0$. Thus, we can use $\mathcal{M}$ instead of N as the variable in calculating the behaviour of functions in the vicinity of $r = r_c$ (of $\mathcal{M} = 0$).

Since $R(t, r_c) = 0$ and $N(r_c) = 0$, we find from (19.45):

$$\lim_{r\to r_c} \frac{R^3}{N} = \lim_{r\to r_c} \frac{3R^2 R_{,r}}{N_{,r}} = \lim_{r\to r_c} \frac{3}{4\pi\epsilon}\left(\Gamma - \frac{QQ_{,N}}{R}\right).$$ (19.90)

This limit will be finite if $\lim_{r \to r_c} (Q Q_{,N}/R) < \infty$. We already know that $Q_{,N}(r_c) \overset{\text{def}}{=} \tilde{q}_0 =$ constant and $Q = \tilde{q}_0 N + O_1(N)$, using the notation of Section 18.4. Thus, $Q = q_0 \mathcal{M} + O_1(\mathcal{M})$ and $\lim_{r \to r_c}(Q Q_{,N}/R) < \infty$ if $R = \text{constant} \times \mathcal{M}^\gamma + O_\gamma(\mathcal{M})$, where $\gamma < 1$. Then, (19.90) imposes the further condition that, in the neighbourhood of r_c,

$$R = \beta(t) \mathcal{M}^{1/3} + O_{1/3}(\mathcal{M}). \tag{19.91}$$

Since $R(t, r_c) = 0$ at all times, we have $R_{,t}(t, r_c) = 0$. All other terms in (19.42) except $(\Gamma^2 - \varepsilon)$ vanish at $r = r_c$, so, in the spherically symmetric case $\varepsilon = +1$, we must have

$$\lim_{r \to r_c} \Gamma^2(r) = 1 \implies \lim_{r \to r_c} E(r) = 0. \tag{19.92}$$

Note that this does not exclude $\Gamma < 0$.

19.3.6 * Shell crossings in charged dust

As can be seen from (19.75) and (19.72), $F = 0$ is equivalent to $R_{,r} = 0$, so $F = 0$ is a locus of shell crossings. Then, from (19.82) we see that $\Gamma F u^R$ must be negative for the density of dust to be positive. Since $u^R = dR/ds < 0$ during collapse, ΓF must then be positive.

Now let us write the solution of (19.89) as follows:

$$F = -\int_{R_1}^{R} \frac{1}{(u^R(\mathcal{M}, x))^3} \left[\Gamma_{,\mathcal{M}} + \frac{1}{x\Gamma} \left(1 - \frac{c^4}{G} \left(Q_{,N}{}^2 + Q Q_{,NN} \right) \right) \right] dx + g(\mathcal{M}), \tag{19.93}$$

where $g(\mathcal{M})$ is an arbitrary function and R_1 is the initial value of R. When approaching the bounce in the central region, we have $R_1 > R$, so the integral term is positive when the expression in square brackets is negative.

We saw in Section 19.3.3 that there are dust particles with all values of R, including arbitrarily small values. At the turning point in collapse $u^R \to 0$. The integrand is of the form $(ax^2 + bx + c)^{-3/2}(a_2 x^3 + a_3 x^2)$, and the trinomial has real zeros, so the integral is unbounded (which shows that $u^R = 0$ is a coordinate singularity). From (19.93) it is seen that as we approach near to $\{R = 0, \mathcal{M} = 0\}$, the coefficient $1/(x\Gamma)$ will dominate. We know from the regularity conditions that $\lim_{\mathcal{M} \to 0} \Gamma^2 = 1$, $\lim_{\mathcal{M} \to 0} Q = 0$ and $\lim_{\mathcal{M} \to 0} Q_{,N} = \text{constant} < \infty$. Thus, as long as $Q_{,N}{}^2 < G/c^4$, the term containing $1/(x\Gamma)$ will dominate in the vicinity of $R = 0$ and will determine the sign of the infinity in F. Now it turns out that the sign of F will necessarily change to the opposite during collapse. If $\Gamma > 0$, then $F > 0$ can be achieved at $R = R_1$, as is necessary, by a choice of $\Gamma_{,\mathcal{M}}$ and $g(\mathcal{M})$, but $F \to -\infty$ as $R \to 0$. If $\Gamma < 0$, then $F < 0$ can be achieved at $R = R_1$, but $F \to +\infty$ as $R \to 0$. This means that there is necessarily a shell crossing somewhere at $R > 0$ if $Q_{,N}{}^2 < G/c^4$ holds all the way down to $\mathcal{M} = 0$. This is the theorem proven by Ori (1990, 1991).[1]

---

[1] We thank Amos Ori for an extended correspondence on this point. The discussion helped to clarify several other points in this section.

The infinity in F can be avoided altogether if the term in square brackets in (19.93) is zero at the same x at which $u^R = 0$. The zero of u^R is given by (19.60); it is $R = R_+$. In that case F is finite at $R = R_+$ and, by (19.82), ϵ becomes infinite, i.e. $R = R_+$ becomes a true curvature singularity. Thus, also in this case the charged dust cannot tunnel through the Reissner–Nordström throat.

The only situations in which both the BB/BC and shell crossing singularities can be avoided are the following:

1. When $\lim_{\mathcal{M}\to 0} Q,_N{}^2 = G/c^4$. Then, because of $\lim_{\mathcal{M}\to 0} Q/R = 0$, the term $\Gamma,_{\mathcal{M}}$ in (19.93) has a chance to outbalance the other one and secure the right sign of F everywhere.
2. When $Q,_N{}^2 > G/c^4$, $E > 0$ and $M < 0$. Then, as shown in Section 19.3.3, a nonsingular bounce is possible, and shell crossings will not occur, either.

However, at the time of writing this book, no examples of such solutions are known.

19.4 The Datt–Ruban solution

Now we will deal with the special case $R,_r = 0$ in (19.11)–(19.12). The Einstein–Maxwell equations and the equations of motion, (19.22)–(19.32), are still valid. Equations (19.23) and (19.27) imply $R,_t C,_r = 0$. With $R,_t = 0$, taking the combination $\mathrm{e}^{-C}G_{00} + \mathrm{e}^{-A}G_{11}$, we obtain $\kappa\epsilon = 0$, i.e. an (electro)vacuum solution, which we will not consider. Thus $C,_r = 0$, which means that the dust is moving on geodesics. A transformation of time can then be used to achieve $C = 0$. Using the notation $Q^2 \overset{\text{def}}{=} Q_e{}^2 + Q_m{}^2$, we find from (19.24) and (19.28)

$$\frac{\varepsilon}{R^2} + \frac{R,_t{}^2}{R^2} + 2\frac{R,_{tt}}{R} - \frac{GQ^2}{c^4 R^4} + \Lambda = 0. \tag{19.94}$$

Since $R,_r = 0$, Q must be constant. According to (19.18) and (19.20), this means that the densities of electric and magnetic charge must be zero. Thus, the only kind of electromagnetic field that is compatible with the geometry we are now considering is the free field; the dust is uncharged and moves in the exterior electric field.

Multiplying (19.94) by $R^2 R,_t$ and integrating we obtain

$$R,_t{}^2 = -\varepsilon + \frac{2M}{R} - \frac{GQ^2}{c^4 R^2} - \frac{1}{3}\Lambda R^2, \tag{19.95}$$

where M is a constant. Then, Eqs. (19.25) and (19.29) imply

$$RR,_{tt} + \frac{1}{2}RA,_t R,_t + \frac{1}{4}R^2 A,_t{}^2 + \frac{1}{2}R^2 A,_{tt} + \frac{GQ^2}{c^4 R^2} + \Lambda R^2 = 0. \tag{19.96}$$

This is much simplified by the substitution $A = 2 \ln u(t, r)$ and then $u = R_{,t} w(t, r)$. For the function w the following equation results:

$$2RR_{,tt} w_{,t} + RR_{,t} w_{,tt} + R_{,t}^2 w_{,t} = 0. \tag{19.97}$$

Its general solution is $w = X(r) \int \left[1 / \left(RR_{,t}^2 \right) \right] dt + Y(r)$, where $X(r)$ and $Y(r)$ are arbitrary functions. Thus

$$e^{A/2} = R_{,t} \left(X(r) \int \frac{1}{RR_{,t}^2} dt + Y(r) \right). \tag{19.98}$$

The expression for the metric is

$$ds^2 = dt^2 - e^{A/2} dr^2 - R^2(t) \left[d\vartheta^2 + (1/\varepsilon) \sin^2 \left(\sqrt{\varepsilon} \vartheta \right) d\varphi^2 \right], \tag{19.99}$$

and the expression for matter density is found from (19.22) and (19.26) with use of (19.95) and (19.98):

$$\kappa \epsilon = \frac{2X}{R^2 e^{A/2}}. \tag{19.100}$$

This solution of the Einstein–Maxwell equations was first semi-published by Ruban (1972), and then mentioned in a later paper (Ruban, 1983).

The Ruban spacetime is globally a T-region:[1] the gradient of R is timelike everywhere. Consequently, as mentioned in Section 14.1, the curvature coordinates do not exist in this case. Since R depends only on t, the spaces $t = $ constant do not contain their centres of symmetry: R is not zero at any point of such a space (except when $R = 0$ at the given value of t, but this is a singularity analogous to the Big Bang). In fact, Eq. (19.95) shows that with $\Lambda = 0$ and $Q \neq 0$ no Big-Bang-like singularity can occur: the term with the electric charge is negative and tends to $-\infty$ when $R \to 0$, while the right-hand side of this equation must be positive. (Note that with $\Lambda = 0$ and $\varepsilon = +1$ Eq. (19.95) has no solutions at all if $M^2 < GQ^2/c^4$, and has only the static solution $R = M$ when $M^2 = GQ^2/c^4$. However, we have shown at the beginning of this section that the solution with $R_{,t} = 0$ is electrovacuum.) The radius R is constant in the 3-space $t = $ constant. The geometry of this 3-space with $\varepsilon = +1$ is that of a 3-dimensional cylinder whose sections $r = $ constant are spheres, all of the same radius, and the coordinate r measures the position along the generator. The space is inhomogeneous along the r-direction, and the electric field has its only component also in the r-direction. The radius of the cylinder evolves with time according to Eq. (19.95).

The subcase $Q = 0, \varepsilon = +1$ of this solution, i.e. the case of zero electric field and spherical symmetry, was found by Ruban in an earlier paper (Ruban, 1968) and discussed in an illuminating way in yet another paper (Ruban, 1969); we will report on that discussion below. The further subcase $\Lambda = 0$ appeared in a paper by Datt published as early as 1938 (Datt, 1938), but the author arbitrarily dismissed it as being of 'little physical significance'. This subcase is the first in which explicit formulae for R and $e^{A/2}$

---

[1] For the definition of the T- and R-regions, see Section 14.1.

may be given. The solution for $R(t)$ is in fact the same as for the $k = +1$ Friedmann model, while

$$e^{A/2} = 2X(r)(1 - Z \cot Z) + Y(r) \cot Z, \qquad Z \stackrel{\text{def}}{=} \arcsin \sqrt{R/(2M)}. \qquad (19.101)$$

From (19.98) and (19.100) it can be seen that the solution becomes spatially homogeneous ($\epsilon_{,r} = 0$) when $X/Y = C = \text{constant}$. Then, by the coordinate transformation $r' = \int Y(r) dr$ the metric component $e^{A/2}$ is made independent of r, which shows that in this case an additional Killing field exists. This subcase has the Kantowski–Sachs geometry introduced in Section 10.7. When, further, $C = 0$, the solution becomes vacuum. In the case $\Lambda = 0$ this vacuum solution corresponds to that part of the Schwarzschild manifold that is not covered by the curvature coordinates, i.e. inside the event horizon, as mentioned in Section 14.4.

Various generalisations to these subcases were found by several authors; see Krasiński (1997) for a complete list. Among these generalisations are, for example, solutions with the Kantowski–Sachs geometry and with various sources that are more general than just perfect fluid. Here we will mention only one generalisation: Korkina and Martinenko (1975) worked out the case when the source in the Einstein equations for the metric (19.99) (with $\varepsilon = +1$) is a general perfect fluid, with nonzero, time-dependent, pressure. With no specific equation of state, the Einstein equations cannot be solved to the end, and are then just reduced to a single ordinary differential equation that contains an arbitrary function of time (the pressure).

Now we come to the geometrical/physical interpretation of the general Ruban solution in the spherically symmetric case, $\varepsilon = +1$. Note that the matter-density in this solution, given by (19.100), depends on r and is everywhere positive if $X > 0$. Thus, the amount of rest mass contained inside a sphere $r = r_0 = \text{constant}$ does depend on the value of r_0, and is an increasing function of r. Nevertheless, as seen from (19.95), the active gravitational mass M that drives the evolution is constant. It looks as if all matter added to the source loses all its ability to gravitate, and the active gravitational mass is just a parameter of the space on which infalling matter has no influence. Ruban (1969) interpreted this property as follows: the gravitational mass defect of any matter added exactly cancels its contribution to the active mass.

Equation (19.95) with $Q = 0$ is the same that governs the evolution of the Friedmann models, Eq. (17.28). Note that the constant $\varepsilon = \pm 1, 0$ in (19.95) that determines the type of symmetry automatically fixes the type of evolution. Thus with $\varepsilon = +1$ (spherical symmetry), the model is necessarily the recollapsing one ($k = +1$). Now compare (19.95) with the law of evolution of the L–T model, (18.14) – the Datt–Ruban (D–R) model has $E = -1/2$ and $R_{,r} = 0$ permanently. Thus, it behaves like a neck in the L–T model (see Section 18.10).

Finally, compare (19.95) with $Q = 0$ with the Schwarzschild solution in the Lemaître–Novikov coordinates, (14.116)–(14.117). It is clear that the two solutions can be matched across $r = r_b$ if $E(r_b) = -1/2$ and $m = M$; see Exercise 7.

In the special case $Q = \Lambda = 0$, the D–R model explodes out of a singularity at $R = 0$ at $t = t_B$, expands until $R = 2M$ is reached at $t = t_B + \pi M$, and then collapses back to $R = 0$ at $t = t_B + 2\pi M$ (Exercise 8). Thus, the hypersurface across which the D–R model is matched to the Schwarzschild solution is hidden inside the Schwarzschild event horizon, except at the instant of maximal expansion, when the D–R sphere touches the Schwarzschild horizon from inside (Ruban, 1983).

The D–R model and its generalisations have no analogues in the Newtonian theory and do not show up in linear approximations to Einstein's theory.

19.5 The Szekeres–Szafron family of solutions

Consider the metric

$$ds^2 = dt^2 - e^{2\alpha} dz^2 - e^{2\beta} (dx^2 + dy^2),$$ (19.102)

where α and β are functions of (t, x, y, z) to be determined from the Einstein equations. The source is taken to be a perfect fluid, and the coordinates of (19.102) are assumed to be comoving so that $u^\mu = \delta^\mu{}_0$. This implies that $\dot{u}^\mu = 0$ and that pressure depends only on time.

Several different parametrisations are in use for the solutions of the Einstein equations with the metric (19.102), and some notations are conflicting. We shall follow Szafron's (1977) exposition. The orthonormal tetrad components of the Einstein tensor for the metric (19.102) are

$$G_{00} = e^{-2\beta} \left(-\alpha,_x{}^2 - \alpha,_y{}^2 - \alpha,_{xx} - \beta,_{xx} - \alpha,_{yy} - \beta,_{yy} \right)$$
$$+ e^{-2\alpha} \left(-3\beta,_z{}^2 - 2\beta,_{zz} + 2\alpha,_z \beta,_z \right) + 2\alpha,_t \beta,_t + \beta,_t{}^2,$$ (19.103)

$$G_{01} = e^{-\alpha} \left(-2\beta,_{tz} + 2\alpha,_t \beta,_z - 2\beta,_t \beta,_z \right),$$ (19.104)

$$G_{02} = e^{-\beta} \left(-\alpha,_{tx} - \beta,_{tx} - \alpha,_t \alpha,_x + \beta,_t \alpha,_x \right),$$ (19.105)

$$G_{03} = e^{-\beta} \left(-\alpha,_{ty} - \beta,_{ty} - \alpha,_t \alpha,_y + \beta,_t \alpha,_y \right),$$ (19.106)

$$G_{11} = e^{-2\alpha} \beta,_z{}^2 + e^{-2\beta} \left(\beta,_{xx} + \beta,_{yy} \right) - 3\beta,_t{}^2 - 2\beta,_{tt},$$ (19.107)

$$G_{12} = e^{-\alpha-\beta} \left(-\beta,_{zx} + \beta,_z \alpha,_x \right),$$ (19.108)

$$G_{13} = e^{-\alpha-\beta} \left(-\beta,_{zy} + \beta,_z \alpha,_y \right),$$ (19.109)

$$G_{22} = e^{-2\beta} \left(\alpha,_y{}^2 + \alpha,_x \beta,_x + \alpha,_{yy} - \alpha,_y \beta,_y \right) + e^{-2\alpha} \left(\beta,_z{}^2 + \beta,_{zz} - \alpha,_z \beta,_z \right)$$
$$- \alpha,_t \beta,_t - \alpha,_t{}^2 - \beta,_t{}^2 - \alpha,_{tt} - \beta,_{tt},$$ (19.110)

$$G_{23} = e^{-2\beta} \left(-\alpha,_{xy} - \alpha,_x \alpha,_y + \alpha,_x \beta,_y + \beta,_x \alpha,_y \right),$$ (19.111)

$$G_{33} = e^{-2\beta} \left(\alpha,_x{}^2 + \alpha,_{xx} \quad \alpha,_x \beta,_x + \alpha,_y \beta,_y \right) + e^{-2\alpha} \left(\beta,_z{}^2 + \beta,_{zz} - \alpha,_z \beta,_z \right)$$
$$- \alpha,_t \beta,_t - \alpha,_t{}^2 - \beta,_t{}^2 - \alpha,_{tt} - \beta,_{tt}.$$ (19.112)

With a perfect fluid source, the Einstein equations are $G_{00} = \kappa\epsilon$ and $G_{11} = G_{22} = G_{33} = -\kappa p$ (the cosmological constant can be taken into account as the special case $\kappa p = -\Lambda$, $\kappa\epsilon \to \kappa\tilde{\epsilon} + \Lambda$).

The equations $G_{01} = G_{12} = G_{13} = 0$ say, respectively,

$$\left(e^{\beta-\alpha}\beta_{,z}\right)_{,t} = 0, \qquad \left(e^{-\alpha}\beta_{,z}\right)_{,x} = \left(e^{-\alpha}\beta_{,z}\right)_{,y} = 0. \qquad (19.113)$$

The last two equations imply that

$$\beta_{,z} = u(t,z)e^{\alpha}. \qquad (19.114)$$

The cases $u = 0$ and $u \neq 0$ have to be considered separately because the integration proceeds in a different way in each case, and the limit $\beta_{,z} \to 0$ of the solution for $\beta_{,z} \neq 0$ is singular (see Section 19.6.3). The relation between these two cases is closely analogous to the relation between the Datt–Ruban model and the L–T model. We will see later that the L–T and Datt–Ruban models are the spherically symmetric limits of the $\beta_{,z} \neq 0$ subfamily and of the $\beta_{,z} = 0$ subfamily, respectively.

19.5.1 The $\beta_{,z} = 0$ subfamily

The equations $G_{01} = G_{12} = G_{13} = 0$ are fulfilled identically. In solving the other equations, we can assume that $\beta_{,tx} = 0 = \beta_{,ty}$ because otherwise the equations have no solutions; the proof is given in Section 19.10. Then

$$\beta = \ln \Phi(t) + \nu(x, y), \qquad (19.115)$$

where Φ and ν are unknown functions, while $G_{02} = G_{03} = 0$ imply that

$$e^{\alpha-\beta}\alpha_{,x} = \tilde{\sigma}_1(z, x, y), \qquad e^{\alpha-\beta}\alpha_{,y} = \tilde{\sigma}_2(z, x, y), \qquad (19.116)$$

where $\tilde{\sigma}_1$ and $\tilde{\sigma}_2$ are other unknown functions. Using (19.115) for β and denoting $\sigma_i(z, x, y) \overset{\text{def}}{=} \tilde{\sigma}_i e^{\nu}$, $i = 1, 2$, we obtain

$$e^{\alpha}\alpha_{,x} = \sigma_1(z, x, y)\Phi(t), \qquad e^{\alpha}\alpha_{,y} = \sigma_2(z, x, y)\Phi(t). \qquad (19.117)$$

The integrability condition $\left(e^{\alpha}\alpha_{,x}\right)_{,y} = \left(e^{\alpha}\alpha_{,y}\right)_{,x}$ implies $\sigma_{1,y} = \sigma_{2,x}$. This means that a function $\sigma(z, x, y)$ exists such that $\sigma_1 = \sigma_{,x}$, $\sigma_2 = \sigma_{,y}$. Knowing this, (19.117) can be integrated to give

$$e^{\alpha} = \Phi(t)\sigma(z, x, y) + \lambda(t, z), \qquad (19.118)$$

where $\lambda(t, z)$ is an unknown function. Substituting for β and α from (19.115) and (19.118) in (19.107) we obtain in the equation $G_{11} = \kappa p$

$$e^{-2\nu}\left(\nu_{,xx} + \nu_{,yy}\right) = 2\Phi\Phi_{,tt} + \Phi_{,t}^{2} + \kappa p\Phi^{2} \overset{\text{def}}{=} -k = \text{constant}, \qquad (19.119)$$

because ν depends only on x and y while Φ and p depend only on t.

Now it is convenient to introduce the complex variables

$$\xi \overset{\text{def}}{=} x + iy, \qquad \bar{\xi} = x - iy,$$ (19.120)

in which the first part of (19.119) becomes

$$-4e^{-2\nu} \nu_{,\xi\bar{\xi}} = k.$$ (19.121)

Differentiating this by ξ we obtain an equation equivalent to $\left(\nu_{,\xi\xi} - \nu_{,\xi}^{2}\right)_{,\bar{\xi}} = 0$, whose solution is

$$\nu_{,\xi\xi} - \nu_{,\xi}^{2} = \tau(\xi),$$ (19.122)

$\tau(\xi)$ being an arbitrary function. Now, $-4e^{-2\nu} \nu_{,\xi\bar{\xi}}$ is the curvature scalar of the 2-dimensional metric

$$ds_2{}^2 = e^{2\nu} d\xi d\bar{\xi},$$ (19.123)

and Eq. (19.121) says that the curvature is constant. Thus, depending on the sign of k, the metric (19.123) is equivalent to one of the 2-metrics in (19.11)–(19.12). We will now transform it to a simpler form.

A transformation of the form $\xi = f(\xi')(\to \bar{\xi} = \bar{f}(\bar{\xi}'))$ is a conformal symmetry of (19.123); the new metric has a different ν:

$$\tilde{\nu}(\xi', \bar{\xi}') = \nu(\xi, \bar{\xi}) + \frac{1}{2} \ln \left(f_{,\xi'}\right) + \frac{1}{2} \ln \left(\bar{f}_{,\bar{\xi}'}\right).$$ (19.124)

After such a transformation, the new function τ in (19.122) becomes

$$\tilde{\tau}(\xi') = \tau(f(\xi')) f_{,\xi'}{}^2 + \frac{1}{2} \left(\ln f_{,\xi'}\right)_{,\xi'\xi'} - \frac{1}{4} \left(\ln f_{,\xi'}\right)_{,\xi'}{}^2.$$ (19.125)

We can choose f so that $\tilde{\tau} = 0$, after which the new ν will obey

$$\nu_{,\xi\xi} - \nu_{,\xi}^{2} = 0.$$ (19.126)

This implies $(e^{-\nu})_{,\xi\xi} = 0 = (e^{-\nu})_{,\bar{\xi}\bar{\xi}}$ since ν is real. Hence

$$e^{-\nu} = a\xi\bar{\xi} + B\xi + \bar{B}\bar{\xi} + c,$$ (19.127)

where $a, B, \bar{B}$ and c are arbitrary constants (a and c being real) that, in consequence of (19.121), must obey

$$ac - B\bar{B} = k/4.$$ (19.128)

Let us note a few properties of the metric (19.123) with (19.127).

Lemma 19.1 *If $k \neq 0$, then (19.123) may be transformed to*

$$ds_2{}^2 = \frac{dx^2 + dy^2}{\left[1 + \frac{1}{4}k(x^2 + y^2)\right]^2}.$$ (19.129)

Proof:

Suppose that $k > 0$. Then $ac > 0$ from (19.128). Carry out the following chain of transformations: (1) $\xi = \xi' - \overline{B}/a$; (2) $\xi' = 1/(a\xi'')$; (3) $\xi'' = x + iy$. The resulting metric is (19.129).

If $k < 0$ and $a \neq 0$, then the same chain of transformations will do the job. If $a = 0$ and $c \neq 0$, then the transformation $\xi = 1/\xi'$ restores $a \neq 0$, with B and $\overline{B}$ interchanged. If $a = c = 0$, then $a \neq 0$ is restored by the Haantjes transformation (8.79) which, in the variables $(\xi, \overline{\xi})$, is

$$\xi = \frac{\xi' - C\xi'\overline{\xi'}}{1 - \overline{C}\xi' - C\overline{\xi'} + C\overline{C}\xi'\overline{\xi'}}, \tag{19.130}$$

where $C = C_1 + iC_2$. The transformed metric is

$$ds_2^{\,2} = \frac{d\xi\,d\overline{\xi}}{\left[B\xi' + \overline{B}\overline{\xi'} - (BC + \overline{BC})\xi'\overline{\xi'}\right]^2}, \tag{19.131}$$

and (19.129) is achieved in the already-known way. Since $c = 0$ in (19.131), Eq. (19.128) is fulfilled with the same B and $\overline{B}$. $\square$

Lemma 19.2 *Equation (19.129) applies also when $k = 0$.*

Proof:

If $a \neq 0$, then $c = B\overline{B}/a$ from (19.128), and the job is done by the chain of transformations (1) $\xi = \xi' - \overline{B}/a$; (2) $\xi' = 1/(a\xi'')$.

If $a = 0$, then, with $k = 0$, Eq. (19.128) implies $B = \overline{B} = 0$, and then the metric is reduced to (19.129) by $\xi = c\xi'$. $\square$

For the proof that (19.129) is equivalent to the 2-dimensional metrics of constant curvature contained in (19.11)–(19.12) see Exercise 9.

Conclusion:

The coordinates (x, y) may be chosen so that $B = 0$, $c = 1$, $a = k/4$:

$$e^\beta = \frac{\Phi(t)}{1 + \frac{1}{4}k(x^2 + y^2)} = \frac{\Phi(t)}{1 + \frac{1}{4}k\xi\overline{\xi}} \overset{\text{def}}{=} \Phi e^\nu. \tag{19.132}$$

We will continue to use the complex coordinates (19.120). As shown in Section 19.10, the equations $G_{22} - G_{33} - G_{23} = 0$ and $G_{22} - G_{33} + G_{23} = 0$ then become (19.334) and its complex conjugate. After substituting from (19.118) and (19.132) they change to

$$(e^{-\nu}\sigma)_{\xi\xi} = (e^{-\nu}\sigma)_{\overline{\xi}\overline{\xi}} = 0. \tag{19.133}$$

Thus, similarly to $e^{-\nu}$ itself, $e^{-\nu}\sigma$ is of first degree in ξ and of first degree in $\bar{\xi}$. Since σ depends on z, the integration constants will be arbitrary functions of z. In the (x, y) coordinates, the solution of (19.133) is[1]

$$\sigma = e^{\nu}\left[\frac{1}{2}U(z)\left(x^2 + y^2\right) + V_1(z)x + V_2(z)y + 2W(z)\right]. \qquad (19.134)$$

With this, the only equations that are not yet satisfied are $G_{22} = \kappa p$ and $G_{00} = \kappa\epsilon$. Substituting (19.134), (19.118), (19.132) and (19.113) in (19.110) and taking $G_{22} = \kappa p$ we obtain

$$\lambda,_{tt}\Phi + \lambda,_t\Phi,_t + \lambda\Phi,_{tt} + \lambda\Phi\kappa p = U(z) + kW(z). \qquad (19.135)$$

Substituting all the results in (19.103) we obtain for the matter density

$$\kappa\epsilon = 2\left(\lambda\Phi,_{tt}/\Phi - \lambda,_{tt}\right)e^{-\alpha} + 3\Phi,_t^2/\Phi^2 + 3k/\Phi^2. \qquad (19.136)$$

With this, all the Einstein equations are solved. Collecting all the information together, the metric thus found is

$$ds^2 = dt^2 - e^{2\alpha}\,dz^2 - e^{2\beta}\left(dx^2 + dy^2\right),$$

$$e^{\beta} = \Phi(t)e^{\nu}, \qquad e^{\alpha} = \Phi(t)\sigma(z, x, y) + \lambda(t, z),$$

$$\sigma = e^{\nu}\left[\frac{1}{2}U(z)\left(x^2 + y^2\right) + V_1(z)x + V_2(z)y + 2W(z)\right], \qquad (19.137)$$

$$e^{-\nu} = 1 + \frac{1}{4}k\left(x^2 + y^2\right),$$

where k is an arbitrary constant, $\Phi(t)$ is determined by the equation

$$2\Phi\Phi,_{tt} + \Phi,_t^2 + \kappa p\Phi^2 + k = 0, \qquad (19.138)$$

while $\lambda(t, z)$ and the matter-density are given by (19.135) and (19.136).

This is as far as one can get without assuming any equation of state, which is necessary to determine p and then solve (19.138). However, since the pressure depends only on t, the barotropic equation of state $\epsilon = \epsilon(p)$ makes the matter density spatially homogeneous, and the resulting model ceases to be interesting for advanced cosmology.

The model (19.137)–(19.138) is a generalisation of the Datt–Ruban class, which results from here as the limit $V_1 = V_2 = 0$, $U = kW$.

All the Robertson–Walker models are also contained here as the limit $\lambda = 0$, $U = -kW$, in the form (17.83)–(17.84). They result in a simpler form if $U = W = V_2 = 0$, $V_1 = 1$ in addition (which just means a more specific choice of coordinates in the R–W limit):

$$ds^2 = dt^2 - \frac{\Phi^2(t)}{\left[1 + \frac{1}{4}k\left(x^2 + y^2\right)\right]^2}\left(x^2\,dz^2 + dx^2 + dy^2\right), \qquad (19.139)$$

---

[1] The coefficients at U and W were added in order to make the subsequent formulae consistent with those of Szafron (1977). There is an inconsistency between Szafron's Eqs. (2.23) and (2.24).

and then the following transformation will reduce (19.139) to (17.1):

$$\Phi = R, \qquad x = r \sin \vartheta, \qquad y = r \cos \vartheta, \qquad z = \varphi. \tag{19.140}$$

Equation (19.138) is identical to (17.25) that governs the evolution of the R–W models (the cosmological constant can be taken into account here by redefining p and ϵ).

Note that $W = 0$ can be assumed without loss of generality because $W = 0$ results by the reparametrisation $U = \tilde{u} + kW$, $\lambda = \tilde{\lambda} - 2\Phi W$; $\tilde{u}$ then replaces U and $\tilde{\lambda}$ replaces λ in the equations. Note also that if $k \neq 0$, then the reparametrisation $\lambda = \tilde{\lambda} - \Phi(U/k + W)$, $W = 2\tilde{W} + U/k$ leads to the same result as if $U = -kW$, i.e. as if the right-hand side of (19.135) were zero. Both these parametrisations have been used in various papers; see Krasiński (1997) and Section 19.6.3.

In general, this family of spacetimes has no symmetry (Bonnor, Sulaiman and Tomimura (1977) and Exercise 10). However, when $U = kW$ and $V_1 = V_2 = 0$, it acquires a 3-dimensional symmetry group acting on 2-dimensional orbits; the symmetry is spherical, plane or hyperbolic when $k > 0$, $k = 0$ or $k < 0$, respectively. In the general case, a certain quasi-symmetry is present: the surfaces given by $\{t = \text{constant}, z = \text{constant}\}$ have constant curvature proportional to k. The lack of symmetry in the spaces $t = \text{constant}$ is due to the spheres being placed non-concentrically (when $k > 0$) and to the 'planes' being nonparallel (when $k = 0$).

The orbits of the O(3) group of (19.139) have nothing to do with the 2-surfaces of constant curvature present in the general Szafron spacetime: the former are the spheres on which $x^2 + y^2 = \text{constant}$, the latter are surfaces of constant z. In contrast to this, the Datt–Ruban limit results in a natural way: the 2-surfaces of constant curvature become orbits of the symmetry group; for example, the spheres become concentric when $k > 0$ and the 'planes' become parallel when $k = 0$.

An explicit solution of (19.138) and (19.135) corresponding to $\kappa p = -\Lambda$ (i.e. to dust with a cosmological constant) was given by Barrow and Stein-Schabes (1984); it involves elliptic functions.

When $p = 0$, Eq. (19.138) has the same solutions as those given in Section 17.4. Equation (19.135) can then be solved explicitly, too. We shall come back to these solutions in Section 19.6.

19.5.2 The $\beta_{,z} \neq 0$ subfamily

We go back to Eq. (19.114) and consider the case when $u(t, z) \neq 0$. Then

$$e^{\alpha} = \frac{\beta_{,z}}{u(t, z)}, \tag{19.141}$$

and the first of (19.113) becomes $(e^{\beta} u)_{,t} = 0$, whose solution is

$$e^{\beta} = \Phi(t, z) e^{\nu(z,x,y)}, \tag{19.142}$$

where $\Phi \overset{\text{def}}{=} 1/u$. It follows that $\beta_{,tx} = \beta_{,ty} = 0$. Equation (19.141) implies $\mathrm{e}^{\alpha} = \Phi(t, z)\beta_{,z}$, and an arbitrary factor dependent on z can be introduced in e^{α} by a transformation of the form $z = f(z')$. This will simplify the limiting transition to the Robertson–Walker models. Thus,

$$\mathrm{e}^{\alpha} = h(z)\Phi(t, z)\beta_{,z}, \tag{19.143}$$

where $h(z)$ is an arbitrary function.

The equation $G_{11} = \kappa p$, in consequence of (19.141), (19.142) and (19.107), becomes

$$\mathrm{e}^{-2\nu}\left(\nu_{,xx} + \nu_{,yy}\right) + 1/h^2 = 2\Phi\Phi_{,tt} + \Phi_{,t}{}^2 + \kappa p\Phi^2 \overset{\text{def}}{=} -k(z); \tag{19.144}$$

the last part of the equation follows because the left-hand side does not depend on t, while the middle part does not depend on x and y. The function $k(z)$ is unknown, and it will remain arbitrary.

We can now repeat the reasoning in Eqs. (19.120)–(19.128), with the modification that the functions ν, τ, $\tilde{\nu}$, $\tilde{\tau}$, f and $\bar{f}$ now depend on z. The metric (19.123) and its curvature were only an auxiliary construction that helped us guess the substitution (19.124) that reduced Eq. (19.122) to (19.126). However, the reasoning that led to (19.129) is no longer valid because it involved coordinate transformations in the (x, y) subspace. Thus, $\mathrm{e}^{-\nu}$ will have here a form analogous to (19.127), but with the coefficients being arbitrary functions of z:

$$\mathrm{e}^{-\nu} = A(z)\left(x^2 + y^2\right) + 2B_1(z)x + 2B_2(z)y + C(z), \tag{19.145}$$

while $\Phi(t, z)$ is defined by the equation

$$2\frac{\Phi_{,tt}}{\Phi} + \frac{\Phi_{,t}{}^2}{\Phi^2} + \kappa p(t) + \frac{k(z)}{\Phi^2} = 0. \tag{19.146}$$

This shows that $\Phi(t, z)$ can be redefined by $\Phi \to \Phi f(z)$, which will result only in rescaling $k(z)$. In consequence of (19.144), the functions $A(z)$, $B_1(z)$, $B_2(z)$, $C(z)$, $h(z)$ and $k(z)$ must obey

$$AC - B_1{}^2 - B_2{}^2 = \frac{1}{4}\left[1/h^2(z) + k(z)\right]. \tag{19.147}$$

All the Einstein equations are now satisfied, and $G_{00} = \kappa\epsilon$ defines the matter-density. In order to make the transition to the L–T limit easier, we will represent the solution of (19.146) by the formal integral

$$\Phi_{,t}{}^2 = \frac{2M(z)}{\Phi} - k(z) - \frac{\kappa}{3\Phi}\int p\left(\frac{\partial\Phi^3}{\partial t}\right)\mathrm{d}t, \tag{19.148}$$

and then the matter-density is given by

$$\kappa\epsilon = \mathrm{e}^{\nu}\left[(\mathrm{e}^{\beta})_{,z}\right]^{-1}\left[\frac{2M_{,z}}{\Phi^2}\right.$$

$$\left. + \frac{6M\nu_{,z}}{\Phi^2} - \frac{\kappa}{3\Phi^2}\int p\left(\frac{\partial^2\Phi^3}{\partial t\partial z}\right)\mathrm{d}t - \frac{\kappa\nu_{,z}}{\Phi^2}\int p\left(\frac{\partial\Phi^3}{\partial t}\right)\mathrm{d}t\right], \tag{19.149}$$

where the first line is all that remains in the L–T limit; see further on. Collecting all the information together, the metric we obtained is given by the formulae

$$ds^2 = dt^2 - e^{2\alpha}\,dz^2 - e^{2\beta}\left(dx^2 + dy^2\right),$$

$$e^\beta = \Phi(t, z)e^{\nu(z,x,y)},$$

$$e^\alpha = h(z)\Phi(t, z)\beta_{,z} \equiv h(z)\left(\Phi_{,z} + \Phi\nu_{,z}\right),$$

$$e^{-\nu} = A(z)\left(x^2 + y^2\right) + 2B_1(z)x + 2B_2(z)y + C(z),$$

(19.150)

with the matter-density defined by (19.149), $\Phi(t, z)$ obeying (19.146) and the arbitrary functions obeying (19.147).

Equation (19.146) can be integrated once $p(t)$ has been specified.

This subfamily, like the preceding one, has in general no symmetry (Exercise 10), and acquires a G_3 with 2-dimensional orbits when A, B_1, B_2 and C are all constant (that is, when $\nu_{,z} = 0$). If $\kappa p = -\Lambda$ in addition, then the L–T model results, with $z = r$ and $\Phi = R$; the metric of the spheres ($t = $ constant, $z = $ constant) can then be transformed to the standard form as explained in Lemmas 19.1 and 19.2 and Exercise 9; for the interpretation of the coordinates in (19.150) see the next section. The sign of $g(z) \overset{\text{def}}{=} AC - B_1{}^2 - B_2{}^2$ determines here the geometry of the surfaces ($t = $ constant, $z = $ constant) in the same way as the sign of k did for (19.136)–(19.138). However here, with A, B_1, B_2 and C being functions of z, the surfaces $z = $ constant within a single space $t = $ constant may have different geometries (i.e. they can be spheres in one part of the space and the surfaces of constant negative curvature elsewhere, the curvature being zero at the boundary). Moreover, the sign of $g(z)$ is here independent of the sign of $k(z)$, so the geometries of these surfaces are independent of the type of evolution.

The Robertson–Walker limit in the form (17.80)–(17.82) follows when $\Phi(t, z) = f(z)R(t)$, $k = k_0 f^2$ and $k_0 = $ constant. When $B_1 = B_2 = 0$, $C = 4A = 1$ and $f = z$ in addition, the R–W limit is 'natural'; its O(3) orbits are the spheres from the Szafron $\beta_{,z} \neq 0$ spacetime, made concentric in the limit. (The additional specialisations of the arbitrary functions amount to a more specific choice of coordinates in the R–W limit.)

19.5.3 Interpretation of the Szekeres–Szafron coordinates

The transition from the metric of the 2-surface $\{t = $ constant, $r = $ constant$\}$ in (19.11) to the metric (19.129) is called a **stereographic projection**. With $\varepsilon > 0$, the surface ($t = $ constant, $r = $ constant, $R = 1$) in (19.11) has the metric of a sphere of unit radius. With $\varepsilon = 0$, the same surface has the metric of a plane. With $\varepsilon < 0$, the metric can be visualised as that of a two-sheeted hyperboloid, but in a 3-space with indefinite metric (Exercise 11). In the $\beta_{,z} = 0$ subfamily, where the metric of the 2-surface is (19.129), the transformation is as in Exercise 9:

$$(x, y) = \begin{cases} (2/\sqrt{-k})\coth(\vartheta/2)(\cos\varphi, \sin\varphi) & \text{for } k < 0, \\ (2/\sqrt{k})\cot(\vartheta/2)(\cos\varphi, \sin\varphi) & \text{for } k > 0. \end{cases}$$

(19.151)

In the $\beta_{,z} \neq 0$ subfamily, the metric in the 2-surface $\{t = \text{constant}, r = \text{constant}\}$ is not as simple, since the functions A, B_1, B_2 and C depend on z and cannot be transformed away. By the same method as used in Lemma 19.1 it may be verified that $A \neq 0$ can be restored in (19.150) by coordinate transformations in the (x, y) surface, even if $A = 0$ initially. Suppose that $\ell^2 \overset{\text{def}}{=} 1/h^2 + k \neq 0$. Then, writing $A = |\ell(z)|/(2S)$, $B_1 = -|\ell(z)|P/(2S)$, $B_2 = -|\ell(z)|Q/(2S)$, $\varepsilon \overset{\text{def}}{=} \ell/|\ell|$ and redefining Φ by $\Phi = \widetilde{\Phi}|\ell(z)|$, we can represent the metric (19.150) as[1]

$$e^{-\nu} \overset{\text{def}}{=} \mathcal{E} = \frac{S}{2}\left[\left(\frac{x-P}{S}\right)^2 + \left(\frac{y-Q}{S}\right)^2 + \varepsilon\right],$$

$$ds^2 = dt^2 - \frac{(\Phi_{,z} - \Phi\mathcal{E}_{,z}/\mathcal{E})^2}{\varepsilon - k(z)}\,dz^2 - \frac{\Phi^2}{\mathcal{E}^2}\left(dx^2 + dy^2\right). \tag{19.152}$$

When $\ell = 0$, the transition from (19.150) to (19.152) is $A = 1/(2S)$, $B_1 = -P/(2S)$, $B_2 = -Q/(2S)$ and Φ is unchanged. Then (19.152) applies with $\varepsilon = 0$.

In all cases, within each single $\{t = \text{constant}, r = \text{constant}\}$ surface, the transformation from the (ϑ, φ) coordinates to the (x, y) coordinates is

$$\varepsilon = +1: \qquad (x - P, y - Q)/S = \cot(\vartheta/2)(\cos\varphi, \sin\varphi),$$

$$\varepsilon = 0: \qquad (x - P, y - Q)/S = (2/\vartheta)(\cos\varphi, \sin\varphi),$$

$$\varepsilon = -1: \qquad (x - P, y - Q)/S = \coth(\vartheta/2)(\cos\varphi, \sin\varphi). \tag{19.153}$$

The quantity ε determines whether the (x, y) 2-surfaces are spherical ($\varepsilon = +1$), pseudospherical ($\varepsilon = -1$), or planar ($\varepsilon = 0$). The geometric interpretation of the stereographic projection (19.153) in the cases $\varepsilon = \pm 1$ is shown in Fig. 19.1. With $\varepsilon = 0$, the interpretation of the coordinates in (19.150) is most easily seen when the (ϑ, φ) coordinates are transformed to the Cartesian coordinates $X = \vartheta\cos\varphi$, $Y = \vartheta\sin\varphi$. Then, using (19.153), we find that the transformation $(x, y) \to (X, Y)$ is

$$(X, Y) = \frac{2S}{(x-P)^2 + (y-Q)^2}(x - P, y - Q), \tag{19.154}$$

which is an inversion with respect to the sphere of radius $\sqrt{2S}$ centred at $(x, y) = (P, Q)$.

The (x, y) coordinates in the cases $\varepsilon = +1$ and $\varepsilon = 0$ have the range $(-\infty, +\infty)$. With $\varepsilon = -1$, x and y vary in that range in which $\mathcal{E}$ has a constant sign. Coming back to the representation (19.150), we see that, for $\varepsilon = -1$ and $A \neq 0$, $\mathcal{E}$ is zero when

$$(x + B_1/A)^2 + (y + B_2/A)^2 = 1/(4A)^2.$$

$\mathcal{E}$ is positive for x and y outside this circle, and is negative for x and y inside it. Figure 19.1 suggests that with $\varepsilon = -1$ we should rather take $(-\mathcal{E})$ as the metric function so that x and y have finite rather than semi-infinite ranges. However, both the $\mathcal{E} > 0$ and the

---

[1] The tildes were dropped in Eq. (19.152) for better readability. The Φ in (19.152) is in fact $\widetilde{\Phi}$, the $e^{-\nu}$ is $e^{-\nu} \overset{\text{def}}{=} e^{-\nu}/|\ell|$, and the $k(z)$ is $\widetilde{k}(z) \overset{\text{def}}{=} k(z)/\ell^2$.

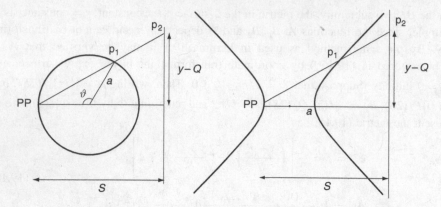

Fig. 19.1. The stereographic projection from the (ϑ, φ) coordinates to the Szekeres–Szafron coordinates (x, y) for a sphere (left) and for a two-sheeted hyperboloid (right). The radius of the sphere is $a = 1$; a is also half the distance between the vertices of the hyperboloid. A point p_1 in the surface with the (ϑ, φ) coordinates is mapped on the point p_2 in the plane with the (x, y) coordinates related to (ϑ, φ) by (19.153). The point PP is the pole of projection; it lies on the straight line p_1–p_2. S is the distance from the pole to the plane. On the hyperboloid, the coordinate ϑ has no simple metric interpretation, but, similarly to the spherical ϑ-coordinate, it determines the distance of the point p_1 from the point antipodal to PP. The figure is the cross-section of the setup with the $x = P$ plane. The coordinate φ in both surfaces is the azimuthal angle around the axis of symmetry that passes through PP.

$\mathcal{E} < 0$ regions are Szekeres spacetimes because they are mapped onto one another by $(x, y) = (x', y')/(x'^2 + y'^2)$, which interchanges the roles of A and C.

The surface area of a $\{t = \text{constant}, r = \text{constant}\}$ surface is finite only in the $\varepsilon = +1$ case, for which it equals $4\pi R^2$. In the other two cases, it is infinite.

19.5.4 Common properties of the two subfamilies

Two common properties have already been mentioned: the lack of any symmetry in general and the existence of the surfaces of constant curvature $\{t = \text{constant}, z = \text{constant}\}$. Other properties in common are the following:

The Weyl tensor of the Szafron spacetimes has its magnetic part with respect to the velocity field of the source equal to zero (Szafron and Collins, 1979; Barnes and Rowlingson, 1989) and is in general of Petrov type D (Szafron, 1977). It degenerates to zero in the R–W limit only.

Rotation and acceleration of the fluid source are zero, the expansion is nonzero, and the shear tensor is

$$\sigma^\alpha{}_\beta = \frac{1}{3} \sum \text{diag}(0, 2, -1, -1), \qquad \text{where}$$

$$\sum = \frac{\Phi_{,tz} - \Phi_{,t}\,\Phi_{,z}/\Phi}{\Phi_{,z} - \Phi\mathcal{E}_{,z}/\mathcal{E}} \qquad \text{for } \beta_{,z} \neq 0,$$

$$\sum = \frac{\lambda_{,t} - \lambda\Phi_{,t}/\Phi}{\lambda + \Phi\sigma} \qquad \text{for } \beta_{,z} = 0. \tag{19.155}$$

Two eigenvalues of the 3-dimensional Ricci tensor, $^{(3)}R_{AB}$, of the slices $t = \text{constant}$ are also equal, and its eigenframe coincides with that of shear (Collins and Szafron, 1979). The eigenspaces corresponding to the degenerate eigenvalues of shear and of $^{(3)}R_{AB}$ are the surfaces of constant curvature mentioned above.

The slices $t = \text{constant}$ of these spacetimes are conformally flat (Berger, Eardley and Olson, 1977), i.e. their Cotton–York tensor (7.50) is zero.

The curvature of the (x, y)-surfaces is a global constant only in the $\beta_{,z} = 0$ subfamily, and there it is equal to the curvature index of the $t = \text{constant}$ slices in the resulting R–W limit. In the $\beta_{,z} \neq 0$ subfamily, the curvature of the (x, y)-surfaces is determined by $g(z) = (AC - B_1{}^2 - B_2{}^2)$ and is independent of the curvature index of the R–W limit, which is determined by $k(z)$. Both $k(z)$ and $g(z)$ are constant only within each (x, y)-surface and can vary within a $t = \text{constant}$ slice. Even the sign of k is not a global property of a general Universe. Its global constancy is a peculiarity of the R–W class.

In the Szafron spacetimes in their full generality, with arbitrary $p(t)$, the dynamic equations are not yet solved. The missing element is the equation of state. The favourite equation of state for physicists and astronomers used to be the barotropic one, $f(\epsilon, p) = 0$. It produces nontrivial results when $p = \text{constant}$, in particular when $p = 0$, and this case will be discussed in Section 19.6. When $\epsilon = \epsilon(p)$ and $p = p(t)$, the Szafron spacetimes trivialise: those with $\beta_{,z} \neq 0$ become R–W, and those with $\beta_{,z} = 0$ acquire either an R–W or a K–S geometry, or the plane and hyperbolic counterparts of the latter (Spero and Szafron, 1978).

The Szafron solutions are an example to which the discussion of the thermodynamical interpretation and of the existence of a thermodynamical scheme (see Section 15.5) applies. As shown by Krasiński, Quevedo and Sussman (1997), in the $\beta_{,z} \neq 0$ subfamily Eqs. (15.52) and (15.55) necessarily either make the thermodynamics trivial (because $p = \text{constant}$) or else imply a symmetry group with at least 2-dimensional orbits. Hence, the general $\beta_{,z} \neq 0$ Szafron spacetimes require an interpretation in terms of a more complicated source than a single-component perfect fluid – it could be a mixture in which chemical reactions occur, or a mixture of two fluids, such as was first introduced by Letelier (1980).

In the $\beta_{,z} = 0$ subfamily, there is a subset of solutions that obey (15.52) and (15.55) while still having no symmetry.[1]

19.5.5 * The invariant definitions of the Szekeres–Szafron metrics

The results reported in this subsection will be quoted without proofs. While the results are important and worth knowing, the proofs are rather technical and laborious, and would lead us too far from the subject of cosmology. Interested readers are advised to consult the original papers.

---

[1] We omit the derivation of these results because the formulae involved are rather complicated, while the subcase that admits a thermodynamical scheme without a symmetry has no clear physical interpretation. We refer the reader to the original paper (Krasiński, Quevedo and Sussman, 1997).

In Section 19.5 we began with the historically earliest definition, through the metric (19.102). Later, a few different coordinate-independent definitions were discovered.

The first invariant definition of the whole family was given by Wainwright (1977) for the Szekeres limit $p = 0$, and then extended by Szafron (1977) to the general case. The general Szafron spacetimes follow when the following properties are assumed:

1. The velocity field of the fluid is geodesic and irrotational.
2. The Weyl tensor is of type D, and the velocity vector of the fluid at every point of the spacetime lies in the 2-plane spanned by the two principal null directions.
3. Any vector orthogonal to both repeated principal null directions is an eigenvector of shear.
4. The 2-surfaces generated by the principal null directions admit orthogonal 2-surfaces.

It follows from here that the repeated principal null directions are orthogonal to the surfaces of constant curvature, $\{t = \text{constant}, z = \text{constant}\}$.

The principal null directions are not in general geodesic. They become geodesic if and only if the spacetimes acquire local rotational symmetry, that is, a G_3/S_2 symmetry group.

Szafron and Collins (1979) worked out another invariant definition. The following conditions must be obeyed simultaneously:

1. The metric is a solution of the Einstein equations with a perfect fluid source.
2. The flow lines of the fluid source are geodesic and nonrotating (the last property implies that the flow lines are orthogonal to a family S_t of spacelike hypersurfaces).
3. Each hypersurface of the family S_t is conformally flat.
4. Two of the eigenvalues of the Ricci tensor $^{(3)}R_{AB}$ of the hypersurfaces S_t are equal.
5. Two of the eigenvalues of the shear tensor $\sigma_{\alpha\beta}$ are equal.

For all spacetimes that have a G_3/S_2 symmetry group properties 4, 5 and $\omega = 0$ are automatically fulfilled, and property 3 results from $\dot{u}^\alpha = 0$ via the field equations. Hence, these subcases of the Szekeres–Szafron spacetimes are defined only by property 1 and $\dot{u}^\alpha = 0$.

If conditions 1–5 are fulfilled, then there exist coordinates in which the metric has the form (19.102). These coordinates are defined up to the following transformations:

$$t = t' + C, \qquad z = f(z'), \qquad u = g(u') + \overline{g(u')},$$
$$u \stackrel{\text{def}}{=} x + iy, \qquad u' \stackrel{\text{def}}{=} x' + iy', \tag{19.156}$$

where C is an arbitrary constant, f is an arbitrary real function and g is an arbitrary complex function.

The proper Szekeres solutions result from the two definitions when the additional requirement $p = 0$ is made.

Yet one more invariant definition was provided by Barnes and Rowlingson (1989): the Szafron spacetimes result from the Einstein equations if the following assumptions are made:

1. The source is a geodesic and rotation-free perfect fluid.
2. The Weyl tensor is purely electric and of Petrov type D.

3. The shear tensor has two equal eigenvalues and its degenerate eigensurface coincides with that of the Weyl tensor.

19.6 The Szekeres solutions and their properties

19.6.1 The $\beta_{,z} = 0$ subfamily

When $\kappa p = -\Lambda = $ constant, Eq. (19.138) is integrated to

$$\Phi_{,t}^2 = -k + \frac{2M}{\Phi} + \frac{1}{3}\Lambda\Phi^2 \overset{\text{def}}{=} \mathcal{U}(\Phi) \tag{19.157}$$

and, using (19.138), Eq. (19.135) can be integrated to

$$\lambda_{,t}\,\Phi\Phi_{,t} + \frac{\lambda M}{\Phi} - \frac{1}{3}\Lambda\lambda\Phi^2 = (U + kW)\Phi + X(z), \tag{19.158}$$

where M is an arbitrary constant and $X(z)$ is an arbitrary function (the integral (19.158) is valid only when $\Phi_{,t} \neq 0$, but this will be assumed always to be the case, otherwise the solution is static).

The solution of (19.158) can be formally represented as

$$\lambda = \sqrt{\mathcal{U}}\left(\int \frac{(U+kW)\Phi + X(z)}{\Phi\mathcal{U}^{3/2}}\,\mathrm{d}\Phi + Y(z)\right), \tag{19.159}$$

where $Y(z)$ is another arbitrary function.

If we want to interpret the constant Λ as the cosmological constant, then the energy-density has to be redefined by $\epsilon = \widetilde{\epsilon} - \Lambda$. Then, from (19.136), (19.157) and (19.158), we obtain

$$\kappa\widetilde{\epsilon} = \frac{2X + 6M\sigma}{\Phi^2 e^\alpha}. \tag{19.160}$$

With $\Lambda \neq 0$, the solutions of (19.157) involve elliptic functions. Each solution will contain an additional arbitrary constant which defines the initial instant of evolution; it can be assumed zero with no loss of generality. The evolution may or may not begin with a Big Bang singularity (a sufficiently large positive Λ would prevent it), but if it does, then the singularity is necessarily simultaneous in the comoving time t.

An additional singularity of infinite density occurs where (and if) $e^\alpha = 0$. This is a shell crossing singularity, discussed for the $\beta_{,z} \neq 0$ subfamily in Subsection 19.7.4 and for the L–T model in Section 18.10.

Equation (19.157) is identical to the Friedmann equation, and so are its solutions. Note that M becomes the mass integral in the R–W limit, so in fact it should be assumed positive for 'physical reasons'. However, a solution exists when $M < 0$ and $\{k < 0$ or $\Lambda > 0\}$, and it should not be discarded without deeper consideration – which is still missing.

The solution (19.159) contains one more arbitrary function $Y(z)$, but one of the functions $\{U, V_1, V_2, W, X, Y\}$ can be specified by a choice of the coordinate z. Hence, the general solution depends on two arbitrary constants (k and M) and five arbitrary functions of z.

In the parametrisation in which $U = -kW$ (see the text after (19.140)), Eq. (19.159) reproduces (19.98), with $(\Phi, z, Q, k) = (R, r, 0, \varepsilon)$.

The solution of (19.157)–(19.158) with $\Lambda \neq 0 = k$ involves elementary functions only and was given by Barrow and Stein-Schabes (1984). When $M \neq 0$, one can redefine $\lambda = \tilde{\lambda} + X\Phi/(3M)$, $W = \tilde{W} - X/(3M)$ and $U = \tilde{U} - kX/(3M)$, and the result is the same as if $X = 0$. Thus $X = 0$ can be assumed without loss of generality whenever $M \neq 0$, and the Barrow–Stein-Schabes result is given in such a parametrisation. (We shall keep X for the sake of comparison with other solutions later.) That solution tends asymptotically to the de Sitter spacetime.

Both (19.157) and (19.158) have elementary solutions when $\Lambda = 0$; they define the Szekeres (1975a) models with $\beta_{,z} = 0$. The solutions of (19.157) with $\Lambda = 0$ are identical to the Friedmann solutions (17.33)–(17.35). Given Φ and $\Lambda = 0$, Eq. (19.158) can be integrated to find $\lambda(t, z)$. The results are presented in the papers by Szekeres (1975a) and by Bonnor and Tomimura (1976). We shall avoid displaying the explicit solutions because then the various possible signs of k and M have to be considered separately (but see Section 19.8). Bonnor and Tomimura discussed the evolution of each model. Depending on the case, the metric or the matter-density may or may not approach homogeneity as the initial singularity is approached and as the model evolves to infinity or to a final singularity. The growing and decaying inhomogeneities are both present, just as in the L–T model; see Section 18.19.

In the model with $k = +1$, the spheres $\{t = \text{constant}, z = \text{constant}\}$ evidently expand and recollapse just as in the Friedmann $k > 0$ model – because the function R is here exactly the same. Along the z-direction the space first has infinite extent at the instant of the Big Bang, then collapses to a minimum size, and expands to infinite extent as the final singularity is reached (Barrow and Silk, 1981). The maximum of expansion of the spheres and the minimum of collapse in the z-direction do not in general occur simultaneously.

19.6.2 The $\beta_{,z} \neq 0$ subfamily

When $\kappa p = -\Lambda = \text{constant}$, Eq. (19.146) has the integral

$$\Phi_{,t}^{2} = -k(z) + \frac{2M(z)}{\Phi} + \frac{1}{3}\Lambda\Phi^{2}; \tag{19.161}$$

the only difference from (19.157) being the dependence of k, M and Φ on z. As before, the density is modified to $\tilde{\epsilon} = \epsilon - \Lambda$, and

$$\kappa\tilde{\epsilon} = \frac{(2Me^{3\nu})_{,z}}{e^{2\beta}(e^{\beta})_{,z}}. \tag{19.162}$$

Again, with $\Lambda \neq 0$ the solutions of (19.161) involve elliptic functions. A general formal integral of (19.161) was presented by Barrow and Stein-Schabes (1984). Any solution of (19.161) will contain one more arbitrary function of z that will be denoted $t_B(z)$, and will enter the solution in the combination $[t - t_B(z)]$. The instant $t = t_B(z)$ defines the initial moment of evolution; when $\Lambda = 0$ it is necessarily a singularity corresponding to $\Phi = 0$, and it goes over into the Big Bang singularity in the Friedmann limit. When $t_{B,z} \neq 0$ (that is, in general) the instant of singularity is position-dependent, just as it was in the L–T model.

As before, another singularity may occur where $(e^{\beta})_{,z} = 0$ (if this equation has solutions). This is a shell crossing, but now it is qualitatively different from that in the L–T model. If this singularity is present in the L–T model, then whole spherical shells stick together at it. Here, as can be seen from (19.150), the equation $(e^{\beta})_{,z} = 0$ may define a subset of the $\{x, y\}$ surface. When a shell crossing exists, its intersection with a $t = $ constant space will be a circle, or, in exceptional cases, a single point, not a sphere (see Section 19.7.4).

Equation (19.161) is formally identical with the Friedmann equation, but, just as in the L–T limit, with k and M depending on z, each surface $z = $ constant evolves independently of the others.

The models defined by (19.150) and (19.161) contain eight functions of z, but only six of them are arbitrary; one can be specified by a choice of z (still arbitrary up to now), the other is determined by (19.147).

The elementary solution of (19.161) that results when $k(z) = 0$ was presented by Barrow and Stein-Schabes (1984).

The solutions of (19.161) with $\Lambda = 0$ are formally identical with Friedmann's, and define the Szekeres (1975a) models with $\beta_{,z} \neq 0$. We will present them in the Goode–Wainwright representation in Section 19.8.

The subcase $AC - B_1{}^2 - B_2{}^2 > 0$, in which the $\{t = $ constant, $z = $ constant$\}$ surfaces are spheres, is the most important one physically. It is called the **quasi-spherical Szekeres model**. Its properties are discussed in Section 19.7, but two little curiosities are listed below.

The mass within the sphere $z = r$ at the time $t = t_0$ is

$$m(t_0, r) = \int_{t=t_0} \epsilon(-g)^{1/2} \, d_3 x = 4\pi \int_0^r 2M_{,z} \, h(z) \, dz = m(r), \qquad (19.163)$$

i.e. it does not depend on time and is given by the same formula as in the spherically symmetric case (Szekeres, 1975a). The density may die off with increasing r so fast that $\lim_{r \to \infty} m(r) < \infty$.

The Szekeres spacetime with $AC - B_1{}^2 - B_2{}^2 > 0$ can be matched to the Schwarzschild solution across a $z = $ constant hypersurface, even though the interior metric still has no symmetry (Bonnor, 1976).

19.6.3 * The $\beta_{,z} = 0$ family as a limit of the $\beta_{,z} \neq 0$ family

The integration of the Einstein equations required that the cases $\beta_{,z} = 0$ and $\beta_{,z} \neq 0$ be considered separately. Having derived the solutions, we can show now that the $\beta_{,z} = 0$

family is a limit of the $\beta_{,z} \neq 0$ family. The consideration below follows that of Hellaby (1996b), with notation adapted to ours and with some corrections.[1]

Equation (19.147) does not in fact put any limitations on the functions A, B_1, B_2 and C; given A, B_1, B_2, C and k it simply defines the function $h(z)$. Choose the arbitrary functions in (19.150) and (19.161) as follows:

$$A = \int \frac{1}{2} U(z) \sqrt{\varepsilon - k} \, \mathrm{d}z + \frac{1}{4} k,$$

$$B_1 = \int \frac{1}{2} V_1(z) \sqrt{\varepsilon - k} \, \mathrm{d}z, \qquad B_2 = \int \frac{1}{2} V_2(z) \sqrt{\varepsilon - k} \, \mathrm{d}z,$$

$$C = \int 2W(z) \sqrt{\varepsilon - k} \, \mathrm{d}z + 1,$$

$$M = \int X(z) \sqrt{\varepsilon - k} \, \mathrm{d}z + M_0, \tag{19.164}$$

$$h = \frac{1}{\sqrt{\varepsilon - k}}, \qquad t_\mathrm{B} = t_1 \int \sqrt{\varepsilon - k} \, \mathrm{d}z + t_0,$$

$$\mathcal{E}_1 \stackrel{\mathrm{def}}{=} 1 + \frac{1}{4} \varepsilon(x^2 + y^2),$$

$$\mathcal{E}_0 \stackrel{\mathrm{def}}{=} \frac{1}{2} U(z)(x^2 + y^2) + V_1(z)x + V_2(z)y + 2W(z),$$

where M_0, t_1 and t_0 are constants. We will take the limit $k \to \varepsilon$ of all these quantities, but first we must check what happens with the other quantities entering the metric under such a reparametrisation.

Assuming that each integral becomes zero in the limit $k \to \varepsilon$ (it is up to us to choose the arbitrary constants appropriately) we have

$$\lim_{k \to \varepsilon} \mathrm{e}^{-\nu} = \mathcal{E}_1, \qquad \lim_{k \to \varepsilon} \left(-\frac{\nu_{,z}}{\sqrt{\varepsilon - k}} \right) = \mathcal{E}_0. \tag{19.165}$$

The limits of the various other quantities are

$$M \to M_0, \qquad t_\mathrm{B} \to t_0, \qquad k \to \varepsilon,$$

$$\frac{M_{,z}}{\sqrt{\varepsilon - k}} \to X(z), \qquad \frac{t_{\mathrm{B},z}}{\sqrt{\varepsilon - k}} \to t_1, \tag{19.166}$$

$$\frac{k_{,z}}{\sqrt{\varepsilon - k}} \to -2(U + \varepsilon W).$$

Knowing this, we define

$$\lim_{k \to \varepsilon} \Phi(t, z) \stackrel{\mathrm{def}}{=} \Phi_1(t), \qquad \lim_{k \to \varepsilon} \frac{\Phi_{,z}}{\sqrt{\varepsilon - k}} \stackrel{\mathrm{def}}{=} \lambda(t, z). \tag{19.167}$$

---

[1] As given, the limit in Hellaby (1996b) reproduces only the subcase of the $\beta_{,z} = 0$ Szekeres solution in which U, V_1, V_2 and W are all constant.

This Φ_1 obeys (19.157) with $k = \varepsilon$ and $M = M_0$. Then we differentiate (19.161) by z, divide the result by $\sqrt{\varepsilon - k}$, take the limit $k \to \varepsilon$ and use (19.166)–(19.167), and find that $\lambda(t, z)$ obeys (19.158).

Using (19.164)–(19.166) in (19.150) we finally obtain

$$ds^2 = dt^2 - (\lambda - \Phi_1 \mathcal{E}_1 / \mathcal{E}_0)^2 dz^2 - (\Phi_1 / \mathcal{E}_0)^2 (dx^2 + dy^2), \tag{19.168}$$

which coincides with (19.137). Thus the $\beta_{,z} = 0$ Szekeres solutions are a limit of the $\beta_{,z} \neq 0$ solutions, also when $\Lambda \neq 0$.

Since (19.161) is the same as (18.14), we can use here Eq. (18.104), with $z = r$, $R = \Phi$ and $E = -k/2$, to write in the $\Lambda = 0$ case:

$$\Phi_{,z} = \left(\frac{M_{,z}}{M} - \frac{k_{,z}}{k} \right) \Phi + \left[\left(\frac{3k_{,z}}{2k} - \frac{M_{,z}}{M} \right) (t - t_B) - t_{B,z} \right] \Phi_{,t}. \tag{19.169}$$

Using this and (19.167), we can verify that the explicit solution for λ that results from (19.169) in the limit $k \to \varepsilon$ coincides with (19.101).

19.7 Properties of the quasi-spherical Szekeres solutions with $\beta_{,z} \neq 0 = \Lambda$

This subclass of the Szekeres models is physically the most important and has been best investigated. With the exception of Subsection 19.7.8, the material in this section is repeated after Hellaby and Krasiński (2002).

We will represent the quasi-spherical Szekeres solution with $\beta_{,z} \neq 0$ in the parametrisation introduced in (19.152). The metric is

$$ds^2 = dt^2 - \frac{(\Phi_{,z} - \Phi \mathcal{E}_z / \mathcal{E})^2}{\varepsilon - k} dz^2 - \frac{\Phi^2}{\mathcal{E}^2} (dx^2 + dy^2), \tag{19.170}$$

and the formula for density in these variables is

$$\kappa \epsilon = \frac{2(M_{,z} - 3M\mathcal{E}_{,z}/\mathcal{E})}{\Phi^2 (\Phi_{,z} - \Phi \mathcal{E}_{,z}/\mathcal{E})}. \tag{19.171}$$

19.7.1 Basic physical restrictions

We choose $\Phi \geq 0$ ($\Phi = 0$ is an origin, the bang or the crunch; in no case is a continuation to negative Φ possible) and $M(z) \geq 0$, so that any vacuum exterior has positive Schwarzschild mass.

We require the metric to be non-degenerate and nonsingular, except at the bang or crunch. Since $(dx^2 + dy^2)/\mathcal{E}^2$ maps to the unit sphere, plane or pseudosphere, $|S(r)| \neq 0$ is needed for a sensible mapping, so $S > 0$ is a reasonable choice. In the cases $\varepsilon = 0$ and $\varepsilon = -1$, $\mathcal{E}$ necessarily goes to zero at certain (x, y) values where the mapping is badly behaved. For a well-behaved z-coordinate, we need to require

$$\infty > \frac{(\Phi_{,z} - \Phi \mathcal{E}_{,z}/\mathcal{E})^2}{\varepsilon - k} > 0, \tag{19.172}$$

i.e. $\varepsilon - k > 0$ except where $(\Phi,_z - \Phi \mathcal{E},_z /\mathcal{E})^2 = 0$. Quasi-pseudo-spherical regions, $\varepsilon = -1$, then require $k \leq -1$, hence they may exist only in hyperbolic spatial sections. Similarly, quasi-planar regions, $\varepsilon = 0$, may exist only in parabolic or hyperbolic spatial sections, $k \leq 0$, whereas quasi-spherical regions are possible for all $k \leq 1$. In the Lemaître–Tolman limit $(\mathcal{E},_z = 0, \varepsilon = 1)$, the equality $1 - k = 0 = (\Phi,_z)^2$ can occur in closed models where $\Phi = R$ on a spatial section is at a maximum, or in wormhole models where it is minimum, $\Phi,_z (t, r_m) = 0 \,\forall t$. These can only occur at constant z and must hold for all (x, y) values. We will consider maxima and minima in Section 19.7.5.

The density must be positive, and the Kretschmann scalar $R_{\alpha\beta\gamma\delta} R^{\alpha\beta\gamma\delta}$ must be finite, which adds

$$\text{either} \qquad M,_z - 3M\mathcal{E},_z /\mathcal{E} \geq 0 \qquad \text{and} \qquad \Phi,_z - \Phi\mathcal{E},_z /\mathcal{E} \geq 0 \qquad (19.173)$$

$$\text{or} \qquad M,_z - 3M\mathcal{E},_z /\mathcal{E} \leq 0 \qquad \text{and} \qquad \Phi,_z - \Phi\mathcal{E},_z /\mathcal{E} \leq 0. \qquad (19.174)$$

If $(\Phi,_z - \Phi\mathcal{E},_z /\mathcal{E})$ passes through 0 anywhere other than at a regular extremum, we have a shell crossing.

19.7.2 The significance of $\mathcal{E}$

The Szekeres metric is covariant with the transformations $z = g(\tilde{z})$, where g is an arbitrary function. Hence, if $\Phi,_z < 0$ in the neighbourhood of some value $z = z_0$, we can take $g = 1/\tilde{z}$ and obtain $\mathrm{d}\Phi/\mathrm{d}\tilde{z} > 0$. Therefore, $\Phi,_z > 0$ can always be assumed to hold in some neighbourhood of any $z = z_0$. However, if $\Phi,_z$ changes sign somewhere, then this is a coordinate-independent property.

As seen from Eq. (19.152), with $\varepsilon = +1$, $\mathcal{E}$ is always nonzero. Since the sign of $\mathcal{E}$ is not defined by the metric, we can assume that $\mathcal{E} > 0$.

Can $\mathcal{E},_z$ change sign?

$$\mathcal{E},_z = \frac{1}{2} S,_z \left\{ -\left[(x - P)^2 + (y - Q)^2 \right] / S^2 + \varepsilon \right\}$$
$$- \left[(x - P)P,_z + (y - Q)Q,_z \right] / S. \qquad (19.175)$$

The discriminant of this with respect to $(x - P)$ is

$$\Delta_x = \frac{1}{S^2} \left(-\frac{S,_z^2}{S^2} (y - Q)^2 - 2\frac{S,_z}{S} (y - Q)Q,_z + P,_z^2 + \varepsilon S,_z^2 \right). \qquad (19.176)$$

The discriminant of Δ_x with respect to $(y - Q)$ is

$$\Delta_y = 4S,_z^2 \left(P,_z^2 + Q,_z^2 + \varepsilon S,_z^2 \right) / S^6. \qquad (19.177)$$

Since, with $\varepsilon = +1$, this is never negative, the equation $\mathcal{E},_z = 0$ will always have at least one solution (in exceptional situations) and in general will have two. The two exceptional situations are when $\Delta_y = 0$. They are as follows:

(i) $S_{,z} = 0$ at $z = z_0$. Then $\mathcal{E}_{,z}(z_0) = 0$ has a family of solutions anyway, but the solutions define a straight line in the (x, y) plane. This will be dealt with below (see after Eq. (19.187)).

(ii) $S_{,z} = P_{,z} = Q_{,z} = 0$ at $z = z_0$. Then $\mathcal{E}_{,z} \equiv 0$ at $z = z_0$, and we see from Eq. (19.171) that ϵ will be spherically symmetric there.

When $\Delta_y > 0$, Δ_x will change sign at the following two values of y:

$$y_{1,2} = Q + S\left(-Q_{,z} \pm \sqrt{P_{,z}^2 + Q_{,z}^2 + \varepsilon S_{,z}^2}\right) / S_{,z}. \tag{19.178}$$

For every y such that $y_1 < y < y_2$ there will be two values of x (and one value of x when $y = y_1$ or $y = y_2$) such that $\mathcal{E}_{,z} = 0$. They are

$$x_{1,2} = P - P_{,z}\frac{S}{S_{,z}} \pm S\sqrt{-\left(\frac{y-Q}{S} + \frac{Q_{,z}}{S_{,z}}\right)^2 + \frac{P_{,z}^2 + Q_{,z}^2}{S_{,z}^2} + \varepsilon}. \tag{19.179}$$

The regions where $\mathcal{E}_{,z}$ is positive and negative depend on the sign of $S_{,z}$. If $S_{,z} > 0$, then $\mathcal{E}_{,z} > 0$ for $x < x_1$ and for $x > x_2$; if $S_{,z} < 0$, then $\mathcal{E}_{,z} > 0$ for $x_1 < x < x_2$. $\mathcal{E}_{,z} = 0$ for $x = x_1$ and $x = x_2$, but note that x_1 and x_2 are members of a continuous family labelled by y. All the values of x and y from (19.178)–(19.179) lie on the circle

$$\left[x - (P - P_{,z} S/S_{,z})\right]^2 + \left[y - (Q - Q_{,z} S/S_{,z})\right]^2 = S^2\left[(P_{,z}^2 + Q_{,z}^2)/S_{,z}^2 + \varepsilon\right]. \tag{19.180}$$

The centre of this circle is at the point

$$(x, y) = (P - P_{,z} S/S_{,z}, Q - Q_{,z} S/S_{,z}), \tag{19.181}$$

and its radius is

$$L_{(\mathcal{E}_{,z}=0)} = S\sqrt{(P_{,z}^2 + Q_{,z}^2)/S_{,z}^2 + \varepsilon}. \tag{19.182}$$

Thus, with $S_{,z} > 0$, the function $\mathcal{E}_{,z}$ is negative inside the circle, zero on the circle and positive outside it.

We will consider the variation of $\mathcal{E}(z, x, y)$ around the spheres of constant t and z. Setting $\varepsilon = +1$ and applying the transformation (19.153) to (19.152) and to its derivative gives

$$\mathcal{E} = S/(1 - \cos\vartheta), \tag{19.183}$$

$$\mathcal{E}_{,z} = -\frac{S_{,z}\cos\vartheta + \sin\vartheta\,(P_{,z}\cos\varphi + Q_{,z}\sin\varphi)}{1 - \cos\vartheta}, \tag{19.184}$$

$$\mathcal{E}_{,zz} = -\frac{S_{,zz}\cos\vartheta + \sin\vartheta\,(P_{,zz}\cos\varphi + Q_{,zz}\sin\varphi)}{1 - \cos\vartheta}$$
$$+ 2\frac{S_{,z}}{S}\frac{S_{,z}\cos\vartheta + \sin\vartheta\,(P_{,z}\cos\varphi + Q_{,z}\sin\varphi)}{1 - \cos\vartheta}$$
$$+ (S_{,z}^2 + P_{,z}^2 + Q_{,z}^2)/S. \tag{19.185}$$

The locus of $\mathcal{E}_{,z} = 0$ is

$$S_{,z}\cos\vartheta + P_{,z}\sin\vartheta\cos\varphi + Q_{,z}\sin\vartheta\sin\varphi = 0. \qquad (19.186)$$

Writing $z = \cos\vartheta$, $y = \sin\vartheta\cos\varphi$ and $x = \sin\vartheta\sin\varphi$, we see that (x, y, z) is on a unit sphere through $(0, 0, 0)$, and (19.186) becomes $S_{,z}z + P_{,z}x + Q_{,z}y = 0$, which is the equation of an arbitrary plane through $(0, 0, 0)$. Such planes all intersect the unit sphere along great circles, therefore $\mathcal{E}_{,z} = 0$ is a great circle, with locus

$$\tan\vartheta = -S_{,z}/(P_{,z}\cos\varphi + Q_{,z}\sin\varphi). \qquad (19.187)$$

The plane has unit normal $(P_{,z}, Q_{,z}, S_{,z})/\sqrt{P_{,z}^2 + Q_{,z}^2 + S_{,z}^2}$.

Now it is easy to understand the meaning of the special case $S_{,z} = 0$ mentioned after Eq. (19.177). As seen from (19.187), with $S_{,z} = 0$ we have $\vartheta = 0$, which means that the great circle defined by $\mathcal{E}_{,z} = 0$ passes through the pole of stereographic projection. In this case, the image of the circle $\mathcal{E}_{,z} = 0$ on the (x, y) plane is a straight line passing through $(x, y) = (P, Q)$, as indeed follows from (19.175). $\mathcal{E}_{,z}$ has a different sign on each side of the straight line. Compare this also with Figs. 19.2 and 19.3. The other special case, $\mathcal{E}_{,z} \equiv 0$, corresponds to the spherically symmetric subcase. Then, the positions of the great circle from Eq. (19.186) and of the poles from Eq. (19.193) are undetermined.

From (19.184) and (19.183) we find

$$\mathcal{E}_{,z}/\mathcal{E} = -\left[S_{,z}\cos\vartheta + \sin\vartheta\,(P_{,z}\cos\varphi + Q_{,z}\sin\varphi)\right]/S, \qquad (19.188)$$

thus

$$\mathcal{E}_{,z}/\mathcal{E} = \text{constant} \Rightarrow S_{,z}z + P_{,z}x + Q_{,z}y = S \times \text{constant}, \qquad (19.189)$$

which is a plane parallel to the $\mathcal{E}_{,z} = 0$ plane, implying that all loci of $\mathcal{E}_{,z}/\mathcal{E} = \text{constant}$ are circles parallel to the $\mathcal{E}_{,z} = 0$ great circle.

The locations of the extrema of $\mathcal{E}_{,z}/\mathcal{E}$ are found as follows:

$$\frac{\partial(\mathcal{E}_{,z}/\mathcal{E})}{\partial\varphi} = \frac{\sin\vartheta\,(P_{,z}\sin\varphi - Q_{,z}\cos\varphi)}{S} = 0 \Rightarrow \tan\varphi_e = \frac{Q_{,z}}{P_{,z}}$$

$$\Rightarrow \cos\varphi_e = \varepsilon_1\frac{P_{,z}}{\sqrt{P_{,z}^2 + Q_{,z}^2}}, \qquad \varepsilon_1 = \pm 1, \qquad (19.190)$$

$$\frac{\partial(\mathcal{E}_{,z}/\mathcal{E})}{\partial\vartheta} = \frac{S_{,z}\sin\vartheta - P_{,z}\cos\vartheta\cos\varphi - Q_{,z}\cos\vartheta\sin\varphi}{S} = 0 \Rightarrow$$

$$\tan\vartheta_e = \frac{P_{,z}\cos\varphi_e + Q_{,z}\sin\varphi_e}{S_{,z}} = \varepsilon_1\frac{\sqrt{P_{,z}^2 + Q_{,z}^2}}{S_{,z}}, \qquad (19.191)$$

$$\cos\vartheta_e = \varepsilon_2\frac{S_{,z}}{\sqrt{S_{,z}^2 + P_{,z}^2 + Q_{,z}^2}}, \qquad \varepsilon_2 = \pm 1. \qquad (19.192)$$

The extreme value is then

$$(\mathcal{E},_z/\mathcal{E})_{\text{extreme}} = -\varepsilon_2\sqrt{S,_z^2 + P,_z^2 + Q,_z^2}/S. \tag{19.193}$$

Since

$$(\sin\vartheta_e\cos\varphi_e, \sin\vartheta_e\sin\varphi_e, \cos\vartheta_e) = \frac{\varepsilon_2(P,_z, Q,_z, S,_z)}{\sqrt{P,_z^2 + Q,_z^2 + S,_z^2}},$$

Eq. (19.184) shows that the extreme values of $\mathcal{E},_z/\mathcal{E}$ are poles to the great circles of $\mathcal{E},_z = 0$.

Clearly $\mathcal{E},_z/\mathcal{E}$ has a dipole variation around each constant-z sphere, changing sign when we go over to the antipodal point: $(\vartheta, \varphi) \to (\pi - \vartheta, \varphi + \pi)$. Writing

$$\Phi,_z - \Phi\mathcal{E},_z/\mathcal{E} = \Phi,_z + \Phi[S,_z\cos\vartheta + \sin\vartheta(P,_z\cos\varphi + Q,_z\sin\varphi)]/S \tag{19.194}$$

we see that $\Phi\mathcal{E},_z/\mathcal{E}$ is the correction to the radial separation $\Phi,_z$ of constant-z shells, due to their not being concentric. In particular, $\Phi S,_z/S$ is the forward ($\vartheta = 0$) displacement, and $\Phi P,_z/S$ and $\Phi Q,_z/S$ are the two sideways displacements ($\vartheta = \pi/2, \varphi = 0$) and ($\vartheta = \pi/2, \varphi = \pi/2$). The shortest 'radial' distance is where $\mathcal{E},_z/\mathcal{E}$ is maximum.

It will be shown in Subsection 19.7.4 that, where $\Phi,_z > 0$, $\mathcal{E},_z/\mathcal{E} \le M,_z/(3M)$ and $\mathcal{E},_z/\mathcal{E} \le \Phi,_z/\Phi$ are required to avoid shell crossings, and also $\Phi,_z/\Phi > M,_z/(3M)$ in Eq. (19.234). These inequalities, together with $M,_z > 0$, imply that the density given by (19.171), as a function of $x \stackrel{\text{def}}{=} \mathcal{E},_z/\mathcal{E}$

$$\epsilon = \frac{2M,_z}{\Phi^2\Phi,_z}\frac{1 - 3Mx/M,_z}{1 - \Phi x/\Phi,_z} \tag{19.195}$$

has a negative derivative by x

$$\epsilon,_x = \frac{\Phi/\Phi,_z - 3M/M,_z}{(1 - \Phi x/\Phi,_z)^2}\cdot\frac{2M,_z}{\Phi^2\Phi,_z} < 0, \tag{19.196}$$

so the density is minimum where $\mathcal{E},_z/\mathcal{E}$ is maximum.

19.7.3 Conditions of regularity at the origin

For reference, we write out the evolution equations of the Szekeres models, even though they are identical to those of the L–T and Friedmann models.

The function $\Phi = \Phi(t, z)$ satisfies the Friedmann equation for dust that follows from (19.161) as the limit $\Lambda = 0$:

$$\Phi,_t^2 = 2M(z)/\Phi - k(z), \tag{19.197}$$

where $M(z)$ and $k(z)$ are arbitrary functions. It follows that the acceleration of Φ is always negative,

$$\Phi,_{tt} = -M/\Phi^2. \tag{19.198}$$

The function $M(z)$ plays the role of the active gravitational mass within the sphere of coordinate radius z. We assume that $M \geq 0$ and $\Phi \geq 0$. In (19.197) $-k(z)$ represents twice the energy per unit mass of the particles in the shells of matter at constant z, but in the metric (19.170) it also determines the geometry of the spatial sections $t = \text{constant}$. The solution of (19.197) depends on the sign of k; it can be

(1) hyperbolic (when $k < 0$):

$$\Phi = -\frac{M}{k}(\cosh \eta - 1), \qquad \sinh \eta - \eta = \frac{(-k)^{3/2}\sigma(t - t_{\text{B}})}{M}; \qquad (19.199)$$

(2) parabolic (when $k = 0$):

$$\Phi = \left[9M(t - t_{\text{B}})^2/2\right]^{1/3}; \qquad (19.200)$$

(3) elliptic (when $k > 0$):

$$\Phi = M(1 - \cos \eta)/k, \qquad \eta - \sin \eta = k^{3/2}\sigma(t - t_{\text{B}})/M, \qquad (19.201)$$

where $t_{\text{B}}(z)$ is an arbitrary function, giving the local time of the Big Bang or Crunch $\Phi = 0$, and $\sigma = \pm 1$ permits time reversal. More correctly, the three types of evolution hold for $k/M^{2/3} < 0, = 0$ and > 0, since $k = 0$ at a spherical type origin for all three evolution types (see further on in this subsection). The behaviour of $\Phi(t, z)$ is identical to that of $R(t, r)$ in the L–T model, and is unaffected by the dependence of the Szekeres solutions on the (x, y) coordinates.

When $\varepsilon = +1$, $\Phi = 0$ occurs at an origin of spherical coordinates (which we assume to be at $z = 0$; this can always be satisfied by a transformation of z), i.e. $\Phi(t, 0) = 0$, $\forall t$, where the 2-spheres have no size. Similarly, $\Phi_{,t}(t, 0) = 0 = \Phi_{,tt}(t, 0)$, etc. $\forall t$. There will be a second origin, at $z = z_{\text{O}}$ say, in any closed, regular, $k > 0$ model. Thus, by (19.201), (19.199) and (19.197)–(19.198), for each constant η

$$\lim_{z \to 0}(M/k) = 0, \qquad \lim_{z \to 0} k = 0, \qquad \lim_{z \to 0}(k^2/M) = 0. \qquad (19.202)$$

The type of time evolution at the origin must be the same as that of its neighbourhood, i.e., along a constant-t slice away from the bang or crunch, by (19.201) and (19.199),

$$0 < \lim_{z \to 0}\left[|k|^{3/2}(t - t_{\text{B}})/M\right] < \infty. \qquad (19.203)$$

Clearly, we need $M \to 0$, $k \to 0$ and

$$0 < \lim_{z \to 0}\left(|k|^{3/2}/M\right) < \infty. \qquad (19.204)$$

Using de l'Hôpital's rule, this gives

$$\lim_{z \to 0}\left[3Mk_{,z} / (2M_{,z}k)\right] = 1. \qquad (19.205)$$

The density and the Kretschmann scalar

$$\mathcal{K} \stackrel{\text{def}}{=} R^{\alpha\beta\gamma\delta} R_{\alpha\beta\gamma\delta} = \kappa^2\left(4\bar{\rho}^2/3 - 8\bar{\rho}\rho/3 + 3\rho^2\right), \qquad (19.206)$$

where

$$\kappa \overline{\rho} \stackrel{\text{def}}{=} 6M/R^3 \tag{19.207}$$

is the mean density within the 'radius' z, must be well behaved. We do not consider a vacuum region of finite size at the origin (since that is just Minkowski space) or a single vacuum point at the origin. Because ϵ and $\overline{\rho}$ evolve differently, we also need

$$0 < \lim_{z \to 0} \frac{6M}{\Phi^3} = \lim_{z \to 0} \frac{2M,_z}{\Phi^2 \Phi,_z} < \infty \quad \Rightarrow \quad \lim_{z \to 0} \frac{3\Phi,_z M}{\Phi M,_z} = 1 \tag{19.208}$$

and

$$0 < \lim_{z \to 0} \frac{2(M,_z - 3M\mathcal{E},_z/\mathcal{E})}{\Phi^2 (\Phi,_z - \Phi \mathcal{E},_z/\mathcal{E})} = \lim_{z \to 0} \frac{2M,_z}{\Phi^2 \Phi,_z} \cdot \frac{1 - 3M\mathcal{E},_z/(M,_z \mathcal{E})}{1 - \Phi\mathcal{E},_z/(\Phi,_z \mathcal{E})} < \infty. \tag{19.209}$$

We have

$$\Phi^2 \frac{\Phi,_z}{M,_z} = \frac{M^2}{k^3}(1 - \cos\eta)\left[-\left(1 - \frac{3Mk,_z}{2M,_z k}\right)\sin\eta(\eta - \sin\eta)\right.$$

$$\left. + \left(1 - \frac{Mk,_z}{M,_z k}\right)(1 - \cos\eta)^2 - \frac{k^{3/2}t_{\text{B},z}}{M,_z}\sin\eta\right], \tag{19.210}$$

where, in consequence of (19.204) and (19.205), the first term is zero and the second finite non-zero at an origin for all $0 < \eta < 2\pi$. Thus, we need only $\lim_{z \to 0} Mt_{\text{B},z}/M,_z < \infty$.

Lastly, the metric must be well behaved, so $\mathcal{E}$ should have no unusual behaviour, such as $S = 0$, that would compromise a valid mapping of $(dx^2 + dy^2)/\mathcal{E}^2$ to the unit sphere. Also, to ensure that the rate of change of the proper radius with respect to the areal radius is that of an origin, $g_{rr}/(\Phi,_z)^2$ should be finite:

$$0 < \lim_{z \to 0} \frac{(\Phi,_z - \Phi\mathcal{E},_z/\mathcal{E})^2}{(1 - k)\Phi,_z^2} < \infty \quad \Rightarrow \tag{19.211}$$

$$0 < \lim_{z \to 0} \left(1 - \frac{3M\mathcal{E},_z}{M,_z \mathcal{E}}\right)^2 < \infty \quad \Rightarrow \tag{19.212}$$

$$-\infty \leq \lim_{z \to 0} \left|\frac{M\mathcal{E},_z}{M,_z \mathcal{E}}\right| < \infty \quad \text{and} \quad \lim_{z \to 0} \left|\frac{M\mathcal{E},_z}{M,_z \mathcal{E}}\right| \neq \frac{1}{3}, \tag{19.213}$$

where the last of (19.208) has been used. This should hold for all (x, y), i.e. all (ϑ, φ). Thus (19.188) gives

$$-\infty \leq \lim_{z \to 0} \left|\frac{MS,_z}{M,_z S}\right| < \infty, \quad -\infty \leq \lim_{z \to 0} \left|\frac{MP,_z}{M,_z S}\right| < \infty,$$

$$-\infty \leq \lim_{z \to 0} \left|\frac{MQ,_z}{M,_z S}\right| < \infty, \tag{19.214}$$

all three limits being different from $1/3$.

All of the above suggests that, near an origin,

$$M \sim \Phi^3, \quad k \sim -\Phi^2, \quad S \sim \Phi^n, \quad P \sim \Phi^n, \quad Q \sim \Phi^n, \quad n \geq 0. \tag{19.215}$$

The condition $\mathcal{E}_{,z}/\mathcal{E} \le M_{,z}/(3M)$ that will be obtained in the next subsection implies that $n \le 1$ near an origin.

19.7.4 Shell crossings

A shell crossing, if it exists, is the locus of zeros of the function $\chi \overset{\text{def}}{=} \mathcal{E}\Phi_{,z}/\Phi - \mathcal{E}_{,z}$. Suppose that $\chi = 0$ holds for all z at some $t = t_0$. This leads to $S_{,z} = P_{,z} = Q_{,z} = \Phi_{,z} = 0$. Since P, Q and S depend only on z, this means that they are constant throughout the spacetime. As seen from (19.171), the Szekeres metric reduces then to the L–T metric, so this case need not be considered.

Suppose that $\chi = 0$ holds for all t at some $z = z_0$. This is an algebraic equation in x and y whose coefficients depend on t and z. Taking the coefficients of different powers of x and y we find $P_{,z} = Q_{,z} = S_{,z} = \Phi_{,z} = 0$, but this time these functions vanish only at $z = z_0$, while $\Phi_{,z}(t, z_0)$ will vanish for all t. This will either be a singularity (when $M_{,z}(z_0) \ne 0$) or a neck (when $M_{,z}(z_0) = 0$, see Section 18.10). Hence, $\chi \ne 0$ except at a shell crossing or at special locations.

Now $\Phi_{,z} > 0$ and $\chi < 0$ cannot hold for all x and y. This would lead to $\mathcal{E}_{,z} > \mathcal{E}\Phi_{,z}/\Phi > 0$, and we know that $\mathcal{E}_{,z}$ cannot be positive at all x and y. Hence, with $\Phi_{,z} > 0$, there must be a region in which $\chi > 0$. By a similar argument, $\Phi_{,z} < 0$ and $\chi > 0$ cannot hold for all x and y, so with $\Phi_{,z} < 0$ there must be a region in which $\chi < 0$.

Assuming $\Phi_{,z} > 0$, can χ be positive for all x and y? Writing

$$\chi = \frac{1}{2S}\left(\frac{S_{,z}}{S} + \frac{\Phi_{,z}}{\Phi}\right)[(x-P)^2 + (y-Q)^2]$$

$$-\frac{1}{2}\varepsilon S\left(\frac{S_{,z}}{S} - \frac{\Phi_{,z}}{\Phi}\right) + \frac{1}{S}[(x-P)P_{,z} + (y-Q)Q_{,z}], \tag{19.216}$$

the discriminants of this with respect to $(x-P)$ and $(y-Q)$ are

$$\Delta_x = \frac{P_{,z}^2}{S^2} - \frac{1}{S^2}\left(\frac{S_{,z}}{S} + \frac{\Phi_{,z}}{\Phi}\right)$$

$$\times\left[\left(\frac{S_{,z}}{S} + \frac{\Phi_{,z}}{\Phi}\right)(y-Q)^2 + 2(y-Q)Q_{,z} - \varepsilon\left(\frac{S_{,z}}{S} - \frac{\Phi_{,z}}{\Phi}\right)\right], \tag{19.217}$$

$$\Delta_y = 4\frac{1}{S^2}\left(\frac{S_{,z}}{S} + \frac{\Phi_{,z}}{\Phi}\right)^2\left(\frac{P_{,z}^2 + Q_{,z}^2 + \varepsilon S_{,z}^2}{S^2} - \varepsilon\frac{\Phi_{,z}^2}{\Phi^2}\right). \tag{19.218}$$

Thus χ will have the same sign for all x and y (i.e. there will be no shell crossings) when $\Delta_y < 0$, that is, if and only if

$$\frac{\Phi_{,z}^2}{\Phi^2} > \varepsilon\frac{P_{,z}^2 + Q_{,z}^2 + \varepsilon S_{,z}^2}{S^2} \overset{\text{def}}{=} \Psi^2(z). \tag{19.219}$$

When $\varepsilon = 0$, this can fail only at those points where $\Phi_{,z} = 0$.

If $\Phi,_z^2/\Phi^2 = \Psi^2$, then $\Delta_y = 0$, so $\Delta_x = 0$ at just one value of $y = y_{SS}$. With this value of y, $\chi = 0$ at one value of $x = x_{SS}$. In this case, the shell crossing is a single point in the constant-(t, z)-surface, i.e. a curve in a space of constant t and a 2-surface in spacetime.

If $\Phi,_z^2/\Phi^2 < \Psi^2$, then the locus of $\chi = 0$ is in general a circle (a straight line in the special case $S,_z/S = -\Phi,_z/\Phi$) in the (x, y) plane. The straight line is a projection onto the (x, y) plane of a circle on the sphere of constant t and z, and hence is not really any special case.

When $\Delta_y > 0(\Phi,_z^2/\Phi^2 < \Psi^2)$, the two limiting values of y at which Δ_x changes sign are

$$y_{1,2} = Q + \frac{-Q,_z \pm \sqrt{\delta}}{S,_z/S + \Phi,_z/\Phi},$$

$$\delta \overset{\text{def}}{=} P,_z^2 + Q,_z^2 + \varepsilon\left(S,_z^2 - S^2\Phi,_z^2/\Phi^2\right), \tag{19.220}$$

and then for every y such that $y_1 < y < y_2$ there are two values of x (only one if $y = y_1$ or $y = y_2$) such that $\chi = 0$. These are

$$x_{1,2} = P + \frac{-P,_z \pm \sqrt{-\left[(S,_z/S + \Phi,_z/\Phi)(y - Q) - Q,_z\right]^2 + \delta}}{S,_z/S + \Phi,_z/\Phi}. \tag{19.221}$$

The x and y obeying (19.221) lie on a circle with the centre at

$$(x_{SC}, y_{SC}) = \left(P - \frac{P,_z}{S,_z/S + \Phi,_z/\Phi}, Q - \frac{Q,_z}{S,_z/S + \Phi,_z/\Phi}\right), \tag{19.222}$$

and with the radius $L_{SC} = \sqrt{\delta}/(S,_z/S + \Phi,_z/\Phi)$. This is in general a different circle from the one defined by $\mathcal{E},_z = 0$. As seen from (19.216), the shell crossing set intersects with the surface of constant t and z along the line $\mathcal{E},_z/\mathcal{E} = \Phi,_z/\Phi = $ constant. As noted after Eq. (19.189), this is a circle that lies in a plane parallel to the $\mathcal{E},_z = 0$ great circle. It follows that the $\mathcal{E},_z = 0$ and SC circles cannot intersect unless they coincide.

Now we will consider the conditions for avoiding shell crossings. These were worked out by Szekeres (1975b) and improved upon by Hellaby and Krasiński (2002). The account here is based on the latter reference.

For positive density, (19.171) shows that $(M,_z - 3M\mathcal{E},_z/\mathcal{E})$ and χ must have the same sign. Consider the case when both are positive. When $M,_z - 3M\mathcal{E},_z/\mathcal{E} \leq 0$ and $\chi < 0$, the inequalities in all the following should be reversed.

Both $(M,_z - 3M\mathcal{E},_z/\mathcal{E})$ and χ can be zero for a particular (x, y) value if $M,_z/3M = \Phi,_z/\Phi$, but the latter cannot hold for all time. This case can hold for all (x, y) only if $M,_z = 0$, $\mathcal{E},_z = 0$ and $\Phi,_z = 0$, which requires all of $M,_z$, $k,_z$, $t_{B,z}$, $S,_z$, $P,_z$ and $Q,_z$ to be zero at some z value.

Consider the inequality $M,_z - 3M\mathcal{E},_z/\mathcal{E} \geq 0$. It must hold for all values of $\mathcal{E},_z/\mathcal{E}$, including the extreme value (19.193), for which

$$\frac{M,_z}{3M} \geq \left.\frac{\mathcal{E},_z}{\mathcal{E}}\right|_{\text{max}} = \frac{\sqrt{(S,_z)^2 + (P,_z)^2 + (Q,_z)^2}}{S} \qquad \forall z. \tag{19.223}$$

It is obvious that this is sufficient, and also that $M_{,z} \geq 0$ for all z.

We will now consider $\chi > 0$ for all three types of evolution.

Hyperbolic evolution, $k < 0$.

For hyperbolic models, we can write:

$$\frac{\Phi_{,z}}{\Phi} = \frac{M_{,z}}{M}(1 - \phi_4) + \frac{k_{,z}}{k}\left(\frac{3}{2}\phi_4 - 1\right) - \frac{(-k)^{3/2}t_{B,z}}{M}\phi_5, \qquad (19.224)$$

where

$$\phi_4 \overset{\text{def}}{=} \frac{\sinh\eta(\sinh\eta - \eta)}{(\cosh\eta - 1)^2}, \qquad \phi_5 \overset{\text{def}}{=} \frac{\sinh\eta}{(\cosh\eta - 1)^2}. \qquad (19.225)$$

At early times, when $\eta \to 0$, we have $\Phi \to 0$, $\phi_5 \to +\infty$ and $\phi_4 \to 2/3$. Thus the term with ϕ_5 dominates and goes to $\pm\infty$, its sign being determined by $-(-k)^{3/2}t_{B,z}/M$. Consequently, $\chi > 0$ implies

$$t_{B,z} < 0 \qquad \forall z. \qquad (19.226)$$

Similarly, at late times, where $\eta \to \infty$, we have $\Phi \to \infty$, $\phi_5 \to 0$, $\phi_4 \to 1$ and $\Phi_{,z}/\Phi \to \frac{1}{2}k_{,z}/k$, so that

$$\Phi_{,z}/\Phi - \mathcal{E}_{,z}/\mathcal{E} > 0 \qquad \Rightarrow \qquad k_{,z}/(2k) - \mathcal{E}_{,z}/\mathcal{E} > 0. \qquad (19.227)$$

Following the above analysis of $M_{,z} - 3M\mathcal{E}_{,z}/\mathcal{E} \geq 0$ we obtain

$$k_{,z}/(2k) > \sqrt{(S_{,z})^2 + (P_{,z})^2 + (Q_{,z})^2}/S \qquad \forall z, \qquad (19.228)$$

which obviously implies $k_{,z} < 0 \,\forall z$. Since we already have $M_{,z} \geq 0$, this is sufficient, and implies $\Phi_{,z} > 0$.

Parabolic evolution, $k = 0$.

The easiest way to obtain the conditions for the case $k = 0$, $k_{,z} \neq 0$ is to put $\tilde{\eta} = \eta/\sqrt{-k} > 0$ in the hyperbolic case, and take the limit $k \to 0$, $\eta \to 0$. All terms involving $k_{,z}/k$ cancel out and we retain conditions (19.226) and (19.223). Naturally, (19.228) ceases to impose any limit.

Elliptic evolution, $k > 0$.

For elliptic models, the following holds:

$$\frac{\Phi_{,z}}{\Phi} = \frac{M_{,z}}{M}(1 - \phi_1) + \frac{k_{,z}}{k}\left(\frac{3}{2}\phi_1 - 1\right) - \frac{k^{3/2}t_{B,z}}{M}\phi_2, \qquad (19.229)$$

where

$$\phi_1 \overset{\text{def}}{=} \frac{\sin\eta(\eta - \sin\eta)}{(1 - \cos\eta)^2}, \qquad \phi_2 \overset{\text{def}}{=} \frac{\sin\eta}{(1 - \cos\eta)^2}. \qquad (19.230)$$

At early times $\eta \to 0$ and $\Phi \to 0$, $\phi_2 \to +\infty$, $\phi_1 \to 2/3$. Thus the term containing ϕ_2 dominates and, by the same argument as in the hyperbolic case, (19.226) is needed to keep $\Phi(\Phi_{,z}/\Phi - \mathcal{E}_{,z}/\mathcal{E}) > 0$. Similarly, at late times, $\eta \to 2\pi$, $\Phi \to 0$,

$\phi_2 \to -\infty$, $\phi_1 \to -\infty$ and $\phi_2/\phi_1 \to 1/(2\pi)$. Consequently, $\Phi^{3/2}(\Phi_{,z}/\Phi - \mathcal{E}_{,z}/\mathcal{E}) > 0$ now gives

$$\frac{2\pi M}{k^{3/2}}\left(\frac{M_{,z}}{M} - \frac{3k_{,z}}{2k}\right) + t_{B,z} > 0 \qquad \forall z, \tag{19.231}$$

which says that the crunch time must increase with z. Since we already have $M_{,z} \geq 0$, these conditions are sufficient to keep $\Phi_{,z} > 0$ for all η.

We now show that the above also ensure $\Phi\chi > 0$. Since the crunch time is

$$t_C = t_B + 2\pi M/k^{3/2}, \tag{19.232}$$

we can rewrite (19.229) as

$$\frac{\Phi_{,z}}{\Phi} = \frac{M_{,z}}{3M} + \frac{t_{C,z}}{(t_C - t_B)}\left(\frac{2}{3} - \phi_1\right) - \frac{t_{B,z}}{t_C - t_B}\left(\frac{2}{3} - \phi_1 + 2\pi\phi_2\right). \tag{19.233}$$

The derivative of $(2/3 - \phi_1)$ is $(2\eta - 3\sin\eta + \eta\cos\eta)/(1 - \cos\eta)^2$, and the third derivative of the numerator of the latter is $\eta\sin\eta$. It follows that $2/3 - \phi_1 \geq 0$ and declines monotonically from $+\infty$ to 0 as η goes from 2π to 0. Since $(2/3 - \phi_1 + 2\pi\phi_2)$ is the mirror image in $\eta = \pi$ of $(2/3 - \phi_1)$, we have that

$$\Phi_{,z}/\Phi > M_{,z}/(3M), \tag{19.234}$$

so that (19.223) guarantees that for each given z, the maximum of $\mathcal{E}_{,z}/\mathcal{E}$ as (x, y) are varied is no more than the minimum of $\Phi_{,z}/\Phi$ as η varies.

Although (19.231) implies $k_{,z}/(2k) < M_{,z}/(3M)$, a condition such as (19.228) is not needed in this case.

19.7.5 Regular maxima and minima

Certain topologies necessarily have extrema in Φ. For example, closed spatial sections have a maximum areal radius, and wormholes have a minimum areal radius, i.e. $\Phi_{,z}(t, z_m) = 0, \forall t$.

Suppose that $\varepsilon - k = 0$ at some $z = z_m$. We need $\chi_1 \stackrel{\text{def}}{=} \Phi_{,z} - \Phi\mathcal{E}_{,z}/\mathcal{E} = 0$ to keep g_{rr} finite, and hence $M_{,z} - 3M\mathcal{E}_{,z}/\mathcal{E} = 0$ (from (19.171)) to keep ϵ finite, both holding $\forall(t, x, y)$ at that z_m. More specifically, along any given spatial slice away from the bang or crunch, we want

$$\chi_1/\sqrt{\varepsilon - k} \to L, \qquad 0 < L < \infty, \tag{19.235}$$

$$(M_{,z} - 3M\mathcal{E}_{,z}/\mathcal{E})/\chi_1 \to N, \qquad 0 \leq N < \infty. \tag{19.236}$$

In consequence of (19.188), (19.224) and (19.229), $\chi_1 = 0$ implies

$$M_{,z} = k_{,z} = t_{B,z} = S_{,z} = P_{,z} = Q_{,z} = 0. \tag{19.237}$$

The limits (19.235) and (19.236) must hold good for all t and for all (x, y), so using (19.152), (19.224) and (19.229) with $\Phi > 0$, $M > 0$ and $S > 0$ shows that

$$\frac{M,_z}{\sqrt{\varepsilon - k}}, \qquad \frac{k,_z}{\sqrt{\varepsilon - k}}, \qquad \frac{t_{B,z}}{\sqrt{\varepsilon - k}}, \qquad \frac{\Phi,_z}{\sqrt{\varepsilon - k}}, \tag{19.238}$$

$$\frac{S,_z}{\sqrt{\varepsilon - k}}, \qquad \frac{P,_z}{\sqrt{\varepsilon - k}}, \qquad \frac{Q,_z}{\sqrt{\varepsilon - k}}, \qquad \frac{\mathcal{E},_z}{\sqrt{\varepsilon - k}} \tag{19.239}$$

must all have finite limits at $z \to z_m$. Using de l'Hôpital's rule, each of the above limits can be expressed in the form

$$L_{(M,_z)} \stackrel{\text{def}}{=} \lim_{k \to \varepsilon} \frac{M,_z}{\sqrt{\varepsilon - k}} = -\frac{2M,_{zz}\sqrt{\varepsilon - k}}{k,_z} \stackrel{\text{def}}{=} -\frac{2M,_{zz}}{L_{(k,_z)}}. \tag{19.240}$$

19.7.6 The apparent horizons

With $\varepsilon = +1$, a general null vector field k^α obeys

$$\frac{\chi^2}{1 - k}\left(\frac{dz}{dt}\right)^2 = 1 - \frac{\Phi^2}{\mathcal{E}^2}\left[\left(\frac{dx}{dt}\right)^2 + \left(\frac{dy}{dt}\right)^2\right]. \tag{19.241}$$

At each event dz/dt is maximal where $k^x = 0 = k^y$. Since Φ is independent of (x, y), this also gives the direction of maximum $d\Phi/dt|_{\text{null}}$ at any event. We will call this 'radial' motion, and its paths 'rays'. Thus, the differential equation $dt/dz|_n = j\chi_1/\sqrt{1-k}$, $j = \pm 1$, when solved, would give $t = t_n(z)$ along the 'ray'. In general, this is not a geodesic, but we regard it as the limit of a sequence of accelerating timelike paths, and thus the boundary to possible motion through a wormhole. The areal radius along a 'ray' is $\Phi_n = \Phi(t_n(z), z)$. These rays stop diverging when $(\Phi_n),_z = 0$, where, using the previous formulae,

$$(\Phi_n),_z = \Phi,_t\,(t_n),_z + \Phi,_z = \ell j \chi_1 \frac{\sqrt{2M/\Phi - k}}{\sqrt{1 - k}} + \Phi,_z, \qquad \ell = \pm 1. \tag{19.242}$$

Light rays initially moving along constant x and y will not remain so. However, since these 'radial' directions are at each point the fastest possible escape route, we define this locus to be the apparent horizon (AH).

At an ordinary spacetime point the metric components will be nonzero. Assuming that Φ is increasing with z on constant-t slices, i.e. that $\Phi,_z > 0$ and $\chi_1 > 0$, for a solution of $(\Phi_n),_z = 0$ to exist, we must require $\ell j = -1$, i.e. either $\{j = +1, \ell = -1\}$ (outgoing rays in a collapsing phase $\Longleftrightarrow$ future AH, denoted AH$^+$) or $\{j = -1, \ell = +1\}$ (incoming rays in an expanding phase $\Longleftrightarrow$ past AH, denoted AH$^-$). By 'outgoing' we mean moving away from the neck at $z = 0$. A ray passing through the neck would change from incoming to outgoing at $z = 0$, and, since $\Phi,_z$ flips sign there, j would also have to flip there.

Now define $D \overset{\text{def}}{=} \sqrt{1-k} - \sqrt{2M/\Phi - k}$. Then $(D > 0) \Longleftrightarrow (\Phi > 2M)$. Since $M/\Phi > 0$, we see that $D \geq 1$ leads to a contradiction, and hence $D < 1$, but $D < -1$ is not prohibited. We have

$$(D < -1) \Longrightarrow \left[\Phi < M / \left(1 + \sqrt{1-k} \right) \right]. \tag{19.243}$$

This will always occur when Φ is close to the Big Bang/Big Crunch.

Using D, the equation of the AH, Eq. (19.242), is

$$\Phi \mathcal{E}_{,z} + D \Phi_{,z} \mathcal{E} = 0, \tag{19.244}$$

and in terms of x and y this equation is

$$(S_{,z}/S - D\Phi_{,z}/\Phi) \left[(x-P)^2 + (y-Q)^2 \right]$$
$$+ 2 \left[(x-P)P_{,z} + (y-Q)Q_{,z} \right] - S^2 \left(\frac{S_{,z}}{S} + D\frac{\Phi_{,z}}{\Phi} \right) = 0. \tag{19.245}$$

The discriminant of this with respect to $(x - P)$ is

$$\Delta_x = 4P_{,z}{}^2 - 4 \left(\frac{S_{,z}}{S} - D\frac{\Phi_{,z}}{\Phi} \right)$$
$$\times \left[\left(\frac{S_{,z}}{S} - D\frac{\Phi_{,z}}{\Phi} \right)(y-Q)^2 + 2(y-Q)Q_{,z} - S^2 \left(\frac{S_{,z}}{S} + D\frac{\Phi_{,z}}{\Phi} \right) \right]. \tag{19.246}$$

The discriminant of this with respect to $(y - Q)$ is

$$\Delta_y = 64 \left(\frac{S_{,z}}{S} - D\frac{\Phi_{,z}}{\Phi} \right)^2 \left[P_{,z}{}^2 + Q_{,z}{}^2 + S^2 \left(\frac{S_{,z}{}^2}{S^2} - D^2\frac{\Phi_{,z}{}^2}{\Phi^2} \right) \right]. \tag{19.247}$$

Now, if $\Delta_y < 0$ everywhere, then $\Delta_x < 0$ for all y, in which case there is no x obeying (19.244), i.e. the apparent horizon does not intersect this particular surface of constant (t, z).

If $\Delta_y = 0$, then $\Delta_x < 0$ for all y except one value $y = y_0$, at which $\Delta_x = 0$. At this value of $x = x_0$, (19.244) has a solution, so the intersection of the apparent horizon with this one constant-(t, z) surface is a single point. Note that the situation when the apparent horizon touches the whole 3-dimensional $t = $ constant hypersurface at a certain value of t is exceptional; this requires, from (19.244), that $P_{,z} = Q_{,z} = S_{,z} = \Phi_{,z} = 0$ at this value of t. The first three functions being zero mean just spherical symmetry, but the fourth one defines a special location, as mentioned at the beginning of Section 19.7.4. These equations hold in the Datt–Ruban solution; see Section 19.4.

If $\Delta_y > 0$, then $\Delta_x > 0$ for every y such that $y_1 < y < y_2$, where

$$y_{1,2} = Q + \frac{-Q_{,z} \pm \sqrt{\lambda}}{S_{,z}/S - D\Phi_{,z}/\Phi}, \tag{19.248}$$

$$\lambda \overset{\text{def}}{=} P_{,z}{}^2 + Q_{,z}{}^2 + S^2 \left(S_{,z}{}^2/S^2 - D^2\Phi_{,z}{}^2/\Phi^2 \right)$$

and then a solution of (19.244) exists given by

$$x_{1,2} = P + \frac{-P,_z \pm \sqrt{-\left[(S,_z/S - D\Phi,_z/\Phi)(y - Q) + Q,_z\right]^2 + \lambda}}{S,_z/S - D\Phi,_z/\Phi}. \tag{19.249}$$

Except for the special case when $S,_z/S = D\Phi,_z/\Phi$, these values lie on a circle in the (x, y) plane, with the centre at

$$(x_{AH}, y_{AH}) = \left(P - \frac{P,_z}{S,_z/S - D\Phi,_z/\Phi}, Q - \frac{Q,_z}{S,_z/S - D\Phi,_z/\Phi}\right), \tag{19.250}$$

and with the radius $L_{AH} = \sqrt{\lambda}/(S,_z/S - D\Phi,_z/\Phi)$. The special case $S,_z/S = D\Phi,_z/\Phi$ (when the locus of AH in the (x, y) plane is a straight line) is again an artefact of the stereographic projection because this straight line is an image of a circle on the sphere.

In summary, the intersection of AH with the (x, y) plane is

(1) nonexistent when $\Phi,_z^2/\Phi^2 > \Psi^2/D^2$ (this is the Ψ defined in (19.219) for the shell crossing);
(2) a single point when $\Phi,_z^2/\Phi^2 = \Psi^2/D^2$;
(3) a circle or a straight line when $\Phi,_z^2/\Phi^2 < \Psi^2/D^2$.

The condition $\Phi,_z^2/\Phi^2 < \Psi^2/D^2$ is consistent with the condition for no shell crossings, Eq. (19.219), when $|D| < 1$. We already know that necessarily $D < 1$, but $D < -1$ is not excluded.

With $|D| < 1$, when the intersection of the AH with $(t = \text{constant}, r = \text{constant})$ is a single point, a shell crossing is automatically excluded.

From (19.244) and from $\Phi > 0$, $\mathcal{E} > 0$ and $\Phi,_z > 0$ we have

$$(D > 0) \Longrightarrow (\mathcal{E},_z < 0), \quad (D < 0) \Longrightarrow (\mathcal{E},_z > 0). \tag{19.251}$$

But $D > 0$ and $D < 0$ define regions independent of x and y. Hence, on that surface on which $D > 0$, $\mathcal{E},_z < 0$ on the whole of the AH. Where $D < 0$, $\mathcal{E},_z > 0$ on the whole of the AH. This implies that the $\mathcal{E},_z = 0$ circle and the AH cannot intersect unless they coincide. Indeed, these circles lie in parallel planes, by the same argument that was used following Eq. (19.189): the line on the $(t, z) = \text{constant}$ surface defined by (19.244) has the property $\mathcal{E},_z/\mathcal{E} = -D\Phi,_z/\Phi = \text{constant}$, so it must be a circle in a plane parallel to the $\mathcal{E},_z = 0$ great circle. It follows that of the three circles ($\mathcal{E},_z = 0$, SC and AH) no two can intersect unless they coincide.

When the $\mathcal{E},_z = 0$ and AH circles are disjoint, they may be either one inside the other or each one outside the other. However, when projected back onto the sphere, these two situations turn out to be topologically equivalent: depending on the position of the point of projection, the same two circles may project on the plane either as one circle inside the other or as two separate circles; see Figs. 19.2 and 19.3.

Along $\Phi = 2M$

$$(\Phi_n),_z = \Phi,_z(1 + \ell j) - \Phi \mathcal{E},_z/\mathcal{E}, \tag{19.252}$$

so $\Phi = 2M$ does not coincide with the AH except where $\mathcal{E},_z = 0$.

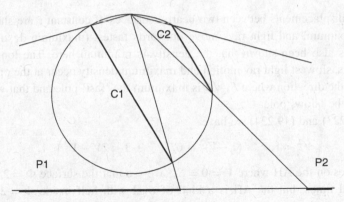

Fig. 19.2. The circles C1 and C2 on a sphere (seen here edge on) will project onto the plane (seen here as the horizontal line) as the circles P1 and P2 that are outside each other. Only parts of P1 and P2 are shown here. Circle C1 is the $\mathcal{E}_{,z} = 0$ set; circle C2 is the apparent horizon circle.

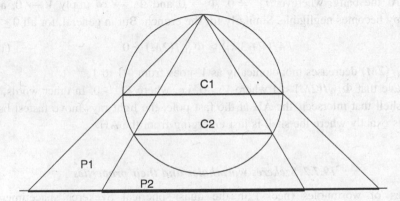

Fig. 19.3. The same circles as in Fig. 19.2 projected onto a plane from a different pole will project as one inside the other. The transition from the situation of Fig. 19.2 to that of Fig. 19.3 is continuous and occurs when the sphere is rotated, but the pole and the plane are not moved. Then one of the circles (C1 when a clockwise rotation is applied to Fig. 19.2) will pass through the pole at one value $\varphi = \varphi_0$ of the rotation angle. Its image on the plane acquires a larger and larger radius with increasing φ, until it becomes a straight line when $\varphi = \varphi_0$. When φ increases further, the straight line bends in the opposite direction and surrounds P2.

Equation (19.242) with $\ell j = -1$ can be written

$$\Phi_{\mathrm{AH}} = \frac{2M\left[1 - \Phi\mathcal{E}_{,z}/(\Phi_{,z}\mathcal{E})\right]^2}{1 - k\left\{1 - \left[1 - \Phi\mathcal{E}_{,z}/(\Phi_{,z}\mathcal{E})\right]^2\right\}} = \frac{2M(1-V)^2}{1 - k(2V - V^2)},$$

$$V \stackrel{\mathrm{def}}{=} \Phi\mathcal{E}_{,z}/(\Phi_{,z}\mathcal{E}). \tag{19.253}$$

The effect of $\mathcal{E}(z, x, y)$ is to create a dipole in the geometry and density around each (t, z) shell, with $\mathcal{E}_{,z} = 0$ on an 'equator', and extreme values given by (19.193) at the poles.

The 'radial' displacements between two nearby surfaces of constant z are shortest where $\mathcal{E}_{,z}/\mathcal{E}$ is maximum, and light rays move outwards fastest (maximum dz/dt, minimum dt/dz). It has also been shown that the density is minimum here. The longest 'radial' displacements, slowest light ray motion and maximum density occur at the opposite pole. We will call the direction where $\mathcal{E}_{,z}/\mathcal{E}$ is maximum the 'fast' pole and that where $\mathcal{E}_{,z}/\mathcal{E}$ is minimum the 'slow' pole.

From (19.223) and (19.234) we have

$$V^2 < 1, \qquad (1-V)^2 > 0, \qquad -3 < 2V - V^2 < 1. \tag{19.254}$$

In those places on the AH where $V = 0 = \mathcal{E}_{,z}$ we see that the surface $\Phi = 2M$ intersects the AH at all times, but the AH is a kind of oval with half inside $\Phi = 2M$ and half outside.

For $k = 1$, $\Phi_{\text{AH}}/(2M) = 1$ regardless of V. Thus the AH$^+$ and AH$^-$ cross in a 2-sphere at the neck ($k = 1$) at the moment of maximum expansion ($\Phi = 2M$), as in the L–T model. At the bang, wherever $t_{\text{B},z} \neq 0$, $\Phi \to 0$ and $\Phi_{,z} \to \infty$ imply $V \to 0$, and the anisotropy becomes negligible. Similarly for the crunch. But in general, for all $0 \leq k < 1$,

$$4/(1 + 3|k|) \geq \Phi_{\text{AH}}/(2M) \geq 0 \tag{19.255}$$

and $\Phi_{\text{AH}}/(2M)$ decreases monotonically as V goes from -1 to 1.

We have that $\Phi_{\text{AH}}/(2M) < 1$ where $V > 0$, i.e. where $\mathcal{E}_{,z} > 0$. In other words, taking a (t, z) shell that intersects the AH at the fast pole, the light rays move fastest between the shells exactly where the shell is just emerging from the AH.

19.7.7 Szekeres wormholes and their properties

Properties of wormholes (necks) in the quasi-spherical Szekeres spacetimes were discussed by Hellaby and Krasiński (2002). However, that discussion requires complicated calculations, so we refer the reader there for details. Here, only the results are briefly reported.

Since the Szekeres spacetime is not spherically symmetric, its wormhole is in general 'bent' – it is shorter along one side and longer along the opposite side. This raises two questions:

1. Can the wormhole be bent so strongly that the regions on its opposite ends are matched together, thus forming a single space, with the wormhole being a handle on it, as in Fig. 19.4?
2. With one side of the wormhole being shorter, can one send a light ray through the wormhole to the region outside the apparent horizons?

Unfortunately, both answers are 'no'. The matching hypersurface would have to coincide with a shell crossing, or else $\Phi_{,t} = 0$.

The paper by Hellaby and Krasiński (2002) also contains numerical examples of light rays propagating through the wormhole.

longer

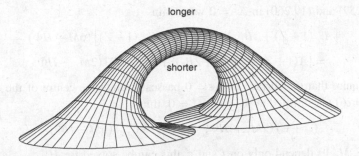

shorter

Fig. 19.4. A conceptual illustration of joining a Szekeres wormhole model to itself, to form a handle on a 3-space. The matching surface (an intersection of the matching hypersurface with the space of the figure) would run between the ends of the tube shown. (Figure provided by C. Hellaby.)

19.7.8 The mass-dipole

In the $\beta,_z \neq 0$ Szekeres solution, the distribution of mass over each single sphere $\{t = \text{constant}, z = \text{constant}\}$ has the form of a mass-dipole superposed with a monopole. This was first noted by Szekeres (1975b) and then explained in much more detail by de Souza (1985). The presentation here is based on the latter reference, but somewhat modified.

The basic idea is to separate the expression for matter density, (19.171), into a spherically symmetric part ϵ_s, depending only on t and z, and a nonsymmetric $\Delta\epsilon$. Without additional requirements, this can be done in an infinite number of ways. Add and subtract $H(t, z)/\Phi^2$ on the right-hand side of (19.171), where $H(t, z)$ is an arbitrary function. The result is

$$\epsilon = \epsilon_s(t, z) + \Delta\epsilon(t, z, x, y), \tag{19.256}$$

where:

$$\kappa\epsilon_s = \frac{H}{\Phi^2}, \qquad \kappa\Delta\epsilon = \frac{\mathcal{E}(2M,_z - H\Phi,_z) - \mathcal{E},_z(6M - H\Phi)}{\Phi^2(\mathcal{E}\Phi,_z - \Phi\mathcal{E},_z)}. \tag{19.257}$$

Now additional requirements on H will make the splitting unique. Transform (x, y) to spherical polar coordinates on a sphere of radius 1 that is tangent to the (x, y) plane at the point $(x, y) = (0, 0)$. The transformation is as in (19.153), but with $S = 1$ and $P = Q = 0$, thus

$$x = \cot(\vartheta/2)\cos\varphi, \qquad y = \cot(\vartheta/2)\sin\varphi. \tag{19.258}$$

This transforms $\mathcal{E} = \mathrm{e}^{-\nu}$, as given by (19.150), to

$$\mathcal{E} = A\cot^2(\vartheta/2) + 2B_1\cot(\vartheta/2)\cos\varphi + 2B_2\cot(\vartheta/2)\sin\varphi + C. \tag{19.259}$$

We substitute this in (19.257) and consider the equation $\Delta\epsilon = 0$. We embed the sphere in a Euclidean 3 space, and express (ϑ, φ) through the Cartesian coordinates in the space

$$X = \sin\vartheta\cos\varphi, \qquad Y = \sin\vartheta\sin\varphi, \qquad Z = \cos\vartheta. \tag{19.260}$$

Using (19.259) and (19.260) in $\Delta\epsilon = 0$ we obtain

$$\left[A_{,z}(1+Z)+2B_{1,z}X+2B_{2,z}Y+C_{,z}(1-Z)\right](6M-H\Phi)$$
$$= [A(1+Z)+2B_1X+2B_2Y+C(1-Z)](2M_{,z}-H\Phi_{,z}). \qquad (19.261)$$

Now we require that the surface of $\Delta\epsilon = 0$ passes through the centre of the sphere, i.e. that Eq. (19.261) is fulfilled at $X = Y = Z = 0$, thus

$$(A+C)(2M_{,z}-H\Phi_{,z}) = (A_{,z}+C_{,z})(6M-H\Phi). \qquad (19.262)$$

Since (A, C, M, Φ) depend only on t and z, this can be solved for H:

$$H = \frac{2M_{,z}(A+C)-6M(A+C)_{,z}}{\Phi_{,z}(A+C)-\Phi(A+C)_{,z}}. \qquad (19.263)$$

This solution makes sense except when $[(A+C)/\Phi]_{,z} \equiv 0$. But then, the Szekeres model would degenerate into the Friedmann model. Hence, (19.263) applies whenever the Szekeres model is inhomogeneous.

With H given by (19.263), Eqs. (19.257) become:

$$\kappa\epsilon_s = \frac{2M_{,z}(A+C)-6M(A+C)_{,z}}{\Phi^2\left[\Phi_{,z}(A+C)-\Phi(A+C)_{,z}\right]}, \qquad (19.264)$$

$$\kappa\Delta\epsilon = \frac{A_{,z}+C_{,z}-(A+C)\mathcal{E}_{,z}/\mathcal{E}}{\Phi^2(\Phi_{,z}-\Phi\mathcal{E}_{,z}/\mathcal{E})} \times \frac{6M\Phi_{,z}-2M_{,z}\Phi}{\Phi^2\left[\Phi_{,z}(A+C)-\Phi(A_{,z}+C_{,z})\right]}. \qquad (19.265)$$

Now, $\Delta\epsilon = 0$ has two solutions:

$$A_{,z}+C_{,z}-(A+C)\mathcal{E}_{,z}/\mathcal{E} = 0 \qquad (19.266)$$

and $(2M/\Phi^3)_{,z} = 0$. The second one defines a hypersurface that depends on t, that is, it is not comoving except when $(2M/\Phi^3)_{,z} \equiv 0$, but then the matter-density becomes spatially homogeneous. The first hypersurface, call it H_1, has its equation independent of t – it is a world-sheet of a comoving surface. Moreover, $\Delta\epsilon$ changes sign when H_1 is crossed, and, in the variables (X, Y, Z), $\Delta\epsilon$ is antisymmetric with respect to H_1. Hence, $\Delta\epsilon$ is a dipole-like contribution to matter density. Although the separation (19.256) is global, the orientation of the dipole axis is different on every sphere $\{t = \text{constant}, r = \text{constant}\}$.

We will now verify that H_1 intersects every $(t = \text{constant}, z = \text{constant})$ sphere along a circle, unless $P_{,z} = Q_{,z} = S_{,z} = 0 \ (= A_{,z} = C_{,z})$, in which case the dipole component of density is simply zero. The intersection of H_1 with any sphere of constant z is a circle parallel to the great circle $\mathcal{E}_{,z} = 0$, as noted after Eq. (19.189). It will coincide with the $\mathcal{E}_{,z} = 0$ circle at those points where $A_{,z}+C_{,z} = 0$ (if they exist). The dipole-like component will be antisymmetric with respect to $\mathcal{E}_{,z}/\mathcal{E}$ only at those values of z where $(A+C)\Phi = 0 = (A_{,z}+C_{,z})\Phi_{,z}$, but such values may exist only at the centre, $\Phi = 0$, because $A+C = 0$ contradicts Eq. (19.147) with $k > 0$.

The solution of (19.266) will exist when

$$\left(\frac{\mathcal{E}_{,z}}{\mathcal{E}}\right)_{\min} \leq \frac{A_{,z}+C_{,z}}{A+C} \leq \left(\frac{\mathcal{E}_{,z}}{\mathcal{E}}\right)_{\max}. \qquad (19.267)$$

Since $(\mathcal{E},_z/\mathcal{E})_{\min} = -(\mathcal{E},_z/\mathcal{E})_{\max}$, Eq. (19.267) is equivalent to

$$\frac{(A,_z+C,_z)^2}{(A+C)^2} \le \left(\frac{\mathcal{E},_z}{\mathcal{E}}\right)^2_{\text{extreme}} = \frac{P,_z^2+Q,_z^2+S,_z^2}{S^2}. \qquad (19.268)$$

We have

$$A+C = \left(1+P^2+Q^2+S^2\right)/(2S),$$

$$A,_z+C,_z = \frac{S,_z\left(S^2-P^2-Q^2-1\right)}{2S^2} + \frac{PP,_z+QQ,_z}{S}. \qquad (19.269)$$

Substituted in (19.268), this leads to

$$4S^2S,_z^2\left(1+P^2+Q^2\right) - 4S^2\left(PP,_z+QQ,_z\right)^2$$
$$-4SS,_z\left(PP,_z+QQ,_z\right)\left(S^2-P^2-Q^2-1\right)$$
$$+\left(P,_z^2+Q,_z^2\right)\left(1+P^2+Q^2+S^2\right) \ge 0. \qquad (19.270)$$

The discriminant of this with respect to $S,_z$ is

$$\Delta = -16S^2\left(1+P^2+Q^2+S^2\right)^2\left[(PQ,_z-QP,_z)^2+P,_z^2+Q,_z^2\right], \qquad (19.271)$$

and is negative unless $P,_z = Q,_z = 0$. Thus, with $(P,_z, Q,_z) \ne (0,0)$, the left-hand side of (19.270) is strictly positive. Even when $P,_z = Q,_z = 0$, it is still strictly positive unless $S,_z = 0$. However, $P,_z = Q,_z = S,_z = 0$ implies $A,_z = C,_z = 0$ and $\mathcal{E},_z = 0$ on the whole sphere, and then $\Delta\epsilon = 0$; i.e. on such a sphere the density is spherically symmetric. Hence, apart from the spherically symmetric subcase, Eq. (19.267) is fulfilled, with sharp inequalities in both places. This means that the $\Delta\epsilon = 0$ hypersurface intersects every $(t = \text{constant}, r = \text{constant})$ sphere along a circle parallel to the $\mathcal{E},_z = 0$ circle (see the remark after Eq. (19.189)).

19.8 * The Goode–Wainwright representation of the Szekeres solutions

Goode and Wainwright (1982) introduced a description of the Szekeres solutions in which many properties of the two subfamilies can be considered at one go.[1] The metric is

$$ds^2 = dt^2 - S^2\left[e^{2\nu}\left(dx^2+dy^2\right)+H^2W^2\,dz^2\right], \qquad (19.272)$$

where $S(t, z)$ is defined by

$$S,_t^2 = -k+2\mathcal{M}/S, \qquad (19.273)$$

$k = 0, \pm1$, and $\mathcal{M}(z)$ is an arbitrary function,

$$H = A(x, y, z) - \beta_+ f_+ - \beta_- f_-, \qquad (19.274)$$

---

[1] The G–W notation became a standard that is partly in conflict with the notation widely used for the L–T models. It is used only in this section.

A, e^ν and W will be defined below (they differ for each subfamily), $\beta_+(z)$ and $\beta_-(z)$ are functions of z defined below, and $f_+(t, z)$ and $f_-(t, z)$ are the two linearly independent solutions of the equation

$$F_{,tt} + 2(S_{,t}/S)F_{,t} - (3\mathcal{M}/S^3)F = 0. \tag{19.275}$$

Those solutions of (19.273) for which $S_{,t} > 0$ are

$$S = \mathcal{M}g_{,\eta}(\eta), \qquad t - T(z) = \mathcal{M}g(\eta), \tag{19.276}$$

where $T(z)$ is an arbitrary function (the bang time) and $g(\eta)$ is

$$g(\eta) = \begin{cases} \eta - \sin\eta & \text{when } k = +1, \\ \sinh\eta - \eta & \text{when } k = -1, \\ \eta^3/6 & \text{when } k = 0. \end{cases} \tag{19.277}$$

With $k = 0$, S can be rescaled so that $\mathcal{M} = $ constant (see later), and this choice of S will be assumed. The corresponding solutions of (19.275) are

$$f_+ = \begin{cases} (6\mathcal{M}/S)\,[1 - (\eta/2)\cot(\eta/2)] - 1 & \text{when } k = +1, \\ (6\mathcal{M}/S)\,[1 - (\eta/2)\coth(\eta/2)] + 1 & \text{when } k = -1, \\ \eta^2/10 & \text{when } k = 0, \end{cases} \tag{19.278}$$

$$f_- = \begin{cases} (6\mathcal{M}/S)\cot(\eta/2) & \text{when } k = +1, \\ (6\mathcal{M}/S)\coth(\eta/2) & \text{when } k = -1, \\ 24/\eta^3 & \text{when } k = 0. \end{cases} \tag{19.279}$$

The coefficients in $f_\pm$ were chosen for later convenience. The solutions with $S_{,t} < 0$ are time reverses of (19.278)–(19.279), and $\mathcal{M} > 0$ is assumed for correspondence of the results with the Friedmann models. With these assumptions, f_+ are the solutions that increase with time and f_- are the solutions that decrease with time.

The two subfamilies are now defined as follows:

$$\beta_{,z} \neq 0$$

$T_{,z}^2 + \mathcal{M}_{,z}^2 \neq 0$, so $S_{,z} \neq 0$, and then

$$e^\nu = \frac{f(z)}{a(z)\,(x^2 + y^2) + 2b(z)x + 2c(z)y + d(z)}, \tag{19.280}$$

where $f(z)$ is arbitrary and a, b, c, d are functions of z subject to:

$$ad - b^2 - c^2 = \varepsilon/4, \qquad \varepsilon = 0, \pm 1, \tag{19.281}$$

$$W^2 = (\varepsilon - kf^2)^{-1}, \tag{19.282}$$

$$\beta_+ = -kf\mathcal{M}_{,z}/(3\mathcal{M}), \qquad \beta_- = fT_{,z}/(6\mathcal{M}), \tag{19.283}$$

$$A = f\nu_{,z} - k\beta_+, \tag{19.284}$$

and it is understood that $\mathcal{M},_z = 0$ when $k = 0$ (see later).

$$\beta,_z = 0$$

$\mathcal{M},_z = T,_z = 0$, so $S,_z = 0$, and:

$$e^{-\nu} = 1 + \frac{1}{4}k\left(x^2 + y^2\right), \qquad W = 1, \tag{19.285}$$

$$A = \begin{cases} e^\nu \{a(z)\left[1 - \frac{1}{4}k\left(x^2 + y^2\right)\right] \\ \quad + b(z)x + c(z)y\} - k\beta_+ & \text{for } k = \pm 1, \\ a(z) + b(z)x + c(z)y \\ \quad - \frac{1}{2}\beta_+\left(x^2 + y^2\right) & \text{for } k = 0, \end{cases} \tag{19.286}$$

a, b, c, β_+ and β_- being arbitrary functions of z.

The remarkable thing about the Goode–Wainwright (G–W) parametrisation is that Eqs. (19.272)–(19.279) all hold for both subfamilies, the $\beta,_z = 0$ subfamily differing from the other one only by $\mathcal{M}$ and T being both constant. In both subfamilies the subcase $\beta_+ = \beta_- = 0$ gives the Friedmann models, but represented in a rarely used coordinate system. That this subcase is indeed a coordinate transform of the Friedmann models can be seen by applying the criteria of Section 17.12.

In the $\beta,_z \neq 0$ subfamily, the G–W representation arises as follows. Define $\phi(z)$ by

$$k(z) = K\phi^2(z), \tag{19.287}$$

where $k(z)$ is the function from (19.144), and $K = 0, \pm 1$. When $k \neq 0$, $\phi = |k|^{1/2}$; when $k = 0$, Eq. (19.287) does not define $\phi(z)$, and in this case we take $\phi(z)$ to be a new arbitrary function. Then define $S(t, z)$ by

$$\Phi = \phi S, \tag{19.288}$$

so that S obeys

$$2\frac{S,_{tt}}{S} + \frac{S,_t^2}{S^2} + \frac{K}{S^2} = 0, \tag{19.289}$$

resulting from (19.146) with (19.287), (19.288) and $p = 0$. Equation (19.273) is the integral of (19.289). Equation (19.147) now becomes

$$AC - B_1^2 - B_2^2 = \frac{1}{4}\left(\frac{1}{h^2(z)} + K\phi^2\right). \tag{19.290}$$

Let us define $G(z)$ by

$$1/h^2(z) + K\phi^2 = \varepsilon G^2, \tag{19.291}$$

where $\varepsilon = 0, \pm 1$ so that $G = \left|1/h^2 + K\phi^2\right|^{1/2}$ when $1/h^2 + K\phi^2 \neq 0$ and $0 \neq G(z)$ is an arbitrary function otherwise. Then let us define a, b, c, d, f and W by

$$(A, B_1, B_2, C, \phi) = (a, b, c, d, f)G, \qquad h = W/G; \tag{19.292}$$

the functions a, b, c, d will then obey (19.281). In the new variables, from (19.142), (19.145) and (19.288), we obtain $e^\beta = Se^{\bar\nu}$, where $\bar\nu$ is the ν given by (19.280) (not equal to the one from (19.145)), and from (19.150), (19.288) and (19.291)–(19.292) we obtain

$$e^\alpha = WS\,(fS_{,z}/S + f\nu_{,z})\tag{19.293}$$

and so $H = fS_{,z}/S + f\nu_{,z}$. When $k = 0$, the function $\phi(z)$ (and with it $f(z)$) is not yet defined. In this case, let us define ϕ by

$$\phi = M^{1/3},\tag{19.294}$$

where M is the function from (19.161). Then, S as defined by (19.288) will obey (19.273) with $k = 0$ and $\mathcal{M} = 1$, all other equations (19.272)–(19.284) remaining unchanged. Hence, when $k = 0$, one can assume $\mathcal{M} = \text{constant}$, and we will assume it in this section.

Now it can be verified, case by case from (19.277)–(19.279), (19.283), (19.284) and (19.294), that

$$F \overset{\text{def}}{=} \beta_+ f_+ + \beta_- f_- = A - H = -fS_{,z}/S - k\beta_+\tag{19.295}$$

obeys (19.275). Note that with $k = 0$ necessarily $\beta_+ = 0$. In the case under consideration, the formula for matter density (19.162) becomes

$$\kappa\epsilon = 6\mathcal{M}(1 + F/H)/S^3 \equiv 6\mathcal{M}S^3 H/A.\tag{19.296}$$

In the $\beta_{,z} = 0$ subfamily, the constant k can be rescaled to $+1$ or -1 when $k \neq 0$ by simple rescalings of x, y, ϕ, U and W. We will therefore assume that $k = 0, \pm1$. The G–W representation is then introduced in a different way when $k = \pm1$ and when $k = 0$.

When $k = \pm1$ we can redefine U, W and λ by

$$U = (u - kw)/2, \qquad W = (ku + w)/2, \qquad \lambda = \widetilde\lambda - ku\Phi,\tag{19.297}$$

and then $\widetilde\lambda$ obeys (19.135) with $p = 0$ and $U + kW = 0$. This can be assumed with no loss of generality – see the comments after (19.140). The G–W representation then arises by

$$\Phi = S, \qquad e^\beta = Se^\nu,$$

$$W = a/2, \qquad U = -ka/2, \qquad V_1 = b, \qquad V_2 = c,\tag{19.298}$$

$$e^\alpha = (A + k\beta_+)S + \lambda \overset{\text{def}}{=} HS, \qquad \beta_+ = -kX(z)/(3M),$$

where e^ν and A are given by (19.285)–(19.286) and $X(z)$ is the function from (19.158). Hence

$$F = A - H = -\lambda/S - k\beta_+,\tag{19.299}$$

and it can be verified, case by case again, with the help of (19.159), that F obeys (19.275) and $F = \beta_+ f_+ + \beta_- f_-$, where f_+ and f_- are given by (19.278)–(19.279) and $\beta_-(z)$ is an arbitrary function.

When $k = 0$, the procedure is different. We first observe that when $\Lambda = k = 0$, the function $X(z)$ in (19.158) can be assumed zero with no loss of generality because $X = 0$ results after the reparametrisation $\lambda = \tilde{\lambda} + X\Phi/3$, $W = w - X/6$ (then $e^\alpha = \tilde{\lambda} + \Phi\sigma$, as before). Then, we take $\Phi = S = e^\beta$ and $e^\alpha = \lambda + S\sigma$, where σ is given by (19.134) and S is given by (19.276)–(19.277). Next, we find λ from (19.159) in the case $k = \Lambda = 0$. Finally, we define

$$F \overset{\text{def}}{=} -\lambda/S = \beta_+ f_+ + \beta_- f_- + A, \tag{19.300}$$

where $\beta_+ = -U$, $b = V_1$, $c = V_2$, $a = 2W$ and $\beta_-(z)$ is arbitrary. The functions f_+ and f_- are as in (19.278)–(19.279), while A is as in (19.286). Using the information given in this paragraph, one can verify that such an F obeys (19.275). The matter-density is given by (19.296) for the $\beta_{,z} = 0$ subfamily as well.

As Goode and Wainwright observed, Eq. (19.275) is linear and has the same form as the equation derived for the density perturbation in the linearised perturbation scheme around the Friedmann background, if the perturbed solution is also dust. This holds both in relativity and in Newtonian theory; see Eqs. (15.9.23) and (15.10.57) in the Weinberg (1972) textbook and Eqs. (12.10) and (12.31) in the textbook by Raychaudhuri (1979). However, this is only a formal similarity. The following differences should be noted:

1. In the corresponding equation of the linearised perturbation scheme, the coefficients $(2S_{,i}/S)$ and $3\mathcal{M}/S^3 = \kappa\rho/2$ are taken from the background Friedmann model, whereas here they come from the perturbed model. Assuming F small, this means, not surprisingly, that the equation of the linearised perturbation is the linear approximation to (19.275).

2. In the linearised perturbation scheme, the solution of (19.275) is $\delta = (\epsilon_p - \epsilon_b)/\epsilon_b$, where ϵ_p is the perturbed density and ϵ_b is the Friedmann background density. In the Szekeres models, assuming F small and linearising about the background, one obtains $F = A(\epsilon_p - \epsilon_b)/\epsilon_b$, that is, $F \cong \delta$ only if $A \cong 1$, which is a limitation imposed on the model. Otherwise, F has no clear physical meaning, although it does define a certain deformation on the Friedmann background.

3. In the linearised perturbation scheme, every solution of (19.275) generates a perturbed model. In the Szekeres models of the $\beta_{,z} \neq 0$ subfamily the coefficients β_+ and β_- are provided by other field equations (see (19.283)), and when $k = 0$ the growing mode is necessarily absent ($\beta_+ \equiv 0$). In the $\beta_{,z} = 0$ subfamily with $k \neq 0$, the coefficients are truly arbitrary, but with $k = 0$ the β_+ is again provided from elsewhere.

Still, the G–W representation is enlightening in several ways. From Eq (19.283) one can see that the growing mode of perturbation is generated by the spatial inhomogeneity in the mass distribution ($\mathcal{M}_{,z} \neq 0$), and the decaying mode is generated by the non-simultaneity of the Big Bang ($T_{,z} \neq 0$) – a result that required much labour to obtain in the standard representation of the L–T model; see Section 18.19.

The expansion and shear of the dust source are in both cases:

$$\theta = 3S_{,t}/S - F/H, \qquad 2\sigma_{11} = 2\sigma_{22} = -\sigma_{33} = 2F/(3H). \tag{19.301}$$

From the above and from (19.278)–(19.279) one can now show that (19.275) is the Raychaudhuri equation (15.43), which, surprisingly, becomes linear for the Szekeres models in the G–W representation.

In general, there can be two scalar polynomial curvature singularities; they occur where $S = 0$ and $H = 0$. The first is the Big Bang. It is pointlike when $\beta_- = 0$ or cigar-type when $\beta_- \neq 0$. In the $\beta_{,z} \neq 0$ subfamily, $\beta_- = 0$ implies a simultaneous Big Bang; in the $\beta_{,z} = 0$ subfamily the Big Bang is always simultaneous. The second singularity is the shell crossing. When $\beta(z_0) > 0$, the shell crossing along the flow lines with $z = z_0$ will occur later than the Big Bang, so it is astrophysically relevant. Whenever it occurs, it is of the pancake type.

The book by Krasiński (1997) contains a list of other papers in which the G–W representation was used.

19.9 Selected interesting subcases of the Szekeres–Szafron family

19.9.1 The Szafron–Wainwright model

Szafron and Wainwright (1977) considered the subcase of the Szafron $\beta' = 0$ class in which $k = 0 = W$ and $p = \alpha/t^2$, $\alpha = $ constant. Then (19.138) is fulfilled by $\Phi(t) = [Q(t)]^{2/3}$, where Q obeys

$$4Q_{,tt}/3 + (\alpha/t^2) Q = 0, \tag{19.302}$$

and (19.135) reduces to

$$(4/3)\chi_{,tt} + (\alpha/t^2) \chi = 4U(z)/(3Q^{1/3}), \tag{19.303}$$

where $\chi \overset{\text{def}}{=} \lambda Q^{1/3}$. Note that (19.302) is the homogeneous part of (19.303). The general solution of (19.302) is

$$Q(t) = C_1 t^{1-q} + C_2 t^q, \tag{19.304}$$

where C_1 and C_2 are arbitrary constants and $\alpha = (4/3)q(1-q)$; q will be real only when $\alpha \leq 1/3$. With $\alpha = 1/3$, the two basis solutions in (19.304) become linearly dependent, and there exists an additional solution. With q complex, the constants C_1 and C_2 must be complex, too, and the real solutions for Q have different properties than those with real q. All these cases have to be considered separately. Szafron and Wainwright considered only the case $\alpha < 1/3$ ($q \neq 1/2$ and real).

A. The case $\alpha < 1/3$

With reference to (19.137), the solution is

$$\Phi(t) = [Q(t)]^{2/3}, \qquad Q(t) = C_1 t^{1-q} + C_2 t^q,$$

$$\lambda(t, z) = \chi/[Q(t)]^{1/3}, \qquad \chi = \lambda_1 t^{1-q} + \lambda_2 t^q,$$

$$\lambda_1 = B(z) + \frac{U(z)}{1-2q} \int \frac{t^q}{Q^{1/3}} \, dt, \tag{19.305}$$

$$\lambda_2 = A(z) - \frac{U(z)}{1-2q} \int \frac{t^{1-q}}{Q^{1/3}} \, dt,$$

where $A(z)$ and $B(z)$ are arbitrary functions.

This model begins with a (simultaneous) Big Bang at $t = 0$ and then expands forever. It reduces to a $k = 0$ Robertson–Walker metric when $U = V_1 = V_2 = 0$, $B = C_1$ and $A = C_2$, but the equation of state in the limit is none of the astrophysicists' favourites. The density in the R–W limit is $\kappa\epsilon = (4/3)(Q_{,t}/Q)^2$, with Q the same as in (19.304).

B. The case $\alpha = 1/3$ $(q = 1/2)$

Only the expressions for Q and χ are different from the previous case:

$$\chi = A(z) \ln t \sqrt{t} + B(z)\sqrt{t} + U(z) \left((\ln t) \int \frac{\sqrt{t}}{Q^{1/3}} \, dt - \int \frac{(\ln t)\sqrt{t}}{Q^{1/3}} \, dt \right),$$

$$Q(t) = C_1 \sqrt{t} + C_2 \sqrt{t} \ln t. \tag{19.306}$$

The evolution of this model is qualitatively similar to the previous one: a simultaneous Big Bang followed by expansion forever. The R–W limit results in the same way as before.

C. The case $\alpha > 1/3$

In this case, q is complex:

$$q_{\pm} = 1/2 \pm iq_2, \qquad q_2 = \sqrt{3\alpha - 1}/2 \tag{19.307}$$

(note that $q_- = 1 - q_+$), and the solution is

$$Q(t) = C_1 \sqrt{t} \cos(q_2 \ln t) + C_2 \sqrt{t} \sin(q_2 \ln t),$$

$$\chi = A(z)\sqrt{t} \sin(q_2 \ln t) + B(z)\sqrt{t} \cos(q_2 \ln t)$$

$$+ \frac{U(z)\sqrt{t}}{q_2} \left(\sin(q_2 \ln t) \int \frac{\sqrt{t}\cos(q_2 \ln t)}{Q^{1/3}} \, dt \right. \tag{19.308}$$

$$\left. - \cos(q_2 \ln t) \int \frac{\sqrt{t}\sin(q_2 \ln t)}{Q^{1/3}} \, dt \right).$$

The function Ψ begins from zero value at $t = t_B$, increases to a maximum and then decreases back to zero a finite time later. Unlike in typical R–W models, the duration

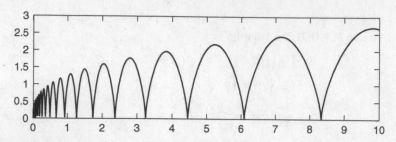

Fig. 19.5.   A graph of the function $\Phi(t)$ for a model of the Szafron–Wainwright class with $\alpha > 1/3$. The graph is self-similar: it looks the same for every scale on the horizontal axis.

of the cycle increases with increasing t. The period between bang and crunch tends to zero as $t \to 0$ and to infinity as $t \to \infty$, and the same is true for the maximal value of Φ in each cycle; see Fig. 19.5. The pressure $\kappa p = \alpha/t^2$ acts like a cosmological constant decreasing with time, and is responsible for this behaviour, which persists also in the R–W limit (achieved in the same way as in Case A).

The value of Φ is the radius of the cylindrical space $t = \text{constant}$. Along the generator of the cylinders, the variation of λ with time is still more complicated. This model has never been investigated from the point of view of cosmology.

The solutions described above have their exact counterparts in the $\beta_{,z} \neq 0$ family. With $p = \alpha/t^2$, the solution for Φ is of the same algebraic form, only its constants become arbitrary functions of z. The one corresponding to $\alpha < 1/3$ was found by Szafron (1977); the other two have never been investigated.

19.9.2 *The toroidal Universe of Senin*

Senin (1982) found a solution whose local geometry (not just topology!) is that of a 3-dimensional torus. We shall re-derive it here.

The parametric equations of a 2-dimensional torus embedded in a Euclidean 3-space, with large radius b and small radius a (Fig. 19.6) are

$$x = (a \cos \psi + b) \cos \varphi, \qquad y = (a \cos \psi + b) \sin \varphi, \qquad z = a \sin \psi, \qquad (19.309)$$

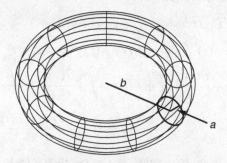

Fig. 19.6.   A 2-torus of large radius b and small radius a.

where φ is the azimuthal angle around the vertical axis in Fig. 19.6, while ψ is the angle around the small circle; both angles vary in the range $[0, 2\pi]$. We will make the torus 3-dimensional by allowing that the subset $\varphi = $ constant is a sphere rather than a circle. The generalisation is straightforward, the appropriate parametric equations being:

$$x = (a\cos\psi + b)\cos\varphi, \qquad y = (a\cos\psi + b)\sin\varphi,$$
$$z = a\sin\psi\sin\zeta, \qquad u = a\sin\psi\cos\zeta, \tag{19.310}$$

the 3-torus being now embedded in a 4-dimensional Euclidean space with the metric $dx^2 + dy^2 + dz^2 + du^2$. The metric of the 3-torus is

$$ds^2 = a^2\left[d\psi^2 + (\cos\psi + b/a)^2\,d\varphi^2 + \sin^2\psi\,d\zeta^2\right]. \tag{19.311}$$

In this 3-space, the surfaces $\zeta = $ constant are 2-dimensional tori, and the surfaces $\varphi = $ constant have the local geometry of 2-spheres. Note, however, that, although on a surface $\varphi = $ constant the coordinate ψ plays the role of the lateral angle (normally denoted by ϑ), it varies from 0 to 2π, not from 0 to π. Moreover, in the 3-space (19.311) the points with $\{\psi = \psi_1 = \pi + \psi_0, \zeta = \zeta_0\}$ *are not* equivalent to points with $\{\psi = \psi_2 = \pi - \psi_0, \zeta = \pi + \zeta_0\}$, as they would be on a sphere, because the coefficient of $d\varphi^2$ does not return to its initial value after ψ is increased by π. We conclude from this that each surface $\varphi = $ constant is actually a pair of spheres that touch each other at one pole, and their opposite poles are identified, as in Fig. 19.7.

The next step is to make the 3-torus (19.311) a subspace $t = $ constant of a 4-dimensional spacetime, and to allow the parameters of the torus to evolve with time. The simplest

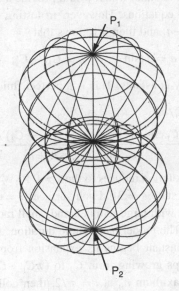

Fig. 19.7. The subspace $\varphi = $ constant of the 3-torus with the metric (19.311). The two spheres are tangent at one pole; the poles P_1 and P_2 are identified. The coordinate ψ has the value 0 at P_1, π at the common point of the spheres and 2π at P_2; it increases along the meridians of the spheres.

thing to do is to assume that the two radii a and b are momentary values of functions of time, so

$$ds^2 = a^2(t) \left\{ dt^2 - d\psi^2 - \sin^2 \psi \, d\xi^2 - [\cos \psi + b(t)/a(t)]^2 \, d\varphi^2 \right\}. \tag{19.312}$$

Somewhat miraculously, it turns out that the metric does contain perfect fluid solutions. The tetrad components of the Einstein tensor, in the orthonormal tetrad defined by (19.312), are

$$G_{00} = \frac{a_{,t}^2}{a^4} + \frac{2\cos \psi \, a_{,t}^2}{a^3 S} + \frac{2a_{,t} \, b_{,t}}{a^3 S} + \frac{2\cos \psi}{aS} + \frac{1}{a^2},$$

$$G_{11} = G_{22} = \frac{a_{,t}^2}{a^4} - \frac{a_{,tt}}{a^3} - \frac{\cos \psi \, a_{,tt}}{a^2 S} - \frac{b_{,tt}}{a^2 S} - \frac{\cos \psi}{aS}, \tag{19.313}$$

$$G_{33} = \frac{a_{,t}^2}{a^4} - \frac{2a_{,tt}}{a^3} - \frac{1}{a^2},$$

where $S \overset{\text{def}}{=} a \cos \psi + b$. To allow a perfect fluid source, only one equation, $G_{22} = G_{33}$, must be imposed on the set (19.313), and it is equivalent to

$$a_{,tt}/a = b_{,tt}/b - 1. \tag{19.314}$$

Hence, one of the radii of the torus can be chosen arbitrarily, while the other one will follow from (19.314). Since the metric depends on just two variables, t and ψ, the existence of a thermodynamical scheme is guaranteed (see Section 15.5). One sensible choice would then be to define an equation of state, which would provide a second equation connecting a and b, and then solve all equations. However, in testing a completely new class of models, Senin chose $a_{,tt} = -a$, and then it follows that

$$a = a_0 \sin t, \qquad b = C_1 t + C_2; \tag{19.315}$$

a_0, C_1 and C_2 being arbitrary constants. This is a perfect fluid solution with

$$u^\alpha = \delta_0^\alpha/a, \qquad \kappa p = a_0^2/a^4,$$

$$\kappa \epsilon = \frac{2C_1 \cos t + 3a_0 \cos \psi + (C_1 t + C_2)/\sin t}{a_0^2 \sin^3 t (a \cos \psi + b)}. \tag{19.316}$$

When $C_1 = C_2 = b = 0$, the solution reduces to the $k > 0$ R–W model in the form (17.77)–(17.78) with the equation of state $\varepsilon = 3p$. Each surface $\{t = \text{constant}, \xi = \text{constant}\}$ has the local geometry of a torus with large radius b and small radius a; the two radii evolve according to different laws. The model has a finite duration. At $t = 0$, it has a Big-Bang singularity; the space $t = \text{constant}$ begins its expansion from a ring of radius C_2. The large radius of the torus keeps growing from C_2 to $(\pi C_1 + C_2)$; the small radius grows from zero at $t = 0$ to the maximum a_0 at $t = \pi/2$, then collapses again to zero at the final singularity $t = \pi$. The final singularity is again a ring, of radius $(2\pi C_1 + C_2)$. With $C_2 = 0$, the Big Bang is pointlike. The energy-density will be positive throughout the evolution when $C_2 = 0$ if C_1/a_0 is sufficiently large.

In order to relate the Senin solution to the others, it is convenient to transform the coordinates as follows:

$$t = \arccos\left[(a_0 - T)/a_0\right], \qquad \psi = 2\arctan(r/2). \tag{19.317}$$

Then the metric becomes

$$ds^2 = dT^2 - \frac{a^2(T)}{(1+r^2/4)^2}\left(dr^2 + r^2\,d\zeta^2\right) - \left(a(T)\frac{1-r^2/4}{1+r^2/4} + \frac{1}{2}b(T)\right)^2 d\varphi^2,$$

$$\tag{19.318}$$

$$a^2(T) = a_0{}^2 - (a_0 - T)^2, \tag{19.319}$$

$$b(T) = C_1 \arccos\left[(a_0 - T)/a_0\right] + C_2,$$

and is the following limit of the Szafron $\beta,_z = 0$ class (19.137):

$$k = 1, \qquad U = -1/2, \qquad V_1 = V_2 = 0, \qquad W = 1/2. \tag{19.320}$$

19.10 * The discarded case in (19.103)–(19.112)

In Section 19.5 we omitted the cases when $\beta,_{tx} \neq 0$ or $\beta,_{ty} \neq 0$ in (19.103)–(19.112). We show here that they contain no perfect fluid solutions. For clarity, we formulate the various statements as lemmas.

Lemma 19.3 *If $\beta,_z = 0$ and $\beta,_{tx} \neq 0$, and the metric (19.102) is to obey the Einstein equations with a perfect fluid source, then coordinates may be chosen so that $\alpha,_z = 0$.*

Proof:
Rewrite Eqs. (19.105), (19.106) and (19.111) as follows:

$$\alpha,_{tx} = -\beta,_{tx} - \alpha,_t\,\alpha,_x + \beta,_t\,\alpha,_x, \tag{19.321}$$

$$\alpha,_{ty} = -\beta,_{ty} - \alpha,_t\,\alpha,_y + \beta,_t\,\alpha,_y, \tag{19.322}$$

$$\alpha,_{xy} = -\alpha,_x\,\alpha,_y + \alpha,_x\,\beta,_y + \beta,_x\,\alpha,_y. \tag{19.323}$$

Apply the integrability conditions $\alpha,_{tx,y} - \alpha,_{ty,x} = 0$, $\alpha,_{ty,x} - \alpha,_{xy,t} = 0$ and $\alpha,_{tx,y} - \alpha,_{xy,t} = 0$. The resulting equations are, respectively,

$$\alpha,_x\,\beta,_{ty} - \alpha,_y\,\beta,_{tx} = 0, \tag{19.324}$$

$$-\beta,_{txy} = 2\alpha,_x\,\beta,_{ty} - \beta,_y\,\beta,_{tx} + \beta,_x\,\beta,_{ty}, \tag{19.325}$$

$$-\beta,_{txy} = 2\alpha,_y\,\beta,_{tx} - \beta,_y\,\beta,_{tx} + \beta,_x\,\beta,_{ty}. \tag{19.326}$$

Now differentiate (19.326) by z. Since $\beta,_z = 0$, the result is $\alpha,_{yz}\,\beta,_{tx} = 0$. Then, since $\beta,_{tx} \neq 0$ by assumption, the implication is

$$\alpha,_{yz} = 0. \tag{19.327}$$

But from (19.324) with $\beta,_{tx} \neq 0$ we have

$$\alpha,_y = \beta,_{ty}\, \alpha,_x / \beta,_{tx}. \tag{19.328}$$

Differentiating this by z and using $\beta,_z = 0$ and (19.327) we obtain $\alpha,_{xz}\, \beta,_{ty} = 0$. We thus have to consider two cases separately.

Case 1: $\beta,_{ty} = 0$.

Then from (19.324) we have immediately $\alpha,_y = 0$, and from (19.325) $\beta,_y = 0$. Substitute this in (19.110) and (19.112) and take the equation $G_{22} - G_{33} = 0$; the result is

$$\alpha,_{xx} - \alpha,_x{}^2 - 2\alpha,_x\, \beta,_x = 0. \tag{19.329}$$

Differentiate this by t and use (19.321) to eliminate $\alpha,_{tx}$ and $\alpha,_{txx}$. The result is

$$-\beta,_{txx} - 2\alpha,_x\, \beta,_{tx} + 2\beta,_x\, \beta,_{tx} = 0. \tag{19.330}$$

Differentiate this by z. Since $\beta,_{tx} \neq 0$, the conclusion $\alpha,_{xz} = 0$ follows anyway. Thus we go back to consider

Case 2: $\alpha,_{xz} = 0$.

Equations (19.327) and the above imply that

$$\alpha = \alpha_1(t, x, y) + \alpha_2(t, z). \tag{19.331}$$

Substitute this in (19.321) and differentiate the result by z. The resulting equation is $\alpha_{1,x}\alpha_{2,tz} = 0$. But with $\alpha_{1,x} = 0$ we would have $\alpha,_x = 0$, and then $\beta,_{tx} = 0$ from (19.321) – contrary to the assumption. Hence, $\alpha_{2,tz} = 0$, which means that $\alpha_2 = \alpha_3(t) + \alpha_4(z)$. Then, the coordinate transformation $z' = \int e^{\alpha_4}\, dz$ will make the metric independent of z. $\square$

Lemma 19.4 *If $\beta,_z = 0$ and $\beta,_{tx} \neq 0$, and the metric (19.102) is to obey the Einstein equations with a perfect fluid source, then coordinates may be chosen so that $\alpha,_y = \alpha,_z = \beta,_y = 0$, i.e. so that the metric depends only on t and x.*

Proof:

We have already proven that no generality is lost on assuming that $\alpha,_z = 0$.

Substitute (19.328) in (19.322) and use (19.321) to eliminate $\alpha,_{tx}$. The result is $\alpha,_x (\beta,_{ty} / \beta,_{tx}),_t = 0$. Since $\alpha,_x = 0$ would lead to $\beta,_{tx} = 0$ (from (19.321)), this implies that

$$\beta,_{ty} / \beta,_{tx} = f(x, y) \Longrightarrow \alpha,_y = f(x, y)\alpha,_x, \tag{19.332}$$

where $f(x, y)$ is an arbitrary function independent of t.

With $\alpha,_z = \beta,_z = 0$, do the complex coordinate transformation in Eqs. (19.110)–(19.112):

$$\xi = x + iy, \qquad \bar{\xi} = x - iy, \tag{19.333}$$

and consider the equations $G_{22} - G_{33} - G_{23} = 0$ and $G_{22} - G_{33} + G_{23} = 0$. They are

$$\alpha,_{\xi\xi} + \alpha,_\xi{}^2 - 2\alpha,_\xi\, \beta,_\xi = 0 \tag{19.334}$$

and its complex conjugate. The solution of (19.334) is

$$e^{\alpha - 2\beta}\alpha_{,\xi} = h(t, \overline{\xi}), \tag{19.335}$$

where $h(t, \overline{\xi})$ is an arbitrary function, independent of ξ. Since α and β are real functions, by taking the complex conjugate of (19.335) we obtain

$$e^{\alpha - 2\beta}\alpha_{,\overline{\xi}} = \overline{h}(t, \xi), \tag{19.336}$$

which happens to be the general solution of the complex conjugate equation to (19.334). Note that $h \neq 0$, or else α becomes a function of t only, which leads to $\beta_{,tx} = 0$ via (19.321) – contrary to the assumption.

Now do the transformation (19.333) in (19.332) and calculate

$$f(x, y) = \frac{\alpha_{,y}}{\alpha_{,x}} \equiv i\frac{\alpha_{,\xi} - \alpha_{,\overline{\xi}}}{\alpha_{,\xi} + \alpha_{,\overline{\xi}}} \equiv i\frac{\alpha_{,\xi}/\alpha_{,\overline{\xi}} - 1}{\alpha_{,\xi}/\alpha_{,\overline{\xi}} + 1}. \tag{19.337}$$

This shows that $\alpha_{,\xi}/\alpha_{,\overline{\xi}} = h/\overline{h}$ does not depend on t. Hence

$$\frac{\alpha_{,\xi}}{\alpha_{,\overline{\xi}}} = \frac{h(t, \overline{\xi})}{\overline{h}(t, \xi)} = \frac{h(t_0, \overline{\xi})}{\overline{h}(t_0, \xi)}, \tag{19.338}$$

where t_0 is some fixed instant. Now define the function $g(\xi)$ by

$$1/\overline{h}(t_0, \xi) = dg/d\xi. \tag{19.339}$$

Then Eq. (19.338) may be written as

$$\frac{\alpha_{,\xi}}{\alpha_{,\overline{\xi}}} = \frac{dg/d\xi}{d\overline{g}/d\overline{\xi}}. \tag{19.340}$$

Let us take $g(\xi)$ and $\overline{g}(\overline{\xi})$ as the new variables. Then Eq. (19.340) is equivalent to $\partial\alpha/\partial g = \partial\alpha/\partial\overline{g}$, whose solution is

$$\alpha = \alpha(t, g + \overline{g}). \tag{19.341}$$

Now use (19.339) in (19.335) and (19.336), obtaining

$$e^{2\beta} = e^{\alpha}\alpha'\left(\overline{g}_{,\overline{\xi}}\right)^2 = e^{\alpha}\alpha'\left(g_{,\xi}\right)^2, \tag{19.342}$$

where the prime denotes the derivative with respect to $(g + \overline{g})$. Since ξ and $\overline{\xi}$ are independent variables, Eq. (19.342) implies that

$$g_{,\xi} = \pm\overline{g}_{,\overline{\xi}} = C = \text{constant}, \tag{19.343}$$

the constant being real in order that β given by (19.342) is real. We choose the sign $+$ to avoid having $g + \overline{g} = 0$. Equations (19.341), (19.343) and (19.342) then imply that α and β are functions of the one argument $g + \overline{g} = C(\xi + \overline{\xi})$. Then choose

$$g + \overline{g} \stackrel{\text{def}}{=} x', \qquad -i(g - \overline{g}) \stackrel{\text{def}}{=} y' \tag{19.344}$$

as the new coordinates. Clearly, $\alpha = \alpha(t, x')$ and $\beta = \beta(t, x')$, and, applying the chain of transformations $(x, y) \to (\xi, \bar{\xi}) \to (g, \bar{g}) \to (x', y')$ to the metric $e^{2\beta}(dx^2 + dy^2)$, we obtain

$$e^{2\beta}\left(dx^2 + dy^2\right) = e^{2\beta}\,d\xi\,d\bar{\xi} = e^{-2\beta}\,e^{2\alpha}\alpha_{,z}{}^2\,dg\,d\bar{g}$$

$$= e^{2\alpha - 2\beta}\alpha_{,z}{}^2\left(dx'^2 + dy'^2\right)/4, \tag{19.345}$$

where $\tilde{\beta} = \alpha - \beta + \ln \alpha_{,z} - \ln 2$ depends only on t and x'. $\square$

Conclusion. *If $\beta_{,ty} \neq 0$ and $\beta_{,z} = 0$, and the metric (19.102) is to obey the Einstein equations with a perfect fluid source, then coordinates may be chosen so that the metric depends only on t and y.*

Proof:

The transformation $(x, y) = (y', x')$ is an isometry of (19.102). The rest of the proof follows by interchanging x and y in the proof above. $\square$

Since the cases $\beta_{,tx} \neq 0$ and $\beta_{,ty} \neq 0$ are thus equivalent, we will follow only the case $\beta_{,tx} \neq 0 \Longrightarrow \alpha = \alpha(t, x)$, $\beta = \beta(t, x)$.

Lemma 19.5 *If the functions α and β in the metric (19.102) depend only on t and x, while $\beta_{,tx} \neq 0$, then the metric does not obey the Einstein equations with a perfect fluid source.*

Proof:

Substitute $\alpha_{,y} = \alpha_{,z} = \beta_{,y} = \beta_{,z} = 0$ in (19.103)–(19.112) and take the equation $G_{22} - G_{33} = 0$. It is

$$\alpha_{,xx} + \alpha_{,x}{}^2 - 2\alpha_{,x}\beta_{,x} = 0. \tag{19.346}$$

Its solution is

$$e^{\alpha}\alpha_{,x} = h e^{2\beta} \Longrightarrow 2\beta = \alpha + \ln \alpha_{,x} - \ln h, \tag{19.347}$$

where $h(t)$ is an arbitrary function. We assume $h \neq 0$, or else (19.321) would imply $\beta_{,tx} = 0$. Use (19.347) to eliminate β from $G_{02} = 0$; the result is

$$\alpha_{,tx} + \frac{\alpha_{,txx}}{2\alpha_{,x}} - \frac{\alpha_{,xx}\alpha_{,tx}}{2\alpha_{,x}{}^2} + \frac{1}{2}\alpha_{,t}\,\alpha_{,x} + \frac{h_{,t}}{2h}\alpha_{,x} = 0. \tag{19.348}$$

This is integrated with respect to x, yielding

$$e^{\alpha}\left(\alpha_{,t} + h_{,t}/h\right) = k_1(t)\alpha + k_2(t), \tag{19.349}$$

where $k_1(t)$ and $k_2(t)$ are arbitrary functions. Substituting (19.347) and (19.349) into the integrability condition $(e^{\alpha}\alpha_{,t})_{,x} = (e^{\alpha}\alpha_{,x})_{,t}$ we obtain

$$\beta_{,t} = k_1 e^{-\alpha}/2 - h_{,t}/h. \tag{19.350}$$

Now take the equation $G_{11} - G_{22} = 0$, use (19.347), (19.349) and (19.350) to eliminate β, $\alpha_{,t}$, $\alpha_{,tt}$, $\beta_{,t}$ and $\beta_{,tt}$, and multiply the result by $\alpha_{,x}$, obtaining

$$he^{-\alpha}\left(\beta_{,xx} - \alpha_{,x}\beta_{,x}\right) + \left(-\frac{1}{2}k_{1,t}e^{-\alpha} + (k_{1,t}\alpha + k_{2,t})e^{-\alpha}\right.$$

$$\left. - 2\frac{h_{,t}}{h}(k_1\alpha + k_2)e^{-\alpha} - \frac{1}{2}k_1{}^2 e^{-2\alpha} + 2k_1(k_1\alpha + k_2)e^{-2\alpha}\right)\alpha_{,x} = 0.$$

$$(19.351)$$

The integral of this is

$$he^{-\alpha}\beta_{,x} - \frac{1}{2}k_{1,t}e^{-\alpha} - (k_{1,t}\alpha + k_{2,t})e^{-\alpha} + 2\frac{h_{,t}}{h}(k_1\alpha + k_2)e^{-\alpha}$$

$$+ 2k_1\frac{h_{,t}}{h}e^{-\alpha} - \frac{1}{4}k_1{}^2 e^{-2\alpha} - k_1(k_1\alpha + k_2)e^{-2\alpha} + \ell_1(t) = 0. \quad (19.352)$$

Substitute here for β from (19.347) and multiply the resulting equation by $e^{2\alpha}\alpha_{,x}$. This can be integrated with respect to x again; the result is

$$\frac{1}{2}he^{\alpha}\alpha_{,x} + \frac{1}{2}k_{1,t}e^{\alpha} - (k_{1,t}\alpha + k_{2,t})e^{\alpha} + 2\frac{h_{,t}}{h}(k_1\alpha + k_2)e^{\alpha}$$

$$- \frac{1}{4}k_1{}^2\alpha - k_1\left(\frac{1}{2}k_1\alpha^2 + k_2\alpha\right) + \frac{1}{2}\ell_1(t)e^{2\alpha} + \ell_2(t) = 0. \quad (19.353)$$

On this equation, the following sequence of operations should now be carried out: (1) replace $e^{\alpha}\alpha_{,x}$ from (19.347); (2) differentiate the result by t; (3) use (19.350) to eliminate $\beta_{,t}$; (4) use (19.349) to eliminate $\alpha_{,t}$; (5) multiply (19.353) by $k_1 e^{-\alpha}$ and subtract the product from the equation obtained in the previous step. The term containing β will cancel out, and what will remain will be a polynomial in α, $e^{-\alpha}$ and e^{α}. Since $\alpha_{,x} \neq 0$ and the other functions do not depend on x, the coefficients of independent functions of α must vanish separately. The coefficient of $\alpha^2 e^{-\alpha}$ will vanish only when $k_1 = 0$. However, then we have $\beta_{,t} = -h_{,t}/h$ in (19.350), which makes $\beta_{,tx} = 0$, contrary to our assumption. Consequently, no perfect fluid solutions exist for (19.102) when $\alpha = \alpha(t, x)$, $\beta = \beta(t, x)$ and $\beta_{,tx} \neq 0$. $\square$

19.11 Exercises

1. Find the Killing vector fields for the transformation (19.1). Then solve the Killing equations for the components of the metric tensor in spacetime and verify that the solution is indeed given by (19.3).
2. Verify that $-\delta^{\alpha\beta}_{\gamma\delta}/\Phi^2$ is the curvature tensor for the metric (19.5).
3. Find the Killing fields for the metric (19.5) and verify that they are given by (19.6). Then solve the Killing equations in four dimensions with the Killing fields (19.6) and verify that the resulting metric is (19.7).

4. Calculate the Riemann tensor for (19.9) and verify that it can vanish only with $\varepsilon = -1$. Then, (19.9) is flat when $\mathrm{d}f(r)/\mathrm{d}r = 1$.

5. Find the Bianchi types for the algebras of Killing fields given by (8.41) and (19.6) and for the algebra defined by the transformations (19.1).

6. Verify that by equating the two expressions for G_{22}, Eqs. (19.25) and (19.29), we do indeed obtain (19.44). Hints are given in the text.

7. Show that the Datt–Ruban solution, given by (19.95) with $Q = 0$ and (19.99), can be matched to the Schwarzschild solution in the Lemaître–Novikov coordinates, (14.116)–(14.117).

8. Solve the equation of evolution of the Datt–Ruban model, (19.95), in the case $Q = \Lambda = 0$. Show that the model expands from a singularity $\Phi = 0$ at $t = t_\mathrm{B}$ to the maximal size $\Phi = 2M$ at $t = t_\mathrm{B} + \pi M$, and then recollapses in a time-symmetric manner.

9. Show that the 2-dimensional metric (19.129) is a coordinate transform of the 2-dimensional metrics resulting from (19.11)–(19.12) when $t = $ constant, $r = $ constant, with $k = \varepsilon$ and $\Phi = e^\nu$. **Hint.** With $k > 0$, apply the transformation $x = (2/\sqrt{k}) \cot(\vartheta/2) \cos\varphi$, $y = (2/\sqrt{k}) \cot(\vartheta/2) \sin\varphi$ to (19.129). With $k < 0$, the transformation is $x = (2/\sqrt{-k}) \coth(\vartheta/2) \cos\varphi$, $y = (2/\sqrt{-k}) \coth(\vartheta/2) \sin\varphi$.

10. Prove that, in the most general case, the Szafron metrics given by (19.136)–(19.138) and (19.149)–(19.150) have no symmetry. **Hint.** Solve the Killing equations and verify that with no further limitations on the metric the solution is $k^\mu \equiv 0$. **Note.** this is a laborious exercise. You may use the paper by Bonnor, Sulaiman and Tomimura (1977) as an aid.

11. Consider a surface given by the equation $r = a = $ constant in the coordinates (r, ϑ, φ) related to the Cartesian coordinates (x, y, z) by

$$x = r \sinh\vartheta \cos\varphi, \qquad y = r \sinh\vartheta \sin\varphi, \qquad z = r \coth\vartheta \qquad (19.354)$$

in the pseudo-Euclidean flat 3-space with the metric $\mathrm{d}s^2 = \mathrm{d}x^2 + \mathrm{d}y^2 - \mathrm{d}z^2$. Show that the surface obeys the equation of a two-sheeted hyperboloid $(z/a)^2 - (x^2 + y^2)/a^2 = 1$ and that its internal metric coincides with that of the surface $(t = $ constant, $\Phi = a)$ in the metric (19.11) with $f(\vartheta) = \sinh\vartheta$.

12. Verify that the magnetic part of the Weyl tensor with respect to the velocity field in the Szafron spacetimes (19.150) and (19.137) is zero.

13. Prove that (1) two eigenvalues of the 3-dimensional Ricci tensor, $^{(3)}\Phi_{AB}$, of the slices $t = $ constant are equal (in both classes of Szafron's spacetimes); (2) the eigenframe of this Ricci tensor coincides with that of shear; (3) the eigenspaces corresponding to the degenerate eigenvalues of shear and of $^{(3)}\Phi_{AB}$ are the surfaces of constant curvature defined by the Szekeres geometry.

14. Prove that the slices $t = $ constant of the Szafron spacetimes are conformally flat, i.e. that their Cotton–York tensor (7.50) is zero.

15. Verify that the barotropic equation of state $\varepsilon = \varepsilon(p)$ reduces the $\beta_{,z} \neq 0$ Szafron spacetimes to the Robertson–Walker ones, and the $\beta_{,z} = 0$ spacetimes to either R–W or Kantowski–Sachs-type spacetimes, or the plane and hyperbolically symmetric counterparts of the latter. **Note.** this is a very laborious exercise. It is advisable to use the paper by Spero and Szafron (1978) as an aid.

16. Calculate the integral in (19.159) when $\Lambda = 0$ and verify that along the z-direction the space has infinite extent at the instant of the Big Bang, then collapses to a minimum size, and expands to

infinite extent as the final singularity is reached. Show that the maximum of expansion of the spheres and the minimum of collapse in the z-direction do not in general occur simultaneously.

17. Verify Equation (19.163).
18. Verify that the Szekeres solution with $AC - B_1^2 - B_2^2 > 0$ can be matched to the Schwarzschild solution.
19. Verify that the metric (19.152) results from (19.150) after the reparametrisation indicated is carried out.
20. Verify that the explicit solution for λ that results from (19.169) in the limit $k \to \varepsilon$ coincides with (19.101).

20

The Kerr solution

20.1 The Kerr–Schild metrics

Most of the known solutions of Einstein's equations resulted from a purposeful search related to the later application. The Kerr metric was discovered rather accidentally, in the course of formal mathematical investigations not obviously related to physics, and only afterwards was it found to describe the exterior gravitational field of a rotating body or black hole. Actually, no explicit solution of Einstein's equations with a perfect fluid source has been found until today that could be matched to the Kerr metric. Its main application is to the description of rotating black holes, which earned it great importance in relativistic astrophysics.

In the first two sections we shall briefly describe the way in which Kerr (1963) first chanced upon his solution. In the next sections, we shall introduce the quasi-derivation by Carter (1973), and will derive its most important properties. Sections 20.1 and 20.2 are based on the papers by Boyer and Lindquist (1967) and by Kerr and Schild (1965).

The starting point is the consideration of metrics of the form

$$g_{\mu\nu} = \eta_{\mu\nu} - l_\mu l_\nu, \tag{20.1}$$

where $\eta_{\mu\nu}$ is the flat (Minkowski) metric in any coordinates and l_μ is a null vector field. The reasons given for studying this metric vary from paper to paper. Kerr and Schild (1965) justified it by saying that it allows for an easy conversion between covariant and contravariant components (see below). Boyer and Lindquist (1967) quoted another reason: the Schwarzschild solution has this form, which inspired the search for further metrics with the same property. The vector field l_μ is required to be null with respect to the metric $g_{\mu\nu}$, but then it follows that it is null with respect to $\eta_{\mu\nu}$ as well and that, consequently, it does not matter which metric is used to raise and lower the index of l. It follows also that the inverse metric has the form

$$g^{\mu\nu} = \eta^{\mu\nu} + l^\mu l^\nu. \tag{20.2}$$

Note that the sign of $l_\mu l_\nu$ cannot be changed by coordinate transformations, so it is relevant. Depending on it, one or another coordinate will become time. For example, take coordinates in which the flat background metric is

$$\begin{bmatrix} 0 & 1 & 0 & 0 \\ 1 & 0 & 0 & 0 \\ 0 & 0 & -1 & 0 \\ 0 & 0 & 0 & -1 \end{bmatrix}$$

and take $l^\mu = (1,0,0,0)$. Consider both possible signs of $l_\mu l_\nu$ in the metric, thus take $\varepsilon l_\mu l_\nu$ with $\varepsilon = \pm 1$. The full metric, with coordinates (u, v, y, z), is then

$$ds^2 = 2\,du\,dv + \varepsilon\,dv^2 - dy^2 - dz^2.$$

With $\varepsilon = +1$ we have

$$ds^2 = d(u+v)^2 - du^2 - dy^2 - dz^2,$$

whereas with $\varepsilon = -1$ we have

$$ds^2 = du^2 - d(u-v)^2 - dy^2 - dz^2.$$

To be consistent with the existing literature, we must take the sign as in (20.1) (note that we use a different signature from that in the papers cited).

Now consider the vacuum Einstein equations, $R_{\mu\nu} = 0$, and take the component $R_{\mu\nu}l^\mu l^\nu = 0$. The following formulae help in the calculation:

$$\begin{Bmatrix} \alpha \\ \beta\gamma \end{Bmatrix} l^\beta l^\gamma = \begin{Bmatrix} \alpha \\ \beta\gamma \end{Bmatrix}(\eta) l^\beta l^\gamma,$$

$$\begin{Bmatrix} \alpha \\ \beta\gamma \end{Bmatrix} l_\alpha l^\gamma = \begin{Bmatrix} \alpha \\ \beta\gamma \end{Bmatrix}(\eta) l_\alpha l^\gamma, \qquad (20.3)$$

$$\begin{Bmatrix} \alpha \\ \beta\alpha \end{Bmatrix} = \begin{Bmatrix} \alpha \\ \beta\alpha \end{Bmatrix}(\eta),$$

$$\begin{Bmatrix} \alpha \\ \beta\gamma \end{Bmatrix} l^\gamma = \begin{Bmatrix} \alpha \\ \beta\gamma \end{Bmatrix}(\eta) l^\gamma - \frac{1}{2} l^\rho \left(l^\alpha l_\beta \right)_{;\rho}, \qquad (20.4)$$

$$\begin{Bmatrix} \alpha \\ \beta\gamma \end{Bmatrix} l_\alpha = \begin{Bmatrix} \alpha \\ \beta\gamma \end{Bmatrix}(\eta) l_\alpha + \frac{1}{2} l^\rho \left(l_\beta l_\gamma \right)_{;\rho},$$

where $\begin{Bmatrix} \alpha \\ \beta\gamma \end{Bmatrix}(\eta)$ denotes the Christoffel symbols corresponding to the flat metric $\eta_{\alpha\beta}$ (recall that $\eta_{\alpha\beta}$ may be expressed in non-Lorentzian coordinates, so the Christoffel symbols do not have to be zero). The covariant derivatives in the formulae can be taken with respect to any of the two metrics, so for definiteness let us assume that they are with respect to $\eta_{\alpha\beta}$. Using these formulae, after a lot of algebra, we find that $R_{\mu\nu}l^\mu l^\nu = 0$ implies

$$l^\beta l_\beta = 0, \qquad (20.5)$$

where

$$i^\alpha \overset{\text{def}}{=} l^\mu l^\alpha{}_{;\mu}. \qquad (20.6)$$

In consequence of Eqs. (20.3) (20.4), it does not matter which of the two metrics is used to calculate the covariant derivative in (20.6), both definitions of i^α give the same result.

Now, in consequence of $\eta_{\mu\nu}l^{\mu}l^{\nu} = 0$ we have $\eta_{\mu\nu}l^{\mu}\dot{l}^{\nu} = 0$. Thus, $\dot{l}^{\mu}$ is orthogonal to the null vector l^{μ} and is itself null, by (20.5). One null vector can be orthogonal to another null vector only when they are collinear. Consequently

$$l^{\mu}l^{\alpha}{}_{;\mu} = \sigma l^{\alpha}. \tag{20.7}$$

Thus, the vacuum Einstein equations imply that l^{α} must be tangent to geodesics, but the geodesics are not necessarily affinely parametrised. To obtain an affine parametrisation, we define

$$k^{\mu} \stackrel{\text{def}}{=} l^{\mu}/\sqrt{2H}, \tag{20.8}$$

where the function H is defined by $H_{,\nu}l^{\nu}/(2H) = \sigma$. Then

$$k^{\mu}{}_{;\nu}k^{\nu} = 0; \qquad g_{\mu\nu} = \eta_{\mu\nu} - 2Hk_{\mu}k_{\nu}, \qquad g^{\mu\nu} = \eta^{\mu\nu} + 2Hk^{\mu}k^{\nu}. \tag{20.9}$$

In terms of k^{α}, Eqs. (20.3) do not change, whereas, in consequence of k^{α} being geodesic and affinely parametrised, Eqs. (20.4) simplify to

$$
\begin{aligned}
\left\{ \begin{matrix} \alpha \\ \beta\gamma \end{matrix} \right\} k^{\gamma} &= \left\{ \begin{matrix} \alpha \\ \beta\gamma \end{matrix} \right\} (\eta) k^{\gamma} - k^{\rho} H_{,\rho} k^{\alpha} k_{\beta}, \\[2mm]
\left\{ \begin{matrix} \alpha \\ \beta\gamma \end{matrix} \right\} k_{\alpha} &= \left\{ \begin{matrix} \alpha \\ \beta\gamma \end{matrix} \right\} (\eta) k_{\alpha} + k^{\rho} H_{,\rho} k_{\beta} k_{\gamma}.
\end{aligned}
\tag{20.10}
$$

In the next step, the following formula will be needed

$$
\begin{aligned}
\left\{ \begin{matrix} \alpha \\ \beta\gamma \end{matrix} \right\} = &\left\{ \begin{matrix} \alpha \\ \beta\gamma \end{matrix} \right\} (\eta) - k^{\alpha} k_{\beta} H_{,\gamma} - H k^{\alpha} k_{\beta;\gamma} - H k^{\alpha}{}_{;\gamma} k_{\beta} - H_{,\beta} k^{\alpha} k_{\gamma} \\[2mm]
& - H k^{\alpha} k_{\gamma,\beta} - H k^{\alpha}{}_{;\beta} k_{\gamma} + \eta^{\alpha\rho} H_{,\rho} k_{\beta} k_{\gamma} + \eta^{\alpha\rho} H k_{\beta;\rho} k_{\gamma} \\[2mm]
& + \eta^{\alpha\rho} H k_{\beta} k_{\gamma,\rho} + 2 H k^{\rho} H_{,\rho} k^{\alpha} k_{\beta} k_{\gamma}.
\end{aligned}
\tag{20.11}
$$

The covariant derivatives are with respect to $\eta_{\alpha\beta}$. Note that the difference of two Christoffel symbols is a tensor, as follows from (4.23).

A direct calculation with use of Eqs. (20.3) rewritten in terms of k^{μ} and of (20.10) gives for the Riemann tensor

$$R_{\alpha\beta\gamma\delta}k^{\beta}k^{\delta} = H_{;\mu\nu}k^{\mu}k^{\nu}k_{\alpha}k_{\gamma}. \tag{20.12}$$

In vacuum, the Weyl tensor will obey the same equation. But then, it will obey also (11.55), and so we have derived the following

Lemma 20.1 *The vacuum Kerr–Schild metrics cannot be of the most general Petrov type, they are* **algebraically special***, that is, of Petrov type II or simpler. The Kerr–Schild null geodesic vector field k^{α} is at the same time the degenerate Debever vector of the Weyl tensor.*

From the Goldberg–Sachs theorem 16.4 it follows then immediately that the field k^α is not only geodesic, but also shearfree.

20.2 The derivation of the Kerr solution by the original method

The Kerr solution has no clear-cut invariant definition. Therefore, in order to understand how it came about, it is most instructive to follow the historical path. The presentation in this section is based on Kerr and Schild (1965). (Another derivation can be found in Debney, Kerr and Schild (1969), but in that paper a tetrad different from the one used here is employed and a different notation is used.)

In the Minkowski space, let us choose the real null coordinates (u, v) and the complex null coordinates $(\xi, \overline{\xi})$, related to the Cartesian orthogonal coordinates (t, x, y, z) by

$$(u, v, \xi, \overline{\xi}) = (t+x, t-x, y+iz, y-iz)/\sqrt{2}. \tag{20.13}$$

In these coordinates, the Minkowski metric is

$$[\eta_{\alpha\beta}] = [\eta^{\alpha\beta}] = \begin{bmatrix} 0 & 1 & 0 & 0 \\ 1 & 0 & 0 & 0 \\ 0 & 0 & 0 & -1 \\ 0 & 0 & -1 & 0 \end{bmatrix}. \tag{20.14}$$

A general real null direction field in the Minkowski spacetime is in these coordinates given by

$$k_\mu \, dx^\mu = du + Y\overline{Y} \, dv + \overline{Y} \, d\xi + Y \, d\overline{\xi} \Longleftrightarrow k^\mu = (Y\overline{Y}, 1, -Y, -\overline{Y}), \tag{20.15}$$

where $Y(u, v, \xi, \overline{\xi})$ is an arbitrary complex function. If this field is to be geodesic, then the function Y must obey

$$k^\rho Y_{,\rho} = k^\rho \overline{Y}_{,\rho} = 0. \tag{20.16}$$

With (20.16) fulfilled, (20.15) is already affinely parametrised.

Let us now consider a Kerr–Schild metric (20.9) with (20.14) as the background Minkowski metric and the geodesic field (20.15)–(20.16) as the field k^α. Let us introduce a double-null tetrad such as was defined in Section 16.5 for the metric (20.9), with $e_0{}^\alpha = k^\alpha$. The simplest choice of the other tetrad vectors is[1]

$$e_1{}^\alpha \overset{\text{def}}{=} \ell^\alpha = \delta^\alpha{}_0 + Hk^\alpha, \qquad \ell_\alpha = dv - Hk_\alpha \, dx^\alpha,$$

$$e_2{}^\alpha \overset{\text{def}}{=} m^\alpha = (\overline{Y}, 0, -1, 0), \qquad m_\alpha = \overline{Y} \, dv + d\overline{\xi}, \tag{20.17}$$

$$e_3{}^\alpha \overset{\text{def}}{=} \overline{m}^\alpha = (Y, 0, 0, -1), \qquad \overline{m}_\alpha = Y \, dv + d\xi.$$

---

[1] The tetrad (20.17) obeys the orthogonality relations (16.45) with respect to both $g_{\alpha\beta}$ and $\eta_{\alpha\beta}$.

Then, by virtue of (16.51), the field (20.15) will be shearfree if Y and $\overline{Y}$ obey in addition

$$m^\rho \overline{Y}_{,\rho} = \overline{m}^\rho Y_{,\rho} = 0. \tag{20.18}$$

By (16.54) we also have

$$Z = \theta + i\omega \equiv \Gamma^1{}_{23} = \overline{m}^\sigma \overline{Y}_{,\sigma}. \tag{20.19}$$

In what follows we will be assuming that $\theta + i\omega \neq 0$.

The tetrad (20.15)–(20.17) is not the special one we used in Section 16.6. Nevertheless, the equations[1] $R_{00} = R_{02} = R_{03} = R_{22} = R_{33} = 0$ that were used in proving the Goldberg–Sachs theorem are now all fulfilled identically in consequence of the special form (20.17) and of the geodesic – shearfree properties (20.16) and (20.18). Some of the Ricci rotation coefficients vanish independently of the property $\Gamma^1{}_{21} = 0$; they are listed in the last line of (20.20). The other ones vanish in consequence of the form of the tetrad (20.15)–(20.17):

$$\Gamma^a{}_{b0} = \Gamma^0{}_{02} = \Gamma^1{}_{12} = \Gamma^0{}_{03} = \Gamma^1{}_{13} = \Gamma^a{}_{22} = \Gamma^a{}_{33} = \Gamma^2{}_{23} = \Gamma^3{}_{32} = 0,$$

$$\Gamma^a{}_{00} = \Gamma^1{}_{a0} = \Gamma^1{}_{22} = \Gamma^1{}_{33} = 0. \tag{20.20}$$

Since we are working in coordinates in which the background metric is constant, the background Christoffel symbols vanish and the covariant derivatives in (20.11) become partial derivatives. Using (20.11) thus simplified, we find (the list does not include those Γs that are obtained by complex conjugation)

$$\Gamma^0{}_{01} = -\Gamma^1{}_{11} = -k^\rho H_{,\rho}, \qquad \Gamma^1{}_{23} = \Gamma^3{}_{03} = Z,$$

$$\Gamma^0{}_{21} = \Gamma^3{}_{11} = -m^\rho H_{,\rho} - H\overline{Y}_{,u}, \qquad \Gamma^1{}_{21} = \Gamma^3{}_{01} = \overline{Y}_{,u}, \tag{20.21}$$

$$\Gamma^0{}_{23} = \Gamma^3{}_{13} = H\overline{Z}, \qquad \Gamma^2{}_{21} = H(Z - \overline{Z}).$$

Substituting all these in the equation $0 = R_{23} = R^0{}_{203} + R^1{}_{213} + R^2{}_{223}$ and using (9.21), (16.104) and (20.16)–(20.18) we find

$$k^\rho H_{,\rho}(Z + \overline{Z}) + H\left(Z^2 + \overline{Z}^2\right) = 0. \tag{20.22}$$

Since Z obeys the equation $k^\rho Z_{,\rho} = -Z^2$ (which follows from (16.84)), the above says that $k^\rho[H/(Z + \overline{Z})]_{,\rho} = 0$, whose solution is

$$H = e^{3P}(Z + \overline{Z}), \tag{20.23}$$

where P is a real function such that $k^\rho P_{,\rho} = 0$.

---

[1] In this section, the indices of the Riemann and Ricci tensors are tetrad indices, with the tetrad vectors being labelled as in Section 16.6: $(e_0, e_1, e_2, e_3) = (k, \ell, m, \overline{m})$.

In consequence of (20.23), the equation $R_{10} = 0$ is fulfilled identically, while the equations $R_{21} = 0$ and $R_{31} = 0$ give

$$m^\rho P,_\rho = -\frac{\overline{Z}}{Z}\overline{Y},_u \equiv -\frac{\overline{Z}}{Z}\,\ell^\rho \overline{Y},_\rho, \qquad \overline{m}^\rho P,_\rho = -\frac{Z}{\overline{Z}}Y,_u \equiv -\frac{Z}{\overline{Z}}\ell^\rho Y,_\rho. \tag{20.24}$$

These two equations imply the integrability condition $\overline{m}^\rho(m^\sigma P,_\sigma),_\rho - m^\rho(\overline{m}^\sigma P,_\sigma),_\rho = -2\Gamma^s{}_{23}e_s{}^\rho P,_\rho$ (see (16.104)). Explicitly, it reads

$$S \overset{\text{def}}{=} \frac{Z}{\overline{Z}^2}Y,_u\, m^\rho \overline{Z}_\rho - \frac{Z}{\overline{Z}}m^\rho Y_{u\rho} - \frac{\overline{Z}}{Z^2}\overline{Y},_u\, \overline{m}^\rho Z_\rho + \frac{\overline{Z}}{Z}\overline{m}^\rho \overline{Y}_{u\rho}$$

$$+ \left(\frac{Z}{\overline{Z}} - \frac{\overline{Z}}{Z}\right)Y,_u\, \overline{Y},_u - \ell^\rho P,_\rho(\overline{Z} - Z) = 0. \tag{20.25}$$

The equation $R_{11} = 0$ is the most difficult to handle. The following identities are useful in calculating R_{11}; they follow from the equations established so far, with use of the commutator formula (16.104):

$$m^\rho Z,_\rho = (\overline{Z} - Z)\overline{Y},_u, \tag{20.26}$$

$$\ell^\rho Z,_\rho = Y,_u\, \overline{Y},_u - HZ^2 + \overline{m}^\rho \overline{Y}_{u\rho}, \tag{20.27}$$

and their complex conjugates. Also, it is important to note that $R_{11} = R^2{}_{121} + R^3{}_{131} \equiv -2R_{1213} = -2R^0{}_{213}$. Although R_{11} must be real, the equation $R_{1213} = R_{1312}$ does not hold identically (see the remark after Eq. (16.54)),[1] so $2R_{1213}$ is not identically equal to $R_{1213} + R_{1312}$. Calculating $R^0{}_{213} = 0$ with use of (9.21) and (20.20)–(20.27) we find

$$\overline{Z}\ell^\rho(e^{3P}),_\rho(Z + \overline{Z}) + 3H\frac{Z + \overline{Z}}{Z}Y,_u\, \overline{Y},_u + 3Z\,e^{3P}m^\rho Y_{u\rho}$$

$$-3\frac{\overline{Z}^2}{Z}e^{3P}\overline{m}^\rho \overline{Y}_{u\rho} + 3\frac{\overline{Z}^2}{Z^2}e^{3P}\overline{Y},_u\, \overline{m}^\rho Z_\rho - 3\frac{Z}{\overline{Z}}e^{3P}Y,_u\, m^\rho \overline{Z}_\rho = 0. \tag{20.28}$$

Taking the imaginary part of the above, we obtain $(Z + \overline{Z})S = 0$, where S is the expression defined in (20.25). Thus, the imaginary part of (20.28) vanishes by virtue of the integrability condition of (20.24). Using now (20.25) in (20.28) to eliminate the $m^\rho Y,_{u\rho}$ terms and recalling that $Y,_u = \ell^\rho Y,_\rho$, we obtain the last equation of the $R_{ij} = 0$ set:

$$\ell^\rho P,_\rho = -(1/Z + 1/\overline{Z})Y,_u\, \overline{Y},_u. \tag{20.29}$$

The functions Y and $\overline{Y}$ are independent, i.e. $dY \wedge d\overline{Y} \neq 0$ (because otherwise $Z = 0$; see Exercise 7). Each of $(P, Y, \overline{Y})$ obeys

$$k^\rho \phi,_\rho = 0, \qquad v^\rho \phi,_\rho = 0, \qquad v^\rho \overset{\text{def}}{=} \ell^\rho - \frac{\ell^\sigma Y,_\sigma}{Z}m^\rho - \frac{\ell^\sigma \overline{Y},_\sigma}{\overline{Z}}\overline{m}^\rho, \tag{20.30}$$

---

[1] It holds by virtue of (20.25).

where ϕ stands for any of $(P, Y, \overline{Y})$. The vectors k^α and v^α are linearly independent. This means that the gradients of P, Y and $\overline{Y}$ are orthogonal to two linearly independent vectors, i.e. that they lie in the same 2-dimensional plane within the tangent vector space at each point of the manifold. Consequently, the gradients $P_{,\alpha}$, $Y_{,\alpha}$ and $\overline{Y}_{,\alpha}$ are linearly dependent, so there exists a functional relation of the form $\psi(P, Y, \overline{Y}) = 0$, and hence P is a function of Y and $\overline{Y}$. Then, Eqs. (20.24) become

$$P_{,Y} = -\overline{Y}_{,u}/Z, \qquad P_{,\overline{Y}} = -Y_{,u}/\overline{Z}. \tag{20.31}$$

With this, and with $P = P(Y, \overline{Y})$, (20.29) is satisfied identically. Calculating the directional derivative of the first equation in (20.31) along m^μ, and of the second equation along $\overline{m}^\mu$, and using the commutator equation (16.104) to simplify the right-hand sides, we obtain $\overline{Z}P_{,Y Y} = \overline{Z}P_{,Y}{}^2$ and $ZP_{,\overline{Y}\overline{Y}} = ZP_{,\overline{Y}}{}^2$, which is equivalent to

$$(e^{-P})_{,YY} = 0 = (e^{-P})_{,\overline{Y}\overline{Y}}. \tag{20.32}$$

The solution of this is $e^{-P} = B + A_1\overline{Y} + A_2 Y + CY\overline{Y}$, where A_1, A_2, B and C are arbitrary constants. Since e^{-P} must be real, it follows that B and C must be real, while $A \overset{\text{def}}{=} \overline{A}_1 = A_2$ can be complex. Thus

$$e^{-P} = B + \overline{A}\overline{Y} + AY + CY\overline{Y}. \tag{20.33}$$

Each of the functions Y, $\overline{Y}$, Z and $\overline{Z}$ satisfies the equation

$$\mathcal{K}^\rho \phi_{,\rho} \overset{\text{def}}{=} B\phi_u + C\phi_{,v} + A\phi_{,\overline{\xi}} + \overline{A}\phi_{,\xi} = 0, \tag{20.34}$$

where ϕ is any of $(Y, \overline{Y}, Z, \overline{Z})$. It follows that P and H obey the same equation, so $\mathcal{K}^\rho g_{\alpha\beta,\rho} = 0$ for the whole metric. The components of $\mathcal{K}^\alpha$ are constant (in the coordinates currently used), so Eqs. (8.12) are satisfied. Consequently, $\mathcal{K}^\alpha$ is a Killing field – simultaneously for the metric $g_{\alpha\beta}$ and for the flat background metric $\eta_{\alpha\beta}$.

The coordinates $(u, v, \xi, \overline{\xi})$ have not yet been defined uniquely – we are still free to transform them so as to preserve the form of the flat background metric (i.e. by the Lorentz transformations). Since $\eta^{\mu\nu}$ does not change under a Lorentz transformation, the equations $k^\mu Y_{,\mu} = g^{\mu\nu}k_\nu Y_{,\mu} \equiv \eta^{\mu\nu}k_\nu Y_{,\mu} = 0 = \overline{m}^\mu Y_{,\mu} = g^{\mu\nu}\overline{m}_\nu Y_{,\mu} \equiv \eta^{\mu\nu}\overline{m}_\nu Y_{,\mu}$ still hold in unchanged form after the transformation. Let us then assume that the Killing field $\mathcal{K}^\alpha$ is timelike with respect to $\eta^{\mu\nu}$,[1] and let us carry out a Lorentz transformation after which $\mathcal{K}^\alpha$ becomes collinear with the time axis of the background Minkowski space, thus

$$A = 0, \qquad B = C > 0 \Longrightarrow \mathcal{K}^\alpha = B(1, 1, 0, 0). \tag{20.35}$$

---

[1] Kerr and Schild (1965) considered also the cases when $\mathcal{K}^\alpha$ is spacelike or null; see also Debney, Kerr and Schild (1969).

Then, Equations (20.16), (20.18) and (20.34) imply

$$Y,_{\bar{\xi}} = YY,_u, \qquad Y,_\xi = Y,_v/Y, \qquad Y,_u + Y,_v = 0. \tag{20.36}$$

Each of these is a quasi-linear partial differential equation of first order and can be solved by standard methods (Courant and Hilbert, 1965). The general solution of the whole set may be represented as

$$F(Y, \xi, \bar{\xi}, u - v) \overset{\text{def}}{=} Y^2\bar{\xi} - \xi + (u - v)Y + \Phi(Y) = 0, \tag{20.37}$$

where Φ is an arbitrary complex function of one variable.

Calculating $0 = m^\rho F,_\rho \equiv \bar{Y}F,_u - F,_\xi$ we find

$$\bar{Z} = -(1 + Y\bar{Y})/F,_Y \tag{20.38}$$

(where $F,_Y = 2Y\bar{\xi} + u - v + \Phi,_Y$). Then, calculating $0 = F,_u \equiv F,_Y Y,_u + Y$ we find, with use of (20.31) and (20.38),

$$Y,_u = -Y/F,_Y = -\bar{Z}P,_{\bar{Y}} = (1 + Y\bar{Y})P,_{\bar{Y}}/F,_Y. \tag{20.39}$$

Hence $P,_{\bar{Y}} = -Y/(1 + Y\bar{Y})$, so $e^{-P} = \text{constant} \times (1 + Y\bar{Y})$ and

$$e^{3P} = \sqrt{2}m/(1 + Y\bar{Y})^3, \tag{20.40}$$

where m is an arbitrary constant.

Equations (20.9), (20.15), (20.23), (20.40), (20.37) and (20.38) implicitly define the most general vacuum Kerr–Schild metric in which the (necessarily existing) Killing field is timelike. The definition is implicit because Y has not been given as a function of the coordinates. In order to find it, we must specify the function $\Phi(Y)$ in (20.37).

It is here that we shall make an arbitrary non-covariant assumption, in order to find an explicit solution. We assume that $\Phi(Y)$ is a polynomial of second degree of the form $\Phi(Y) = \alpha Y^2 + \beta Y - \bar{\alpha}$.

We adapted the coordinates to the Killing field $\mathcal{K}^\alpha$, but we are still free to carry out transformations preserving the direction of $\mathcal{K}^\alpha$. These are the translations and rotations in the 3-space orthogonal to $\mathcal{K}^\alpha$. By the translation $\bar{\xi} = \bar{\xi}' - \alpha$ we can then cancel the term $\alpha Y^2 - \bar{\alpha}$ in $\Phi(Y)$, and by one of the translations $u = u' - \text{Re}\,\beta$ or $v = v' + \text{Re}\,\beta$ we can cancel the real part of β. Finally, we obtain $\Phi = \sqrt{2}iaY$, where $\text{Im}\,\beta = \sqrt{2}ia$ was assumed in this form for later convenience.

With this Φ, we transform back to the real coordinates (t, x, y, z) by

$$\xi = \frac{1}{\sqrt{2}}(x + iy), \qquad u = \frac{1}{\sqrt{2}}(t + z), \qquad v = \frac{1}{\sqrt{2}}(t - z), \tag{20.41}$$

and F becomes

$$F = (x - iy)Y^2/\sqrt{2} + \sqrt{2}(z - ia)Y - (x + iy)/\sqrt{2}, \tag{20.42}$$

the function $Y(x, y, z)$ being defined by $F = 0$. In order to simplify the final expression, we introduce the real function $r(x, y, z) > 0$ by[1]

$$\frac{x^2 + y^2}{r^2 + a^2} + \frac{z^2}{r^2} = 1.$$ (20.43)

The solution of $F = 0$ is now[2]

$$Y = \frac{(r - z)(r + ia)}{r(x - iy)}.$$ (20.44)

Substituting this where necessary, we obtain

$$F_{,Y} = \frac{\sqrt{2}}{r}\left(r^2 - iaz\right), \qquad Y\overline{Y} = \frac{r - z}{r + z},$$

$$\overline{Z} = -\frac{\sqrt{2}r^2\left(r^2 + iaz\right)}{(r + z)\left(r^4 + a^2z^2\right)}, \qquad e^{3P} = \frac{\sqrt{2}m(r + z)^3}{8r^3},$$ (20.45)

$$H = -\frac{mr(r + z)^2}{2\left(r^4 + a^2z^2\right)},$$

and then substituting (20.41) in (20.15) we obtain in (20.9)

$$ds^2 = dt^2 - dx^2 - dy^2 - dz^2 - \frac{2mr^3}{r^4 + a^2z^2}\left(dt + \frac{z}{r}dz\right.$$

$$\left. + \frac{r}{r^2 + a^2}(x\,dx + y\,dy) + \frac{a}{r^2 + a^2}(x\,dy - y\,dx)\right)^2.$$ (20.46)

This is the form in which the Kerr metric first appeared in the literature,[3] but Kerr (1963) did not give any hint on how it had been derived.

Nothing in this derivation presaged the prominence that the result acquired later. The Kerr solution turned out to be extremely important from the point of view of physics. It is still the simplest exact solution of Einstein's equations that describes the exterior field of a rotating body and the spacetime around a stationary rotating black hole. Because of this, it became a basis of hundreds of papers (if not more) discussing astrophysical aspects of black holes. Also, the Kerr solution is believed to be the universal asymptotic state towards which all nonstationary uncharged black holes should evolve. This last statement is based on the still unproven **cosmic censorship conjecture** (see Section 18.16), and we shall not consider the arguments in favour of it.

---

[1] If x, y and z are interpreted as Cartesian coordinates, then the surfaces $r = $ constant are confocal ellipsoids of revolution, with a being the radius of the focal ring and $2r$ being the smallest diameter of the ellipsoid.

[2] The other solution of $F = 0$ is obtained by the replacement $r \to -r$. We discard it because we assumed $r > 0$. The value of r is the semiminor axis of the ellipsoid $r = $ constant. The resulting metric *can* be extended to the region of negative values of r, but the extension involves nontrivial problems. We will discuss it in Section 20.8.

[3] Actually, the signature in Kerr's paper was different from ours.

20.3 Basic properties

The Kerr metric becomes the Minkowski metric when $m = 0$. Also, it becomes approximately flat when r becomes large – the Kerr–Schild term then becomes negligible compared with the Lorentzian part. Thus, at large values of r the metric (20.46) obeys the assumptions made in Section 12.18, where we discussed the weak-field limit of relativity. The component

$$g_{00} = 1 - \frac{2mr^3}{r^4 + a^2 z^2} \approx 1 - \frac{2m}{r} + \frac{2ma^2 z^2}{r^5} + O(1/r^5), \qquad (20.47)$$

when compared with (12.153), allows us to recognise m as the mass of the source. The g_{0I} components, however, are not exactly of the form (12.154) because they contain terms of order $1/r$, namely

$$g_{tx} = -\frac{2mr^4 x - 2mar^3 y}{(r^2 + a^2)(r^4 + a^2 z^2)}, \qquad g_{ty} = -\frac{2mr^4 y + 2mar^3 x}{(r^2 + a^2)(r^4 + a^2 z^2)},$$

$$g_{tz} = -\frac{2mr^2 z}{r^4 + a^2 z^2}. \qquad (20.48)$$

The unwanted terms are $2mr^4 x$ and $2mr^4 y$ in g_{tx} and g_{ty}, and the whole of g_{tz}. We guess that the coordinates of (20.46) are not those that were used in deriving (12.153)–(12.155), so we need a transformation of the form (12.131) that would transform the metric as in (12.152). The simplest choice is to take $b_I = 0$ and a time-independent b_0. The following b_0 will do the job:

$$b_{0,x} = -\frac{2mr^4 x}{(r^2 + a^2)(r^4 + a^2 z^2)}, \qquad b_{0,y} = -\frac{2mr^4 y}{(r^2 + a^2)(r^4 + a^2 z^2)},$$

$$b_{0,z} = -\frac{2mr^2 z}{r^4 + a^2 z^2}. \qquad (20.49)$$

The partial derivatives of b_0 determined above identically obey the integrability conditions $b_{0,xy} = b_{0,yx}$ and $b_{0,xz} = b_{0,zx}$ (by virtue of (20.43)), so such a b_0 does exist, and the transformed g_{0I} are:

$$\tilde{g}_{tx} = \frac{2mar^3 y}{(r^2 + a^2)(r^4 + a^2 z^2)} \approx \frac{2may}{r^3} + O(1/r^3),$$

$$\tilde{g}_{ty} = -\frac{2mar^3 x}{(r^2 + a^2)(r^4 + a^2 z^2)} \approx -\frac{2max}{r^3} + O(1/r^3), \qquad (20.50)$$

$$\tilde{g}_{tz} = 0.$$

These are exactly of the form (12.154), with the angular momentum $P^I = (0, 0, -ma)$ (where $(x^1, x^2, x^3) = (x, y, z)$). Thus the parameter a is the total angular momentum of the source per unit mass. This much was said in the original paper of Kerr (1963), and the physical interpretation became immediately clear.

The coordinates of (20.46), although convenient for interpreting the parameters, are not convenient in most other calculations because of the complicated form of the function $r(x, y, z)$. Often, it is more practical to use r as one of the coordinates. One such coordinate system was introduced by Kerr (1963). It is related to the coordinates of (20.46) by

$$x = (r\cos\varphi + a\sin\varphi)\sin\vartheta \equiv \sqrt{r^2 + a^2}\sin\vartheta\cos[\varphi - \arctan(a/r)],$$

$$y = (r\sin\varphi - a\cos\varphi)\sin\vartheta \equiv \sqrt{r^2 + a^2}\sin\vartheta\sin[\varphi - \arctan(a/r)],$$

$$z = r\cos\vartheta, \tag{20.51}$$

where $r \geq 0$, $\vartheta \in [0, \pi]$ and $\varphi \in [0, 2\pi]$. The transformed metric is

$$ds^2 = dt^2 - dr^2 - 2a\sin^2\vartheta\, dr\, d\varphi - \Sigma\, d\vartheta^2 - (r^2 + a^2)\sin^2\vartheta\, d\varphi^2$$

$$- \frac{2mr}{\Sigma}\left(dt + dr + a\sin^2\vartheta\, d\varphi\right)^2,$$

$$\Sigma \stackrel{\text{def}}{=} r^2 + a^2\cos^2\vartheta. \tag{20.52}$$

In these coordinates, since the metric is independent of φ, it is seen that one more Killing vector field exists, in addition to the timelike Killing field $k^\alpha_{(1)} = \delta^\alpha_0$. It is $k^\alpha_{(2)} = \delta^\alpha_3$, with $x^3 = \varphi$. The group generated by $k^\alpha_{(1)}$ and $k^\alpha_{(2)}$ is Abelian, and it is a complete symmetry group of the Kerr metric, as can be verified by solving the Killing equations. In the Minkowski limit $m = 0$, the coordinate φ becomes the azimuthal angle, so we say that the Kerr metric is axially symmetric.

Since the (x, y, z) coordinates of (20.46) become Cartesian rectangular at infinity, they can be imagined as 'approximately Cartesian' at finite distances. Then, as already stated, the surfaces of constant r are confocal ellipsoids of revolution whose foci lie on the ring $\{z = 0, x^2 + y^2 = a^2\}$. On this ring $r = 0$ and $\vartheta = \pi/2$; it is a singularity of the coordinate system of (20.52), and, as we shall see below, a singularity of the spacetime as well. It is easily verified, using (20.51), that the following holds

$$\frac{x^2 + y^2}{a^2\sin^2\vartheta} - \frac{z^2}{a^2\cos^2\vartheta} = 1, \tag{20.53}$$

which shows that the surfaces $\vartheta = $ constant are one-sheeted hyperboloids of revolution, with foci on the same ring. The hyperboloid corresponding to $\vartheta = \pi/2$ is degenerated to the plane $z = 0$ with the ring $r = 0$ and its interior removed. Figure 20.1 shows the axial cross-section through the family of the $r = $ constant and $\vartheta = $ constant hypersurfaces.

The geometry of the $\varphi = $ constant surfaces is more complicated. To see it, let us change the variables in (20.51) as follows:

$$\xi = x\cos\varphi + y\sin\varphi, \qquad \eta = -x\sin\varphi + y\cos\varphi. \tag{20.54}$$

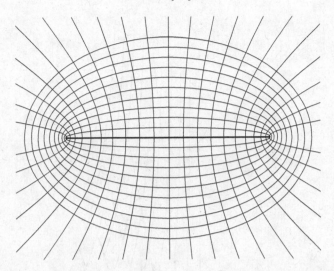

Fig. 20.1. A cross-section through the space $t = $ constant by a plane containing the axis of symmetry. The figure shows the surfaces of constant r (the ellipses) and the surfaces of constant ϑ (the hyperbolae). The ellipsoids and the hyperboloids all have their foci on the ring of radius a, seen here as the thick horizontal line. The (r, ϑ, φ) coordinates are singular on this ring; it has the equation $r = 0$, $\vartheta = \pi/2$. When $\vartheta \to \pi/2$, the hyperboloid degenerates to the $z = 0$ plane with the disc $r = 0$ removed.

It follows then from (20.51) that

$$(\xi = r \sin \vartheta, \eta = -a \sin \vartheta) \Longleftrightarrow (a/\eta)^2 - (z/\xi)^2 = 1. \tag{20.55}$$

Since $r \geq 0$ and $\vartheta \in [0, \pi]$, we have $\xi \geq 0$ and $\eta \leq 0$. The coordinates ξ and η are obtained by rotating the x- and y-axes by the angle φ. The surface (20.55) consists of two sheets that are mirror images of each other; the sheet corresponding to $z \geq 0$ is shown in Fig. 20.2. Other $\varphi = $ constant surfaces are obtained by rotating the one in Fig. 20.2 around the z-axis. The surface contains the straight half-lines $\{z = \xi \cot \vartheta, \vartheta = \text{constant}\}$ along which it intersects the $\vartheta = $ constant hyperboloids.

A still more readable form of the Kerr metric results when the **Boyer–Lindquist** (1967) **(B–L) coordinates** are introduced. They are related to those of (20.52) by[1]

$$r = r', \qquad \vartheta = \vartheta',$$

$$t = t' + 2m \int \frac{r \, dr}{\Delta_r(r)}, \qquad \varphi = -\varphi' - a \int \frac{dr}{\Delta_r(r)}, \tag{20.56}$$

$$\Delta_r(r) \overset{\text{def}}{=} r^2 - 2mr - a^2.$$

---

[1] The φ' in (20.56) and the term with $dr\,d\varphi$ in (20.37) have opposite signs to those implied by Boyer and Lindquist (1967). We have made the change in order to be consistent with the papers referred to in subsequent sections.

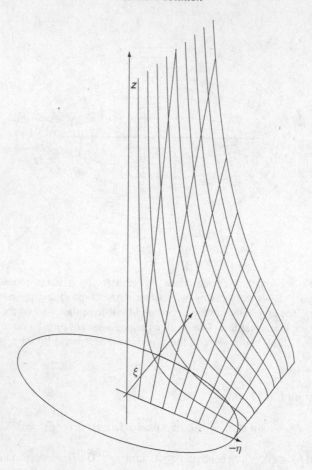

Fig. 20.2.   A surface of constant φ in the coordinates of (20.52). The tangent planes of this surface become vertical at the right edge and at $z \to \infty$. The distance from the z-axis to the right edge of the surface is a. The surface contains the straight half-lines $\{z = \xi \cot \vartheta, \xi \geq 0\}$ seen in the figure, they are intersections with the surfaces of constant ϑ. The figure shows only one half of this surface, for better clarity. The other half is the mirror reflection in the $z = 0$ plane; other surfaces of constant φ are obtained by rotating this one around the z-axis. The η-axis is drawn in reverse, so the actual coordinate on it is $(-\eta)$. The circle is the singular ring $\{r = a, \vartheta = \pi/2\}$.

The resulting metric, with primes dropped, is

$$\mathrm{d}s^2 = \left(1 - \frac{2mr}{\Sigma}\right)\mathrm{d}t^2 + \frac{4mra\sin^2\vartheta}{\Sigma}\,\mathrm{d}t\,\mathrm{d}\varphi - \frac{\Sigma}{\Delta_r}\,\mathrm{d}r^2 - \Sigma\,\mathrm{d}\vartheta^2$$

$$- \left(\frac{2mra^2\sin^2\vartheta}{\Sigma} + r^2 + a^2\right)\sin^2\vartheta\,\mathrm{d}\varphi^2. \qquad (20.57)$$

The limit $a = 0$ of this is seen to be the Schwarzschild solution.

Since the B–L coordinates (r, ϑ, φ) are orthogonal in the hypersurface $t = $ constant, the surfaces $\varphi = $ constant can be imagined as planes orthogonal to the ellipsoids and to the hyperboloids. (In fact, they are only approximately planes – they are clearly nonflat.)

The Schwarzschild metric has a spurious singularity at $r = 2m$, at which $g_{rr} \to \infty$ and $g_{00} = 0$. The coordinate t in the Schwarzschild metric (and also in the Kerr metric in the B–L coordinates) coincides with the proper time of an observer at infinity. In consequence of $g_{00} = 0$, the ratio of the proper time interval of the observer at rest in the set $g_{00} = 0$ to the corresponding proper time interval of infinity is zero, which means that the light emitted from $g_{00} = 0$ arrives to a distant observer with an infinite redshift. In the Kerr metric, the set $g_{rr} = \infty$ is not the same as the set $g_{00} = 0$. The first one, given by $r^2 - 2mr + a^2 = 0$, exists only when $a^2 \leq m^2$ and will be seen later to be a spurious singularity, removable by a coordinate transformation. Actually, when $a^2 < m^2$, the spurious singularity consists of two disjoint sets

$$r = r_{\pm} \overset{\text{def}}{=} m \pm \sqrt{m^2 - a^2}. \tag{20.58}$$

The set $r^2 - 2mr + a^2 \cos^2 \vartheta = 0$ is sometimes called an *infinite redshift hypersurface*, by analogy with the Schwarzschild metric. However, as rightly pointed out by Carter (1973), this name is misleading. An observer at rest in the B–L coordinates has the coordinates r, ϑ and φ all constant. Now, on the hypersurface $g_{00} = 0$ and inside it (where $g_{00} < 0$), for such an 'observer' the interval $ds^2 \leq 0$, which means that he/she would have to be moving with the velocity of light or faster just to remain at rest. This means that the state of rest with respect to infinity is simply impossible where $g_{00} \leq 0$ – stationary observers do not exist in that region. This is why this hypersurface has also a second, more appropriate name: the **stationary limit hypersurface**.

It is seen from (20.57) that when $a^2 < m^2$, r becomes time in those regions where $\Delta_r < 0$. Consequently, just as in the Schwarzschild solution, it is impossible to stay at constant r there. In those regions where $\Delta_r > 0$ and $g_{00} \leq 0$, timelike vectors must necessarily have a nonzero φ-component. The minimal value of the φ-component is found from $ds^2 = 0$ at $dr = d\vartheta = 0$ using (20.57); it is:

$$v_{\min}^{\varphi} = v^0 \frac{2mra - \Sigma\sqrt{\Delta_r}/\sin\vartheta}{2mra^2 \sin^2\vartheta + \Sigma(r^2 + a^2)},$$

where v^0 is the t-component of that vector. (Note that, as expected, $v_{\min}^{\varphi} \to 0$ as $g_{00} \to 0$, i.e. as $\Sigma \to 2mr$ and $\Delta_r \to a^2 \sin^2 \vartheta$.) This is an example of the phenomenon of **frame dragging** in the gravitational field of rotating bodies, which was mentioned in Section 12.18; see after Eq. (12.156).

When ϑ is near $\pi/2$, while r is negative and $|r|$ sufficiently small, the term $(2mra^2 \sin^2 \vartheta/\Sigma)$ in the component $g_{\varphi\varphi}$ of (20.57) becomes negative and large in absolute value, which means that φ becomes a timelike coordinate. Hence, the curves of constant t, ϑ and r in that region are timelike. If we require that, by continuity, these lines are closed with period 2π also in that region, then *closed timelike curves exist in the* $r < 0$ *sheet of the extended Kerr manifold.* This applies to all three varieties of the Kerr solution. This observation was reported by Boyer and Lindquist (1967), and attributed to a remark by Carter.

The existence or nonexistence of regions with $\Delta_r \leq 0$, their geometry and their relation to the regions with $g_{00} \leq 0$ will be discussed in Section 20.5. Their physical meaning will emerge gradually later.

The singularities of the Kerr metric are best recognised when, following Carter (1973), one calculates the tetrad components of the Riemann tensor in the orthonormal tetrad defined by

$$e^0 = \sqrt{\frac{\Delta_r(r)}{\Sigma(r)}} \left(dt - a \sin^2 \vartheta \, d\varphi \right), \qquad e^1 = \sqrt{\frac{\Sigma(r)}{\Delta_r(r)}} \, dr,$$

$$e^2 = \sqrt{\Sigma(r)} d\vartheta, \qquad e^3 = \sin \vartheta \frac{a \, dt - \left(r^2 + a^2 \right) d\varphi}{\sqrt{\Sigma(r)}}. \tag{20.59}$$

The tetrad components of the Riemann tensor are then

$$R_{0101} = -R_{2323} = 2I_1,$$

$$R_{0123} = 2R_{0213} = -2R_{0312} = -2I_2,$$

$$R_{0202} = R_{0303} = -R_{1212} = -R_{1313} = -I_1,$$

$$I_1 \stackrel{\text{def}}{=} mr \frac{r^2 - 3a^2 \cos^2 \vartheta}{\Sigma^3},$$

$$I_2 \stackrel{\text{def}}{=} ma \cos \vartheta \frac{3r^2 - a^2 \cos^2 \vartheta}{\Sigma^3}. \tag{20.60}$$

The singularity is located on the ring $\{r = a, \vartheta = \pi/2\}$, where $\Sigma = 0$. Since the interior of the ring is nonsingular, one can extend the metric through this set, to negative values of r. From (20.51) we see that the (t, x, y, z) coordinates cover both sides of this surface sheet. With $a^2 \leq m^2$, the spurious singularities lie in the $r > 0$ region, so the extension through the interior of the singular ring does not simply result in another copy of the same set. We will come back to this in Section 20.8.

Equations (20.60) imply that the Kerr metric is of Petrov type D.

20.4 * Derivation of the Kerr metric by Carter's method – from the separability of the Klein–Gordon equation

The Kerr metric emerged in Section 20.2 after imposing coordinate-dependent assumptions of unknown interpretation. It would be good to have an invariant definition that would single out the Kerr metric by a set of geometric properties. Such a definition is still lacking. Carter (1973) came one step closer to it by considering generalisations of the Schwarzschild metric in which the Klein–Gordon equation, in appropriately chosen coordinates, is still separable. Although this method is not unique and not invariant either, and makes ample use of the expected result, its bonus is that it allows one to generalise the Kerr metric for the cosmological constant and the electric/magnetic charge in a rather simple way. We will present this derivation now.[1]

The Klein–Gordon equation in curved space has the form

$$\frac{1}{\psi} \frac{\partial}{\partial x^\alpha} \left(\sqrt{-g} g^{\alpha\beta} \frac{\partial \psi}{\partial x^\beta} \right) - m_0^{\ 2} \sqrt{-g} = 0, \tag{20.61}$$

where m_0 is not to be confused with the mass parameter m of the Kerr metric. Equation (20.61) is called separable when the function ψ is a product of one-variable functions, $\psi = \prod_i \psi_i(x^i)$. Such a form of ψ causes the left-hand side of (20.61) to become a sum of four ordinary differential equations, each involving just one coordinate.

In considering the Klein–Gordon equation, it is convenient to use the metric co-form $(\partial/\partial s)^2 = g^{\alpha\beta} (\partial/\partial x^\alpha) (\partial/\partial x^\beta)$ instead of the metric form. For the Schwarzschild solution, introducing the coordinate $\mu \overset{\text{def}}{=} \cos \vartheta$ (to simplify the determinant of the metric), the metric co-form is

$$\left(\frac{\partial}{\partial s} \right)^2 = \frac{1}{r^2} \left[-(1 - \mu^2) \left(\frac{\partial}{\partial \mu} \right)^2 - \frac{1}{1 - \mu^2} \left(\frac{\partial}{\partial \varphi} \right)^2 \right.$$
$$\left. - \Delta_r \left(\frac{\partial}{\partial r} \right)^2 + \frac{Z_r^{\ 2}}{\Delta_r} \left(\frac{\partial}{\partial t} \right)^2 \right], \tag{20.62}$$

where $\Delta_r \overset{\text{def}}{=} r^2 - 2mr$ and $Z_r \overset{\text{def}}{=} r^2$, these quantities being introduced for later convenience. The corresponding Klein–Gordon equation is

$$\frac{r^2}{Z_r} \left\{ -\frac{1}{\psi} \frac{\partial}{\partial \mu} \left[(1 - \mu^2) \frac{\partial \psi}{\partial \mu} \right] - \frac{1}{\psi (1 - \mu^2)} \frac{\partial^2 \psi}{\partial \varphi^2} - m_0^{\ 2} r^2 \right\}$$
$$- \frac{1}{\psi} \frac{\partial}{\partial r} \left(\frac{r^2 \Delta_r}{Z_r} \frac{\partial \psi}{\partial r} \right) + \frac{1}{\psi} \frac{r^2 Z_r}{\Delta_r} \frac{\partial^2 \psi}{\partial t^2} = 0. \tag{20.63}$$

The substitution $\psi = \prod_i \psi_i(x^i)$ and multiplication by the appropriate functions allows one now to separate out the term that depends only on φ, then the term that depends only on μ, and then the terms depending only on r and t.

Now we would like to generalise the co-form (20.62) so as to preserve the separability property, but allow the Kerr metric as a subcase. We thus keep the assumption that the metric functions are independent of t and φ. We also keep the property that the co-form is proportional to the sum of a term independent of r and a term independent of μ.

---

[1] Carter (1968a) showed that the separability of the Klein–Gordon equation implies the separability of the Hamilton–Jacobi equation for the geodesics. We shall not quote the proof, since this implication will not be exploited here.

We must take into account that the Kerr metric is not diagonal and contains the $g_{t\varphi}$ terms. Consequently, we take as our first hypothesis

$$\left(\frac{\partial}{\partial s}\right)^2 = \frac{1}{Z}\left[-\Delta_\mu\left(\frac{\partial}{\partial \mu}\right)^2 - \frac{1}{\Delta_\mu}\left(Z_\mu\frac{\partial}{\partial t} + Q_\mu\frac{\partial}{\partial \varphi}\right)^2\right]$$
$$+ \frac{1}{Z}\left[-\Delta_r\left(\frac{\partial}{\partial r}\right)^2 + \frac{1}{\Delta_r}\left(Z_r\frac{\partial}{\partial t} + Q_r\frac{\partial}{\partial \varphi}\right)^2\right], \qquad (20.64)$$

where $\Delta_\mu(\mu)$, $Z_\mu(\mu)$, $Q_\mu(\mu)$, $\Delta_r(r)$, $Z_r(r)$ and $Q_r(r)$ are arbitrary functions, and Z will be determined later. The factors Δ_μ and Δ_r could be made equal to 1 by coordinate transformations and renaming the other functions, but it is more convenient to keep them and make use of the freedom later in another way. The determinant of the metric is

$$\sqrt{-g} = Z^2/|Z_r Q_\mu - Z_\mu Q_r|. \qquad (20.65)$$

The Klein–Gordon equation in the metric determined by (20.64) is

$$\frac{1}{\psi}\left[-\frac{\partial}{\partial \mu}\left(\frac{\sqrt{-g}}{Z}\Delta_\mu\frac{\partial \psi}{\partial \mu}\right) - \frac{\partial}{\partial r}\left(\frac{\sqrt{-g}}{Z}\Delta_r\frac{\partial \psi}{\partial r}\right) - \frac{\sqrt{-g}}{Z}m_0^2 Z\psi\right]$$
$$-\frac{1}{\psi}\left(Z_\mu\frac{\partial}{\partial t} + Q_\mu\frac{\partial}{\partial \varphi}\right)\left[\frac{1}{\Delta_\mu}\frac{\sqrt{-g}}{Z}\left(Z_\mu\frac{\partial \psi}{\partial t} + Q_\mu\frac{\partial \psi}{\partial \varphi}\right)\right]$$
$$+\frac{1}{\psi}\left(Z_r\frac{\partial}{\partial t} + Q_r\frac{\partial}{\partial \varphi}\right)\left[\frac{1}{\Delta_r}\frac{\sqrt{-g}}{Z}\left(Z_r\frac{\partial \psi}{\partial t} + Q_r\frac{\partial \psi}{\partial \varphi}\right)\right] = 0. \qquad (20.66)$$

The factor $\sqrt{-g}/Z$ present in each term may depend only on r and μ in consequence of the assumed symmetry. But, to achieve separation, it must be a product of a function of r by a function of μ. Then, after dividing Eq. (20.66) by $\sqrt{-g}/Z$, we can transform $r = f_1(r')$, $\mu = f_2(\mu')$, and redefine Δ_μ, Δ_r, Z_μ, Q_μ, Z_r and Q_r so as to achieve the result as if $\sqrt{-g}/Z = 1$, which, by (20.65), means that we choose

$$Z = Z_r Q_\mu - Z_\mu Q_r. \qquad (20.67)$$

However, this is only a necessary, not yet a sufficient, condition for separability; the term that may still cause problems is the one containing m_0. It will allow separability if Z has the form $Z = U_\mu(\mu) + U_r(r)$, where U_μ and U_r are functions of one variable, as yet undetermined. The derivative $Z_{,r\mu}$ must then vanish, which means, from (20.67)

$$\frac{dZ_r}{dr}\frac{dQ_\mu}{d\mu} - \frac{dZ_\mu}{d\mu}\frac{dQ_r}{dr} = 0. \qquad (20.68)$$

This can be satisfied in three ways: (1) at least one of (Z_r, Q_μ, Z_μ, Q_r) is zero; (2) Z_r and Z_μ are constant; (3) Q_r and Q_μ are constant. We shall believe Carter (1973) that case (1) leads to subcases of (2) and (3). Cases (2) and (3) are clearly equivalent, so, for better correspondence with the Schwarzschild limit, we will take

$$Q_r = C_r = \text{constant}, \qquad Q_\mu = C_\mu = \text{constant}. \qquad (20.69)$$

In this way, we have arrived at the following metric form:

$$ds^2 = \frac{\Delta_r \left(C_\mu\, dt - Z_\mu\, d\varphi\right)^2 - \Delta_\mu \left(C_r\, dt - Z_r\, d\varphi\right)^2}{C_\mu Z_r - C_r Z_\mu}$$

$$- \left(C_\mu Z_r - C_r Z_\mu\right)\left(\frac{dr^2}{\Delta_r} + \frac{d\mu^2}{\Delta_\mu}\right). \tag{20.70}$$

Now we take into account the electromagnetic field. The Klein–Gordon equation with the electromagnetic 4-potential included is

$$\frac{1}{\psi}\left(\frac{\partial}{\partial x^\alpha} - ieA_\alpha\right)\left[\sqrt{-g}\,g^{\alpha\beta}\left(\frac{\partial\psi}{\partial x^\beta} - ieA_\beta\right)\right] - m_0^2\sqrt{-g} = 0. \tag{20.71}$$

For simplicity, we assume that only the t- and φ-components of the 4-potential are nonzero, just as in the Schwarzschild case, and we assume the same symmetry as for the Kerr metric, so the nonzero components depend only on r and μ. Thus

$$A_\alpha\, dx^\alpha = A_0(r,\mu)dt + A_3(r,\mu)d\varphi. \tag{20.72}$$

Substituting this in (20.71), and taking into account the simplifications achieved in (20.70), we obtain

$$\frac{1}{\psi}\left[-\frac{\partial}{\partial\mu}\left(\Delta_\mu \frac{\partial\psi}{\partial\mu}\right) - \frac{\partial}{\partial r}\left(\Delta_r \frac{\partial\psi}{\partial r}\right) - m_0^2 Z\psi\right]$$

$$- \frac{1}{\psi}\left(Z_\mu\frac{\partial}{\partial t} + C_\mu\frac{\partial}{\partial\varphi} + ieX_\mu\right)\left[\frac{1}{\Delta_\mu}\left(Z_\mu\frac{\partial\psi}{\partial t} + C_\mu\frac{\partial\psi}{\partial\varphi} + ieX_\mu\psi\right)\right]$$

$$+ \frac{1}{\psi}\left(Z_r\frac{\partial}{\partial t} + C_r\frac{\partial}{\partial\varphi} - ieX_r\right)\left[\frac{1}{\Delta_r}\left(Z_r\frac{\partial\psi}{\partial t} + C_r\frac{\partial\psi}{\partial\varphi} - ieX_r\psi\right)\right] = 0, \tag{20.73}$$

where the following abbreviations were introduced:

$$A_0 Z_r + A_3 C_r \stackrel{\text{def}}{=} X_r, \qquad A_0 Z_\mu + A_3 C_\mu \stackrel{\text{def}}{=} -X_\mu. \tag{20.74}$$

Equation (20.73) will be separable if X_μ depends only on μ and X_r depends only on r. Assuming this, the 4-potential is

$$ZA_\alpha\, dx^\alpha = X_r\left(C_\mu\, dt - Z_\mu\, d\varphi\right) + X_\mu\left(C_r\, dt - Z_r\, d\varphi\right). \tag{20.75}$$

The constants C_μ and C_r, if they are nonzero, can be rescaled by coordinate transformations of the form $t = \alpha t'$, $\varphi = \beta\varphi'$, accompanied by redefinitions of Δ_r and Δ_μ, and by a similar rescaling of r. Both these constants could be rescaled to 1 in this way. However, in anticipation of the result we aim at, we rescale only C_μ and rename C_r as follows: $C_\mu = 1$, $C_r = a$. We thus leave out the case $C_\mu = 0$.

With this, we now proceed to the Einstein–Maxwell equations. We will include the cosmological constant and the electromagnetic field of an electric and a magnetic charge.

We will calculate the Einstein tensor using the following orthonormal tetrad of differential forms:

$$e^0 = \sqrt{\frac{\Delta_r}{Z}}\left(C_\mu\, dt - Z_\mu\, d\varphi\right), \qquad e^1 = \sqrt{\frac{Z}{\Delta_r}}\, dr, \qquad e^2 = \sqrt{\frac{Z}{\Delta_\mu}}\, d\mu,$$

$$e^3 = \sqrt{\Delta_\mu/Z}\,(C_r\, dt - Z_r\, d\varphi). \tag{20.76}$$

Using (20.75)–(20.76) and (13.10) we find

$$T_{00} = -T_{11} = T_{22} = T_{33} = \left(F_{01}{}^2 + F_{23}{}^2\right)/2, \tag{20.77}$$

where the F_{ij} are the only two tetrad components of the electromagnetic field tensor that do not vanish:

$$F_{01} = -\left[X_{r,r}\left(Z_r - aZ_\mu\right) - Z_{r,r}\left(X_r + aX_\mu\right)\right]/Z^2,$$
$$F_{23} = \left[X_{\mu,\mu}\left(Z_r - aZ_\mu\right) + Z_{\mu,\mu}\left(X_r + aX_\mu\right)\right]/Z^2. \tag{20.78}$$

For the Einstein tensor we find

$$G_{00} = -\frac{1}{2Z}\Delta_{\mu,\mu\mu} - \frac{a}{2Z^2}\Delta_{\mu,\mu}Z_{\mu,\mu} - \frac{a^2}{4Z^3}\Delta_\mu Z_{\mu,\mu}{}^2 + \frac{3}{4Z^3}\Delta_r Z_{\mu,\mu}{}^2$$
$$- \frac{a^2}{4Z^3}\Delta_\mu Z_{r,r}{}^2 - \frac{1}{Z^2}\Delta_r Z_{r,rr} + \frac{3}{4Z^3}\Delta_r Z_{r,r}{}^2 - \frac{1}{2Z^2}\Delta_{r,r}Z_{r,r}, \tag{20.79}$$

$$G_{03} = \frac{1}{2Z^2}\sqrt{\Delta_r\Delta_\mu}\left(aZ_{r,rr} + Z_{\mu,\mu\mu}\right), \tag{20.80}$$

$$G_{11} = \frac{1}{2Z}\Delta_{\mu,\mu\mu} + \frac{a}{2Z^2}\Delta_{\mu,\mu}Z_{\mu,\mu} + \frac{a^2}{4Z^3}\Delta_\mu Z_{\mu,\mu}{}^2 - \frac{1}{4Z^3}\Delta_r Z_{\mu,\mu}{}^2$$
$$+ \frac{a^2}{4Z^3}\Delta_\mu Z_{r,r}{}^2 - \frac{1}{4Z^3}\Delta_r Z_{r,r}{}^2 + \frac{1}{2Z^2}\Delta_{r,r}Z_{r,r}, \tag{20.81}$$

$$G_{22} = -\frac{a^2}{4Z^3}\Delta_\mu Z_{\mu,\mu}{}^2 + \frac{1}{4Z^3}\Delta_r Z_{\mu,\mu}{}^2 - \frac{a}{2Z^2}\Delta_{\mu,\mu}Z_{\mu,\mu}$$
$$- \frac{a^2}{4Z^3}\Delta_\mu Z_{r,r}{}^2 + \frac{1}{2Z}\Delta_{r,rr} - \frac{1}{2Z^2}\Delta_{r,r}Z_{r,r} + \frac{1}{4Z^3}\Delta_r Z_{r,r}{}^2, \tag{20.82}$$

$$G_{33} = -\frac{a}{Z^2}\Delta_\mu Z_{\mu,\mu\mu} - \frac{a}{2Z^2}\Delta_{\mu,\mu}Z_{\mu,\mu} - \frac{3a^2}{4Z^3}\Delta_\mu Z_{\mu,\mu}{}^2$$
$$+ \frac{1}{4Z^3}\Delta_r Z_{\mu,\mu}{}^2 - \frac{3a^2}{4Z^3}\Delta_\mu Z_{r,r}{}^2 + \frac{1}{2Z}\Delta_{r,rr} - \frac{1}{2Z^2}\Delta_{r,r}Z_{r,r}$$
$$+ \frac{1}{4Z^3}\Delta_r Z_{r,r}{}^2. \tag{20.83}$$

We are now going to solve the equations $G_{ij} + \Lambda g_{ij} = \kappa T_{ij}$.

The G_{03} equation is a sum of two terms, one of which depends only on μ and the other only on r. The solution is easily found to be

$$Z_r = Cr^2 + C_1 r + C_2, \qquad Z_\mu = -aC\mu^2 + C_3\mu + C_4, \tag{20.84}$$

where the C and C_i are arbitrary constants. Substituting these in the equation $G_{22} - G_{33} = 0$ we obtain

$$C_4 = C_2/a - (C_1{}^2 + C_3{}^2)/(4aC). \tag{20.85}$$

Substituting this in (20.84) we see that the transformation $\mu = \mu' + C_3/(2aC)$ and the redefinition $C_2 = aC_2' + C_1{}^2/(4C)$ have the same result as if $C_3 = 0$. Thus

$$Z_r = C[r + C_1/(2C)]^2 + aC_2', \qquad Z_\mu = C_2' - aC\mu^2. \tag{20.86}$$

The transformation $t = t' + C_2'\varphi - aC\varphi$ does not change the combination $(Z_r - aZ_\mu)$ and has the same result as if $C_2' = aC$. Then, a translation of r makes $C_1 = 0$, and the rescaling $\varphi = \varphi'/C$, $\Delta_\mu = C\Delta_\mu'$, $\Delta_r = C\Delta_r'$ has the same result as if $C = 1$. Thus finally, we have

$$Z_\mu = a(1 - \mu^2), \qquad Z_r = r^2 + a^2. \tag{20.87}$$

We now solve the Maxwell equations. The equations $F_{[\alpha\beta,\gamma]} = 0 = (\sqrt{-g}F^{1\beta})_{,\beta} = (\sqrt{-g}F^{2\beta})_{,\beta}$ (coordinate indices) are already fulfilled. The ones that remain to be solved are

$$(Z_r F_{01})_{,r} - (Z_\mu F_{23})_{,\mu} = 0, \tag{20.88}$$

$$aF_{01,r} - F_{23,\mu} = 0, \tag{20.89}$$

the indices of F_{ij} above being tetrad indices. The solution of (20.89) is

$$X_r = Dr^4 + er - aD_1, \qquad X_\mu = -a^3 D\mu^4 + q\mu + D_1, \tag{20.90}$$

where e and q are arbitrary constants. Equation (20.88) results in $D = 0$. The constant D_1 does not enter the electromagnetic field tensor, and can thus be assumed zero with no loss of generality. The final solution of the Maxwell equations is then $X_r = er$, $X_\mu = q\mu$.

We now go back to the Einstein equations. The equation $G_{00} - G_{22} = -2\Lambda$ is equivalent to a function of μ being equal to a function of r, and thus being constant. Calling this constant E and integrating, we obtain

$$\Delta_r = \frac{1}{3}\Lambda r^4 + Er^2 + 2mr + E_2, \qquad \Delta_\mu = \frac{1}{3}\Lambda a^2\mu^4 - E\mu^2 + E_3\mu + E_4, \tag{20.91}$$

where m and E_i are arbitrary constants. The last equation to solve is

$$G_{22} - \Lambda = (e^2 + q^2)/(r^2 + a^2\mu^2)^2, \tag{20.92}$$

and it leads to $E_2 - a^2 E_4 = e^2 + q^2$.

With this, the Einstein–Maxwell equations are solved. However, a few conditions have to be imposed on the solution in order that the metric is physically reasonable. One of them is $E_3 = 0$ – otherwise there would be a term proportional to $\mu = \cos \vartheta$ in the metric and the field would not be mirror-symmetric with respect to the equatorial plane. To avoid a singularity at the axis $\vartheta = 0$, we require $E = 1 + \Lambda a^2/3$, and to obtain the correct Schwarzschild limit we require $E_4 = 1$. Thus

$$\Delta_\mu = \left(1 - \frac{1}{3}\Lambda a^2 \mu^2\right)\left(1 - \mu^2\right),$$

$$\Delta_r = \frac{1}{3}\Lambda r^2 (r^2 + a^2) + r^2 - 2mr + a^2 + e^2 + q^2. \tag{20.93}$$

The final result for the metric is

$$\begin{aligned} \mathrm{d}s^2 = {} & \frac{\Delta_r}{r^2 + a^2 \cos^2 \vartheta} \left(\mathrm{d}t - a \sin^2 \vartheta \mathrm{d}\varphi\right)^2 \\ & - \frac{\sin^2 \vartheta \left(1 - \frac{1}{3}\Lambda a^2 \cos^2 \vartheta\right)}{r^2 + a^2 \cos^2 \vartheta} \left[a\,\mathrm{d}t - \left(r^2 + a^2\right)\mathrm{d}\varphi\right]^2 \\ & - \left(r^2 + a^2 \cos^2 \vartheta\right)\left(\frac{\mathrm{d}r^2}{\Delta_r} + \frac{\mathrm{d}\vartheta^2}{1 - \frac{1}{3}\Lambda a^2 \cos^2 \vartheta}\right). \end{aligned} \tag{20.94}$$

This agrees with the result of Carter (1973) up to a constant factor in the first two terms[1] that can be recovered by a simple rescaling of t and φ. A result equivalent to (20.94) under a coordinate transformation was obtained by Frolov (1974).[2]

It may be verified that the algebraic structure of the Weyl tensor for the metric (20.94) is still the same as in (20.60), so the generalised Kerr metric is still of Petrov type D.

Comparing (20.94) with the spherically symmetric limit ($a = 0$, Section 14.4) and the proper Kerr limit ($\Lambda = 0 = e = q$, Eq. (20.57)) we recognise e and q as the electric and magnetic charges, respectively,[3] and a as the angular momentum per unit mass (while, of course, m is the mass of the source and Λ is the cosmological constant).

The farthest-reaching generalisation (so far) of the Kerr solution was obtained by Debever, Kamran and McLenaghan (1983, 1984). It contains 13 arbitrary constants, for most of which no physical or geometrical interpretation has been provided. Among them, in addition to the parameters contained in (20.94), it includes the acceleration of the source (Stephani et al., 2003). It is still of Petrov type D, and has the same Abelian symmetry group. The paper by Debever, Kamran and McLenaghan (1984) and the book by Stephani et al. (2003) contain lists of specialisations of this metric that recover the earlier-known solutions.

Historically, the generalisation for electric charge was discovered by Newman et al. (1965) by a procedure equally mysterious as the derivation of the Kerr metric itself. They allowed the constants and coordinates in the Reissner–Nordström solution to be complex

---

[1] Another difference is that Carter used the signature $(+++-)$.
[2] We are grateful to Valeri Frolov for demonstrating the equivalence in a letter to A. K., long ago.
[3] As remarked in Section 14.4, the magnetic charge may be removed by a duality rotation, so the generalisation with respect to the case $q = 0$ is rather illusory.

and applied such a complex coordinate transformation that the result of it was real – and was a generalisation of the Kerr metric for electric charge. The trick is known to work also in other cases, leading from given solutions of the Einstein or Einstein–Maxwell equations to other solutions, but nobody knows why it works, and why only in these particular cases – see Stephani *et al.* (2003) for more information. This metric (the limit $\Lambda = 0$ of (20.94)) is called the **Kerr–Newman solution**. The first author to generalise the Kerr metric for the cosmological constant was Carter (1968a); that paper in fact contains a 10-parameter electrovacuum solution and a list of possible specialisations of it.

20.5 The event horizons and the stationary limit hypersurfaces

We come back now to the proper Kerr solution, with $\Lambda = e = q = 0$, and we will consider this case to the end of this chapter.

We have already noted that the hypersurfaces given, in the B–L coordinates, by (20.58), if they exist, play a special role. We will call them **event horizons** – we will see later that they are indeed event horizons. We also noted that the stationary limit hypersurfaces, given by

$$r^2 - 2mr\cos\vartheta + a^2\cos^2\vartheta = 0 \implies r = m \pm \sqrt{m^2 - a^2\cos^2\vartheta} \tag{20.95}$$

play a special role. We will now consider the shapes of these two families of hypersurfaces and their relation to each other.

When $a^2 < m^2$, there are two event horizons, the one at $r = r_-$ is contained inside the one at $r = r_+$. There are also two stationary limit hypersurfaces in this case. The outer one, at $r = m + \sqrt{m^2 - a^2\cos^2\vartheta}$, envelops the outer event horizon, and is tangent to it only at the axis, $\vartheta = 0$ and $\vartheta = \pi$. The inner stationary limit surface, at $r = m - \sqrt{m^2 - a^2\cos^2\vartheta}$, lies all within the inner event horizon and is tangent to it also only at the axis. It is tangent to the disk $r = 0$ at its singular edge. The geometry of these four hypersurfaces is shown in Fig. 20.3.

As $a \to 0$, the Kerr solution tends to the Schwarzschild solution. Then, the inner stationary limit surface and the inner event horizon both shrink to a point, together with the disc $r = 0$. The outer stationary limit surface and the outer event horizon coalesce and go over into the Schwarzschild horizon at $r = 2m$.

As $a \to m$, the two event horizons approach each other to meet at $r = m$ when $a^2 = m^2$. The concave regions of the outer stationary limit surface shrink to points, and the surface becomes conical in their neighbourhoods, the vertices of the cones touching the event horizon. Similarly, the neighbourhoods of those points of the inner stationary limit surface that lie on the axis of symmetry become conical, the vertices being common with the outer surface. The geometry of these surfaces in the case $a^2 = m^2$ is shown in Fig. 20.4.

When $a^2 > m^2$, the event horizon disappears completely. This case, similarly to the previous one, has no Schwarzschild limit. The conical points of the stationary limit surface change to open holes, and the two stationary limit surfaces become parts of one surface

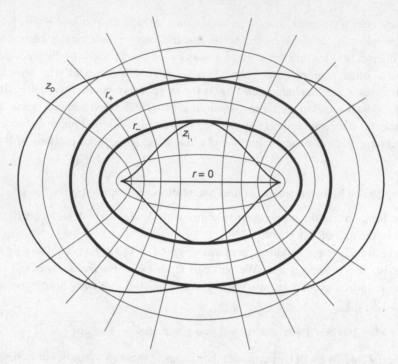

Fig. 20.3.   An axial cross-section through the space $t = \text{constant}$ in the Kerr metric, in the Boyer–Lindquist coordinates, in the case $a^2 < m^2$. The surfaces $r = r_+$ and $r = r_-$ are disjoint; the outer stationary limit surface ('z_o' in the figure) is tangent to $r = r_+$ at the axis of symmetry. Likewise, the inner stationary limit surface (z_i) is tangent to $r = r_-$ at the axis. The disc $r = 0$ is seen as the horizontal line; the surface z_i has a sharp edge, tangent to the disc. When $|a|/m$ decreases, the surface $r = r_-$ approaches the disc $r = 0$, whereas the surface $r = r_+$ recedes from the disc and becomes more spherical. When $a \leftrightarrow 0$, the disc $r = 0$ and the surfaces z_i and $r = r_-$ all collapse to a single point, while the surfaces z_o and $r = r_+$ coalesce at $r = 2m$ and become spherical.

that has the topology of a torus. The hole in the surface around the axis of symmetry is the larger the greater the difference $a^2 - m^2$. The surface still has an edge at $r = 0$, where it is tangent to the disc $r = 0$, but now the 'outer' and the 'inner' parts of it join smoothly at two rings, given by $(r = m, \vartheta = \arccos(m/a))$ and $(r = m, \vartheta = \pi - \arccos(m/a))$. The geometry of this case is shown in Fig. 20.5.

A comment must be added here. We have already alluded to the possibility of extending the Kerr spacetime to negative values of the Boyer–Lindquist r-coordinate (in the footnote to Eq. (20.44)), and such an extension will be discussed in Section 20.8. The following question then arises: where should the region with $r < 0$ be placed in Figs. 20.3–20.5? In the theory of analytic functions, one deals with the problem of multi-valued mappings by introducing *Riemann surfaces*: multi-sheeted manifolds such that to each value of a given function there corresponds only a single point of the Riemann surface (see, for example, Knopp (1996)). In our present case, one has to imagine that each of the planes of Figs. 20.3–20.5 branches out into two sheets at the disc $r = 0$. By going through the interior of the ring $\{r = 0, \vartheta = \pi/2\}$ from above, one does not reach the lower half of the figure, but a sheet with $r < 0$. There are in fact two such sheets. To enter the second one from the first, one has to go back up through the interior of the ring, go around the ring on the right or on the left, and then go through the disc $r = 0$ from below. Note that by going around the ring in this way one has to intersect the inner horizon $r = r_-$ (in Fig. 20.3) or *the* horizon $r = m$ (in Fig. 20.4) twice, each time a different half of the horizon, and the two halves meet only at the ring singularity. This agrees with the picture of the maximally extended Kerr manifold shown in Fig. 20.14: there are two disjoint $r < 0$ sheets, and to proceed from one to the other one has to go through the $r = r_-$ horizon at least twice.

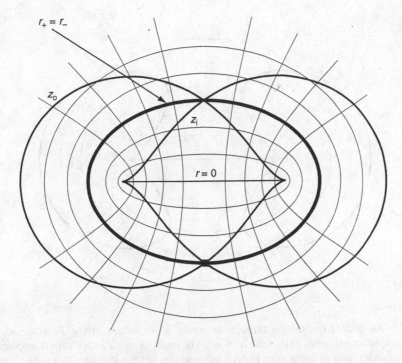

Fig. 20.4. An axial cross-section through the space $t = $ constant in the Kerr metric, in the Boyer–Lindquist coordinates, in the case $a^2 = m^2$. There is now a single surface $r = r_+ = r_-$, the surfaces z_o and z_i both touch it at the axis (but are not tangent to it – they approach it at nonzero angle). As $a \to m$, the concave areas on the z_o surface from Fig. 20.3 shrink to points, and the neighbourhoods of those points become conical. The z_i surface becomes conical at the axis as well.

Figure 20.6 shows the 3-dimensional subspace of the Kerr spacetime with $a^2 < m^2$ given by setting $\vartheta = \pi/2$ in (20.57). The vertical axis in Fig. 20.6 is time. The top part of Fig. 20.6 shows a perspective view; the bottom part shows the view from above. The direction of rotation is clockwise ($a < 0$), so that the X-axis in the top figure would move towards the viewer. The outermost ring is the stationary limit surface at $r = 2m$, the two middle rings are the event horizons $r_\pm$ and the innermost ring is the inner stationary limit surface that, because of $\vartheta = \pi/2$, coincides with the ring singularity at $r = 0$. Several future light cones are shown. For $r \gg 2m$ they look like slightly deformed Minkowski light cones. At $r = 2m$, the light cone has one generator parallel to the t-axis: no timelike vector at that point can have a zero φ-component.[1] As we move from $r = 2m$ towards $r = r_+$, the cones lean forward further and become thinner in all directions. The limit $r \to r_+$ is discontinuous. As $r \to r_+$ from the $r > r_+$ side, the cones tend to a single beam

---

[1] Because of $g_{t,\varphi} \neq 0$, the time axis is not orthogonal to the t – constant hypersurface, but it is drawn as orthogonal for better readability.

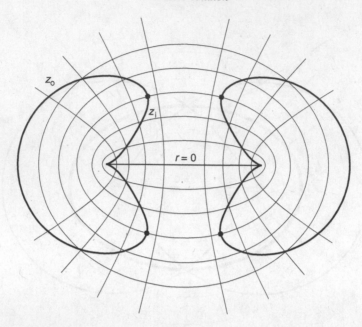

Fig. 20.5. An axial cross-section through the space $t = \text{constant}$ in the Kerr metric, in the Boyer–Lindquist coordinates, in the case $a^2 > m^2$. The spurious singularities have disappeared, and the two stationary limit surfaces have merged into one that has the topology of a torus. The points mark the places where z_o (the surface with the plus sign in (20.95)) and z_i meet; this happens at $(r = m, \vartheta = \vartheta_0 \overset{\text{def}}{=} \arccos(m/a))$ and $(r = m, \vartheta = \pi - \vartheta_0)$.

along the Y-direction in the $T = 0$ plane. As $r \to r_+$ from the $r < r_+$ side, the cones tend to the whole plane $X = \sqrt{r_+^2 + a^2} = \text{constant}$ (not shown in Fig. 20.6 for the sake of better clarity). With $r > r_+$, the intersections of the cones with a $T = \text{constant}$ plane are ellipses that recede to $Y \to +\infty$ as $r \to r_+$. With $r_- < r < r_+$, the intersections of the cones with the $X = \text{constant}$ planes are ellipses elongated and rotated as shown in the inset; their axes both become infinite as $r \to r_+$. A similar discontinuity exists at $r = r_-$; the intersections of the cones with a $T = \text{constant}$ plane again become ellipses in the region $r < r_-$.

Since at $r = r_+$ the cone degenerates to the plane $x = \sqrt{r_+^2 + a^2} = \text{constant}$, no timelike vector attached there can have a zero r-component; in fact r takes over as the time coordinate here. Inside $r = r_+$, it is impossible for an ingoing timelike or null curve to turn back without becoming spacelike along an arc or having a non-differentiable reflection. This shows that $r = r_+$ is indeed an event horizon. In Fig. 20.6, ingoing and outgoing null curves cross at $r = r_+$, but the B–L coordinates give a false picture here. In reality, there are two event horizons at $r = r_+$. The Kruskal diagram (Fig. 14.6) gives a good picture of the situation, and we will deal with the corresponding extension for the Kerr manifold in Section 20.8. For $r < r_-$, the cones become similar to those in the sector $(r_+, 2m)$. As

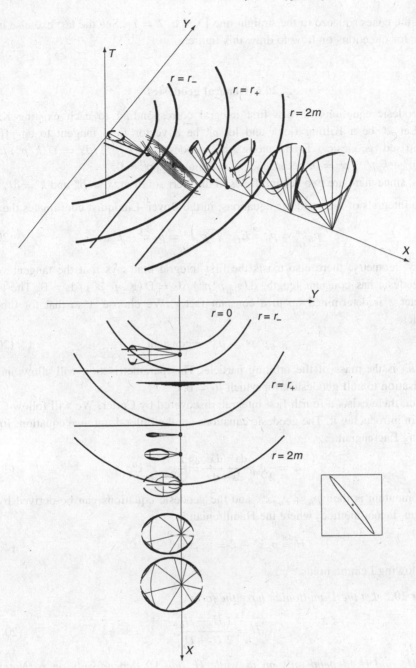

Fig. 20.6. A general perspective view (top graph) and a view along the time axis from above (bottom graph) of the Kerr subspace $\vartheta = \pi/2$ in the coordinates of (20.57). Explanation is given in the text. Large dots mark the positions of the vertices of the cones in the $T = 0$ plane. The inset is an $X = $ constant intersection of the light cone from the sector $r \subset (r_-, r_+)$.

$r \to 0$, the cones squeeze to the straight line $\{X = a, T = Y\}$. See the last exercise in this chapter for directions on how to draw this figure.

20.6 General geodesics

The geodesic equations have a first integral corresponding to each existing Killing field. Let k^α be a Killing field, and let p^α be a vector field tangent to an affinely parametrised geodesic. Then, along the geodesic, $d(k_\alpha p^\alpha)/ds = D(k_\alpha p^\alpha)/ds = k_{\alpha;\beta} p^\alpha p^\beta + k_\alpha p^\alpha{}_{;\beta} p^\beta = 0$, because $k_{(\alpha;\beta)} = 0$ and $p^\alpha{}_{;\beta} p^\beta = 0$.

Thus, since there are two Killing fields in the Kerr solution, $k^\alpha = \delta^\alpha_0$ and $k^\alpha = \delta^\alpha_3$, there are two integrals of the geodesic equations; in the Boyer–Lindquist coordinates they are

$$p_\alpha \underset{(t)}{k^\alpha} = p_0 \overset{\text{def}}{=} E, \qquad p_\alpha \underset{(\varphi)}{k^\alpha} = p_3 \overset{\text{def}}{=} -L_z. \tag{20.96}$$

In every geometry, there also exists the first integral that says that the tangent vector to a geodesic has constant length, $d\left(g_{\alpha\beta} p^\alpha p^\beta\right)/ds = D\left(g_{\alpha\beta} p^\alpha p^\beta\right)/ds = 0$. The affine parameter s is determined up to a constant factor. We choose it so that for timelike geodesics

$$g_{\alpha\beta} p^\alpha p^\beta = \mu_0{}^2 = \text{constant}, \tag{20.97}$$

where μ_0 is the mass of the orbiting particle. This parametrisation will allow an easy specialisation to null geodesics, for which $\mu_0 = 0$.

In fact, there exists a fourth first integral, discovered by Carter. We will follow Carter (1973) in introducing it. The geodesic equations are the Euler–Lagrange equations for the following Lagrangian:

$$\mathcal{L} = \frac{1}{2} g_{\alpha\beta} \frac{dx^\alpha}{ds} \frac{dx^\beta}{ds} \overset{\text{def}}{=} \frac{1}{2} g_{\alpha\beta} \dot{x}^\alpha \dot{x}^\beta. \tag{20.98}$$

The momentum is then $p_\alpha = g_{\alpha\beta} \dot{x}^\beta$, and the geodesic equations can be derived by the Hamilton–Jacobi method, where the Hamiltonian is

$$H \overset{\text{def}}{=} p_\alpha \dot{x}^\alpha - \mathcal{L} = g_{\alpha\beta} \dot{x}^\alpha \dot{x}^\beta / 2 = \mathcal{L}. \tag{20.99}$$

The following Lemma holds:

Lemma 20.2 *Let the Hamiltonian have the form*

$$H = \frac{1}{2} \frac{H_r + H_\mu}{U_r + U_\mu}, \tag{20.100}$$

where H_μ and U_μ depend only on μ, while H_r and U_r depend only on r. (Note: the functions depend also on the components of momentum, which are variables independent of the coordinates.) Then

$$K \overset{\text{def}}{=} -\frac{U_r H_\mu - U_\mu H_r}{U_r + U_\mu} \tag{20.101}$$

commutes (in the sense of vanishing Poisson bracket) with the Hamiltonian, and thus is a constant of the motion.

The proof follows by simple calculation.

The Kerr Hamiltonian with $\Lambda = 0$ is

$$H = \frac{-\Delta_\mu p_\mu^{\,2} - \Delta_\mu^{-1} \left(Z_\mu p_t + C_\mu p_\varphi\right)^2}{2\left(C_\mu Z_r - C_r Z_\mu\right)}$$

$$+ \frac{-\Delta_r p_r^{\,2} + \Delta_r^{-1} \left(Z_r p_t + C_r p_\varphi\right)^2}{2\left(C_\mu Z_r - C_r Z_\mu\right)}. \tag{20.102}$$

Substituting whatever necessary in (20.102) and (20.101), and using (20.52), we obtain for the fourth constant of the motion

$$K = a^2 \sin^2 \vartheta \left(\frac{\Sigma}{\Delta_r} \dot{r}^2 - \frac{\Delta_r}{\Sigma} \left(a \sin^2 \vartheta \dot{\varphi} - \dot{t}\right)^2 \right)$$

$$+ \left(r^2 + a^2\right) \left(\Sigma \dot{\vartheta}^2 + \frac{\sin^2 \vartheta}{\Sigma} \left[\left(r^2 + a^2\right) \dot{\varphi} - a \dot{t}\right]^2 \right). \tag{20.103}$$

The constants defined in (20.96) and (20.97) are

$$E = \frac{r^2 - 2mr + a^2 \cos^2 \vartheta}{\Sigma} \dot{t} + \frac{2mra \sin^2 \vartheta}{\Sigma} \dot{\varphi}, \tag{20.104}$$

$$L_z = -\frac{2mra \sin^2 \vartheta}{\Sigma} \dot{t} + \frac{\left[-\Delta_r a^2 \sin^2 \vartheta + \left(r^2 + a^2\right)^2\right] \sin^2 \vartheta}{\Sigma} \dot{\varphi}, \tag{20.105}$$

$$\mu_0^{\,2} = -\Sigma \left(\frac{\dot{r}^2}{\Delta_r} + \dot{\vartheta}^2 \right) - \frac{\sin^2 \vartheta}{\Sigma} \left[\left(r^2 + a^2\right) \dot{\varphi} - a \dot{t}\right]^2 + \frac{\Delta_r}{\Sigma} \left(a \sin^2 \vartheta \dot{\varphi} - \dot{t}\right)^2. \tag{20.106}$$

These equations can now be solved for the velocities, which allows some general discussion. We introduce the following new functions:

$$R(r) \stackrel{\text{def}}{=} -K \Delta_r - \mu_0^{\,2} \left(r^2 + a^2\right) \Delta_r + \left[\left(r^2 + a^2\right) E - a L_z\right]^2, \tag{20.107}$$

$$\Theta(\vartheta) \stackrel{\text{def}}{=} K - \mu_0^{\,2} a^2 \sin^2 \vartheta - \sin^2 \vartheta \left(aE - \frac{L_z}{\sin^2 \vartheta} \right)^2. \tag{20.108}$$

The components of the velocity are then given by

$$\Sigma^2 \dot{r}^2 = R(r), \tag{20.109}$$

$$\Sigma^2 \dot{\vartheta}^2 = \Theta(\vartheta), \tag{20.110}$$

$$\Sigma \dot{\varphi} = \left(\frac{1}{\sin^2 \vartheta} - \frac{a^2}{\Delta_r} \right) L_z + \frac{2mra}{\Delta_r} E, \tag{20.111}$$

$$\Sigma \dot{t} = -\frac{2mra}{\Delta_r} L_z + \left(\frac{\left(r^2 + a^2\right)^2}{\Delta_r} - a^2 \sin^2 \vartheta \right) E. \tag{20.112}$$

This can be disentangled still further. From (20.109) and (20.110) we see that $\Sigma\dot\vartheta/\sqrt{\Theta(\vartheta)} = \Sigma\dot r/\sqrt{R(r)} = 1$. We take the quotient of these two equations to obtain (20.113), and their sum to obtain (20.114) below. In deriving (20.114) we substitute for Σ and group the terms so that each integrand depends on just one coordinate. The result is

$$\int \frac{d\vartheta}{\sqrt{\Theta}} = \int \frac{dr}{\sqrt{R}} = Q = \text{constant}, \tag{20.113}$$

$$\lambda - \lambda_0 = \int \frac{a^2 \cos^2\vartheta d\vartheta}{\sqrt{\Theta}} + \int \frac{r^2\, dr}{\sqrt{R}}, \tag{20.114}$$

where λ is the affine parameter on the geodesic. The two equations above determine $r(\lambda)$ and $\vartheta(\lambda)$ implicitly.

In order to disentangle (20.111)–(20.112) in a similar way, we note from (20.109)–(20.110) that $1/\Sigma = \dot\vartheta/\sqrt{\Theta(\vartheta)} = \dot r/\sqrt{R(r)}$, and use one or the other of these equations to obtain an integrand that is a function of r only, or one that is a function of ϑ only. The result is

$$\varphi - \varphi_0 = \int \frac{L_z\, d\vartheta}{\sin^2\vartheta\sqrt{\Theta}} + \int \frac{2mraE - a^2 L_z}{\Delta_r\sqrt{R}}\, dr, \tag{20.115}$$

$$t - t_0 = \int \frac{\left(r^2 + a^2\right)^2 E - 2mraL_z}{\Delta_r\sqrt{R}}\, dr - \int \frac{a^2\sin^2\vartheta E}{\sqrt{\Theta}}\, d\vartheta. \tag{20.116}$$

In this form, the geodesic equations can be solved and investigated numerically. The equations of null geodesics follow as the subcase $\mu_0 = 0$.

Without numerical integration, only a few qualitative conclusions may be drawn. For example, if an orbit lies in the equatorial plane ($\vartheta = \pi/2$ and $\dot\vartheta = 0$ on the whole orbit) or is tangent to that plane at one point ($\vartheta = \pi/2$ and $\dot\vartheta = 0$ only at that point), then

$$C \overset{\text{def}}{=} K + \mu_0^2 a^2 - (aE - L_z)^2 = 0. \tag{20.117}$$

For the orbits that cross the equatorial plane at a nonzero angle, $C > 0$. (For the equations to be well defined, R and Θ must be non-negative.) For the orbits that never go through $\vartheta = \pi/2$, C may be negative. For the orbits that run along the symmetry axis, $L_z = 0$.

A detailed investigation of arbitrary timelike geodesics in the Kerr solution, and also in its generalisation for the cosmological constant, was published recently by Kraniotis (2004). The discussion includes physical effects, such as frame dragging, in the proper ($\Lambda = 0$) Kerr metric, and it has been announced that a similarly detailed discussion of the physical effects in the presence of Λ will be published in the future.

20.7 Geodesics in the equatorial plane

This section is based on the papers by Bardeen (1973) and by Boyer and Lindquist (1967).

In the spherically symmetric case each timelike or null geodesic lies in a plane, and then the coordinates are chosen so that it is the equatorial plane. In the Kerr spacetime, which is mirror-symmetric with respect to the plane $\vartheta = \pi/2$ (in the B–L coordinates), but not spherically symmetric, we do not have that freedom. The equatorial plane is $\vartheta = \pi/2$, and a given geodesic either lies in it or does not. Equations (20.109)–(20.116) simplify a lot when it does, and we will consider this case now.

Consider Eq. (20.109) in the case $\vartheta = \pi/2$, $\dot\vartheta = 0$. The regions accessible for motion are determined by $R(r) \geq 0$. It is more convenient to treat $R(r)$ as a function of E. For motions in the equatorial plane, $C = 0$ in (20.117), which allows us to express K through the other constants. Substituting such K in (20.107) we obtain

$$\frac{1}{r}R(r) = \left[r\left(r^2 + a^2\right) + 2ma^2\right]E^2 - 4amL_zE - (r - 2m)L_z^2 - r\Delta_r\mu_0^2. \tag{20.118}$$

The discriminant of this, with respect to E, is

$$\delta = 4\Delta_r\left\{r^2L_z^2 + \mu_0^2r\left[r\left(r^2 + a^2\right) + 2ma^2\right]\right\}, \tag{20.119}$$

and for each $r > 0$ its sign is the same as the sign of Δ_r. Hence, R has zeros only in those regions where $\Delta_r \geq 0$. Before we conclude what this means, note that the component p^0 of the tangent vector to the geodesic must be positive: since $p^0 \propto dx^0/ds$, the opposite would mean that we follow the motion backward in time. From (20.96) and (20.57) we find

$$p^0 = g^{00}E + g^{03}L_z = B(E - \omega L_z)/(\Delta_r\Sigma), \tag{20.120}$$

where

$$B \overset{\text{def}}{=} \left(r^2 + a^2\right)^2 - \Delta_r a^2 \sin^2\vartheta, \qquad \omega \overset{\text{def}}{=} 2amr/B. \tag{20.121}$$

Hence, $p^0 > 0 \Longleftrightarrow E > \omega L_z$, which, in the plane $\vartheta = \pi/2$, reduces to

$$E > 2amL_z/D, \qquad D \overset{\text{def}}{=} r\left(r^2 + a^2\right) + 2ma^2. \tag{20.122}$$

The two roots of the equation $R(r) = 0$ are

$$E_\pm = \frac{1}{D}\left(2amL_z \pm \sqrt{\Delta_r\left(r^2L_z^2 + \mu_0^2rD\right)}\right). \tag{20.123}$$

We see that the root E_- does not obey (20.122). Thus, the motion in the equatorial plane can occur only at values of $E > E_+$.

Now consider the sign of Δ_r. If $a^2 > m^2$, then $\Delta_r > 0$ for all values of r, and there exists no hypersurface analogous to the Schwarzschild horizon at $r = 2m$. This means that there are no obstacles to communication with the singular ring at $\{r = 0, \vartheta = \pi/2\}$. This is the case of **naked singularity**. The prevailing opinion in the astrophysical community is that real astronomical objects, when they are about to collapse to a black hole state, must somehow get rid of the excess angular momentum to achieve $a^2 < m^2$. Nevertheless, our familiar celestial bodies, like the Sun and the Earth, have $a^2 > m^2$ (verification of this is left as an exercise for the reader).

If $a^2 < m^2$, then $\Delta_r > 0$ for $r < r_-$ and for $r > r_+$, where the $r_\pm$ are given by (20.58). Between r_- and r_+, the discriminant (20.119) is negative, which means that $R(r)$ cannot have any zeros there, which in turn means that $dr/ds \neq 0$. Hence, there exist no circular orbits in that region and no turning points for other orbits. An orbiting body that enters the region (r_-, r_+) from the side of $r > r_+$ must keep decreasing its r until it flies through the hypersurface $r = r_-$. Unlike in the Schwarzschild case, however, it does not have to hit the singularity, since the orbit can have a turning point at $r < r_-$. Conversely, a body that entered the region (r_-, r_+) from the side of $r < r_-$ must keep increasing its r until it flies through the hypersurface $r = r_+$. Thus, the region $r_- < r < r_+$ is analogous to the region $r < 2m$ in the Schwarzschild spacetime and to the respective region in the Reissner–Nordström spacetime. The analogy with the R–N spacetime is far-reaching; see Section 20.8.

More information about the orbits follows from the graph of the function $E_{\min}(r) = E_+(r)$; see Fig. 20.7. We consider only the case $a^2 < m^2$ and the region $r > r_+$, so the curves begin at $r = r_+$. Each orbit has a constant $E \geq E_{\min}(r) = E_+(r)$. Thus, the area accessible for motion is above the graph of $E_{\min}(r)$, and the value of E determines the allowed range of r on the orbit. Note that $E_{\min}/\mu_0 \underset{r \to \infty}{\to} 1$, independently of the value of L_z. When $L_z = 0$, the function $E_{\min}/\mu_0$ has a positive derivative for all $r \geq r_+$, and so is smaller than 1 for all $r \geq r_+$. However, if $|L_z|$ is sufficiently large, then there exists an interval (r_1, r_2), with $r_+ \leq r_1 < r_2$, in which $E_{\min}/\mu_0 \geq 1$, and $E_{\min}/\mu_0 < 1$ for $r > r_2$ (see Exercise 11), independently of the sign of L_z. Thus, with $|L_z|$ sufficiently large, $E_{\min}/\mu_0$ necessarily has a local maximum at some $r = r_u > r_+$ and a local minimum at $r = r_s > r_u$, just as is shown in Fig. 20.7. The region where $E_{\min}/\mu_0 < 1$ is the locus of

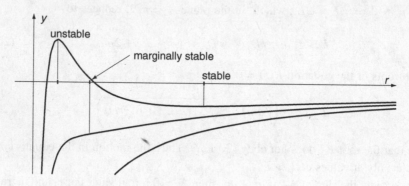

Fig. 20.7.   Graphs of the function $y(r) \overset{\text{def}}{=} E_{\min}(r)/\mu_0 - 1$ for different values of L_z. For every value of L_z, $y(r) \to 0$ as $r \to \infty$. The lower graph corresponds to $L_z = 0$; then the function is all monotonic. The upper graph corresponds to $|L_z|$ being large – the function $y(r)$ then has a maximum that determines the position of the unstable circular orbit and a minimum that determines the stable circular orbit. The middle graph approximately corresponds to such an intermediate value of L_z, at which the maximum and the minimum coalesce into a single point where $d^2y/dr^2 = 0$. This point then determines the position of the marginally stable circular orbit. The curves begin at $r = r_+$; the y-axis is drawn at approximately this value of r.

bound orbits, and $r = r_s$ is the radius of the stable circular orbit (i.e. there exists one for each sufficiently large value of $|L_z|$, for each sign of L_z). A circular orbit also exists at $r = r_u$, but it is unstable.

Note in (20.123) that a and L_z may have opposite signs. When $aL_z > 0$, the orbital angular momentum of the orbiting body and the internal angular momentum of the source of the gravitational field have the same sense. Such orbits are called **direct**. When $aL_z < 0$, the two angular momenta have opposite senses. Such orbits are called **retrograde**. The value of E_{min} at a given r is different for each of these orbits. The difference between them is a relativistic effect: the Newtonian gravitational field does not 'feel' in which direction the central body is rotating. It is only sensitive to the asymmetries in the gravitational field created by asymmetries (polar flattening) in the central body caused by the centrifugal force. Retrograde orbits in Newton's theory are exactly the same as direct orbits.

The qualitative difference between them in relativity is in fact quite pronounced. Note from (20.123) that with $aL_z < 0$ and r sufficiently close to r_+ (i.e. with Δ_r close to zero) $E_{min} < 0$, so E can be negative. This is the total energy 'at infinity', with the energy equivalent of the rest mass included.[1] Thus, $E < 0$ means that the whole energy contained in the particle is insufficient to actually send it to infinity. In other words, the energy that the particle must have lost on entering such an orbit was greater than the rest energy. This effect does not appear for direct orbits, or in the Schwarzschild limit $a = 0$.

Where exactly can such negative-energy orbits occur? Note that the momentum of a physical particle must be a timelike vector, $g^{\alpha\beta} p_\alpha p_\beta > 0$. Let us choose an orthonormal tetrad, in which $g^{ij} = \text{diag}(+1, -1, -1, -1)$; then for the tetrad components of the momentum we have

$$p_{\hat{0}}^2 - p_{\hat{1}}^2 - p_{\hat{2}}^2 - p_{\hat{3}}^2 = \mu_0^2 > 0 \implies |p_{\hat{3}}| < p_{\hat{0}}. \tag{20.124}$$

For the metric (20.57), we choose the following orthonormal tetrad:

$$e^0 = e^\nu \, dt, \qquad e^1 = e^\lambda \, dr, \qquad e^2 = \sqrt{\Sigma} \, d\vartheta, \qquad e^3 = e^\psi (d\varphi - \omega \, dt), \tag{20.125}$$

where

$$e^{2\nu} = \Delta_r \Sigma / B, \qquad e^{2\lambda} = \Sigma / \Delta_r, \qquad e^{2\psi} = B \sin^2 \vartheta / \Sigma, \tag{20.126}$$

and B and ω were defined in (20.121).[2] The tetrad components of the momentum, defined in (20.96), are then

$$p_{\hat{0}} = e^{-\nu} (E - \omega L_z), \qquad p_{\hat{3}} = e^{-\psi} L_z. \tag{20.127}$$

From the first equation we have $E = e^\nu p_{\hat{0}} + \omega e^\psi p_{\hat{3}}$. But $p_{\hat{0}} = p^{\hat{0}} > 0$, as explained after (20.119). Hence, for E to be negative, $\omega e^\psi p_{\hat{3}}$ must be negative. Since $|p_{\hat{3}}| < p_{\hat{0}}$, it follows that $\omega^2 e^{2\psi} > e^{2\nu}$, which means that $g_{00} < 0$. Thus, orbits with negative energy can exist only between the stationary limit hypersurfaces.

---

[1] The gravitational field 'at infinity' vanishes, hence special relativity applies there, and in special relativity $p^0 = E$ is the total energy of the orbiting particle.

[2] The verification that the metric defined by such a tetrad coincides with (20.57) is rather laborious. The first thing to verify is that $\Delta_r \Sigma^2 - 4a^2 m^2 r^2 \sin^2 \vartheta = B(r^2 - 2mr + a^2 \cos^2 \vartheta)$.

This was a necessary condition for the existence of orbits of negative energy. What is a sufficient condition? Suppose that $p_{\hat{1}} = 0$ at an initial point of the orbit. Then $p_{\hat{0}}$ is bounded because $p_0{}^2 = \mu_0{}^2 + p_2{}^2 + p_3{}^2$, while $p_2{}^2 = \Sigma^2 \dot\vartheta^2$ and $p_3{}^2 = \mathrm{e}^{-2\psi} L_z{}^2$ are both bounded in the region $r > r_+$; see (20.110) and (20.121). Consequently, $\mathrm{e}^\nu p_{\hat{0}}$ can be made as small as we wish by taking r sufficiently close to r_+. Then, since $E = \mathrm{e}^\nu p_{\hat{0}} + \omega L_z$, and ω is seen to have a positive value at $r = r_+$, it suffices to take $aL_z < 0$ and such r that $\mathrm{e}^\nu p_{\hat{0}} < |\omega L_z|$ to achieve $E < 0$.

Equation (20.123) simplifies for photon orbits, for which $\mu_0 = 0$:

$$E_{\min}^\nu = \pm L_z \left(r\sqrt{\Delta_r} \pm 2am \right) \Big/ D, \tag{20.128}$$

where '$+$' corresponds to $L_z > 0$ and '$-$' to $L_z < 0$. One can show that, for any sign, the function $F(r) \overset{\text{def}}{=} E_{\min}^\nu / |L_z|$ has exactly one maximum and no minima for $r \in (r_+, \infty)$ (see Exercise 12). Thus, a photon orbit in the equatorial plane can have only one turning point and there are no stable circular orbits. For each value of L_z, there exists a circular orbit that lies on the maximum of the function $F(r)$, and so is unstable. The same conclusion follows in the Schwarzschild limit $a = 0$ (see Misner, Thorne and Wheeler (1973) for more details on this subcase). A representative graph of the function $F(r)$ is shown in Fig. 20.8.

Still more information about equatorial null geodesics can be drawn from the equation of radial motion. As seen from (20.109) and (20.118), for null geodesics (where $\mu_0 = 0$) the affine parameter may be redefined by $s = Es'$, so that only the ratio L_z/E enters the equation. Consequently, it may be assumed without loss of generality that $E = 1$. We redefine some of the variables as follows:

$$\rho \overset{\text{def}}{=} r/(2m), \qquad \lambda \overset{\text{def}}{=} - L_z/(2m), \qquad \alpha \overset{\text{def}}{=} a/(2m). \tag{20.129}$$

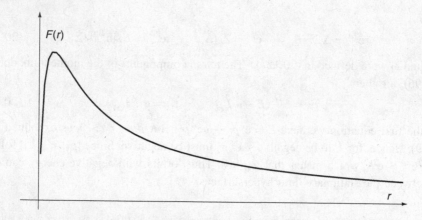

Fig. 20.8. The analogue of the graph from Fig. 20.7 for null geodesics in the equatorial plane. The curve begins at $r = r_+$. The graph has the same general shape for all values of L_z.

Then, for equatorial geodesics, for which $C = 0$, we obtain

$$\dot{r}^2 = \left[\rho^3 + (\alpha^2 - \lambda^2)\rho + (\alpha + \lambda)^2\right]/\rho^3 \overset{\text{def}}{=} \psi(\rho)/\rho^3. \qquad (20.130)$$

The turning points of the null geodesics are at the zeros of the polynomial $\psi(\rho)$, and the motion can take place in those ranges of ρ where $\psi(\rho)/\rho^3 > 0$. Since we will extend the Kerr manifold in the next section to include also negative values of r, we will take them into account here.

Apart from the special arrangement in which $\alpha = -\lambda$, i.e. $a = L_z$, the last term in ψ is always positive, so $\psi(0) > 0$, while ψ becomes negative for sufficiently large negative ρ. Thus, ψ typically has a real root at $\rho = \rho_1 < 0$ and is positive for some $\rho > \rho_1$. Then any ray sent in from $r = -\infty$ will eventually return to $r = -\infty$ (the ray with $\alpha = -\lambda$ hits the $r = 0$ singularity). If $\lambda^2 \leq \alpha^2$ while $\lambda \neq -\alpha$, then there are no positive roots of ψ, and any ray sent in from the side of positive ρ will necessarily hit the singularity. If λ^2 is smaller than some $\lambda_1{}^2$, then ψ is positive for all $\rho > 0$ and such rays will hit the singularity, too. However, if λ^2 is sufficiently large, then ψ will have two positive roots. In that case, there will be rays that come in from $r = +\infty$ and turn back at a $\rho_2 > 0$, possibly after circling the central body several times in a spiral, and rays that come from the side of $r = 0$ and turn back towards $r = 0$ after reaching a maximal distance $0 < \rho_1 < \rho_2$. The situation is shown in Fig. 20.9. Figures 20.10 and 20.11 show the situations when $a^2 = m^2$ and $a^2 > m^2$, respectively. The contours separating the allowed and prohibited areas are graphs of the $\lambda(\rho)$ function found by solving the equation $\psi(\rho) = 0$ for λ; from the definition (20.130) the solution is

$$\lambda = \left(-\alpha \pm |\rho|\sqrt{\rho^2 - \rho + \alpha^2}\right) \bigg/ (1 - \rho) \overset{\text{def}}{=} \lambda_\pm(\rho). \qquad (20.131)$$

In comparing Eq. (20.131) with the figures note well that the relation between the various branches of $\lambda(\rho)$ depends on the sign of ρ, in consequence of the $|\rho|$ in front of the square root. Namely, in the $\rho < 0$ area of Fig. 20.9, the upper branch corresponds to $\lambda_-(\rho)$, and the lower branch to $\lambda_+(\rho)$, while the opposite is true in the $\rho > 0$ area. Also, note well that it is ψ/ρ^3 that must be positive in the allowed region; hence in the $\rho < 0$ area the allowed region is where $\psi < 0$.

For timelike geodesics, we can choose the affine parameter so that $\mu_0 = 1$. We define $\Gamma \overset{\text{def}}{=} E^2 - 1$, and, using (20.129), we obtain from (20.109) and (20.118) (with $\vartheta = \pi/2$):

$$\dot{r}^2 = \frac{1}{\rho^3}\left[\Gamma\rho^3 + \rho^2 + (\alpha^2\Gamma - \lambda^2)\rho + (\alpha E + \lambda)^2\right] \overset{\text{def}}{=} \frac{1}{\rho^3}\overline{\psi}(\rho). \qquad (20.132)$$

The allowed range of r is given by $\overline{\psi}(\rho)/\rho^3 \geq 0$. The zeros of $\overline{\psi}(\rho)$ exist only for those values of ρ at which the discriminant of $\overline{\psi}(\rho)$ treated as a function of λ is non-negative, thus

$$\delta = \rho(\rho - \rho_-)(\rho - \rho_+)(\Gamma\rho + 1) \geq 0, \qquad (20.133)$$

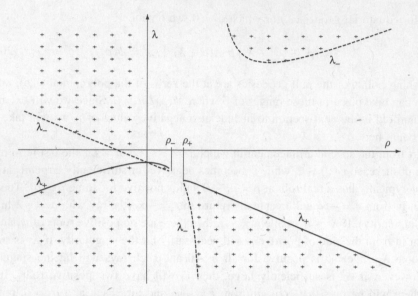

Fig. 20.9. The allowed ranges of λ and ρ for null geodesics in the case $a^2 < m^2$ (the areas covered with crosses are prohibited). For λ sufficiently small or sufficiently large, the rays can either emanate from $\rho = 0$ and turn back at a finite distance, or come in from $r = +\infty$ and turn back to infinity at a finite distance. In between, on the $\rho > 0$ side, the rays mostly hit $\rho = 0$, except for a small range of λ, in which there is the little prohibited 'peninsula'. On the $\rho < 0$ side, all rays that come in from the direction of $r = -\infty$ turn back at a finite distance, except the rays with $\lambda = -\alpha$ that hit the singularity. The right tip of the peninsula is at $\rho = \rho_-$, the radius of the inner event horizon. The left tip of the prohibited wedge in the $\rho > 0$ region is at $\rho = \rho_+$, the outer event horizon. With decreasing α, the peninsula shrinks and its left tip moves towards the ρ-axis (and so does the tip of the white wedge on the left), the upper prohibited area on the right moves down and left, while the lower prohibited area moves down and right. In the limit $a = 0$ the peninsula disappears, and the whole graph becomes mirror-symmetric with respect to the ρ-axis. For what happens when a increases, see Figs. 20.10 and 20.11.

where $\rho_\pm$ are the event horizons; $\rho_- < 1/2, \rho_- < \rho_+ < 1$. Note that $\Gamma \geq -1$ by definition. The following cases have to be considered:

Case 1: $\Gamma < 0$.
Then $|\Gamma| \leq 1$, and $\delta \geq 0$ for $0 \leq \rho \leq \rho_-$ and for $\rho_+ \leq \rho \leq 1/|\Gamma|$. By considering the signs of δ and of $(1 - \rho)$ (the coefficient of λ^2), the following subcases are discerned:

(a) $\rho < 0$ – this area is all prohibited;
(b) $0 \leq \rho \leq \rho_-$ – orbits exist for $\lambda \leq \lambda_-$ and for $\lambda \geq \lambda_+$, where

$$\lambda_\pm(\rho) = \left(-\alpha E \pm \sqrt{\delta}\right) \Big/ (1 - \rho) \qquad (20.134)$$

(note that $\lambda_- < \lambda_+$ when $\rho < 1$ and $\lambda_+ < \lambda_-$ when $\rho > 1$);
(c) $\rho_- < \rho < \rho_+$ – all values of λ are allowed;
(d) $\rho_+ \leq \rho < 1$ – orbits exist for $\lambda \leq \lambda_-$ and for $\lambda \geq \lambda_+$;

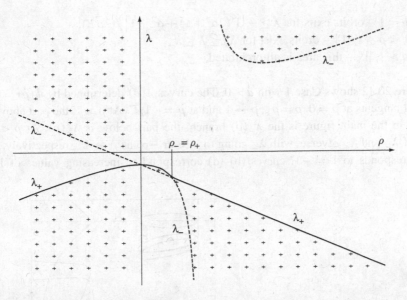

Fig. 20.10.   The allowed ranges of λ and ρ for null geodesics in the critical case $a^2 = m^2$. The little prohibited peninsula from Fig. 20.9 has extended to touch the tip of the lower prohibited area. The contact point has $\rho = \rho_- = \rho_+$.

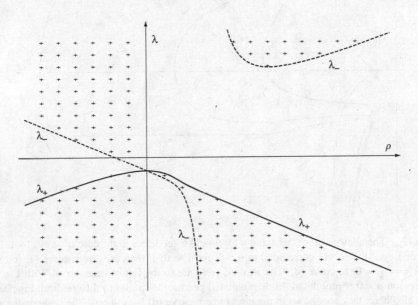

Fig. 20.11.   The allowed ranges of λ and ρ for null geodesics in the case $a^2 > m^2$. The two prohibited areas at $\{\rho > 0, \lambda < 0\}$ merged into one. As a^2 increases, the lower prohibited wedge becomes gradually wider and its tip moves down, while the upper prohibited area on the right moves up and farther to the right.

(e) $\rho = 1$ – orbits exits for $\lambda \geq -\left[\Gamma\left(2\alpha^2+1\right)+\alpha^2+1\right]/(2\alpha E)$;
(f) $1 < \rho \leq 1/|\Gamma|$ – orbits exist for $\lambda_+ \leq \lambda \leq \lambda_-$;
(g) $\rho > 1/|\Gamma|$ – this area is all prohibited.

Figure 20.12 shows Case 1 with $a > 0$. The curves $\lambda(\rho)$ determined by $\overline{\psi}(\rho) = 0$ have vertical tangents at $\rho = 0$, $\rho = \rho_\pm$, $\rho \to 1$ and at $\rho = -1/\Gamma$. At $\rho > 1$, the part above each 'belly' in the main figure is the $\lambda_-(\rho)$ branch; the part below is $\lambda_+(\rho)$. At $\rho < 1$, the roles of λ_- and λ_+ reverse, with λ_- going to $+\infty$ or $-\infty$ as $\rho \to 1^\pm$, respectively. Curve (a) corresponds to $\Gamma = -1$; curves (b)–(d) correspond to increasing values of Γ. The

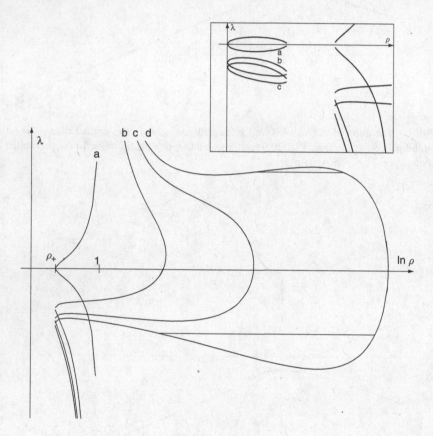

Fig. 20.12.   The allowed ranges of λ and ρ for timelike geodesics in the case $a^2 < m^2$, for different values of Γ. All the curves correspond to $a > 0$ and $\Gamma < 0$; Γ increases from curve (a) (on which $\Gamma = -1 \iff E = 0$) to curve (d). The allowed areas are to the left of each curve. With $\Gamma < 0$, the whole region $\rho < 0$ is prohibited. Little prohibited peninsulas similar to the one from Fig. 20.9 are present also here; they are shown in the inset (where curve (d) is omitted). The scale on the ρ-axis is logarithmic in the main figure and linear in the inset. Horizontal lines show the allowed ranges of ρ on two bound orbits. Since $a > 0$ for all curves, $\lambda > 0$ corresponds to retrograde orbits and $\lambda < 0$ to direct orbits; the figure clearly shows their inequivalence. There exist allowed regions also with $\rho < 0$ when $\Gamma > 0$; see the text.

information about the allowed ranges of λ can be briefly summarised by saying that the allowed area is to the left of each curve. The inset shows a magnified view of the region $0 \leq \rho \leq \rho_-$, where small prohibited 'peninsulas', similar to the one seen in Fig. 20.9, exist.

Case 2: $\Gamma > 0$.
Then $\delta \geq 0$ and there are the following ranges of ρ to consider: (a) $\rho \leq -1/\Gamma$, (b) $-1/\Gamma < \rho < 0$, (c) $0 \leq \rho \leq \rho_-$, (d) $\rho_- < \rho < \rho_+$, (e) $\rho_+ \leq \rho \leq 1$ and (f) $\rho > 1$.

In cases (a), (c) and (e) the allowed range is $\lambda \leq \lambda_-$ and $\lambda \geq \lambda_+$, in case (f) the allowed range is $\lambda_- \leq \lambda \leq \lambda_+$, in the remaining two cases all values of λ are allowed.

Case 3: $\Gamma = 0$.
Details of Case 2 and the whole Case 3 are left as an exercise for the readers.

Bound orbits exist in those cases where a line $\lambda = \text{constant}$ in the graph has a segment that runs in the allowed region and has both ends on the same $\lambda(\rho)$ curve. In Fig. 20.12 such bound states exist for curves c and d; two of them are shown.

20.8 * The maximal analytic extension of the Kerr spacetime

This section is based on the paper by Boyer and Lindquist (1967), who were the first to construct a complete extension of the case $a^2 < m^2$. Important contributions to this subject were published also by Carter, who first constructed such an extension along the symmetry axis (Carter, 1966b) and then extended the Boyer–Lindquist (B–L) construction to the charged case (Carter, 1968b) and to the critical case $a^2 = m^2$ (Carter, 1973).

It can be predicted that the maximal extension of the Kerr spacetime will be qualitatively similar to the corresponding extension of the Reissner–Nordström (R–N) spacetime (see Section 14.15). In the Schwarzschild spacetime, the singularity at $r = 0$ lies in the region where the hypersurfaces $r = \text{constant}$ are spacelike, i.e. transversal to all causal curves. Hence, any causal curve running in that region must hit the singularity after a finite proper time (or, for null geodesics, at a finite value of the affine parameter). In the R–N and Kerr spacetimes, the singularity at $r = 0$ lies in the region where the hypersurfaces $r = \text{constant}$ are timelike. Consequently, it is possible for a causal curve in that region to steer clear of the singularity. Any such curve in the R–N spacetime can be continued infinitely far to the future, but, in consequence of the geometric arrangement of the event horizons, it cannot return to the asymptotically flat region from which it entered – this would require travelling backward in time. This is how the infinite chain structure arises. We will find a similar thing to happen in the Kerr spacetime.

The B–L coordinates are useful for many calculations, but not for considering the extensions across spurious singularities, since they had in fact *introduced* these singularities. Therefore, we will have to begin with other coordinates.

Take the Kerr metric in the form (20.52) and find its inverse. It is

$$\left(\frac{\partial}{\partial s}\right)^2 = \left(\frac{\partial}{\partial t}\right)^2 - \frac{1}{\Sigma}\left[\left(r^2+a^2\right)\left(\frac{\partial}{\partial r}\right)^2 - 2a\frac{\partial}{\partial r}\frac{\partial}{\partial \varphi}\right.$$

$$\left. + \frac{1}{\sin^2\vartheta}\left(\frac{\partial}{\partial \varphi}\right)^2 + \left(\frac{\partial}{\partial \vartheta}\right)^2\right] + \frac{2mr}{\Sigma}\left(\frac{\partial}{\partial t} - \frac{\partial}{\partial r}\right)^2. \tag{20.135}$$

This has the Kerr–Schild form (20.9), in which the flat metric (the limit $m = 0$ of (20.135)) is expressed in spheroidal coordinates. The Kerr–Schild vector field k^α is seen to be proportional to

$$k^\alpha = \left(k^t, k^r, k^\vartheta, k^\varphi\right) = (1, -1, 0, 0). \tag{20.136}$$

It turns out that this is exactly *the* Kerr–Schild field – it obeys $k^\rho k^\alpha;_\rho = 0$, i.e. it is affinely parametrised. In fact, since $k^\alpha = dx^\alpha/ds$, the coordinate t or r can be chosen as the affine parameter. Since $k^t > 0$, the vector field k^α is future-pointing, and, since $k^r < 0$, it is ingoing. It is tangent to the surfaces ($\vartheta = $ constant, $\varphi = $ constant). With $\vartheta \neq \pi/2$, the null geodesics tangent to k^α proceed towards smaller r until they cross the $r = 0$ disc and continue into the $r < 0$ region. The null geodesics lying in $\vartheta = \pi/2$ hit the ring singularity at $\{r = 0, \vartheta = \pi/2\}$.

However, since the Kerr metric is of Petrov type D, it defines a second degenerate Debever congruence (call it ℓ^α), which, by virtue of the Goldberg–Sachs theorem 16.4, must also be geodesic and shearfree. There is no reason why one of these congruences should be preferred, so it should be possible to express the Kerr metric in the Kerr–Schild form (20.9) also with respect to ℓ^α. The B–L coordinates of (20.57) are helpful in identifying ℓ^α. The metric expressed in them is invariant under the substitution $dr \to -dr$, which changes the ingoing field (20.136) into an outgoing one, and so puts an outgoing field into the position of k^α. That outgoing field must thus be a Kerr–Schild field, too. To find that second field we transform k^α to the B–L coordinates, obtaining

$$k^\alpha = \left[\left(r^2+a^2\right)/\Delta_r, -1, 0, a/\Delta_r\right], \tag{20.137}$$

which implies that

$$\ell^\alpha = \left[\left(r^2+a^2\right)/\Delta_r, +1, 0, a/\Delta_r\right], \tag{20.138}$$

and then we transform ℓ^α back to the coordinates of (20.52):

$$\ell^\alpha = \left[\left(r^2+a^2+2mr\right)/\Delta_r, +1, 0, -2a/\Delta_r\right]. \tag{20.139}$$

The coordinate r is still an affine parameter on the geodesics tangent to ℓ^α, but t is not. The expressions (20.138) and (20.139) become singular where $\Delta_r = 0$. To find the expression for ℓ^α on the horizons, we note that it is proportional to the vector field (in the B–L coordinates)

$$\widetilde{\ell}^\alpha = \left(r^2+a^2, \Delta_r, 0, a\right). \tag{20.140}$$

At $r = r_{\pm}$, the field $\tilde{\ell}^{\alpha}$, while being still null, becomes $\tilde{\ell}^{\alpha}_{\pm} = N(2mr_{\pm}, 0, 0, a)$, i.e. it becomes tangent to the $r = r_{\pm}$ hypersurfaces. Thus, the hypersurfaces $r = r_{\pm}$ are tangent to the light cones, i.e. matter and light can cross them in one direction only (they are crossed by the null curves tangent to k^{α}, which is an ingoing field, so the allowed direction is from outside in). This confirms that they are event horizons. Moreover, note that $\tilde{\ell}^{\alpha}_{\pm}$ is geodesic and affinely parametrised.

In the coordinates of (20.52), the fields k^{α} and ℓ^{α} appear non-symmetrically, while they play similar roles in the metric. This is because the coordinates of (20.52) are adapted to the field k^{α}. Since in the B–L coordinates the roles of these fields are interchanged by the substitution $dr \to -dr$, we find the coordinates adapted to ℓ^{α} in the following way: (1) carry out the substitution $dr \to -dr$ in (20.57); (2) carry out the transformation inverse to (20.56) on the result of step (1) – the result will be (20.52) with dr replaced by $-dr'$ (or, equivalently, with (t, φ) replaced by $(-t, -\varphi)$); (3) carry out the same two transformations on the components of the fields k^{α} and ℓ^{α}. The result will be

$$k^{\alpha} = \left[(r^2 + a^2 + 2mr)/\Delta_r, -1, 0, -2a/\Delta_r\right], \qquad \ell^{\alpha} = (1, +1, 0, 0), \qquad (20.141)$$

i.e. the roles of k^{α} and ℓ^{α} are now interchanged, except that k^{α} is still ingoing, while ℓ^{α} is still outgoing. In these coordinates it is seen that k^{α} at the horizons is tangent to them.

This looks mysterious, but the coordinates of (20.136) (call them, after Boyer and Lindquist (1967), an E frame) and of (20.141) (call them an E' frame) do not have the same domain, and the horizon tangent to k^{α} is not the same one that is tangent to ℓ^{α}. To see this, note that both vector fields are future-pointing, k^{α} is ingoing and ℓ^{α} is outgoing. Begin at a point in the $r > r_+$ region and proceed along the null geodesics tangent to k^{α}. Since it is ingoing, by moving to the future you proceed towards smaller r, and you will eventually cross the surface $r = r_+$ (at which k^{α} has no singularity). Not so with the ℓ^{α} field – if you follow its integral lines to the future from any point in $r > r_+$, then you proceed towards larger r and you will never reach any of the horizons. In order to meet the $r = r_+$ hypersurface, you would have to move along ℓ^{α} to the past.

The transformation from E to E' is regular for all values of r only when $a^2 > m^2$, but then no horizons exist. When $a^2 < m^2$, the domains of the two frames overlap only partially. The transformation is

$$t' = t - \frac{2m}{\sqrt{m^2 - a^2}} \left(r_+ \ln\left|\frac{r - r_+}{2m}\right| - r_- \ln\left|\frac{r - r_-}{2m}\right| \right),$$

$$\varphi' = \varphi - \frac{a}{\sqrt{m^2 - a^2}} \ln\left|\frac{r - r_+}{r - r_-}\right|, \qquad (20.142)$$

and it is singular at both horizons. Thus, the extension of the region $r > r_+$ along the k field, obtained by transforming to the E frame, is not the same as the extension along the ℓ field obtained by transforming to the E' frame. This is an analogy to the Kruskal extension of the Schwarzschild manifold.

These inequivalent extensions now have to be put together into an extended manifold. Let us begin with the region $r > r_+$ and the E' frame, and let us move back in time along the ℓ field. We can carry out the transformation from the E$'$ frame to the B–L coordinates separately in each of the regions $r > r_+$, $r_- < r < r_+$ and $-\infty < r < r_-$. Hence, this E$'$ frame can be understood to be the common extension of these three regions; we will call them regions (0), (-1) and (-2), respectively. Now let us go back to the $r > r_+$ region, let us transform the coordinates to the E frame and let us move forward in time along the k field. Again, we can carry out the transformation from the E frame to the B–L coordinates in each of the regions $r > r_+$, $r_- < r < r_+$ and $-\infty < r < r_-$. The first one will be the same (0) as before, but the other two will be different from (-1) and (-2) since they lie to the future of (0). We shall call them ($+1$) and ($+2$), respectively. The result of these extensions is schematically shown in Fig. 20.13 on the left.

But then, in the (-1) region the field k exists together with ℓ. Since $\Delta_r < 0$ there, we see from (20.141) that a future-pointing field is actually ($-k^\alpha$), not k^α itself, and ($-k^\alpha$) is *outgoing* there, just like ℓ^α. We will thus hit the $r = r_+$ horizon irrespective of whether we move along ($-k^\alpha$) or along ℓ^α. However, the horizon that we meet when moving along ($-k^\alpha$) is not the same as that which we meet when moving along ℓ^α. At the first one, ℓ^α becomes tangent to it, while ($-k^\alpha$) crosses it smoothly (as can be seen by transforming to the E frame); at the second one ($-k^\alpha$) becomes tangent and ℓ^α crosses it. Thus, just like we did in the Kruskal extension of the Schwarzschild spacetime, we must assume that there are two ways back to the future from region (-1) and that there exists

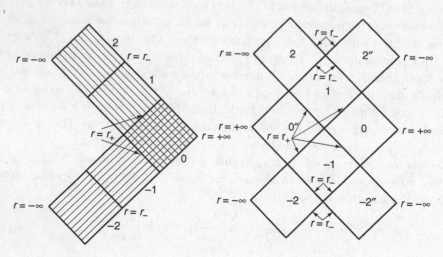

Fig. 20.13. **Left:** The coordinate patches -1 and -2 obtained by extending the region $r > r_+$ along the ℓ field (in the E$'$ frame), and the coordinate patches 1 and 2 obtained by extending the same region along the k field (in the E frame). **Right:** The coordinate patches $0''$ and $-2''$ were added by extending the -1 patch along the k field, and then the coordinate patches $+1''$ and $+2''$ were added by extending the $0''$ patch along the ℓ field. At this point there is no justification yet for identifying the patches $1''$ and 1, and not even for placing $0''$ side by side with 1, but a proof that this is correct is provided further on in the text. The straight segments of the edge of the figure are null infinities. Spatial infinities are the extreme points on the left and on the right. The extreme points at mid-height are $+\infty$; the others are $-\infty$.

a second copy of region (0) that we will call (0″). Similarly, when we move from region (−1) along (−k) back in time across $r = r_-$, we obtain a second copy of region (−2) that we will call (−2″).

Let us then move from region (−1) to (0″) along the (−k) field.[1] We are then back in a region where $\Delta_r > 0$, so k^α itself is future-pointing and ingoing, while ℓ^α is future-pointing and outgoing. By moving along k^α to the future across $r = r_+$ and then across $r = r_-$, we create second copies of regions (+1) and (+2) that we call (+1″) and (+2″), respectively. By analogy with the Kruskal extension of the Schwarzschild spacetime, we expect that we can identify regions (+1) and (+1″) – after all, there are timelike curves inside the light cone tangent to the k and ℓ fields with vertex in the region (−1), and their future should be continuously predictable. This (at this point still hypothetical) identification is shown in Fig. 20.13 on the right. However, in order to prove that such an identification is correct, we must find a coordinate map that will cover regions (0), (+1), (+1″), (0″) and (−1) simultaneously and will show explicitly that (+1) and (+1″) coincide.

There is a difficulty here. In the Schwarzschild case, the $a \to 0$ limits of k^α and ℓ^α are tangent to the surfaces $\{\vartheta = \text{constant}, \varphi = \text{constant}\}$, and the Kruskal diagram shows that surface. Here, the fields k^α and ℓ^α are not surface-forming, so we cannot adapt coordinates to them. We will choose such coordinates u and v that the integral curves of k^α and ℓ^α will lie in the hypersurfaces $u + v = \text{constant}$ and $u - v = \text{constant}$, respectively.

Take the B–L coordinates, since in them both null fields are treated on a nearly equal footing – see Eqs. (20.137) and (20.138). As we approach the $r = r_+$ hypersurface along these fields, they approach the direction $(2mr_+, 0, 0, a)$, which is a tangent vector to the helix in spacetime with the equation $d\varphi/dt = a/(2mr_+)$. In the first step, we transform the φ-coordinate so that the fields approach the horizon along the hypersurface $\varphi' = \text{constant}$. We call the new φ-coordinate w and transform

$$w = \varphi - at/(2mr_+) \tag{20.143}$$

(this is not the only possible transformation that 'untwists' the fields at the horizon, but it is the simplest one). This makes $k^w(r_+) = 0$, as desired.

The k and ℓ fields obey

$$dt/dr = \pm \left(r^2 + a^2\right)/\Delta_r \tag{20.144}$$

(plus the equation that determines $w(r)$). The solution of (20.144) is

$$\exp\left(\frac{r \pm t}{\sigma_+}\right) \frac{r - r_+}{2m} \left(\frac{r - r_-}{2m}\right)^{-r_-/r_+} \stackrel{\text{def}}{=} F(u \pm v) = \text{constant}, \tag{20.145}$$

$$\sigma_\pm \stackrel{\text{def}}{=} mr_\pm/\sqrt{m^2 - a^2}.$$

[1] Actually, after we have crossed the $r = r_+$ hypersurface, we have to move on along k again.

We want to cancel the spurious singularity at $r = r_+$ (we will deal with the one at $r = r_-$ later). Using the extension of the R–N metric as a guide (see Section 14.15), we guess

$$F(u \pm v) = (u \pm v)^2. \tag{20.146}$$

Then we obtain

$$u \pm v = \sqrt{\frac{r - r_+}{2m}} \left(\frac{r - r_-}{2m} \right)^{-r_-/(2r_+)} \exp \left(\frac{r \pm t}{2\sigma_+} \right), \tag{20.147}$$

$$\Psi(r) \stackrel{\text{def}}{=} \frac{r - r_+}{2m} \left(\frac{r - r_+}{2m} \right)^{-r_-/r_+} e^{r/\sigma_+} = u^2 - v^2, \tag{20.148}$$

$$t = \sigma_+ \text{arctanh} \left(\frac{2uv}{u^2 + v^2} \right).$$

This is well defined in the (u, v) plane except on the straight lines $u^2 = v^2$, where t becomes infinite. We have $\Psi(r) \underset{r \to r_-}{\to} -\infty$, $\Psi(r_+) = 0$ and $\Psi(r) \underset{r \to +\infty}{\to} +\infty$. The function $\Psi(r)$ is well defined and analytic on the whole plane, so $r(u, v)$ exists for all values of u and v. The function $t(u, v)$ is analytic everywhere except at the two straight lines. In spite of that, the Kerr metric transformed to the (u, v, ϑ, w) coordinates is analytic for *all* values of u and v, including the two straight lines. To see this, verify that the result of the transformation is

$$ds^2 = \frac{1}{\Sigma \Delta_r} \left[\left(\frac{2\sigma_+ f \Sigma_+}{r_+^2 + a^2} (v \, du - u \, dv) - a \sin^2 \vartheta \Delta_r \, dw \right)^2 \right.$$

$$\left. - \left(\frac{2\sigma_+ f \Sigma}{r^2 + a^2} (v \, du - u \, dv) \right)^2 \right] + \frac{4\sigma_+^2 f \Sigma}{(r^2 + a^2)^2} (dv^2 - du^2)$$

$$- \Sigma \, d\vartheta^2 - \frac{\sin^2 \vartheta}{\Sigma} \left(\frac{af (r + r_+)}{\sqrt{m^2 - a^2} (r - r_-)} (v \, du - u \, dv) + (r^2 + a^2) \, dw \right)^2, \tag{20.149}$$

$$\Sigma_\pm \stackrel{\text{def}}{=} \Sigma (r_\pm, \vartheta),$$

$$f(r) \stackrel{\text{def}}{=} \frac{\Delta_r}{\Psi(r)} = 4m^2 \left(\frac{r - r_-}{2m} \right)^{2m/r_+} e^{-r/\sigma_+}.$$

The only term that might cause problems is the one with Δ_r in the denominator, which has a zero at $r = r_+$. However, the numerator has a zero of the same (first) order there, so the ratio has a finite limit and the metric is analytic for all values of u and v in the range $(-\infty, +\infty)$, of $w \in [0, 2\pi]$ and of $\vartheta \in [0, \pi]$. (But recall that $u^2 - v^2 = -\infty$ corresponds to $r = r_-$, so the extension we have just found has dealt with the spurious singularity at $r = r_+$, but not with the one at $r = r_-$. We will deal with that one below.) Since $r = r_+$ is no longer a singularity for (20.149), this is the metric that applies throughout regions (0), (+1), (0'') and (−1) in Fig. 20.13, so we have shown that the identifications made in that figure can indeed be made.

In order to remove the spurious singularities at $r = r_-$, we would have to choose $F(u \pm v) = (u \pm v)^{-2r_-/r_+}$ in (20.146), with a result analogous to (20.148). Then the range $u \in (-\infty, +\infty)$, $v \in (-\infty, +\infty)$ would cover the region $r \in (-\infty, r_+)$, but the true singularity at $\{r(u, v) = 0, \vartheta = \pi/2\}$ would still be there. The coordinate system thus obtained would apply throughout the regions $\{-2, -1, -2''\}$ and $\{+2, +1, +2''\}$, plus in the regions $(+3)$ and (-3) that we are going to construct now.[1]

We can continue the process of extending the spacetime in the same way both upwards and downwards, see Fig. 20.14. Thus, we can generate regions $(+3)$ and $(+4)$ by adapting the coordinates to the ℓ field in region $(+2)$ and proceeding to the future, and we can generate regions $(+3)$ and $(+4'')$ by adapting coordinates to the k field in region $(+2'')$ and proceeding to the future. Likewise, we can extend region (-2) along the k field to the past and thus generate regions (-3) and (-4), and extend region $(-2'')$ along the ℓ field to the past, thus generating regions (-3) and $(-4'')$. Exactly as in the case of the R–N spacetime (see Section 14.15) we must then decide whether we wish to continue the process *ad infinitum*, or to identify two regions that are isometric, for example (-1) and $(+3)$.

Like in the R–N case, the lines $r = $ constant are hyperbolae in the surface $\{\vartheta = $ constant, $\varphi = $ constant$\}$ (compare Eqs. (14.149) and (20.148)), so the remark after Eq. (14.156) applies also here – they will still be hyperbolae after the surface has been compactified by (14.154)–(14.156). They are drawn accordingly in Fig. 20.14. Again in analogy to the R–N case, they are timelike for $r > r_+$ and $r < r_-$, and spacelike for $r_- < r < r_+$.

The arcs of hyperbolae in Fig. 20.14 represent the $r = $ constant hypersurfaces; they are timelike in even-numbered areas and spacelike in odd-numbered areas. The thick arcs represent the sets $r = 0$; they are nonsingular except at $\vartheta = \pi/2$. This is the only difference from the corresponding diagram for the R–N spacetime (Fig. 14.10): here, null and timelike curves can be continued through the open disc $\{r = 0, \vartheta \neq \pi/2\}$ to $r \to -\infty$. The left and right edges of Fig. 20.14 are null infinities; those with numbers divisible by 4 are $+\infty$, the remaining ones are $-\infty$. The extreme points of the edges are the corresponding spatial infinities. Thick straight segments are the event horizons, alternately r_- and r_+. Identifications can be made so that an odd-numbered area with number n is identified with the area numbered $(n + 4k)$, where k is a natural number.

There remains now the question of *geodesic completeness* of the mosaic manifold represented by Fig. 20.14. We call a manifold **geodesically complete** if every geodesic in it can be continued to an arbitrarily large absolute value of the affine parameter. We already know that the extended Kerr manifold is not geodesically complete in the strict sense: there are null geodesics that hit the ring singularity $\{r = 0, \vartheta = \pi/2\}$ at a finite value of the affine parameter and cannot be continued further. But we can consider the geodesics that do not run into these singularities and then ask whether they can be continued indefinitely. It turns out that in the extended Kerr geometry they can.

---

[1] In fact, these coordinates could be used to extend regions (-2) and $(-2'')$ through $r = r_-$ to cover the not-yet-constructed region (-3) between them, and likewise to extend $(+2)$ and $(+2'')$ to $(+3)$, but we will do it by the previous method.

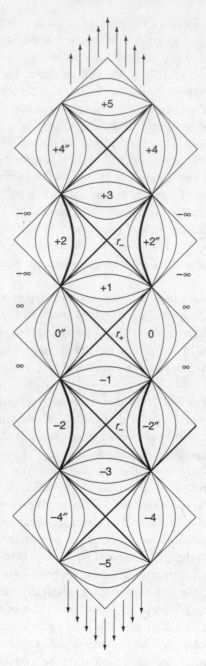

Fig. 20.14.   The compactified (u, v) surface of the maximally extended Kerr spacetime. The arcs of hyperbolae are the $r = $ constant lines, the thick arcs are the $r = 0$ lines. More explanation is given in the text.

Consider the geodesic equations (20.109)–(20.112). We applied them to timelike and null geodesics only, but spacelike geodesics are automatically included if we replace the parameter μ_0^2 with a negative quantity. The infinite values that appear as $r \to \pm\infty$ are no problem, since we know already that the spacetime becomes approximately flat there, and so complete. The components of the tangent vector to a geodesic become infinite at $\Sigma = 0$, $\sin\vartheta = 0$ and $\Delta_r = 0$. The set $\Sigma = 0$ is just the singularity that we know exists. The set $\sin\vartheta = 0$ is the axis of symmetry. As seen from (20.110), a geodesic can run only in the regions where $\Theta \geq 0$, and $\Theta \to -\infty$ as $\vartheta \to 0$, unless $L_z = 0$. Thus the axis cannot be intersected by any geodesic that has $L_z \neq 0$, and for those with $L_z = 0$ the set $\sin\vartheta = 0$ is nonsingular. Thus, we have to take care only about the infinities in $\dot{t}$ and $\dot{\varphi}$ that appear on the horizons $\Delta_r = 0$.

Before we continue, note that a timelike or null geodesic can properly be called ingoing only when along it, simultaneously, either $(\mathrm{d}t/\mathrm{d}s > 0$ and $\mathrm{d}r/\mathrm{d}s < 0)$ or $(\mathrm{d}t/\mathrm{d}s < 0$ and $\mathrm{d}r/\mathrm{d}s > 0)$. This can be summarised in the single inequality $\mathrm{d}r/\mathrm{d}t < 0$. By looking at Eqs. (20.107), (20.109) and (20.112) we see that a geodesic is ingoing at $r_\pm$ if and only if the quantities $\dot{r}(r_\pm)$ and $(2mr_\pm E - aL_z)$ have opposite signs.

Now take the geodesic equations (20.111) and (20.112) and transform them by the transformation inverse to (20.56) (in which, recall, t' and φ' are the B–L coordinates). The result is (the non-primed coordinates are now those of (20.52)):

$$\Sigma\dot{t} = -Ea^2\sin^2\vartheta + \left\{\left(r^2 + a^2\right)^2 E - 2mraL_z + 2mr\dot{r}\right\}/\Delta_r,$$

$$\Sigma\dot{\varphi} = L_z/\sin^2\vartheta + a\left\{2mrE - aL_z + \dot{r}\right\}/\Delta_r. \tag{20.150}$$

If $\dot{r}$ and $(2mr_\pm E - aL_z)$ have opposite signs (i.e. if the geodesic is ingoing), then the terms in braces vanish at both horizons, and their quotients by Δ_r have finite limits, which means that we have removed the singularities by a coordinate transformation. But why does this work only for ingoing geodesics? This is because the coordinates of (20.52) are adapted to the k field, which is ingoing, so they apply only in those regions where k^α is finite. In order to deal with outgoing geodesics, we must employ coordinates adapted to the outgoing ℓ field, i.e. coordinates related to those of (20.150) and (20.52) by (20.142) (t' and φ' are the new coordinates, while t and φ are those of (20.150)). Indeed, after the transformation the terms containing $\dot{r}$ in (20.150) change signs, so these coordinates remove the singularities at $\Delta_r = 0$ for outgoing geodesics. Thus, by switching between coordinate systems we can carry all geodesics through both horizons. Consequently, if a geodesic does not strike the true singularity at $\Sigma = 0$, it can be continued to arbitrarily large absolute values of its affine parameter.

The foregoing discussion concerned geodesics intersecting $r = r_\pm$, but in fact the conclusion applies also to geodesics that have turning points at $r = r_\pm$ or approach $r = r_\pm$ asymptotically (by winding around the horizon). The key point was that $\dot{t}$ and $\dot{\varphi}$ remain finite as $r \to r_\pm$, asymptotically or not.

What remains are the null geodesics that run along the horizons – we left them after (20.140) for separate consideration. For such geodesics $r = r_\pm = \text{constant}(\dot{r} = 0)$, $\Delta_r = 0$

and, as seen from (20.140), $\dot{\vartheta} = 0$. It follows further from (20.140) that on the horizon

$$\dot{\varphi} = a\dot{t}/(2mr_{\pm}).   \tag{20.151}$$

With this, (20.104) and (20.105) imply $E = L_z = 0$, and then (20.108) implies that also $K = 0$. Substituting (20.140) and (20.151) in the geodesic equation we find[1]

$$t = s + s_1, \qquad \varphi = a(s + s_2)/(2mr_{\pm}),   \tag{20.152}$$

where s is an affine parameter and s_1 and s_2 are arbitrary constants. Such a geodesic can obviously be continued to $s \to \pm\infty$.

Finally, then, all geodesics that do not strike the ring singularity $\Sigma = 0$ can be continued to infinite absolute values of their affine parameters.

We could draw figures similar to Figs. 14.7 and 14.11 for the surface $\{t = \text{constant}, \vartheta = \pi/2\}$ in (20.57). This surface has the metric

$$ds_2{}^2 = \frac{r^2}{r^2 - 2mr + a^2}\, dr^2 + \left(\frac{2ma^2}{r} + r^2 + a^2\right) d\varphi^2.   \tag{20.153}$$

If we want to embed it in a flat space, then the Cartesian coordinates in the flat space must be given in terms of r and φ as

$$x = F(r)\cos\varphi, \qquad y = F(r)\sin\varphi, \qquad z = \pm G(r),   \tag{20.154}$$

with F and G obeying

$$F = \sqrt{2ma^2/r + r^2 + a^2},$$

$$G_{,r}{}^2 = \chi(r) \stackrel{\text{def}}{=} \frac{r^2}{r^2 - 2mr + a^2} - \frac{(r - ma^2/r^2)^2}{2ma^2/r + r^2 + a^2}.   \tag{20.155}$$

Just as in the R–N case, the embedding can occur in a Euclidean space when $\chi(r) \geq 0$ and in a pseudo-Euclidean space otherwise. It may be verified that $\chi(r) > 0$ for $r \geq r_{\pm}$ (see Exercise 14). However, for $r < r_{+}$, $\chi(r)$ can change sign several times. Thus, several embeddings would be necessary to represent the full collection of shapes. Figure 20.15 shows the embedding of the region $r \geq r_{+}$.

In the extreme case $a^2 = m^2$ we use as a guide the $e^2 = m^2$ case of R–N Eqs. (14.163)–(14.166)) and the procedure applied earlier in this section. The considerations about the null Kerr–Schild fields k and ℓ do not depend on whether $a^2 < m^2$ or $a^2 = m^2$. So, in the present case, one of the fields passes through the hypersurface $r = m$ smoothly, while the other becomes tangent to it. The transformation (20.143) that 'untwists' the helical field is here $w = \varphi - t/(2m)$, and the solution of (20.144) is

$$\pm t = \xi(r) \stackrel{\text{def}}{=} r - m - \frac{2m^2}{r - m} + 2m\ln\left(\frac{r}{m} - 1\right).   \tag{20.156}$$

---

[1] The result obtained by Boyer and Lindquist (1967) in a strangely roundabout way is equivalent to (20.152), but their parameter is non-affine.

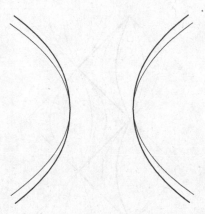

Fig. 20.15. An axial cross-section through the surface (20.154) for the region $r \geq r_+$ (thicker line) and the corresponding cross-section for the Schwarzschild metric (thin line; this is a cross-section through the surface from Fig. 14.7). For the Kerr metric, the quantity on the horizontal axis is $F(r)$ – the geometric radius of a circle $r =$ constant in (20.153). The surfaces are tangent at the equator (verify that $F(r_+) = 2m$). Compare this figure with Fig. 14.11 for the Reissner–Nordström spacetime.

We now introduce the null coordinates p and q such that the fields k^α and ℓ^α obeying (20.144) lie in the hypersurfaces $p \pm q =$ constant; they are $p = t + \xi$, $q = t - \xi$. The transformed metric is

$$ds^2 = \frac{1}{4}\left(1 - \frac{2mr}{\Sigma}\right)(dp + dq)^2 - \frac{\Sigma(r-m)^2}{4(r^2 + m^2)}(dp - dq)^2 - \Sigma\, d\vartheta^2$$

$$+ \frac{2m^2 r \sin^2 \vartheta}{\Sigma}(dp + dq)\left(dw + \frac{dp}{2m} + \frac{dq}{2m}\right)$$

$$- \left(\frac{2m^3 r \sin^2 \vartheta}{\Sigma} + r^2 + m^2\right)\left(dw + \frac{dp}{2m} + \frac{dq}{2m}\right), \tag{20.157}$$

and it is non-singular everywhere except at the ring $\Sigma = 0$. The coordinates (p, q) have infinite ranges, so in order to compactify the space and be able to draw a diagram like Fig. 20.14, we transform (p, q) just like we did in (14.166), $p = \tan P$, $q = \tan Q$. The resulting picture is again easy to construct, by analogy with Fig. 14.14. We can extend the region $r > m$ across $r = m$ either along the ingoing k field to the future, or along the outgoing ℓ field to the past. Once we have crossed the $r = m$ hypersurface, there is nothing to prevent us from going to $r \to -\infty$, unless we hit the $\{r = 0, \vartheta = \pi/2\}$ singularity. But on the other side of the horizon, when we used the k field to cross it, we can use the ℓ field to go to the past; and when we crossed the horizon along the ℓ field, we can use the k field to go to the future. What results looks like Fig. 20.14, but with the odd-numbered regions all squeezed to lines, together with their lower-left and upper-right neighbours; see Fig. 20.16. All $r =$ constant surfaces are now either timelike or null. This

Fig. 20.16. The (P, Q) surface of the maximally extended Kerr spacetime with $a^2 = m^2$. The whole left edge represents $r = -\infty$, the whole right edge corresponds to $r = +\infty$. The thick straight segments form the horizon $r = m$; the thick hyperbolae are the $r = 0$ sets. More explanation is given in the text.

extension was first constructed by Carter; see Carter (1973) for a brief presentation and references to earlier work.

Just like in the extremal R–N case, the invariant spatial distance from any point in the $r > m$ region to the horizon at $r = m$ along a curve of constant (t, ϑ, φ) is infinite, and the embedding of the $r > m$ part of the surface $\{t = \text{constant}, \vartheta = \pi/2\}$ looks very much like the left part of Fig. 14.15.

20.9 * The Penrose process

Penrose (1969) contemplated a process by which, *in principle*, the rotational energy of a Kerr black hole can be extracted. The idea is based on the observation we made after Eq. (20.127), that a body on a retrograde orbit inside the stationary limit hypersurface can have a negative total energy if it is close enough to the event horizon $r = r_+$. In brief, Penrose's idea was this: put two masses at the ends of a sufficiently strong spring, then squeeze the spring and bind the masses together. Then send the composite on an orbit that enters the region between the stationary limit hypersurface and the event horizon $r = r_+$. That region is called the **ergosphere**, and the meaning of this name will become clear at the end of this section. Design the orbit so that it has its turning point very close to $r = r_+$ (how close will become clear below). When the composite object is at the turning point, release the spring in such a direction that one of the masses is sent, with $\dot{r} = 0$ at the initial point, on a retrograde orbit with a negative energy and with $|L_z|$ sufficiently small that it falls through the event horizon. It follows from the reasoning between Eqs. (20.127) and (20.128) that this is possible: we direct the ejected mass so that $aL_z < 0$, we make $|L_z|$ small enough that the mass is sure to go through $r = r_+$ (see Fig. 20.12), and the orbit has to be pre-designed so that at the turning point $e^\nu p_{\hat{0}} < |\omega L_z|$.

Since the mass dropped into the black hole carried away some negative energy, the other mass acquires some additional energy, and additional momentum by recoil so that it is sure to return to the outside of the stationary limit hypersurface – and it does so having a greater energy than it had at the beginning of the journey. This trick can be applied for as long as the stationary limit hypersurface and the horizon exist. The logical conclusion is that the extra energy of the returning mass was gained at the expense of the rotational energy of the black hole: the slower the black hole rotates, the smaller $|a|$ becomes, and the smaller the volume of the ergosphere. Strictly speaking, however, this is a speculation that goes beyond the area of applicability of the Kerr metric. In order to discuss this energy-extraction process in a perfectly correct way, we would have to use a nonstationary solution in which the angular momentum of the source of the gravitational field can depend on time.

The ergosphere was named after the Greek word $\varepsilon\rho\gamma o$, meaning 'work' – because, as shown, thanks to its existence, a rotating black hole is in principle able to do some work.

20.10 Stationary–axisymmetric spacetimes and locally nonrotating observers

The Kerr spacetime, with its two commuting Killing fields, is an example of a **stationary–axisymmetric spacetime**. We call a spacetime **stationary** when its metric tensor allows a timelike Killing field (if the Killing field is hypersurface-orthogonal, then the spacetime is called **static**). We call it **axisymmetric** when there exists a Killing field whose integral lines are closed, and there exists a location in the spacetime at which the length of the lines goes to zero. (Take a look at (20.57) – the length of the lines of constant (t, r, ϑ) is zero at $\vartheta = 0$.) For the stationary–axisymmetric spacetimes it is assumed in addition that the two Killing fields commute; in consequence they are surface-forming.

Then, coordinates can be chosen so that the metric is independent of $x^0 = t$ (where $\mathrm{d}x^\alpha/\mathrm{d}t = \underset{(0)}{k^\alpha}$ – the timelike Killing field) and of $x^3 = \varphi$ (where $\mathrm{d}x^\alpha/\mathrm{d}\varphi = \underset{(3)}{k^\alpha}$ – the Killing field connected with axial symmetry). Let us call the other two coordinates x^1 and x^2.

At this point, it is assumed in addition that the surfaces generated by the Killing fields admit orthogonal surfaces, i.e. that in the coordinates adapted to the Killing fields $g_{01} = g_{02} = g_{13} = g_{23} = 0$ (this property is called **orthogonal transitivity**). This is equivalent to the requirement that the metric, and the motion of matter if any is present, is invariant under the discrete transformation $(t, \varphi) \rightarrow (-t, -\varphi)$.[1] Several theorems were proven in which the property $\left[\underset{(0)}{k}, \underset{(3)}{k}\right] = 0$ and the orthogonal transitivity follow from other assumptions. The intention of those theorems (see Stephani *et al.* (2003) for a brief listing) was to show that spacetimes that do not possess these properties are rare, unimportant or weird in some sense. The fact is, though, that not much is known about the cases left out – no exact solutions of the Einstein equations have been found in those cases.

In a stationary–axisymmetric spacetime that is orthogonally transitive, coordinates in the (x^1, x^2) surfaces can be chosen so that $g_{12} = 0$ – since we know that every 2-dimensional metric is conformally flat, and the property $g_{12} = 0$ is even weaker than that. The metric is thus

$$ds^2 = g_{00}\,dt^2 + 2g_{03}\,dt\,d\varphi + g_{33}\,d\varphi^2 + g_{11}\,d(x^1)^2 + g_{22}\,d(x^2)^2. \tag{20.158}$$

When two Killing fields exist and nothing else is assumed about them, the basis of the space of Killing fields can be chosen arbitrarily. The transformation of the basis of the form

$$\underset{(0)}{k'} = C_0 \underset{(0)}{k} + C_3 \underset{(3)}{k}, \qquad \underset{(3)}{k'} = D_0 \underset{(0)}{k} + D_3 \underset{(3)}{k}$$

induces a transformation of the coordinates adapted to the Killing fields; the coordinates (t', φ') adapted to $\underset{(0)}{k'}$ and $\underset{(3)}{k'}$ are related to (t, φ) by

$$t' = C_0 t + C_3 \varphi, \qquad \varphi' = D_0 t + D_3 \varphi. \tag{20.159}$$

Such a transformation preserves the independence of the metric tensor of t and φ and the orthogonal transitivity property; it only reshuffles the components g_{00}, g_{03} and g_{33} among themselves. Note, however, that the Killing fields are in fact *unique* if they correspond to stationarity and axial symmetry, and it is assumed in addition that the spacetime is asymptotically flat. Then the integral lines of $\underset{(3)}{k}$ are closed, and the coordinate φ defined by this field is periodic with the period 2π. This excludes those transformations (20.159) in which $C_3 \neq 0$, or else the strange behaviour illustrated in Fig. 20.17 would occur: after increasing φ' by $2\pi C_0 D_3 / (C_0 D_3 - C_3 D_0)$ we would land at the same t' line from which we started, but with the t'-coordinate increased by $\Delta t' \overset{\text{def}}{=} -2\pi C_3 D_3 / (C_0 D_3 - C_3 D_0)$. The orbits of the $\underset{(3)}{k}$ field would thus be disrupted and changed into infinite helices, while on two sides of the initial t line adjacent points would exist whose time coordinate would differ by $\Delta t'$. The time coordinate would thus fail to be a continuous function of the spacetime point.[2]

---

[1] An example of a configuration that does not obey this is a rotating gaseous body, inside which the gas circulates in the meridional planes.

[2] However, exactly this kind of time coordinate is used on the surface of the Earth – the discontinuity occurs across the line of change of date that runs through the middle of the Pacific.

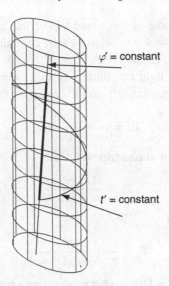

Fig. 20.17. A coordinate transformation of the form (20.159) would result in disrupting the orbits of the k field and in a discontinuous time coordinate. The thick helix segment shows the jump in t' along the $\varphi' = 0$ line (given in the text).

The constant D_0 has to be zero for a different reason. Suppose that the coordinates of the stationary–axisymmetric spacetime that is asymptotically flat go over, asymptotically, into the cylindrical coordinates:

$$ds^2 = dt^2 - r^2\,d\varphi^2 - dr^2 - dz^2. \tag{20.160}$$

This means that asymptotically g_{03} must go to zero while g_{33}/r^2 goes to -1. After a transformation (20.159) with $D_0 \neq 0$, in the coordinates adapted to the transformed Killing fields, the g_{03} component becomes

$$g'_{03} = (g_{03} - D_0 g_{33}) / (D_3 C_0),$$

and would fail to go to zero asymptotically at infinity. Consequently, $D_0 - 0$ in consequence of the assumed asymptotic flatness.

Finally, $D_3 = 1$ in order that the period of φ is 2π. Thus, the only freedom remaining is the choice of the unit of time, connected with C_0.

This uniqueness allows us to introduce the notion of *locally nonrotating observers* (Bardeen, 1970). Write the metric (20.158) in the form

$$ds^2 = e^{2\nu}\,dt^2 - e^{2\psi}\,(d\varphi - \omega\,dt)^2 - e^{2\lambda}\,dr^2 - e^{2\mu}\,d\vartheta^2. \tag{20.161}$$

Now imagine an observer moving with the 4-velocity U^α such that $U^r = U^\vartheta = 0$, i.e. circulating within the (t, φ) surface. Define $\Omega \overset{\text{def}}{=} U^\varphi/U^t \equiv d\varphi/dt$ – this is the angular velocity of the observer. Imagine that she has set up mirrors all along her orbit so that light rays emitted by her source can travel around the same circle in both directions and come back to her. Let the time of the round-trip of the light ray be T_ε, with $\varepsilon = +1$ for the ray rotating forward and $\varepsilon = -1$ for the ray rotating backward with respect to the observer. The observer will during the time T_ε move from the position φ_1 to the position $\varphi_2 = \varphi_1 + \Omega T_\varepsilon$. The light ray, during its round-trip, will move from φ_1 to $\varphi_3 = \varphi_1 + \Omega T_\varepsilon + 2\pi\varepsilon$. Along the ray $ds^2 = 0$, so $\varepsilon e^\nu dt = e^\psi (d\varphi - \omega dt)$, and consequently

$$dt = \frac{e^\psi}{\omega e^\psi + \varepsilon e^\nu} d\varphi. \tag{20.162}$$

Integrating this over the whole round-trip we obtain

$$T_\varepsilon = \frac{\Omega T_\varepsilon + 2\pi\varepsilon}{\omega + \varepsilon e^{\nu - \psi}}. \tag{20.163}$$

Solving this for T_ε we obtain

$$T_\varepsilon = \frac{2\pi}{e^{\nu - \psi} - \varepsilon(\Omega - \omega)}. \tag{20.164}$$

Along the observer's path $d\varphi = \Omega dt$, so her proper time s is related to her time coordinate by

$$ds^2 = \left[e^{2\nu} - e^{2\psi}(\Omega - \omega)^2 \right] dt^2. \tag{20.165}$$

Hence, the proper time of the round-trip of the ray measured by the observer is

$$S_\varepsilon = \sqrt{e^{2\nu} - e^{2\psi}(\Omega - \omega)^2} \, T_\varepsilon \equiv 2\pi e^\psi \sqrt{\frac{1 + \varepsilon e^{\psi - \nu}(\Omega - \omega)}{1 - \varepsilon e^{\psi - \nu}(\Omega - \omega)}}. \tag{20.166}$$

This time is different for the forward-sent ray and for the backward-sent ray ($S_1 \neq S_{-1}$), except when $\Omega = \omega$. Since, as has been shown, in an asymptotically flat spacetime the (t, φ) coordinates of (20.161) are unique, ω is uniquely determined by the geometry, so *there is a uniquely determined set of observers for whom the effect of rotation, $(S_1 - S_{-1})$, disappears.* They are called **locally nonrotating observers**, after Bardeen (1970). Their worldlines are orthogonal to the $t = $ constant hypersurfaces, so their rotation tensor defined by (15.30) is zero.

The quantities present in (20.161) were defined for the Kerr metric in Eqs. (20.121) and (20.126). The Kerr metric is stationary, axisymmetric and asymptotically flat, with the period of φ equal to 2π, so it is an example of a spacetime that allows the existence of locally nonrotating observers.

20.11 * Ellipsoidal spacetimes

Together with locally non-rotating observers, one can define another set of observers in the Kerr spacetime – those for whom the locally orthogonal subspaces are composed of confocal ellipsoids of revolution.

In order to define these observers, we first introduce the notion of an *ellipsoidal spacetime*. It is defined in analogy to the spherically symmetric spacetimes – that are composed of concentric spheres. The ellipsoidal spacetimes are composed of concentric ellipsoids in a similar manner.[1]

Consider a Euclidean space E_3 with rectangular Cartesian coordinates (x, y, z) and the metric $ds^2 = dx^2 + dy^2 + dz^2$. Introduce such (r, ϑ, Φ) coordinates that $r = \text{constant}$ defines an ellipsoid of revolution:

$$x = g(r) \sin \vartheta \cos \Phi, \qquad y = g(r) \sin \vartheta \sin \Phi, \qquad z = r \cos \vartheta. \tag{20.167}$$

They will be orthogonal when $g^2(r) - r^2 = \text{constant}$, i.e. when

$$g(r) = \sqrt{r^2 + a^2}, \qquad a = \text{constant}. \tag{20.168}$$

This implies that the ellipsoids corresponding to different values of r are confocal, and a is the radius of their common focal ring $\{r = 0, \vartheta = \pi/2\}$. The surfaces $\vartheta = \text{constant}$ are one-sheeted hyperboloids of revolution with their foci on the same ring. The metric in these coordinates becomes

$$ds^2 = \frac{r^2 + a^2 \cos^2 \vartheta}{r^2 + a^2} dr^2 + \left(r^2 + a^2 \cos^2 \vartheta\right) d\vartheta^2 + \left(r^2 + a^2\right) \sin^2 \vartheta \, d\Phi^2. \tag{20.169}$$

The space with the metric (20.169) is still flat. We make it curved by assuming that $g_{rr} = f^2(r, \vartheta)$ is an arbitrary function – in analogy to a curved spherically symmetric space that results when the Cartesian $g_{rr} = 1$ is replaced by $g_{rr} = f^2(r)$.

We now define an **ellipsoidal spacetime** as follows: there should exist a congruence of timelike lines in it, with the tangent vector field

$$u^\alpha = U\delta^\alpha{}_0 + V\delta^\alpha{}_3 \tag{20.170}$$

(where $x^0 = t$ and $x^3 = \varphi$) such that the metric

$$ds_3{}^2 = f^2(r, \vartheta)dr^2 + \left(r^2 + a^2 \cos^2 \vartheta\right) d\vartheta^2 + \left(r^2 + a^2\right) \sin^2 \vartheta \, d\Phi^2 \tag{20.171}$$

coincides with the metric of the space locally orthogonal to u^α, i.e.

$$h_{\alpha\beta} = g_{\alpha\beta} - u_\alpha u_\beta. \tag{20.172}$$

Since we intend to apply this construction to the Kerr metric, we assume that the ellipsoidal spacetime is stationary and axially symmetric, and thus has the metric (20.158). Equating (20.171) with (20.172) we obtain

$$-h_{\alpha\beta} \, dx^\alpha \, dx^\beta = -g_{11} \, dr^2 - g_{22} \, d\vartheta^2 - \left(g_{00}g_{33} - g_{03}{}^2\right) U^2 \left(d\varphi - \frac{V}{U} dt\right)^2$$

$$= f^2(r, \vartheta)dr^2 + \left(r^2 + a^2 \cos^2 \vartheta\right) d\vartheta^2 + \left(r^2 + a^2\right) \sin^2 \vartheta \, d\Phi^2. \tag{20.173}$$

---

[1] In this section, we will consider confocal ellipsoids of revolution. In principle, it should be possible to generalise this construction to other types and arrangements of the ellipsoids, but so far those other possibilities have not been explored.

Thus

$$g_{00}g_{33} - g_{03}{}^2 = -\left(r^2 + a^2\right)\sin^2\vartheta / U^2, \qquad g_{11} = -f^2(r, \vartheta),$$
$$g_{22} = -\left(r^2 + a^2\cos^2\vartheta\right), \qquad d\Phi = d\varphi - (V/U)dt, \tag{20.174}$$

and from the normalisation $g_{\alpha\beta}u^\alpha u^\beta = 1$ we have

$$g_{00}U^2 + 2g_{03}UV + g_{33}V^2 = 1. \tag{20.175}$$

The set (20.174)–(20.175) cannot be uniquely solved for $g_{\alpha\beta}$ because we defined only the 3-dimensional projection of the metric, $h_{\alpha\beta}$. Hence, $g_{\alpha\beta}$ will contain one arbitrary function, which we choose to be

$$k^2 \overset{\text{def}}{=} g_{33} + \left(r^2 + a^2\right)\sin^2\vartheta. \tag{20.176}$$

The resulting 4-dimensional metric is then

$$ds^2 = \left(\frac{1 \mp kV}{U}\,dt \pm k\,d\varphi\right)^2 - f^2(r, \vartheta)dr^2 - \left(r^2 + a^2\cos^2\vartheta\right)d\vartheta^2$$
$$- \left(r^2 + a^2\right)\sin^2\vartheta\left[d\varphi - (V/U)dt\right]^2. \tag{20.177}$$

The conditions (20.174)–(20.175) define metric components, so they are coordinate-dependent. Thus, the condition of ellipsoidality in fact says that *coordinates should exist in which the metric obeys (20.174)–(20.175)*. In order to show that the Kerr metric is ellipsoidal, the Boyer–Lindquist coordinates of (20.57) have to be transformed as follows

$$t = t' + a\varphi, \qquad (r, \vartheta, \varphi) = (r', \vartheta', \varphi'). \tag{20.178}$$

This transformation is of the 'forbidden' type (20.159) that leads to a discontinuous time coordinate. The resulting metric is

$$ds^2 = \left(1 - \frac{2mr}{\Sigma}\right)dt^2 + 2a\left(1 - \frac{2mr\cos^2\vartheta}{\Sigma}\right)dt\,d\varphi$$
$$+ \left(a^2 - \frac{2mra^2\cos^4\vartheta}{\Sigma} - \left(r^2 + a^2\right)\sin^2\vartheta\right)d\varphi^2 - \frac{\Sigma}{\Delta_r}\,dr^2$$
$$- \Sigma\,d\vartheta^2. \tag{20.179}$$

This is seen to coincide with (20.177) when

$$k^2 = a^2\left(1 - 2mr\cos^4\vartheta/\Sigma\right), \qquad U^2 = \left(r^2 + a^2\right)/\Delta_r,$$
$$V = \left(\pm k - g_{03}U\right)/g_{33}, \tag{20.180}$$

where the metric components refer to (20.179).

The ellipsoidal spacetimes were introduced by Krasiński (1978) with the hope that the construction would help in the search for a perfect fluid source of the Kerr metric. It did not (see Section 20.13), and the ellipsoidal form turned out to be rather difficult to use in calculations. So far, it has not found any practical application.

20.12 A Newtonian analogue of the Kerr solution

We found in Section 12.11 that the Newtonian theory results as the $c \to \infty$ limit of relativity if the metric, when developed in power series with respect to $1/c^2$, has the form (12.46). The corresponding expression for the Kerr metric is given in Eq. (20.47). Since the $r =$ constant surfaces were found to be confocal ellipsoids of revolution, the Newtonian limit of the Kerr metric might be a gravitational field whose equipotential surfaces are such ellipsoids. However, the potential implied in (20.47) is not necessarily $-2m/r$; this may be just an approximation to it. Thus, in order to find the Newtonian potential that is constant on the surfaces of confocal ellipsoids of revolution, we have to consider the complete set of Newtonian field equations and equations of motion.

In fact, an example of a source of such a potential has been known for a long time; it was found by Chasles in 1840.[1] The source was a homoeoid, i.e. an ellipsoid of revolution with a uniform 2-dimensional distribution of matter over it. The equipotential surfaces are ellipsoids confocal to the source. In the spheroidal coordinates, related to the Cartesian ones by (20.167)–(20.168), the exterior potential is

$$V_e(r) = -(GM/a) \arctan(a/r), \tag{20.181}$$

where M is the total mass of the homoeoid. (See Krasiński (1980) for details. The proof that the potential of the homoeoid is independent of ϑ is tricky; hints for it may be found in Kellogg (1929).) Knowing that the potential depends only on r, it is actually quite easy to guess a continuous density distribution that will generate the same exterior potential; it is (Krasiński, 1980)

$$\rho(r, \vartheta) = f(r) / \left(r^2 + a^2 \cos^2 \vartheta \right), \tag{20.182}$$

where $f(r)$ is an arbitrary function. The total mass of the source is then

$$M = 4\pi \int_0^{r_0} f(r') \mathrm{d}r', \tag{20.183}$$

where $r = r_0$ is the equation of the surface of the source. The field (20.181) may be verified to be the unique solution of the Laplace equation that depends only on r (recall that this is the spheroidal r!) and obeys the boundary conditions $\lim_{r \to \infty} V = 0$ and $\lim_{r \to \infty} r^2 \, \mathrm{d}V/\mathrm{d}r = GM$.

Some properties of this source were investigated by Krasiński (1980) and by Bażański, Kaczyńska and Krasiński (1985). Among other things, the interior potential was found, and the equations of motion were solved to find the distribution of velocity of matter

---

[1] This information is based on Chandrasekhar (1969) and Webster (1949); the authors had no access to the original paper by Chasles.

and of pressure. The equipotential surfaces are ellipsoids of the same family (confocal) also inside the source. If $\rho_{,r} < 0$, then the surfaces of constant matter-density are more oblate than the equipotential surfaces; they do not coincide in any case. The pressure is not constant on the equipotential surfaces either. The focal ring of the ellipsoids, $\{r = 0, \vartheta = \pi/2\}$, is in general a singularity; in order to make it nonsingular the density would have to *increase* with increasing r, at least in some neighbourhood of the ring.

We refer the reader to Krasiński (1980) and Bażański, Kaczyńska and Krasiński (1985) for a more extended list of properties of this configuration. Let us only note a few curiosities here:

1. The exterior field is determined only by the mass and angular momentum of the body; it does not depend on the body's size.
2. At a point P inside the body, the gravitational potential is determined only by the mass contained inside the equipotential ellipsoid passing through P; the matter exterior to that ellipsoid gives a zero contribution to the potential.
3. Although the configuration is rather special, in the spherically symmetric limit $a \to 0$ it reproduces *all possible* spherically symmetric configurations (because of the arbitrary function $f(r)$).

This Newtonian solution is reported here for a certain reason. It is fashionable in some sectors of the astronomical community to claim that relativity is just too difficult to be used, and ad-hoc simplified methods were introduced in order to mimic relativistic effects within Newtonian gravitation. One such artificial crutch is the so-called 'pseudo-Newtonian potential'. A pseudo-Newtonian potential is a 'potential' pulled out of somebody's mind without verifying whether it obeys the Laplace equation (usually it does not), for example $V(r) = \text{constant}/(r - r_0)$, where r is the ordinary spherical coordinate and r_0 is a constant. The orbits in such a potential are not ellipses; they display the 'relativistic effect', the perihelion shift, which is supposed to justify using this potential in place of the Schwarzschild solution.[1] Never mind that the potential is not vacuum.

The field (20.181) is an example showing that, with a little extra effort, one can have 'relativistic effects' in a potential that is at least a correct Newtonian vacuum potential.

20.13 A source of the Kerr field?

The Kerr solution has been known for more than 40 years now, and from the very beginning its existence provoked the simple question: what material body could generate such a vacuum field around it? Several authors have tried very hard to find a model of the source, but so far without success. The most promising positive result is that of Roos (1976), who investigated the Einstein equations with a perfect fluid or an anisotropic non-dissipating fluid source, with the boundary condition that the Kerr metric is matched

---

[1] It has been known since time immemorial that *any* non-Coulombian spherically symmetric potential has the same property.

to the solution. Roos' result was that the equations form an integrable set. All attempts so far to find an explicit example of a solution have failed.[1] The sources known are all rather artificial (see Krasiński (1978) for references): a 2-dimensional disc spanned on the singular ring of the Kerr metric; a body with anisotropic stresses, sometimes enveloped in a crust of another kind of anisotropic matter. The continuing lack of success prompted some authors to spread the suspicion that a perfect fluid source might not exist; rumours about this suspicion were then taken as a serious suggestion. The opinion of one of the present authors (A. K.) is that a bright new idea is needed, as opposed to routine standard tricks tested so far. Simply having a look at the Newtonian models that, even in the apparently simple cases of homogeneous density distribution, can be very difficult (see Chandrasekhar (1969), for example) allows one to realise that the corresponding problem in relativity must be at least as difficult.

20.14 Exercises

1. Prove that the Schwarzschild solution has the property (20.1).
2. Prove that the vector field l_μ in (20.1) is null with respect to $g_{\mu\nu}$ if and only if it is null with respect to $\eta_{\mu\nu}$. Prove then that $l^\mu \overset{\text{def}}{=} g^{\mu\nu} l_\nu \equiv \eta^{\mu\nu} l_\nu$ and verify that (20.2) holds.
3. Verify Eqs. (20.3) and (20.4). **Note.** verifying (20.4) is rather laborious; it requires using the definition of the covariant derivative and observing that a symmetric tensor doubly contracted with an antisymmetric one gives zero identically.
4. Verify Eq. (20.12) (it is rather laborious and requires multiple changes of names of indices in order to detect simplifications).
5. Verify that Eqs. (20.16) and (20.18) are consistent with each other and do admit nonzero Ys. **Hint.** Solve (20.16) and (20.18) for $Y,_\xi$ and $Y,_{\bar\xi}$, and then check the integrability condition $Y,_{\xi\bar\xi} = Y,_{\bar\xi\xi}$.
6. Verify Eqs. (20.25) and (20.26).
7. Prove that the assumption $Z \neq 0$ in (20.19) implies that Y and $\overline{Y}$ in (20.15) obey $dY \wedge d\overline{Y} \neq 0 (\Longleftrightarrow Y,_{[\alpha} \overline{Y},_{\beta]} \neq 0)$.
   Hint. Assume that $Y,_{[\alpha} \overline{Y},_{\beta]} = 0$, which implies $\overline{Y},_\alpha = \mathcal{A} Y,_\alpha$, where $\mathcal{A}$ is a function, then write out explicitly the invariance equations (20.16) and the definition of Z, Eq. (20.19). It will follow that $Z = 0$.
8. Verify that the functions Y, $\overline{Y}$, Z and $\overline{Z}$ obey (20.34).
   Hint. To prove that Y obeys (20.34), solve $k^\rho Y,_\rho = 0 = m^\rho Y,_\rho$ for $Y,_\xi$ and $Y,_{\bar\xi}$, then find $Y,_v$ from $\overline{Z} = YY,_u - Y,_\xi$ and substitute for $Y,_u$ from (20.31). To prove (20.34) for Z note that the directional derivatives along $\mathcal{K}^\alpha$ and along $\overline{m}^\alpha$ commute, and use the result that $\mathcal{K}^\rho \overline{Y},_\rho = 0$.
9. Verify that if x, y and z are interpreted as Cartesian coordinates, then the surfaces $r = $ constant defined by (20.43) are confocal ellipsoids of revolution. Since they are axially symmetric, the foci of cross-sections through the axis all lie on a circle. Verify that the radius of that circle is a and that $2r$ is the smallest diameter of the ellipsoid $r = $ constant.
10. Verify, by any method, that the Kerr metric is of Petrov type D.

---

[1] See Krasiński (1978) for a systematic overview of the attempts prior to 1976; all the later attempts can be classified according to the same scheme and none has moved the matter forward.

11. Prove that with $a^2 < m^2$ and a sufficiently large value of $|L_z|$, there exists a range (r_1, r_2), with $r_+ \leq r_1 < r_2$, such that $E_{\min}(r)/\mu_0 > 1$ for $r_1 < r < r_2$ and $E_{\min}(r)/\mu_0 < 1$ for $r > r_2$ in (20.123).
Hint. Substitute (20.123) in the inequality $E_{\min}(r)/\mu_0 > 1$ and simplify. Note that $r^2\Delta_r - 4a^2m^2 = (r - 2m)D$. You will see that, if L_z^2 is large enough, then the discriminant of the resulting square trinomial in r is positive, hence $E_{\min}(r)/\mu_0 > 1$ for $r_1 < r < r_2$, with

$$r_{1,2} = \frac{1}{4m}\left[\left(\frac{L_z}{\mu_0}\right)^2 \mp \sqrt{\left(\frac{L_z}{\mu_0}\right)^4 - 16m^2\left(\frac{L_z}{\mu_0} - a\right)^2}\right],$$

and $E_{\min}(r)/\mu_0 < 1$ for $r > r_2$. Then verify that $r_1 \geq r_+$ for any L_z.

12. Prove that the function $F(r) \overset{\text{def}}{=} E_{\min}^v/|L_z|$ defined in (20.128) has only one maximum and no minima in the range $r \in (r_+, \infty)$.
Hint. One method to do it is as follows: 1. Note that $F = (1 - 2m/r)/\left(\sqrt{\Delta_r} \mp 2am\right)$. Note also that $F \neq 0$ at $r = 2m$ with the lower sign because the denominator vanishes there, too, and the limit of the whole expression is finite. This remark applies also to the derivatives of F. 2. Calculate dF/dr, equate it to zero, and observe that it is $+\infty$ at $r = r_+$ and goes to zero from below as $r \to \infty$. Hence it has at least one zero in the range $r \in (r_+, \infty)$. The derivative has the form $\sqrt{\Delta_r} +$ (polynomial in r) $= 0$. 3. Solve this for $\sqrt{\Delta_r}$ and square the result. The resulting fifth order equation looks bad, but it factorises and becomes $(r - 2m)^2\left(r^3 - 6mr^2 + 9m^2r - 4ma^2\right) \overset{\text{def}}{=} (r - 2m)^2 H(r) = 0$. (The $r = 2m$ is not a zero of the derivative: with the minus sign, the limit of $F'(r)$ at $r \to 2m$ is finite; with the plus sign, $F'(r)$ has a nonzero value there.) 4. Observe that $H(r)$ has in fact two zeros in the range $r \in (r_+, \infty)$; however, only one of them is also a zero of $F'(r)$. This is done as follows: $H(r)$ is positive at $r = r_+$ and at $r = 4m$, but negative at its minimum at $r = 3m$. Thus one of its zeros is in $(r_+, 3m)$ and the other in $(3m, 4m)$. Now, with the minus sign, $F'(r)$ changes sign an odd number of times in $(r_+, 3m)$, but an even number of times in $(3m, 4m)$, hence the second zero of H is not a zero of F'. With the plus sign, the opposite is true: $F'(r)$ changes sign an even number of times in $(r_+, 3m)$, but an odd number of times in $(3m, 4m)$, so the first zero of H is not a zero of F'. Consequently, F' has only one zero in (r_+, ∞). That this corresponds to a maximum of F follows from point 2 above: F is increasing in the neighbourhood of $r = r_+$ and decreasing as $r \to \infty$.

13. Draw graphs analogous to Fig. 20.12 for the cases $\Gamma > 0$ and $\Gamma = 0$ (preferably by using a good plotting program).

14. Prove that $\chi(r) > 0$ in (20.155) for all $r \geq r_+$.
Hint. Note that if $f(r_0) > 0$ and $df/dr > 0$ for all $r > r_0$, then $f(r) > 0$ for all $r > r_0$. Write the condition $\chi(r) > 0$, multiply this inequality by whatever is necessary to obtain an inequality $P(r) > 0$, where $P(r)$ is a polynomial, then differentiate $P(r)$ repeatedly until you find out that one of the derivatives is seen to have the property $f(r) > 0$ for all $r \geq r_+$. Then use the observation to deduce the same about all the lower-order derivatives.

15. Prove that for the locally non-rotating observers defined in Section 20.10 the rotation tensor is zero and that the worldlines of these observers are orthogonal to the $r = $ constant hypersurfaces.

16. Find the equations of the light cones in Fig. 20.6.
Hint. The metric of the subspace $\vartheta = \pi/2$ in (20.57) is

$$ds^2 = \left(1 - \frac{2m}{r}\right)dt^2 + \frac{4ma}{r}dt\,d\varphi - \left(\frac{2ma^2}{r} + r^2 + a^2\right)d\varphi^2 - \frac{r^2}{\Delta_r}dr^2.$$

The equation of the light cones is $ds^2 = 0$. The relation between the B–L coordinates (r, φ) and the (X, Y) spatial coordinates in the figure is $X = \sqrt{r^2 + a^2} \cos \varphi$, $Y = \sqrt{r^2 + a^2} \sin \varphi$. The cones are drawn at the points of the plane $\varphi = 0$, where $dr = \left(\sqrt{r^2 + a^2}/r \right) dX$ and $d\varphi = dY/\sqrt{r^2 + a^2}$. Thus the equation of a cone with vertex at $(T, X, Y) = (0, 0, X_0)$ is

$$\left(1 - \frac{2m}{r} \right) T^2 + \frac{4ma}{r\sqrt{r^2 + a^2}} TY - \left(\frac{2ma^2}{r(r^2 + a^2)} + 1 \right) Y^2$$

$$- \frac{r^2 + a^2}{\Delta_r} (X - X_0)^2 = 0. \tag{20.184}$$

The vertical axis in the figure is T, the (X, Y) axes are as shown. The equations for drawing the graphs are obtained by solving the above for T or for X, but a few special cases must be treated separately. The limit $r = 2m$ (the outer stationary limit hypersurface) in the above equation is nonsingular. Before taking the limits $r \to r_{\pm}$, the equation must be multiplied by Δ_r. These limits are discontinuous, as described at the end of Section 20.5. On approaching $r = r_{\pm}$ from inside the segment (r_-, r_+), the equation becomes $X = X_0(r_{\pm})$, which is a vertical plane. On approaching these values from outside that segment, the equation becomes singular, but by calculating the curves along which the cones intersect with a $T = $ constant plane one finds that they are ellipses whose both axes shrink to a point, while the ellipses themselves recede to $Y \to -\infty$ as $r \to r_{\pm}$. Before taking the limit $r \to 0$, the equation must be multiplied by r; in the limit it becomes $-2m [T - (a/|a|)Y]^2 = 0$, which is one of the straight lines $T = Y$ (when $a > 0$) or $T = -Y$ (when $a < 0$, the case shown in the figure).

21

Subjects omitted from this book

As stated in the introduction, it would not be possible to include the whole of relativity in a book of manageable size. We chose to go into several selected topics in depth, but omitted some other topics completely or nearly so. This short chapter is a list of the topics we omitted, with some suggestions to the reader for further reading.

The following topics were covered inadequately or not at all:

1. **Gravitational waves.** Speaking most generally, a gravitational wave is any gravitational field that propagates through space independently of matter. It may, but need not, be periodic. There exists a large collection of exact solutions of Einstein's equations describing waves, for these see Stephani *et al.* (2003). There exists also an elaborate theory of nearly-linear waves, a relatively good source for it is the book by Ohanian and Ruffini (1994), and also the classic MTW course (Misner, Thorne and Wheeler, 1973). The theory of generation and detection of gravitational waves is worked out rather well, but progress in it is still going on, so current knowledge can be gained only from papers. A sophisticated and elaborate experimental technology is already in place, but to keep up with this one has to attend conferences in addition to reading the literature. The pioneer of the search for gravitational waves was Weber (1961); his small book can be recommended to readers interested in the history of the subject.

2. **The Cauchy problem.** In each coordinate system, the set of Einstein's equations can be separated into those equations that contain at most the first-order time derivatives of the metric components and those that are of second order in time. The former are limitations imposed on the initial data, the latter are the dynamical evolution equations. This approach makes it possible to discuss such problems as the global existence or nonexistence of solutions of Einstein's equations, horizons in general spacetimes and general principles of propagation of gravitational waves. The pioneering paper in this field was that by Arnowitt, Deser and Misner (1962), which gave the name ADM to the whole approach. Misner, Thorne and Wheeler (1973) and Joshi (1993) are also good sources for this topic, and Hawking and Ellis (1973) make elaborate use of it.

3. **Generating new stationary–axisymmetric solutions out of known solutions.** This is a very large field of activity. We saw in Section 20.2 that the Kerr solution resulted from the Einstein equations in consequence of a set of assumptions, not all of which

had a clear interpretation. By changing these assumptions one can obtain other solutions. The first important step in this direction was made by Ernst (1968a), who reduced the Einstein equations for stationary–axisymmetric vacuum spacetimes to one equation for one complex function, now called the **Ernst potential** (in a subsequent paper, this approach was extended to the Einstein–Maxwell equations (Ernst, 1968b)). Then it was discovered that these equations admit transformations of variables that lead from one solution to another, with a geometry that is not a coordinate transform of the initial one. On the basis of this finding, several **generating techniques** for obtaining new solutions were introduced. Meanwhile, the most famous among the next-generation stationary–axisymmetric metrics, the Tomimatsu–Sato (1972) solution, was found. An overview of the generating techniques can be found in Stephani *et al.* (2003); a more extended overview is the recent book by Belinskii and Verdaguer (2001).

4. **The Penrose transform.** We mentioned this approach when discussing the maximal analytic extensions of the Reissner–Nordström spacetime in Section 14.15, and of the Kerr metric in Section 20.8. More on it can be found in the books by Misner, Thorne and Wheeler (1973) and by Hawking and Ellis (1973). The main advantage of this approach is that it maps points at infinity into finite points of another manifold, which in turn allows one to discuss values of functions instead of limits. This is a powerful tool, even though few spacetimes are known for which the Penrose transform has been constructed explicitly.

5. **Cosmic censorship.** We mentioned this subject in Section 18.16, but our brief account does not fairly represent the activity behind it. This is a lively paradigm, on which perhaps the best source is the book by Joshi (1993).

6. **Experimental tests.** Apart from the very few classical and most basic tests, we did not really do justice to this subject. This is now a science in itself, with large groups of physicists involved in projects lasting many years. Apparently, there exists no up-to-date book on it. As a historical introduction to the subject, the old volume of proceedings of the Fermi school from 1972 can be recommended (Bertotti, 1974); a discussion of results and their meaning for the theory can be found in the book by Will (1981).

7. **Spinor methods.** Our Chapter 11 is a very concise introduction to this subject, but spinors can do much more than we described there. Fortunately, a large monograph (Penrose and Rindler, 1984) is available.

8. **Relativistic astrophysics.** We did not give a fair representation of the classical applications of relativity to astrophysics because we concentrated on the conceptual basis of relativity. The most extended course on this subject is still the two volumes by Zel'dovich and Novikov (1971, 1974); briefer accounts can be found in Misner, Thorne and Wheeler (1973) and Weinberg (1972).

9. **The history of relativity.** This subject is treated rather superficially in most textbooks, and ours is no exception. That history is important and can be exciting is best attested

by the classic book by Pais (1982) – a detailed account of Einstein's life and scientific activities. Other important sources are the book by Mehra (1974) that explains how relativity had been taking shape step by step, and the collection of original papers in which special and general relativity were created (Einstein *et al.*, 1923). Some important bits of history can be found in the monograph by Dicke (1964).

10. **Special relativity.** We omitted this subject altogether because we assumed that special relativity is now part of all courses on electrodynamics and should be familiar to anyone setting out to study general relativity. Should any reader need to learn, we recommend the following sources: Synge's (1965) book is an expert-level complete textbook. The book by Kopczyński and Trautman (1992) is only in a small part devoted to special relativity, but it presents an enlightening geometric approach that simplifies many problems. Equally enlightening is the textbook by Rindler (1980). Finally, special relativity in the context of electrodynamics is presented in a readable way by Jackson (1975).

References

Allen, C. W. (1973). *Astrophysical Quantities*. London, The Athlone Press. [181]

Alpher, R. A. and Herman, R. C. (1948). Evolution of the Universe, *Nature* **162**, 774. [288]

Appelquist, T. and Chodos, A. (1983). Quantum effects in Kaluza–Klein theories, *Phys. Rev. Lett.* **50**, 141. [164]

Arnau, J. V., Fullana, M., Monreal, L. and Saez, D. (1993). On the microwave background anisotropies produced by nonlinear voids, *Astrophys. J.* **402**, 359. [330, 331]

Arnau, J. V., Fullana, M. and Saez, D. (1994). Great attractor-like structures and large-scale anisotropy, *Mon. Not. Roy. Astr. Soc.* **268**, L17. [330, 331]

Arnowitt, R., Deser, S. and Misner, C. W. (1962). The dynamics of general relativity, in *Gravitation: An Introduction to Current Research*. Edited by L. Witten. New York: Wiley Interscience. [498]

Bardeen, J. M. (1970). A variational principle for rotating stars in general relativity, *Astrophys. J.* **162**, 71. [489, 490]

Bardeen, J. M. (1973). Timelike and null geodesics in the Kerr metric, in: *Black Holes – les astres occlus*. Edited by C. de Witt and B. S. de Witt. New York, London, Paris: Gordon and Breach, pp. 219–239. [466]

Barnes, A. (1970). On gravitational collapse against a cosmological background, *J. Phys.* **A3**, 653. [313, 325]

Barnes, A. (1973). On shearfree normal flows of a perfect fluid, *Gen. Rel. Grav.* **4**, 105. [230]

Barnes, A. (1984). Shear-free flows of a perfect fluid, in *Classical General Relativity*. Edited by W. B. Bonnor, J. N. Islam and M. A. H. MacCallum. Cambridge: Cambridge University Press, pp. 15–23. [70]

Barnes, A. and Rowlingson, R. R. (1989). Irrotational perfect fluids with a purely electric Weyl tensor, *Class. Quant. Grav.* **6**, 949. [396, 398]

Barrow, J. D. and Silk, J. (1981). The growth of anisotropic structures in a Friedmann Universe, *Astrophys. J.* **250**, 432. [400]

Barrow, J. D. and Stein-Schabes, J. A. (1984). Inhomogeneous cosmologies with cosmological constant, *Phys. Lett.* **A103**, 315. [392, 400, 401]

Bażański, S., Kaczyńska, R. and Krasiński, A. (1985). Physical properties of the extended Chasles equilibrium figure, *Phys. Lett.* **A115**, 33. [493, 494]

Belinskii, V. and Verdaguer, E. (2001). *Gravitational Solitons*. Cambridge: Cambridge University Press. [499]

Berger, B. K., Eardley, D. M. and Olson, D. W. (1977). Note on the spacetimes of Szekeres, *Phys. Rev.* **D16**, 3086. [397]

Bergmann, P. G. (1968). Comments on the scalar–tensor theory, *Int. J. Theor. Phys.* **1**, 25. [150]

Bertotti, B. (1959). Uniform electromagnetic field in the theory of general relativity, *Phys. Rev.* **116**, 1331. **[171]**

Bertotti, B. (editor) (1974). *Experimental Gravitation. Proceedings of the International School of Physics 'Enrico Fermi', Course 56*. New York and London: Academic Press. **[499]**

Bianchi, L. (1898). Sugli spazi a tre dimensioni che ammettono un gruppo continuo di movimenti [On the three-dimensional spaces which admit a continuous group of motions], *Memorie di Matematica e di Fisica della Società Italiana delle Scienze* **11**, 267. English translation, with historical comments: *Gen. Rel. Grav.* **33**, 2171 (2002). **[99, 101]**

Boesgaard, A. M. and Steigman, G. (1985). Big bang nucleosynthesis – theories and observations, *Ann. Rev. Astron. Astrophys.* **23**, 319. **[288]**

Bona, C. and Coll, B. (1985). On the Stephani Universe, *C. R. Acad. Sci. Paris* **301**, 613. **[233]**

Bona, C. and Coll, B.(1988). On the Stephani Universes, *Gen. Rel. Grav.* **20**, 297. **[233]**

Bondi, H. (1947). Spherically symmetrical models in general relativity *Mon. Not. Roy. Astr. Soc.* **107**, 410; reprinted, with historical comments, in *Gen. Rel. Grav.* **31**, 1777 (1999). **[297, 298, 313, 316, 328]**

Bonnor, W. B. (1956). The formation of the nebulae, *Z. Astrophysik* **39**, 143; reprinted, with historical comments, in *Gen. Rel. Grav.* **30**, 1111 (1998). **[303, 353]**

Bonnor, W. B. (1976a). Non-radiative solutions of Einstein's equations for dust, *Commun. Math. Phys.* **51**, 191. **[401]**

Bonnor, W. B. (1976b). Do freely falling bodies radiate? *Nature* **263**, 301. **[401]**

Bonnor, W. B. (1985). An open recollapsing cosmological model with $\Lambda = 0$, *Mon. Not. Roy. Astr. Soc.* **217**, 597. **[317]**

Bonnor, W. B., Sulaiman, A. H. and Tomimura, N. (1977). Szekeres's space-times have no Killing vectors, *Gen. Rel. Grav.* **8**, 549. **[392, 436]**

Bonnor, W. B. and Tomimura, N. (1976). Evolution of Szekeres's cosmological models, *Mon. Not. Roy. Astr. Soc.* **175**, 85. **[400]**

Boyer, R. H. and Lindquist, R. W. (1967). Maximal analytic extension of the Kerr metric, *J. Math. Phys.* **8**, 265. **[438, 449, 452, 466, 475, 477, 484]**

Brans, C. and Dicke, R. H. (1961). Mach's principle and a relativistic theory of gravitation, *Phys. Rev.* **124**, 925. **[149]**

Bronnikov, K. A. (1983). Some exact models of nonspherical collapse. II, *Gen. Rel. Grav.* **15**, 823. **[372]**

Bronnikov, K. A. and Pavlov, N. V. (1979). Relyativistskiye raspredeleniya zaryazhenoy pyli s ploskoy, sfericheskoy i psevdosfericheskoy simmetriyami [Relativistic distributions of charged dust with flat, spherical and pseudospherical symmetries], in: *Diskusyonnye voprosy teorii otnositelnosti i gravitatsii [Controversial Questions of the Theory of Relativity and Gravitation]*. Moscow: Nauka, p. 59. **[369, 374]**

Cahen, M. and Defrise, L. (1968). Lorentzian 4-dimensional manifolds with 'local isotropy', *Commun. Math. Phys.* **11**, 56. **[175]**

Canuto, V., Adams, P. J., Hsieh, S.-H. and Tsiang, E. (1977). Scale-covariant theory of gravitation and astrophysical applications, *Phys. Rev.* **D16**, 1643. **[150]**

Carter, B. (1966a). The complete analytic extension of the Reissner–Nordström metric in the special case $e^2 = m^2$, *Phys. Lett.* **21**, 423. **[212, 216]**

Carter, B. (1966b). Complete analytic extension of the symmetry axis of Kerr's solution of Einstein's equations, *Phys. Rev.* **141**, 1242. **[475]**

Carter, B. (1968a). Hamilton–Jacobi and Schrödinger separable solutions of Einstein's equations, *Commun. Math. Phys.* **10**, 280. **[453, 459]**

Carter, B. (1968b). Global structure of the Kerr family of gravitational fields, *Phys. Rev.* **174**, 1559. **[475]**

Carter, B. (1973). Black hole equilibrium states. Part I: Analytic and geometric properties of the Kerr solutions, in: *Black Holes – les astres occlus*. Edited by C. de Witt and B. S. de Witt. New York, London, Paris: Gordon and Breach, p. 61. **[212, 216, 438, 451, 452, 454, 458, 464, 475, 486]**

Celerier, M. N. (2000). Do we really see a cosmological constant in the supernovae data?, *Astron. Astrophys.* **353**, 63. **[287]**

Celerier, M. N. and Schneider, J. (1998). A solution to the horizon problem: a delayed big bang singularity, *Phys. Lett.* **A249**, 37. **[347]**

Celerier, M. N. and Szekeres, P. (2002). Timelike and null focusing singularities in spherical symmetry: A solution to the cosmological horizon problem and a challenge to the cosmic censorship hypothesis, *Phys. Rev.* **D** 65, 123516. **[347]**

Chandrasekhar, S. (1969). *Ellipsoidal Figures of Equilibrium*. New Haven and London: Yale University Press, p. 46. **[493, 495]**

Christodoulou, D. (1984). Violation of cosmic censorship in the gravitational collapse of a dust cloud, *Commun. Math. Phys.* **93**, 171. **[333, 336, 343, 346]**

Clarke, C. J. S. (1993). A review of cosmic censorship, *Class. Quant. Grav.* **10**, 1375. **[346]**

Clarke, C. J. S. and O'Donnell, N. (1992). Dynamical extension through a space-time singularity, *Rendiconti del Seminario Matematico della Università e Politecnico di Torino* **50**(1), 39. **[338]**

Coles P. and Ellis, G. F. R. (1997). *Is the Universe Open or Closed? The density of matter in the Universe.* Cambridge: Cambridge University Press. **[268, 286]**

Coll, B. and Ferrando, J. J. (1989). Thermodynamic perfect fluid. Its Rainich theory, *J. Math. Phys.* **30**, 2918. **[233]**

Collins, C. B. (1979). Intrinsic symmetries in general relativity, *Gen. Rel. Grav.* **10**, 925. **[105]**

Collins, C. B. and Ellis, G. F. R. (1979). Singularities in Bianchi cosmologies, *Phys. Rep.* **56** no 2, 65. **[105]**

Collins, C. B. and Szafron, D. A. (1979). A new approach to inhomogeneous cosmologies: intrinsic symmetries. I, *J. Math. Phys.* **20**, 2347. **[397]**

Collins, C. B. and Wainwright, J. (1983). On the role of shear in general relativistic cosmological and stellar models, *Phys. Rev.* **D27**, 1209. **[105]**

Courant, R. and Hilbert, D. (1965). *Methods of Mathematical Physics, vol. II: Partial Differential Equations*. New York: Wiley Interscience. **[445]**

Datt, B. (1938). Über eine Klasse von Lösungen der Gravitationsgleichungen der Relativität [On a class of solutions of the gravitation equations of relativity], *Z. Physik* **108**, 314; English translation, with historical comments: *Gen. Rel. Grav.* **31**, 1615 (1999). **[110, 385]**

Dautcourt, G. (1980). The cosmological problem as initial value problem on the observer's past light cone, in: *9th International Conference on General Relativity and Gravitation*, University of Jena, p. 315. **[363]**

Dautcourt, G. (1983a). The cosmological problem as initial value problem on the observer's past light cone: observations, *Astron. Nachr.* **304**, 153. **[363]**

Dautcourt, G. (1983b). The cosmological problem as initial value problem on the observer's past light cone: geometry, *J. Phys.* **A16**, 3507. **[363]**

Dautcourt, G. (1985). Observer dependence of past light cone initial data in cosmology, *Astron. Nachr.* **306**, 1. **[363]**

de Sitter, W. (1916). On Einstein's theory of gravitation and its astronomical consequences. Second paper, *Mon. Not. Roy. Astr. Soc.* **77**, 155 (1916–17). **[175]**

de Sitter, W. (1917). On Einstein's theory of gravitation and its astronomical consequences. Third paper, *Mon. Not. Roy. Astr. Soc.* **78**, 3. **[93]**

de Souza, M. M. (1985). Hidden symmetries of Szekeres quasi-spherical solutions, *Revista Brasileira de Física* **15**, 379. **[419]**

Debever, R. (1959). Sur le tenseur de super-énergie [On the super-energy tensor], *C. R. Acad. Sci. Paris* **249**, 1324; Tenseur de super-énergie, tenseur de Riemann: cas singuliers [The super-energy tensor, the Riemann tensor: the singular cases], *C. R. Acad. Sci Paris* **249**, 1744. **[70, 122]**

Debever, R. (1964). Le rayonnement gravitationnel: le tenseur de Riemann en relativité génerale [Gravitational radiation: the Riemann tensor in general relativity], *Cahiers Phys.* **18**, 303. **[122]**

Debever, R., Kamran N. and McLenaghan, R. G. (1983). A single expression for the general solution of Einstein's vacuum and electrovac field equations with cosmological constant for Petrov type D admitting a nonsingular aligned Maxwell field, *Phys. Lett.* **93A**, 399; also reported in *10th International Conference on General Relativity and Gravitation, Padova 1983 – contributed papers*. Edited by B. Bertotti, F. de Felice and A. Pascolini. Rome: Consiglio Nazionale delle Ricerche, pp. 216–218. **[458]**

Debever, R., Kamran, N. and McLenaghan, R. G. (1984). Exhaustive integration and a single expression for the general solution of the type D vacuum and electrovac field equations with cosmological constant for a nonsingular aligned Maxwell field, *J. Math. Phys.* **25**, 1955. **[458]**

Debney, G. C., Kerr, R. P. and Schild, A. (1969). Solutions of the Einstein and Einstein–Maxwell equations, *J. Math. Phys.* **10**, 1842. **[441,444]**

Deshingkar, S. S., Joshi, P. S. and Dwivedi, I. H. (1999). Physical nature of the central singularity in spherical collapse, *Phys. Rev.* **D59**, 044018. **[346]**

Dicke, R. H. (1964), *The Theoretical Significance of Experimental Relativity*. New York: Gordon and Breach. **[3, 149, 182, 500]**

Dicke, R. H., Peebles, P. J. E., Roll, P. G. and Wilkinson, D. T. (1965). Cosmic black-body radiation, *Astrophys. J.* **142**, 414. **[288]**

Droste, J. (1917). The field of a single centre in Einstein's theory of gravitation, and the motion of a particle in that field, *Koninklijke Nederlandsche Akademie van Wetenschappen Proceedings* **19**, 197; reprinted, with historical comments, in *Gen. Rel. Grav.* **34**, 1545 (2002). **[175]**

Dwivedi, I. H. and Joshi, P. S. (1992). Cosmic censorship violation in non-self-similar Tolman–Bondi models, *Class. Quant. Grav.* **9**, L69. **[346]**

Dyer, C. C. (1979). A spherically symmetric self-similar universe, *Mon. Not. Roy. Astr. Soc.* **189**, 189. **[344, 345]**

Dyson, F. W., Eddington, A. S. and Davidson, C. (1920). A determination of the deflection of light by the Sun's gravitational field, from observations made at the total eclipse of May 29, 1919, *Phil. Trans. Roy. Soc. London* A220, 291; cited after Will (1981, p. 5). **[186]**

Eardley, D. M. (1974a). Self-similar spacetimes: geometry and dynamics, *Commun. Math. Phys.* **37**, 287. **[105]**

Eardley, D. M. (1974b). Death of white holes in the early Universe, *Phys. Rev. Lett.* **33**, 442. **[344]**

Eardley, D. M. and Smarr, L. (1979). Time functions in numerical relativity: Marginally bound dust collapse, *Phys. Rev.* **D19**, 2239. **[333, 344, 346]**

Ehlers, J. (1961). Beiträge zur relativistischen Mechanik kontinuierlicher Medien [Contributions to the relativistic mechanics of continuous media], *Abhandlungen der Mathematisch-Naturwissenschaftlichen Klasse der Akademie der Wissenschaften und Literatur Mainz*, No 11; English translation: *Gen. Rel. Grav.* **25**, 1225 (1993). **[227, 229]**

Ehlers, J., and Kundt, W. (1962). Exact solutions of the gravitational field equations, in *Gravitation, an introduction to current research*. Edited by L. Witten. New York: Wiley Interscience, pp. 49–101. **[70, 119]**

Einstein, A. (1916). Die Grundlage der allgemeinen Relativitätstheorie [The foundation of the general theory of relativity], *Ann. Physik* **49**, 769; English translation: Einstein et al. (1923, pp. 109–164). **[125]**

Einstein, A. (1925). Einheitliche Feldtheorie von Gravitation und Elektrizität [Unified field theory of gravitation and electricity], *Sitzungsberichte Preuss. Akad. Wiss.* p. 414. **[136]**

Einstein, A., Lorentz, H. A., Weyl, H. and Minkowski, H. (1923). *The Principle of Relativity. A Collection of Original Papers on the Special and General Theory of Relativity*. Dover Publications. **[4, 132, 500]**

Einstein, A. and Straus, E. G. (1945). The influence of the expansion of space on the gravitation fields surrounding the individual stars, *Rev. Mod. Phys.* **17**, 120 (1945); Corrections and additional remarks to our paper: The influence of the expansion of space on the gravitation fields surrounding the individual stars, *Rev. Mod. Phys.* **18**, 148 (1946). **[309]**

Eisenhart, L. P. (1940). *An Introduction to Differential Geometry with Use of Tensor Calculus*. Princeton: Princeton University Press. **[48, 54]**

Eisenhart, L. P. (1964). *Riemannian Geometry*. Princeton University Press: Princeton. **[48, 63]**

Ellis, G. F. R. (1967). Dynamics of pressure-free matter in general relativity, *J. Math. Phys.* **8**, 1171. **[369, 374]**

Ellis, G. F. R. (1971). Relativistic cosmology, in: *Proceedings of the International School of Physics 'Enrico Fermi', Course 47: General Relativity and Cosmology*. Edited by R. K. Sachs. New York and London: Academic Press, pp. 104–182. **[228, 229, 237, 253, 255, 259]**

Ellis, G. F. R. (1973). Relativistic cosmology, in: *Cargèse Lectures in Physics*, vol. 6. Edited by E. Schatzman. New York: Gordon and Breach, p. 1. **[231, 237]**

Ellis, G. F. R., Hellaby, C. and Matravers, D. R. (1990). Density waves in cosmology. *Astrophys. J.* **364**, 400. **[366]**

Ellis, G. F. R. and MacCallum, M. A. H. (1969). A class of homogeneous cosmological models, *Commun. Math. Phys.* **12**, 108. **[99, 112]**

Ernst, F. J. (1968a). New formulation of the axially symmetric gravitational field problem, *Phys. Rev.* **167**, 1175. **[499]**

Ernst, F. J. (1968b). New formulation of the axially symmetric gravitational field problem II, *Phys. Rev.* **168**, 1415. **[499]**

Estabrook, F. B., Wahlquist, H. D. and Behr, C. G. (1968). Dyadic analysis of spatially homogeneous world models, *J. Math. Phys.* **9**, 497. **[99]**

Etherington, I. M. H. (1933). On the definition of distance in general relativity. *Phil. Mag., ser. 7* **15**, 761. **[253, 255]**

Ferraris, M., Francaviglia, M. and Reina, C. (1982). Variational formulation of general relativity from 1915 to 1925 'Palatini's method' discovered by Einstein in 1925, *Gen. Rel. Grav.* **14**, 243. **[136]**

Finkelstein, D. (1958). Past–future asymmetry of the gravitational field of a point particle, *Phys. Rev.* **110**, 965. [191]

Flanders, H. (1963). *Differential Forms with Applications to Physical Sciences.* New York and London: Academic Press. [294]

Fomalont, E. B. and Sramek, R. (1975). A confirmation of Einstein's general theory of relativity by measuring the bending of microwave radiation in the gravitational field of the Sun, *Astrophys. J.* **199**, 749 (1975); Measurements of the solar gravitational deflection of radio waves in agreement with general relativity, *Phys. Rev. Lett.* **36**, 1475 (1976); The deflection of radio waves by the Sun, *Comments Astrophys.* **7**, 19 (1977). [188]

Fowler, W. A. (1967). *Nuclear Astrophysics.* Philadelphia: American Philosophical Society. [288]

Foyster, J. M. and McIntosh, C. B. G. (1972). A class of solutions of Einstein's equations which admit a 3-parameter group of isometries, *Commun. Math. Phys.* **27**, 241. [171]

Friedmann, A. A. (1922). Über die Krümmung des Raumes [On the curvature of space], *Z. Physik* **10**, 377 (1922); Über die Möglichkeit einer Welt mit konstanter negativer Krümmung des Raumes [On the possibility of a world with constant negative curvature of space], *Z. Physik* **21**, 326 (1924); English translation of both papers with historical comments: *Gen. Rel. Grav.* **31**, 1985 (1999); + addendum: *Gen. Rel. Grav.* **32**, 1937 (2000). [111, 236, 262, 267, 276, 287]

Frolov, V. P. (1974). Resheniya tipa Kerra – Nyumana – Unti – Tamburino uravneniy Eynshteina s kosmologicheskim chlenom [Solutions of Einstein's equations of the Kerr – Newman – Unti – Tamburino type with the cosmological term], *Teor. Mat. Fiz.* **21**, 213; English translation: *Theor. Math. Phys.* **21**, 1088 (1974). [458]

Frolov, V. P. and Novikov, I. D. (1998). *Black Hole Physics: Basic Concepts & New Developments.* Amsterdam: Kluwer, 770 pp. [201]

Fronsdal, C. (1959). Completion and embedding of the Schwarzschild solution, *Phys. Rev.* **116**, 778. [197]

Gamow, G. (1948). The evolution of the Universe, *Nature* **162**, 680. [288]

Gautreau, R. (1984). Imbedding a Schwarzschild mass into cosmology, *Phys. Rev.* **D29**, 198. [300, 310]

Gödel, K. (1949). An example of a new type of cosmological solutions of Einstein's field equations of gravitation, *Rev. Mod. Phys.* **21**, 447; reprinted, with historical comments, in *Gen. Rel. Grav.* **32**, 1409 (2000). [99]

Goldberg, J. and Sachs, R. K. (1962). A theorem on Petrov types, *Acta Phys. Polon* **22** (**supplement**), 13. [253]

Goode, S. W. and Wainwright, J. (1982). Singularities and evolution of the Szekeres cosmological models, *Phys. Rev.* **D26**, 3315. [421]

Gorini, V., Grillo, G. and Pelizza, M. (1989). Cosmic censorship and Tolman–Bondi spacetimes, *Phys. Lett.* **A135**, 154. [344]

Graves, J. C. and Brill, D. R. (1960). Oscillatory character of Reissner–Nordström metric for an ideal charged wormhole, *Phys. Rev.* **120**, 1507. [194, 207, 212, 217]

Gravity Probe B mission update (2004). http://einstein.stanford.edu/. [160]

Gregory S. A. and Thompson L. A. (1978). The Coma/A1367 supercluster and its environs, *Astrophys. J.* **222**, 784. [301]

Grillo, G. (1991). On a class of naked strong-curvature singularities, *Class. Quant. Grav.* **8**, 739. [344]

Grishchuk, L. P. (1967). Cosmological models and spatial homogeneity criteria, *Astron. Zh.* **44**, 1097; English translation: *Sov. Astron. A. J.* **11**, 881 (1968). [112]

Guth, A. H. (1981). Inflationary Universe: a possible solution to the horizon and flatness problems. *Phys. Rev.* **D23**, 347. **[282, 284, 285]**

Haantjes, J. (1937). Conformal representations of an *n*-dimensional euclidean space with a non-definite fundamental form on itself, *Koninklijke Nederlandsche Akademie van Wetenschappen Proceedings* **40**, 700 (1937); Die Gleichberechtigung gleichförmig beschleunigter Beobachter für die elektromagnetischen Erscheinungen [Equivalence of uniformly accelerated observers with respect to electromagnetic phenomena], *Koninklijke Nederlandsche Akademie van Wetenschappen Proceedings* **43**, 1288 (1940). **[92]**

Harness, R. S. (1982). Space-times homogeneous on a time-like hypersurface, *J. Phys.* **A15**, 135. **[105]**

Hawking, S. W. and Ellis, G. F. R. (1973). *The Large-scale Structure of Spacetime*. Cambridge: Cambridge University Press. **[231, 498, 499]**

Hellaby, C. (1987). A Kruskal-like model with finite density, *Class. Quant. Grav.* **4**, 635. **[325, 357, 359]**

Hellaby, C. (1996a). The nonsimultaneous nature of the Schwarzschild $R = 0$ singularity, *J. Math. Phys.* **37**, 2892. **[321]**

Hellaby, C. (1996b). The null and KS limits of the Szekeres model, *Class. Quant. Grav.* **13**, 2537. **[402]**

Hellaby, C. and Krasiński, A. (2002). You cannot get through Szekeres wormholes or regularity, topology and causality in quasi-spherical Szekeres models, *Phys. Rev.* **D66**, 084011. **[325, 403, 411, 418]**

Hellaby, C. and Lake, K. (1984). The redshift structure of the Big Bang in inhomogeneous cosmological models. I. Spherical dust solutions, *Astrophys. J.* **282**, 1; + erratum *Astrophys. J.* **294**, 702 (1985). **[337]**

Hellaby, C. and Lake, K. (1985). Shell crossings an the Tolman model, *Astrophys. J.* **290**, 381 + erratum *Astrophys. J.* **300**, 461 (1986). **[296, 317, 324, 354]**

Hellaby, C. and Lake, K. (1988). *The singularity of Eardley, Smarr and Christodoulou*. Preprint 88/7, Cape Town: Institute of Theoretical Physics and Astrophysics, University of Cape Town. **[333]**

Hu, W. (2004). webpage: http://background.uchicago.edu/w̃hu/physics/tourpage.html **[290]**

Hubble, E. P. (1929). A relation between distance and radial velocity among extragalactic nebulae, *Proc. Nat. Acad. Sci. USA* **15**, 169. **[145, 262, 287]**

Hubble, E. P. (1936). *The Realm of the Nebulae* (New Haven: Yale University Press, 1982), 207 (Republication; First published 1936). **[262, 287]**

Hubble, E. P. (1953). The law of the redshifts, *Mon. Not. Roy. Astr. Soc.* **113**, 658. **[145, 262]**

Jackson, J. D. (1975). *Classical Electrodynamics*. Second edition. New York: J. Wiley & Sons, Inc. **[500]**

Jantzen, R. (2001). Editor's note: On the three-dimensional spaces which admit a continuous group of motions, *Gen. Rel. Grav.* **33**, 2157. **[99]**

Joshi, P. S. (1993). *Global Aspects in Gravitation and Cosmology*. Oxford: Clarendon Press. **[324, 341, 346, 498, 499]**

Joshi, P. S. and Dwivedi, I. H. (1993). Naked singularities in spherically symmetric inhomogeneous Tolman–Bondi dust cloud collapse, *Phys. Rev.* **D47**, 5357. **[346]**

Kaluza, T. (1921). Zum Unitätsproblem der Physik [On the problem of unity of physics], *Sitzungsber. Preuss. Akad. Wiss. Berlin, Math. Phys. Kl.* p. 966. **[164]**

Kantowski, R. (1965). Some relativistic cosmological models (PhD Thesis), reprinted, with historical comments, in *Gen. Rel. Grav.* **30**, 1665 (1998). **[110]**

Kantowski, R. and Sachs, R. K. (1966). Some spatially homogeneous anisotropic relativistic cosmological models, *J. Math. Phys.* **7**, 443. **[86, 110, 192]**

Kasner, E. (1921). Geometrical theorems on Einstein's cosmological equations, *Amer. J. Math.* **43**, 217. **[148]**

Kellogg, O. D. (1929). *Foundations of potential theory.* New York: Frederick Ungar, pp. 184–191. **[493]**

Kerr, R. P. (1963). Gravitational field of a spinning mass as an example of algebraically special metrics, *Phys. Rev. Lett.* **11**, 237. **[438, 446, 448]**

Kerr, R. P. and Schild, A. (1965). A new class of vacuum solutions of the Einstein field equations, in *Atti del Convegno sulla Relatività Generale: Problemi dell'Energia e Onde Gravitazionali.* Florence: G. Barbèra, pp. 1–12. **[438, 441, 444]**

Kinnersley, W. (1969). Type D vacuum metrics, *J. Math. Phys.* **10**, 1195. **[175]**

Klapdor-Kleingrothaus, H. V. (Editor) (2001). *Dark Matter in Astro- and Particle Physics. Proceedings of the International Conference DARK 2000, Heidelberg, Germany, 2000.* Berlin – Heidelberg: Springer. **[287]**

Klein, O. (1926). Quantentheorie und fünfdimensionale Relativitätstheorie [Quantum theory and five-dimensional relativity], *Z. Physik* **37**, 895 (1926); The atomicity of electricity as a quantum theory law, *Nature* **118**, 516 (1926). **[164]**

Knopp, K. (1996). *Theory of Functions, Part II.* New York: Dover. **[460]**

Kompaneets, A. S. and Chernov, A. S. (1964). Solutions of the gravitation equations for a homogeneous anisotropic model, *Zh. Eksper. Teor. Fiz.* **47**, 1939; English translation: *Sov. Phys. JETP* **20**, 1303 (1965). **[110]**

Kopczyński, W. and Trautman, A. (1992). *Spacetime and Gravitation.* Warsaw: Państwowe Wydawnictwo Naukowe, and Chichester – New York – Brisbane – Toronto – Singapore: J. Wiley. **[500]**

Korkina, M. P. and Martinenko, V. G. (1975). Dovilni 'T-kuli' v zagalniy teorii vidnosti [Arbitrary 'T-spheres' in the general theory of relativity], *Ukr. Fiz. Zh.* **20**, 626. **[386]**

Kottler, F. (1918). Über die physikalischen Grundlagen der Einsteinschen Gravitationstheorie [On the physical foundations of Einstein's gravitation theory], *Ann. Physik* **56**, 410. **[175]**

Kraniotis, G. V. (2004). Precise relativistic orbits in Kerr space-time with a cosmological constant. Preprint, gr-qc 0405095. **[466]**

Krasiński, A. (1974). Solutions of the Einstein field equations for a rotating perfect fluid. Part I – Presentation of the flow-stationary and vortex-homogeneous solutions, *Acta. Phys. Polon.* **B5**, 411 (1974); Part II – Properties of the flow-stationary and vortex-homogeneous solutions, *Acta. Phys. Polon.* **B6**, 223 (1975). **[105]**

Krasiński, A. (1978). Ellipsoidal spacetimes, sources for the Kerr metric, *Ann. Phys.* **112**, 22. **[492, 495]**

Krasiński, A. (1980). A Newtonian model of the source of the Kerr metric, *Phys. Lett.* **A80**, 238. **[493, 494]**

Krasiński, A. (1981). Spacetimes with spherically symmetric hypersurfaces, *Gen. Rel. Grav.* **13**, 1021. **[105]**

Krasiński, A (1983). Symmetries of the Riemann tensor, in *10th International Conference on General Relativity and Gravitation.* Abstracts of contributed papers. Edited by F. de Felice and A. Pascolini. Padua: University of Padua, p. 290. **[78, 79]**

Krasiński, A. (1997). *Inhomogeneous Cosmological Models.* Cambridge: Cambridge University Press. **[xviii, 146, 202, 291, 298, 302, 304, 309, 353, 374, 386, 392, 426]**

Krasiński, A. (1998). Rotating dust solutions of Einstein's equations with 3-dimensional symmetry groups; Part 1: Two Killing fields spanned on u^α and w^α, *J. Math. Phys.*

39, 380; Part 2: One Killing field spanned on u^α and w^α, *J. Math. Phys.* **39**, 401; Part 3: All Killing fields linearly independent of u^α and w^α, *J. Math. Phys.* **39**, 2148. **[105]**

Krasiński, A. (1999). [Editorial note to the Nariai 1950 papers] *Gen. Rel. Grav.* **31**, 945. **[171]**

Krasiński, A. (2001a). Rotating Bianchi type V dust models generalizing the $k = -1$ Friedmann models, *J. Math. Phys.* **42**, 355; Friedmann limits of hypersurface-homogeneous rotating dust models, *J. Math. Phys.* **42**, 3628. **[105]**

Krasiński, A. (2001b). The newest release of the Ortocartan set of programs for algebraic calculations in general relativity, *Gen. Rel. Grav.* **33**, 145. **[98]**

Krasiński, A., Behr, C. G., Schücking, E., Estabrook, F. B., Wahlquist, H. D., Ellis, G. F. R., Jantzen, R. and Kundt, W. (2003). The Bianchi classification in the Schücking–Behr approach, *Gen. Rel. Grav.* **35**, 475. **[99]**

Krasiński, A. and Ellis, G. F. R. (1999). Editor's note: On the curvature of space; On the possibility of a world with constant negative curvature of space, *Gen. Rel. Grav.* **31**, 1985. **[145, 262, 263]**

Krasiński, A. and Hellaby, C. (2002) Structure formation in the Lemaître–Tolman model, *Phys. Rev.* **D65**, 023501. **[303, 304, 305, 306, 308, 332]**

Krasiński, A. and Hellaby, C. (2004a). More examples of structure formation in the Lemaître–Tolman model, *Phys. Rev.* **D69**, 023502. **[303, 304, 307, 308, 332]**

Krasiński, A. and Hellaby, C. (2004b). Formation of a galaxy with a central black hole in the Lemaître–Tolman model, *Phys. Rev.* **D69**, 043502. **[304, 308, 313, 325, 332]**

Krasiński, A. and Hellaby, C. (2005) Structure formation in the Lemaître–Tolman cosmological model (a non-perturbative approach), in: Topics in mathematical physics, general relativity and cosmology. Proceedings of the symposium on the occasion of the 75th birthday of J. Plebański, Mexico 2002. In press. **[303, 304, 308, 332]**

Krasiński, A. and Plebański, J. (1980). N-dimensional complex Riemann–Einstein spaces with $O(n-1, \mathbb{C})$ as the symmetry group, *Rep. Math. Phys.* **17**, 217. **[171]**

Krasiński, A., Quevedo, H. and Sussman, R. (1997). On thermodynamical interpretation of perfect fluid solutions of the Einstein equations with no symmetry, *J. Math. Phys.* **38**, 2602; more detailed version available as a preprint. **[233, 397]**

Kristian, J. and Sachs, R. K. (1966). Observations in cosmology, *Astrophys. J.* **143**, 379. **[237]**

Kruskal, M. (1960). Maximal extension of Schwarzschild metric, *Phys. Rev.* **119**, 1743. **[191]**

Kundt, W. (2003). The spatially homogeneous cosmological models, *Gen. Rel. Grav.* **35**, 491. **[99]**

Kurki-Suonio, H. and Liang, E. (1992). Relation of redshift surveys to matter distribution in spherically symmetric dust Universes, *Astrophys. J.* **390**, 5. **[360]**

Lake, K. *General Relativity Database.* (An interactive program to identify manifolds with given properties.) Available at www.grdb.org. **[71]**

Lake, K. and Pim, R. (1985). Development of voids in the thin-wall approximation. I General characteristics of spherical vacuum voids, *Astrophys. J.* **298**, 439. **[310]**

Lang, K. (1974). *Astrophysical Formulae.* Berlin – Heidelberg – New York: Springer. **[182, 268]**

Lang, K. (1999). *Astrophysical Formulae. Volume I: Radiation, Gas Processes and High Energy Astrophysics; Volume II: Space, Time, Matter and Cosmology.* Berlin – Heidelberg – New York: Springer. **[283, 284, 289]**

Laplace, P. S. (1795). *Exposition du Système du Monde [Presentation of the World System]*; cited after Schneider, Ehlers and Falco (1992). **[201]**

Lemaître, G. (1927). Un univers homogène de masse constante et de rayon croissant, rendant compte de la vitesse radiale de nébuleuses extra-galactiques [A homogeneous Universe of constant mass and increasing radius accounting for the radial velocity of the extra-galactic nebulae], *Ann. Soc. Sci. Bruxelles* **A47**, 49; English translation (somewhat updated): *Mon. Not. Roy. Astr. Soc.* **91**, 483 (1931). **[236, 262, 287]**

Lemaître, G. (1931). A homogeneous Universe of constant mass and increasing radius accounting for the radial velocity of the extra-galactic nebulae, *Mon. Not. R. Astr. Soc.* **91**, 483. **[262, 263]**

Lemaître, G. (1933a). L'Univers en expansion [The expanding Universe], *Ann. Soc. Sci. Bruxelles* **A53**, 51; English translation, with historical comments: *Gen. Rel. Grav.* **29**, 637 (1997). **[191, 203, 296, 297, 298]**

Lemaître, G. (1933b). La formation des nébuleuses dans l'univers en expansion [Formation of nebulae in the expanding Universe], *C. R. Acad. Sci. Paris* **196**, 1085. **[303]**

Lemos, J. P. S. (1991). On naked singularities in self-similar Tolman–Bondi spacetimes, *Phys. Lett.* **A158**, 279. **[344]**

Lense, J. and Thirring, H. (1918). Über den Einfluß der Eigenrotation der Zentralkörper auf die Bewegung der Planeten und Monde nach der Einsteinschen Gravitationstheorie [On the influence of the proper rotation of central bodies on the motions of planets and moons according to Einstein's theory of gravitation], *Phys. Zeitschr.* **19**, 156. English translation with comments: *Gen. Rel. Grav.* **16**, 711 (1984). **[160]**

Letelier, P. S. (1980). Anisotropic fluids with two perfect-fluid components, *Phys. Rev.* **D22**, 807. **[397]**

Li, J. Z. and Liang, C. B. (1985). A new plane-symmetric solution of Einstein–Maxwell equations, *Chin. Phys. Lett.* **2**, 23. **[172]**

MacCallum, M. A. H. (1979). Anisotropic and inhomogeneous relativistic cosmologies, in *General Relativity, an Einstein Centenary Survey*, Edited by S. W. Hawking and W. Israel, Cambridge: Cambridge University Press, pp. 552–553. **[136]**

MacCallum, M. A. H. and Ellis, G. F. R. (1970). A class of homogeneous cosmological models, *Commun. Math. Phys.* **19**, 31. **[99]**

Maeda, K., Sasaki, M. and Sato, H. (1983). Voids in the closed Universe, *Progr. Theor. Phys.* **69**, 89. **[303]**

Maeda, K. and Sato, H. (1983a). Expansion of a thin shell around a void in expanding Universe, *Progr. Theor. Phys.* **70**, 772. **[303]**

Maeda, K. and Sato, H. (1983b). Expansion of a thin shell around a void in expanding Universe. II, *Progr. Theor. Phys.* **70**, 1276. **[303]**

Markov, M. A. and Frolov, V. P. (1970). Metrika zakrytogo mira Fridmana, vozmushchennaya elektricheskim zaryadom (k teorii elektromagnitnykh 'Fridmonov' [The metric of a closed Friedman world perturbed by an electric charge (towards a theory of electromagnetic 'Friedmons')], *Teor. Mat. Fiz.* **3**, 3; English translation: *Theor. Math. Phys.* **3**, 301 (1970). **[374]**

Mather, J. C., Bennett, C. L., Boggess, N. W., Hauser, M. G., Smoot, G. F. and Wright, E. L. (1993). Recent results from COBE, in: *General Relativity and Gravitation 1992. Proceedings of the 13th International Conference on General Relativity and Gravitation at Cordoba 1992*. Edited by R. J. Gleiser, C. N. Kozameh and O. M. Moreschi. Boston and Philadelphia: Institute of Physics Publishing, pp. 151–158. **[331]**

Mattig, W. (1958) Über den Zusammenhang zwischen Rotverschiebung und scheinbarer Helligkeit [On the connection between redshift and apparent luminosity]. *Astron. Nachr.* **284**, 109. **[271]**

Mehra, J. (1974). *Einstein, Hilbert and the Theory of Gravitation*. Dordrecht: D. Reidel. **[4, 132, 500]**

Mészáros, A. (1986). New model of the metagalaxy, *Acta Phys. Hung.* **60**, 75. **[357]**

Miller, B. D. (1976). Negative-mass lagging cores of the Big Bang, *Astrophys. J.* **208**, 275. **[353]**

Milne, E. A. (1934). A Newtonian expanding Universe, *Quart. J. Math. Oxford* **5**, 64 (1934); McCrea, W. H. and Milne, E. A. Newtonian Universes and the curvature of space, *Quart. J. Math. Oxford* **5**, 73 (1934); both papers reprinted, with historical comments, in *Gen. Rel. Grav.* **32**, 1933 (2000). **[145, 273]**

Milne, E. A. (1948). *Kinematic Relativity*. Oxford: Clarendon Press. **[273]**

Misner, C. W. and D. H. Sharp (1964). Relativistic equations for adiabatic, spherically symmetric gravitational collapse, *Phys. Rev.* **B136**, 571. **[296]**

Misner, C. W., Thorne, K. S. and Wheeler, J. A. (1973). *Gravitation*. San Francisco: Freeman. **[140, 145, 289, 470, 498, 499]**

Mustapha, N. and Hellaby, C. (2001). Clumps into voids, *Gen. Rel. Grav.* **33**, 455. **[303]**

Nariai, H. (1950). On some static solutions of Einstein's gravitational field equations in a spherically symmetric case, *Scientific Reports of the Tôhoku University* **34**, 160 (1950); On a new cosmological solution of Einstein's field equations of gravitation, *Scientific Reports of the Tôhoku University* **35**, 46 (1951); both papers reprinted, with historical comments, in *Gen. Rel. Grav.* **31**, 951 (1999). **[171, 296]**

Neeman, Y. and Tauber, G. (1967). The lagging-core model for quasi-stellar sources, *Astrophys. J.* **150**, 755. **[201, 353]**

Newman, E. T., Couch, E., Chinnapared, K., Exton, A., Prakash, A. and Torrence, R. (1965). Metric of a rotating charged mass, *J. Math. Phys.* **6**, 918. **[458]**

Newman, R. P. A. C. (1986). Strengths of naked singularities in Tolman–Bondi space-times, *Class. Quant. Grav.* **3**, 527. **[336, 337, 344]**

Nordström, G. (1918). On the energy of the gravitational field in Einstein's theory, *Koninklijke Nederlandsche Akademie van Wetenschappen Proceedings* **20**, 1238. **[175]**

Novikov, I. D. (1962a). O nekotorykh svoystvakh resheniy uravneniy Eynshteina dlya sfericheski simmetrichnykh polyey tyagotyeniya (I) [On some properties of solutions of Einstein's equations for spherically symmetric gravitational fields (I)], *Vestn. Mosk. Univ.* no. 5, 90. **[353, 373]**

Novikov, I. D. (1962b). O povedenii sfericheski-symmetrichnykh raspredeleniy mass v obshchey teorii otnositelnosti [On the behaviour of spherically symmetric mass distributions in the general relativity theory], *Vestn. Mosk. Univ.* no. 6, 66. **[171, 362]**

Novikov, I. D. (1964a). Zaderzhka vzryva chasti Fridmanovskogo mira i sverkhzvezdy [Delayed explosion of a part of the Fridman Universe and quasars], *Astron. Zh.* **41**, 1075; English translation: *Sov. Astr. A. J.* **8**, 857 (1965). **[201, 353]**

Novikov, I. D. (1964b). R- i T-oblasti v prostranstve-vremeni so sfericheski-simetrichnym prostranstvom [R- and T-regions in a spacetime with a spherically symmetric space], *Soobshcheniya GAISh* **132**, 3; English translation, with historical comments: *Gen. Rel. Grav.* **33**, 2255 (2001). **[191, 203, 363]**

Ohanian, H. C. and Ruffini, R. (1994). *Gravitation and Spacetime*. New York – London: W. W. Norton & Company. **[498]**

Omer, G. C. (1965). Spherically symmetric distributions of matter without pressure, *Proc. Nat. Acad. Sci. USA* **53**, 1. **[298]**

Oppenheimer, J. R. and Snyder, H. (1939). On continued gravitational contraction, *Phys. Rev.* **56**, 455. **[316]**

Ori, A., (1990). The general solution for spherical charged dust, *Class. Quant. Grav.* **7**, 985. **[379, 380, 381, 382, 383]**

Ori, A. (1991). Inevitability of shell crossing in the gravitational collapse of weakly charged dust spheres, *Phys. Rev.* **D44**, 2278. **[379, 383]**

Padmanabhan, T. (1993). *Structure Formation in the Universe*, Cambridge: Cambridge University Press. **[268, 285, 286, 288]**

Padmanabhan, T. (1996). *Cosmology and Astrophysics Through Problems*, Cambridge: Cambridge University Press. **[303]**

Pais, A. (1982). *Subtle is the Lord... The Science and the Life of Albert Einstein*. Oxford: Oxford University Press. **[500]**

Palatini, A. (1919). Deduzione invariantiva delle equazioni gravitazionali dal principio di Hamilton [Invariant deduction of the gravitational equations from Hamilton's principle], *Rendiconti del Circolo Matematico di Palermo* **43**, 203 (1919). **[136]**

Partovi, H. H. and Mashhoon, B. (1984). Toward verification of large-scale homogeneity in cosmology, *Astrophys. J.* **276**, 4. **[359]**

Peebles, P. J. E. (1993). *Principles of Physical Cosmology*. Princeton: Princeton University Press. **[268, 285, 286]**

Penrose, R. (1960). A spinor approach to general relativity, *Ann. Phys.* **10**, 171. **[70, 113, 117, 119]**

Penrose, R. (1964). Conformal treatment of infinity, in *Relativity, groups and cosmology*. Edited by B. and C. deWitt. New York – London: Gordon and Breach, pp. 565–584. **[321]**

Penrose, R. (1969). Gravitational Collapse: the Role of General Relativity. *Rivista del Nuovo Cimento, Numero Speziale* **I**, 252; reprinted, with historical comments, in *Gen. Rel. Grav.* **34**, 1135 (2002). **[339, 486]**

Penrose, R. and Rindler, W. (1984). *Spinors and Space-Time*. Cambridge: Cambridge University Press. **[113, 499]**

Penzias, A. A. and Wilson, R. W. (1965). A measurement of excess antenna temperature at 4080 Mc/s, *Astrophys. J.* **142**, 419. **[288]**

Perlmutter, S., Aldering, G., Goldhaber, G., Knop, R. A., Nugent, P., Castro, P. G., Deustua, S., Fabbro, S., Goobar, A., Groom, D. E., Hook, I. M., Kim, A. G., Kim, M. Y., Lee, L. C., Nunes, N. J., Pain, R., Pennypacker, C. R., Quimby, R., Lidman, C., Ellis, R.S., Irwin, M., McMahon, R. G., Ruiz-Lapuente, P. , Walton, N., Schaefer, B., Boyle, B. J., Filippenko, A. V., Matheson, T., Fruchter, A. S., Panagia, N., Newberg, H. J. M. and Couch, W. J. (1999). Measurements of Omega and Lambda from 42 High-Redshift Supernovae. *Astrophys. J.* **517**, 565. **[286, 289]**

Petrov, A. Z. (1954). Klassifikacya prostranstv opredelyaushchikh polya tyagoteniya [The classification of spaces defining gravitational fields]. *Uchenye Zapiski Kazanskogo Gosudarstvennogo Universiteta im. V. I. Ulyanovicha-Lenina* **114**(8), 55. English translation, with historical comments: *Gen. Rel. Grav.* **32**, 1665 (2000). **[70]**

Pirani, F. A. E. (1957). Invariant formulation of gravitational radiation theory. *Phys. Rev.* **105**, 1089. **[70]**

Plebański, J. (1964). *Notes from Lectures on General Relativity. Part I: Mathematical Introduction*. CINVESTAV, Mexico. [Note: this is a xeroxed typescript. Part II and following may have never appeared even in this form.] **[xvii]**

Plebański, J. (1967). *On Conformally Equivalent Riemannian Spaces*. CINVESTAV, Mexico. [Note: this is a xeroxed typescript.] **[69, 92]**

Plebański, J. (1974). *Spinors, Tetrads and Forms. Part I*. CINVESTAV, Mexico 1974, Part II: CINVESTAV, Mexico 1975. [Note: these are xerox copies of handwritten notes. This work has apparently never been published.] **[113]**

Podurets, M. A. (1964). Ob odnoy forme uravneniy Eynshteyna dlya sfericheski simmet-richnogo dvizheniya sploshnoy sredy [On one form of Einstein's equations for spherically symmetrical motion of a continuous medium], *Astron. Zh.* **41**, 28; English translation: *Sov. Astr. A. J.* **8**, 19 (1964). [296]

Quevedo, H. and Sussman, R. (1995). On the thermodynamics of simple non-isentropic perfect fluids in general relativity, *Class. Quant. Grav.* **12**, 859; Thermodynamics of the Stephani Universe, *J. Math. Phys.* **36**, 1365. [233]

Raine, D. J. and Thomas, E. G. (1981). Large-scale inhomogeneity in the Universe and the anisotropy of the microwave background, *Mon. Not. Roy. Astr. Soc.* **195**, 649. [331]

Raszewski, P. K. (1958). *Geometria Riemanna i analiza tensorowa [Riemann Geometry and Tensor Analysis]*. (Polish translation from Russian). Warsaw: Państwowe Wydawnictwo Naukowe. [58]

Raychaudhuri, A. K. (1953). Arbitrary concentrations of matter and the Schwarzschild singularity, *Phys. Rev.* **89**, 417. [191]

Raychaudhuri, A. K. (1955). Relativistic cosmology. I *Phys. Rev.* **98**, 1123 (1955); reprinted, with historical comments: *Gen. Rel. Grav.* **32**, 743 (2000); Singular state in relativistic cosmology, *Phys. Rev.* **106**, 172 (1957). [229]

Raychaudhuri, A. K. (1979). *Theoretical Cosmology*. Oxford: Clarendon Press. [425]

Reissner, H. (1916). Über die Eigengravitation des elektrischen Feldes nach der Einstein-schen Theorie [On the self-gravitation of the electric field according to Einstein's theory], *Ann. Physik* **50**, 106. [175]

Ribeiro, M. B. (1992a). On modelling a relativistic hierarchical (fractal) cosmology by Tolman's spacetime. I. Theory, *Astrophys. J.* **388**, 1. [332, 362]

Ribeiro, M. B. (1992b). On modelling a relativistic hierarchical (fractal) cosmology by Tolman's spacetime. II. Analysis of the Einstein–de Sitter model, *Astrophys. J.* **395**, 29. [362]

Ribeiro, M. B. (1993). On modelling a relativistic hierarchical (fractal) cosmology by Tolman's spacetime. III. Numerical results, *Astrophys. J.* **415**, 469. [362]

Rindler, W. (1956). Visual horizons in world models, *Mon. Not. Roy. Astr. Soc.* **116**, 662; reprinted, with historical comments: *Gen. Rel. Grav.* **34**, 131 (2002). [277]

Rindler, W. (1980). *Essential Relativity. Special, General and Cosmological*. Revised 2nd ed. Berlin: Springer. [276, 500]

Robertson, H. P. (1929). On the foundations of relativistic cosmology *Proc. Nat. Acad. Sci. USA* **15**, 822. [111, 262]

Robertson, H. P. (1933). Relativistic cosmology, *Rev. Mod. Phys.* **5**, 62. [111, 263]

Robertson, H. P. and T. W. Noonan (1968). *Relativity and Cosmology*. Philadelphia – London – Toronto: W. B. Saunders Company, p. 374–378. [276]

Robinson, I. (1959). A solution of the Einstein–Maxwell equations, *Bull. Acad. Polon. Sci., Ser. Mat. Fis. Astr.* **7**, 351. [171]

Roos, W. (1976). On the existence of interior solutions in general relativity, *Gen. Rel. Grav.* **7**, 431. [494]

Rosen, N. (1973). A bi-metric theory of gravitation, *Gen. Rel. Grav.* **4**, 435. [151]

Rothman, T. (2002). [Editorial note to the Droste 1917 paper] *Gen. Rel. Grav.* **34**, 1541. [175]

Rothman, T. and Ellis, G. F. R. (1987). Metaflation? *Astronomy* **15** no 2, 6; see also the University of Cape Town preprint 85/18. [286]

Royal Astronomical Society Discussion (1930). *Observatory* **53**, 39. [262]

Ruban, V. A. (1968). *T*-modeli 'shara' v obshchey teorii otnositelnosti [*T*-models of a 'sphere' in general relativity theory], *Pis'ma v Red. ZhETF* **8**, 669; English translation: *Sov. Phys. JETP Lett.* **8**, 414 (1968); reprinted, with historical comments: *Gen. Rel. Grav.* **33**, 363 (2001). **[110, 385]**

Ruban, V. A. (1969). Sfericheski-symmetrichnye *T*-modeli v obshchey teorii otnositel- nosti [Spherically symmetric *T*-models in the general theory of relativity], *Zh. Eksper. Teor. Fiz.* **56**, 1914; English translation: *Sov. Phys. JETP* **29**, 1027 (1969); reprinted, with historical comments: *Gen. Rel. Grav.* **33**, 375 (2001). **[110, 385, 386]**

Ruban, V. A. (1972). Neodnorodnye kosmologicheskiye modeli s ploskoy i psevdos- fericheskoy simmetriyami [Inhomogeneous cosmological models with flat and pseu- dospherical symmetries], in: *Tezisy dokladov 3-y Sovetskoy Gravitatsyonnoy Konfer- entsii* [*Theses of Lectures of the 3rd Soviet Conference on Gravitation*]. Izdatel'stvo Erevanskogo Universiteta, Erevan, pp. 348–351. **[385]**

Ruban, V. A. (1983). T-modeli shara w obshchey teorii otnositelnosti (II) [T-models of a sphere in the general theory of relativity (II)], *Zh. Eksper. Teor. Fiz.* **85**, 801; English translation: *Sov. Phys. JETP* **58**, 463 (1983). **[385, 387]**

Saez, D., Arnau, J. V. and Fullana, M. J. (1993). The imprints of the Great Attractor and the Virgo cluster on the microwave background, *Mon. Not. Roy. Astr. Soc.* **263**, 681. **[330, 331]**

Sato, H. (1984). Voids in expanding Universe, in: *General Relativity and Gravitation*. Edited by B. Bertotti, F. de Felice and A. Pascolini. Dordrecht: D. Reidel, pp. 289– 312. **[303, 310]**

Sato, H. and Maeda, K. (1983). The expansion law of the void in the expanding Universe, *Progr. Theor. Phys.* **70**, 119. **[303]**

Schneider, P., Ehlers, J. and Falco, E. E. (1992). *Gravitational Lenses*. Berlin: Springer. **[187, 189, 190, 191, 220]**

Schouten, J. A. and Struik, D. J. (1935). *Einführung in die neueren Methoden der Differentialgeometrie*, Band 1 [*Introduction to the newer methods of differential geometry, Vol. 1*]. Groningen - Batavia: P. Noordhoff N. V., p. 142. **[80]**

Schouten, J. A. and van Kampen, E. R. (1934). Beiträge zur Theorie der Deformation [Contributions to the deformation theory], *Prace Matematyczno-Fizyczne* **41**, 1. [Note: in citations of this paper, even by Schouten himself, the title of the journal and the date of publication are distorted. The citation here is correct.] **[80]**

Schwarzschild, K. (1916a). Über das Gravitationsfeld eines Massenpunktes nach der Einsteinschen Theorie [On the gravitational field of a point mass according to Einstein's theory], *Sitzungsber. Preuss. Akad. Wiss.* p. 189. **[175]**

Schwarzschild, K. (1916b). Über das Gravitationsfeld einer Kugel aus inkompressibler Flüssigkeit nach der Einsteinschen Theorie [On the gravitational field of a sphere of incompressible fluid according to the Einstein theory], *Sitzungsber. Preuss. Akad. Wiss.* p. 424. **[206]**

Sen, N. R. (1934). On the stability of cosmological models *Z. Astrophysik* **9**, 215; reprinted, with historical comments, in *Gen. Rel. Grav.* **29**, 1473 (1997). **[301]**

Senin, Yu. E. (1982). Kosmologiya s toroidalnym raspredeleniyem zhidkosti [Cosmology with a toroidal distribution of a fluid], in: *Problemy teorii gravitatsii i elementarnykh chastits* [*Problems of the Theory of Gravitation and Elementary Particles*], 13th issue. Edited by K. P. Stanyukovich. Moscow: Energoizdat, pp. 107–111. **[428]**

Senovilla, J. M. M. (1998). Singularity theorems and their consequences *Gen. Rel. Grav.* **30**, 701. **[231]**

Shikin, I. S. (1972). Gravitational fields with groups of motions on two-dimensional transitivity hypersurfaces in a model with matter and a magnetic field, *Commun. Math. Phys.* **26**, 24. **[374, 378]**

Shikin, I. S. (1974). Issledovaniye klassa polyey tyagotyeniya dlya zaryazhennoy pylev-idnoy sredy [Investigation of a class of gravitational fields of charged dust medium], *Zh. Eksper. Teor. Fiz.* **67**, 433; English translation: *Sov. Phys. JETP* **40**, 215 (1975). **[374]**

Silk, J. (1977). Large-scale inhomogeneity of the Universe, *Astron. Astrophys.* **59**, 53. **[299, 349]**

Ślebodziński, W. (1931). Sur les équations canoniques de Hamilton [On Hamilton's canonical equations], *Bulletins de la Classe des Sciences, Acad. Royale de Belg.* (5) **17**, 864. **[80]**

Soldner, J. (1804). Über die Ablenkung eines Lichtstrahls von seiner geradelinigen Bewe-gung durch die Attraktion eines Weltkörpers, an welchem er nahe vorbeigeht [On the deflection of a light ray from its rectilinear motion by the attraction of a heavenly body, nearby which it is passing], *Berliner Astronomisches Jahrbuch* 1804, p. 161; cited after Schneider, Ehlers and Falco (1992). **[187, 220]**

Spero, A. and Szafron, D. A. (1978). Spatial conformal flatness in homogeneous and inhomogeneous cosmologies, *J. Math. Phys.* **19**, 1536. **[397, 436]**

Stephani, H. (1967a). Über Lösungen der Eisteinschen Feldgleichungen, die sich in einen fünfdimensionalen flachen Raum einbetten lassen [On solutions of Einstein's equations that can be embedded in a five-dimensional flat space], *Commun. Math. Phys.* **4**, 137. **[230, 233, 263, 291]**

Stephani, H. (1967b). Konform flache Gravitationsfelder [Conformally flat gravitational fields], *Commun. Math. Phys.* **5**, 337. **[69]**

Stephani, H. (1990). *General Relativity.* Second edition. Cambridge: Cambridge Univer-sity Press. **[78, 154, 156, 159, 266]**

Stephani, H., Kramer, D., MacCallum, M., Hoenselaers, C. and Herlt, E. (2003). *Exact Solutions of Einstein's Field Equations.* 2nd Edition. Cambridge: Cambridge Univer-sity Press. **[70, 71, 105, 108, 145, 146, 154, 171, 230, 245, 253, 263, 291, 375, 458, 459, 488, 498, 499]**

Stoeger, R. W., Ellis, G. F. R. and Nel, S. D. (1992). Observational cosmology: III. Exact spherically symmetric dust solutions, *Class. Quant. Grav.* **9**, 509. **[300, 321, 363]**

Suto, Y., Sato, K. and Sato, H. (1984). Expansion of voids in a matter-dominated Universe, *Progr. Theor. Phys.* **71**, 938; Nonlinear evolution of negative density perturbations in a radiation-dominated Universe, *Progr. Theor. Phys.* **72**, 1137. **[325]**

Synge, J. L. (1965). *Relativity: The Special Theory.* 2nd ed. North-Holland, Amster-dam. **[500]**

Szafron, D. A. (1977). Inhomogeneous cosmologies: New exact solutions and their evolu-tion, *J. Math. Phys.* **18**, 1673. **[233, 387, 391, 396, 398, 428]**

Szafron, D. A. and Collins, C. B. (1979). A new approach to inhomogeneous cosmologies. II. Conformally flat slices and an invariant classification, *J. Math. Phys.* **20**, 2354. **[396, 398]**

Szafron, D. A. and Wainwright, J. (1977). A class of inhomogeneous perfect fluid cosmologies, *J. Math. Phys.* **18**, 1668. **[426]**

Szekeres, G. (1960). On the singularities of a Riemannian manifold, *Publicationes Math-ematicae Debrecen* **7**, 285; reprinted, with historical comments: *Gen. Rel. Grav.* **34**, 1995 (2002). **[191]**

Szekeres, P. (1975a). A class of inhomogeneous cosmological models, *Commun. Math. Phys.* **41**, 55. **[400, 401]**

Szekeres, P. (1975b). Quasispherical gravitational collapse, *Phys. Rev.* **D12**, 2941. **[311, 411, 419]**

Szekeres, P. (1980). Naked singularities, in: *Gravitational Radiation, Collapsed Objects and Exact Solutions*. Edited by C. Edwards. New York: Springer (Lecture Notes in Physics, vol. 124), pp. 477–487. **[337]**

Taub, A. H. (1951). Empty space-times admitting a three parameter group of motions, *Ann. Math.*, **53**, 472; reprinted, with historical comments: *Gen. Rel. Grav.* **36**, 2689 (2004). **[99]**

Tolman, R. C. (1934). Effect of inhomogeneity on cosmological models, *Proc. Nat. Acad. Sci. USA* **20**, 169; reprinted, with historical comments: *Gen. Rel. Grav.* **29**, 931 (1997). **[297, 301]**

Tomimatsu, A. and Sato, H. (1972). New exact solution for the gravitational field of a spinning mass, *Phys. Rev. Lett.* **29**, 1344. **[499]**

Trautman, A. (1972). On the Einstein–Cartan equations, *Bull. Acad. Polon. Sci., sér. sci. math., astr. et phys.* **20** Part I: 185–190, Part II: 503–506, Part III: 895–896 (1972); **21** Part IV: 345–346 (1973); On the structure of the Einstein-Cartan equations, *Symposia Mathematica* **12** 139 (1973). **[150]**

Vickers, P. A. (1973). Charged dust spheres in general relativity, *Ann. Inst. Poincaré* **A18**, 137. **[374, 376, 377, 380]**

Wagoner, R. V. (1970). Scalar–tensor theory and gravitational waves, *Phys. Rev.* **D1**, 3209. **[150]**

Wagoner, R. V., Fowler, W. A. and Hoyle, F. (1967). On the synthesis of elements at very high temperatures, *Astrophys. J.* **148**, 3. **[288]**

Wainwright, J. (1977). Characterization of the Szekeres inhomogeneous cosmologies as algebraically special space-times, *J. Math. Phys.* **18**, 672. **[398]**

Wainwright, J. and Yaremovicz, P. E. A. (1976). Killing vector fields and the Einstein–Maxwell field equations with perfect fluid source, *Gen. Rel. Grav.* **7**, 345; Symmetries of the Einstein–Maxwell field equations: the null field case, *Gen. Rel. Grav.* **7**, 595. **[172]**

Walker, A. G. (1935). On Riemannian spaces with spherical symmetry about a line, and the conditions for isotropy in general relativity, *Quart. J. Math. Oxford*, ser. 6, 81. **[111, 263]**

Waugh, B. and Lake, K. (1988). Strengths of shell-focusing singularities in marginally bound collapsing self-similar Tolman spacetimes, *Phys. Rev.* **D38**, 1315. **[344]**

Waugh, B. and Lake, K. (1989). Shell-focusing singularities in spherically symmetric self-similar Tolman spacetimes, *Phys. Rev.* **D40**, 2137. **[344]**

Weber, J. (1961). *General Relativity and Gravitational Waves*. New York: Wiley Interscience. **[498]**

Webster, A. G. (1949). *The Dynamics of Particles and of Rigid, Elastic and Fluid Bodies*. New York: Hafner, p. 413. **[493]**

Weinberg, S. (1972). *Gravitation and Cosmology*. Wiley: New York. **[425, 499]**

Werle, J. (1957). *Termodynamika fenomenologiczna [Phenomenological Thermodynamics]*. Warsaw: Państwowe Wydawnictwo Naukowe, pp. 78–80 (in Polish). **[232]**

Will, C. M. (1981). *Theory and Experiment in Gravitational Physics*. Cambridge: Cambridge University Press. **[2, 140, 149, 182, 187, 499]**

Will, C. M. (1988). Henry Cavendish, Johann von Soldner, and the deflection of light, *Amer. J. Phys.* **56**, 413; cited after Schneider, Ehlers and Falco (1992). **[187, 220]**

WMAP (2004). Fluctuations in the cosmic microwave background, http://map.gsfc.nasa.gov/m_uni/uni_101Flucts.html **[331]**

Wolf, T. (1985). *Exakte Lösungen der Einsteinschen Feldgleichungen mit flachen Schnitten* [*Exact solutions of the Einstein field equations with flat sections.*] Ph. D. Thesis, University of Jena. **[105]**

Yodzis, P., Seifert, H. J. and Müller zum Hagen, H. (1973). On the occurrence of naked singularities in general relativity, *Commun. Math. Phys.* **34**, 135. **[341, 343]**

Zel'dovich, Ya. B. and Novikov, I. D. (1971). *Relativistic Astrophysics: Volume I: Stars and Relativity*. Chicago: University of Chicago Press. **[499]**

Zel'dovich, Ya. B. and Novikov, I. D. (1974). *Relativistic Astrophysics: Volume II: The Universe and Relativity*. Chicago: University of Chicago Press. **[499]**

Index

Printed in the United States
By Bookmasters